D1130961

Stochastic Mechanics
Random Media
Signal Processing
and Image Synthesis
Mathematical Economics and Finance
Stochastic Optimization
Stochastic Control

Applications of Mathematics

Stochastic Modelling
and Applied Probability

33

Edited by I. Karatzas
M. Yor

Springer
Berlin
Heidelberg
New York
Barcelona
Hong Kong
London
Milan
Paris
Singapore
Tokyo

Applications of Mathematics

1 Fleming/Rishel, **Deterministic and Stochastic Optimal Control** (1975)
2 Marchuk, **Methods of Numerical Mathematics,** Second Edition (1982)
3 Balakrishnan, **Applied Functional Analysis,** Second Edition (1981)
4 Borovkov, **Stochastic Processes in Queueing Theory** (1976)
5 Liptser/Shiryayev, **Statistics of Random Processes I: General Theory** (1977)
6 Liptser/Shiryayev, **Statistics of Random Processes II: Applications** (1978)
7 Vorob'ev, **Game Theory: Lectures for Economists and Systems Scientists** (1977)
8 Shiryayev, **Optimal Stopping Rules** (1978)
9 Ibragimov/Rozanov, **Gaussian Random Processes** (1978)
10 Wonham, **Linear Multivariable Control: A Geometric Approach,** Third Edition (1985)
11 Hida, **Brownian Motion** (1980)
12 Hestenes, **Conjugate Direction Methods in Optimization** (1980)
13 Kallianpur, **Stochastic Filtering Theory** (1980)
14 Krylov, **Controlled Diffusion Processes** (1980)
15 Prabhu, **Stochastic Storage Processes: Queues, Insurance Risk, and Dams** (1980)
16 Ibragimov/Has'minskii, **Statistical Estimation: Asymptotic Theory** (1981)
17 Cesari, **Optimization: Theory and Applications** (1982)
18 Elliott, **Stochastic Calculus and Applications** (1982)
19 Marchuk/Shaidourov, **Difference Methods and Their Extrapolations** (1983)
20 Hijab, **Stabilization of Control Systems** (1986)
21 Protter, **Stochastic Integration and Differential Equations** (1990)
22 Benveniste/Métivier/Priouret, **Adaptive Algorithms and Stochastic Approximations** (1990)
23 Kloeden/Platen, **Numerical Solution of Stochastic Differential Equations** (1992)
24 Kushner/Dupuis, **Numerical Methods for Stochastic Control Problems in Continuous Time** (1992)
25 Fleming/Soner, **Controlled Markov Processes and Viscosity Solutions** (1993)
26 Baccelli/Brémaud, **Elements of Queueing Theory** (1994)
27 Winkler, **Image Analysis, Random Fields and Dynamic Monte Carlo Methods** (1995)
28 Kalpazidou, **Cycle Representations of Markov Processes** (1995)
29 Elliott/Aggoun/Moore, **Hidden Markov Models: Estimation and Control** (1995)
30 Hernández-Lerma/Lasserre, **Discrete-Time Markov Control Processes** (1995)
31 Devroye/Györfi/Lugosi, **A Probabilistic Theory of Pattern Recognition** (1996)
32 Maitra/Sudderth, **Discrete Gambling and Stochastic Games** (1996)
33 Embrechts/Klüppelberg/Mikosch, **Modelling Extremal Events for Insurance and Finance** (1997)
34 Duflo, **Random Iterative Models** (1997)
35 Kushner/Yin, **Stochastic Approximation Algorithms and Applications** (1997)
36 Musiela/Rutkowski, **Martingale Methods in Financial Modelling** (1997)
37 Yin, **Continuous-Time Markov Chains and Applications** (1998)
38 Dembo/Zeitouni, **Large Deviations Techniques and Applications** (1998)
39 Karatzas, **Methods of Mathematical Finance** (1998)
40 Fayolle/Iasnogorodski/Malyshev, **Random Walks in the Quarter-Plane** (1999)

Paul Embrechts
Claudia Klüppelberg
Thomas Mikosch

Modelling Extremal Events

for Insurance and Finance

With 100 Figures

Springer

Authors

Paul Embrechts
Department of Mathematics
ETH Zurich
CH-8092 Zurich, Switzerland

Thomas Mikosch
Department of Mathematics
University of Groningen
P.O. Box 800
NL-9700 Groningen, The Netherlands

Claudia Klüppelberg
Center for Mathematical Sciences
Munich University of Technology
D-80290 Munich, Germany

Managing Editors

I. Karatzas
Departments of Mathematics and Statistics, Columbia University
New York, NY 10027, USA

M. Yor
Laboratoire de Probabilités, Université Pierre et Marie Curie
4 Place Jussieu, Tour 56, F-75230 Paris Cedex, France

Corrected Second Printing 1999

Cover picture: Flooding of the Lower Rhine in 1988 (dpa)

Mathematics Subject Classification (1991):
60-00, 60-01, 62-01, 60G70, 62P05

Library of Congress Cataloging-in-Publication Data
Embrechts, Paul, 1953 –
Modelling extremal events for insurance and finance / Paul Embrechts, Claudia Klüppelberg, Thomas Mikosch. p. cm. – (Applications of mathematics, ISSN 0172-4568 ; 33)
Includes bibliographical references and index.
ISBN 3-540-60931-8 (hc : alk. paper)
1. Business mathematics. 2. Insurance–Mathematics.
I. Klüppelberg, Claudia, 1953 – . II. Mikosch, Thomas. III. Title. IV. Series.
HF5691. E46 1997
650'.01'513 – dc21 97-12308 CIP

ISSN 0172-4568
ISBN 3-540-60931-8 Springer-Verlag Berlin Heidelberg New York

Printed in Italy

Typeset from the authors' LaTEX files using Springer-TEX style files
Cover design: *design & production* GmbH, Heidelberg
SPIN: 10723391 41/3143 - 5 4 3 2 1 0 – Printed on acid-free paper

Voor Gerda, Krispijn, Eline en Frederik.
Na al dit werk blijft één vraag onbeantwoord:
"Hoe kan ik jullie ooit danken voor de opoffering en steun?"

Paul

Meinen Eltern

Thomas

Preface

Qui scribunt libros caveant a judice multo
Cum multus judex talibus immineat

Abelard (1121)

Writers of books should beware of the verdict of the crowd
For the verdict of the crowd is prejudiced against them

Translation by Christopher Platt

In a recent issue, *The New Scientist* ran a cover story under the title: "Mission improbable. How to predict the unpredictable"; see Matthews [448]. In it, the author describes a group of mathematicians who claim that extreme value theory (EVT) is capable of doing just that: predicting the occurrence of rare events, outside the range of available data. All members of this group, the three of us included, would immediately react with: "Yes, but, ...", or, "Be aware ...". Rather than at this point trying to explain what EVT can and cannot do, we would like to quote two members of the group referred to in [448]. Richard Smith said, "There is always going to be an element of doubt, as one is extrapolating into areas one doesn't know about. But what EVT is doing is making the best use of whatever data you have about extreme phenomena." Quoting from Jonathan Tawn, "The key message is that EVT cannot do magic – but it can do a whole lot better than empirical curve–fitting and guesswork. My answer to the sceptics is that if people aren't given well–founded methods like EVT, they'll just use dubious ones instead."

These two quotes set the scene for the book you are holding. Over many years we have been in contact with potential users of EVT, such as actuaries, risk managers, engineers, ... Whatever theory can or cannot predict about extremal events, in practice the problems are there! As scientists, we cannot duck the question of the height of a sea dyke to be built in Holland, claiming that this is an inadmissible problem because, to solve it, we would have to extrapolate beyond the available data. Likewise, reinsurers have for a long time known a great deal about extremal events; in their case, premiums have to be set which both cover the insured in case of a claim, and also are calculated in such a way that in the event of a catastrophe, the company stays solvent. Finally, recent developments in the financial markets create products such as catastrophe–linked bonds where the repayment value is contingent on the occurrence of some well–defined catastrophe. These and many more examples benefit from a well–established body of theory which is now referred to as EVT. Our book gives you an introduction to the mathematical and statistical theory underlying EVT. It is written with a broad audience of potential users in mind. From the subtitle however, it is clear that the main target group is in the financial industry. A reason for this emphasis is that the latter have been less exposed to EVT methodology. This is in contrast to hydrologists and reliability engineers, for instance, where for a long time EVT has belonged to the standard toolkit.

While our readership is expected to be broad, we do require a certain mathematical level. Through the availability of standardised software, EVT can be at the fingertips of many. However, a clear understanding of its capabilities and limitations demands a fair amount of mathematical knowledge. Basic courses in linear algebra, calculus, probability and statistics are essential. We have tried hard to keep the technical level minimal, stressing the understanding of new concepts and results rather than their detailed discussions and proofs. Plentiful examples and figures should make the introduction of new methodology more digestible.

Those who have no time to read the book from cover to cover, and rather want a fairly streamlined introduction to EVT in practice, could immediately start with Chapter 6. Do however read the Guidelines first. From the applied techniques presented in Chapter 6, you will eventually discover relevant material from other chapters.

A long list of references, together with numerous sections of Notes and Comments should guide the reader to a wealth of available material. Though our list of references is long, as always it reflects our immediate interest. Many important papers which do not fit our presentation have been omitted. Even in more than 600 pages, one cannot achieve completeness; the biggest gap is doubtless multivariate extreme value theory. This is definitely a shortcoming! We feel that mathematical theory has to go hand in hand with statistical theory and computer software before it can safely be presented to the end–user, but for the multivariate case, despite important recent progress, we do not feel that the theory has reached a stage as well–established as the one–dimensional one.

As with any major project, we owe thanks to lots of people. First of all, there are those colleagues and friends who have helped us in ways which go far beyond what normally can be hoped for. Charles Goldie was a constant source of inspiration and help, both on mathematical issues, as well as on stylistic ones. He realised early on that three authors who are not native English speakers, when left alone, will produce a Flemish–Dutch–German–Swiss version of the English language which is bound to bemuse many. In his typical diplomatic manner, Charles constructed confidence bands around proper English which he hoped we would not overstep too often. The fine tuning and final decisions were of course always in our hand, hence also the full responsibility for the final outcome.

Gabriele Baltes, Jutta Gonska and Sigrid Hoffmann made an art out of producing numerous versions in LaTeX of half readable manuscripts at various stages. They went far beyond the support expected from a secretary. The many computer graphs in the book show only the tip of the iceberg. For each one produced, numerous were proposed, discussed, altered, ... We owe many thanks, also for various other support throughout the project, to Franco Bassi, Klemens Binswanger, Milan Borkovec, Hansjörg Furrer, Natascha Jung, Anne Kampovsky, Alexander McNeil, Patricia Müller, Annette Schärf and Alfred Schöttl. For the software used we thank Alexander McNeil, John Nolan and Richard Smith.

Many colleagues helped in proofreading parts of the book at various stages: Gerd Christoph, Daryl Daley, Rüdiger Frey, Jan Grandell, Maria Kafetzakis, Marcin Kotulski, Frank Oertel, Sid Resnick, Chris Rogers, Gennady Samorodnitsky, Hanspeter Schmidli and Josef Steinebach. Their critical remarks kept us on our toes! Obviously there has been an extensive exchange with the finance industry as potential end–user, in the form of informal discussions, seminars or lectures. Moreover, many were generous in sharing their data with us. We hope that the final outcome will also help them in their everyday handling of extremal events: Alois Gisler (Winterthur Versicherungen), René Held and Hans Fredo List (Swiss Reinsurance), Richard Olsen (Olsen and Associates), Mette Rytgaard (Copenhagen Reinsurance) and Wolfgang Schmidt (Deutsche Bank).

All three of us take pleasure in thanking our respective home institutions and colleagues for their much appreciated support. One colleague means something special to all three of us: Hans Bühlmann. His stimulating enthusiasm for the beauty and importance of actuarial mathematics provided the ideal environment for our project to grow. We have benefitted constantly from his scholarly advice and warm friendship.

The subtitle of the book "For Insurance and Finance" hints at the potential financial applications. The "real thing", be it either Swiss Francs, German Marks or Dutch Guilders, was provided to us through various forms of support. Both the Forschungsinstitut für Mathematik (ETH) and the Mathematisches Forschungsinstitut Oberwolfach provided opportunities for face–to–face meetings at critical stages.

PE recalls fondly the most stimulating visit he had, as part of his sabbatical in the autumn of 1996, at the School of ORIE at Cornell University. The splendid social and academic environment facilitated the successful conclusion of the book. CK worked on this project partly at ETH Zürich and partly at the Johannes Gutenberg University of Mainz. During most of the time she spent on the book in Zürich she was generously supported by the Schweizerische Lebensversicherungs– und Rentenanstalt, the Schweizerische Rückversicherungs–Gesellschaft (Swiss Re), Winterthur–Versicherungen, and the Union Rückversicherungs–Gesellschaft. Her sincere thanks go to these companies. TM remembers with nostalgia his time in New Zealand where he wrote his first parts of the book. The moral support of his colleagues at ISOR of the Victoria University of Wellington allowed him to concentrate fully on writing. He gratefully acknowledges the financial support of a New Zealand FRST Grant.

Last but not least, we thank our students! One of the great joys of being an academic is being able to transfer scientific knowledge to young people. Their questions, projects and interest made us feel we were on the right track. We hope that their eagerness to learn and enthusiasm to communicate is felt throughout the pages of this book.

March, 1997

PE, Zürich
CK, Mainz
TM, Groningen

Table of Contents

Reader Guidelines

The basic question each author should pose him/herself, preferably in the future tense before starting, is

Why have we written this book?

In our case the motivation came from many discussions we had with mathematicians, economists, engineers and physicists, mainly working in insurance companies, banks or other financial institutions. Often, these people had as students learnt the more classical theory of stochastics (probability theory, stochastic processes and statistics) and were interested in its applications to insurance and finance. In these discussions notions like *extremes, Pareto, divergent moments, leptokurtosis, tail events, Hill estimator* and many, many more would appear. Invariably, a question would follow, "Where can I read more on this?" An answer would usually involve a relatively long list of books and papers with instructions like "For this, look here, for that, perhaps you may find those papers useful, concerning the other, why not read ...". You see the point! After years of frustration concerning the non–existence of a relevant text we decided to write one ourselves. You now hold the fruit of our efforts: a book on the modelling of extremal events with special emphasis on applications to insurance and finance. The latter fields of application were mainly motivated by our joint research and teaching at the ETH where various chapters have been used for many years as *Capita Selecta* in the ETH programme on insurance mathematics. Parts of the book have also formed the basis for a Summer School of the Swiss Society of Actuaries (1994) and the Master's Programme in Insurance and Finance at ESSEC, Paris (1995). These trials have invariably led to an increase in the size of the book, due to questions like "Couldn't you include this or that?". Therefore, dear reader, you are holding a rather hefty volume. However, as in insurance and finance where everything is about "operational time" rather than real time, we hope that you will judge the "operational volume" of this book, i.e. measure its value not in physical weight but in "information" weight.

For whom have we written this book?

As already explained in the previous paragraph, in the first place for all those working in the broader financial industry faced with questions concerning extremal or rare events. We typically think of the actuarial student, the professional actuary or finance expert having this book on a corner of the desk ready for a quick freshen–up concerning a definition, technique, estimator or example when studying a particular problem involving extremal events. At the same time, most of the chapters may be used in teaching a special–topics course in insurance or mathematical finance. As such both undergraduate as well as graduate students interested in insurance and/or finance related subjects will find this text useful: the former because of its development of specific techniques in analysing extremal events, the latter because of its comprehensive review of recent research in the larger area of extreme value theory. The extensive list of references will serve both. The emphasis on economic applications does not imply that the intended readership is restricted to those working on such problems. Indeed, most of the material presented is of a much more general nature so that anyone with a keen interest in extreme value theory, say, or more generally interested in how classical probabilistic results change if the underlying assumptions allow for larger shocks in the system, will find useful material in it. However, the reader should have a good background in mathematics, including stochastics, to benefit fully. This brings us to the key question

What is this book about?

Clearly about extremal events! But what do we mean by this?

In the introduction to their book on *Outliers in Statistics*, Barnett and Lewis [51], the authors write: "When all is said and done, the major problem in outlier study remains the one that faced the very earliest research workers in the subject – what is an outlier?" One could safely repeat this sentence for our project, replacing *outlier* by *extremal event*. In their case, they provide methodology which allows for a possible description of outliers (influential observations) in statistical data. The same will be true for our book: we will mainly present those models and techniques that allow a precise mathematical description of certain notions of extremal events. The key question to what extent these theoretical notions correspond to specific events in practice is of a much more general (and indeed fundamental) nature, not just restricted to the methodology we present here. Having said that, we will not shy away from looking at data and presenting applied techniques designed for the user. It is all too easy for the academic to hide constantly behind the screen of theoretical research: the actuary or finance expert facing the real problems has to take important decisions based on the data at hand. We shall provide him or her with the necessary language, methods, techniques and examples which will allow for a more consistent handling of questions in the area of extremal events.

Whatever definition one takes, most will agree that Table 1, taken from *Sigma* [582] contains extremal events. When looked upon as single events, each of them exhibits some common features.

- *Their (financial) impact on the (re)insurance industry is considerable.* As stated in *Sigma* [582], at $US 150 billion, the total estimated losses in 1995 amounted to ten times the cost of insured losses – an exceptionally high amount, more than half of which was accounted for by the Kobe earthquake. Natural catastrophes alone caused insured losses of $US 12.4 billion, more than half of which were accounted for by four single disasters costing some billion dollars each; the Kobe earthquake, hurricane "Opal", a hailstorm in Texas and winter storms combined with floods in Northern Europe. Natural catastrophes also claimed 20 000 of the 28 000 fatalities in the year of the report.
- *They are difficult to predict a long time ahead.* It should be noted that 28 of the insurance losses reported in Table 1 are due to natural events and only 2 are caused by man–made disasters.
- If looked at within the larger context of all insurance claims, *they are rare events.*

Extremal events in insurance and finance have (from a mathematical point of view) the advantage that they are mostly quantifiable in units of money. However most such events have a non–quantifiable component which more and more economists are trying to take into account. Going back to the data presented in Table 1, extremal events may clearly correspond to individual (or indeed grouped) claims which by far exceed the capacity of a single insurance company; the insurance world's reaction to this problem is the creation of a reinsurance market. One does not however have to go to this grand scale. Even looking at standard claim data within a given company one is typically confronted with statements like "In this portfolio, 20% of the claims are

Losses	Date	Event	Country
16 000	08/24/92	Hurricane "Andrew"	USA
11 838	01/17/94	Northridge earthquake in California	USA
5 724	09/27/91	Tornado "Mireille"	Japan
4 931	01/25/90	Winterstorm "Daria"	Europe
4 749	09/15/89	Hurricane "Hugo"	P. Rico
4 528	10/17/89	Loma Prieta earthquake	USA
3 427	02/26/90	Winter storm "Vivian"	Europe
2 373	07/06/88	Explosion on "Piper Alpha" offshore oil rig	UK
2 282	01/17/95	Hanshin earthquake in Kobe	Japan
1 938	10/04/95	Hurricane "Opal"	USA
1 700	03/10/93	Blizzard over eastern coast	USA
1 600	09/11/92	Hurricane "Iniki"	USA
1 500	10/23/89	Explosion at Philips Petroleum	USA
1 453	09/03/79	Tornado "Frederic"	USA
1 422	09/18/74	Tornado "Fifi"	Honduras
1 320	09/12/88	Hurricane "Gilbert"	Jamaica
1 238	12/17/83	Snowstorms, frost	USA
1 236	10/20/91	Forest fire which spread to urban area	USA
1 224	04/02/74	Tornados in 14 states	USA
1 172	08/04/70	Tornado "Celia"	USA
1 168	04/25/73	Flooding caused by Mississippi in Midwest	USA
1 048	05/05/95	Wind, hail and floods	USA
1 005	01/02/76	Storms over northwestern Europe	Europe
950	08/17/83	Hurricane "Alicia"	USA
923	01/21/95	Storms and flooding in northern Europe	Europe
923	10/26/93	Forest fire which spread to urban area	USA
894	02/03/90	Tornado "Herta"	Europe
870	09/03/93	Typhoon "Yancy"	Japan
865	08/18/91	Hurricane "Bob"	USA
851	02/16/80	Floods in California and Arizona	USA

Table 1 *The* 30 *most costly insurance losses 1970–1995. Losses are in million $US at 1992 prices. For a precise definition of the notion of catastrophic claim in this context see* Sigma [582].

responsible for more than 80% of the total portfolio claim amount". This is an extremal event statement as we shall discuss more in detail in Section 8.2.

By stating above that the quantifiability of insurance claims in monetary units makes the mathematical modelling more tractable, we do not want to trivialise the enormous human suffering underlying such events. It is indeed striking that, when looking at the 30 worst catastrophes, in terms of fatalities over the same period in Table 2 only one event (the Kobe earthquake) figures

Fatalities	Date/start	Event	Country
300 000	11/14/70	Hurricane	Bangladesh
250 000	07/28/76	Earthquake in Tangshan	China
140 000	04/29/91	Hurricane "Gorky"	Bangladesh
60 000	05/31/70	Earthquake	Peru
50 000	06/21/90	Earthquake	Iran
25 000	12/07/88	Earthquake in Armenia	former USSR
25 000	09/16/78	Earthquake	Iran
23 000	11/13/85	Volcanic eruption "Nevado del Ruiz"	Columbia
22 000	02/04/76	Earthquake	Guatemala
15 000	09/19/85	Earthquake in Mexico City	Mexico
15 000	08/11/79	Damburst	India
15 000	09/01/78	Flood	India
10 800	10/31/71	Flood	India
10 000	05/25/85	Hurricane	Bangladesh
10 000	11/20/77	Tornado	India
9 500	09/30/93	Earthquake in Marashtra state	India
8 000	08/16/76	Earthquake on Mindanao	Philippines
6 304	11/05/91	Typhoons "Thelma" and "Uring"	Philippines
6 000	01/17/95	Great Hanshin earthquake in Kobe	Japan
5 300	12/28/74	Earthquake	Pakistan
5 000	04/10/72	Earthquake in Fars	Iran
5 000	12/23/72	Earthquake in Managua	Nicaragua
5 000	06/30/76	Earthquake in Westirian	Indonesia
4 800	11/23/80	Earthquake	Italy
4 500	10/10/80	Earthquake	Algeria
4 000	02/15/72	Storm; snow	Iran
4 000	11/24/76	Earthquake in Van	Turkey
3 800	09/08/92	Floods in Punjab	Pakistan
3 200	04/16/78	Tornado	Reunion
3 000	08/01/88	Flood	Bangladesh

Table 2 *The* 30 *worst catastrophes in terms of fatalities 1970–1995, taken from* Sigma [582].

on both lists. Also, Table 1 mainly involves industrialised nations, whereas Table 2 primarily concerns Third World countries.

Within the finance context, extremal events present themselves spectacularly whenever major stock market crashes like the one in 1987 occur. Or recent casualties within the realm of derivatives such as the collapse of Barings Bank, the losses of the Metallgesellschaft, Proctor & Gamble, Kashima Oil, Orange County, or Sumitomo. The full analysis of events of such grand scale again goes well beyond the prime content of this book, and any claim that the managements of financial institutions will find the means of avoiding such disasters in our book would be absurd. In most of the above cases the setting–up (both in structure as well as people) of a well–functioning risk management and control system was called for. On a much smaller scale however, questions related to the estimation of Profit–and–Loss distributions or Value–at–Risk measures have to be answered with techniques presented in some of the following chapters. Though not providing a risk manager in a bank with the final product he or she can use for monitoring financial risk on a global scale, we will provide that manager with stochastic methodology needed for the construction of various components of such a global tool.

Events that concern both branches are to be found in credit insurance, mortgage–backed securities, the recent developments around catastrophic insurance futures or indeed more generally the problem of securitisation of risk. In all of these areas, there is an increasing need for modelling of events that cause larger shocks to the underlying financial system. As an example of how knowledge of basic underlying stochastic methodology may be used, consider the problem of potential increases in both the frequency as well as (inflation–adjusted) sizes of well–defined catastrophic claims. A simple, but at the same time intuitively clear method, is to plot the successive records in the data. In Figure 3 we have plotted such records for yearly frequency and insured loss data both for man–made as well as natural catastrophes over the period 1970–1995. For a precise definition of the underlying data see *Sigma* [582]. If the data were independent and identically distributed (iid), what sort of picture would one expect? An answer to this question is given in Section 6.2.4. Intuition tells us that successive records for iid data should become more and more rare as time goes by: it becomes more and more difficult to exceed all past observations.

By now, the reader should have some idea of the type of problems we are interested in. The next step would be to dig a bit deeper and explain which mathematical models we plan to discuss and what methodology we want to introduce. Before doing so, some general comments on the format of the chapters is called for.

How is new material to be presented, and indeed how should one read this book?

As stated before, we typically think of an actuary, a finance expert or a student, working on a problem in which a technique related to rare though potentially influential events is to be used. Take as an example a finance expert in the area of risk management, concerned with Value–at–Risk estimation for a specific portfolio. The Value–at–Risk may for instance be defined as the left 5% quantile of the portfolio Profit–Loss distribution. The latter is

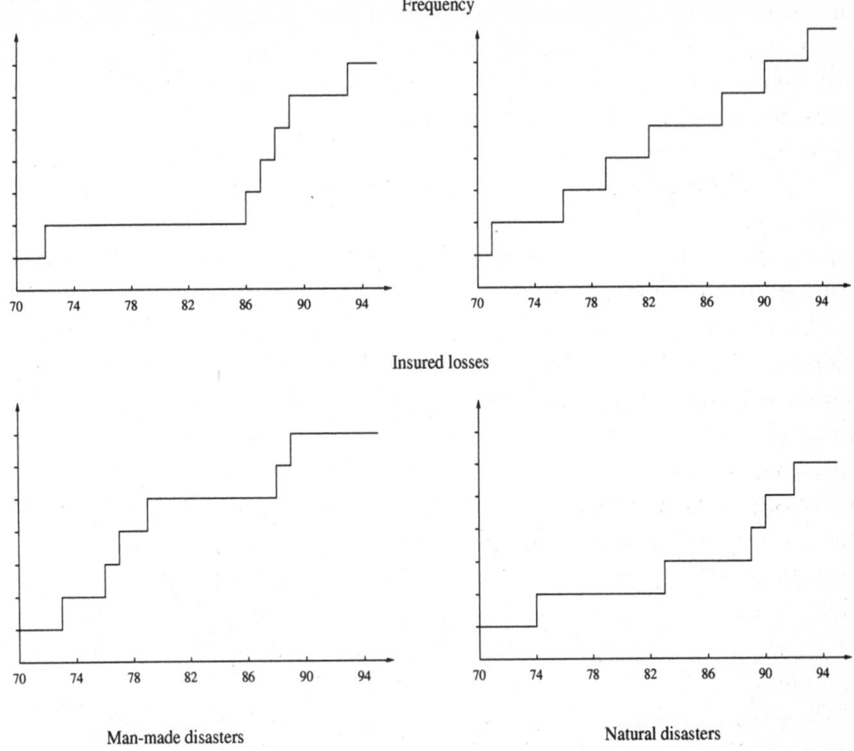

Figure 3 *Record years of catastrophic insurance claims 1970–1995: frequency and insured losses (in 1992 prices) both for man–made and natural disasters, taken from* Sigma [582]. *The graphs show a jump for each year in which a new record occurred. For instance, one observes* 8 *records for the frequency of natural disasters and* 6 *records for the insured losses.*

typically skewed with heavy tails both at left (losses) and right (gains); see Figure 4. So we end up with questions that concern finding relevant classes of Profit–Loss distributions, as well as statistical fitting and tail estimation. It is exactly for this type of problems that our book will provide the necessary background material or indeed specific techniques.

A typical chapter will introduce the new methodology in a rather intuitive (though always mathematically correct) way, stressing more the understanding of new techniques rather than following the usual theorem–proof path. We do, however, usually state theorems in their most general form, provided that this form is practically relevant. Proofs are usually given either as a sketch of the main ideas, or as a way of showing how new methods can be used in technical calculations. Sometimes we use them to highlight the instances in

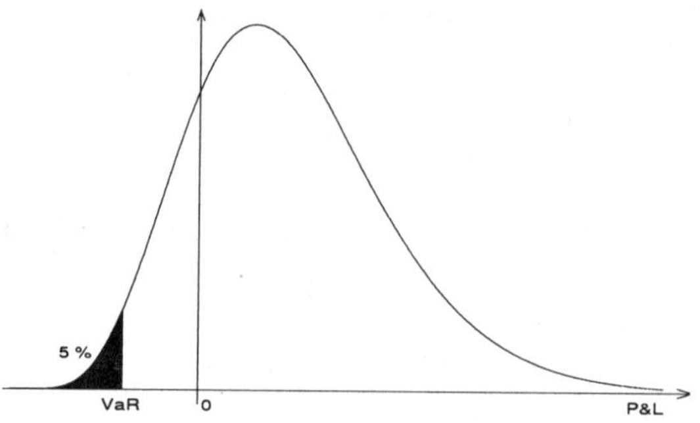

Figure 4 *Profit–Loss (P&L) density function with related Value–at–Risk* (VaR).

the argument where classical techniques break down (explaining why), and how arguments relating to extremal events have to be handled. Each section ends with Notes and Comments giving the reader further guidance towards relevant literature on related topics. Various examples, tables and graphs have been included for illustrative purposes, but at the same time for reasons of making the text (at least optically) easier to digest. Few readers will want to read the text from cover to cover; the ideal way would be to read those sections that are necessary for the problems at hand.

Which basic models in insurance and finance do we consider?

Our main motivation comes from insurance, and consequently a bias towards problems (and topics) from that field of applications is certainly to be found in the text. On the other hand, except for Chapters 1 and 8, all chapters are aimed at a much larger audience than workers in insurance.

Mathematical modelling in finance and insurance can be traced back many centuries. For our purposes, however, history starts around the beginning of the 20th century. In 1900, Louis Bachelier showed in his thesis [35] that Brownian motion lies at the heart of any model for asset returns. Around the same time, Filip Lundberg introduced in his thesis [431] the collective risk model for insurance claim data. Lundberg showed that the homogeneous Poisson process, after a suitable time transformation, is the key model for insurance liability data. Of course, both Brownian motion and the homogeneous Poisson process are the prime examples of the wider class of stochastic processes with stationary and independent increments. We shall treat both examples more in detail and provide techniques concerning extremal events

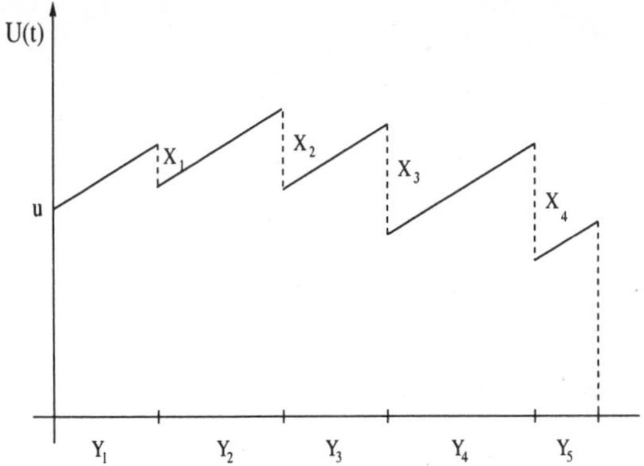

Figure 5 *One realisation of the risk process* $(U(t))$.

useful in either case. Embedded in these processes is the structure of a random walk, i.e. the sum of iid random variables. So a more profound study of extremal events in the iid case is called for. This forms the basis for classical statistical theory and classical extreme value theory. More general models can often be transformed to the iid case; this allows us for instance to analyse general (linear) time series.

In Chapter 1 we study *the* classical model for insurance risk,

$$U(t) = u + ct - S(t)\,, \quad S(t) = \sum_{i=1}^{N(t)} X_i\,, \quad t \geq 0\,, \tag{1}$$

where u stands for initial capital, c for loaded premium rate and the total claim amount $S(t)$ consists of a random sum of iid claims X_i. Here $N(t)$ stands for the number of claims until time t. It is common to simplify this model further by assuming (as Lundberg did) that $(N(t))$ is a homogeneous Poisson process, independent of (X_i). For a realisation of $(U(t))$ see Figure 5. The process $(S(t))$ and its ramifications have been recognised as a very tractable (and reasonably realistic) model and a vast amount of literature in risk theory has been devoted to it. An important question concerns the influence of individual extremal events, i.e. large claims, on the global behaviour of $(U(t))$. In Chapter 1 the latter question will be answered via a detailed analysis of ruin probabilities associated with the process $(U(t))$. Under a condition of "small claims" (see for instance Theorem 1.2.2), the traditional Cramér–Lundberg estimate for the ruin probability yields bounds which are exponential in the initial capital u. However, in reality claims are mostly modelled by heavy–tailed distributions like Pareto, loggamma, lognormal, or heavy–tailed Weibull. See for instance Figure 6, where the left–hand picture shows those

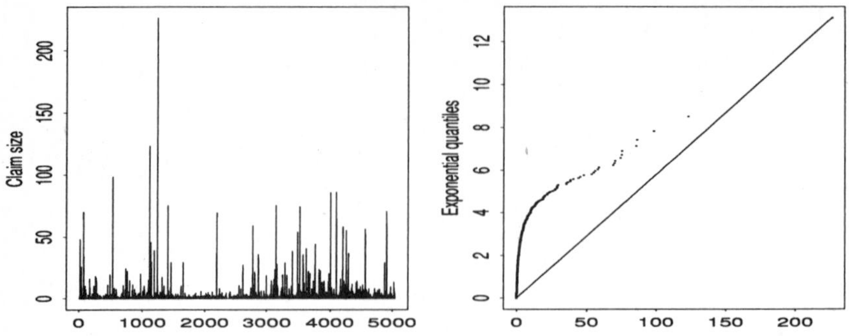

Figure 6 *Fire insurance data and corresponding exponential QQ–plot. The claim sizes are in* 1 000 *SFr.*

claim sizes of a portfolio of fire insurance that are larger than a given franchise (1 000 SFr). In the right–hand picture one finds a so–called QQ–plot of the data, measuring the fit achieved by an exponential distribution function (df). The curvature (i.e. departure from a straight line) present in the QQ–plot implies that the tails of the df of the fire data are much heavier than exponential. For a detailed discussion of these and related plotting techniques see Section 6.2.1.

Chapter 1 mainly deals with the mathematical analysis of ruin estimation under precise heavy–tailed model assumptions. Whereas Poisson processes form the basic building block underlying insurance liability processes, within finance the basic models can be transformed back to simple random walks. This is certainly true for the Cox–Ross–Rubinstein and the Black–Scholes models; see for instance Föllmer and Schweizer [242] for a nice account of the economic whys and wherefores concerning these processes.

The skeleton model in finance, corresponding to the homogeneous Poisson process in insurance, is without doubt geometric Brownian motion, i.e. the stochastic process

$$\exp\left\{\left(c-\sigma^2/2\right)t+\sigma B_t\right\}\,,\quad t\geq 0\,,$$

with (B_t) Brownian motion. Here c stands for the mean rate of return and σ for the *volatility* (riskiness). It is the solution to an Itô stochastic differential equation and provides the basis of the Black–Scholes option pricing formula and many other parts of financial theory. One of the attractions of the above model is its simplicity; indeed, as a consequence it follows that logarithmic returns are iid, normally distributed. At this point, as in insurance,

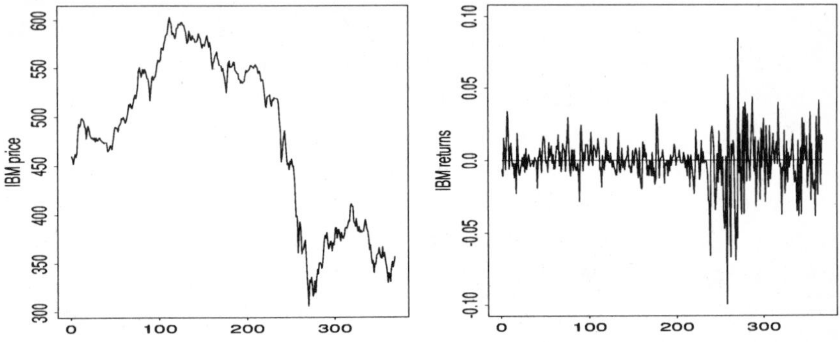

Figure 7 *Daily IBM common stock closing prices: May 17, 1961–Nov. 2, 1962.*

one should ask the question "what do the data tell us?" An answer to this question would, and indeed does, fill a book. A summary answer, fit for this introduction, is that, on the whole, geometric Brownian motion is a good first model. If however one looks more closely at data, one often finds situations as in Figure 7. In it we observe a clear change in volatility possibly triggered by some extreme returns. A multitude of models for such phenomena has been introduced including α–stable processes (as heavy–tailed alternatives to Brownian motion), and heavy–tailed time series models, for instance ARCH and GARCH models. The basic characteristics of such models will be discussed in later chapters, for instance Chapter 7, Sections 8.4 and 8.8.

From a naive point of view both fields, insurance and finance, have in common that we can observe certain financial or actuarial phenomena such as prices, exchange rates, interest rates, insurance claims, claim arrival times etc. We will later classify these observations or data, but we first want to consider them simply as a *time series* or a *continuous–time stochastic process*, i.e. we assign to each instant of time t a real random variable X_t. One of our usual requirements is that (X_t) itself or a transformed version of it (for instance the first–order differences or the log–differences) forms a *stationary* process (strictly stationary or stationary in the wide sense). In particular, this includes the important case of iid observations which provides the basis for classical fluctuation and extreme value theory, as well as for statistical estimation.

In Chapter 2 we give a general asymptotic theory for sums of iid random variables (random walk), and in Sections 2.5.1 and 2.5.3 we especially emphasize random sums like $S(t)$ in (1). This theory includes classical results such as the central limit theorem, the law of large numbers, the law of the iterated logarithm, the functional central limit theorem and their ramifications and refinements. They are important building blocks for the asymptotic theory which is a basic tool of this book. We also introduce two important classes of continuous–time stochastic processes: *Brownian motion* and *α–stable motion.* Both are continuous–time limits of appropriate partial sum processes. As such, they can be understood as *random walks in continuous time.*

After having recalled the basic partial sum theory, in Chapters 3 and 4 we turn to the analogous theory for partial maxima and order statistics. These chapters are conceived in such a way that the reader can compare and contrast results for maxima with similar ones for sums. Special attention will also be given to those questions where both theories complement one another. As a start we first present extreme value theory for iid sequences, thereby paving the way for similar results in the case of stationary sequences (X_t). In particular, we will describe and study *maxima, minima, records, record times, excesses over thresholds, the frequency of exceedances* and many other features of such sequences which are related to their extremal behaviour.

Though most of the material of this book can be found scattered over various textbooks and/or research papers, some material is presented here for the first time in textbook form. One such example is the study of linear processes

$$X_t = \sum_{j=-\infty}^{\infty} \psi_j \, Z_{t-j} \,, \quad t \in \mathbb{Z} \,, \tag{2}$$

for iid innovations Z_t with infinite variance. Over the past 20 years methods have been developed to deal with these objects, and Chapter 7 contains a survey of the relevant results. The proofs are mostly very technical and accessible only to the specialist. This is the reason why we omitted them, but we give a very detailed reference list where the interested reader will find a wealth of extra reading material. The extreme value theory for the process (2) is dealt with in Section 5.5 under different assumptions on the innovations which include the heavy–tailed case. The extremes of more general stationary sequences are treated in Sections 4.4 and 5.3.2.

In summary, the stochastic processes of main interest can be roughly classified as follows:

- Discrete time sequences $(X_t)_{t\in\mathbb{Z}}$, in particular stationary and iid sequences as models for log–returns of prices, for exchange rates, for individual claim sizes, for inter–arrival times of claims.
- Random walk models, i.e. sums of the X_t or continuous–time models such as Brownian motion $(B_t)_{t\geq 0}$ and α–stable motion, as models for the total claim amount, aggregated returns or building blocks for price processes etc.

- Random sum processes $(S(t))_{t\geq 0}$ (see (1)) as models for the total claim amount in an insurance portfolio.
- The risk process $(U(t))_{t\geq 0}$; see (1).
- Poisson processes and Poisson random measures as means to describe rare events in space and time. The homogeneous Poisson process also serves as a basic model for claim arrival times.

After having introduced our basic models we may ask

Which distributions and stochastic processes typically describe extremal events in these models?

When we are interested in the extremal behaviour of the models described above we have to ask *how extremal events occur*. This means we have to find appropriate mathematical methods in order to explain events that occur with relatively small probability but have a significant influence on the behaviour of the whole model. For example, we may ask about the inter–relation between the iid individual claim sizes X_i and the total claim amount $S(t)$ in (1). In particular, under what assumptions and how do the values of the largest claims determine the value $S(t)$? A natural class of large claim distributions is given by the ***subexponential distributions***. They are extensively treated in Chapter 1 and Appendix A3.2. Their defining property is:

$$\lim_{x\to\infty} \frac{P\left(X_1+\cdots+X_n>x\right)}{P\left(\max\left(X_1,\ldots,X_n\right)>x\right)}=1$$

for every $n\geq 2$. Thus the tails of the distribution of the sum and of the maximum of the first n claims are asymptotically of the same order. This clearly indicates the strong influence of the largest claim on the total claim amount.

Whereas in insurance heavy–tailed (i.e. subexponential) distributions are well recognised as standard models for individual claim sizes, the situation in finance is much more complicated. The latter is partly due to the fact that one often works with near continuous–time observed (so–called high–density) data. At the same time, marginal distributions are heavy–tailed and return data exhibit clustering of extremes and long–range dependence. There is no universally accepted nor indeed easy model that explains all these phenomena. In Section 2.4, for instance, we introduce α–stable motion $(0<\alpha<2)$ as a limit of partial sum processes with infinite variance. For a realisation of a 1.5–stable motion see Figure 8, where also a plot of Brownian motion is given. The α–stable processes form fundamental building blocks within more

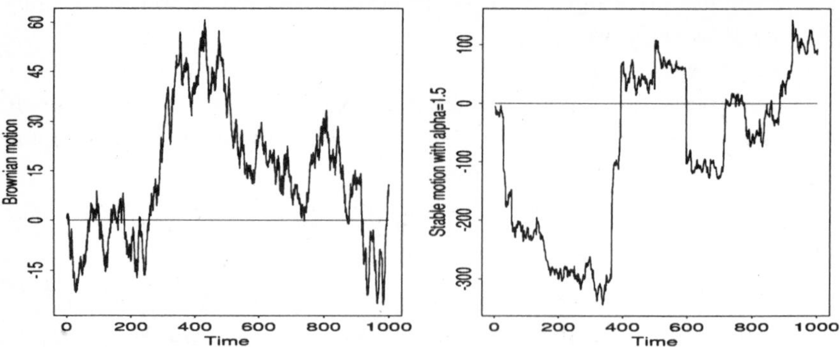

Figure 8 *Paths of Brownian motion and of a 1.5–stable motion.*

general model constructions and anyone interested in rare events ought to know them.

Many distributions of interest in extreme value theory turn out to be closely related to α–stable distributions. The α–stable laws are the *only* possible limit distributions for properly normalised and centred sums of iid random variables. The case $\alpha = 2$ corresponds to the normal limit; we know that a finite second moment is sufficient for the application of the central limit theorem. The case $\alpha < 2$ arises for infinite–variance iid summands. The infinite variance property has not prevented practitioners in insurance from working with such models. A quick simulation of a scenario of the total claim amount under these heavy–tailed assumptions is helpful for making a decision about the insurability of such claims. In that sense, α–stable or other heavy–tailed distributions often can be used as a worst–case scenario.

Extreme value theory is one of the main objectives of this book, and so when talking about relevant distributions in that context, we have to mention the extreme value distributions, the Gumbel law Λ, the Fréchet law Φ_α and the Weibull law Ψ_α. They are the *only* possible limit distributions for maxima of properly normalised and centred iid random variables. As such they essentially play the same role as the α–stable distributions for sums of iid random variables. Sections 3.2 and 3.3 are devoted to their study. Furthermore, in Sections 4.1 and 4.2 the theory is extended from maxima to upper order statistics.

There are of course many more distributions of interest which are somehow related to extremes. Above we have mentioned the essential ones and the way they enter applied modelling in the presence of extremal events. We will provide lists and examples of particular distributions, densities and tails in the corresponding sections.

We have already encountered the Poisson distribution in the context of risk theory. Both the Poisson distribution as well as the Poisson process are key tools in the analysis of extremal events, as we shall see on various occasions.

In sum, the following classes of distributions are of main importance in the context of extremal events:

- the subexponential distributions as realistic models for heavy–tailed random variables,
- the α–stable distributions for $\alpha < 2$ as the limit laws for sums of infinite–variance iid random variables,
- the Fréchet, the Weibull, and the Gumbel distributions, as limit laws for maxima of iid random variables,
- the normal distribution as limit law for sums of iid, finite–variance random variables,
- the Poisson distribution as limit law of binomial distributions which represent a counting measure of rare events.

As important stochastic processes we would like to mention:

- Poisson processes,
- α–stable processes $(0 < \alpha < 2)$ and Brownian motion,
- more general processes using the above as input.

What are the main probabilistic tools?

Besides standard introductory probability theory and the theory of stochastic processes, many results presented will be based upon a deeper understanding of relevant asymptotic methods. One of the main tools falling into the latter category is the theory of *weak convergence of probability distributions*, both on the real line and in certain function spaces. A short summary of the methodological background is given in Appendices A1 and A2. Abstract weak–convergence techniques are needed in order to prove that suitable partial sum processes converge towards Brownian motion or α–stable processes. The strength of this process convergence is illustrated by various examples in Chapter 2. This theory allows us to characterise those distributions and processes that may arise as useful stochastic models for certain insurance and finance data.

The analysis of extremes further requires the framework of *point processes*. The general theory for the latter is rather involved, though the benefit for applications, especially those towards extremal event modelling, is considerable once the whole machinery has been set up. In Chapter 5 we give an ample number of examples of this. We have tried hard to avoid unnecessary technical details. Point process techniques are by now an unavoidable tool in modern extreme value theory, and the results are convincing and give a deep insight into the structure and occurrence of extremes.

The basic idea of weak convergence of point processes is analogous to Poisson's classical limit theorem. Weak limits of the point processes under consideration (as analogues of binomial random variables) are quite often (general) *Poisson processes* or *Poisson random measures* (as analogues to the Poisson distribution). These notions will be made precise in Sections 5.1 and 5.2.

Limit theory for sums, maxima or point processes is closely related to the power law behaviour of tails, of normalising constants, of characteristic functions in the neighbourhood of the origin etc. Exact power laws mainly occur in the very limit, but if, for instance, we discuss domains of attraction of stable laws or of extreme value distributions, power laws do not appear in "pure" form, but slightly disturbed by *slowly varying functions.* A power law times a slowly varying function is called *regularly varying.* The theory of regularly varying functions and their generalisations and extensions are important analytical tools throughout this book. Their basic properties are given in Appendix A3.1.

In Chapter 7 we provide an analysis of time series with heavy tails. A lean introduction to the relevant notions of time series analysis is given, but the reader without the necessary background will certainly have to consult some of the standard textbooks. The main objects in Chapter 7 are linear processes with heavy–tailed innovations. That chapter and Section 5.5, where extreme value theory for linear processes is treated, give quite a complete picture about this kind of process with heavy–tailed innovations.

To sum up, besides the basic classical techniques and facts from probability theory our main probabilistic tools are the following:

- weak convergence of distributions of random variables such as sums, random sums and maxima of random variables,
- weak convergence of sum processes and maximum processes to their limits in appropriate function spaces,
- point processes for describing the random distribution of points in space and time with applications to extreme value theory.

What are the appropriate statistical tools?

Insurers and bankers are interested in assessing, pricing and hedging their risks. They calculate premiums and price financial instruments including coverage against major risks. The probable maximal loss of a risk or investment portfolio is determined by extremal events. The problem we want to solve may therefore be described in its broadest terms as how to make statistical inference about the extreme values in a population or a random process. Quantities like the following may serve as indicators:

- the distribution of the annual extremes,
- the distribution of the largest values in a portfolio,
- the return period of some rare event,
- the frequency of extremal events,
- the mean excess over a given threshold,
- the distribution of the excesses,
- the time development of records.

Every piece of knowledge we can acquire about these quantities from our data helps us to predict extremal events, and hence potentially protect ourselves against adverse effects caused by them. In Chapter 6 we present a collection of methods for statistical inference based on extreme values in a sample.

Some simple exploratory data–analytical methods can be extremely useful at a descriptive stage. An example has been given in Figure 3 where a plot of the records manifests a trend in the frequency of natural disasters. Methods based on probability plots, estimated return periods or empirical mean excess functions provide first information about the extremes of a data set.

For iid data the classical extreme value distributions, the Gumbel Λ, the Fréchet Φ_α and the Weibull distribution Ψ_α, are the obvious candidates to model the largest values of a sample. We review parameter estimation methods for extreme value distributions, investigate their asymptotic properties and discuss their different merits and weaknesses. Extensions to upper order statistics of a sample are also treated.

Our interest focusses on extremal events of the form $\{X > x\}$ for some random variable X and large x, i.e. we want to estimate tails in their far regions and, also, high quantiles. We survey various tail and quantile estimators which are only to be found rather scattered through the literature. We also describe a variety of statistical methods based on upper order statistics and on so–called threshold methods.

Before you start!

We think it a bad idea for a methodological book like this one to distinguish too strongly between those readers working in insurance and those working more in finance. It would be especially bad to do so at the present time, when experts from both fields are increasingly collaborating either on questions of related interest (risk management say) or on new product development involving both insurance and finance features (for instance index–linked life insurance, catastrophe futures and options, securitisation of insurance risk). It is important for both sides to learn more about each other's basic models and tools. We therefore hope that a broad spectrum of readers will find various interesting facts in this book.

We start with a somewhat specialised chapter on *risk theory*; however, the basic model treated in it reappears in many fields of applications as for instance queueing theory, dam theory, inventory systems, shock models etc. Its main purpose is that it provides an ideal vehicle for the introduction of the important class of *subexponential distributions*. At the same time, the liability model that is fundamental to insurance is also discussed. From Chapter 2 onwards, standard theory is first of all reviewed (Chapter 2 on *sums*) before the core material on *probabilistic modelling of extremes* together with their *statistical analysis* are treated in Chapters 3–6. A mathematically more demanding, though with respect to applications rewarding, excursion to *point process methods* is presented in Chapter 5. Typically you would start with Chapters 2 and 3 and embark first on the *statistical methods* in Chapter 6 before coming back for a more detailed analysis of some of the techniques from Chapter 5. Chapter 7 treats the more specialised topic of *heavy–tailed time series* models. It fits into the framework of *extremes for dependent data* which earlier appears in Sections 4.4, 5.3 and 5.5. Together, Chapters 1 through 7 give a sound introduction to one–dimensional extremal event modelling. Having this methodology at our finger tips, we may start using it for understanding and solving various related problems. This is exactly what is presented in Chapter 8 on *special topics*. In it, we have brought together various problems, all of which use the foregoing theory in some form or another. Take for instance Section 8.2 where a *large claim index* is discussed, describing mathematically the 20–80 rule of thumb used by actuaries to specify the dangerousness of certain portfolios. Chapter 8 is also used to discuss briefly those extensions of the theory which should come next, such as for instance Sections 8.1 (on the *extremal index*), 8.4 (on *perpetuities and ARCH processes*) and 8.7 (on *reinsurance treaties*). This chapter could have grown considerably; somewhere however we had to stop. Therefore, most of the sections presented reflect somehow our own teaching, research and/or consulting experience. We have based an extreme value theory course for mathematics students specialising in actuarial mathematics on most of the material presented in Chapters 3 to 6, together with some sections in Chapter 8. Naturally, the Appendix is there for reviewing those tools from mathematics used most often throughout the text and which may not belong to everybody's basic toolkit.

Epilogue

You are now ready to start: good luck!

P.E., C.K. and T.M.

1

Risk Theory

For most of the problems treated in insurance mathematics, risk theory still provides the quintessential mathematical basis. The present chapter will serve a similar purpose for the rest of this book. The basic risk theory models will be introduced, stressing the instances where a division between small and large claims is relevant. Nowadays, there is a multitude of textbooks available treating risk theory at various mathematical levels. Consequently, our treatment will not be encyclopaedic, but will focus more on those aspects of the theory where we feel that, for modelling extremal events, the existing literature needs complementing. Readers with a background in finance rather than insurance may use this chapter as a first introduction to the stochastic modelling of claim processes.

After the introduction of the basic risk model in Section 1.1, we derive in Section 1.2 the classical Cramér–Lundberg estimate for ruin probabilities in the infinite horizon case based on a small claim condition. Using the Cramér–Lundberg approach, a first estimation of asymptotic ruin probabilities in the case of regularly varying claim size distributions is obtained in Section 1.3.1. The natural generalisation to subexponentially distributed claim sizes is given in Sections 1.3.2, 1.3.3 and further discussed in Section 1.4. The latter section, together with Appendix A3, contains the basic results on regular variation and subexponentiality needed further in the text.

1.1 The Ruin Problem

The basic insurance risk model goes back to the early work by Filip Lundberg [431] who in his famous Uppsala thesis of 1903 laid the foundation of actuarial risk theory. Lundberg realised that Poisson processes lie at the heart of non–life insurance models. Via a suitable time transformation (so–called operational time) he was able to restrict his analysis to the homogeneous Poisson process. This "discovery" is similar to the recognition by Bachelier in 1900 that Brownian motion is the key building block for financial models. It was then left to Harald Cramér and his Stockholm School to incorporate Lundberg's ideas into the emerging theory of stochastic processes. In doing so, Cramér contributed considerably to laying the foundation of both non–life insurance mathematics as well as probability theory. The basic model coming out of these first contributions, referred to in the sequel as *the Cramér–Lundberg model*, has the following structure:

Definition 1.1.1 (The Cramér–Lundberg model, the renewal model)
The Cramér–Lundberg model *is given by conditions* (*a*)–(*e*)*:*

(*a*) *The* claim size process:
the claim sizes $(X_k)_{k\in\mathbb{N}}$ *are positive iid rvs having common non–lattice df* F, *finite mean* $\mu = EX_1$, *and variance* $\sigma^2 = \mathrm{var}(X_1) \le \infty$.

(*b*) *The* claim times:
the claims occur at the random instants of time

$$0 < T_1 < T_2 < \cdots \quad \text{a.s.}$$

(*c*) *The* claim arrival process:
the number of claims in the interval $[0,t]$ *is denoted by*

$$N(t) = \sup\{n \ge 1 : T_n \le t\}\,, \quad t \ge 0\,,$$

where, by convention, $\sup\emptyset = 0$.

(*d*) *The* inter–arrival times

$$Y_1 = T_1\,, \quad Y_k = T_k - T_{k-1}\,, \quad k = 2,3,\ldots, \tag{1.1}$$

are iid exponentially distributed with finite mean $EY_1 = 1/\lambda$.

(*e*) *The sequences* (X_k) *and* (Y_k) *are independent of each other.*

The renewal model *is given by* (*a*)–(*c*), (*e*) *and*

(*d′*) *the inter–arrival times* Y_k *given in* (1.1) *are iid with finite mean* $EY_1 = 1/\lambda$. □

Remarks. 1) A consequence of the above definition is that $(N(t))$ is a homogeneous Poisson process with intensity $\lambda > 0$ (for a definition we refer to Example 2.5.2). Hence

$$P(N(t) = k) = e^{-\lambda t} \frac{(\lambda t)^k}{k!}, \quad k = 0, 1, 2, \ldots .$$

2) The renewal model is a slight generalisation of the Cramér–Lundberg model which allows for renewal counting processes (see Section 2.5.2). The latter are more general than the Poisson process for the claim arrivals. □

The *total claim amount process* $(S(t))_{t \geq 0}$ of the underlying portfolio is defined as

$$S(t) = \begin{cases} \sum_{i=1}^{N(t)} X_i\,, & N(t) > 0\,, \\ 0\,, & N(t) = 0\,. \end{cases} \tag{1.2}$$

The general theory of random sums will be discussed in Section 2.5. It is clear that in the important case of the Cramér–Lundberg model more detailed information about $(S(t))$ can be obtained. We shall henceforth treat this case as a basic example on which newly introduced methodology can be tested. An important quantity in this context is the *total claim amount distribution* (or *aggregate claim (size) distribution*)

$$G_t(x) = P(S(t) \leq x) = \sum_{n=0}^{\infty} e^{-\lambda t} \frac{(\lambda t)^n}{n!} F^{n*}(x)\,, \qquad x \geq 0\,, \quad t \geq 0\,, \tag{1.3}$$

where $F^{n*}(x) = P(\sum_{i=1}^{n} X_i \leq x)$ is the n–fold convolution of F. Throughout the text, for a general df H on $(-\infty, \infty)$,

$$H^{0*}(x) = \begin{cases} 1 & x \geq 0\,, \\ 0 & x < 0\,. \end{cases}$$

The resulting *risk process* $(U(t))_{t \geq 0}$ is now defined as

$$U(t) = u + ct - S(t)\,, \qquad t \geq 0\,. \tag{1.4}$$

In (1.4), $u \geq 0$ denotes the *initial capital* and $c > 0$ stands for the *premium income rate.* The choice of c is discussed below; see (1.7). For an explanation on why in this case a deterministic (linear) income rate makes sense from an actuarial point of view; see for instance Bühlmann [98]. In Figure 1.1.2 some realisations of $(U(t))$ are given in the case of exponentially distributed claim sizes.

In the classical Cramér–Lundberg set–up, the following quantities are relevant for various insurance–related problems.

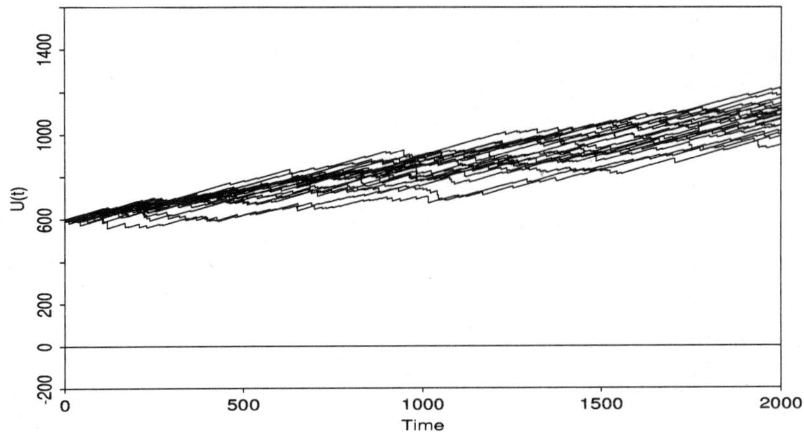

Figure 1.1.2 *Some realisations of $(U(t))$ for exponential claim sizes.*

Definition 1.1.3 (Ruin)
The ruin probability in finite time (*or* with finite horizon) :

$$\psi(u,T) = P(U(t) < 0 \quad \textit{for some } t \le T)\,, \quad 0 < T < \infty\,, \quad u \ge 0\,.$$

The ruin probability in infinite time (*or* with infinite horizon) :

$$\psi(u) = \psi(u,\infty)\,, \quad u \ge 0\,.$$

The ruin times*:*

$$\tau(T) = \inf\{t : 0 \le t \le T\,,\ U(t) < 0\}\,, \qquad 0 < T \le \infty\,,$$

where, by convention, $\inf \emptyset = \infty$. *We usually write* $\tau = \tau(\infty)$ *for the* ruin time with infinite horizon. □

The following result is elementary.

Lemma 1.1.4 *For the renewal model,*

$$EU(t) = u + ct - \mu EN(t)\,. \tag{1.5}$$

For the Cramér–Lundberg model,

$$EU(t) = u + ct - \lambda\mu t\,. \tag{1.6}$$

Proof. Since $EU(t) = u + ct - ES(t)$, and

$$\begin{aligned} ES(t) &= \sum_{k=0}^{\infty} E(S(t) \mid N(t) = k)\, P(N(t) = k) \\ &= \sum_{k=1}^{\infty} E\left(\sum_{i=1}^{N(t)} X_i \middle| N(t) = k \right) P(N(t) = k) \\ &= \sum_{k=1}^{\infty} E\left(\sum_{i=1}^{k} X_i \right) P(N(t) = k) \\ &= \mu \sum_{k=1}^{\infty} k\, P(N(t) = k) \\ &= \mu EN(t)\,, \end{aligned}$$

relation (1.5) follows. Because $EN(t) = \lambda t$ for the homogeneous Poisson process, (1.6) follows immediately. □

This elementary lemma yields a first guess of the premium rate c in (1.1). The latter is a major problem in insurance to which, at least for more general models, a vast amount of literature has been devoted; see for instance Goovaerts, De Vylder and Haezendonck [279]. We shall restrict our discussion to the above models. The determination of a suitable insurance premium rate obviously depends on the criteria used in order to define "suitable". It all depends on the measure of solvency we want to optimise over a given time period. The obvious (but by no means the only) measures available to us are the ruin probabilities $\psi(u, T)$ for $T \leq \infty$. The premium rate c should be chosen so that a small $\psi(u, T)$ results for given u and T. A first step in this direction would be to require that $\psi(u) < 1$, for all $u \geq 0$. However, since $\psi(u) = P(\tau < \infty)$, this is equivalent to $P(\tau = \infty) > 0$: the company is given a strictly positive probability of infinitely long survival. Clearly, adjustments to this strategy have to be made before real premiums can be cashed. Anyhow, to set the stage, the above criterion is a useful one.

It follows immediately from (1.5) and Proposition 2.5.12 that in the renewal model, for $t \to \infty$,

$$\begin{aligned} EU(t) &= u + (c - \lambda\mu)\, t\, (1 + o(1)) \\ &= u + \left(\frac{c}{\lambda\mu} - 1 \right) \lambda\mu t\, (1 + o(1))\,. \end{aligned}$$

Therefore, $EU(t)/t \to c - \lambda\mu$, and an obvious condition towards solvency is $c - \lambda\mu > 0$, implying that $(U(t))$ has a positive drift for large t. This leads to the basic *net profit condition* in the renewal model:

$$\rho = \frac{c}{\lambda\mu} - 1 > 0\,. \tag{1.7}$$

The constant ρ is called the *safety loading*, which can be interpreted as a *risk premium rate*; indeed, the premium income over the period $[0,t]$ equals $ct = (1+\rho)\lambda\mu t$.

By definition of the risk process, ruin can occur only at the claim times T_i, hence for $u \geq 0$,

$$\begin{aligned}
\psi(u) &= P\left(u + ct - S(t) < 0 \quad \text{for some } t \geq 0\right) \\
&= P\left(u + cT_n - S\left(T_n\right) < 0 \quad \text{for some } n \geq 1\right) \\
&= P\left(u + \sum_{k=1}^{n} \left(cY_k - X_k\right) < 0 \quad \text{for some } n \geq 1\right) \\
&= P\left(\sup_{n\geq 1} \sum_{k=1}^{n} \left(X_k - cY_k\right) > u\right).
\end{aligned}$$

Therefore, $\psi(u) < 1$ is equivalent to the condition

$$1 - \psi(u) = P\left(\sup_{n\geq 1} \sum_{k=1}^{n} \left(X_k - cY_k\right) \leq u\right) > 0\,, \qquad u \geq 0\,. \tag{1.8}$$

From (1.8) it follows that, in the renewal model, the determination of the non–ruin probability $1-\psi(u)$ is reduced to the study of the df of the ultimate maximum of a random walk. Indeed, consider the iid sequence

$$Z_k = X_k - cY_k\,, \quad k \geq 1\,,$$

and the corresponding random walk

$$R_0 = 0\,, \quad R_n = \sum_{k=1}^{n} Z_k\,, \quad n \geq 1\,. \tag{1.9}$$

Notice that $EZ_1 = \mu - c/\lambda < 0$ is just the net profit condition (1.7). Then the non–ruin probability is given by

$$1 - \psi(u) = P\left(\sup_{n\geq 1} R_n \leq u\right).$$

This probability can for instance be determined via Spitzer's identity (cf. Feller [235], p. 613), which, for a general random walk, gives the distribution of its ultimate supremum. An application of the latter result allows us to express the non–ruin probability as a compound geometric df, i.e.

$$1 - \psi(u) = (1 - \alpha) \sum_{n=0}^{\infty} \alpha^n H^{n*}(u) \tag{1.10}$$

for some constant $\alpha \in (0, 1)$ and a df H. As before, H^{n*} denotes the nth convolution of H. Both α and H can in general be determined via the classical Wiener–Hopf theory; see again Feller [235], Sections XII.3 and XVIII.3, and Resnick [531].

Estimation of $\psi(u)$ can be worked out for a large category of models by applying a variety of (mostly analytic) techniques to functional relationships like (1.10). It is beyond the scope of this text to review these methods in detail. Besides the Wiener–Hopf methodology for the calculation of $\psi(u)$, renewal theory also yields relevant estimates, as we shall show in the next section. In doing so we shall concentrate on the Cramér–Lundberg model, first showing what typical estimates in a "small claim regime" look like. We then discuss what theory may be used to yield estimates for "large claims".

Notes and Comments

In recent years a multitude of textbooks on risk theory has been published. The interested reader may consult for instance Bowers et al. [85], Bühlmann [97], Gerber [256], Grandell [282], Straub [609], or Beard, Pentikäinen and Pesonen [54]. The latter book has recently appeared in a much updated form as Daykin, Pentikäinen and Pesonen [167]. In the review paper Embrechts and Klüppelberg [211] further references are to be found. A summary of Cramér's work on risk theory is presented in Cramér [140]; see also the recently published collected works of Cramér [141, 142] edited by Anders Martin-Löf. For more references on the historical background to this earlier work, together with a discussion on "where risk theory is evolving to" see Embrechts [202]. A proof of Spitzer's identity, which can be used in order to calculate the probability in (1.8), can be found in any basic textbook on stochastic processes; see for instance Chung [120], Karlin and Taylor [371], Prabhu [505], Resnick [531]. A classical source on Wiener–Hopf techniques is Feller [235], see also Asmussen [27]. An elementary proof of the Wiener–Hopf factorisation, relating the so-called ladder–height distributions of a simple random walk to the step distribution, is to be found in Kennedy [376]. A detailed discussion, including the estimation of ruin probabilities as an application, is given in Prabhu [507]; see also Prabhu [506, 508]. A comment on the relationship between the net profit condition and the asymptotic behaviour of the random walk (1.9) is to be found in Rogozin [548]. For a summary of the Wiener–Hopf theory relevant for risk theory see for instance Asmussen [28], Bühlmann [97] or Embrechts and Veraverbeke [218].

In Section 8.3 we come back to ruin probabilities. There we describe $\psi(u)$ via the distribution of the ultimate supremum of a random walk. Moreover, we characterise a sample path of the risk process leading to ruin.

1.2 The Cramér–Lundberg Estimate

In the previous section we mentioned a general method for obtaining estimates of the ruin probability $\psi(u)$ in the renewal model. If we further restrict ourselves to the Cramér–Lundberg model we can obtain a formula for $\psi(u)$ involving the claim size df F explicitly. Indeed, for the Cramér–Lundberg model under the net profit condition $\rho = c/(\lambda\mu) - 1 > 0$ one can show that

$$1 - \psi(u) = \frac{\rho}{1+\rho} \sum_{n=0}^{\infty} (1+\rho)^{-n}\, F_I^{n*}(u)\,, \tag{1.11}$$

where

$$F_I(x) = \frac{1}{\mu} \int_0^x \overline{F}(y)\, dy\,, \quad x \ge 0\,, \tag{1.12}$$

denotes *the integrated tail distribution* and

$$\overline{F}(x) = 1 - F(x)\,, \quad x \ge 0\,,$$

denotes the tail of the df F. Later we shall show that formula (1.11) is the key tool for estimating ruin probabilities under the assumption of large claims. Also a proof of (1.11) will be given in Theorem 1.2.2 below.

In the sequel, the notion of Laplace–Stieltjes transform plays a crucial role.

Definition 1.2.1 (Laplace–Stieltjes transform)
Let H be a df concentrated on $(0, \infty)$, then

$$\widehat{h}(s) = \int_0^\infty e^{-sx}\, dH(x)\,, \quad s \in \mathbb{R}\,,$$

denotes the Laplace–Stieltjes transform of H. □

Remark. 1) Depending on the behaviour of $\overline{H}(x)$ for x large, $\widehat{h}(s)$ may be finite for a larger set of s–values than $s \ge 0$. In general, $\widehat{h}(s) < \infty$ for $s > -\gamma$ say, where $0 \le \gamma < \infty$ is the abscissa of convergence for $\widehat{h}(s)$. □

The following Cramér–Lundberg estimates of the ruin probability $\psi(u)$ are fundamental in risk theory.

Theorem 1.2.2 (Cramér–Lundberg theorem)
Consider the Cramér–Lundberg model including the net profit condition $\rho > 0$. *Assume that there exists a* $\nu > 0$ *such that*

$$\widehat{f}_I(-\nu) = \int_0^\infty e^{\nu x}\, dF_I(x) = \frac{c}{\lambda\mu} = 1 + \rho\,. \tag{1.13}$$

Then the following relations hold.

(*a*) *For all* $u \geq 0$,

$$\psi(u) \leq e^{-\nu u}\,. \tag{1.14}$$

(*b*) *If, moreover,*

$$\int_0^\infty x\, e^{\nu x}\, \overline{F}(x)\, dx < \infty\,, \tag{1.15}$$

then

$$\lim_{u\to\infty} e^{\nu u}\, \psi(u) = C < \infty\,, \tag{1.16}$$

where

$$C = \left[\frac{\nu}{\rho\mu} \int_0^\infty x\, e^{\nu x}\, \overline{F}(x)\, dx\right]^{-1}\,. \tag{1.17}$$

(*c*) *In the case of an exponential df* $F(x) = 1 - e^{-x/\mu}$, (1.11) *reduces to*

$$\psi(u) = \frac{1}{1+\rho}\, \exp\left\{-\frac{\rho}{\mu(1+\rho)}\, u\right\}, \quad u \geq 0\,. \tag{1.18}$$

Remarks. 2) The fundamental, so–called *Cramér–Lundberg condition* (1.13), can also be written as

$$\int_0^\infty e^{\nu x} \overline{F}(x)\, dx = \frac{c}{\lambda}\,.$$

3) It follows immediately from the definition of Laplace–Stieltjes transform that, whenever ν in (1.13) exists, it is uniquely determined; see also Grandell [282], p. 58.

4) Although the above results can be found in any basic textbook on risk theory, it is useful to discuss the proof of (b) in order to indicate how renewal–theoretic arguments enter (we have summarised the necessary renewal theory in Appendix A4). More importantly, we want to explain why the condition (1.13) has to be imposed. Very readable accounts of the relevant arguments are Feller [235], Sections VI.5, XI.7a, and Grandell [282]. □

Proof of (b). Denote $\delta(u) = 1 - \psi(u)$. Recall from (1.8) that $\delta(u)$ can be expressed via the random walk generated by $(X_i - cY_i)$. Then

$$\begin{aligned}
&\delta(u)\\
&= P(S(t) - ct \le u \text{ for all } t > 0)\\
&= P\left(\sum_{k=1}^{n}(X_k - cY_k) \le u \text{ for all } n \ge 1\right)\\
&= P\left(\sum_{k=2}^{n}(X_k - cY_k) \le u + cY_1 - X_1 \text{ for all } n \ge 2\,,\ X_1 - cY_1 \le u\right)\\
&= P\left(S'(t) - ct \le u + cY_1 - X_1 \text{ for all } t > 0\,,\ X_1 - cY_1 \le u\right)\,,
\end{aligned}$$

where S' is an independent copy of S. Hence

$$\begin{aligned}
&\delta(u)\\
&= E\left(P\left(S'(t) - ct \le u + cY_1 - X_1 \text{ for all } t > 0\,,\ X_1 - cY_1 \le u \mid Y_1\,,\ X_1\right)\right)\\
&= \int_0^\infty \int_0^{u+cs} P\left(S'(t) - ct \le u + cs - x \text{ for all } t > 0\right) dF(x) \lambda e^{-\lambda s} ds\\
&= \int_0^\infty \lambda e^{-\lambda s} \int_0^{u+cs} \delta(u + cs - x) dF(x) ds\\
&= \frac{\lambda}{c}\, e^{u\lambda/c} \int_u^\infty e^{-\lambda z/c} \left[\int_0^z \delta(z - x)\, dF(x)\right] dz\,, \qquad (1.19)
\end{aligned}$$

where we used the substitution $u + cs = z$. The reader is urged to show explicitly where the various conditions in the Cramér–Lundberg model were used in the above calculations! This shows that δ is absolutely continuous with density

$$\delta'(u) = \frac{\lambda}{c}\,\delta(u) - \frac{\lambda}{c} \int_0^u \delta(u - x)\, dF(x)\,. \qquad (1.20)$$

From this equation for $1 - \psi(u)$ the whole theory concerning ruin in the classical Cramér–Lundberg model can be developed. A key point is that the integral in (1.20) is of *convolution type*; this opens the door to renewal theory. Integrate (1.20) from 0 to t with respect to Lebesgue measure to find

$$\begin{aligned}
\delta(t) &= \delta(0) + \frac{\lambda}{c} \int_0^t \delta(u)\, du - \frac{\lambda}{c} \int_0^t\!\!\int_0^u \delta(u - x)\, dF(x)\, du\\
&= \delta(0) + \frac{\lambda}{c} \int_0^t \delta(t - u)\, du - \frac{\lambda}{c} \int_0^t \delta(t - x)\, F(x)\, dx\,.
\end{aligned}$$

We finally arrive at the solution,

$$\delta(t) = \delta(0) + \frac{\lambda}{c} \int_0^t \delta(t-x)\overline{F}(x)\,dx\,. \tag{1.21}$$

Note that $\delta(0)$ is still unknown. However, letting $t \uparrow \infty$ in (1.21) and using the net profit condition (yielding $\delta(\infty) = 1 - \psi(\infty) = 1$) one finds $1 = \delta(0) + \mu\lambda/c$, hence $\delta(0) = 1 - \mu\lambda/c = \rho/(1+\rho)$. Consequently,

$$\delta(t) = \frac{\rho}{1+\rho} + \frac{1}{1+\rho} \int_0^t \delta(t-x)\,dF_I(x)\,, \tag{1.22}$$

where the integrated tail distribution F_I is defined in (1.12). Note that from (1.22), using Laplace–Stieltjes transforms, formula (1.11) immediately follows. The reader is advised to perform this easy calculation as an exercise and also to derive at this point formula (1.18). Equation (1.22) *looks* like a renewal equation; there is however one crucial difference and *this is exactly the point in the proof where a small claim condition of the type* (1.13) *enters.* First, rewrite (1.22) as follows in terms of $\psi(u) = 1 - \delta(u)$, setting $\alpha = 1/(1+\rho) < 1$,

$$\psi(u) = \alpha\overline{F}_I(u) + \int_0^u \psi(u-x)\,d\left(\alpha F_I(x)\right)\,. \tag{1.23}$$

Because $0 < \alpha < 1$, this is a so-called *defective renewal equation* (for instance Feller [235], Section XI.7). In order to cast it into the standard renewal set-up of Appendix A4, we define the following *exponentially tilted* or *Esscher transformed* df $F_{I,\nu}$:

$$dF_{I,\nu}(x) = e^{\nu x} d\left(\alpha F_I(x)\right)\,,$$

where ν is the exponent appearing in the condition (1.13). Using this notation, (1.23) becomes

$$e^{\nu u}\psi(u) = \alpha\, e^{\nu u}\,\overline{F}_I(u) + \int_0^u e^{\nu(u-x)}\psi(u-x)\,dF_{I,\nu}(x)$$

which, by condition (1.13), is a standard renewal equation. A straightforward application of the key renewal theorem (Theorem A4.3(b)) yields

$$\lim_{u\to\infty} e^{\nu u}\psi(u) = \left[\frac{\nu}{\rho\mu}\int_0^\infty x\,e^{\nu x}\overline{F}(x)\,dx\right]^{-1}$$

which is exactly (1.16)–(1.17). The conditions needed for applying Theorem A4.3 are easily checked. By partial integration and using (1.15),

$$\alpha e^{\nu u}\overline{F}_I(u) = \int_u^\infty e^{\nu x} d\left(\alpha F_I(x)\right) - \nu \int_u^\infty \alpha\overline{F}_I(x)e^{\nu x}\,dx\,.$$

Hence $\alpha e^{\nu u}\overline{F}_I(u)$ is the difference of two non–increasing Riemann integrable functions, and therefore it is directly Riemann integrable. Moreover,

$$\int_0^\infty \alpha\, e^{\nu u}\overline{F}_I(u)\, du = \alpha\, \frac{1-\widehat{f}_I(-\nu)}{-\nu} = \frac{\rho}{\nu(1+\rho)} < \infty\,,$$

and

$$\int_0^\infty x\, dF_{I,\nu}(x) = \frac{1}{\mu(1+\rho)} \int_0^\infty x\, e^{\nu x}\overline{F}(x)\, dx < \infty\,,$$

by (1.15). □

Because of the considerable importance for insurance the solution ν of (1.13) gained a special name:

Definition 1.2.3 (Lundberg exponent)
Given a claim size df F, the constant $\nu > 0$ satisfying

$$\int_0^\infty e^{\nu x}\overline{F}(x)\, dx = \frac{c}{\lambda}\,,$$

is called the Lundberg exponent *or* adjustment coefficient *of the underlying risk process.* □

Returning to (1.13), clearly the existence of ν implies that $\widehat{f}_I(s)$ has to exist in a non–empty neighbourhood of 0, implying that the tail $\overline{F}_I$ of the integrated claim size df, and hence also the tail $\overline{F}$, is exponentially bounded. Indeed, it follows from Markov's inequality that

$$\overline{F}(x) \le e^{-\nu x}\, E\, e^{\nu X_1}\,, \quad x > 0\,.$$

This inequality means that large claims are very unlikely (exponentially small probabilities!) to occur. For this reason (1.13) is often called a *small claim condition.*

The Cramér–Lundberg condition can easily be discussed graphically. The existence of ν in (1.13) crucially depends on the left abscissa of convergence $-\gamma$ of $\widehat{f}_I$. Various situations can occur as indicated in Figure 1.2.4. The most common case, and indeed the one fully covered by Theorem 1.2.2, corresponds to Figure 1.2.4(1). Typical claim size dfs and densities (denoted by f) covered by this regime are given in Table 1.2.5. We shall not discuss in detail the intermediate cases, unimportant for applications, of Figure 1.2.4(2) and (3).

If one scans the literature with the following question in mind:

Which distributions do actually fit claim size data?

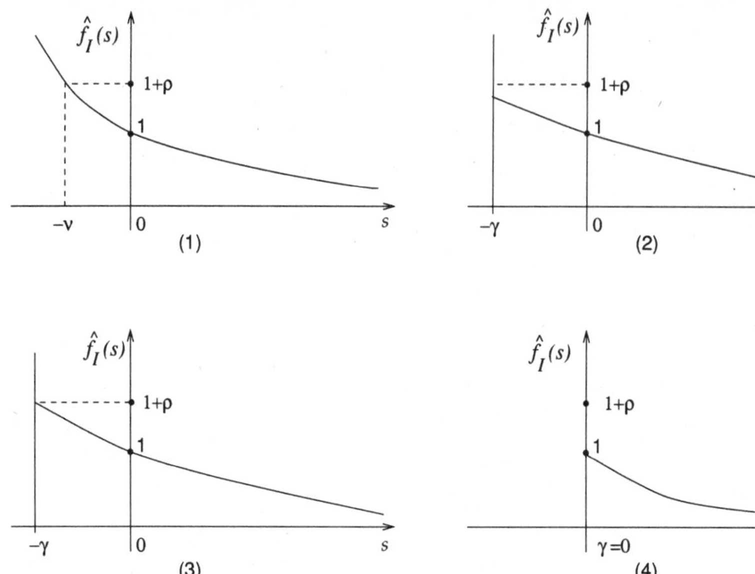

Figure 1.2.4 *Special cases in the Cramér–Lundberg condition.*

then most often one will find one of the dfs listed in Table 1.2.6. All the dfs in Table 1.2.5 allow for the construction of the Lundberg exponent. For the ones in Table 1.2.6 however, this exponent does not exist. Indeed, the case (4) in Figure 1.2.4 applies. For that reason we have labelled the two tables with "small claims", respectively "large claims". A more precise discussion of these distributions follows in Section 1.4. A detailed study of the properties of the distributions listed in Table 1.2.6 with special emphasis on insurance is to be found in Hogg and Klugman [330]. A wealth of material on these and related classes of dfs is presented in Johnson and Kotz [358, 359, 360].

For the sake of argument, assume that we have a portfolio following the Cramér–Lundberg model for which individual claim sizes can be modelled by a Pareto df

$$\overline{F}(x) = (1+x)^{-\alpha}\,, \qquad x \geq 0\,, \quad \alpha > 1\,.$$

It then follows that $EX_1 = \int_0^\infty (1+x)^{-\alpha}\,dx = (\alpha-1)^{-1}$ and the net profit condition amounts to $\rho = c(\alpha-1)/\lambda - 1 > 0$. Question:

Can we work out the exponential Cramér–Lundberg estimate in this case, for a given premium rate c satisfying the above condition?

The answer to this question is *no*. Indeed, in this case, for every $\nu > 0$

$$\int_0^\infty e^{\nu x}(1+x)^{-\alpha}\,dx = \infty\,,$$

Name	Tail $\overline{F}$ or density f	Parameters
Exponential	$\overline{F}(x) = e^{-\lambda x}$	$\lambda > 0$
Gamma	$f(x) = \dfrac{\beta^\alpha}{\Gamma(\alpha)}\, x^{\alpha-1}\, e^{-\beta x}$	$\alpha,\, \beta > 0$
Weibull	$\overline{F}(x) = e^{-cx^\tau}$	$c > 0,\, \tau \geq 1$
Truncated normal	$f(x) = \sqrt{\frac{2}{\pi}}\, e^{-x^2/2}$	—
Any distribution with bounded support		

Table 1.2.5 *Claim size dfs: "small claims". All dfs have support* $(0, \infty)$.

i.e. there is *no* exponential Cramér–Lundberg estimate in this case. We are in the regime of Figure 1.2.4(4): zero is an essential singularity of $\widehat{f}_I$, this means that $\widehat{f}_I(-\varepsilon) = \infty$ for every $\varepsilon > 0$.

However, it turns out that most individual claim size data are modelled by such dfs; see for instance Hogg and Klugman [330] and Ramlau–Hansen [522, 523] for very convincing empirical evidence on this. In Chapter 6 we shall analyse insurance data and come to the conclusion that also in these cases (1.13) is violated. So clearly, classical risk theory has to be adjusted to take this observation into account. In the next section we discuss in detail the class of *subexponential distributions* which will be *the* candidates for loss distributions in the heavy–tailed case. A detailed discussion of the theory of subexponential distributions is rather technical, so we content ourselves with an overview of that part of the theory which is most easily applicable within risk theory in particular and insurance and finance in general. In Section 1.3 we present the large–claims equivalent of the Cramér–Lundberg estimate; see for instance Theorem 1.3.6.

Notes and Comments

The reader interested in the various mathematical approaches to calculating ruin probabilities should consult any of the standard textbooks on risk theory; see the Notes and Comments of Section 1.1. A short proof of (1.14) based on martingale techniques is for instance discussed in Grandell [283]; see also Gerber [255, 256]. An excellent review on the subject is Grandell [283]. Theo-

Name	Tail $\overline{F}$ or density f	Parameters
Lognormal	$f(x) = \frac{1}{\sqrt{2\pi}\,\sigma x} e^{-(\ln x-\mu)^2/(2\sigma^2)}$	$\mu \in \mathbb{R},\ \sigma > 0$
Pareto	$\overline{F}(x) = \left(\frac{\kappa}{\kappa + x}\right)^{\alpha}$	$\alpha, \kappa > 0$
Burr	$\overline{F}(x) = \left(\frac{\kappa}{\kappa + x^{\tau}}\right)^{\alpha}$	$\alpha, \kappa, \tau > 0$
Benktander–type–I	$\overline{F}(x) = (1 + 2(\beta/\alpha)\ln x)$ $e^{-\beta(\ln x)^2-(\alpha+1)\ln x}$	$\alpha, \beta > 0$
Benktander–type–II	$\overline{F}(x) = e^{\alpha/\beta} x^{-(1-\beta)} e^{-\alpha\, x^{\beta}/\beta}$	$\alpha > 0$ $0 < \beta < 1$
Weibull	$\overline{F}(x) = e^{-cx^{\tau}}$	$c > 0$ $0 < \tau < 1$
Loggamma	$f(x) = \frac{\alpha^{\beta}}{\Gamma(\beta)} (\ln x)^{\beta-1}\, x^{-\alpha-1}$	$\alpha, \beta > 0$
Truncated α–stable	$\overline{F}(x) = P(\lvert X\rvert > x)$ where X is an α–stable rv (see Definition 2.2.1)	$1 < \alpha < 2$

Table 1.2.6 *Claim size dfs: "large claims". All dfs have support* $(0,\infty)$ *except for the Benktander cases and the loggamma with* $(1,\infty)$.

rem 1.2.2 can also be formulated for the renewal model; a detailed analysis of the Wiener–Hopf technique together with relevant renewal–type arguments can be worked out; see for instance Embrechts and Veraverbeke [218] for details and further references. Useful textbooks containing a discussion on the link between the asymptotic behaviour of the tail of a measure to properties of its Laplace–Stieltjes transform in a neighbourhood of zero are Bingham, Goldie and Teugels [72] (see for instance Section 1.7 in the latter), Feller [235], Section XIII.5, and Widder [642].

Using Wiener–Hopf theory, a theorem similar to Theorem 1.2.2 can be proved in the renewal model with F supported by $(-\infty,+\infty)$. For details see

Embrechts and Veraverbeke [218] and Thorin [622]. An interesting survey paper is Thorin [624].

Exponential–type ruin estimates hold for much wider classes of risk processes; see for instance Embrechts, Grandell and Schmidli [208], Grandell [282] and the references therein. The latter references also concentrate in detail on ruin estimation in finite time. For an approach based on diffusion approximations see Example 2.5.18.

A detailed discussion of ruin estimation under the various regimes given in Figure 1.2.4 is to be found in Embrechts and Veraverbeke [218]; see also Embrechts [201] for an example based on the generalised inverse Gaussian distribution. A useful review of the various claim size models used in non–life insurance is Hogg and Klugman [330]. The reader should be aware that for most models there is no standard notation or indeed parametrisation. We shall on some occasions say that "under the assumption of a Pareto distribution", meaning that the exact parametrisation is not important for that particular discussion. If however the specific parameter values are of interest, we will always make this clear, in many cases by explicitly stating which functional form of the density f or the df F is used.

1.3 Ruin Theory for Heavy–Tailed Distributions

Throughout this section, all rvs are positive with infinite support, i.e. $F(x) < 1$ for all $x > 0$. We have already seen that Pareto distributions violate the Cramér–Lundberg condition (1.13) so that Theorem 1.2.2 is not applicable for such claim size distributions. What alternative methodology can be used? The answer lies in the representation (1.11), together with Lemma 1.3.1 below. As from Section 1.3.1 onwards, we shall extensively use the theory of regular variation. The reader unfamiliar with the latter theory is urged first to read Appendix A3.1 before proceeding. We denote by $\mathcal{R}_\alpha$ the class of regularly varying functions with index $\alpha \in \mathbb{R}$. The case $\alpha = 0$ corresponds to the so–called slowly varying functions.

From Section 1.3.2 onwards, the class of subexponential distributions will play a fundamental role. For the latter, no complete textbook treatment exists. Because of their importance for the modelling of large claims, we have included a more detailed analysis of their properties. The results immediately needed for proving ruin estimates in the heavy–tailed case are presented in this chapter. For some of the more technical theorems, the reader is referred to Appendix A3.2. The main ideas underlying subexponentiality are presented in Section 1.3.2; Section 1.4 may be skipped upon first reading.

1.3.1 Some Preliminary Results

We start our discussion with a convolution closure property for regularly varying dfs. From Appendix A3.1, recall that L belongs to $\mathcal{R}_0$, i.e. L is slowly varying, whenever for all $t > 0$

$$\lim_{x\to\infty} \frac{L(tx)}{L(x)} = 1 .$$

The following result is to be found in Feller [235], p. 278.

Lemma 1.3.1 (Convolution closure of dfs with regularly varying tails)
If F_1, F_2 are two dfs such that $\overline{F}_i(x) = x^{-\alpha}L_i(x)$ for $\alpha \geq 0$ and $L_i \in \mathcal{R}_0$, $i = 1, 2$, then the convolution $G = F_1 * F_2$ has a regularly varying tail such that

$$\overline{G}(x) \sim x^{-\alpha}\left(L_1(x) + L_2(x)\right) , \qquad x \to \infty .$$

Proof. Let X_1, X_2 be independent rvs with dfs F_1, respectively F_2. Using $\{X_1 + X_2 > x\} \supset \{X_1 > x\} \cup \{X_2 > x\}$ one easily checks that

$$\overline{G}(x) \geq \left(\overline{F}_1(x) + \overline{F}_2(x)\right)(1 - o(1)) .$$

If $0 < \delta < 1/2$, then from

$$\{X_1 + X_2 > x\} \subset \{X_1 > (1-\delta)x\} \cup \{X_2 > (1-\delta)x\} \cup \{X_1 > \delta x, X_2 > \delta x\} ,$$

it follows that

$$\begin{aligned} \overline{G}(x) &\leq \overline{F}_1((1-\delta)x) + \overline{F}_2((1-\delta)x) + \overline{F}_1(\delta x)\,\overline{F}_2(\delta x) \\ &= \Big(\overline{F}_1((1-\delta)x) + \overline{F}_2((1-\delta)x)\Big)(1 + o(1)) . \end{aligned}$$

Hence

$$1 \leq \liminf_{x\to\infty} \frac{\overline{G}(x)}{\overline{F}_1(x) + \overline{F}_2(x)} \leq \limsup_{x\to\infty} \frac{\overline{G}(x)}{\overline{F}_1(x) + \overline{F}_2(x)} \leq (1-\delta)^{-\alpha} ,$$

which proves the result upon letting $\delta \downarrow 0$. One easily shows that $L_1 + L_2$ is slowly varying. □

An alternative proof of this result can be based upon Karamata's Tauberian theorem (Theorem A3.9). An important corollary obtained via induction on n is the following:

Corollary 1.3.2 *If $\overline{F}(x) = x^{-\alpha}L(x)$ for $\alpha \geq 0$ and $L \in R_0$, then for all $n \geq 1$,*

$$\overline{F^{n*}}(x) \sim n\overline{F}(x) , \quad x \to \infty . \qquad \square$$

Suppose now that $X_1, \ldots, X_n$ are iid with df F as in the above corollary. Denote the partial sum of $X_1, \ldots, X_n$ by $S_n = X_1 + \cdots + X_n$ and their maximum by $M_n = \max(X_1, \ldots, X_n)$. Then for all $n \geq 2$,

$$\begin{aligned}
P(S_n > x) &= \overline{F^{n*}}(x), \\
P(M_n > x) &= \overline{F^n}(x) \\
&= \overline{F}(x) \sum_{k=0}^{n-1} F^k(x) \\
&\sim n\overline{F}(x), \quad x \to \infty. \qquad (1.24)
\end{aligned}$$

Therefore, with the above notation, Corollary 1.3.2 can be reformulated as

$$\overline{F} \in \mathcal{R}_{-\alpha}, \quad \alpha \geq 0,$$

implies

$$P(S_n > x) \sim P(M_n > x), \quad x \to \infty.$$

This implies that for dfs with regularly varying tails, the tail of the df of the sum S_n is mainly determined by the tail of the df of the maximum M_n. This is exactly one of the intuitive notions of heavy–tailed distribution or large claims. Hence, stated in a somewhat vague way:

Under the assumption of regular variation, the tail of the maximum determines the tail of the sum.

Recall that in the Cramér–Lundberg model the following relation holds; see (1.11):

$$\psi(u) = \frac{\rho}{1+\rho} \sum_{n=0}^{\infty} (1+\rho)^{-n} \, \overline{F_I^{n*}}(u), \quad u \geq 0,$$

where $F_I(x) = \mu^{-1} \int_0^x \overline{F}(y)\, dy$ is the integrated tail distribution. Under the condition $\overline{F}_I \in \mathcal{R}_{-\alpha}$ for some $\alpha \geq 0$, *we might hope that* the following asymptotic estimate holds:

$$\begin{aligned}
\frac{\psi(u)}{\overline{F}_I(u)} &= \frac{\rho}{1+\rho} \sum_{n=0}^{\infty} (1+\rho)^{-n} \frac{\overline{F_I^{n*}}(u)}{\overline{F}_I(u)} \qquad (1.25) \\
&\to \frac{\rho}{1+\rho} \sum_{n=0}^{\infty} (1+\rho)^{-n} n \quad = \quad \rho^{-1}, \quad u \to \infty. \qquad (1.26)
\end{aligned}$$

The key problem left open in the above calculation is the step from (1.25) to (1.26).

Can one safely interchange limits and sums?

The answer is *yes*; see Theorem 1.3.6. Consequently, (1.26) is the natural ruin estimate whenever $\overline{F}_I$ is regularly varying. Below we shall show that a similar estimate holds true for a much wider class of dfs. In its turn, (1.26) can be reformulated as follows.

> For claim size distributions with regularly varying tails, ultimate ruin $\psi(u)$ for large initial capital u is essentially determined by the tail $\overline{F}(y)$ of the claim size distribution for large values of y, i.e.
>
> $$\psi(u) \sim \frac{1}{\rho\mu} \int_u^\infty \overline{F}(y)\, dy\,, \quad u \to \infty\,.$$

From Table 1.2.6 we obtain the following typical claim size distributions covered by the above result:

- Pareto
- Burr
- loggamma
- truncated stable distributions.

1.3.2 Cramér–Lundberg Theory for Subexponential Distributions

As stated above, the crucial step in obtaining (1.26) was the property $\overline{F_I^{n*}}(x) \sim n\overline{F}_I(x)$ for $x \to \infty$ and $n \geq 2$. This naturally leads us to a class of dfs which allows for a very general theory of ruin estimation for large claims. The main result in this set–up is Theorem 1.3.6 below.

Definition 1.3.3 (Subexponential distribution function)
A df F with support $(0,\infty)$ *is* subexponential, *if for all $n \geq 2$,*

$$\lim_{x\to\infty} \frac{\overline{F^{n*}}(x)}{\overline{F}(x)} = n\,. \tag{1.27}$$

The class of subexponential dfs will be denoted by $\mathcal{S}$. □

Remark. 1) Relation (1.27) yields the following intuitive characterisation of subexponentiality; see (1.24).

$$\text{For all } n \geq 2, \quad P(S_n > x) \sim P(M_n > x)\,, \quad x \to \infty\,. \tag{1.28}$$

□

In order to check for subexponentiality, one does not need to show (1.27) for all $n \geq 2$.

Lemma 1.3.4 (A sufficient condition for subexponentiality)
If $\limsup_{x\to\infty} \overline{F^{2*}}(x)/\overline{F}(x) \leq 2$, *then* $F \in \mathcal{S}$.

Proof. As F stands for the df of a positive rv, it follows immediately that $F^{2*}(x) \leq F^2(x)$, i.e. $\overline{F^{2*}}(x) \geq \overline{F^2}(x)$ for all $x \geq 0$. Therefore, $\liminf_{x\to\infty} \overline{F^{2*}}(x)/\overline{F}(x) \geq 2$, so that because of the condition of the lemma, the limit relation (1.27) holds for $n = 2$. The proof is then by induction on n. For $x \geq y > 0$,

$$\begin{aligned}
\frac{\overline{F^{(n+1)*}}(x)}{\overline{F}(x)} &= 1 + \frac{F(x) - F^{(n+1)*}(x)}{\overline{F}(x)} \qquad (1.29)\\
&= 1 + \int_0^x \frac{\overline{F^{n*}}(x-t)}{\overline{F}(x)}\, dF(t)\\
&= 1 + \left(\int_0^{x-y} + \int_{x-y}^x\right)\left(\frac{\overline{F^{n*}}(x-t)}{\overline{F}(x-t)}\,\frac{\overline{F}(x-t)}{\overline{F}(x)}\right) dF(t)\\
&= 1 + I_1(x) + I_2(x)\,.
\end{aligned}$$

By inserting $-n+n$ in I_1 and noting that $(\overline{F^{n*}}(x-t)/\overline{F}(x-t) - n)$ can be made arbitrarily small for $0 \leq t \leq x-y$ and y sufficiently large, it follows that

$$I_1(x) = (n + o(1)) \int_0^{x-y} \frac{\overline{F}(x-t)}{\overline{F}(x)}\, dF(t)\,.$$

Now

$$\begin{aligned}
\int_0^{x-y} \frac{\overline{F}(x-t)}{\overline{F}(x)}\, dF(t) &= \frac{F(x) - F^{2*}(x)}{\overline{F}(x)} - \int_{x-y}^x \frac{\overline{F}(x-t)}{\overline{F}(x)}\, dF(t)\\
&= \frac{F(x) - F^{2*}(x)}{\overline{F}(x)} - J(x,y)\\
&= (1 + o(1)) - J(x,y)\,,
\end{aligned}$$

where $J(x,y) \leq (F(x) - F(x-y))/\overline{F}(x) \to 0$ as $x \to \infty$ by Lemma 1.3.5 (a) below. Therefore $\lim_{x\to\infty} I_1(x) = n$.

Finally, since $\overline{F^{n*}}(x-t)/\overline{F}(x-t)$ is bounded for $x-y \le t \le x$ and $\lim_{x\to\infty} J(x,y)=0$, $\lim_{x\to\infty} I_2(x)=0$, completing the proof. $\square$

Remarks. 2) The condition in Lemma 1.3.4 is trivially necessary for $F \in \mathcal{S}$.

3) In the beginning of the above proof we used that for the df F of a positive rv, always $\liminf_{x\to\infty} \overline{F^{2*}}(x)/\overline{F}(x) \ge 2$. One easily shows in this case that, for all $n \ge 2$,

$$\liminf_{x\to\infty} \overline{F^{n*}}(x)/\overline{F}(x) \ge n\,.$$

Indeed $S_n \ge M_n$, hence $\overline{F^{n*}}(x) = P(S_n > x) \ge P(M_n > x) = \overline{F^n}(x)$. Therefore

$$\liminf_{x\to\infty} \frac{\overline{F^{n*}}(x)}{\overline{F}(x)} \ge \lim_{x\to\infty} \frac{\overline{F^{n}}(x)}{\overline{F}(x)} = n\,. \qquad \square$$

The following lemma is crucial if we want to derive (1.26) from (1.25) for subexponential F_I.

Lemma 1.3.5 (Some basic properties of subexponential distributions)

(a) *If $F \in \mathcal{S}$, then uniformly on compact y–sets of* $(0,\infty)$,

$$\lim_{x\to\infty} \frac{\overline{F}(x-y)}{\overline{F}(x)} = 1\,. \tag{1.30}$$

(b) *If* (1.30) *holds then, for all* $\varepsilon > 0$,

$$e^{\varepsilon x}\overline{F}(x) \to \infty\,, \quad x \to \infty\,.$$

(c) *If $F \in \mathcal{S}$ then, given $\varepsilon > 0$, there exists a finite constant K so that for all $n \ge 2$,*

$$\frac{\overline{F^{n*}}(x)}{\overline{F}(x)} \le K(1+\varepsilon)^n\,, \quad x \ge 0\,. \tag{1.31}$$

Proof. (a) For $x \ge y > 0$, by (1.29),

$$\begin{aligned}\frac{\overline{F^{2*}}(x)}{\overline{F}(x)} &= 1 + \int_0^y \frac{\overline{F}(x-t)}{\overline{F}(x)}\,dF(t) + \int_y^x \frac{\overline{F}(x-t)}{\overline{F}(x)}\,dF(t)\\ &\ge 1 + F(y) + \frac{\overline{F}(x-y)}{\overline{F}(x)}\,(F(x)-F(y))\,.\end{aligned}$$

Thus, for x large enough so that $F(x)-F(y) \ne 0$,

$$1 \le \frac{\overline{F}(x-y)}{\overline{F}(x)} \le \left(\frac{\overline{F^{2*}}(x)}{\overline{F}(x)} - 1 - F(y)\right)(F(x)-F(y))^{-1}\,.$$

In the latter estimate, the rhs tends to 1 as $x \to \infty$. Uniformity immediately follows from monotonicity in y. Alternatively, use that the property (1.30) can be reformulated as $\overline{F} \circ \ln \in \mathcal{R}_0$, then apply Theorem A3.2.

(b) By (a), $\overline{F} \circ \ln \in \mathcal{R}_0$. But then the conclusion that $x^\varepsilon \overline{F}(\ln x) \to \infty$ as $x \to \infty$ follows immediately from the representation theorem for $\mathcal{R}_0$; see Theorem A3.3.

(c) Let $\alpha_n = \sup_{x \ge 0} \overline{F^{n*}}(x)/\overline{F}(x)$. Using (1.29) we obtain, for every $T < \infty$,

$$\begin{aligned}\alpha_{n+1} &\le 1 + \sup_{0 \le x \le T} \int_0^x \frac{\overline{F^{n*}}(x-y)}{\overline{F}(x)} \, dF(y) \\ &\quad + \sup_{x \ge T} \int_0^x \frac{\overline{F^{n*}}(x-y)}{\overline{F}(x-y)} \frac{\overline{F}(x-y)}{\overline{F}(x)} \, dF(y) \\ &\le 1 + A_T + \alpha_n \sup_{x \ge T} \frac{F(x) - F^{2*}(x)}{\overline{F}(x)} ,\end{aligned}$$

where $A_T = (\overline{F}(T))^{-1} < \infty$. Now since $F \in \mathcal{S}$ we can, given any $\varepsilon > 0$, choose T such that

$$\alpha_{n+1} \le 1 + A_T + \alpha_n (1 + \varepsilon) .$$

Hence

$$\alpha_n \le (1 + A_T) \, \varepsilon^{-1} \, (1 + \varepsilon)^n ,$$

implying (1.31). □

Remark. 4) Lemma 1.3.5(b) justifies the name subexponential for $F \in \mathcal{S}$; indeed $\overline{F}(x)$ decays to 0 slower than any exponential $e^{-\varepsilon x}$ for $\varepsilon > 0$. Furthermore, since for any $\varepsilon > 0$:

$$\int_y^\infty e^{\varepsilon x} \, dF(x) \ge e^{\varepsilon y} \, \overline{F}(y) , \quad y \ge 0 ,$$

it follows from Lemma 1.3.5(b) that for $F \in \mathcal{S}$, $\widehat{f}(-\varepsilon) = \infty$ for all $\varepsilon > 0$. Therefore the Laplace–Stieltjes transform of a subexponential df has an essential singularity at 0. This result was first proved by Chistyakov [115], Theorem 2. As follows from the proof of Lemma 1.3.5(b) the latter property holds true for the larger class of dfs satisfying (1.30). For a further discussion on these classes see Section 1.4. □

Recall that for a df F with finite mean μ, $F_I(x) = \mu^{-1} \int_0^x \overline{F}(y) \, dy$. An immediate, important consequence from the above result is the following.

Theorem 1.3.6 (The Cramér–Lundberg theorem for large claims, I)
Consider the Cramér–Lundberg model with net profit condition $\rho > 0$ and $F_I \in \mathcal{S}$, then

$$\psi(u) \sim \rho^{-1}\, \overline{F}_I(u)\,, \qquad u \to \infty\,. \tag{1.32}$$

Proof. Since $(1+\rho)^{-1} < 1$, there exists an $\varepsilon > 0$ such that $(1+\rho)^{-1}(1+\varepsilon) < 1$. Hence because of (1.31),

$$(1+\rho)^{-n}\,\frac{\overline{F_I^{n*}}(u)}{\overline{F}_I(u)} \le (1+\rho)^{-n}\,K(1+\varepsilon)^n\,, \quad u \ge 0\,,$$

which allows by dominated convergence the interchange of limit and sum in (1.25), yielding the desired result. □

In Figure 1.3.7 realisations of the risk process $(U(t))$ are given in the case of lognormal and Pareto claims. These realisations should be compared with the ones in Figure 1.1.2 (exponential claims).

This essentially finishes our task of finding a Cramér–Lundberg type estimate in the heavy-tailed case.

For claim size distributions with subexponential integrated tail distribution, ultimate ruin $\psi(u)$ is given by (1.32).

In addition to dfs with regularly varying tails, the following examples from Table 1.2.6 yield the estimate (1.32). This will be shown in Section 1.4.

- lognormal
- Benktander–type–I
- Benktander–type–II
- Weibull $(0 < \tau < 1)$.

From a mathematical point of view, the result in Theorem 1.3.6 can be substantially improved. Indeed, Corollary A3.21 yields the following result.

Theorem 1.3.8 (The Cramér–Lundberg theorem for large claims, II)
Consider the Cramér–Lundberg model with net profit condition $\rho > 0$. Then the following assertions are equivalent:

(a) $F_I \in \mathcal{S}$,

(b) $1 - \psi(u) \in \mathcal{S}$,

(c) $\lim_{u\to\infty} \psi(u)/\overline{F}_I(u) = \rho^{-1}$. □

Consequently, the estimate (1.32) is *only* possible under the condition $F_I \in \mathcal{S}$. In the case of the Cramér–Lundberg theory, $\mathcal{S}$ is therefore the natural class when it comes to ruin estimates whenever the Cramér–Lundberg condition

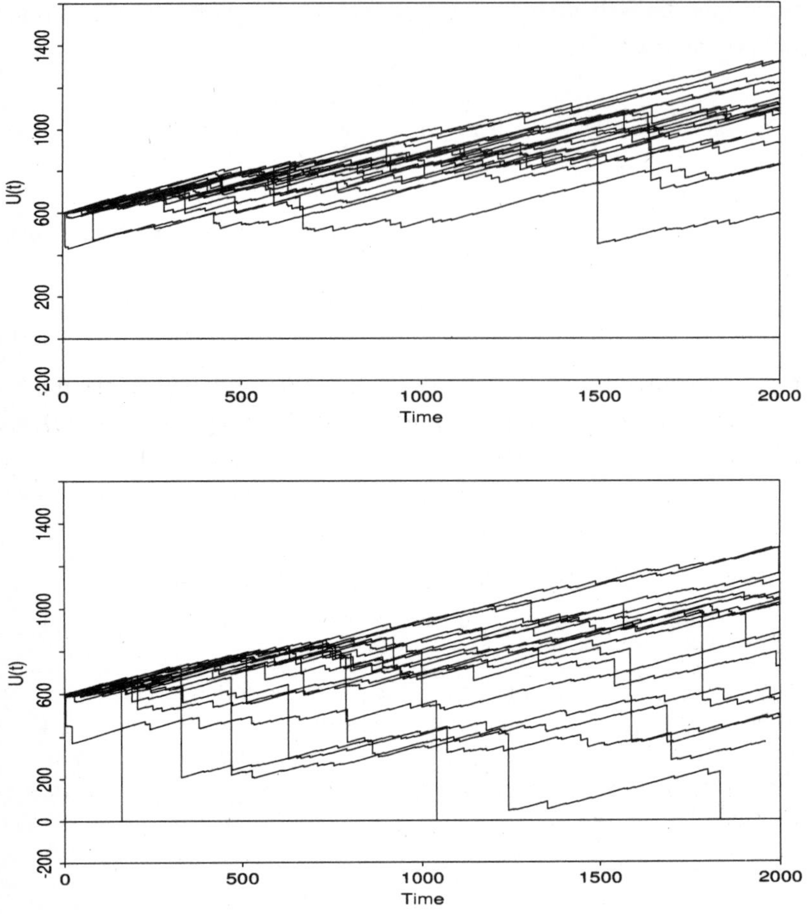

Figure 1.3.7 *Some realisations of the risk process $(U(t))$ for lognormal (top) and Pareto (bottom) claim sizes.*

(1.13) is violated. In Section 1.4 we shall come back to the condition $F_I \in \mathcal{S}$, relating it to conditions on F itself.

1.3.3 The Total Claim Amount in the Subexponential Case

In Section 1.3.2 we have stressed the importance of $\mathcal{S}$ for the estimation of ruin probabilities for large claims. From a mathematical point of view it is important that in the Cramér–Lundberg model, $1 - \psi(u)$ can be expressed as a compound geometric sum; see (1.11). The same methods used for proving

Theorem 1.3.6 yield an estimate of the total claim amount distribution for large claims. Indeed, in Section 1.1 we observed that, within the Cramér–Lundberg model, for all $t \geq 0$,

$$G_t(x) = P(S(t) \leq x) = \sum_{n=0}^{\infty} e^{-\lambda t} \frac{(\lambda t)^n}{n!} F^{n*}(x) , \quad x \geq 0 , \tag{1.33}$$

where $S(t) = \sum_{k=1}^{N(t)} X_k$ is the total (or aggregate) claim amount up to time t. The claim arrival process $(N(t))_{t\geq 0}$ in (1.33) is a homogeneous Poisson process with intensity $\lambda > 0$, hence

$$P(N(t) = n) = e^{-\lambda t} \frac{(\lambda t)^n}{n!} , \quad n \geq 0 . \tag{1.34}$$

The same calculation leading up to (1.33) would, for a general claim arrival process (still assumed to be independent of the claim size process (X_k)), yield the formula

$$G_t(x) = \sum_{n=0}^{\infty} p_t(n) F^{n*}(x) , \quad x \geq 0 , \tag{1.35}$$

where

$$p_t(n) = P(N(t) = n) , \quad n \in \mathbb{N}_0 ,$$

defines a probability measure on $\mathbb{N}_0$. In the case of a subexponential df F the same argument as given for the proof of Theorem 1.3.6 yields the following result.

Theorem 1.3.9 (The total claim amount in the subexponential case)
Consider (1.35) *with* $F \in \mathcal{S}$, *fix* $t > 0$, *and suppose that* $(p_t(n))$ *satisfies*

$$\sum_{n=0}^{\infty} (1+\varepsilon)^n p_t(n) < \infty \tag{1.36}$$

for some $\varepsilon > 0$. *Then* $G_t \in \mathcal{S}$ *and*

$$\overline{G}_t(x) \sim EN(t) \overline{F}(x) , \quad x \to \infty . \tag{1.37}$$

□

Remarks: 1) Condition (1.36) is equivalent to the fact that the probability generating function $\sum_{n=0}^{\infty} p_t(n) s^n$ is analytic in a neighbourhood of $s = 1$.

2) The most general formulation of Theorem 1.3.9 is to be found in Cline [124], Theorem 2.13. □

Example 1.3.10 (The total claim amount in the Cramér–Lundberg model)
Suppose $(N(t))$ is a homogeneous Poisson process with individual probabilities (1.34) so that trivially $p_t(n)$ satisfies (1.36). Then, for $F \in \mathcal{S}$,

$$\overline{G}_t(x) \sim \lambda t \overline{F}(x) , \quad x \to \infty .$$

□

Example 1.3.11 (The total claim amount in the negative binomial case) The negative binomial process is a claim arrival process satisfying

$$p_t(n) = \binom{\gamma + n - 1}{n} \left(\frac{\beta}{\beta + t}\right)^\gamma \left(\frac{t}{\beta + t}\right)^n, \quad n \in \mathbb{N}_0, \quad \beta, \gamma > 0. \quad (1.38)$$

Seal [572, 573] stresses that, apart from the homogeneous Poisson process, this process is the main realistic model for the claim number distribution in insurance applications. One easily verifies that

$$EN(t) = \gamma t/\beta, \quad \mathrm{var}(N(t)) = \gamma t(1 + t/\beta)/\beta.$$

Denoting $q = \beta/(\beta + t)$ and $p = t/(\beta + t)$, one obtains from (1.38), by using Stirling's formula $\Gamma(x + 1) \sim \sqrt{2\pi x}\,(x/e)^x$ as $x \to \infty$, that

$$p_t(n) \sim p^n\, n^{\gamma - 1}\, q^\gamma / \Gamma(\gamma), \quad n \to \infty.$$

Therefore the condition (1.36) is fulfilled, so that for $F \in \mathcal{S}$,

$$\overline{G}_t(x) \sim \frac{\gamma t}{\beta}\, \overline{F}(x), \quad x \to \infty.$$

Recall that in the homogeneous Poisson case, $EN(t) = \lambda t = \mathrm{var}(N(t))$. For the negative binomial process,

$$\mathrm{var}(N(t)) = \left(1 + \frac{t}{\beta}\right) EN(t) > EN(t), \quad t > 0. \quad (1.39)$$

The condition (1.39) is referred to as *over–dispersion* of the process $(N(t))$; see for instance Cox and Isham [134], p. 12. As discussed in McCullagh and Nelder [449], p. 131, over–dispersion may arise in a number of different ways, for instance

(*a*) by observing a homogeneous Poisson process over an interval whose length is random rather than fixed,
(*b*) when the data are produced by a clustered Poisson process, or
(*c*) in behavioural studies and in accident–proneness when there is inter–subject variability.

It is mainly (c) which is often encountered in the analysis of insurance data. The features mentioned under (c) can be modelled by *mixed Poisson processes*. Their precise definition given below is motivated by the following example. Suppose Λ is a rv which is $\Gamma(\gamma, \beta)$ distributed with density

$$f_\Lambda(x) = \frac{\beta^\gamma}{\Gamma(\gamma)}\, x^{\gamma - 1} e^{-\beta x}, \quad x > 0.$$

Then

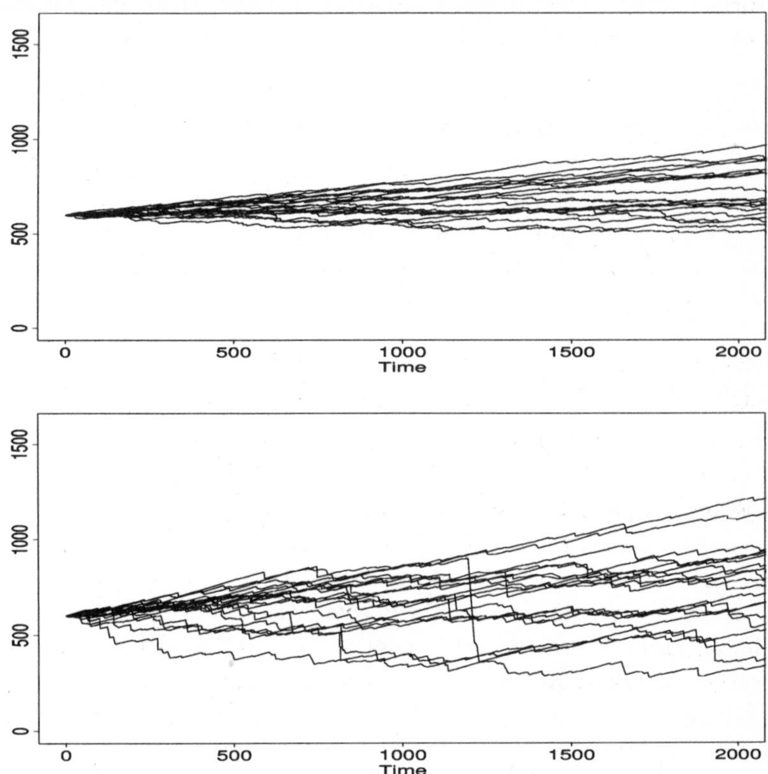

Figure 1.3.12 *Realisations of the risk process $(U(t))$ with linear premium income and total claim amount process $S(t) = \sum_{i=1}^{N(t)} X_i$, where $(N(t))$ is a negative binomial process. The claim size distribution is either exponential (top) or lognormal (bottom). Compared with Figure 1.1.2, the top figure clearly shows the over–dispersion effect. If we compare the bottom graph with the corresponding Figure 1.3.7 we notice the accumulation of many small claims.*

$$\int_0^\infty e^{-xt}\frac{(xt)^n}{n!} f_\Lambda(x)\, dx = \frac{\Gamma(n+\gamma)}{n!\Gamma(\gamma)}\left(\frac{\beta}{\beta+t}\right)^\gamma \left(\frac{t}{\beta+t}\right)^n, \quad n = 0,1,2,\ldots.$$

The latter formula equals $p_t(n)$ in (1.38). Hence we have obtained the negative binomial probabilities by *randomising* the Poisson parameter λ over a gamma distribution. This is exactly an example of what is meant by a mixed Poisson process. □

Definition 1.3.13 (Mixed Poisson process)
Suppose Λ is a rv with $P(\Lambda > 0) = 1$, and suppose $N = (N(t))_{t\geq 0}$ is a ho-

mogeneous Poisson process with intensity 1, *independent of* Λ. *The process* $(N(\Lambda t))_{t\geq 0}$ *is called* mixed Poisson. □

The rv Λ in the definition above can be interpreted as a random time change. Processes more general than mixed Poisson, for instance Cox processes, have belonged to the toolkit of actuaries for a long time. Mixed Poisson processes are treated in every standard text on risk theory. Recent textbook treatments are Grandell [284] and Panjer and Willmot [489]. The homogeneous Poisson process with intensity $\lambda > 0$ is obtained for Λ degenerate at λ, i.e. $P(\Lambda = \lambda) = 1$.

Notes and Comments

The class of subexponential distributions was independently introduced by Chistyakov [115] and Chover, Ney and Wainger [116], mainly in the context of branching processes. An early textbook treatment is given in Athreya and Ney [34], from which the proof of Lemma 1.3.5 is taken. Lemma 1.3.5(c) is attributed to Kesten. An independent introduction of $\mathcal{S}$ through questions in queueing theory is to be found in Borovkov [83, 84]; see also Pakes [488]. The importance of $\mathcal{S}$ as a useful class of heavy–tailed dfs in the context of applied probability in general, and insurance mathematics in particular, was realised early on by Teugels [617]. A recent survey paper is Goldie and Klüppelberg [274].

In the next section we shall prove that the condition $F_I \in \mathcal{S}$ is also satisfied for F lognormal. Whenever F is Pareto, it immediately follows that $F_I \in \mathcal{S}$. In these forms (i.e. Pareto, lognormal F), Theorem 1.3.6 has an interesting history, kindly communicated to us by Olof Thorin. In Thorin [623] the estimate

$$\psi(u) \sim k \int_u^\infty \overline{F}(x)\,dx\,, \qquad u \to \infty\,,$$

for some constant k was obtained for a wide class of distributions F assuming certain regularity conditions:

$$F(y) = \int_0^\infty \left(1 - e^{-yt}\right) V'(t)\,dt\,, \qquad y \geq 0\,,$$

with V' continuous, positive for $t > 0$, and having

$$V'(0) = 0 \quad \text{and} \quad \int_0^\infty V'(t)\,dt = 1\,.$$

An interesting special case is obtained by choosing $V'(t)$ as a gamma density with shape parameter greater than 1, giving the Pareto case. It was pointed out in Thorin and Wikstad [625] that the same method also works for the lognormal distribution. The Pareto case was obtained independently by von Bahr [633] and early versions of these results were previously discussed by Thorin at the 19th International Congress of Actuaries in Oslo (1972) and at the Wisconsin Actuarial Conference in 1971. Thorin also deals with the renewal case. Embrechts and Veraverbeke [218] obtained the full answer as presented in Theorem 1.3.8. In the latter paper these results were also formulated in the most general form for the renewal model allowing for real–valued, not necessarily positive claims. It turns out that also in that case, under the assumption $F_I \in \mathcal{S}$, the estimate $\psi(u) \sim \rho^{-1}\overline{F}_I(u)$ holds. In the renewal model however, we do not have the full equivalence relationships as in the Cramér–Lundberg case.

A recent overview concerning approximation methods for $\overline{G}_t(x)$ is given in Buchwalder, Chevallier and Klüppelberg [95]. The use of the fast Fourier transform method with special emphasis on heavy tails is highlighted in Embrechts, Grübel and Pitts [209]. A particularly important methodology for application is the so–called Panjer recursion method. For a discussion and further references on this topic; see Dickson [180] or Panjer and Willmot [489]. A light–tailed version of Example 1.3.11 is to be found in Embrechts, Maejima and Teugels [214]. An especially useful method in the light–tailed case is the so–called saddlepoint approximation; see Jensen [356] for an introduction including applications to risk theory. A very readable textbook treatment on approximation methods is Hipp and Michel [328]; see also Feilmeier and Bertram [232].

There are much fewer papers on statistical estimation of $\overline{G}_t$ than there are on asymptotic expansions. Clearly, one could work out a parametric estimation procedure or use non–parametric methods. The latter approach looks especially promising, as can be seen from Pitts [502, 503] and references therein.

1.4 Cramér–Lundberg Theory for Large Claims: a Discussion

1.4.1 Some Related Classes of Heavy–Tailed Distributions

In order to get direct conditions on F so that the heavy–tailed Cramér–Lundberg estimate in Theorem 1.3.6 holds, we first study some related classes of heavy–tailed dfs.

The class of *dominatedly varying distributions* is defined as

$$\mathcal{D} = \left\{ F \text{ df on } (0,\infty) : \limsup_{x\to\infty} \overline{F}(x/2)/\overline{F}(x) < \infty \right\} .$$

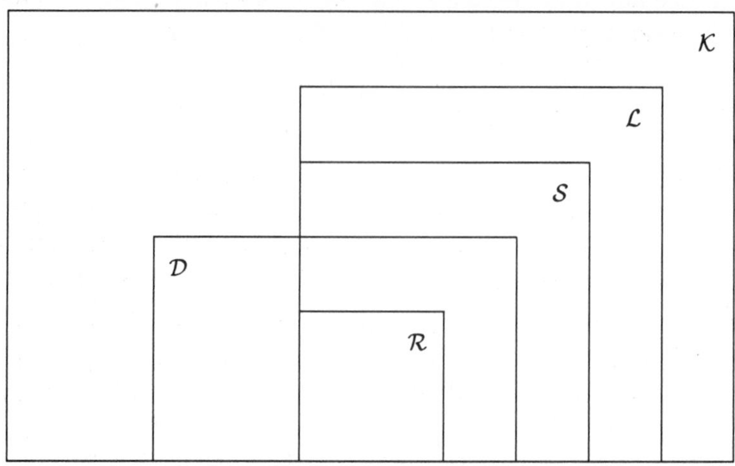

Figure 1.4.1 *Classes of heavy-tailed distributions.*

We have already encountered members of the following three families:

$$\mathcal{L} = \left\{ F \text{ df on } (0,\infty) : \lim_{x\to\infty} \overline{F}(x-y)/\overline{F}(x) = 1 \quad \text{for all } y > 0 \right\},$$

$$\mathcal{R} = \left\{ F \text{ df on } (0,\infty) : \overline{F} \in \mathcal{R}_{-\alpha} \quad \text{for some } \alpha \geq 0 \right\},$$

$$\mathcal{K} = \left\{ F \text{ df on } (0,\infty) : \widehat{f}(-\varepsilon) = \int_0^\infty e^{\varepsilon x} dF(x) = \infty \quad \text{for all } \varepsilon > 0 \right\}.$$

From the definition of slowly varying functions, it follows that $F \in \mathcal{L}$ if and only if $\overline{F} \circ \ln \in \mathcal{R}_0$. The following relations hold:

(a) $\mathcal{R} \subset \mathcal{S} \subset \mathcal{L} \subset \mathcal{K}$ and $\mathcal{R} \subset \mathcal{D}$,
(b) $\mathcal{L} \cap \mathcal{D} \subset \mathcal{S}$,
(c) $\mathcal{D} \not\subset \mathcal{S}$ and $\mathcal{S} \not\subset \mathcal{D}$.

The situation is summarised in Figure 1.4.1. A detailed discussion of these interrelationships is to be found in Embrechts and Omey [216]; see also Klüppelberg [387]. Most of the implications are easy, and indeed some of them we have already proved ($\mathcal{R} \subset \mathcal{S}$ in Corollary 1.3.2, $\mathcal{L} \subset \mathcal{K}$ in Remark 4 after the proof of Lemma 1.3.5). A mistake often encountered in the literature is the claim that $\mathcal{D} \subset \mathcal{S}$; the following df provides a counterexample.

Example 1.4.2 (The Peter and Paul distribution)
Consider a game where the first player (Peter) tosses a fair coin until it falls

head for the first time, receiving from the second player (Paul) 2^k Roubles, if this happens at trial k. The df of Peter's gain is

$$F(x) = \sum_{k:2^k \le x} 2^{-k}, \quad x \ge 0.$$

The problem underlying this game is the famous St. Petersburg paradox; see for instance Feller [234], Section X.4. It immediately follows that for all $k \in \mathbb{N}$, $\overline{F}(2^k - 1)/\overline{F}(2^k) = 2$ so that $F \notin \mathcal{L}$ and a fortiori $F \notin \mathcal{S}$. On the other hand, one easily shows that $F \in \mathcal{D}$. For a full analysis see Goldie [271]. □

The result $\mathcal{S} \neq \mathcal{L}$ is non–trivial; relevant examples are to be found in Embrechts and Goldie [204] and Pitman [501]. Concerning the relationship between $\mathcal{L}$ and $\mathcal{S}$, consider for $x \ge 0$,

$$\frac{\overline{F^{2*}}(x)}{\overline{F}(x)} = 1 + \int_0^x \frac{\overline{F}(x-y)}{\overline{F}(x)}\, dF(y).$$

By definition, $F \in \mathcal{L}$ implies that for every y fixed the integrand above converges to 1. By the uniform convergence theorem for slowly varying functions (Theorem A3.2), this convergence holds also uniformly on compact y–intervals. In order however to interchange limits and integrals one needs some sort of uniform integrability condition (dominated convergence, monotonicity in $x, \ldots$). In general (i.e. for $F \in \mathcal{L}$) these conditions fail.

Let us first look at $\mathcal{S}$–membership in general. We have already established $\mathcal{R} \subset \mathcal{S}$ and $\mathcal{S} \subset \mathcal{L}$ (Lemma 1.3.5(a)), the latter implying that for all $\varepsilon > 0$, $\exp\{\varepsilon x\}\overline{F}(x) \to \infty$ as $x \to \infty$. From this it immediately follows that the exponential df $F(x) = 1 - \exp\{-\lambda x\}$ does *not* belong to $\mathcal{S}$. One could of course easily verify this directly, or use the fact that $\mathcal{S} \subset \mathcal{L}$ and immediately note that $F \notin \mathcal{L}$.

So by now we know that the dfs with power law tail behaviour (i.e. $F \in \mathcal{R}$) belong to $\mathcal{S}$. On the other hand, exponential distributions (and indeed dfs F with faster than exponential tail behaviour) do not belong to $\mathcal{S}$. What can be said about classes "in between", such as for example the important class of Weibull type variables where $\overline{F}(x) \sim \exp\{-x^\tau\}$ with $0 < \tau < 1$? Proposition A3.16, formulated in terms of the density f of F, the hazard rate $q = f/\overline{F}$ and the hazard function $Q(x) = \int_0^x q(y)dy$, immediately yields the following examples in $\mathcal{S}$. Note that, using the above notation, $\overline{F}(x) = \exp\{-Q(x)\}$.

Example 1.4.3 (Examples of subexponential distributions)

(a) Take F a Weibull distribution with parameters $0 < \tau < 1$ and $c > 0$, i.e.

$$\overline{F}(x) = e^{-c\,x^\tau}, \quad x \ge 0.$$

Then $f(x) = c\tau x^{\tau-1}e^{-cx^\tau}$, $Q(x) = cx^\tau$ and $q(x) = c\tau x^{\tau-1}$ which decreases to 0 if $\tau < 1$. We can immediately apply Proposition A3.16(b) since

$$x \mapsto e^{x\,q(x)} f(x) = e^{c\,(\tau-1)x^\tau} c\,\tau\,x^{\tau-1}$$

is integrable on $(0,\infty)$ for $0 < \tau < 1$. Therefore $F \in \mathcal{S}$.

(*b*) Using Proposition A3.16, one can also prove for

$$\overline{F}(x) \sim e^{-x(\ln x)^{-\beta}}\,, \quad x \to \infty\,, \quad \beta > 0\,,$$

that $F \in \mathcal{S}$. This example shows that one can come fairly close to exponential tail behaviour while staying in $\mathcal{S}$.

(*c*) At this point one could hope that for

$$\overline{F}(x) \sim e^{-x^\tau L(x)}\,, \quad x \to \infty\,, \quad 0 \le \tau < 1\,, \quad L \in \mathcal{R}_0\,,$$

F would belong to $\mathcal{S}$. Again, in this generality the answer to this question is *no*. One can construct examples of $L \in \mathcal{R}_0$ so that the corresponding F does not even belong to $\mathcal{L}$! An example for $\tau = 0$ was communicated to us by Charles Goldie; see also Cline [123] where counterexamples for $0 < \tau < 1$ are given. □

A particularly useful result is the following.

Proposition 1.4.4 (Dominated variation and subexponentiality)

(*a*) *If* $F \in \mathcal{L} \cap \mathcal{D}$, *then* $F \in \mathcal{S}$.

(*b*) *If* F *has finite mean* μ *and* $F \in \mathcal{D}$, *then* $F_I \in \mathcal{L} \cap \mathcal{D}$. *Consequently, because of* (*a*), $F_I \in \mathcal{S}$.

Proof. (a) Because of (A.17), for $x \ge 0$,

$$\begin{aligned}\frac{\overline{F^{2*}}(x)}{\overline{F}(x)} &= 2\int_0^{x/2} \frac{\overline{F}(x-y)}{\overline{F}(x)}\,dF(y) + \frac{\left(\overline{F}(x/2)\right)^2}{\overline{F}(x)} \\ &= 2\int_0^{x/2} \frac{\overline{F}(x-y)}{\overline{F}(x)}\,dF(y) + o(1)\,,\end{aligned}$$

where the $o(1)$ is a consequence of $F \in \mathcal{D}$. Now for all $0 \le y \le x/2$,

$$\frac{\overline{F}(x-y)}{\overline{F}(x)} \le \frac{\overline{F}(x/2)}{\overline{F}(x)}\,,$$

so that because $F \in \mathcal{D}$, we can apply dominated convergence, yielding for $F \in \mathcal{L}$ the convergence of the integral to 1. Hence $F \in \mathcal{S}$.

(b) For ease of notation, and without loss of generality, we put $\mu = 1$. Since, for all $x \ge 0$,

$$\overline{F}_I(x) = \int_x^\infty \overline{F}(y)\,dy \geq \int_x^{2x} \overline{F}(y)\,dy \geq x\overline{F}(2x)\,, \tag{1.40}$$

it follows from $F \in \mathcal{D}$ that

$$\limsup_{x\to\infty} \frac{x\overline{F}(x)}{\overline{F}_I(x)} \leq \limsup_{x\to\infty} \frac{\overline{F}(x)}{\overline{F}(2x)} < \infty\,.$$

Moreover,

$$\begin{aligned} \frac{\overline{F}_I(x/2)}{\overline{F}_I(x)} &= \frac{\displaystyle\int_{x/2}^\infty \overline{F}(y)\,dy}{\displaystyle\int_x^\infty \overline{F}(y)\,dy} = 1 + \frac{\displaystyle\int_{x/2}^x \overline{F}(y)\,dy}{\displaystyle\int_x^\infty \overline{F}(y)\,dy} \leq 1 + \frac{\overline{F}(x/2)x/2}{\overline{F}_I(x)} \\ &= 1 + 2^{-1}\frac{\overline{F}(x/2)}{\overline{F}(x)}\,\frac{x\overline{F}(x)}{\overline{F}_I(x)}\,, \end{aligned}$$

whence

$$\limsup_{x\to\infty} \frac{\overline{F}_I(x/2)}{\overline{F}_I(x)} < \infty\,,$$

i.e. $F_I \in \mathcal{D}$. Take $y > 0$, then for $x \geq 0$

$$1 = \frac{\displaystyle\int_x^{x+y} \overline{F}(u)\,du + \int_{x+y}^\infty \overline{F}(u)\,du}{\overline{F}_I(x)}\,,$$

hence, by (1.40),

$$1 \leq \frac{y\overline{F}(x)}{\overline{F}_I(x)} + \frac{\overline{F}_I(x+y)}{\overline{F}_I(x)} \leq \frac{y\overline{F}(x)}{x\overline{F}(2x)} + \frac{\overline{F}_I(x+y)}{\overline{F}_I(x)}\,.$$

The first term in the latter sum is $o(1)$ as $x \to \infty$, since $F \in \mathcal{D}$. Therefore,

$$1 \leq \liminf_{x\to\infty} \frac{\overline{F}_I(x+y)}{\overline{F}_I(x)} \leq \limsup_{x\to\infty} \frac{\overline{F}_I(x+y)}{\overline{F}_I(x)} \leq 1\,,$$

i.e. $F_I \in \mathcal{L}$. □

1.4.2 The Heavy–Tailed Cramér–Lundberg Case Revisited

So far we have seen that, from an analytic point of view, the classes $\mathcal{R}$ and $\mathcal{S}$ yield natural models of claim size distributions for which the Cramér–Lundberg condition (1.13) is violated.

In Seal [573], for instance, the numerical calculation of $\psi(u)$ is discussed for various classes of claim size dfs. After stressing the fact that the mixed

Poisson process in general, and the homogeeous Poisson process and the negative binomial process in particular, are the *only* claim arrival processes which fit real insurance data, Seal continues by saying

Types of distributions of independent claim sizes are just as limited, for apart from the Pareto *and* lognormal *distributions, we are* not *aware that any has been fitted successfully to actual claim sizes in actuarial history.*

Although perhaps formulated in a rather extreme form, more than ten years later the main point of this sentence still stands; see for instance Schnieper [568], Benabbou and Partrat [59] and Ramlau–Hansen [522, 523] for a more recent account on this theme. Some examples of insurance data will be discussed in Chapter 6.

In this section we discuss $\mathcal{S}$–membership with respect to standard classes of dfs as given above. We stick to the Cramér–Lundberg model for purposes of illustration on how the new methodology works. Recall in the Cramér–Lundberg set–up the main result of Section 1.3, i.e. Theorem 1.3.6:

$$\boxed{\text{If } F_I \in \mathcal{S} \text{ then } \psi(u) \sim \rho^{-1}\,\overline{F}_I(u),\ u \to \infty\,.}$$

The exponential Cramér–Lundberg estimates (1.14), (1.16) under the small claim condition (1.13) yield surprisingly good estimates for $\psi(u)$, even for moderate to small u. The large claim estimate $\psi(u) \sim \rho^{-1}\overline{F}_I(u)$ is however mainly of theoretical value and can indeed be further improved; see the Notes and Comments. A first problem with respect to applicability concerns the condition $F_I \in \mathcal{S}$. A natural question at this point is:

(1) *Does $F \in \mathcal{S}$ imply that $F_I \in \mathcal{S}$?*

And, though less important for our purposes:

(2) *Does $F_I \in \mathcal{S}$ imply that $F \in \mathcal{S}$?*

It will turn out that, in general, the answer to both questions (unfortunately) is *no*. This leads us immediately to the following task:

Give sufficient conditions for F such that $F_I \in \mathcal{S}$.

Concerning the latter problem, there are numerous answers to be found in the literature. We shall discuss some of them. The various classes of dfs introduced in the previous section play an important role here.

An immediate consequence of Proposition 1.4.4 is the following result.

Corollary 1.4.5 (The Cramér–Lundberg theorem for large claims, III)
Consider the Cramér–Lundberg model with net profit condition $\rho > 0$ and $F \in \mathcal{D}$, then

$$\psi(u) \sim \rho^{-1}\overline{F}_I(u)\,, \quad u \to \infty\,. \qquad \square$$

The condition $F \in \mathcal{D}$ is readily verified for all relevant examples; this is in contrast to the non–trivial task of checking $F_I \in \mathcal{S}$. It is shown in Seneta [575], Appendix A3, that any $F \in \mathcal{D}$ has the property that there exists a $k \in \mathbb{N}$ so that $\int_0^\infty x^k \, dF(x) = \infty$, i.e. there always exist divergent (higher) moments. It immediately follows from Karamata's theorem (Theorem A3.6) that $F \in \mathcal{R}$ implies $F_I \in \mathcal{R}$ and hence $F_I \in \mathcal{S}$. For detailed estimates in the heavy–tailed Cramér–Lundberg model see Klüppelberg [388]. In the latter paper, also, various sufficient conditions for $F_I \in \mathcal{S}$ are given in terms of the *hazard rate* $q(x) = f(x)/\overline{F}(x)$ for F with density f or the *hazard function* $Q = -\ln \overline{F}$; see also Cline [123].

Lemma 1.4.6 (Sufficient conditions for $F_I \in \mathcal{S}$)
If one of the following conditions holds, then $F_I \in \mathcal{S}$.

(*a*) $\limsup_{x\to\infty} x\, q(x) < \infty$,
(*b*) $\lim_{x\to\infty} q(x) = 0$, $\lim_{x\to\infty} x\, q(x) = \infty$, *and one of the following conditions holds*:
  (*1*) $\limsup_{x\to\infty} x\, q(x)/Q(x) < 1$,
  (*2*) $q \in \mathcal{R}_\delta$, $-1 \le \delta < 0$,
  (*3*) $Q \in \mathcal{R}_\delta$, $0 < \delta < 1$, *and q is eventually decreasing,*
  (*4*) *q is eventually decreasing to* 0, $q \in \mathcal{R}_0$ *and* $Q(x) - x\, q(x) \in \mathcal{R}_1$. □

In Klüppelberg [387], Theorem 3.6, a Pitman–type result (see Proposition A3.16) is presented, characterising $F_I \in \mathcal{S}$ for certain absolutely continuous F with hazard rate q decreasing to zero.

Example 1.4.7 (Examples of $F_I \in \mathcal{S}$)
Using Lemma 1.4.6(b)(2), it is not difficult to see that $F_I \in \mathcal{S}$ in the following cases:

- Weibull with parameter $\tau \in (0,1)$
- Benktander–type–I and –II
- lognormal. □

Corollary 1.4.8 (The Cramér–Lundberg theorem for large claims, IV)
Consider the Cramér–Lundberg model with net profit condition $\rho > 0$ and F satisfying one of the conditions (a), (b) of Lemma 1.4.6, then

$$\psi(u) \sim \rho^{-1}\, \overline{F}_I(u)\,, \quad u \to \infty\,. \qquad \square$$

We still have left the questions (1) and (2) above unanswered. Concerning question (2) (does $F_I \in \mathcal{S}$ imply that $F \in \mathcal{S}$?), on using Proposition 1.4.4(b) we find that a straightforward modification of the Peter and Paul distribution yields an example of a df F with finite mean such that $F_I \in \mathcal{S}$ but $F \notin \mathcal{S}$.

For details see Klüppelberg [387]. The latter paper also contains a discussion of question (1) (does $F \in \mathcal{S}$ with finite mean imply that $F_I \in \mathcal{S}$?).

At this point, the reader may have become rather bewildered concerning the properties of $\mathcal{S}$. On the one hand, we have shown that it is the right class of dfs to consider in risk theory under large claim conditions; see Theorem 1.3.8, (c) implies (a). On the other hand, one has to be extremely careful in making general statements concerning $\mathcal{S}$ and its relationship to other classes of dfs.

For our immediate purposes it suffices to notice that for distributions F with finite mean belonging to the families: *Pareto, Weibull* ($\tau < 1$), *lognormal, Benktander–type–I and –II, Burr, loggamma,*

$$\boxed{F \in \mathcal{S} \quad \text{and} \quad F_I \in \mathcal{S}\,.}$$

For further discussions on the applied nature of classes of heavy–tailed distributions, we also refer to Benabbou and Partrat [59], Conti [132], Hogg and Klugman [330] and Schnieper [568]. In the latter paper the reader will find some critical remarks on the existing gap between theoretical and applied usefulness of families of claim size distributions. It also contains some examples of relevant software for the actuary.

Notes and Comments

The results presented so far only give a first, though representative, account of ruin estimation in the heavy–tailed case. The reader should view them also as examples of how the class $\mathcal{S}$, and its various related classes, offer an appropriate tool towards a "heavy–tailed calculus".

Nearly all of the results can be extended. For instance Veraverbeke [630] considers the following model, first introduced by Gerber:

$$U_B(t) = u + ct - S(t) + B_t\,, \quad t \geq 0\,,$$

where u, c and $S(t)$ are defined within the Cramér–Lundberg model, and B is a Wiener process (see Section 2.4), independent of the process S. The process B can be viewed as describing small perturbations (i.e. B_t is distributed as a normal rv with mean 0 and variance $\sigma_B^2 t$) around the risk process U in (1.4). In [630], Theorem 1, it is shown that an estimate similar to the one obtained in the Cramér–Lundberg model for subexponential claim size distributions holds. These results can also be generalised to the renewal model set–up, as noted by Furrer and Schmidli [248]. Subexponentiality is also useful beyond these models as for instance shown by Asmussen, Fløe Henriksen and Klüppelberg [31]. In the latter paper, a risk process, evolving in an environment given by a Markov process with a finite state space, is studied. An appealing example of this type of process with exponential claim sizes is given in Reinhard [524].

Asymptotic estimates for the ruin probability change when the company receives interest on its reserves. For $F \in \mathcal{R}$ and a positive force of interest δ the corresponding ruin probability satisfies

$$\psi_\delta(u) \sim k_\delta \overline{F}(u)\,, \quad u \to \infty\,,$$

i.e. it is tail–equivalent to the claim size df itself. This has been proved in Klüppelberg and Stadtmüller [399]. The case of subexponential claims has been treated in Asmussen [29].

Concerning the definition of $\mathcal{S}$, there is no *a priori* reason for assuming that the limit of $\overline{F^{2*}}(x)/\overline{F}(x)$ equals 2; an interesting class of distributions results from allowing this limit to be any value greater than 2.

Definition 1.4.9 *A df F on $(0,\infty)$ belongs to the class $\mathcal{S}(\gamma)$, $\gamma \geq 0$, if*

(a) $\lim_{x\to\infty} \overline{F^{2*}}(x)/\overline{F}(x) = 2d < \infty$,

(b) $\lim_{x\to\infty} \overline{F}(x-y)/\overline{F}(x) = e^{\gamma y}, \quad y \in \mathbb{R}.$ □

One can show that $d = \int_0^\infty e^{\gamma y}\, dF(y) = \widehat{f}(-\gamma)$, so that $\mathcal{S} = \mathcal{S}(0)$. These classes of dfs turn out to cover exactly the situations illustrated in Figures 1.2.4(2) and (3), in between the light–tailed Cramér–Lundberg case and the heavy–tailed (subexponential) case. A nice illustration of this, using the class of generalised inverse Gaussian distributions, is to be found in Embrechts [201]; see also Klüppelberg [389] and references therein.

For a critical assessment of the approximation $\psi(u) \sim k\overline{F}_I(u)$ for some constant k and $u \to \infty$ see De Vylder and Goovaerts [179]. Further improvements can be obtained only if conditions beyond $F_I \in \mathcal{S}$ are imposed. One such set of conditions is higher–order subexponentiality, or indeed higher–order regular variation. In general $G \in \mathcal{S}$ means that $\overline{G^{2*}}(x) \sim 2\overline{G}(x)$ for $x \to \infty$; higher–order versions of $\mathcal{S}$ involve conditions on the asymptotic behaviour of $\overline{G^{2*}}(x) - 2\overline{G}(x)$ for $x \to \infty$. For details on these techniques the reader is referred to Omey and Willekens [486, 487], and also Bingham et al. [72], p. 185. With respect to the heavy–tailed ruin estimate $\psi(u) \sim \rho^{-1}\overline{F}_I(u)$, second–order assumptions on F lead to asymptotic estimates for $\psi(u) - \rho^{-1}\overline{F}_I(u)$ for $u \to \infty$. A numerical comparison of such results, together with a detailed simulation study of rare events in insurance, is to be found in Asmussen and Binswanger [30] and Binswanger [74].

2

Fluctuations of Sums

In this chapter we consider some basic theory for sums of independent rvs. This includes classical results such as the strong law of large numbers (SLLN) in Section 2.1 and the central limit theorem (CLT) in Section 2.2, but also refinements on these theorems. In Section 2.3 refinements on the CLT are given (asymptotic expansions, large deviations, rates of convergence). Brownian and α–stable motion are introduced in Section 2.4 as weak limits of partial sum processes. They are fundamental stochastic processes which are used throughout this book. This is also the case for the homogeneous Poisson process which occurs as a special renewal counting process in Section 2.5.2. In Sections 2.5.2 and 2.5.3 we study the fluctuations of renewal counting processes and of random sums indexed by a renewal counting process. As we saw in Chapter 1, random sums are of particular interest in insurance; for example, the compound Poisson process is one of the fundamental notions in the field.

The present chapter is the basis for many other results provided in this book. Poisson random measures occur as generalisations of the homogeneous Poisson process in Chapter 5. Since most of the theory given below is classical we only sketch the main ideas of the proofs and refer to some of the relevant literature for details. We also consider extensions and generalisations of the theory for sums in Sections 8.5 and 8.6. There we look at the fine structure of a random walk, in particular at the longest success–run and large deviation results. The latter will find some natural applications in reinsurance (Section 8.7). An introduction to general stable processes is given in Section 8.8.

2.1 The Laws of Large Numbers

Throughout this chapter $X, X_1, X_2, \ldots$ is a sequence of iid non-degenerate rvs defined on a probability space $[\Omega, \mathcal{F}, P]$ with common df F. If we want to get a rough idea about the fluctuations of the X_n we might ask for convergence of the sequence (X_n). Unfortunately, for almost all $\omega \in \Omega$ this sequence does not converge. However, we can obtain some information about how the X_n "behave in the mean". This leads us to the consideration of the cumulative sums

$$S_0 = 0\,, \quad S_n = X_1 + \cdots + X_n, \quad n \ge 1\,,$$

and of the arithmetic (or sample) means

$$\overline{X}_n = n^{-1} S_n\,, \quad n \ge 1\,.$$

Mean values accompany our daily life. For instance, in the newspapers we are often confronted with average values in articles on statistical, actuarial or financial topics. Sometimes they occur in hidden form such as the NIKKEI, DAX, Dow Jones or other indices.

Intuitively, it is clear that an arithmetic mean should possess some sort of "stability" in n. So we expect that for large n the individual values X_i will have less influence on the order of $\overline{X}_n$, i.e. the sequence $(\overline{X}_n)$ stabilises around a fixed value (converges) as $n \to \infty$. This well-known effect is called a *law of large numbers.*

Suppose for the moment that $\sigma^2 = \operatorname{var}(X)$ is finite. Write $\mu = EX$. From Chebyshev's inequality we conclude that for $\epsilon > 0$,

$$P\left(\left|\overline{X}_n - \mu\right| > \epsilon\right) \;\le\; \epsilon^{-2} \operatorname{var}\left(\overline{X}_n\right) = (n\epsilon^2)^{-1}\sigma^2 \to 0\,, \quad n \to \infty\,.$$

Hence

$$\overline{X}_n \stackrel{P}{\to} \mu\,, \quad n \to \infty\,.$$

This relation is called the *weak law of large numbers* (*WLLN*) or simply the *law of large numbers* (*LLN*) for the sequence (X_n). If we interpret the index of X_n as time n then $\overline{X}_n$ is an average over time. On the other hand, the expectation

$$EX = \int_\Omega X(\omega)\, dP(\omega)$$

is a weighted average over the probability space Ω. Hence the LLN tells us that, over long periods of time, the time average $\overline{X}_n$ converges to the space average EX. This is the physical interpretation of the LLN which gained it the special name *ergodic theorem.*

We saw that (X_n) obeys the WLLN if the variance σ^2 is finite. This condition can be weakened substantially:

Theorem 2.1.1 (Criterion for the WLLN)
The WLLN

$$\overline{X}_n \xrightarrow{P} 0$$

holds if and only if the following two conditions are satisfied:

$$\begin{aligned} nP(|X| > n) &\rightarrow 0\,, \\ EXI_{\{|X| \le n\}} &\rightarrow 0\,. \end{aligned}$$ □

Here and throughout we use the notation

$$I_A(\omega) = \begin{cases} 1 & \text{if } \omega \in A\,, \\ 0 & \text{otherwise}\,, \end{cases}$$

for the *indicator function of the event* (of the set) A.

The assumptions of Theorem 2.1.1 are easily checked for the centred sequence $(X_n - \mu)$. Thus we conclude that the WLLN $\overline{X}_n \xrightarrow{P} \mu$ holds provided the expectation of X is finite. But we also see that the existence of a first moment is not necessary for the WLLN in the form $\overline{X}_n \xrightarrow{P} 0$:

Example 2.1.2 Let X be symmetric with tail

$$P(|X| > x) \sim \frac{c}{x \ln x}\,, \quad x \to \infty\,,$$

for some constant $c > 0$. The conditions of Theorem 2.1.1 are easily checked. Hence $\overline{X}_n \xrightarrow{P} 0$. However,

$$E|X| = \int_0^\infty P(|X| > x)\, dx = \infty\,.$$ □

Next we ask what conditions are needed to ensure that $\overline{X}_n$ does not only converge in probability but also *with probability 1* or *almost surely* (*a.s.*). Such a result is then called a *strong law of large numbers* (*SLLN*) for the sequence (X_n). The existence of the first moment is a necessary condition for the SLLN: given that $\overline{X}_n \xrightarrow{\text{a.s.}} a$ for some finite constant a we have that

$$n^{-1} X_n = n^{-1} (S_n - S_{n-1}) \xrightarrow{\text{a.s.}} a - a = 0\,.$$

Hence, for $\epsilon > 0$,

$$P\left(n^{-1} |X_n| > \epsilon \quad \text{i.o.}\right) = 0\,.$$

This and the Borel–Cantelli lemma (see Section 3.5) imply that for $\epsilon > 0$,

$$\sum_{n=1}^{\infty} P\left(\left|n^{-1}X_n\right| > \epsilon\right) = \sum_{n=1}^{\infty} P\left(|X| > \epsilon n\right) < \infty\,,$$

which means that $E|X| < \infty$. This condition is also sufficient for the SLLN:

Theorem 2.1.3 (Kolmogorov's SLLN)
The SLLN

$$\overline{X}_n \overset{\text{a.s.}}{\to} a$$

holds for the sequence (X_n) and some real constant a if and only if $E|X| < \infty$. Moreover, if (X_n) obeys the SLLN then $a = \mu$. □

Formally, Kolmogorov's SLLN remains valid for positive (negative) rvs with infinite mean, i.e. in that case we have

$$\overline{X}_n \overset{\text{a.s.}}{\to} EX = \infty \quad (= -\infty)\,.$$

Example 2.1.4 (Glivenko–Cantelli theorem)
Denote by

$$F_n(x) = \frac{1}{n}\sum_{i=1}^{n} I_{\{X_i \le x\}}\,, \qquad x \in \mathbb{R}\,,$$

the *empirical df* of the iid sample $X_1, \ldots, X_n$. An application of the SLLN yields that

$$F_n(x) \overset{\text{a.s.}}{\to} EI_{\{X \le x\}} = F(x)$$

for every $x \in \mathbb{R}$. The latter can be strengthened (and is indeed equivalent) to

$$\Delta_n = \sup_{x \in \mathbb{R}} |F_n(x) - F(x)| \overset{\text{a.s.}}{\to} 0\,. \tag{2.1}$$

The latter is known as the *Glivenko–Cantelli theorem*. It is one of the fundamental results in non–parametric statistics. In what follows we will frequently make use of it.
We give a proof of (2.1) for a continuous df F. For general F see Theorem 20.6 in Billingsley [70]. Let

$$-\infty = x_0 < x_1 < \cdots < x_k < x_{k+1} = \infty$$

be points such that $F(x_{i+1}) - F(x_i) < \varepsilon$ for a given $\varepsilon > 0$, $i = 0, \ldots, k$. $F(\pm\infty)$ are interpreted in the natural way as limits. By the monotonicity of F and F_n we obtain

$$\begin{aligned}
\Delta_n &= \max_{i=0,\ldots,k} \sup_{x \in (x_i, x_{i+1}]} |F_n(x) - F(x)| \\
&\le \max_{i=0,\ldots,k} \left(F_n(x_{i+1}) - F(x_i), F(x_{i+1}) - F_n(x_i)\right)\,.
\end{aligned}$$

An application of the SLLN to the rhs yields

$$\limsup_{n\to\infty} \Delta_n \le \max_{i=0,\ldots,k} (F(x_{i+1}) - F(x_i)) < \varepsilon \quad \text{a.s.}$$

This concludes the proof of (2.1). The latter remains valid for stationary ergodic sequences (X_n). This is a consequence of Birkhoff's ergodic theorem (for instance Billingsley [68]) which implies that $F_n(x) \stackrel{\text{a.s.}}{\to} F(x)$ for every fixed x. □

The SLLN yields an a.s. first-order approximation of the rv $\overline{X}_n$ by the deterministic quantity μ:

$$\overline{X}_n = \mu + o(1) \quad \text{a.s.}$$

The natural question that arises is:

What is the quality of this approximation?

Refinements of the SLLN are the aim of some of our future considerations. We pose a further question:

What can we conclude about the a.s. fluctuations of the sums S_n if we choose another normalising sequence?

A natural choice of normalising constants is given by the powers of n.

Theorem 2.1.5 (Marcinkiewicz–Zygmund SLLNs)
Suppose that $p \in (0,2)$. The SLLN

$$n^{-1/p}\,(S_n - a\,n) \stackrel{\text{a.s.}}{\to} 0 \tag{2.2}$$

holds for some real constant a if and only if $E|X|^p < \infty$. If (X_n) obeys the SLLN (2.2) then we can choose

$$a = \begin{cases} 0 & \text{if } \; p < 1\,, \\ \mu & \text{if } \; p \in [1,2)\,. \end{cases}$$

Moreover, if $E|X|^p = \infty$ for some $p \in (0,2)$ then for every real a,

$$\limsup_{n\to\infty} n^{-1/p}\,|S_n - a\,n| = \infty \quad \text{a.s.}$$

□

This theorem gives a complete characterisation of the SLLN with normalising power functions of n. Under the conditions of Theorem 2.1.5 we obtain the following refined a.s. first-order approximation of $\overline{X}_n$:

$$\overline{X}_n = \mu + o(n^{1/p-1}) \quad \text{a.s.}\,, \tag{2.3}$$

which is valid if $E|X|^p < \infty$ for some $p \in [1,2)$.

Theorem 2.1.5 allows us to derive an elementary relationship between the large fluctuations of the sums S_n, the summands X_n and the maxima

$$M_1 = |X_1|\,, \quad M_n = \max\left(|X_1|,\ldots,|X_n|\right)\,, \quad n \ge 2\,.$$

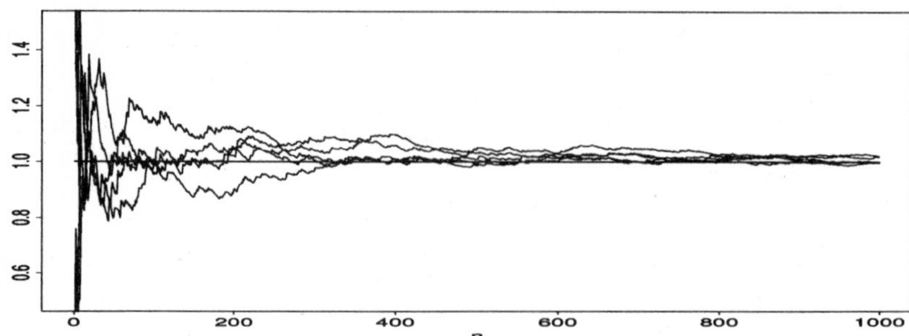

Figure 2.1.6 *Visualisation of the convergence in the SLLN: five sample paths of the process (S_n/n) for iid standard exponential X_n.*

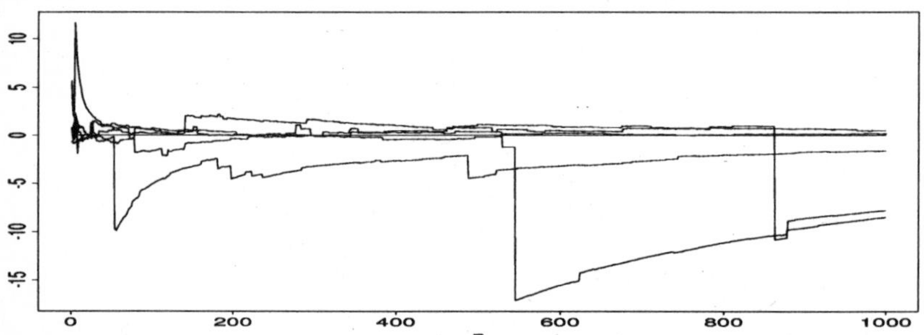

Figure 2.1.7 *Failure of the SLLN: five sample paths of the process (S_n/n) for iid standard symmetric Cauchy rvs X_n with $E|X| = \infty$, hence the wild oscillations of the sample paths.*

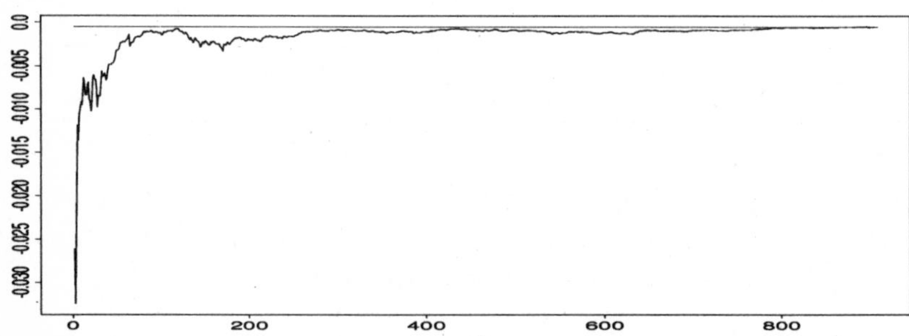

Figure 2.1.8 *The SLLN for daily log–returns of the* NIKKEI *index February 22, 1990 – October 8, 1993. The solid straight line shows the mean -0.000501 of these* 910 *values.*

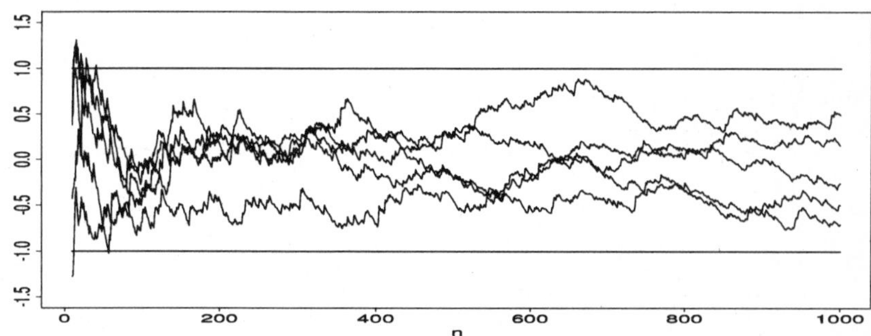

Figure 2.1.9 *Visualisation of the LIL: five sample paths of the process* $((2n \ln \ln n)^{-1/2}(S_n - n))$ *for iid standard exponential rvs* X_n.

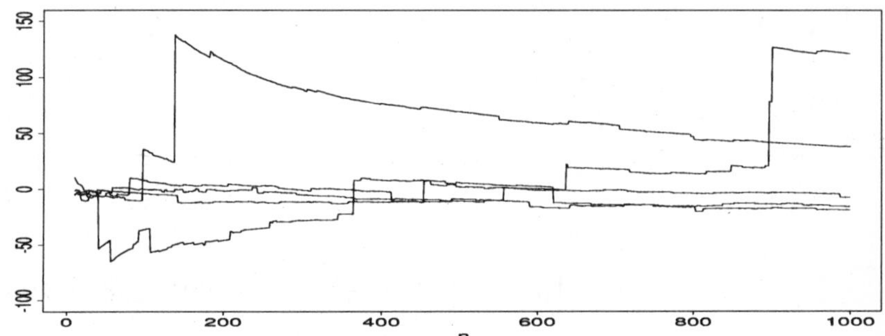

Figure 2.1.10 *Failure of the LIL: five sample paths of the process* $((2n \ln \ln n)^{-1/2} S_n)$ *for iid standard symmetric Cauchy rvs* X_n. *Notice the difference in the vertical scale!*

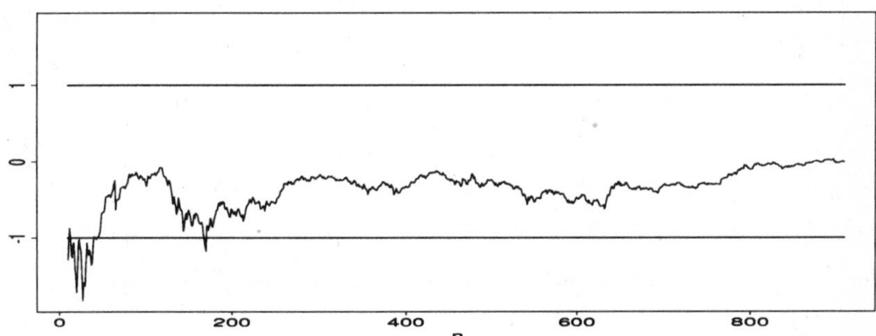

Figure 2.1.11 *The LIL for daily log-returns of the* NIKKEI *index February 22, 1990 – October 8, 1993.*

Corollary 2.1.12 *Suppose that $p \in (0,2)$. Then*

$$E|X|^p < \infty \qquad (= \infty) \tag{2.4}$$

according as

$$\limsup_{n\to\infty} n^{-1/p}|X_n| = 0 \qquad (= \infty) \quad \text{a.s.} \tag{2.5}$$

according as

$$\limsup_{n\to\infty} n^{-1/p} M_n = 0 \qquad (= \infty) \quad \text{a.s.} \tag{2.6}$$

according as

$$\limsup_{n\to\infty} n^{-1/p}|S_n - a\,n| = 0 \qquad (= \infty) \quad \text{a.s.} \tag{2.7}$$

Here a has to be chosen as in Theorem 2.1.5.

Proof. It is not difficult to see that (2.4) holds if and only if

$$\sum_{n=1}^{\infty} P\left(|X| > \epsilon n^{1/p}\right) < \infty \qquad (= \infty) \quad \forall \epsilon > 0\,.$$

A Borel–Cantelli argument yields that this is equivalent to

$$P\left(|X_n| > \epsilon n^{1/p} \quad \text{i.o.}\right) = 0 \qquad (= 1) \quad \forall \epsilon > 0\,.$$

Combining this and Theorem 2.1.5 we see that (2.4), (2.5) and (2.7) are equivalent.

The equivalence of (2.5) and (2.6) is a consequence of the elementary relation

$$\begin{aligned} n^{-1/p}|X_n| &\le n^{-1/p} M_n \\ &\le \max\left(\frac{|X_1|}{n^{1/p}}, \ldots, \frac{|X_{n_0}|}{n^{1/p}}, \frac{|X_{n_0+1}|}{(n_0+1)^{1/p}}, \frac{|X_{n_0+2}|}{(n_0+2)^{1/p}}, \ldots, \frac{|X_n|}{n^{1/p}}\right) \end{aligned}$$

for every fixed $n_0 \le n$. □

This means that the asymptotic order of magnitude of the sums S_n, of the summands X_n and of the maxima M_n is roughly the same.

Another question arises from Theorem 2.1.5:

Can we choose $p = 2$ or even $p > 2$ in (2.2) ?

The answer is unfortunately *no*. More precisely, for all non–degenerate rvs X and deterministic sequences (a_n),

$$\limsup_{n\to\infty} n^{-1/2}\,|S_n - a_n| = \infty \quad \text{a.s.}$$

This is somewhat surprising because we might have expected that the more moments of X exist the smaller the fluctuations of the sums S_n. This is not the case by virtue of the central limit theorem (CLT) which we will consider in more detail in Sections 2.2 and 2.3. Indeed, the CLT requires the normalisation $n^{1/2}$ which makes a result like (2.3) for $p = 2$ impossible. However, a last a.s. refinement can still be done if the second moment of X exists:

Theorem 2.1.13 (Hartman–Wintner law of the iterated logarithm)
If $\sigma^2 = \text{var}(X) < \infty$ *then*

$$\limsup_{n\to\infty} (2n\ln\ln n)^{-1/2}\,(S_n - \mu\, n) \;=\; -\liminf_{n\to\infty} (2n\ln\ln n)^{-1/2}\,(S_n - \mu\, n) \;=\; \sigma \quad \text{a.s.}$$

If $\sigma^2 = \infty$ *then for every real sequence* (a_n)

$$\limsup_{n\to\infty} (2n\ln\ln n)^{-1/2}\,|S_n - a_n| = \infty \quad \text{a.s.} \qquad \square$$

Hence the *law of the iterated logarithm* (*LIL*) as stated by Theorem 2.1.13 gives us the a.s. first–order approximation

$$\overline{X}_n = \mu + O\left(\left(\ln\ln n/n\right)^{-1/2}\right) \quad \text{a.s.},$$

which is the best possible a.s. approximation of $\overline{X}_n$ by its expectation μ. We will see in the next section that we have to change the mode of convergence if we want to derive more information about the fluctuations of the sums S_n. There we will commence with their distributional behaviour.

Notes and Comments

The WLLN for iid sequences (Theorem 2.1.1) can be found in any standard textbook on probability theory; see for instance Breiman [90], Chow and Teicher [118], Feller [235], Loève [427]. The WLLN with other normalisations and for the non–iid case has been treated for instance in Feller [235], Petrov [495, 496], or in the martingale case in Hall and Heyde [312].

More insight into the weak limit behaviour of sums is given by so–called *rates of convergence* in the LLN, i.e. by statements about the order of the probabilities

$$P(|S_n - a_n| > b_n), \quad n \to \infty,$$

for appropriate normalising and centring sequences (b_n), (a_n). We refer to Petrov [495, 496] and the literature cited therein.

The classical Kolmogorov SLLN (Theorem 2.1.3) is part of every standard textbook on probability theory, and the Marcinkiewicz–Zygmund SLLNs can be found for instance in Stout [608]. Necessary and sufficient conditions under non–standard normalisations and for the non–iid case are given for instance in Petrov [495, 496] or in Stout [608]. In Révész [538] and Stout [608] various SLLNs are proved for sequences of dependent rvs. Some remarks on the *convergence rate in the SLLN*, i.e. on the order of the probabilities

$$P\left(\sup_{k \ge n} |S_k - a_k| > b_n\right), \quad n \to \infty,$$

can be found in Petrov [495, 496].

The *ergodic theorem* as mentioned above is a classical result which holds for stationary ergodic (X_n); see for instance Billingsley [68] or Stout [608].

The limit in the SLLN for a sequence (X_n) of independent rvs is necessarily a constant. This is due to the so–called 0–1 *law*; for different versions see for instance Stout [608]. The limit in the SLLN for a sequence of dependent rvs can be a genuine rv.

The Marcinkiewicz–Zygmund SLLNs for an iid sequence exhibit another kind of 0–1 *behaviour:* either the SLLN holds with a constant limit for the normalisation $n^{1/p}$ or, with the same normalisation, the sums fluctuate wildly with upper or lower limit equal to $\pm\infty$. This behaviour is typical for a large class of normalisations (cf. *Feller's SLLN*; see Feller [233], Petrov [495, 496], Stout [608]). Similar behaviour can be observed for a large class of rvs with infinite variance which includes the class of α–stable rvs, $\alpha < 2$, and their domains of attraction; see Section 2.2. To be precise, suppose that for some constant $c > 0$,

$$x^{-2} E X^2 I_{\{|X| \le x\}} \le c\, P(|X| > x), \quad x > 0. \tag{2.8}$$

Let (b_n) be any real sequence such that $b_n \uparrow \infty$ and if $E|X| < \infty$ suppose that $\mu = 0$. Then

$$\limsup_{n \to \infty} b_n^{-1} |S_n| = 0 \qquad (= \infty) \quad \text{a.s.}$$

according as

$$\sum_{n=1}^{\infty} P(|X| > b_n) < \infty \qquad (= \infty).$$

The latter relation is a moment condition. Moreover, the relations between S_n, X_n and M_n (with normalisation b_n) corresponding to Corollary 2.1.12 hold. This SLLN is basically due to Heyde [323]; see Stout [608].

The SLLN can also be extended to sums S_n of *independent but not identically distributed rvs*. There exist results under the condition of finiteness of the second moments of the X_n. The results are typically of the form

$$b_n^{-1}(S_n - ES_n) \stackrel{\text{a.s.}}{\to} 0 ,$$

where b_n is the variance of S_n; see for instance Petrov [495, 496]. However, it seems difficult to use such a model for statistical inference as long as the class of distributions of the X_n is not specified. A more sensitive study is possible for sequences of iid rvs with given deterministic weights. *Weighted sums*

$$T_n = \sum_k w_n(k)\, X_k$$

are standard models in the statistical literature. For example, in time series analysis the linear processes, including the important class of ARMA processes, are weighted sums of iid rvs; see Chapter 7. The rvs T_n can be considered as a mean which, in contrast to $\overline{X}_n$, gives different weight to the observations X_k. Examples are the discounted sums $\sum_{k\geq 0} z^k X_k$ whose asymptotic behaviour (as $z \uparrow 1$) is well studied (so–called Abel summation). There is quite a mass of literature on the a.s. behaviour of the weighted sums T_n. Results of SLLN–type can be found for instance in Mikosch and Norvaiša [461] or in Stout [608]. Overviews of *summability methods* have been given in Bingham and Maejima [71], Maejima [433], Mikosch and Norvaiša [459, 460].

The *Hartman–Wintner LIL* (Theorem 2.1.13) is included in standard textbooks; see for instance Feller [235]. Different proofs and ramifications for non–identically distributed rvs and dependent observations can be found in Csörgő and Révész [145], Hall and Heyde [312], Petrov [495, 496], Stout [608].

There exists a well developed theory about fluctuations of sums of iid rvs with or without normalisation. The latter is also called a *random walk*. For example, necessary and sufficient conditions have been derived for relations of type

$$\begin{aligned} \limsup_{n\to\infty} \quad b_n^{-1}\,|S_n - a_n| &= c_1 \in (0,\infty) \qquad \text{a.s.}\,, \\ \limsup_{n\to\infty} \quad b_n^{-1}\,(S_n - a_n) &= c_2 \in (-\infty,\infty) \qquad \text{a.s.} \end{aligned}$$

(*generalised LIL, one–sided LIL*) for given (a_n), (b_n) and constants c_1, c_2, and also results about the existence of such normalising or centring constants; see for instance Kesten [378], Klass [381, 382], Martikainen [441, 442], Pruitt

[515, 517]. These results give some insight into the complicated nature of the fluctuations of sums. However, they are very difficult to apply: the sequence (b_n) is usually constructed in such a way that one has to know the whole distribution tail of X. Thus these results are very sensitive to changes in the distribution.

A further topic of research has been concentrated around *cluster phenomena* of the sums S_n (normalised or non-normalised) and the general properties of random walks. We refer to Cohen [130], Erickson [222, 223], Kesten [378], Révész [540], Spitzer [604], Stout [608]. The set of *a.s. limit points* of the sequence of normalised sums can be very complicated. However, in many situations the set of a.s. limit points coincides with a closed interval (finite or infinite). The following basic idea from elementary calculus is helpful: let (a_n) be a sequence of real numbers such that $a_n - a_{n-1} \to 0$. Then every point in the interval $[\liminf_{n\to\infty} a_n, \limsup_{n\to\infty} a_n]$ is a limit point of (a_n). Applying this to the Hartman–Wintner LIL with $EX^2 < \infty$ and $A_n = (2n \ln\ln n)^{-1/2}(S_n - \mu\, n)$ we see that $A_n - A_{n-1} \stackrel{\text{a.s.}}{\to} 0$ and hence every point in $[-\sigma, \sigma]$ is a limit point of (A_n) for almost every sample path. This remarkable property means that the points $A_1, A_2, \ldots$ fill the interval $[-\sigma, \sigma]$ densely for almost every sample path. This is somehow counter–intuitive since at the same time $A_n \stackrel{P}{\to} 0$.

2.2 The Central Limit Problem

In the preceding section we saw that the sums S_n of the iid sequence (X_n) diverge a.s. when normalised with $n^{1/2}$. However, we can still get information about the growth of $n^{-1/2} S_n$ if we change to *convergence in distribution* (*weak convergence*).

We will approach the problem from a more general point of view. We ask:

What are the possible (non–degenerate) limit laws for the sums S_n when properly normalised and centred?

This is a classical question in probability theory. Many famous probabilists of this century have contributed to its complete solution: Khinchin, Lévy, Kolmogorov, Gnedenko, Feller,.... It turns out that this question is closely related to another one:

Which distributions satisfy the identity in law

$$c_1 X_1 + c_2 X_2 \stackrel{d}{=} b(c_1, c_2) X + a(c_1, c_2) \tag{2.9}$$

for all non–negative numbers c_1, c_2 and appropriate real numbers $b(c_1, c_2) > 0$ and $a(c_1, c_2)$?

In other words, which classes of distributions are closed (up to changes of location and scale) under convolution and multiplication with real numbers? The possible limit laws for sums of iid rvs are just the distributions which satisfy (2.9) for all non-negative c_1, c_2. Many classes of distributions are closed with respect to convolution but the requirement (2.9) is more stringent. For example, the convolution of two Poisson distributions is a Poisson distribution. However, the Poisson distributions do not satisfy (2.9).

Definition 2.2.1 (Stable distribution and rv)
A rv (a distribution, a df) is called stable *if it satisfies* (2.9) *for iid X, X_1, X_2, for all non-negative numbers c_1, c_2 and appropriate real numbers $b(c_1, c_2) > 0$ and $a(c_1, c_2)$.* □

Now consider the sum S_n of iid stable rvs. By (2.9) we have for some real constants a_n and $b_n > 0$ and $X = X_1$,

$$S_n = X_1 + \cdots + X_n \stackrel{d}{=} b_n X + a_n\,, \quad n \geq 1\,,$$

which we can rewrite as

$$b_n^{-1}(S_n - a_n) \stackrel{d}{=} X\,.$$

We conclude that, if a distribution is stable, then it is a limit distribution for sums of iid rvs. Are there any other possible limit distributions? The answer is NO:

Theorem 2.2.2 (Limit property of a stable law)
The class of the stable (non-degenerate) distributions coincides with the class of all possible (non-degenerate) limit laws for (properly normalised and centred) sums of iid rvs. □

Because of the importance of the class of stable distributions it is necessary to describe them analytically. The most common way is to determine their characteristic functions (chfs):

Theorem 2.2.3 (Spectral representation of a stable law)
A stable distribution has chf

$$\phi_X(t) = E\exp\{iXt\} = \exp\{i\gamma t - c|t|^\alpha(1 - i\beta \operatorname{sign}(t)\, z(t,\alpha))\}\,, \quad t \in \mathbb{R}, \tag{2.10}$$

where γ is a real constant, $c > 0$, $\alpha \in (0,2]$, $\beta \in [-1,1]$, and

$$z(t,\alpha) = \begin{cases} \tan\left(\dfrac{\pi\alpha}{2}\right) & \text{if } \alpha \neq 1\,, \\ -\dfrac{2}{\pi}\ln|t| & \text{if } \alpha = 1\,. \end{cases}$$

□

Remarks. 1) We note that we can formally include the case $c = 0$ which corresponds to a degenerate distribution. Every sequence (S_n) can be normalised and centred in such a way that it converges to a constant (for instance zero) in probability. Thus this trivial limit belongs to the class of the possible limit rvs. However, it is not of interest in the context of weak convergence and therefore excluded from our considerations.

2) The quantity γ is just a location parameter. For the rest of this section we assume $\gamma = 0$.

3) The most important parameter in this representation is α. It determines the basic properties of this class of distributions (moments, tails, asymptotic behaviour of sums, normalisation etc.). □

Definition 2.2.4 *The number α in the chf* (2.10) *is called the* characteristic exponent, *the corresponding distribution* α–stable. □

Remarks. 4) For $\alpha = 2$ we obtain the *normal* or *Gaussian* distributions. In this case, we derive from (2.10) the well known chf

$$\phi_X(t) = \exp\left\{-ct^2\right\}$$

of a Gaussian rv with mean zero and variance $2c$. Thus one of the most important distributions in probability theory and mathematical statistics is a stable law. We also see that the normal law is determined just by two parameters (mean and variance) whereas the other α–stable distributions depend on four parameters. This is due to the fact that a normal distribution is always symmetric (around its expectation) whereas a stable law for $\alpha < 2$ can be asymmetric and even be concentrated on a half–axis (for $\alpha < 1$).

5) Another well–known class of stable distributions corresponds to $\alpha = 1$: the *Cauchy laws* with chf

$$\phi_X(t) = \exp\left\{-c|t|\left(1 + i\beta\frac{2}{\pi}\,\mathrm{sign}\,(t)\ln|t|\right)\right\}.$$

6) For fixed α, the parameters c and β determine the nature of the distribution. The parameter c is a scaling constant which corresponds to $c = \sigma^2/2$ in the Gaussian case and has a similar function as the variance in the non–Gaussian case (where the variance is infinite). The parameter β describes the skewness of the distribution. We see that the chf $\phi_X(t)$ is real–valued if and only if $\beta = 0$. The chf

$$\phi_X(t) = \exp\left\{-c|t|^\alpha\right\} \tag{2.11}$$

corresponds to a symmetric rv X. We will sometimes use the notation $s\alpha s$ for *symmetric α–stable*. If $\beta = 1$ and $\alpha < 1$ the rv X is positive, and for

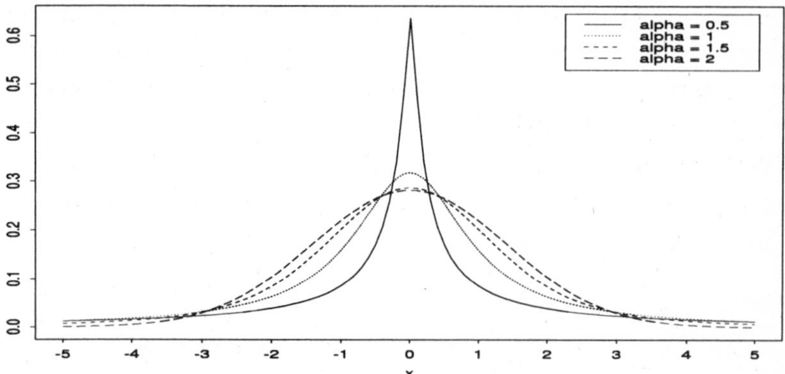

Figure 2.2.5 *Densities of sαs rvs with* $c = 1$.

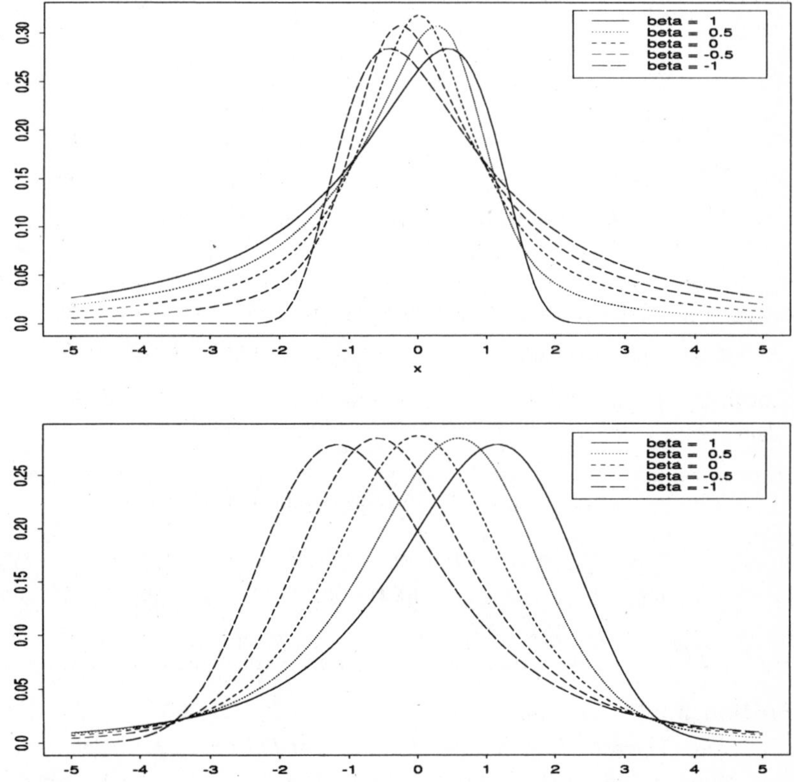

Figure 2.2.6 *Densities of* 1– *and of* 1.5–*stable rvs (top, bottom) with* $c = 1$.

$\beta = -1$ and $\alpha < 1$ it is negative. In the cases $|\beta| < 1$, $\alpha < 1$, or $\alpha \in [1,2]$, the rv X has the whole real axis as support. However, if $\beta = 1$, $\alpha \in [1,2)$ then $P(X \le -x) = o(P(X > x))$ as $x \to \infty$. From the chf (2.10) we also deduce that $-X$ is α–stable with parameters c and $-\beta$. It can be shown that every α–stable rv X with $|\beta| < 1$ is equal in law to $X' - X'' + c_0$ for independent α–stable rvs X', X'' both with parameter $\beta = 1$ and a certain constant c_0.

7) We might ask why we used the inconvenient (from a practical point of view) representation of α–stable rvs via their chf. The answer is simple: it is the best analytic way of characterising all members of this class. Although the α–stable laws are absolutely continuous, their densities can be expressed only by complicated special functions; see Hoffmann–Jørgensen [329] and Zolotarev [646]. Only in a few cases which include the Gaussian ($\alpha = 2$), the symmetric Cauchy ($\alpha = 1, \beta = 0$) and the stable inverse Gaussian ($\alpha = 1/2, \beta = 1$), are these densities expressible explicitly via elementary functions. But there exist asymptotic expansions of the α–stable densities in a neighbourhood of the origin or of infinity; see Ibragimov and Linnik [350] and Zolotarev [645]. Therefore the α–stable distributions (with a few exceptions) are not easy to handle. In particular, they are difficult to simulate; see for instance Janicki and Weron [354]. □

Next we ask:

Given an α–stable distribution G_α, what conditions imply that the normalised and centred sums S_n converge weakly to G_α?

This question induces some further problems:

How must we choose constants $a_n \in \mathbb{R}$ and $b_n > 0$ such that

$$b_n^{-1}(S_n - a_n) \xrightarrow{d} G_\alpha\,? \tag{2.12}$$

Can it happen that different normalising or centring constants imply convergence to different limit laws?

The last question can be answered immediately: the convergence to types theorem (Theorem A1.5) ensures that the limit law is uniquely determined up to positive affine transformations.

Before we answer the other questions we introduce some further notions:

Definition 2.2.7 (Domain of attraction)
We say that the rv X (the df F of X, the distribution of X) belongs to the domain of attraction of the α–stable distribution G_α *if there exist constants $a_n \in \mathbb{R}$, $b_n > 0$ such that* (2.12) *holds. We write $X \in \mathrm{DA}(G_\alpha)$ ($F \in \mathrm{DA}(G_\alpha)$) and say that (X_n)* satisfies the central limit theorem (CLT) with limit G_α. □

If we are interested only in the fact that X (or F) is attracted by some α–stable law whose concrete form is not of interest we will simply write $X \in \mathrm{DA}(\alpha)$ (or $F \in \mathrm{DA}(\alpha)$).

The following result characterises the domain of attraction of a stable law completely. Here and in the remainder of this section we will need some facts about *slowly and regularly varying* functions which are given in Appendix A3.1. We recall that a (measurable) function L is slowly varying if $\lim_{x\to\infty} L(tx)/L(x) = 1$ for all $t > 0$.

Theorem 2.2.8 (Characterisation of domain of attraction)

(a) *The df F belongs to the domain of attraction of a normal law if and only if*

$$\int_{|y|\le x} y^2\, dF(y)$$

is slowly varying.

(b) *The df F belongs to the domain of attraction of an α–stable law for some $\alpha < 2$ if and only if*

$$F(-x) = \frac{c_1 + o(1)}{x^\alpha}\, L(x)\,, \quad 1 - F(x) = \frac{c_2 + o(1)}{x^\alpha}\, L(x)\,, \quad x \to \infty\,,$$

where L is slowly varying and c_1, c_2 are non–negative constants such that $c_1 + c_2 > 0$. □

First we study the case $\alpha = 2$ more in detail. If $EX^2 < \infty$ then

$$\int_{|y|\le x} y^2\, dF(y) \to EX^2\,, \quad x \to \infty\,,$$

hence $X \in \mathrm{DA}(2)$. Moreover, by Proposition A3.8(f) we conclude that slow variation of $\int_{|y|\le x} y^2\, dF(y)$ is equivalent to the tail condition

$$G(x) = P(|X| > x) = o\left(x^{-2} \int_{|y|\le x} y^2\, dF(y)\right)\,, \quad x \to \infty\,. \tag{2.13}$$

Thus we derived

Corollary 2.2.9 (Domain of attraction of a normal distribution)
A rv X is in the domain of attraction of a normal law if and only if one of the following conditions holds:

(a) $EX^2 < \infty$,
(b) $EX^2 = \infty$ *and* (2.13). □

The situation is completely different for $\alpha < 2$: $X \in \mathrm{DA}(\alpha)$ implies that

$$G(x) = x^{-\alpha} L(x)\,, \quad x > 0\,, \tag{2.14}$$

for a slowly varying function L and

$$x^2 G(x) \Big/ \int_{|y| \le x} y^2 \, dF(y) \to \frac{2-\alpha}{\alpha}, \quad x \to \infty\,. \tag{2.15}$$

The latter follows from Proposition A3.8(e). Hence the second moment of X is infinite. Relation (2.14) and Corollary 2.2.9 show that the domain of attraction of the normal distribution is much more general than the domain of attraction of an α–stable law with exponent $\alpha < 2$. We see that DA(2) contains at least all distributions that have a second finite moment.

From Corollary 2.2.9 and from (2.14) we conclude the following about the moments of distributions in $\mathrm{DA}(\alpha)$:

Corollary 2.2.10 (Moments of distributions in $\mathrm{DA}(\alpha)$)
If $X \in \mathrm{DA}(\alpha)$ then

$$\begin{aligned} E|X|^\delta &< \infty \quad \text{for } \delta < \alpha\,, \\ E|X|^\delta &= \infty \quad \text{for } \delta > \alpha \text{ and } \alpha < 2\,. \end{aligned}$$

In particular,

$$\begin{aligned} \operatorname{var}(X) &= \infty \quad \text{for } \alpha < 2\,, \\ E|X| &< \infty \quad \text{for } \alpha > 1\,, \\ E|X| &= \infty \quad \text{for } \alpha < 1\,. \end{aligned}$$

□

Note that $E|X|^\alpha = \int_0^\infty P(|X|^\alpha > x)dx < \infty$ is possible for certain $X \in \mathrm{DA}(\alpha)$, but $E|X|^\alpha = \infty$ for an α–stable X for $\alpha < 2$. Recalling the results of the preceding section we can apply the Marcinkiewicz–Zygmund SLLNs (Theorem 2.1.5) as well as Heyde's SLLN ((2.8) is satisfied in view of (2.15)) to α–stable rvs with $\alpha < 2$, and these theorems show that the sample paths of (S_n) fluctuate wildly because of the non–existence of the second moment.

Next we want to find appropriate normalising and centring constants for the CLT (2.12). Suppose for the moment that X is $s\alpha s$ with chf $\phi_X(t) = \exp\{-c|t|^\alpha\}$; cf. (2.11). We see that

$$\begin{aligned} E\exp\left\{itn^{-1/\alpha}S_n\right\} &= \left(\exp\left\{-c|n^{-1/\alpha}t|^\alpha\right\}\right)^n \qquad (2.16)\\ &= \exp\{-c|t|^\alpha\} \\ &= E\exp\{itX\}\,. \end{aligned}$$

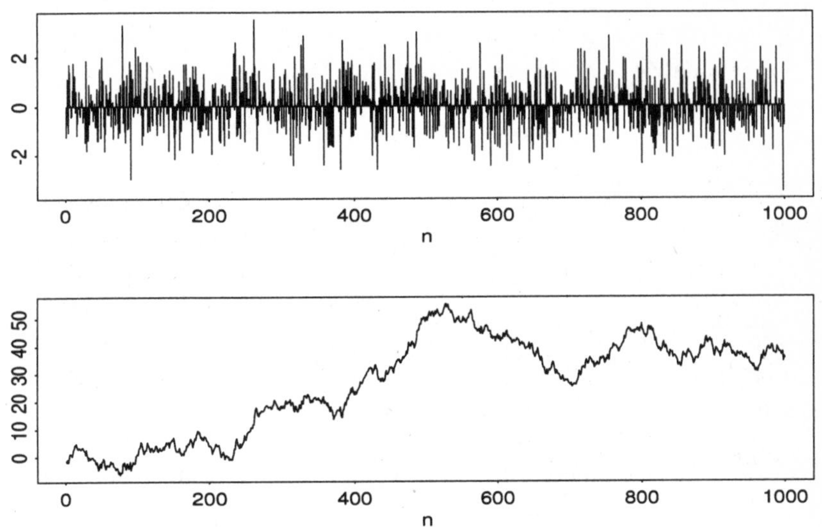

Figure 2.2.11 *One sample path of the process* (X_n) *(top) and of the corresponding path* (S_n) *(bottom) for* X *with a standard normal distribution.*

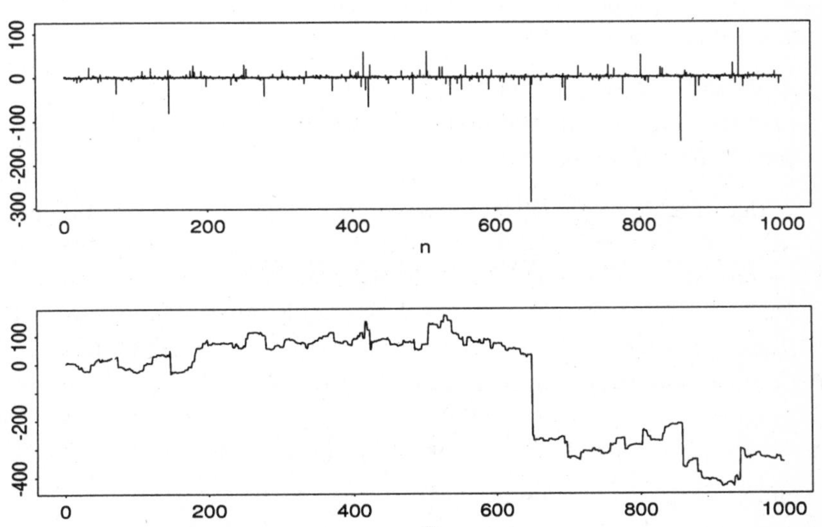

Figure 2.2.12 *One sample path of the process* (X_n) *(top) and of the corresponding path* (S_n) *(bottom) for* X *with a standard symmetric Cauchy distribution.*

Thus

$$n^{-1/\alpha} S_n \stackrel{d}{=} X$$

which gives us a rough impression of the order of the normalising constants.

For symmetric $X \in \mathrm{DA}(\alpha)$ one can show that the relation

$$E\exp\{itX\} = \exp\{-|t|^\alpha L_1(1/t)\}$$

holds in a neighbourhood of the origin, with a slowly varying function L_1 which is closely related to the slowly varying functions which occur in Theorem 2.2.8; see Theorem 2.6.5 in Ibragimov and Linnik [350]. Now we can apply the same idea as in (2.16) although this time we will have to compensate for the slowly varying function L_1. Thus it is not surprising that the normalising constants in the CLT (2.12) are of the form

$$b_n = n^{1/\alpha} L_2(n)$$

for a slowly varying function L_2. To be more precise, introduce the quantities

$$\begin{aligned} K(x) &= x^{-2} \int_{|y|\le x} y^2 \, dF(y)\,, \quad x>0\,, \\ Q(x) &= G(x) + K(x) = P(|X|>x) + K(x)\,, \quad x>0\,. \end{aligned}$$

Note that $Q(x)$ is continuous and decreasing on $[x_0,\infty)$ where x_0 denotes the infimum of the support of $|X|$.

Proposition 2.2.13 (Normalising constants in the CLT)
The normalising constants in the CLT for $F \in \mathrm{DA}(\alpha)$ can be chosen as the unique solution of the equation

$$Q(b_n) = n^{-1}\,, \quad n \ge 1\,. \tag{2.17}$$

In particular, if $\sigma^2 = \mathrm{var}(X) < \infty$ and $EX = 0$ then

$$b_n \sim n^{1/2}\sigma, \quad n \to \infty\,.$$

If $\alpha < 2$ we can alternatively choose (b_n) such that

$$b_n = \inf\{y : G(y) < n^{-1}\}\,, \quad n \ge 1\,. \tag{2.18}$$

We note that (2.18) implies that

$$G(b_n) \sim n^{-1}\,, \quad n \to \infty\,,$$

and that, in view of (2.14),

$$b_n = n^{1/\alpha} L_3(n)\,, \quad n \ge 1\,,$$

for an appropriate slowly varying function L_3.

Sketch of the proof. We omit the calculations leading to (2.17) and restrict ourselves to the particular cases. For a proof we refer to Ibragimov and Linnik [350], Section II.6.
If $EX^2 < \infty$ then

$$G(x) = o\left(x^{-2}\right) , \quad K(x) = x^{-2}\, EX^2(1+o(1)) , \quad x \to \infty .$$

Hence, if $EX = 0$,

$$n^{-1} = Q\left(b_n\right) \sim b_n^{-2}\sigma^2 .$$

If $\alpha < 2$ then, using (2.15), we see immediately that (2.17) and (2.18) yield asymptotically equivalent sequences (b_n) and (b'_n), say, which means that $b_n \sim \; c\, b'_n$ for a positive constant c. □

Proposition 2.2.14 (Centring constants in the CLT)
The centring constants a_n in the CLT (2.12) can be chosen as

$$a_n = n \int_{|y| \le b_n} y\, dF(y) , \tag{2.19}$$

where b_n is given in Proposition 2.2.13. In particular, we can take $a_n = \widetilde{a}\, n$, where

$$\widetilde{a} = \begin{cases} \mu & \text{if} \quad \alpha \in (1,2] , \\ 0 & \text{if} \quad \alpha \in (0,1) , \\ 0 & \text{if} \quad \alpha = 1 \text{ and } F \text{ is symmetric.} \end{cases} \tag{2.20}$$

□

For a proof we refer to Ibragimov and Linnik [350], Section II.6. Now we formulate a general version of the CLT.

Theorem 2.2.15 (General CLT)
Suppose that $F \in \mathrm{DA}(\alpha)$ for some $\alpha \in (0,2]$.

(a) *If $EX^2 < \infty$ then*

$$\left(\sigma n^{1/2}\right)^{-1} \; (S_n - \mu n) \xrightarrow{d} \Phi$$

for the standard normal distribution Φ with mean zero and variance 1.

(b) *If $EX^2 = \infty$ and $\alpha = 2$ or if $\alpha < 2$ then*

$$\left(n^{1/\alpha} L_4(n)\right)^{-1} \; (S_n - a_n) \xrightarrow{d} G_\alpha$$

for an α-stable distribution G_α, an appropriate slowly varying function L_4 and centring constants as in (2.19).
In particular,

$$\left(n^{1/\alpha}L_4(n)\right)^{-1}(S_n - \widetilde{a}n) \xrightarrow{d} G_\alpha\,,$$

where $\widetilde{a}$ is defined in (2.20). □

We notice that it is possible for the normalising constants in the CLT to be of the special form $b_n = cn^{1/\alpha}$ for some constant c. This happens for instance if $EX^2 < \infty$ or if X is α–stable. There is a special name for this situation:

Definition 2.2.16 (Domain of normal attraction)
We say that X (or F) belongs to the domain of normal attraction of an α–stable distribution G_α ($X \in \mathrm{DNA}(G_\alpha)$ or $F \in \mathrm{DNA}(G_\alpha)$) *if $X \in \mathrm{DA}(G_\alpha)$ and if in the CLT we can choose the normalisation $b_n = cn^{1/\alpha}$ for some positive constant c.* □

If we are interested only in the fact that X (or F) belongs to the DNA of some α–stable distribution we write $X \in \mathrm{DNA}(\alpha)$ (or $F \in \mathrm{DNA}(\alpha)$). We recall the characterisation of the domains of attraction via tails; see Theorem 2.2.8. Then (2.17) implies the following:

Corollary 2.2.17 (Characterisation of DNA)

(a) *The relation $F \in \mathrm{DNA}(2)$ holds if and only if $EX^2 < \infty$.*
(b) *For $\alpha < 2$, $F \in \mathrm{DNA}(\alpha)$ if and only if*

$$F(-x) \sim c_1 x^{-\alpha} \quad \textit{and} \quad 1 - F(x) \sim c_2 x^{-\alpha}\,, \quad x \to \infty\,,$$

for non–negative constants c_1, c_2 such that $c_1 + c_2 > 0$.

In particular, every α–stable distribution is in its own DNA. □

So we see that $F \in \mathrm{DNA}(\alpha)$, $\alpha < 2$, actually means that the corresponding tail $G(x)$ has *power law* or *Pareto–like* behaviour. Note that a df F with Pareto–like tail $G(x) \sim cx^{-\alpha}$ for some $\alpha \geq 2$ is in DA(2), and if $\alpha > 2$, then $F \in \mathrm{DNA}(2)$.

Notes and Comments

The theory above is classical and can be found in detail in Araujo and Giné [19], Bingham, Goldie and Teugels [72], Feller [235], Gnedenko and Kolmogorov [267], Ibragimov and Linnik [350], Loève [427] and many other textbooks. For applications of the CLT and related weak convergence results to asymptotic inference in statistics we refer to Ferguson [236] or Serfling [576].

There exists some more specialised literature on stable distributions and stable processes. Mijnheer [456] is one of the first monographs on the topic. Zolotarev [645] covers a wide range of interesting properties of stable distributions, including asymptotic expansions of the stable densities and many useful representations and transformation formulae. Some limit theory for distributions in the domain of attraction of a stable law is given in Christoph and Wolf [119]. An encyclopaedic treatment of stable laws, multivariate stable distributions and stable processes can be found in Samorodnitsky and Taqqu [565]; see also Kwapień and Woyczyński [411] and Janicki and Weron [354]. The latter book also deals with numerical aspects, in particular the simulation of stable rvs and processes. An introduction to stable random vectors and processes will be provided in Section 8.8.

Recently there have been some efforts to obtain efficient methods for the numerical calculation of stable densities. This has been a problem for many years and was one of the reasons that practitioners expressed doubts about the applicability of stable distributions for modelling purposes. McCulloch and Panton [451] and Nolan [480, 481] provided tables and software for calculating stable densities for a large variety of parameters α and β. Their methods allow one to determine those densities for small and moderate arguments with high accuracy; the determination of the densities in their tails needs further investigation. Figures 2.2.5 and 2.2.6 were obtained using software kindly provided to us by John Nolan.

The central limit problem has also been solved for independent non–iid rvs; see for instance Feller [235], Gnedenko and Kolmogorov [267], Ibragimov and Linnik [350], Petrov [495, 496]. To be precise, let $(X_{nk})_{k=1,\ldots,n}$, $n = 1, 2, \ldots$ be a triangular scheme of row–wise independent rvs satisfying the condition of *infinitesimality*:

$$\max_{k=1,\ldots,n} P(|X_{nk}| > \epsilon) \to 0\,, \quad n \to \infty\,, \quad \epsilon > 0\,.$$

The class of possible limit laws for the sums $\sum_{k=1}^{n} X_{nk}$ consists of the *infinitely divisible distributions* including most distributions of interest in statistics. For example, the stable distributions and the Poisson distribution belong to this class. A rv Y (and its distribution) is infinitely divisible if and only if we can decompose it in law:

$$Y \stackrel{d}{=} Y_{n1} + \cdots + Y_{nn}$$

for every n, where $(Y_{nk})_{k=1,\ldots,n}$ are iid rvs with possibly different common distribution for different n. There exist representations of the chf of an infinitely divisible law. Theorem 2.2.3 is a particular case for stable laws.

As in the case of a.s. convergence, see Section 2.1, *weighted sums* are particularly important for applications in statistics. The general limit theory for non–iid rvs can sometimes be applied to weighted sums. However, there exist quite a few results for special *summability methods*; for references see the Notes and Comments of Section 2.1.

2.3 Refinements of the CLT

In this section we consider some refinements of the results of the previous section. We will basically restrict ourselves to the case when $EX^2 < \infty$ and briefly comment on the other ones. It is natural to ask:

How can we determine and improve the quality of the approximation in the CLT?

Berry–Esséen Theorem

Let Φ denote the df of the standard normal distribution and write

$$G_n(x) = P\left(\frac{S_n - n\,\mu}{\sigma\sqrt{n}} \le x\right), \quad x \in \mathbb{R}.$$

From the previous section we know that

$$\Delta_n = \sup_{x\in\mathbb{R}} |G_n(x) - \Phi(x)| \to 0. \tag{2.21}$$

There we formulated only a weak convergence result, i.e. convergence of G_n at every continuity point of Φ. However, Φ is continuous and therefore (2.21) holds; see Appendix A1.1.

One can show that the rate at which Δ_n converges to zero can be arbitrarily slow if we require no more than a finite second moment of X. The typical rate of convergence is $1/\sqrt{n}$ provided the third moment of X exists. We give here a *non–uniform* version of the well–known *Berry–Esséen theorem:*

Theorem 2.3.1 (Berry–Esséen theorem)
Suppose that $E|X|^3 < \infty$. Then

$$|G_n(x) - \Phi(x)| \le \frac{c}{\sqrt{n}(1+|x|)^3}\,\frac{E|X-\mu|^3}{\sigma^3} \tag{2.22}$$

for all x, where c is a universal constant. In particular,

$$\Delta_n \le \frac{c}{\sqrt{n}}\,\frac{E|X-\mu|^3}{\sigma^3}. \tag{2.23}$$

□

From (2.22) we have learnt that the quality of the approximation can be improved substantially for large x. Moreover, the rate in (2.22) and (2.23) is influenced by the order of magnitude of the ratio $E|X-\mu|^3/\sigma^3$ and of the constant c. This is of crucial importance if n is small.

The rates in (2.22) and (2.23) are optimal in the sense that there exist sequences (X_n) such that $\Delta_n \asymp (1/\sqrt{n})$. For example, this is true for symmetric

Bernoulli rvs assuming the values +1 and −1 with equal probability. On the other hand, the Berry–Esséen estimate is rather pessimistic and can be improved when special conditions on X are satisfied, for instance the existence of a smooth density, moment generating function etc.

Results of Berry–Esséen type have been studied for $X \in \mathrm{DNA}(\alpha)$ and $X \in \mathrm{DA}(\alpha)$ with $\alpha < 2$ as well. A unifying result such as Theorem 2.3.1 does not exist and cannot be expected. The results require very special knowledge about the structure of the df in DNA and DA and are difficult to apply.

Notes and Comments

References for the speed of convergence in the CLT are Petrov [495, 496] and Rachev [520].

A proof of the classical Berry–Esséen theorem and its non–uniform version using Fourier methods can be found in Petrov [495, 496]. Also given there are results of Berry–Esséen type in the non–iid situation and for iid rvs under the existence of the $(2+\delta)$th moment for some $\delta \in (0,1]$. The rate of convergence is the slower, the less δ is. In the iid situation the speed is just $n^{-\delta/2}$. The rate can be improved under special conditions on the rv X, although, as mentioned before, an increase of the power of the moments is not sufficient for this.

Attempts have been made to calculate the best constants in the Berry–Esséen theorem; see Petrov [495, 496]. One can take 0.7655 in (2.23) and $0.7655 + 8(1+e)$ in (2.22).

Studies of the rate of convergence in $\mathrm{DA}(\alpha)$ for $\alpha < 2$ can be found in Christoph and Wolf [119] or in Rachev [520]. The former concentrates more on classical methods whereas the latter proposes other techniques for estimating the rate of convergence. For example, appropriate metrics (L^p, Lévy and Lévy–Prokhorov metrics) for weak convergence are introduced and then applied to sums of iid rvs. We also refer to results by de Haan and Peng [298] and Hall [308] who study rates of convergence under second–order regular variation conditions on the tail $\overline{F}$.

The approximation of the df of the cumulative sum by a stable limit distribution and its refinements is not always optimal. There exist powerful direct estimates for these probabilities assuming conditions on the tails, the moments or the bounds of the support of these rvs; see for instance Petrov [495, 496], Shorack and Wellner [579].

Asymptotic Expansions

As mentioned above, the Berry–Esséen estimate (2.23) is optimal for certain dfs F. However, in some cases one can approximate the df G_n by the standard normal df Φ and some additional terms. The approximating function is then *not* a df. A common approximation method is called *Edgeworth* or *asymptotic expansion*: formally we write

$$G_n(x) = \Phi(x) + \sum_{k=1}^{\infty} n^{-k/2} Q_k(x) , \quad x \in \mathbb{R} , \tag{2.24}$$

where the Q_k are expressions involving the Hermite polynomials, the precise form of the expression depending on the moments of X. The expansion (2.24) is derived from a formal Taylor expansion of the logarithm of the corresponding chf. To the latter is then applied a Fourier inversion. This approach does not depend on the specific form of the df G_n and is applicable to much wider classes of distributions, but here we restrict ourselves to G_n for illustrational purposes.

In practice one can take only a finite number of terms Q_k into account. To get an impression we consider the first two terms: let

$$\varphi(x) = (2\pi)^{-1/2} \exp\left\{-x^2/2\right\} , \quad x \in \mathbb{R},$$

denote the density function of the standard normal df Φ. Then, for $x \in \mathbb{R}$,

$$\begin{aligned} Q_1(x) &= -\varphi(x) \frac{H_2(x)}{6} \frac{E(X-\mu)^3}{\sigma^3} , \\ Q_2(x) &= -\varphi(x) \left\{ \frac{H_5(x)}{72} \left(\frac{E(X-\mu)^3}{\sigma^3} \right)^2 + \frac{H_3(x)}{24} \left(\frac{E(X-\mu)^4}{\sigma^4} - 3 \right) \right\} , \end{aligned} \tag{2.25}$$

where H_i denotes the Hermite polynomial of degree i:

$$\begin{aligned} H_2(x) &= x^2 - 1 , \\ H_3(x) &= x^3 - 3x , \\ H_5(x) &= x^5 - 10x^3 + 15x . \end{aligned}$$

Notice that the Q_k in (2.25) vanish if X is Gaussian, and the quantities

$$E(X-\mu)^3/\sigma^3 , \qquad E(X-\mu)^4/\sigma^4$$

are the ***skewness*** and ***kurtosis*** of X, respectively. They measure the "closeness" of the df F to Φ.

If we want to expand G_n with an asymptotically negligible remainder term special conditions on the df F must be satisfied. For example, F must be absolutely continuous or distributed on a lattice. We provide here just one example to illustrate the power of the method.

Theorem 2.3.2 (Asymptotic expansion in the absolutely continuous case)
Suppose that $E|X|^k < \infty$ for some integer $k \geq 3$. If F is absolutely continuous then

$$(1+|x|)^k \left| G_n(x) - \Phi(x) - \sum_{i=1}^{k-2} \frac{Q_i(x)}{n^{i/2}} \right| = o\left(\frac{1}{n^{(k-2)/2}}\right),$$

uniformly in x. In particular,

$$G_n(x) = \Phi(x) + \sum_{i=1}^{k-2} \frac{Q_i(x)}{n^{i/2}} + o\left(\frac{1}{n^{(k-2)/2}}\right),$$

uniformly in x. □

Asymptotic expansions can also be applied to the derivatives of G_n. In particular, if F is absolutely continuous then one can obtain asymptotic expansions for the density of G_n.

Notes and Comments

Results on asymptotic expansions for the iid and non–iid case can be found in Hall [311] or in Petrov [495, 496]. Asymptotic expansions for an arbitrary df have been treated in Field and Ronchetti [239] and Jensen [356]. In Christoph and Wolf [119], Ibragimov and Linnik [350] and Zolotarev [645] one can find some ideas about the construction of asymptotic expansions in the α–stable case.

Large Deviations

The CLT can be further refined if one starts looking at G_n for x taken from certain regions (depending on n) or if $x = x_n \to \infty$ at a given rate. This is the objective of the so–called *large deviation* techniques. Nowadays the theory of large deviations has been developed quite rapidly in different directions with applications in mathematics, statistics, engineering and physics. We will restrict ourselves to large deviations in the classical framework of Cramér [139].

Theorem 2.3.3 (Cramér's theorem on large deviations)
Suppose that the moment generating function $M(h) = E\exp\{hX\}$ exists in a neighbourhood of the origin. Then

$$\frac{1-G_n(x)}{1-\Phi(x)} = \exp\left\{\frac{x^3}{\sqrt{n}}\lambda\left(\frac{x}{n}\right)\right\}\left[1+O\left(\frac{x+1}{\sqrt{n}}\varphi(x)\right)\right],$$

$$\frac{G_n(-x)}{\Phi(-x)} = \exp\left\{\frac{-x^3}{\sqrt{n}}\lambda\left(\frac{-x}{n}\right)\right\}\left[1+O\left(\frac{x+1}{\sqrt{n}}\varphi(x)\right)\right],$$

uniformly for positive $x = o(\sqrt{n})$. *Here* $\lambda(z)$ *is a power series which converges in a certain neighbourhood of the origin and whose coefficients depend only on the moments of* X. □

The power series $\lambda(z)$ is called *Cramér's series.* Instead of determining the general coefficients of this series we consider a particular case of Theorem 2.3.3.

Corollary 2.3.4 *Suppose that the conditions of Theorem 2.3.3 are satisfied. Then*

$$\begin{aligned} 1 - G_n(x) &= (1 - \Phi(x)) \exp\left\{ \frac{x^3}{6\sqrt{n}} \frac{E(X-\mu)^3}{\sigma^3} \right\} + O\left(\frac{1}{\sqrt{n}} \varphi(x) \right), \\ G_n(-x) &= \Phi(-x) \exp\left\{ \frac{-x^3}{6\sqrt{n}} \frac{E(X-\mu)^3}{\sigma^3} \right\} + O\left(\frac{1}{\sqrt{n}} \varphi(x) \right), \end{aligned}$$

for $x \geq 0$, $x = O(n^{1/6})$. *In particular, if* $E(X-\mu)^3 = 0$ *then*

$$G_n(x) - \Phi(x) = O\left(\frac{1}{\sqrt{n}} \varphi(x) \right), \quad x \in \mathbb{R}.$$ □

Large deviation results can be interpreted as refinements of the convergence rates in the CLT. Indeed, let $x = x_n \to \infty$ in such a way that $x_n = o(n^{1/6})$. Then we conclude from Corollary 2.3.4 that

$$P\left(\left| \frac{S_n - \mu n}{\sigma \sqrt{n}} \right| > x_n \right) = 2\left((1 - \Phi(x_n)) + O\left(\frac{1}{\sqrt{n}} \exp\left\{ -\frac{x_n^2}{2} \right\} \right)\right).$$

Note that the x_n are chosen such that

$$\frac{S_n - \mu n}{\sigma \sqrt{n}\, x_n} \xrightarrow{P} 0\,. \tag{2.26}$$

In an analogous way we can also consider large deviation results for $X \in \mathrm{DA}(\alpha)$, $\alpha < 2$. These must be of a completely different nature since the moment generating function $M(h)$ does not exist in any neighbourhood of the origin. However, one can get an impression of the order of decrease for the tail probabilities of S_n. For simplicity we restrict ourselves to symmetric rvs.

Theorem 2.3.5 (Heyde's theorem on large deviations)
Let $X \in \mathrm{DA}(\alpha)$ *be symmetric and* $\alpha \in (0,2)$. *Let* (b_n) *be any sequence such that* $b_n \uparrow \infty$ *and* $P(X > b_n) \sim 1/n$, *and denote by*

$$M_1 = X_1\,, \quad M_n = \max(X_1, \dots, X_n)\,, \quad n \geq 2\,,$$

the sample maxima. Then

$$\lim_{n\to\infty} \frac{P(S_n > b_n x_n)}{nP(X > b_n x_n)} = \lim_{n\to\infty} \frac{P(S_n > b_n x_n)}{P(M_n > b_n x_n)} = 1$$

for every sequence $x_n \to \infty$. □

In view of Theorem 2.2.15, the conditions of Theorem 2.3.5 ensure that $b_n^{-1} S_n \xrightarrow{d} G_\alpha$ for an α–stable law G_α. Thus the relation

$$\frac{S_n}{b_n x_n} \xrightarrow{P} 0$$

is directly comparable with (2.26). Similar results can be established for rvs with regularly varying tails $1 - F(x) \sim x^{-\alpha} L(x)$, where $\alpha \geq 2$, as $x \to \infty$; see Section 8.6. This kind of result is another example of the interplay between sums and maxima of a sample $X_1, \ldots, X_n$ as $n \to \infty$. Notice that Theorem 2.3.5 can be understood as a supplementary result to the limit relation

$$\lim_{x\to\infty} \frac{P(S_n > x)}{P(M_n > x)} = 1 , \quad n = 1, 2, \ldots ,$$

which is a defining property of subexponentiality and is studied in detail in Section 1.3.2 and Appendix A3.

Notes and Comments

Cramér's theorem and other versions of large deviation results (including the non–iid case) can be found in Petrov [495, 496]. Theorem 2.3.5 is due to Heyde [321, 322].

The general theory of large deviations has become an important part of probability theory with applications in different fields, including insurance and finance. By now it has become a theory which can be applied to sequences of arbitrary rvs which do not necessarily have sum structure and which can satisfy very general dependence conditions; see for instance the monographs by Bucklew [96], Dembo and Zeitouni [177], Deuschel and Strook [178], or Ellis [200]. We also mention that large deviation results are closely related to *saddlepoint approximations* in statistics, for instance Barndorff–Nielsen and Cox [48], Field and Ronchetti [239], Jensen [356]. The latter contains various applications to insurance risk theory. It should also be mentioned that, whereas Edgeworth expansions yield good approximations around the mean, they become unreliable in the tails. Saddlepoint approximations remedy this problem.

We give some more specific results on large deviations in Section 8.6. They find immediate applications for the valuation of certain quantities which are closely related to reinsurance problems; see Section 8.7.

2.4 The Functional CLT: Brownian Motion Appears

Let (X_n) be an iid sequence with $0 < \sigma^2 < \infty$. In this section we embed the sequence of partial sums (S_n) in a process on $[0,1]$ and consider the limit process which turns out to be Brownian motion. First consider the process $S_n(\cdot)$ on $[0,1]$ such that

$$S_n\left(n^{-1}k\right) = \frac{1}{\sigma\sqrt{n}}\left(S_k - \mu\, k\right), \quad k = 0,\ldots,n,$$

and define the graph of the process $S_n(\cdot)$ at every point of $[0,1]$ by linear interpolation between the points $(k/n\,, S_n(k/n))$. This graph is just a "broken line" and the sample paths are continuous (polygonal) functions. Suppose for the moment that (X_n) is a sequence of iid standard normal rvs. Then the increments $S_n(k/n) - S_n(l/n)$ for $l < k$ are Gaussian with mean zero and variance $(k-l)/n$. Moreover, the process has independent increments when restricted to the points $(k/n)_{k=0,\ldots,n}$. These properties remind us of one of the most important processes which is used in probability theory, Brownian motion, the definition of which follows:

Definition 2.4.1 (Brownian motion)
Let $(B_t)_{t\in[0,1]}$ be a stochastic process which satisfies the following conditions:

(a) *It starts at zero: $B_0 = 0$ a.s.*
(b) *It has independent increments: for any partition $0 \le t_0 < t_1 < \cdots < t_m \le 1$ and any m the rvs $B_{t_1} - B_{t_0}, \ldots, B_{t_m} - B_{t_{m-1}}$ are independent.*
(c) *For every $t \in [0,1]$, B_t has a Gaussian distribution with mean zero and variance t.*
(d) *The sample paths are continuous with probability* 1.

This process is called (standard) Brownian motion *or* Wiener process on $[0,1]$. □

A consequence of this definition is that the increments $B_t - B_s$, $t > s$, have a $N(0, t-s)$ distribution. Brownian motion on $[0,T]$ and on $[0,\infty)$ is defined in a straightforward way by suitably modifying Definition 2.4.1. We mention that one can give a "minimal" definition of Brownian motion as a process with stationary, independent increments and a.s. continuous sample paths. It can be shown that from these properties alone it follows that the increments *must be* normally distributed.

We write $\mathbb{C}[0,1]$ for the vector space of continuous functions which is equipped with the supremum norm; see Appendix A2.2: for $x \in \mathbb{C}[0,1]$

$$\|x\| = \sup_{0\le t\le 1} |x(t)|\,.$$

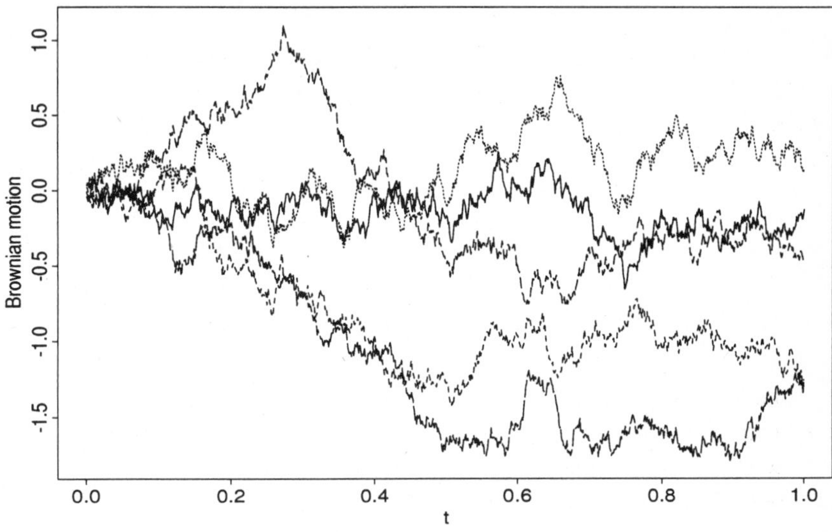

Figure 2.4.2 *Visualisation of Brownian motion: five sample paths of standard Brownian motion on* $[0,1]$.

Notice that the processes $S_n(\cdot)$ and $B.$ assume values in $\mathbb{C}\,[0,1]$.

We introduce still another process on $[0,1]$ which coincides with $S_n(\cdot)$ at the points k/n, $k=0,1,\ldots,n$. It is easier to construct but more difficult to deal with theoretically:

$$\widetilde{S}_n(t) = \frac{1}{\sigma\sqrt{n}}\left(S_{[nt]} - \mu\,[nt]\right)\,,\quad 0 \le t \le 1\,,$$

where $[y]$ denotes the integer part of the real number y. This process has independent increments which are Gaussian (possibly degenerate) if X is Gaussian. Its sample paths are not continuous but have possible jumps at the points k/n. At each point of $[0,1)$ they are continuous from the right and at each point of $(0,1]$ the limit from the left exists. Thus the process $\widetilde{S}_n(\cdot)$ has *cadlag* sample paths, see Appendix A2.2, i.e. they belong to the space $\mathbb{D}\,[0,1]$. The space $\mathbb{D}\,[0,1]$ of cadlag functions can be equipped with different metrics in order to define weak convergence on it. However, our limit process will be Brownian motion which assumes values in $\mathbb{C}\,[0,1]$ so that we are allowed to take the sup–norm as an appropriate metric in $\mathbb{D}\,[0,1]$; see Theorem A2.5.

The following result is known as the *Donsker invariance principle* or *functional CLT* (*FCLT*). In this context it is worthwhile to recall the continuous mapping theorem (Theorem A2.6) and the notion of weak convergence in the

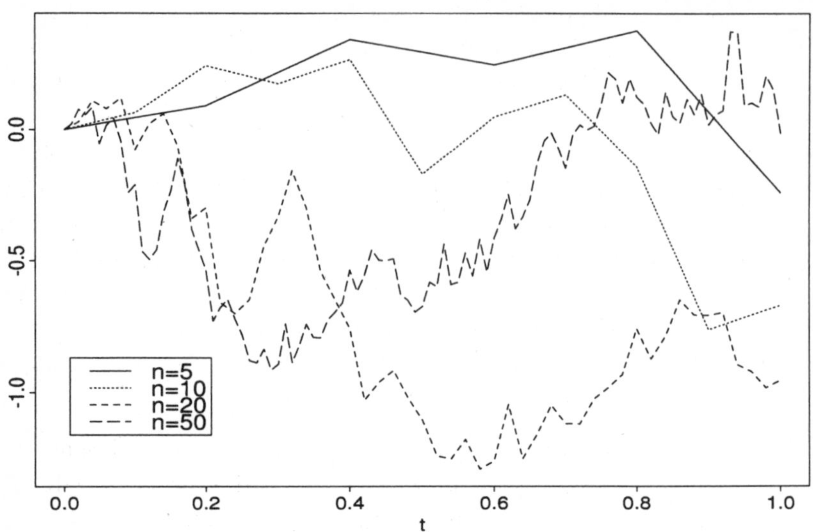

Figure 2.4.3 *Visualisation of the Donsker invariance principle: sample paths of the process* $S_n(\cdot)$ *for the same realisation of* (X_n) *for* $n = 5, 10, 20, 50$.

function spaces $\mathbb{C}\,[0,1]$ and in $\mathbb{D}\,[0,1]$ from Appendix A2.4. To those of you not familiar with this abstract terminology we simply recommend a glance at Figure 2.4.3 which explains how a sample path of Brownian motion is built up from sums of iid rvs.

Theorem 2.4.4 (FCLT, Donsker invariance principle)
Suppose that $EX^2 < \infty$. *Then*

(a) $S_n(\cdot) \xrightarrow{d} B.$ *in* $\mathbb{C}\,[0,1]$ *(equipped with the sup–norm and the* σ*–algebra generated by the open subsets),*

(b) $\widetilde{S}_n(\cdot) \xrightarrow{d} B.$ *in* $\mathbb{D}\,[0,1]$ *(equipped with the sup–norm and the* σ*–algebra generated by the open balls).*

In particular, if f_1 (f_2) *is continuous except possibly on a subset* $A \subset \mathbb{C}\,[0,1]$ $(A \subset \mathbb{D}\,[0,1])$ *for which* $P(B. \in A) = 0$, *then* $f_1(S_n(\cdot)) \xrightarrow{d} f_1(B.)$ *and* $f_2(\widetilde{S}_n(\cdot)) \xrightarrow{d} f_2(B.)$. □

Remarks. 1) The Donsker invariance principle is a very powerful result. It explains why Brownian motion can be taken as a reasonable approximation to many real processes which are in some way related to sums of independent rvs. In finance, the celebrated Black–Scholes model is based on geometric

Brownian motion $X_t = \exp\{ct + \sigma B_t\}$ for constants c and $\sigma > 0$. The rationale of such an approach is that the logarithmic price $\ln X_t$ of a risky asset can be understood as the result of actions and interactions caused by a large number of different independent activities in economy and politics, or indeed individual traders, i.e. it can be understood as a sum process. And indeed, geometric Brownian motion can be viewed as a weak limit of the binomial pricing model of Cox–Ross–Rubinstein; see for instance Lamberton and Lapeyre [412]. In physics a sample path of Brownian motion is often interpreted as the movement of a small particle which is pushed by small independent forces from different directions. Here again the interpretation as a sum process is applicable. As a limit process of normalised and centred random walks, we can consider Brownian motion as a random walk in continuous time.

2) The Donsker invariance principle suggests an easy way of simulating Brownian sample paths by the approximating processes $S_n(\cdot)$ or $\widetilde{S}_n(\cdot)$. They can easily be simulated, for example, if (X_n) is iid Gaussian noise or if (X_n) is a sequence of iid Bernoulli rvs assuming the two values $+1$ and -1 with equal probability. Again, back to the finance world: Donsker explains how to generate from one fair coin the basic process underlying modern mathematical finance. □

The power of a functional limit theorem is considerably increased by the continuous mapping theorem (Theorem A2.6):

Example 2.4.5 (Donsker and continuous mapping theorem)
We may conclude from Theorem 2.4.4 that the finite–dimensional distributions of the processes $S_n(\cdot)$ and $\widetilde{S}_n(\cdot)$ converge. Indeed, consider the mapping $f : \mathbb{D}\,[0,1] \to \mathbb{R}^m$ defined by

$$f(x) = (x_{t_1}, \ldots, x_{t_m})$$

for any $0 \le t_1 < \cdots < t_m \le 1$. It is continuous at elements $x \in \mathbb{C}\,[0,1]$. Then

$$\begin{aligned} f(S_n(\cdot)) &= (S_n(t_1), \ldots, S_n(t_m)) \xrightarrow{d} f(B.) = (B_{t_1}, \ldots, B_{t_m}) \,, \\ f(\widetilde{S}_n(\cdot)) &= (\widetilde{S}_n(t_1), \ldots, \widetilde{S}_n(t_m)) \xrightarrow{d} f(B.) = (B_{t_1}, \ldots, B_{t_m}) \,. \end{aligned}$$

Hence weak convergence of the processes $S_n(\cdot)$ and $\widetilde{S}_n(\cdot)$ implies convergence of the finite–dimensional distributions.

Moreover, the following functionals are continuous on both spaces $\mathbb{C}\,[0,1]$ and $\mathbb{D}\,[0,1]$ when endowed with the sup–norm:

$$f_1(x) = x(1)\,, \quad f_2(x) = \sup_{0 \le t \le 1} x(t)\,, \quad f_3(x) = \inf_{0 \le t \le 1} x(t)\,.$$

In particular,

$$\begin{aligned} f_1(S_n(\cdot)) &= f_1(\widetilde{S}_n(\cdot)) &= \frac{1}{\sigma\sqrt{n}}\,(S_n - n\,\mu)\,, \\ f_2(S_n(\cdot)) &= f_2(\widetilde{S}_n(\cdot)) &= \frac{1}{\sigma\sqrt{n}}\max_{0\le k\le n}\,(S_k - k\,\mu)\,, \\ f_3(S_n(\cdot)) &= f_3(\widetilde{S}_n(\cdot)) &= \frac{1}{\sigma\sqrt{n}}\min_{0\le k\le n}\,(S_k - k\,\mu)\,. \end{aligned}$$

Moreover, the multivariate function (f_1, f_2, f_3) is continuous on both spaces $\mathbb{C}\,[0,1]$ and $\mathbb{D}\,[0,1]$. From Theorem 2.4.4 and the continuous mapping theorem we immediately obtain

$$\frac{1}{\sigma\sqrt{n}}\left(S_n - n\mu\,,\ \max_{0\le k\le n}\,(S_k - k\,\mu)\,,\ \min_{0\le k\le n}\,(S_k - k\,\mu)\right)$$
$$\xrightarrow{d}\left(B_1\,,\ \max_{0\le t\le 1} B_t\,,\ \min_{0\le t\le 1} B_t\right)\,.$$

The joint distribution of B_1, the minimum and maximum of Brownian motion on $[0,1]$ is well known. A derivation is given in Billingsley [69], Chapter 2.11.

At this point it is still worth stressing that, whereas Donsker in conjunction with the continuous mapping theorem offers indeed a very powerful tool, in many applications actually proving that certain functionals on either $\mathbb{C}$ or $\mathbb{D}$ are continuous may be the hard part. Also, once we have a weak convergence result, we may want to use it in two ways. First, in some cases we may derive distributional properties of the limit process through known properties of the approximating process; the latter can for instance be taken to be based on iid Bernoulli rvs. For several examples see Billingsley [69]. However, we may also use the limit process as a useful approximation of a less tangible underlying process; a typical example will be discussed in the diffusion approximation for risk processes, see Example 2.5.18. □

As already stated, Brownian motion is a particular *process with independent, stationary increments*:

Definition 2.4.6 (Process with independent, stationary increments)
Let $\xi = (\xi_t)_{0\le t\le 1}$ be a stochastic process. Then ξ has independent increments *if for any $0 \le t_0 < \cdots < t_m \le 1$ and any $m \ge 1$ the rvs*

$$\xi_{t_1} - \xi_{t_0}, \ldots, \xi_{t_m} - \xi_{t_{m-1}}\,,$$

are independent. This process is said to have stationary increments *if for any $0 \le s < t \le 1$ the rvs $\xi_t - \xi_s$ and ξ_{t-s} have the same distribution.*

A process with independent, stationary increments and sample paths in $\mathbb{D}\,[0,1]$ is also called a Lévy process. □

By a straightforward modification of this definition we can also define processes with independent, stationary increments on $[0,T]$ or on $[0,\infty)$.

We introduce another class of stochastic processes which contains Brownian motion as a special case. Recall from Section 2.2 the definition of an α–stable rv.

Definition 2.4.7 (α–stable motion)
A stochastic process $(\xi_t)_{0\le t\le 1}$ *with sample paths in* $\mathbb{D}\,[0,1]$ *is said to be* α–stable motion *if the following properties hold:*

(a) *It starts at zero:* $\xi_0 = 0$ *a.s.*
(b) *It has independent, stationary increments.*
(c) *For every* $t \in [0,1]$, ξ_t *has an* α*–stable distribution with fixed parameters* $\beta \in [-1,1]$ *and* $\gamma = 0$ *in the spectral representation* (2.10). □

It is straightforward that we can extend this definition to processes on $[0,T]$ or on $[0,\infty)$. We see that Brownian motion (cf. Definition 2.4.1) is just a 2–stable motion. For simplicity, α–stable motions are often called *stable processes* although this might be confusing since in the specialised literature more general stable processes (with dependent or non–stationary stable increments) occur. In Section 8.8 we give an introduction to the world of multivariate stable random vectors and of stable processes.

We need the following elementary relation:

Lemma 2.4.8 *For an* α*–stable motion* $(\xi_t)_{0\le t\le 1}$ *we have*

$$\xi_t - \xi_s \stackrel{d}{=} (t-s)^{1/\alpha}\xi_1\,, \quad 0 \le s < t \le 1\,.$$

Proof. Using the spectral representation (2.10) and the stationary, independent α–stable increments we conclude that

$$\begin{aligned} E\exp\{i\lambda\xi_t\} &= \exp\{-c_t|\lambda|^\alpha(1-i\beta\,\mathrm{sign}(\lambda)z(\lambda,\alpha))\} \\ &= E\exp\{i\lambda\xi_s\}\,E\exp\{i\lambda(\xi_t-\xi_s)\} \\ &= E\exp\{i\lambda\xi_s\}\,E\exp\{i\lambda\xi_{t-s}\} \\ &= \exp\{-(c_s+c_{t-s})\,|\lambda|^\alpha(1-i\beta\,\mathrm{sign}(\lambda)z(\lambda,\alpha))\}\,, \end{aligned}$$

for $\lambda \in \mathbb{R}$ and positive constants c_s, c_t and c_{t-s} which satisfy the relation

$$c_s + c_{t-s} = c_t\,, \quad c_0 = 0\,, \quad 0 \le s < t \le 1\,.$$

The well known measurable solution to this Cauchy functional equation is $c_s = cs$ for a constant c (see Bingham, Goldie and Teugels [72], Theorem 1.1.7), and c must be positive because of the properties of chfs. This proves the lemma. □

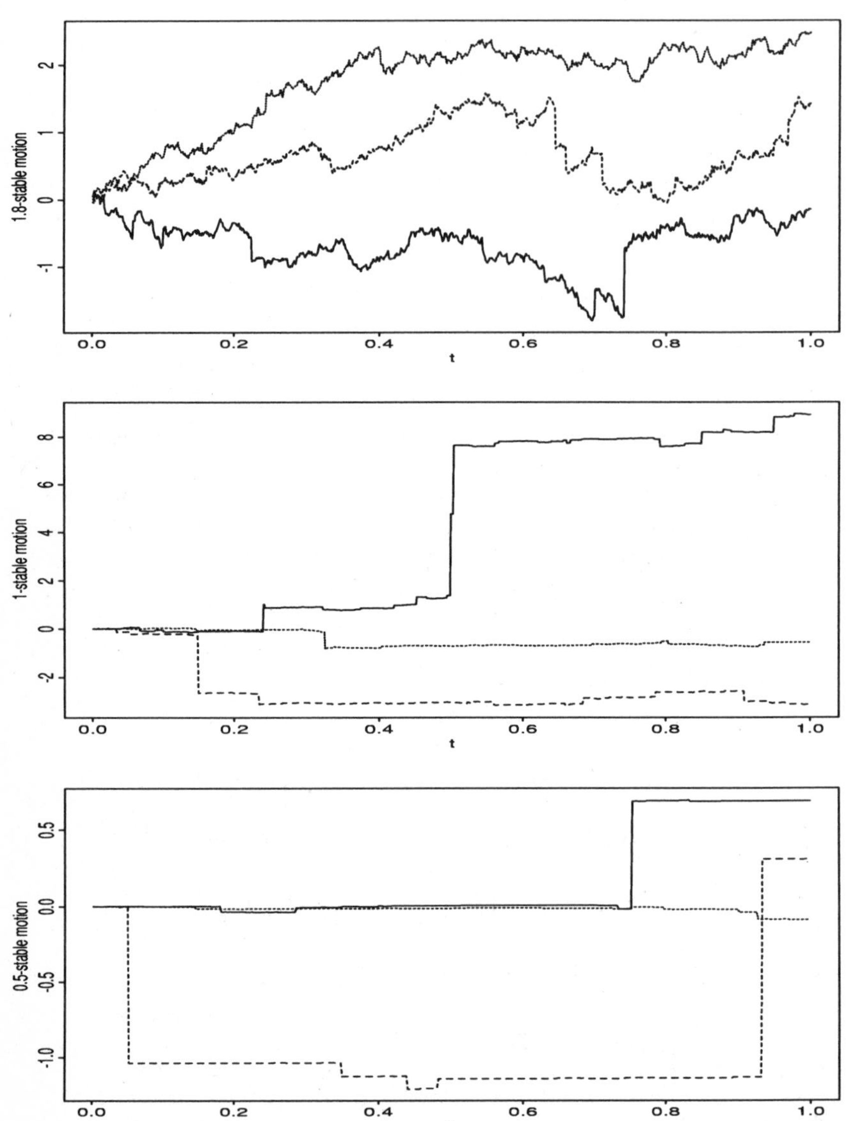

Figure 2.4.9 *Visualisation of symmetric* 1.8–, 1– *and* 0.5*–stable motion (top, middle and bottom): three sample paths of* (ξ_t) *on* $[0,1]$. *The lower two graphs suggest that the sample paths are piecewise constant. This is by no means the case; the set of jumps of almost every sample path is a dense set in* $[0,1]$. *However, the jump heights are in general so tiny that we cannot see them, we only see the large ones.*

From Lemma 2.4.8 we can easily derive the finite–dimensional distributions of an α–stable motion:

$$\begin{aligned}
&(\xi_{t_1}, \xi_{t_2}, \ldots, \xi_{t_m}) \\
&= \left(\xi_{t_1}, \xi_{t_1} + (\xi_{t_2} - \xi_{t_1}), \ldots, \xi_{t_1} + (\xi_{t_2} - \xi_{t_1}) + \cdots + (\xi_{t_m} - \xi_{t_{m-1}})\right) \\
&\stackrel{d}{=} \Big(t_1^{1/\alpha} Y_1, t_1^{1/\alpha} Y_1 + (t_2 - t_1)^{1/\alpha} Y_2, \ldots, \\
&\qquad t_1^{1/\alpha} Y_1 + (t_2 - t_1)^{1/\alpha} Y_2 + \cdots + (t_m - t_{m-1})^{1/\alpha} Y_m\Big)
\end{aligned}$$

for any real numbers $0 \le t_1 < \cdots < t_m \le 1$ and iid α–stable rvs $Y_1, \ldots, Y_m$ such that $Y_1 \stackrel{d}{=} \xi_1$.

Analogously to the Donsker invariance principle we might ask:

Can every α–stable motion be derived as the weak limit of an appropriate sum process?

The answer is YES as the following theorem shows. We refer to Section 2.2 for the definition of domains of attraction and to Appendix A2.4 for the notion of weak convergence of processes.

Theorem 2.4.10 (Stable FCLT)
Let (X_n) be iid rvs in the domain of attraction of an α–stable rv Z_α with parameter $\gamma = 0$ in (2.10). Suppose that

$$\left(n^{1/\alpha} L(n)\right)^{-1} (S_n - a_n) \stackrel{d}{\to} Z_\alpha, \quad n \to \infty,$$

for an appropriate slowly varying function L. Then the process

$$\left(n^{1/\alpha} L(n)\right)^{-1} \left(S_{[nt]} - a_{[nt]}\right), \quad 0 \le t \le 1,$$

converges weakly to an α–stable motion $(\xi_t)_{0 \le t \le 1}$, and $\xi_1 \stackrel{d}{=} Z_\alpha$. Here convergence is understood as weak convergence in $\mathbb{D}[0,1]$ equipped with the J_1–metric and the σ–algebra generated by the open sets. □

We know that Brownian motion has a.s. continuous sample paths. This is not the case for α–stable motions with $\alpha < 2$. Apart from a drift, their sample paths are pure jump processes, and all jumps occur at random instants of time. If we restrict the sample paths of ξ to the interval $[0,1]$ then ξ is a stochastic process which assumes values in $\mathbb{D}[0,1]$, i.e. these sample paths are cadlag. Again we can apply the continuous mapping theorem. For example, the results of Example 2.4.5 remain valid with Brownian motion replaced by a general α–stable motion as limit process.

Notes and Comments

Proofs of the Donsker invariance principle (Theorem 2.4.4) can be found in Billingsley [69] and Pollard [504]. Generalisations to martingales are given in Hall and Heyde [312] and to more general processes in Jacod and Shiryaev [352].

Monographs on Brownian motion and its properties are Hida [325], Karatzas and Shreve [368], Revuz and Yor [541]. An encyclopaedic compendium of facts and formulae for Brownian motion is Borodin and Salminen [82].

FCLTs are applied in insurance mathematics for determining the *probability of ruin* via the so–called *diffusion approximation*; see Grandell [282]. The idea is due to Iglehart [351]. We explain this method in Example 2.5.18.

Methods for simulating Brownian motion are given for instance in Janicki and Weron [354], Kloeden and Platen [385] and the companion book by Kloeden, Platen and Schurz [386].

Definitions of α–stable motion and more general α–stable processes can be found in the literature cited below. A proof of the FCLT in the form of Theorem 2.4.10 follows from the general theory of processes with independent increments; see for instance Gikhman and Skorokhod [262], Jacod and Shiryaev [352], Chapter VII, see also Resnick [529]. Stable motions and processes are treated in various books: Mijnheer [456] concentrates on a.s. properties of the sample paths of α–stable motions. Janicki and Weron [354] discuss various methods for simulating α–stable processes and consider applications. Samorodnitsky and Taqqu [565] give a general theory for α–stable processes including several representations of stable rvs, stable processes and stable integrals. They also develop a theory of stochastic integration with respect to α–stable processes. Kwapień and Woyczyński [411] consider the case of single and multiple stochastic integrals with respect to α–stable processes. Lévy processes are considered in Bertoin [63], Jacod and Shiryaev [352] and Sato [566].

In Section 8.8 we give an introduction to stable processes more general than stable motion.

2.5 Random Sums

2.5.1 General Randomly Indexed Sequences

Random (i.e. randomly indexed) sums are the bread and butter of insurance mathematics. The total claim amount of an insurance portfolio is classically modelled by random sums

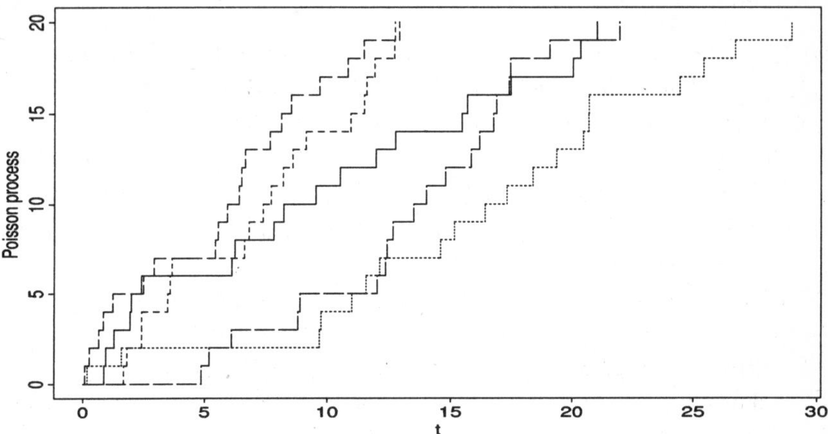

Figure 2.5.1 *Visualisation of a Poisson process with intensity* 1: *five sample paths of* $(N(t))$.

$$S(t) = S_{N(t)} = \begin{cases} 0 & \text{if} \quad N(t) = 0\,, \\ X_1 + \cdots + X_{N(t)} & \text{if} \quad N(t) \geq 1\,, \end{cases} \qquad t \geq 0\,,$$

where $(N(t))_{t\geq 0}$ is a stochastic process on $[0, \infty)$ such that the rvs $N(t)$ are non-negative integer-valued. Usually, $(N(t))$ is assumed to be generated by a sequence $(T_n)_{n\geq 1}$ of non-negative rvs such that

$$0 \leq T_1 \leq T_2 \leq \cdots \quad \text{a.s.}$$

and

$$N(t) = \sup\{n \geq 1 : T_n \leq t\}\,, \quad t \geq 0\,. \tag{2.27}$$

As usual, $\sup A = 0$ if $A = \emptyset$. This is then called a *counting process*. The rv X_n can be interpreted as an individual claim which arrives at the random time T_n, $N(t)$ counts the total number of individual claims and $S(t)$ is the total claim amount in the portfolio up to time t. In the context of finance, $N(t)$ could for instance represent the (random) number of position changes in a foreign exchange portfolio based on tick–by–tick (high frequency) observations. The quantity $S(t)$ then represents the total return over $[0, t]$.

Example 2.5.2 (Homogeneous Poisson process and compound Poisson process)
In the Cramér–Lundberg model (Definition 1.1.1) it is assumed that (X_n) and $(N(t))$ are independent and that $(N(t))$ is a homogeneous *Poisson process with intensity parameter* $\lambda > 0$, i.e. it is a counting process (2.27) with

$$T_n = Y_1 + \cdots + Y_n \,, \quad n \geq 1 \,,$$

and (Y_n) (the inter–arrival times of the claims) are iid exponential rvs with expectation $1/\lambda$. Any counting process which is generated by an iid sum process (T_n) is also called a *renewal counting process.*
Alternatively, a (homogeneous) *Poisson process* is defined by the following three properties:

(a) It starts at zero: $N(0) = 0$.
(b) It has independent, stationary increments.
(c) For every $t > 0$, $N(t)$ is a Poisson rv with parameter λt:

$$P(N(t) = n) = \frac{(\lambda t)^n}{n!} \, e^{-\lambda t} \,, \quad n = 0, 1, 2, \ldots \,.$$

The Poisson process $(N(t))$ is a pure jump process with sample paths in $\mathbb{D}\,[0,\infty)$ which increase to ∞ as $t \to \infty$ and have jumps of height 1 at the random times T_n. It is also a Lévy process; see Definition 2.4.6. A homogeneous Poisson process can be interpreted as a special point process; see Section 5.1.3.
If $(N(t))$ and (X_n) are independent then the process $(S(t))_{t\geq 0}$ is called a *compound Poisson process.*
The Poisson process and Brownian motion and their modifications and generalisations are the most important stochastic processes in probability theory and mathematical statistics. □

The fluctuations of the random sums $S(t)$ for large t can again be described via limit theorems. In what follows we provide some basic tools which show that the asymptotic behaviour of (S_n) and $(S(t))$ is closely linked.

In this section, $(Z_n)_{n\geq 0}$ is a general sequence of rvs and $(N(t))_{t\geq 0}$ is a process of non–negative integer–valued rvs $N(t)$.

Lemma 2.5.3 *Suppose that $Z_n \stackrel{\text{a.s.}}{\to} Z$ as $n \to \infty$ and $N(t) \stackrel{\text{a.s.}}{\to} \infty$ ($N(t) \stackrel{P}{\to} \infty$) as $t \to \infty$. Then*

$$Z_{N(t)} \stackrel{\text{a.s.}}{\to} Z \,, \qquad \left(Z_{N(t)} \stackrel{P}{\to} Z \right) \,, \quad t \to \infty \,.$$

Proof. Suppose $N(t) \stackrel{\text{a.s.}}{\to} \infty$. Set

$$A_1 = \{\omega : N(t)(\omega) \to \infty \,, \quad t \to \infty\} \,, \quad A_2 = \{\omega : Z_n(\omega) \to Z(\omega) \,, \, n \to \infty\} \,,$$

and note that $P(A_1) = P(A_2) = 1$. Then

$$P\left(\{\omega : Z_{N(t)(\omega)}(\omega) \to Z(\omega) \,, \quad t \to \infty\}\right) \geq P\left(A_1 \cap A_2\right) = 1 \,,$$

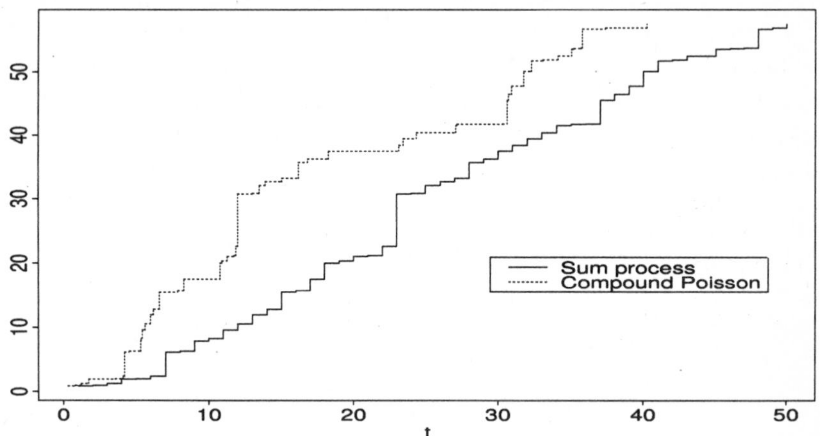

Figure 2.5.4 *One sample path of (S_n) and one sample path of the compound Poisson process $(S(t))$ $(\lambda = 1)$ for the same realisation of iid standard exponential X_n.*

i.e. $Z_{N(t)} \stackrel{\text{a.s.}}{\to} Z$.

Now suppose that $N(t) \stackrel{P}{\to} \infty$ as $t \to \infty$. For every sequence $t_k \to \infty$, $N(t_k) \stackrel{P}{\to} \infty$ as $k \to \infty$, and there exists a subsequence $t_{k_j} \uparrow \infty$ such that $N(t_{k_j}) \stackrel{\text{a.s.}}{\to} \infty$ as $j \to \infty$; see Appendix A1.3. From the first part of the proof, $Z_{N(t_{k_j})} \stackrel{\text{a.s.}}{\to} Z$, hence $Z_{N(t_{k_j})} \stackrel{P}{\to} Z$. Thus every sequence $(Z_{N(t_k)})$ contains a subsequence which converges in probability to Z. Since convergence in probability is metrizable (see Appendix A1.2) this means that $Z_{N(t)} \stackrel{P}{\to} Z$.

□

Combining Theorem 2.1.5 and Lemma 2.5.3 we immediately obtain

Theorem 2.5.5 (Marcinkiewicz–Zygmund SLLNs for random sums)
Suppose that $E|X|^p < \infty$ for some $p \in (0,2)$ and $N(t) \stackrel{\text{a.s.}}{\to} \infty$. Then

$$(N(t))^{-1/p} \ (S(t) - aN(t)) \stackrel{\text{a.s.}}{\to} 0\,, \tag{2.28}$$

where

$$a = \begin{cases} 0 & \text{if} \quad p < 1\,, \\ \mu = EX & \text{if} \quad p \in [1,2)\,. \end{cases}$$

□

We will see in Section 2.5.3 that, if we restrict ourselves to renewal counting processes $(N(t))$, we can replace the random normalising and centring processes in (2.28) by deterministic functions. Moreover, (2.28) can be extended to a LIL.

Now we turn to the case of weak convergence. In particular, we wish to derive the CLT for random sums. The following lemma covers many cases of practical interest, for example, the compound Poisson case as considered in Example 2.5.2.

Lemma 2.5.6 *Suppose that (Z_n) and $(N(t))$ are independent and $N(t) \stackrel{P}{\to} \infty$ as $t \to \infty$. If $Z_n \stackrel{d}{\to} Z$ as $n \to \infty$ then $Z_{N(t)} \stackrel{d}{\to} Z$ as $t \to \infty$.*

Proof. Write $\phi_A(s) = E\exp\{isA\}$ for the chf of any rv A and $f_n(s) = \phi_{Z_n}(s)$ for real s. By independence,

$$E\left(\exp\left\{isZ_{N(t)}\right\}\middle|\, N(t)\right) = f_{N(t)}(s) \quad \text{a.s.}$$

Note that $f_n(s) \to \phi_Z(s)$ as $n \to \infty$ and $N(t) \stackrel{P}{\to} \infty$. By Lemma 2.5.3,

$$f_{N(t)}(s) \stackrel{P}{\to} \phi_Z(s)\,, \quad t \to \infty\,,$$

and since $(f_{N(t)})$ is uniformly integrable,

$$E\exp\left\{isZ_{N(t)}\right\} = E_{N(t)}\left(f_{N(t)}(s)\right) \to E\left(\phi_Z(s)\right) = \phi_Z(s)\,, \quad s \in \mathbb{R}.$$

This proves that $Z_{N(t)} \stackrel{d}{\to} Z$. □

As an immediate consequence we derive an analogue of Theorem 2.2.15 for random sums:

Theorem 2.5.7 (CLT for random sums)
Suppose that (X_n) and $(N(t))$ are independent and that $N(t) \stackrel{P}{\to} \infty$. Assume that $F \in \mathrm{DA}(\alpha)$ for some $\alpha \in (0,2]$. Then Theorem 2.2.15 remains valid if n is everywhere replaced by $N(t)$, i.e. there exist appropriate centring constants a_n and a slowly varying function L such that

$$\left((N(t))^{1/\alpha}L(N(t))\right)^{-1}\left(S(t) - a_{N(t)}\right) \stackrel{d}{\to} G_\alpha\,, \quad t \to \infty\,, \tag{2.29}$$

for an α–stable distribution G_α. □

In Section 2.5.3 we will specify conditions which ensure that the random normalising and centring processes in (2.29) can be replaced by deterministic functions.

The condition that the processes (Z_n) and $(N(t))$ are independent can be relaxed substantially:

Lemma 2.5.8 (Anscombe's theorem)
Suppose that there exists a function $b(t) \uparrow \infty$ such that

$$\frac{N(t)}{b(t)} \xrightarrow{P} 1, \quad t \to \infty, \tag{2.30}$$

and that the following, so–called Anscombe condition *holds:*

$$\forall \epsilon > 0 \, \forall \eta > 0 \, \exists \delta > 0 \, \exists n_0 \quad \text{such that}$$

$$P\left(\max_{m:|m-n|<n\delta} |Z_m - Z_n| > \epsilon \right) < \eta, \quad n > n_0. \tag{2.31}$$

If $Z_n \xrightarrow{d} Z$ as $n \to \infty$ then $Z_{N(t)} \xrightarrow{d} Z$ as $t \to \infty$. □

Roughly speaking, condition (2.30) ensures that the random index $N(t)$ can be replaced by the deterministic function $b(t)$, and if we do so with $Z_{N(t)}$, i.e. if we replace $Z_{N(t)}$ by $Z_{b(t)}$, then (2.31) guarantees that the error $|Z_{N(t)} - Z_{b(t)}|$ is negligible. In other words, Anscombe's condition is a specific stochastic continuity property of the sequence (Z_n).

We note that (2.30) is satisfied for wide classes of renewal counting processes (see Section 2.5.3), including the homogeneous Poisson process. Moreover, (2.31) holds for the (properly normalised and centred) sums S_n. This is the content of the following result which is analogous to Theorems 2.2.15 and 2.5.7. The use of the Anscombe condition in the proof below is not obvious; it is hidden by Kolmogorov's inequality.

Theorem 2.5.9 (Anscombe–type CLT for random sums)
Suppose that

$$\frac{N(t)}{t} \xrightarrow{P} \lambda, \quad t \to \infty, \tag{2.32}$$

for some positive λ and that $F \in \mathrm{DA}(\alpha)$ for some $\alpha \in (0, 2]$ with

$$\left(n^{1/\alpha} L(n) \right)^{-1} (S_n - \widetilde{a} n) \xrightarrow{d} G_\alpha, \tag{2.33}$$

for an α–stable distribution G_α and a slowly varying function L. Here

$$\widetilde{a} = \begin{cases} 0 & \text{if } \alpha \le 1, \\ \mu & \text{if } \alpha \in (1, 2]. \end{cases}$$

Then

$$\left((N(t))^{1/\alpha} L(N(t)) \right)^{-1} (S(t) - \widetilde{a} N(t)) \xrightarrow{d} G_\alpha, \tag{2.34}$$

$$\left((\lambda t)^{1/\alpha} L(t) \right)^{-1} (S(t) - \widetilde{a} N(t)) \xrightarrow{d} G_\alpha. \tag{2.35}$$

In particular, if $\sigma^2 < \infty$ *then*

$$\left(\lambda\sigma^2 t\right)^{-1/2} \left(S(t) - \mu N(t)\right) \xrightarrow{d} \Phi\,,$$

where Φ *is the standard normal distribution.*

In view of Theorem 2.2.15, (2.33) is only a restriction on the distribution of X in the case $\alpha = 1$. It is satisfied for instance for symmetric F. In Section 2.5.3 we will find conditions which ensure that the random centring process in Theorem 2.5.9 can be replaced by a deterministic function.

Sketch of the proof. We restrict ourselves to the case $\alpha = 2$ and $\sigma^2 = \operatorname{var}(X) < \infty$. Without loss of generality we may and do assume that $\sigma^2 = 1$ and $\mu = 0$. We write

$$\frac{S(t)}{(N(t))^{1/2}} = \left(\frac{S_{[\lambda t]}}{(\lambda t)^{1/2}} + \frac{S_{N(t)} - S_{[\lambda t]}}{(\lambda t)^{1/2}}\right)\left(\frac{\lambda t}{N(t)}\right)^{1/2}.$$

By (2.32), the term $(\lambda t/N(t))^{1/2}$ converges to 1 in probability and the classical CLT yields that

$$\frac{S_{[\lambda t]}}{(\lambda t)^{1/2}} \xrightarrow{d} \Phi\,.$$

By virtue of the continuous mapping theorem (Theorem A2.6) it suffices to show that

$$\frac{S_{N(t)} - S_{[\lambda t]}}{(\lambda t)^{1/2}} \xrightarrow{P} 0\,.$$

For every $\epsilon > 0$ and $\delta > 0$ we have that

$$\begin{aligned}
C_\epsilon &= \left\{\frac{|S_{N(t)} - S_{[\lambda t]}|}{(\lambda t)^{1/2}} > \epsilon\right\} \\
&\subset \left\{\left|\frac{N(t)}{t} - \lambda\right| > \delta\right\} \bigcup \left\{\left|\frac{N(t)}{t} - \lambda\right| \le \delta\,,\ \frac{|S_{N(t)} - S_{[\lambda t]}|}{(\lambda t)^{1/2}} > \epsilon\right\} \\
&= A_1 \cup A_2\,.
\end{aligned}$$

By (2.32), $P(A_1) \to 0$ as $t \to \infty$. By Kolmogorov's inequality,

$$\begin{aligned}
P(A_2) &\le P\left(\max_{|n/t-\lambda|\le\delta} \left|S_n - S_{[\lambda t]}\right| > \epsilon(\lambda t)^{1/2}\right) \\
&\le \frac{2}{\epsilon^2 \lambda t} \operatorname{var}\left(S_{[(\lambda+\delta)t]} - S_{[\lambda t]}\right) \\
&\le c\,\frac{\delta}{\epsilon^2}
\end{aligned}$$

for a positive constant c. Thus letting first t tend to ∞ and then δ to 0 we see that $P(C_\epsilon) \to 0$ for every fixed $\epsilon > 0$. This proves (2.34) for $\alpha = 2$.
For $\alpha < 2$ the proof is analogous. Instead of Kolmogorov's inequality one can apply Skorokhod–Ottaviani–type inequalities; see for instance Petrov [496], Theorem 2.3. The details are technical in nature and therefore omitted.
The equivalence of (2.34) and (2.35) is a consequence of (2.32) and of the slow variation of L. □

2.5.2 Renewal Counting Processes

We consider a *renewal counting process* $(N(t))_{t\geq 0}$, i.e.

$$N(t) = \sup\{n \geq 1 : T_n \leq t\}\,, \quad t \geq 0\,, \tag{2.36}$$

and

$$T_n = Y_1 + \cdots + Y_n\,, \quad n \geq 1\,,$$

for iid non–negative (non–zero) rvs $Y, Y_1, Y_2, \ldots$. For applications of this kind of processes to risk theory see Chapter 1. The homogeneous Poisson process (see Example 2.5.2) is such a renewal counting process where Y is exponential with expectation $1/\lambda$.

In this section we answer the question:

What is the order of magnitude of $N(t)$ as $t \to \infty$?

Observe that

$$\{T_n \leq t\} = \{N(t) \geq n\}\,.$$

Kolmogorov's SLLN implies that $T_n \stackrel{\text{a.s.}}{\to} \infty$ and therefore $N(t) \stackrel{\text{a.s.}}{\to} \infty$. However, we can derive much more precise information:

Theorem 2.5.10 (Marcinkiewicz–Zygmund SLLNs/LIL for renewal counting processes)
Suppose that $EY = 1/\lambda \leq \infty$ (if $EY = \infty$ set $\lambda = 0$). Then

$$t^{-1} N(t) \stackrel{\text{a.s.}}{\to} \lambda\,. \tag{2.37}$$

If $EY^p < \infty$ for some $p \in (1,2)$ then

$$t^{-1/p}\,(N(t) - \lambda t) \stackrel{\text{a.s.}}{\to} 0\,. \tag{2.38}$$

If $\sigma_Y^2 = \operatorname{var}(Y) < \infty$ then

$$\begin{aligned}
&\limsup_{t\to\infty}\,(2t\ln\ln t)^{-1/2}\,(N(t) - \lambda t)\\
&= \; -\liminf_{t\to\infty}\,(2t\ln\ln t)^{-1/2}\,(N(t) - \lambda t)\\
&= \; \sigma_Y \lambda^{3/2} \quad \text{a.s.}
\end{aligned}$$

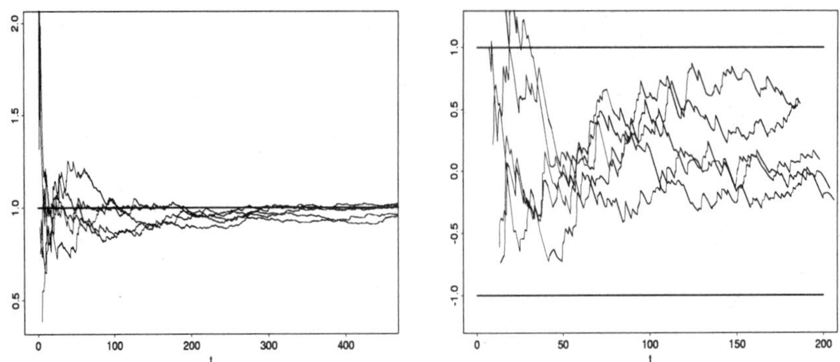

Figure 2.5.11 *Visualisation of the SLLN (left) and of the LIL (right) for the homogeneous Poisson process with intensity* 1*: five sample paths.*

Sketch of the proof. We restrict ourselves to show the SLLNs (2.37) and (2.38). Kolmogorov's SLLN for random sums yields that

$$\frac{T_{N(t)}}{N(t)} \stackrel{\text{a.s.}}{\to} \frac{1}{\lambda}\,.$$

By this and a sandwich argument applied to

$$\frac{T_{N(t)}}{N(t)} \leq \frac{t}{N(t)} \leq \frac{T_{N(t)+1}}{N(t)+1}\,\frac{N(t)+1}{N(t)}\,,$$

we prove (2.37).
Now suppose that $EY^p < \infty$ for some $p \in (1,2)$. Notice that

$$n^{-1/p}(T_{n+1} - T_n) = n^{-1/p} Y_{n+1} \stackrel{\text{a.s.}}{\to} 0\,.$$

This, the Marcinkiewicz–Zygmund SLLNs of Theorem 2.5.5, and (2.37) imply that

$$t^{-1/p}\,\left(T_{N(t)+1} - T_{N(t)}\right) = t^{-1/p} Y_{N(t)+1} \stackrel{\text{a.s.}}{\to} 0\,.$$

This, Theorem 2.5.5 and a sandwich argument applied to

$$\frac{\lambda t - N(t)}{t^{1/p}} \leq \frac{\lambda T_{N(t)+1} - N(t)}{t^{1/p}} \leq \frac{\lambda t - N(t)}{t^{1/p}} + \frac{\lambda Y_{N(t)+1}}{t^{1/p}} \tag{2.39}$$

gives us (2.38). □

Theorem 2.5.10 suggests that $EN(t) \sim \lambda t$ and $\text{var}(N(t)) \sim \sigma_Y^2 \lambda^3 t$. In the case of the Poisson process we even have that $EN(t) = \lambda t$ and $\text{var}(N(t)) = \sigma_Y^2 \lambda^3 t$ since $EY = 1/\lambda$ and $\sigma_Y^2 = 1/\lambda^2$. For the following results see Gut [291], Theorems 5.1 and 5.2.

Proposition 2.5.12 (Moments of renewal counting process)
The following relations hold:

(a) $EN(t) = (\lambda + o(1))t$ *as* $t \to \infty$.
(b) *Suppose that* $\sigma_Y^2 = \mathrm{var}(Y) < \infty$. *Then*

$$\begin{aligned} EN(t) &= \lambda t + O(1)\,, & t \to \infty\,,\\ \mathrm{var}(N(t)) &= \sigma_Y^2 \lambda^3 t + o(t)\,, & t \to \infty\,. \end{aligned}$$

□

From Theorem 2.5.10 and Proposition 2.5.12 we have gained a first impression on the growth of a renewal counting process. Next we study the weak convergence of $(N(t))$.

Theorem 2.5.13 (CLT for renewal counting process)
Suppose that $\sigma_Y^2 < \infty$. *Then*

$$\left(\sigma_Y^2 \lambda^3 t\right)^{-1/2} (N(t) - \lambda t) \xrightarrow{d} \Phi\,, \tag{2.40}$$

where Φ *is the standard normal distribution.*

Recall from Proposition 2.5.12 that $EN(t) \sim \lambda t$ and $\mathrm{var}(N(t)) \sim \sigma_Y^2 \lambda^3 t$. Thus Theorem 2.5.13 is similar to the classical CLT for iid sums. We note that one can prove an analogous result for $(N(t))$ with an α–stable limit.
Proof. We proceed as in (2.39):

$$\frac{\lambda t - N(t)}{\left(\sigma_Y^2 \lambda^3 t\right)^{1/2}} \le \frac{\lambda T_{N(t)+1} - N(t)}{\left(\sigma_Y^2 \lambda^3 t\right)^{1/2}} \le \frac{\lambda t - N(t)}{\left(\sigma_Y^2 \lambda^3 t\right)^{1/2}} + \frac{\lambda Y_{N(t)+1}}{\left(\sigma_Y^2 \lambda^3 t\right)^{1/2}}\,.$$

We have, by independence,

$$\begin{aligned} P\left(Y_{N(t)+1} > \epsilon t^{1/2}\right) &= E\left(P\left(Y_{N(t)+1} > \epsilon t^{1/2} \middle| N(t)\right)\right) \\ &= P\left(Y > \epsilon t^{1/2}\right)\,, \quad \forall \epsilon > 0\,. \end{aligned}$$

Hence

$$\frac{\lambda t - N(t)}{\left(\sigma_Y^2 \lambda^3 t\right)^{1/2}} = \frac{\lambda T_{N(t)} - N(t)}{\left(\sigma_Y^2 \lambda^3 t\right)^{1/2}} + o_P(1)\,. \tag{2.41}$$

In view of Theorem 2.5.9 and by the continuous mapping theorem the rhs converges weakly to Φ. This proves the theorem. □

2.5.3 Random Sums Driven by Renewal Counting Processes

In this section we consider one of the most important models in insurance mathematics. Throughout, we assume that the random sums $S(t) = S_{N(t)}$ are driven by a *renewal counting process* as defined in (2.36). The process $(S(t))$ is a model for the total claim amount of an insurance portfolio. The renewal and the Cramér–Lundberg models (Definition 1.1.1) are included in this setting as particular cases when $(N(t))$ and (X_n) are independent. In general, we do not require this assumption. In what follows we are interested in the asymptotic properties of the process $(S(t))$.

Recall from Section 2.5.2 that $N(t) \stackrel{\text{a.s.}}{\to} \infty$. Hence we may apply the Marcinkiewicz–Zygmund SLLNs for random sums (Theorem 2.5.5),

$$(N(t))^{-1/p}\ (S(t) - a\,N(t)) \stackrel{\text{a.s.}}{\to} 0\,, \tag{2.42}$$

provided $E|X|^p < \infty$ for some $p \in (0,2)$ and

$$a = \begin{cases} 0 & \text{if} \quad p < 1\,, \\ \mu = EX & \text{if} \quad p \in [1,2)\,. \end{cases} \tag{2.43}$$

The following question arises naturally:

May we replace $N(t)$ in (2.42) by a deterministic function, for instance λt?

The answer is

In general: NO.

However, by Theorem 2.5.10, $N(t)/t \stackrel{\text{a.s.}}{\to} \lambda$ provided $EY < \infty$. Hence we may replace the normalising process $(N(t))^{1/p}$ by $(\lambda t)^{1/p}$. The centring process causes some problems. To proceed, suppose $E|X|^p < \infty$ for some $p \in [1,2)$. We write

$$t^{-1/p}\ (S(t) - \lambda\mu t) = t^{-1/p}\ (S(t) - \mu\,N(t)) + \mu t^{-1/p}\ (N(t) - \lambda\,t)\ .$$

In view of (2.42), the first term on the rhs converges to zero a.s. provided the first moment of Y is finite. On the other hand,

$$\mu t^{-1/p}\ (N(t) - \lambda\,t) \stackrel{\text{a.s.}}{\to} 0 \tag{2.44}$$

does not hold in general. But if $EY^p < \infty$ we conclude from Theorem 2.5.10 that (2.44) is satisfied. In summary we obtain:

Theorem 2.5.14 (Marcinkiewicz–Zygmund SLLNs for random sums)
Suppose that $E|X|^p < \infty$ for some $p \in (0,2)$.

(a) *If* $EY < \infty$ *then*

$$t^{-1/p}\ (S(t) - a\,N(t)) \stackrel{\text{a.s.}}{\to} 0\,,$$

where a *is defined by* (2.43).

(b) *If* $p \geq 1$ *and* $EY^p < \infty$ *then*

$$t^{-1/p}\ (S(t) - \mu\lambda\, t) \stackrel{\text{a.s.}}{\to} 0\,. \qquad \square$$

In the weak convergence case we can proceed analogously. Since $N(t) \to \infty$ a.s. for a renewal counting process, the CLT for random sums applies under mild conditions:

Theorem 2.5.15 (Anscombe–type CLT for random sums)
Assume that $F \in \mathrm{DA}(\alpha)$ *for some* $\alpha \in (0,2]$ *and that*

$$\left(n^{1/\alpha}L(n)\right)^{-1}\ (S_n - \widetilde{a}n) \stackrel{d}{\to} G_\alpha\,,$$

for some α*–stable distribution* G_α *and a slowly varying function* L*. Here*

$$\widetilde{a} = \begin{cases} 0 & \text{if} \quad \alpha \leq 1\,, \\ \mu & \text{if} \quad \alpha \in (1,2]\,. \end{cases}$$

(a) *If* $EY < \infty$ *then*

$$\left((\lambda t)^{1/\alpha}L(t)\right)^{-1}\ (S(t) - \widetilde{a}N(t)) \stackrel{d}{\to} G_\alpha\,. \tag{2.45}$$

In particular, if $\sigma^2 = \mathrm{var}(X) < \infty$ *then*

$$\left(\lambda\sigma^2 t\right)^{-1/2}\ (S(t) - \mu N(t)) \stackrel{d}{\to} \Phi\,, \tag{2.46}$$

where Φ *is the standard normal distribution.*

(b) *If* $\alpha \in (1,2)$ *and* $EY^p < \infty$ *for some* $p > \alpha$ *then*

$$\left((\lambda t)^{1/\alpha}L(t)\right)^{-1}\ (S(t) - \lambda\mu\, t) \stackrel{d}{\to} G_\alpha\,. \tag{2.47}$$

Proof. If $EY < \infty$ then, by Theorem 2.5.10, $N(t)/t \stackrel{\text{a.s.}}{\to} \lambda$, and Theorem 2.5.9 applies immediately. This yields (2.45) and (2.46).
If $EY^p < \infty$ and $p \in [1,2)$ then, by Theorem 2.5.10,

$$t^{-1/p}\ (N(t) - \lambda t) \stackrel{\text{a.s.}}{\to} 0\,. \tag{2.48}$$

Hence

$$
\begin{aligned}
&\left((\lambda t)^{1/\alpha} L(t)\right)^{-1} (S(t) - \lambda\mu\, t) \\
&= \left((\lambda t)^{1/\alpha} L(t)\right)^{-1} ((S(t) - \mu\, N(t)) + \mu\, (N(t) - \lambda\, t)) \\
&= \left((\lambda t)^{1/\alpha} L(t)\right)^{-1} (S(t) - \mu\, N(t)) + o(1) \quad \text{a.s.} \qquad (2.49)
\end{aligned}
$$

Here we used (2.48) and the fact that, for $p > \alpha$,

$$\lim_{t\to\infty} \frac{t^{1/\alpha} L(t)}{t^{1/p}} = \infty$$

which is a consequence of the slow variation of L. Relation (2.47) is now immediate from (2.45) and (2.49) by the continuous mapping theorem (Theorem A2.6). □

Note that we excluded the case $\alpha = 2$ from (2.47). In that case the method of proof fails. Indeed, (2.49) is no longer applicable if $\sigma^2 < \infty$. This follows from the CLT in Theorem 2.5.13.

Now we try to combine the CLT for $(S(t))$ and for $(N(t))$. Assume that $\sigma^2 < \infty$ and $\sigma_Y^2 < \infty$. Using (2.41), we obtain

$$
\begin{aligned}
t^{-1/2}\,(S(t) - \lambda\mu\, t) &= t^{-1/2}\,((S(t) - \mu\, N(t)) + \mu(N(t) - \lambda\, t)) \\
&= t^{-1/2}\left((S(t) - \mu\, N(t)) + \mu\left(N(t) - \lambda T_{N(t)}\right)\right) + o_P(1) \\
&= t^{-1/2} \sum_{i=1}^{N(t)} (X_i - \mu\lambda Y_i) + o_P(1)\,. \qquad (2.50)
\end{aligned}
$$

Notice that the rvs $X_i' = X_i - \mu\lambda Y_i$ have mean zero. Moreover, the sequence (X_i') is iid if $((X_n, Y_n))$ is iid. Under the latter condition, Theorem 2.5.9 applies immediately to (2.50) and yields the following result:

Theorem 2.5.16 *Suppose that $((X_n, Y_n))$ is a sequence of iid random vectors and that $\sigma^2 < \infty$ and $\sigma_Y^2 < \infty$. Then*

$$(\operatorname{var}(X - \mu\lambda Y)\lambda t)^{-1/2}\,(S(t) - \mu\lambda t) \stackrel{d}{\to} \Phi\,,$$

where Φ denotes the standard normal distribution.
In particular, if (X_n) and (Y_n) are independent then

$$\left(\left(\sigma^2 + (\mu\lambda\sigma_Y)^2\right)\lambda t\right)^{-1/2}\,(S(t) - \mu\lambda\, t) \stackrel{d}{\to} \Phi\,. \qquad \square$$

Following the idea of proof of Theorem 2.5.16, one can derive results if $\sigma^2 = \infty$ or $\sigma_Y^2 = \infty$ with appropriate stable limit distributions; see for instance Kotulski [405].

It is also possible to derive different versions of FCLTs with Gaussian or α-stable limit processes for $(S(t))$. We state here one standard result, versions of which can be found in Billingsley [69], Section 17, and in Gut [291], Theorem 2.1 in Chapter V. In our presentation we follow Grandell [281], p. 47. Recall the notion of weak convergence from Appendix A2.4 and compare the following result with the Donsker invariance principle (Theorem 2.4.4).

Theorem 2.5.17 (FCLT for random sum process)
Let (X_n) be a sequence of iid rvs such that $\sigma^2 < \infty$. Assume that the renewal counting process $(N(t))$ and (X_n) are independent and that $EY = 1/\lambda$ and $\sigma_Y^2 < \infty$. Let $B.$ be standard Brownian motion on $[0, \infty)$. Then

$$\left(\left(\sigma^2 + (\mu\lambda\sigma_Y)^2\right)\lambda n\right)^{-1/2} \left(S_{N(n\cdot)} - \lambda\mu\, n\cdot\right) \xrightarrow{d} B.$$

in $\mathbb{D}[0,\infty)$ equipped with the J_1-metric and the corresponding σ-algebra of the open sets. □

This theorem has quite an interesting application in insurance mathematics:

Example 2.5.18 (Diffusion approximation of the risk process)
Consider the Cramér–Lundberg model (Definition 1.1.1), i.e. $(S(t))$ is compound Poisson with positive iid claims (X_n) independent of the homogeneous Poisson process $(N(t))$ with intensity $\lambda > 0$. The corresponding risk process with initial capital u and premium income rate $c = (1+\rho)\lambda\mu > 0$ (with safety loading $\rho > 0$) is given in (1.4) as

$$U(t) = u + ct - S(t)\,, \quad t \ge 0\,.$$

In Chapter 1 we mainly studied the ruin probability in infinite time. One method to obtain approximations to the ruin probability $\psi(u,T)$ in finite time T, i.e.

$$\begin{aligned} \psi(u,T) &= P\left(U(t) < 0 \text{ for some } t \le T\right) \\ &= P\left(\inf_{0\le t\le T}(ct - S(t)) < -u\right), \end{aligned}$$

is the so-called *diffusion approximation* which was introduced in insurance mathematics by Iglehart [351]; see also Grandell [282], Appendix A.4, for an extensive discussion of the method. Define

$$\widetilde{\sigma}^2 = \left(\sigma^2 + (\mu\lambda\sigma_Y)^2\right)\lambda = \left(\sigma^2 + \mu^2\right)\lambda\,.$$

Then

$$\begin{aligned}
&\psi(u,T)\\
&= P\left(\inf_{0\le t\le T}((1+\rho)\lambda\mu t - S(t)) < -u\right)\\
&= P\left(\inf_{0\le t\le T/n}((1+\rho)\lambda\mu tn - S(tn)) < -u\right)\\
&= P\left(\inf_{0\le t\le T/n}\left(\widetilde{\sigma}^2 n\right)^{-1/2}((1+\rho)\lambda\mu tn - S(tn)) < -u\left(\widetilde{\sigma}^2 n\right)^{-1/2}\right).
\end{aligned}$$

Now assume that $T_0 = T/n$, $\rho_0 = \rho\lambda\mu\widetilde{\sigma}^{-1}\sqrt{n}$ and $u_0 = u\,(\widetilde{\sigma}^2 n)^{-1/2}$ are constants, i.e. we increase T and u with n, and decrease at the same time the safety loading ρ with n. This means that a small safety loading is compensated by a large initial capital. Then we obtain

$$\psi(u,T) = P\left(\inf_{0\le t\le T_0}\left(\left(\widetilde{\sigma}^2 n\right)^{-1/2}(\lambda\mu tn - S(tn)) + \rho_0 t\right) < -u_0\right).$$

The functional $x(f) = \inf_{0\le t\le T_0} f(t)$ is continuous on $\mathbb{D}[0,T_0]$. Thus we may conclude from Theorem 2.5.17 by the continuous mapping theorem (note that u, ρ and T depend on n) that

$$\begin{aligned}
\psi(u,T) &\to P\left(\inf_{0\le t\le T_0}(\rho_0 t - B_t) < -u_0\right)\\
&= P\left(\sup_{0\le t\le T_0}(B_t - \rho_0 t) > u_0\right).
\end{aligned}\tag{2.51}$$

The latter approach is called a *diffusion approximation* since Brownian motion is a special diffusion process. The distribution of the supremum functional of Brownian motion with linear drift is well known; see for instance Lerche [421], Example 1 on p. 27:

$$P\left(\sup_{0\le t\le T_0}(B_t - \rho_0 t) > u_0\right) = \overline{\Phi}\left(\frac{\rho_0 T_0 + u_0}{\sqrt{T_0}}\right) + e^{-2u_0\rho_0}\Phi\left(\frac{\rho_0 T_0 - u_0}{\sqrt{T_0}}\right).$$

The diffusion approximation has many disadvantages, but also some good aspects. We refer to Grandell [282], Appendix A.4, and Asmussen [28] for a discussion and some recent literature; see also Schmidli [567] and Furrer, Michna and Weron [247]. The latter look at weak approximations of the risk process by α–stable processes. Among the positive aspects of the diffusion approximation is that it is applicable to a wide range of risk processes which deviate from the Cramér–Lundberg model. In that case, the classical methods from renewal theory as developed in Chapter 1 will usually break down, and the diffusion approach is then one of the few tools which work. In such more general models it is usually not possible to choose the premiums as a linear function in time; see for example Klüppelberg and Mikosch [392, 393] for a shot noise risk model. As a contra one might mention that the choice of T_0, ρ_0 and u_0 is perhaps not the most natural one. On the other hand, "large" values of T and u and small values of ρ are relative and down to individual judgement. Notice that in Chapter 1 the probability of ruin in infinite time was approximated for "large" initial capital u. Nevertheless, if one wants to use the diffusion approximation for practical purposes a study of the values of T, ρ and u for which the method yields reasonable results is unavoidable. For example, Grandell [282], Appendix A.4, gives a simulation study. □

Notes and Comments

There are several texts on random sums, renewal counting processes and related questions. They are mainly motivated by renewal theory. A more advanced limit theory, but also the proofs of the standard results above can be found in Gut [291]. The classical theory of random sums relevant for risk theory was reviewed in Panjer and Willmot [489]. Other relevant literature is Asmussen [27] and Grandell [282]. The latter deals with the total claim amount process and related questions of risk and ruin for very general processes. Grandell [284] gives an overview of the corresponding theory for mixed Poisson processes and related risk models. A recent textbook treatment of random sums is Gnedenko and Korolev [268].

3

Fluctuations of Maxima

This chapter is concerned with classical extreme value theory and consequently it is fundamental for many results in this book. The central result is the Fisher–Tippett theorem which specifies the form of the limit distribution for centred and normalised maxima. The three families of possible limit laws are known as extreme value distributions. In Section 3.3 we describe their maximum domains of attraction and derive centring and normalising constants. A short summary is provided in Tables 3.4.2–3.4.4 where numerous examples are to be found.

The basic tool for studying rare events related to the extremes of a sample is the Poisson approximation: a first glimpse is given in Section 3.1. Poisson approximation is also the key to the weak limit theory of upper order statistics (see Section 4.2) and for the weak convergence of point processes (see Chapter 5).

The asymptotic theories for maxima and sums complement and contrast each other. Corresponding results exist for affinely transformed sums and maxima: stable distributions correspond to max–stable distributions, domains of attraction to maximum domains of attraction; see Chapter 2. Limit theorems for maxima and sums require appropriate normalising and centring constants. For maxima the latter are chosen as some quantile (or a closely related quantity) of the underlying marginal distribution. Empirical quantiles open the way for tail estimation. Chapter 6 is devoted to this important statistical problem. In Section 3.4 the mean excess function is introduced.

It will prove to be a useful tool for distinguishing dfs in their right tail and plays an important role in tail estimation; see Chapter 6.

As in Chapters 1 and 2, regular variation continues to play a fundamental role. The maximum domain of attraction of an extreme value distribution can be characterised via regular variation and its extensions; see Section 3.3. We also study the relationship between subexponentiality and maximum domains of attraction. This will have consequences in Section 8.3, where the path of a risk process leading to ruin is characterised.

The theory of Section 3.4 allows us to present various results of the previous sections in a compact way. The key is the generalised extreme value distribution which also leads to the generalised Pareto distribution. These are two crucial notions which turn out to be very important for the statistics of rare events treated in Chapter 6.

The almost sure behaviour of maxima is considered in Section 3.5. These results find applications in Section 8.5, where we study the longest success–run in a random walk.

3.1 Limit Probabilities for Maxima

Throughout this chapter $X, X_1, X_2, \ldots$ is a sequence of iid non–degenerate rvs with common df F. Whereas in Chapter 2 we focussed on cumulative sums, in this chapter we investigate the fluctuations of the *sample maxima*

$$M_1 = X_1\,, \quad M_n = \max\left(X_1, \ldots, X_n\right)\,, \quad n \geq 2\,.$$

Corresponding results for minima can easily be obtained from those for maxima by using the identity

$$\min\left(X_1, \ldots, X_n\right) = -\max\left(-X_1, \ldots, -X_n\right)\,.$$

In Chapter 4 we continue with the analysis of the *upper order statistics* of the sample $X_1, \ldots, X_n$.

There is of course no difficulty in writing down the exact df of the maximum M_n:

$$P\left(M_n \leq x\right) = P\left(X_1 \leq x, \ldots, X_n \leq x\right) = F^n(x)\,, \quad x \in \mathbb{R}, \quad n \in \mathbb{N}.$$

Extremes happen "near" the upper end of the support of the distribution, hence intuitively the asymptotic behaviour of M_n must be related to the df F in its right tail near the right endpoint. We denote by

$$x_F = \sup\{x \in \mathbb{R} : \ F(x) < 1\}$$

the right endpoint of F. We immediately obtain, for all $x < x_F$,

$$P(M_n \le x) = F^n(x) \to 0\,, \quad n \to \infty\,,$$

and, in the case $x_F < \infty$, we have for $x \ge x_F$ that

$$P(M_n \le x) = F^n(x) = 1\,.$$

Thus $M_n \xrightarrow{P} x_F$ as $n \to \infty$, where $x_F \le \infty$. Since the sequence (M_n) is non–decreasing in n, it converges a.s., and hence we conclude that

$$M_n \xrightarrow{\text{a.s.}} x_F\,, \quad n \to \infty\,. \tag{3.1}$$

This fact does not provide a lot of information. More insight into the order of magnitude of maxima is given by weak convergence results for centred and normalised maxima. This is one of the main topics in classical extreme value theory. For instance, the fundamental Fisher–Tippett theorem (Theorem 3.2.3) has the following content: if there exist constants $c_n > 0$ and $d_n \in \mathbb{R}$ such that

$$c_n^{-1}(M_n - d_n) \xrightarrow{d} H\,, \quad n \to \infty\,, \tag{3.2}$$

for some non–degenerate distribution H, then H must be of the type of one of the three so-called standard *extreme value distributions*. This is similar to the CLT, where the stable distributions are the only possible non–degenerate limit laws. Consequently, one has to consider probabilities of the form

$$P\left(c_n^{-1}(M_n - d_n) \le x\right)\,,$$

which can be rewritten as

$$P(M_n \le u_n)\,, \tag{3.3}$$

where $u_n = u_n(x) = c_n x + d_n$. We first investigate (3.3) for general sequences (u_n), and afterwards come back to affine transformations as in (3.2). We ask:

Which conditions on F ensure that the limit of $P(M_n \le u_n)$ for $n \to \infty$ exists for appropriate constants u_n?

It turns out that one needs certain continuity conditions on F at its right endpoint. This rules out many important distributions. For instance, if F has a Poisson distribution, then $P(M_n \le u_n)$ never has a limit in $(0,1)$, whatever the sequence (u_n). This implies that the normalised maxima of iid Poisson distributed rvs do not have a non–degenerate limit distribution. This remark might be slightly disappointing, but it shows the crucial difference between sums and maxima. In the former case, the CLT yields the normal distribution as limit under the very general *moment condition* $EX^2 < \infty$. If $EX^2 = \infty$ the relatively small class of α–stable limit distributions enters. Only in that very heavy–tailed case do *conditions on the tail* $\overline{F} = 1 - F$ guarantee the existence of a limit distribution. In contrast to sums, *we always need rather delicate conditions on the tail* $\overline{F}$ *to ensure that* $P(M_n \le u_n)$ *converges to a non–trivial limit*, i.e. a number in $(0,1)$.

In what follows we answer the question above. We commence with an elementary result which is crucial for the understanding of the weak limit theory of sample maxima. It will become a standard tool throughout this book.

Proposition 3.1.1 (Poisson approximation)
For given $\tau \in [0,\infty]$ *and a sequence* (u_n) *of real numbers the following are equivalent*

$$n\overline{F}(u_n) \to \tau\,, \tag{3.4}$$

$$P(M_n \le u_n) \to e^{-\tau}\,. \tag{3.5}$$

Proof. Consider first $0 \le \tau < \infty$. If (3.4) holds, then

$$P(M_n \le u_n) = F^n(u_n) = \left(1 - \overline{F}(u_n)\right)^n = \left(1 - \frac{\tau}{n} + o\left(\frac{1}{n}\right)\right)^n,$$

which implies (3.5). Conversely, if (3.5) holds, then $\overline{F}(u_n) \to 0$. (Otherwise, $\overline{F}(u_{n_k})$ would be bounded away from 0 for some subsequence (n_k). Then $P(M_{n_k} \le u_{n_k}) = (1 - \overline{F}(u_{n_k}))^{n_k}$ would imply $P(M_{n_k} \le u_{n_k}) \to 0$.) Taking logarithms in (3.5) we have

$$-n \ln\left(1 - \overline{F}(u_n)\right) \to \tau\,.$$

Since $-\ln(1-x) \sim x$ for $x \to 0$ this implies that $n\overline{F}(u_n) = \tau + o(1)$, giving (3.4).

If $\tau = \infty$ and (3.4) holds, but (3.5) does not, there must be a subsequence (n_k) such that $P(M_{n_k} \le u_{n_k}) \to \exp\{-\tau'\}$ as $k \to \infty$ for some $\tau' < \infty$. But then (3.5) implies (3.4), so that $n_k\overline{F}(u_{n_k}) \to \tau' < \infty$, contradicting (3.4) with $\tau = \infty$. Similarly, (3.5) implies (3.4) for $\tau = \infty$. □

Remarks. 1) Clearly, Poisson's limit theorem is the key behind the above proof. Indeed, assume for simplicity $0 < \tau < \infty$ and define $B_n = \sum_{i=1}^n I_{\{X_i > u_n\}}$. This quantity has a binomial distribution with parameters $(n, \overline{F}(u_n))$. An application of Poisson's limit theorem yields $B_n \stackrel{d}{\to} Poi(\tau)$ if and only if $EB_n = n\overline{F}(u_n) \to \tau$ which is nothing but (3.4). Also notice that $P(M_n \le u_n) = P(B_n = 0) \to \exp\{-\tau\}$. This explains why (3.5) is sometimes referred to as *Poisson approximation* to the probability $P(M_n \le u_n)$.

2) Evidently, if there exists a sequence $(u_n^{(\tau)})$ satisfying (3.4) for some fixed $\tau > 0$, then we can find such a sequence for any $\tau > 0$. For instance, if $(u_n^{(1)})$ satisfies (3.4) with $\tau = 1$, $u_n^{(\tau)} = u_{[n/\tau]}^{(1)}$ obeys (3.4) for an arbitrary $\tau > 0$. □

By (3.1), (M_n) converges a.s. to the right endpoint x_F of the df F, hence

$$P(M_n \le x) \to \begin{cases} 0 & \text{if} \quad x < x_F\,, \\ 1 & \text{if} \quad x > x_F\,. \end{cases}$$

The following result extends this kind of 0–1 behaviour.

Corollary 3.1.2 *Suppose that $x_F < \infty$ and*

$$\overline{F}(x_F-) = F(x_F) - F(x_F-) > 0\,.$$

Then for every sequence (u_n) such that

$$P(M_n \le u_n) \to \rho\,,$$

either $\rho = 0$ or $\rho = 1$.

Proof. Since $0 \le \rho \le 1$, we may write $\rho = \exp\{-\tau\}$ with $0 \le \tau \le \infty$. By Proposition 3.1.1 we have $n\overline{F}(u_n) \to \tau$ as $n \to \infty$. If $u_n < x_F$ for infinitely many n we have $\overline{F}(u_n) \ge \overline{F}(x_F-) > 0$ for those n and hence $\tau = \infty$. The other possibility is that $u_n \ge x_F$ for all sufficiently large n, giving $n\overline{F}(u_n) = 0$, and hence $\tau = 0$. Thus $\tau = \infty$ or 0, giving $\rho = 0$ or 1. □

This result shows in particular that *for a df with a jump at its finite right endpoint no non–degenerate limit distribution for M_n exists*, whatever the normalisation.

A similar result is true for certain distributions with infinite right endpoint as we see from the following characterisation, given in Leadbetter, Lindgren and Rootzén [418], Theorem 1.7.13.

Theorem 3.1.3 *Let F be a df with right endpoint $x_F \le \infty$ and let $\tau \in (0,\infty)$. There exists a sequence (u_n) satisfying $n\overline{F}(u_n) \to \tau$ if and only if*

$$\lim_{x \uparrow x_F} \frac{\overline{F}(x)}{\overline{F}(x-)} = 1\,. \tag{3.6}$$

□

The result applies in particular to discrete distributions with infinite right endpoint. If the jump heights of the df do not decay sufficiently fast, then a non–degenerate limit distribution for maxima does not exist. For instance, if X is integer–valued and $x_F = \infty$, then (3.6) translates into $\overline{F}(n)/\overline{F}(n-1) \to 1$ as $n \to \infty$.

These considerations show that some intricate asymptotic behaviour of (M_n) exists. The discreteness of a distribution can prevent the maxima from converging and instead forces "oscillatory" behaviour. Nonetheless, in this situation it is often possible to find a sequence (c_n) of integers such that $(M_n - c_n)$ is *tight*; i.e. every subsequence of $(M_n - c_n)$ contains a weakly convergent subsequence. This is true for the examples to follow; see Aldous [7], Section C2, Leadbetter et al. [418], Section 1.7.

Example 3.1.4 (Poisson distribution)

$$P(X = k) = e^{-\lambda}\, \lambda^k / k!\,, \quad k \in \mathbb{N}_0\,, \quad \lambda > 0\,.$$

Then

$$\begin{aligned}\frac{\overline{F}(k)}{\overline{F}(k-1)} &= 1 - \frac{F(k) - F(k-1)}{\overline{F}(k-1)} \\ &= 1 - \frac{\lambda^k}{k!}\left(\sum_{r=k}^{\infty} \frac{\lambda^r}{r!}\right)^{-1} \\ &= 1 - \left(1 + \sum_{r=k+1}^{\infty} \frac{k!}{r!}\, \lambda^{r-k}\right)^{-1}.\end{aligned}$$

The latter sum can be estimated as

$$\sum_{s=1}^{\infty} \frac{\lambda^s}{(k+1)(k+2)\cdots(k+s)} \le \sum_{s=1}^{\infty} \left(\frac{\lambda}{k}\right)^s = \frac{\lambda/k}{1-\lambda/k}\,, \quad k > \lambda\,,$$

which tends to 0 as $k \to \infty$, so that $\overline{F}(k)/\overline{F}(k-1) \to 0$. Theorem 3.1.3 shows that no non–degenerate limit distribution for maxima exists and, furthermore, that no limit of the form $P(M_n \le u_n) \to \rho \in (0,1)$ exists, whatever the sequence of constants (u_n). □

Example 3.1.5 (Geometric distribution)

$$P(X = k) = p(1-p)^{k-1}\,, \quad k \in \mathbb{N}, \quad 0 < p < 1\,.$$

In this case we have

$$\frac{\overline{F}(k)}{\overline{F}(k-1)} = 1 - (1-p)^{k-1}\left(\sum_{r=k}^{\infty}(1-p)^{r-1}\right)^{-1} = 1 - p \in (0,1)\,.$$

Hence again no limit $P(M_n \le u_n) \to \rho$ exists except for $\rho = 0$ or 1.

Maxima of iid geometrically distributed rvs play a prominent role in the study of the length of the longest success–run in a random walk. We refer to Section 8.5, in particular to Theorem 8.5.13. □

Example 3.1.6 (Negative binomial distribution)

$$P(X=k)=\binom{v+k-1}{k-1}p^v\,(1-p)^{k-1}\,,\quad k\in\mathbb{N}_0\,,\quad 0<p<1\,,v>0\,.$$

For $v\in\mathbb{N}$ the negative binomial distribution generalises the geometric distribution in the following sense: the geometric distribution models the waiting time for the first success in a sequence of independent trials, whereas the negative binomial distribution models the waiting time for the vth success.

Using properties of the binomial coefficients we obtain

$$\frac{\overline{F}(k)}{\overline{F}(k-1)}\le 1-p\in(0,1)\,;$$

i.e. no limit $P(M_n\le u_n)\to\rho$ exists except for $\rho=0$ or 1. □

Notes and Comments

Extreme value theory is a classical topic in probability theory and mathematical statistics. Its origins go back to Fisher and Tippett [240]. Since then a large number of books and articles on extreme value theory has appeared. The interested reader may, for instance, consult the following textbooks: Adler [4], Aldous [7], Beirlant, Teugels and Vynckier [57], Berman [62], Falk, Hüsler and Reiss [225], Galambos [249], Gumbel [290], Leadbetter, Lindgren and Rootzén [418], Pfeifer [497], Reiss [526] and Resnick [530].

Some historical notes concerning the development of extreme value theory starting with Nicolas Bernoulli (1709) can be found in Reiss [526].

Our presentation is close in spirit to Leadbetter, Lindgren and Rootzén [418] and Resnick [530]. The latter book is primarily concerned with extreme value theory of iid observations. Two subjects are central: the main analytic tool of extreme value theory is the theory of regularly varying functions (see Appendix A3.1), and the basic probabilistic tool is point process theory (see Chapter 5). After a brief summary of results for iid observations, Leadbetter et al. [418] focus on extremes of stationary sequences and processes; see Sections 4.4, 5.3 and 5.5. Galambos [249] studies the weak and strong limit theory for extremes of iid observations. Moreover, Galambos [249] and also Resnick [530] include results on multivariate extremes. Beirlant et al. [57], Gumbel [290], Pfeifer [497] and Reiss [526] concentrate more on the statistical aspects; see Chapter 6 for more detailed information concerning statistical methods based on extreme value theory.

Extreme value theory for discrete distributions is treated, for instance, in Anderson [11, 12], Arnold, Balakrishnan and Nagaraja [20] and Gordon, Schilling and Waterman [280].

Adler [4], Berman [62] and Leadbetter et al. [418] study extremes of continuous–time (in particular Gaussian) processes.

3.2 Weak Convergence of Maxima Under Affine Transformations

We come back to the main topic of this chapter, namely to the characterisation of the possible limit laws for the maxima M_n of the iid sequence (X_n) under positive affine transformations; see (3.2). This extreme value problem can be considered as an analogue to the central limit problem. Consequently, the main parts of Sections 3.2 and 3.3 bear some resemblance to Section 2.2 and it is instructive to compare and contrast the corresponding results.

In this section we answer the question:

What are the possible (non–degenerate) limit laws for the maxima M_n when properly normalised and centred?

This question turns out to be closely related to the following:

Which distributions satisfy for all $n \geq 2$ the identity in law

$$\max(X_1, \ldots, X_n) \stackrel{d}{=} c_n X + d_n \tag{3.7}$$

for appropriate constants $c_n > 0$ and $d_n \in \mathbb{R}$?

The question is, in other words, which classes of distributions F are closed (up to affine transformations) for maxima. Relation (3.7) reminds us of the defining properties of a stable distribution; see (2.9) in Chapter 2. Those distributions are the only possible limit laws for sums of normalised and centred iid rvs. A similar notion exists for maxima.

Definition 3.2.1 (Max–stable distribution)
A non–degenerate rv X (the corresponding distribution or df) is called max–stable *if it satisfies* (3.7) *for iid $X, X_1, \ldots, X_n$, appropriate constants $c_n > 0$, $d_n \in \mathbb{R}$ and every $n \geq 2$.* □

Remark. 1) *From now on we refer to the centring constants d_n and the normalising constants c_n jointly as* norming constants. □

Assume for the moment that (X_n) is a sequence of iid max–stable rvs. Then (3.7) may be rewritten as follows

$$c_n^{-1}(M_n - d_n) \stackrel{d}{=} X . \tag{3.8}$$

We conclude that every max–stable distribution is a limit distribution for maxima of iid rvs. Moreover, max–stable distributions are the only limit laws for normalised maxima.

Theorem 3.2.2 (Limit property of max–stable laws)
The class of max–stable distributions coincides with the class of all possible (non–degenerate) limit laws for (properly normalised) maxima of iid rvs.

Proof. It remains to prove that the limit distribution of affinely transformed maxima is max–stable. Assume that for appropriate norming constants,

$$\lim_{n\to\infty} F^n\left(c_n x + d_n\right) = H(x)\,, \quad x \in \mathbb{R},$$

for some non–degenerate df H. We anticipate here (and indeed state precisely in Theorem 3.2.3) that the possible limit dfs H are continuous functions on the whole of $\mathbb{R}$.

Then for every $k \in \mathbb{N}$

$$\lim_{n\to\infty} F^{nk}\left(c_n x + d_n\right) = \left(\lim_{n\to\infty} F^n\left(c_n x + d_n\right)\right)^k = H^k(x)\,, \quad x \in \mathbb{R}.$$

Furthermore,

$$\lim_{n\to\infty} F^{nk}\left(c_{nk} x + d_{nk}\right) = H(x)\,, \quad x \in \mathbb{R}.$$

By the convergence to types theorem (Theorem A1.5) there exist constants $\widetilde{c}_k > 0$ and $\widetilde{d}_k \in \mathbb{R}$ such that

$$\lim_{n\to\infty} \frac{c_{nk}}{c_n} = \widetilde{c}_k \quad \text{and} \quad \lim_{n\to\infty} \frac{d_{nk} - d_n}{c_n} = \widetilde{d}_k\,,$$

and for iid rvs $Y_1, \ldots, Y_k$ with df H,

$$\max\left(Y_1, \ldots, Y_k\right) \stackrel{d}{=} \widetilde{c}_k\, Y_1 + \widetilde{d}_k\,. \qquad \square$$

The following result is *the* basis of classical extreme value theory.

Theorem 3.2.3 (Fisher–Tippett theorem, limit laws for maxima)
Let (X_n) be a sequence of iid rvs. If there exist norming constants $c_n > 0$, $d_n \in \mathbb{R}$ and some non–degenerate df H such that

$$c_n^{-1}\left(M_n - d_n\right) \stackrel{d}{\to} H\,, \tag{3.9}$$

then H belongs to the type of one of the following three dfs:

$$\text{Fréchet:} \quad \Phi_\alpha(x) = \begin{cases} 0\,, & x \le 0 \\ \exp\left\{-x^{-\alpha}\right\}\,, & x > 0 \end{cases} \qquad \alpha > 0.$$

$$\text{Weibull:} \quad \Psi_\alpha(x) = \begin{cases} \exp\left\{-(-x)^{\alpha}\right\}\,, & x \le 0 \\ 1\,, & x > 0 \end{cases} \qquad \alpha > 0.$$

$$\text{Gumbel:} \quad \Lambda(x) = \exp\left\{-e^{-x}\right\}\,, \quad x \in \mathbb{R}.$$

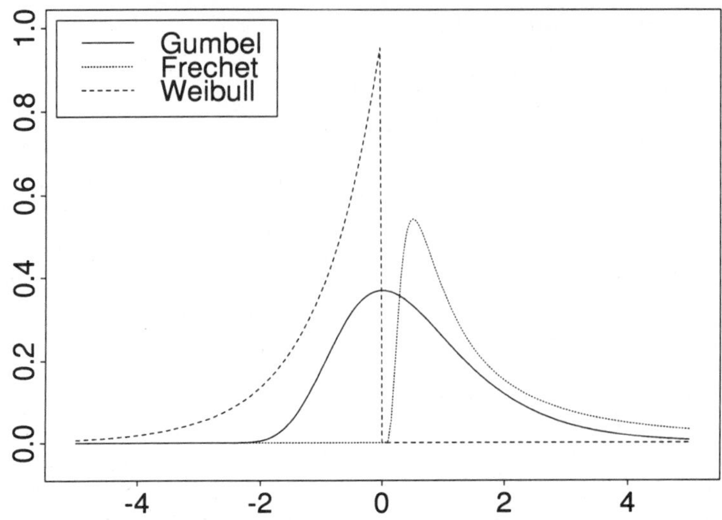

Figure 3.2.4 *Densities of the standard extreme value distributions. We chose* $\alpha = 1$ *for the Fréchet and the Weibull distribution.*

Sketch of the proof. Though a full proof is rather technical, we would like to show how the three limit–types appear; the main ingredient is again the convergence to types theorem, Theorem A1.5. Indeed, (3.9) implies that for all $t > 0$,

$$F^{[nt]}\left(c_{[nt]}x + d_{[nt]}\right) \to H(x)\,, \quad x \in \mathbb{R}\,,$$

where $[\cdot]$ denotes the integer part. However,

$$F^{[nt]}\left(c_n x + d_n\right) = \left(F^n\left(c_n x + d_n\right)\right)^{[nt]/n} \to H^t(x)\,,$$

so that by Theorem A1.5 there exist functions $\gamma(t) > 0$, $\delta(t) \in \mathbb{R}$ satisfying

$$\lim_{n\to\infty} \frac{c_n}{c_{[nt]}} = \gamma(t)\,, \quad \lim_{n\to\infty} \frac{d_n - d_{[nt]}}{c_{[nt]}} = \delta(t)\,, \quad t > 0\,,$$

and

$$H^t(x) = H(\gamma(t)x + \delta(t))\,. \tag{3.10}$$

It is not difficult to deduce from (3.10) that for $s, t > 0$

$$\gamma(st) = \gamma(s)\,\gamma(t)\,, \quad \delta(st) = \gamma(t)\,\delta(s) + \delta(t)\,. \tag{3.11}$$

The solution of the functional equations (3.10) and (3.11) leads to the three types Λ, Φ_α, Ψ_α. Details of the proof are for instance to be found in Resnick [530], Proposition 0.3. □

Remarks. 2) The limit law in (3.9) is unique only up to affine transformations. If the limit appears as $H(cx+d)$, i.e.

$$\lim_{n\to\infty} P\left(c_n^{-1}\left(M_n-d_n\right)\le x\right)=H(cx+d)\,,$$

then $H(x)$ is also a limit under a simple change of norming constants:

$$\lim_{n\to\infty} P\left(\tilde{c}_n^{-1}\left(M_n-\tilde{d}_n\right)\le x\right)=H(x)$$

with $\tilde{c}_n=c_n/c$ and $\tilde{d}_n=d_n-dc_n/c$. The convergence to types theorem shows precisely how affine transformations, weak convergence and types are related.

3) In Tables 1.2.5 and 1.2.6 we defined a Weibull df for $c,\alpha>0$. For $c=1$ it is given by

$$F_\alpha(x)=1-e^{-x^\alpha}\,,\quad x\ge 0\,,$$

which is the df of a positive rv. The Weibull distribution Ψ_α, as a limit distribution for maxima, is concentrated on $(-\infty,0)$:

$$\Psi_\alpha(x)=1-F_\alpha(-x)\,,\quad x<0\,.$$

In the context of extreme value theory we follow the convention and refer to Ψ_α as the Weibull distribution. We hope to avoid any confusion by a clear distinction between the two distributions whose extremal behaviour is completely different. Example 3.3.20 and Proposition 3.3.25 below show that F_α belongs to the maximum domain of attraction of the Gumbel distribution Λ.

4) The proof of Theorem 3.2.3 uses similar techniques as the proof of Theorems 2.2.2 and 2.2.3. Indeed, in the case $S_n=X_1+\cdots+X_n$ we use the characteristic function $\phi_{S_n}(t)=(\phi_X(t))^n$, whereas for partial maxima we directly work with the df $F_{M_n}(x)=(F_X(x))^n$. So not suprisingly do we obtain functions like $\exp\{-c|t|^\alpha\}$ as possible limit chfs in the partial sum case, whereas such functions appear as limits for the dfs of normalised maxima.

5) Though, for modelling purposes, the types of Λ, Φ_α and Ψ_α are very different, from a mathematical point of view they are closely linked. Indeed, one immediately verifies the following properties. Suppose $X>0$, then

$$X \text{ has df } \Phi_\alpha \iff \ln X^\alpha \text{ has df } \Lambda \iff -X^{-1} \text{ has df } \Psi_\alpha\,.$$

These relationships will appear again and again in various disguises throughout the book. □

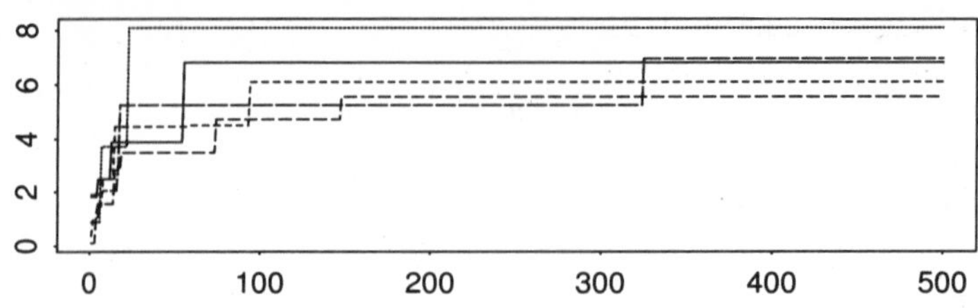

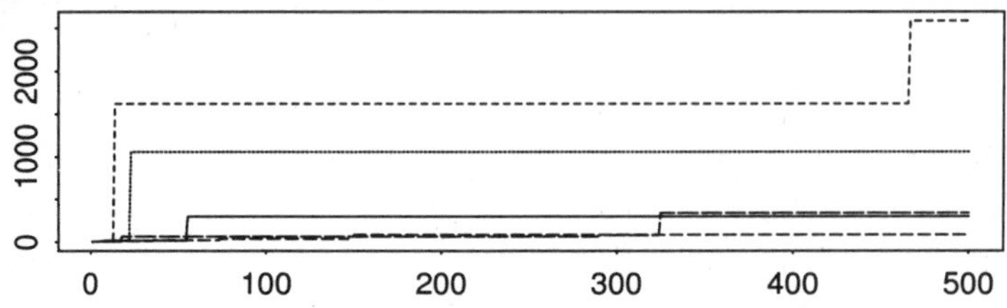

Figure 3.2.5 *Evolution of the maxima M_n of standard exponential (top) and Cauchy (bottom) samples. A sample path of (M_n) has a jump whenever $X_n > M_{n-1}$ (we say that M_n is a record). The graph seems to suggest that there occur more records for the exponential than for the Cauchy rvs. However, the distribution of the number of record times is approximately the same in both cases; see Theorem 5.4.7. The qualitative differences in the two graphs are due to a few large jumps for Cauchy distributed variables. Compared with those the smaller jumps are so tiny that they "disappear" from the computer graph; notice the difference between the vertical scales.*

Definition 3.2.6 (Extreme value distribution and extremal rv)
The dfs Φ_α, Ψ_α and Λ as presented in Theorem 3.2.3 *are called* standard extreme value distributions, *the corresponding rvs* standard extremal rvs. *Dfs of the types of Φ_α, Ψ_α and Λ are* extreme value distributions; *the corresponding rvs* extremal rvs. □

By Theorem 3.2.2, the extreme value distributions are precisely the max–stable distributions. Hence if X is an extremal rv it satisfies (3.8). In particular, the three cases in Theorem 3.2.3 correspond to

Fréchet:	$M_n \stackrel{d}{=} n^{1/\alpha} X$
Weibull:	$M_n \stackrel{d}{=} n^{-1/\alpha} X$
Gumbel:	$M_n \stackrel{d}{=} X + \ln n\,.$

Example 3.2.7 (Maxima of exponential rvs)
See also Figures 3.2.5 and 3.2.9. Let (X_i) be a sequence of iid standard exponential rvs. Then

$$\begin{aligned} P(M_n - \ln n \le x) &= (P(X \le x + \ln n))^n \\ &= \left(1 - n^{-1} e^{-x}\right)^n \\ &\to \exp\left\{-e^{-x}\right\} = \Lambda(x), \quad x \in \mathbb{R}. \end{aligned}$$

For comparison recall that for iid Gumbel rvs X_i,

$$P(M_n - \ln n \le x) = \Lambda(x), \quad x \in \mathbb{R}.$$

□

Example 3.2.8 (Maxima of Cauchy rvs)
See also Figures 3.2.5 and 3.2.10. Let (X_i) be a sequence of iid standard Cauchy rvs. The standard Cauchy distribution is absolutely continuous with density

$$f(x) = \left(\pi\left(1 + x^2\right)\right)^{-1}, \quad x \in \mathbb{R}.$$

By l'Hospital's rule we obtain

$$\lim_{x\to\infty} \frac{\overline{F}(x)}{(\pi x)^{-1}} = \lim_{x\to\infty} \frac{f(x)}{\pi^{-1}x^{-2}} = \lim_{x\to\infty} \frac{\pi x^2}{\pi\left(1 + x^2\right)} = 1,$$

giving $\overline{F}(x) \sim (\pi x)^{-1}$. This implies

$$\begin{aligned} P\left(M_n \le \frac{nx}{\pi}\right) &= \left(1 - \overline{F}\left(\frac{nx}{\pi}\right)\right)^n \\ &= \left(1 - \frac{1}{n}\left(\frac{1}{x} + o(1)\right)\right)^n \\ &\to \exp\left\{-x^{-1}\right\} = \Phi_1(x), \quad x > 0. \end{aligned}$$

□

Notes and Comments

Theorem 3.2.3 marked the beginning of extreme value theory as one of the central topics in probability theory and statistics. The limit laws for maxima were derived by Fisher and Tippett [240]. A first rigorous proof is due to Gnedenko [266]. De Haan [292] subsequently applied regular variation as an analytical tool. His work has been of great importance for the development of modern extreme value theory. Weissman [636] provided a simpler version of de Haan's proof, variations of which are now given in most textbooks on extreme value theory.

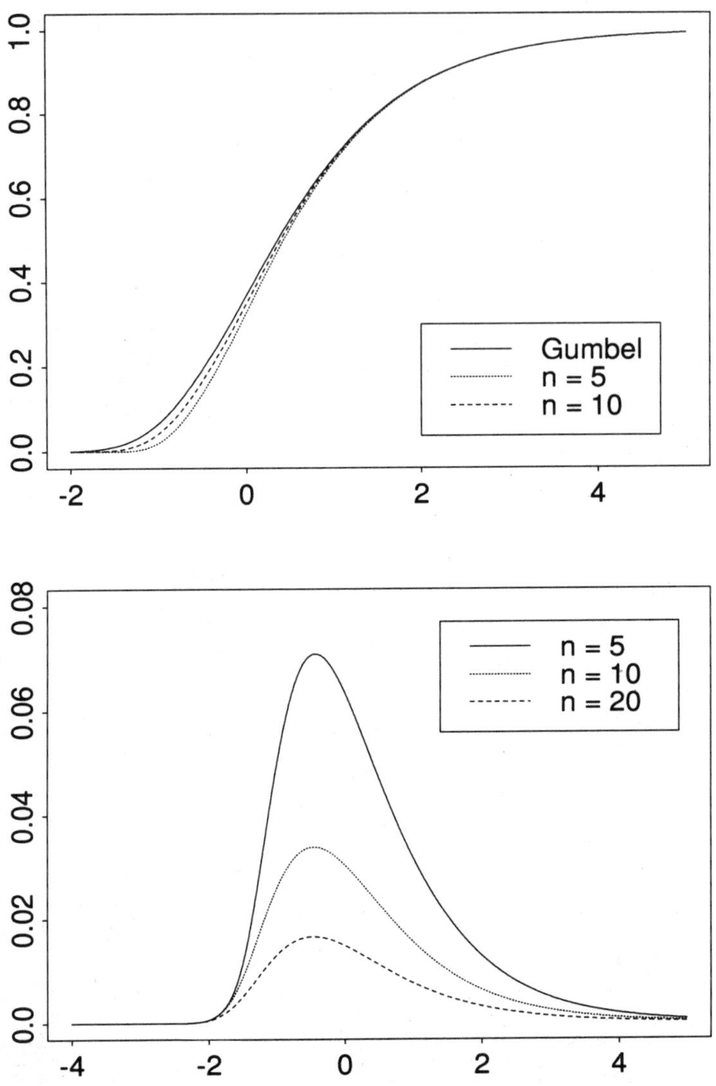

Figure 3.2.9 *The df $P(M_n - \ln n \le x)$ for n iid standard exponential rvs and the Gumbel df (top). In the bottom figure the relative error $(P(M_n - \ln n > x)/\overline{\Lambda}(x)) - 1$ of this approximation is illustrated.*

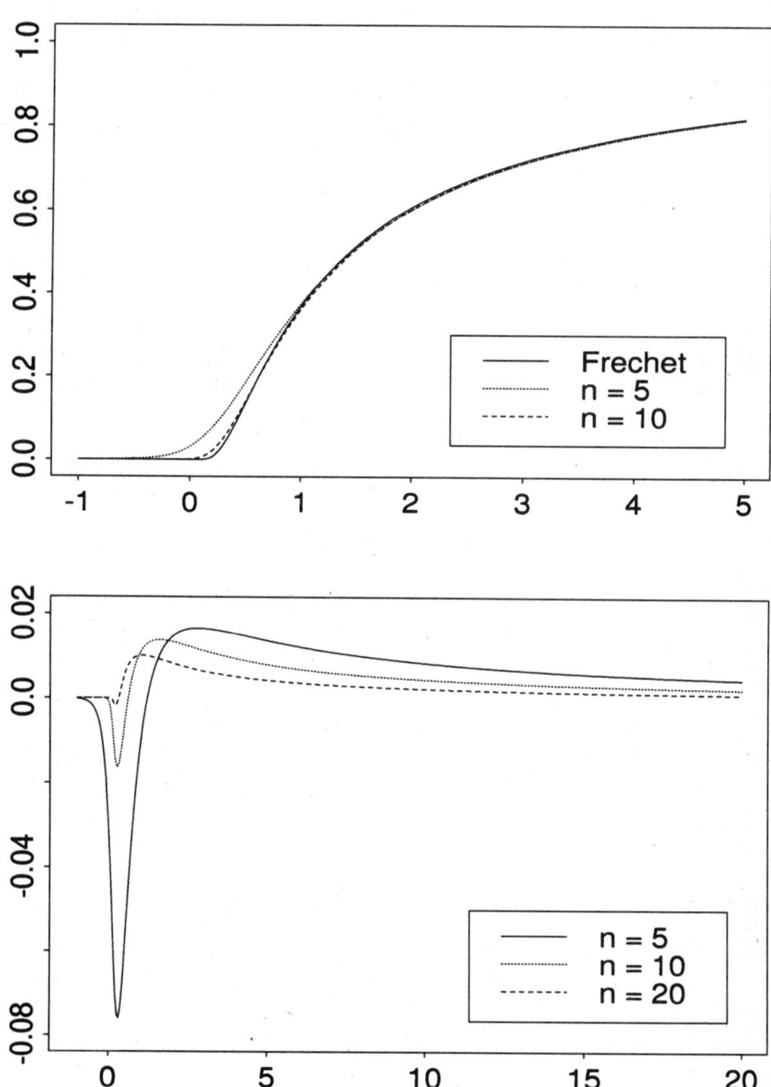

Figure 3.2.10 *The df $P(\pi n^{-1} M_n \leq x)$ of n iid standard Cauchy rvs and the Fréchet df Φ_1 (top). In the bottom figure the relative error $(P(\pi n^{-1} M_n > x)/\overline{\Phi}_1(x)) - 1$ of this approximation is illustrated.*

3.3 Maximum Domains of Attraction and Norming Constants

In the preceding section we identified the extreme value distributions as the limit laws for normalised maxima of iid rvs; see Theorem 3.2.3. This section is devoted to the question:

Given an extreme value distribution H, what conditions on the df F imply that the normalised maxima M_n converge weakly to H?

Closely related to this question is the following:

How may we choose the norming constants $c_n > 0$ and $d_n \in \mathbb{R}$ such that

$$c_n^{-1}(M_n - d_n) \xrightarrow{d} H\,? \tag{3.12}$$

Can it happen that different norming constants imply convergence to different limit laws?

The last question can be answered immediately: the convergence to types theorem (Theorem A1.5) ensures that the limit law is uniquely determined up to affine transformations.

Before we answer the other questions recall from Section 2.2 how we proceeded with the sums $S_n = X_1 + \cdots + X_n$ of iid rvs: we collected all those dfs F in a common class for which the normalised sums S_n had the same stable limit distribution. Such a class is then called a domain of attraction (Definition 2.2.7). For maxima we proceed analogously.

Definition 3.3.1 (Maximum domain of attraction)
We say that the rv X (the df F of X, the distribution of X) belongs to the maximum domain of attraction of the extreme value distribution *H if there exist constants $c_n > 0$, $d_n \in \mathbb{R}$ such that* (3.12) *holds. We write $X \in \mathrm{MDA}(H)$ ($F \in \mathrm{MDA}(H)$).* □

Remark. Notice that the extreme value dfs are continuous on $\mathbb{R}$, hence $c_n^{-1}(M_n - d_n) \xrightarrow{d} H$ is equivalent to

$$\lim_{n\to\infty} P(M_n \le c_n x + d_n) = \lim_{n\to\infty} F^n(c_n x + d_n) = H(x)\,, \quad x \in \mathbb{R}\,. \quad \square$$

The following result is an immediate consequence of Proposition 3.1.1 and will be used throughout the following sections.

Proposition 3.3.2 (Characterisation of $\mathrm{MDA}(H)$)
The df F belongs to the maximum domain of attraction of the extreme value distribution H with norming constants $c_n > 0$, $d_n \in \mathbb{R}$ if and only if

$$\lim_{n\to\infty} n\overline{F}(c_n x + d_n) = -\ln H(x)\,, \quad x \in \mathbb{R}\,.$$

When $H(x) = 0$ the limit is interpreted as ∞. □

For every standard extreme value distribution we characterise its maximum domain of attraction. Using the concept of regular variation this is not too difficult for the Fréchet distribution Φ_α and the Weibull distribution Ψ_α; see Sections 3.3.1 and 3.3.2. Recall that a distribution tail $\overline{F}$ is regularly varying with index $-\alpha$ for some $\alpha \geq 0$, we write $\overline{F} \in \mathcal{R}_{-\alpha}$, if

$$\lim_{x\to\infty} \frac{\overline{F}(xt)}{\overline{F}(x)} = t^{-\alpha}, \quad t > 0.$$

The definition of regularly varying functions and those of their properties most important for our purposes can be found in Appendix A3.1. The interested reader may consult the monograph by Bingham, Goldie and Teugels [72] for an encyclopaedic treatment of regular variation. The maximum domain of attraction of the Gumbel distribution Λ is not so easily characterised; it consists of dfs whose right tail decreases to zero faster than any power function. This will be made precise in Section 3.3.3. If F has a density, simple sufficient conditions for F to be in the maximum domain of attraction of some extreme value distribution are due to von Mises. We present them below for practical (and historical) reasons.

The following concept defines an equivalence relation on the set of all dfs.

Definition 3.3.3 (Tail–equivalence)
Two dfs F and G are called tail–equivalent *if they have the same right end-point, i.e. if $x_F = x_G$, and*

$$\lim_{x\uparrow x_F} \overline{F}(x)/\overline{G}(x) = c$$

for some constant $0 < c < \infty$. □

We show that every maximum domain of attraction is closed with respect to tail–equivalence, i.e. for tail–equivalent F and G, $F \in \text{MDA}(H)$ if and only if $G \in \text{MDA}(H)$. Moreover, for any two tail–equivalent distributions one can take the same norming constants. This will prove to be of great help for calculating norming constants which, in general, can become a rather tedious procedure.

Theorem 3.2.2 identifies the max–stable distributions as limit laws for affinely transformed maxima of iid rvs. The corresponding Theorem 2.2.2 for sums identifies the stable distributions as limit laws for centred and normalised sums. Sums are centred by their medians or by truncated means; see Proposition 2.2.14. The sample maximum M_n is the empirical version of the $(1 - n^{-1})$–quantile of the underlying df F. Therefore the latter is an appropriate centring constant. Quantiles correspond to the "inverse" of a df, which is not always well–defined (dfs are not necessarily strictly increasing). In the following definition we fix upon a left–continuous version.

Definition 3.3.4 (Generalised inverse of a monotone function)
Suppose h is a non-decreasing function on $\mathbb{R}$. The generalised inverse of h *is defined as*

$$h^{\leftarrow}(t) = \inf\{x \in \mathbb{R} : h(x) \geq t\}\,.$$

(We use the convention that the infimum of an empty set is ∞.) □

Definition 3.3.5 (Quantile function)
The generalised inverse of the df F

$$F^{\leftarrow}(t) = \inf\{x \in \mathbb{R} :\ F(x) \geq t\}\,, \quad 0 < t < 1\,,$$

is called the quantile function *of the df F. The quantity $x_t = F^{\leftarrow}(t)$ defines the t–*quantile *of F.* □

We have summarised some properties of generalised inverse functions in Appendix A1.6.

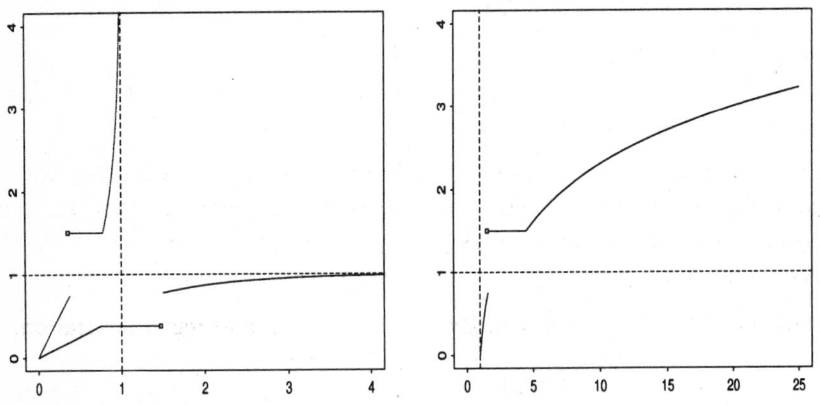

Figure 3.3.6 *An "interesting" df F, its quantile function $F^{\leftarrow}$ (left) and the corresponding function $F^{\leftarrow}(1 - x^{-1})$ (right).*

3.3.1 The Maximum Domain of Attraction of the Fréchet Distribution $\Phi_\alpha(x) = \exp\left\{-x^{-\alpha}\right\}$

In this section we characterise the maximum domain of attraction of Φ_α for $\alpha > 0$. By Taylor expansion,

$$1-\Phi_\alpha(x) = 1-\exp\left\{-x^{-\alpha}\right\} \sim x^{-\alpha}\,, \quad x\to\infty\,,$$

hence the tail of Φ_α decreases like a power law. We ask:

How far away can we move from a power tail and still remain in MDA(Φ_α)*?*

We show that the maximum domain of attraction of Φ_α consists of dfs F whose right tail is regularly varying with index $-\alpha$. For $F\in \text{MDA}(\Phi_\alpha)$ the constants d_n can be chosen as 0 (centring is not necessary) and the c_n by means of the quantile function, more precisely by

$$\begin{aligned} c_n = F^{\leftarrow}(1-n^{-1}) &= \inf\left\{x\in\mathbb{R} : F(x)\ge 1-n^{-1}\right\} \\ &= \inf\left\{x\in\mathbb{R} : \left(1/\overline{F}\right)(x)\ge n\right\} \\ &= \left(1/\overline{F}\right)^{\leftarrow}(n)\,. \end{aligned} \tag{3.13}$$

Theorem 3.3.7 (Maximum domain of attraction of Φ_α)
The df F belongs to the maximum domain of attraction of Φ_α, $\alpha>0$, if and only if $\overline{F}(x)=x^{-\alpha}L(x)$ for some slowly varying function L.
If $F\in$ MDA(Φ_α), then

$$c_n^{-1}\, M_n \xrightarrow{d} \Phi_\alpha\,, \tag{3.14}$$

where the norming constants c_n can be chosen according to (3.13).

Notice that this result implies in particular that every $F\in$ MDA(Φ_α) has an infinite right endpoint $x_F=\infty$. Furthermore, the norming constants c_n form a regularly varying sequence, more precisely, $c_n = n^{1/\alpha}L_1(n)$ for some slowly varying function L_1.

Proof. Let $\overline{F}\in\mathcal{R}_{-\alpha}$ for $\alpha>0$. By the choice of c_n and regular variation,

$$\overline{F}\left(c_n\right)\sim n^{-1}\,, \qquad n\to\infty\,, \tag{3.15}$$

and hence $\overline{F}(c_n)\to 0$ giving $c_n\to\infty$. For $x>0$,

$$n\overline{F}\left(c_n x\right) \sim \frac{\overline{F}\left(c_n x\right)}{\overline{F}\left(c_n\right)} \to x^{-\alpha}\,, \quad n\to\infty\,.$$

For $x<0$, immediately $F^n(c_n x)\le F^n(0)\to 0$, since regular variation requires $F(0)<1$. By Proposition 3.3.2, $F\in$ MDA(Φ_α).

Conversely, assume that $\lim_{n\to\infty}F^n(c_n x+d_n) = \Phi_\alpha(x)$ for all $x>0$ and appropriate $c_n>0$, $d_n\in\mathbb{R}$. This leads to

$$\lim_{n\to\infty} F^n(c_{[ns]}x+d_{[ns]}) = \Phi_\alpha^{1/s}(x) = \Phi_\alpha(s^{1/\alpha}x)\,, \quad s>0\,, x>0\,.$$

By the convergence to types theorem (Theorem A1.5)

$$c_{[ns]}/c_n \to s^{1/\alpha} \quad \text{and} \quad \left(d_{[ns]} - d_n\right)/c_n \to 0 .$$

Hence (c_n) is a regularly varying sequence in the sense of Definition A3.13, in particular $c_n \to \infty$. Assume first that $d_n = 0$, then $n\overline{F}(c_n x) \to x^{-\alpha}$ so that $\overline{F} \in \mathcal{R}_{-\alpha}$ because of Proposition A3.8(a). The case $d_n \neq 0$ is more involved, indeed one has to show that $d_n/c_n \to 0$. If the latter holds one can repeat the above argument by replacing d_n by 0. For details on this, see Bingham et al. [72], Theorem 8.13.2, or de Haan [292], Theorem 2.3.1. Resnick [530], Proposition 1.11 contains an alternative argument. □

We have found the answer to the above question:

$$\boxed{F \in \mathrm{MDA}(\Phi_\alpha) \quad \Longleftrightarrow \quad \overline{F} \in \mathcal{R}_{-\alpha}}$$

Thus we have a simple characterisation of $\mathrm{MDA}(\Phi_\alpha)$. Notice that this class of dfs contains "very heavy–tailed distributions" in the sense that $E(X^+)^\delta = \infty$ for $\delta > \alpha$. Thus they may be appropriate distributions for modelling large insurance claims and large fluctuations of prices, log–returns etc.

Von Mises found some easily verifiable conditions on the density of a distribution for it to belong to some maximum domain of attraction. The following is a consequence of Proposition A3.8(b).

Corollary 3.3.8 (Von Mises condition)
Let F be an absolutely continuous df with density f satisfying

$$\lim_{x\to\infty} \frac{x\, f(x)}{\overline{F}(x)} = \alpha > 0 , \tag{3.16}$$

then $F \in \mathrm{MDA}(\Phi_\alpha)$. □

The class of dfs F with regularly varying tail $\overline{F}$ is obviously closed with respect to tail–equivalence (Definition 3.3.3). The following result gives us some insight into the structure of $\mathrm{MDA}(\Phi_\alpha)$. Besides this theoretical aspect, it will turn out to be a useful tool for calculating norming constants.

Proposition 3.3.9 (Closure property of $\mathrm{MDA}(\Phi_\alpha)$)
Let F and G be dfs and assume that $F \in \mathrm{MDA}(\Phi_\alpha)$ with norming constants $c_n > 0$, i.e.

$$\lim_{n\to\infty} F^n\left(c_n x\right) = \Phi_\alpha(x) , \quad x > 0 . \tag{3.17}$$

Then

$$\lim_{n\to\infty} G^n\left(c_n x\right) = \Phi_\alpha(cx) , \quad x > 0 ,$$

for some $c > 0$ if and only if F and G are tail–equivalent with

$$\lim_{x\to\infty} \overline{F}(x)/\overline{G}(x) = c^\alpha .$$

Proof of the sufficiency. For the necessity part see Resnick [530], Proposition 1.19. Suppose that $\overline{F}(x) \sim q\,\overline{G}(x)$ as $x \to \infty$ for some $q > 0$. By Proposition 3.3.2 the limit relation (3.17) is equivalent to

$$\lim_{n\to\infty} n\overline{F}\,(c_n x) = x^{-\alpha}$$

for all $x > 0$. For such x, $c_n x \to \infty$ as $n \to \infty$ and hence, by tail–equivalence,

$$n\overline{G}\,(c_n x) \sim nq^{-1}\,\overline{F}\,(c_n x) \to q^{-1}\,x^{-\alpha}\,,$$

i.e. again by Proposition 3.3.2,

$$\lim_{n\to\infty} G^n\,(c_n x) = \exp\left\{-\left(q^{1/\alpha}x\right)^{-\alpha}\right\} = \Phi_\alpha\left(q^{1/\alpha}x\right)\,.$$

Now set $c = q^{1/\alpha}$. □

By Theorem 3.3.7, $F \in \mathrm{MDA}(\Phi_\alpha)$ if and only if $\overline{F} \in \mathcal{R}_{-\alpha}$. The representation theorem for regularly varying functions (Theorem A3.3) implies that every $F \in \mathrm{MDA}(\Phi_\alpha)$ is tail–equivalent to an absolutely continuous df satisfying (3.16). We can summarize this as follows:

> MDA (Φ_α) consists of dfs satisfying the von Mises condition (3.16) and their tail–equivalent dfs.

We conclude this section with some examples.

Example 3.3.10 (Pareto–like distributions)

- Pareto
- Cauchy
- Burr
- Stable with exponent $\alpha < 2$.

The respective densities or dfs are given in Table 1.2.6; for stable distributions see Definition 2.2.1. All these distributions are Pareto–like in the sense that their right tails are of the form

$$\overline{F}(x) \sim Kx^{-\alpha}\,, \quad x \to \infty\,,$$

for some K, $\alpha > 0$. Obviously $\overline{F} \in \mathcal{R}_{-\alpha}$ which implies that $F \in \mathrm{MDA}(\Phi_\alpha)$ and as norming constants we can choose $c_n = (Kn)^{1/\alpha}$; see Theorem 3.3.7. Then

$$(Kn)^{-1/\alpha}\,M_n \xrightarrow{d} \Phi_\alpha\,.$$

The Cauchy distribution was treated in detail in Example 3.2.8. □

Example 3.3.11 (Loggamma distribution)
The loggamma distribution has tail

$$\overline{F}(x) \sim \frac{\alpha^{\beta-1}}{\Gamma(\beta)}(\ln x)^{\beta-1}x^{-\alpha}, \quad x\to\infty, \quad \alpha,\beta>0. \tag{3.18}$$

Hence $\overline{F} \in \mathcal{R}_{-\alpha}$ which is equivalent to $F \in \mathrm{MDA}(\Phi_\alpha)$. According to Proposition 3.3.9 we choose c_n by means of the tail–equivalent right–hand side of (3.18). On applying (3.13) and taking logarithms we find we have to solve

$$\alpha \ln c_n - (\beta-1)\ln\ln c_n - \ln(\alpha^{\beta-1}/\Gamma(\beta)) = \ln n. \tag{3.19}$$

The solution satisfies

$$\ln c_n = \alpha^{-1}(\ln n + \ln r_n),$$

where $\ln r_n = o(\ln n)$ as $n\to\infty$. We substitute this into equation (3.19) and obtain

$$\ln r_n = (\beta-1)\ln(\alpha^{-1}\ln n(1+o(1))) + \ln\left(\alpha^{\beta-1}/\Gamma(\beta)\right).$$

This gives the norming constants

$$c_n \sim \left((\Gamma(\beta))^{-1}(\ln n)^{\beta-1}n\right)^{1/\alpha}.$$

Hence

$$\left((\Gamma(\beta))^{-1}(\ln n)^{\beta-1}n\right)^{-1/\alpha} M_n \quad \xrightarrow{d} \quad \Phi_\alpha. \qquad \square$$

3.3.2 The Maximum Domain of Attraction of the Weibull Distribution $\Psi_\alpha(x) = \exp\{-(-x)^\alpha\}$

In this section we characterise the maximum domain of attraction of Ψ_α for $\alpha>0$. An important, though not at all obvious fact is that all dfs F in $\mathrm{MDA}(\Psi_\alpha)$ have finite right endpoint x_F. As was already indicated in Remark 5 of Section 3.2, Ψ_α and Φ_α are closely related, indeed

$$\Psi_\alpha\left(-x^{-1}\right) = \Phi_\alpha(x), \quad x>0.$$

Therefore we may expect that also $\mathrm{MDA}(\Psi_\alpha)$ and $\mathrm{MDA}(\Phi_\alpha)$ will be closely related. The following theorem confirms this.

Theorem 3.3.12 (Maximum domain of attraction of Ψ_α)
The df F belongs to the maximum domain of attraction of Ψ_α, $\alpha > 0$, if and only if $x_F < \infty$ and $\overline{F}(x_F - x^{-1}) = x^{-\alpha} L(x)$ for some slowly varying function L.
If $F \in \mathrm{MDA}(\Psi_\alpha)$, then

$$c_n^{-1}\left(M_n - x_F\right) \xrightarrow{d} \Psi_\alpha\,, \tag{3.20}$$

where the norming constants c_n can be chosen as $c_n = x_F - F^{\leftarrow}(1 - n^{-1})$ and $d_n = x_F$.

Sketch of the proof. The necessity part is difficult; see Resnick [530], Proposition 1.13. Sufficiency can be shown easily by exploiting the link between Φ_α and Ψ_α; see Remark 5 in Section 3.2. So suppose $x_F < \infty$ and $\overline{F}(x_F - x^{-1}) = x^{-\alpha} L(x)$ and define

$$F_*(x) = F\left(x_F - x^{-1}\right), \quad x > 0\,, \tag{3.21}$$

then $\overline{F}_* \in \mathcal{R}_{-\alpha}$ so that by Theorem 3.3.7, $F_* \in \mathrm{MDA}(\Phi_\alpha)$ with norming constants $c_n^* = F_*^{\leftarrow}(1 - n^{-1})$ and $d_n^* = 0$. The remaining part of the proof of sufficiency is now straightforward. Indeed, $F_* \in \mathrm{MDA}(\Phi_\alpha)$ implies that for $x > 0$,

$$F_*^n\left(c_n^* x\right) \to \Phi_\alpha(x)\,,$$

i.e.

$$F^n\left(x_F - (c_n^* x)^{-1}\right) \to \exp\left\{-x^{-\alpha}\right\}.$$

Substitute $x = -y^{-1}$, then

$$F^n\left(x_F + y/c_n^*\right) \to \exp\left\{-(-y)^\alpha\right\}, \quad y < 0\,. \tag{3.22}$$

Finally,

$$\begin{aligned} c_n^* &= F_*^{\leftarrow}\left(1 - n^{-1}\right) \\ &= \inf\left\{x \in \mathbb{R} : F(x_F - x^{-1}) \geq 1 - n^{-1}\right\} \\ &= \inf\left\{(x_F - u)^{-1} : F(u) \geq 1 - n^{-1}\right\} \\ &= \left(x_F - \inf\left\{u : F(u) \geq 1 - n^{-1}\right\}\right)^{-1} \\ &= \left(x_F - F^{\leftarrow}(1 - n^{-1})\right)^{-1}, \end{aligned}$$

completing the proof because of (3.22). □

Consequently,

$$F \in \mathrm{MDA}\left(\Psi_\alpha\right) \quad \Longleftrightarrow \quad x_F < \infty, \quad \overline{F}\left(x_F - x^{-1}\right) \in \mathcal{R}_{-\alpha}.$$

Thus MDA(Ψ_α) consists of dfs F with support bounded to the right. They may not be the best choice for modelling extremal events in insurance and finance, precisely because $x_F < \infty$. Though clearly in all circumstances in practice there is a (perhaps ridiculously high) upper limit, we may not want to incorporate this extra parameter x_F in our model. Often distributions with $x_F = \infty$ should be preferred since they allow for arbitrarily large values in a sample. Such distributions typically belong to MDA(Φ_α) or MDA(Λ). In Chapter 6 we shall discuss various such examples.

In the previous section we found it convenient to characterise membership in MDA(Φ_α) via the density of a df; see Corollary 3.3.8. Having in mind the transformation (3.21), Corollary 3.3.8 can be translated for $F \in$ MDA(Ψ_α).

Corollary 3.3.13 (Von Mises condition)
Let F be an absolutely continuous df with density f which is positive on some finite interval (z, x_F). If

$$\lim_{x \uparrow x_F} \frac{(x_F - x)\, f(x)}{\overline{F}(x)} = \alpha > 0\,, \tag{3.23}$$

then $F \in$ MDA(Ψ_α). □

Applying the transformation (3.21), Proposition 3.3.9 can be reformulated as follows.

Proposition 3.3.14 (Closure property of MDA (Ψ_α))
Let F and G be dfs with right endpoints $x_F = x_G < \infty$ and assume that $F \in$ MDA(Ψ_α) with norming constants $c_n > 0$; i.e.

$$\lim_{n \to \infty} F^n\left(c_n x + x_F\right) = \Psi_\alpha(x)\,, \quad x < 0\,.$$

Then

$$\lim_{n \to \infty} G^n\left(c_n x + x_F\right) = \Psi_\alpha(cx)\,, \quad x < 0\,,$$

for some $c > 0$ if and only if F and G are tail–equivalent with

$$\lim_{x \uparrow x_F} \overline{F}(x)/\overline{G}(x) = c^{-\alpha}\,.$$ □

Notice that the representation theorem for regularly varying functions (Theorem A3.3) implies that every $F \in$ MDA(Ψ_α) is tail–equivalent to an absolutely continuous df satisfying (3.23). We summarize this as follows:

MDA(Ψ_α) consists of dfs satisfying the von Mises condition (3.23) and their tail–equivalent dfs.

We conclude this section with some examples of prominent MDA(Ψ_α)–members.

Example 3.3.15 (Uniform distribution on $(0,1)$)
Obviously, $x_F = 1$ and $\overline{F}(1-x^{-1}) = x^{-1} \in \mathcal{R}_{-1}$. Then by Theorem 3.3.12 we obtain $F \in \text{MDA}(\Psi_1)$. Since $\overline{F}(1-n^{-1}) = n^{-1}$, we choose $c_n = n^{-1}$. This implies in particular

$$n\,(M_n - 1) \xrightarrow{d} \Psi_1\,. \qquad \square$$

Example 3.3.16 (Power law behaviour at the finite right endpoint)
Let F be a df with finite right endpoint x_F and distribution tail

$$\overline{F}(x) = K\,(x_F - x)^\alpha\,, \quad x_F - K^{-1/\alpha} \le x \le x_F\,, \quad K,\alpha > 0\,.$$

By Theorem 3.3.12 this ensures that $F \in \text{MDA}(\Psi_\alpha)$. The norming constants c_n can be chosen such that $\overline{F}(x_F - c_n) = n^{-1}$, i.e. $c_n = (nK)^{-1/\alpha}$ and, in particular,

$$(nK)^{1/\alpha}\,(M_n - x_F) \xrightarrow{d} \Psi_\alpha\,. \qquad \square$$

Example 3.3.17 (Beta distribution)
The beta distribution is absolutely continuous with density

$$f(x) = \frac{\Gamma(a+b)}{\Gamma(a)\,\Gamma(b)}\, x^{a-1}(1-x)^{b-1}\,, \quad 0 < x < 1\,, \quad a\,,b > 0\,.$$

Notice that $f(1-x^{-1})$ is regularly varying with index $-(b-1)$ and hence, by Karamata's theorem (Theorem A3.6),

$$\overline{F}(1-x^{-1}) = \int_{1-x^{-1}}^{1} f(y)\,dy = \int_x^\infty f(1-y^{-1})y^{-2}\,dy \sim x^{-1} f(1-x^{-1})\,.$$

Hence $\overline{F}(1-x^{-1})$ is regularly varying with index $-b$ and

$$\overline{F}(x) \sim \frac{\Gamma(a+b)}{\Gamma(a)\,\Gamma(b+1)}\,(1-x)^b\,, \quad x \uparrow 1\,.$$

Thus the beta df is tail–equivalent to a df with power law behaviour at $x_F = 1$. By Proposition 3.3.14 the norming constants can be determined by this power law tail which fits into the framework of Example 3.3.16 above. $\square$

3.3.3 The Maximum Domain of Attraction of the Gumbel Distribution $\Lambda(x) = \exp\{-\exp\{-x\}\}$

Von Mises Functions

The maximum domain of attraction of the Gumbel distribution Λ covers a wide range of dfs F. Although there is no direct link with regular variation as in the maximum domains of attraction of the Fréchet and Weibull distribution, we will find extensions of regular variation which allow for a complete characterisation of MDA(Λ).

A Taylor expansion argument yields

$$1 - \Lambda(x) \sim e^{-x}, \quad x \to \infty,$$

hence $\overline{\Lambda}(x)$ decreases to zero at an exponential rate. Again the following question naturally arises:

How far away can we move from an exponential tail and still remain in MDA(Λ)*?*

We will see in the present and the next section that MDA(Λ) contains dfs with very different tails, ranging from *moderately heavy* (such as the lognormal distribution) to *light* (such as the normal distribution). Also both cases $x_F < \infty$ and $x_F = \infty$ are possible. Before we give a general answer to the above question, we restrict ourselves to some absolutely continuous $F \in$ MDA(Λ) which have a simple representation, proposed by von Mises. These distributions provide an important building block of this maximum domain of attraction, and therefore we study them in detail. We will see later (Theorem 3.3.26 and Remark 4) that one only has to consider a slight modification of the von Mises functions in order to characterise MDA(Λ) completely.

Definition 3.3.18 (Von Mises function)
Let F be a df with right endpoint $x_F \le \infty$. Suppose there exists some $z < x_F$ such that F has representation

$$\overline{F}(x) = c \exp\left\{-\int_z^x \frac{1}{a(t)}\,dt\right\}, \quad z < x < x_F, \tag{3.24}$$

where c is some positive constant, $a(\cdot)$ is a positive and absolutely continuous function (with respect to Lebesgue measure) with density a' and $\lim_{x \uparrow x_F} a'(x) = 0$.

Then F is called a von Mises function, *the function $a(\cdot)$ the* auxiliary function *of F.* □

Remark. 1) Relation (3.24) should be compared with the Karamata representation of a regularly varying function; see Theorem A3.3. Substituting into (3.24) the function $a(x) = x/\delta(x)$ such that $\delta(x) \to \alpha \in [0, \infty)$ as $x \to \infty$, (3.24) becomes a regularly varying tail with index $-\alpha$. We will see later (see Remark 2 below) that the auxiliary function of a von Mises function with $x_F = \infty$ satisfies $a(x)/x \to 0$. It immediately follows that $\overline{F}(x)$ decreases to zero much faster than any power law $x^{-\alpha}$. □

We give some examples of von Mises functions.

Example 3.3.19 (Exponential distribution)

$$\overline{F}(x) = e^{-\lambda x}\,, \quad x \geq 0\,, \quad \lambda > 0\,.$$

F is a von Mises function with auxiliary function $a(x) = \lambda^{-1}$. □

Example 3.3.20 (Weibull distribution)

$$\overline{F}(x) = \exp\left\{-c\,x^{\tau}\right\}\,, \quad x \geq 0\,, \quad c\,,\tau > 0\,.$$

F is a von Mises function with auxiliary function

$$a(x) = c^{-1}\tau^{-1}x^{1-\tau}\,, \quad x > 0\,.$$ □

Example 3.3.21 (Erlang distribution)

$$\overline{F}(x) = e^{-\lambda x}\sum_{k=0}^{n-1}\frac{(\lambda x)^k}{k!}\,, \quad x \geq 0\,, \quad \lambda > 0\,, n \in \mathbb{N}\,.$$

F is a von Mises function with auxiliary function

$$a(x) = \sum_{k=0}^{n-1}\frac{(n-1)!}{(n-k-1)!}\,\lambda^{-(k+1)}\,x^{-k}\,, \quad x > 0\,.$$

Notice that F is the $\Gamma(n, \lambda)$ df. □

Example 3.3.22 (Exponential behaviour at the finite right endpoint)
Let F be a df with finite right endpoint x_F and distribution tail

$$\overline{F}(x) = K\exp\left\{-\frac{\alpha}{x_F - x}\right\}\,, \quad x < x_F\,, \quad \alpha\,, K > 0\,.$$

F is a von Mises function with auxiliary function

$$a(x) = \frac{(x_F - x)^2}{\alpha}\,, \quad x < x_F\,.$$

For $x_F = 1$, $\alpha = 1$ and $K = e$ we obtain for example

$$\overline{F}(x) = \exp\left\{-\frac{x}{1-x}\right\}\,, \quad 0 \leq x < 1\,.$$ □

Example 3.3.23 (Differentiability at the right endpoint)
Let F be a df with right endpoint $x_F \leq \infty$ and assume there exists some $z < x_F$ such that F is twice differentiable on (z, x_F) with positive density $f = F'$ and $F''(x) < 0$ for $z < x < x_F$. Then it is not difficult to see that F is a von Mises function with auxiliary function $a = \overline{F}/f$ if and only if

$$\lim_{x \uparrow x_F} \overline{F}(x)\, F''(x)/f^2(x) = -1\,. \tag{3.25}$$

Indeed, let $z < x < x_F$ and set $Q(x) = -\ln \overline{F}(x)$ and $a(x) = 1/Q'(x) = \overline{F}(x)/f(x) > 0$. Hence F has representation (3.24). Furthermore,

$$a'(x) = -\frac{\overline{F}(x)\, F''(x)}{f^2(x)} - 1$$

and (3.25) is equivalent to $a'(x) \to 0$ as $x \uparrow x_F$.
Condition (3.25) applies to many distributions of interest, including the normal distribution; see Example 3.3.29. □

In Remark 1 above we gained some indication that regular variation does not seem to be the right tool for describing von Mises functions. Recall the notion of *rapidly varying function* from Definition A3.11. In particular, $\overline{F} \in \mathcal{R}_{-\infty}$ means that

$$\lim_{x \to \infty} \frac{\overline{F}(xt)}{\overline{F}(x)} = \begin{cases} 0 & \text{if } t > 1\,, \\ \infty & \text{if } 0 < t < 1\,. \end{cases}$$

It is mentioned in Appendix A3 that some of the important results for regularly varying functions can be extended to $\mathcal{R}_{-\infty}$ in a natural way; see Theorem A3.12.

Proposition 3.3.24 (Properties of von Mises functions)
Every von Mises function F is absolutely continuous on (z, x_F) with positive densitiy f. The auxiliary function can be chosen as $a(x) = \overline{F}(x)/f(x)$. Moreover, the following properties hold.

(a) If $x_F = \infty$, then $\overline{F} \in \mathcal{R}_{-\infty}$ and

$$\lim_{x \to \infty} \frac{x f(x)}{\overline{F}(x)} = \infty\,. \tag{3.26}$$

(b) If $x_F < \infty$, then $\overline{F}(x_F - x^{-1}) \in \mathcal{R}_{-\infty}$ and

$$\lim_{x \uparrow x_F} \frac{(x_F - x) f(x)}{\overline{F}(x)} = \infty\,. \tag{3.27}$$

Remarks. 2) It follows from (3.26) that $\lim_{x\to\infty} x^{-1}a(x) = 0$, and from (3.27) that $a(x) = o(x_F - x) = o(1)$ as $x \uparrow x_F$.

3) Note that $a^{-1}(x) = f(x)/\overline{F}(x)$ is the *hazard rate* of F. □

Proof. From representation (3.24) we obtain

$$\frac{d}{dx}\left(-\ln \overline{F}(x)\right) = \frac{f(x)}{\overline{F}(x)} = \frac{1}{a(x)}\,, \quad z < x < x_F\,.$$

(a) Since $a'(x) \to 0$ as $x \to \infty$ the Cesàro mean of a' also converges:

$$\lim_{x\to\infty} \frac{a(x)}{x} = \lim_{x\to\infty} \frac{1}{x}\int_z^x a'(t)\,dt = 0\,. \tag{3.28}$$

This implies (3.26). $\overline{F} \in \mathcal{R}_{-\infty}$ follows from an application of Theorem A3.12(b).

(b) We have

$$\begin{aligned}\lim_{x\uparrow x_F} \frac{a(x)}{x_F - x} &= \lim_{x\uparrow x_F} -\int_x^{x_F} \frac{a'(t)}{x_F - x}\,dt \\ &= \lim_{s\downarrow 0} \frac{1}{s}\int_0^s a'\left(x_F - t\right)\,dt\end{aligned}$$

by change of variables. Since $a'(x_F - t) \to 0$ as $t \downarrow 0$, the last limit tends to 0. This implies (3.27). $\overline{F}(x_F - x^{-1}) \in \mathcal{R}_{-\infty}$ follows as above. □

Now we can show that von Mises functions belong to the maximum domain of attraction of the Gumbel distribution. Moreover, the specific form of $\overline{F}$ allows to calculate the norming constants c_n from the auxiliary function.

Proposition 3.3.25 (Von Mises functions and MDA(Λ))
Suppose the df F is a von Mises function. Then $F \in \text{MDA}(\Lambda)$. A possible choice of norming constants is

$$d_n = F^{\leftarrow}(1 - n^{-1}) \quad \text{and} \quad c_n = a(d_n)\,, \tag{3.29}$$

where a is the auxiliary function of F.

Proof. Representation (3.24) implies for $t \in \mathbb{R}$ and x sufficiently close to x_F that

$$\frac{\overline{F}(x + t\,a(x))}{\overline{F}(x)} = \exp\left\{-\int_x^{x+t\,a(x)} \frac{1}{a(u)}\,du\right\}.$$

We set $v = (u - x)/a(x)$ and obtain

$$\frac{\overline{F}(x + t\,a(x))}{\overline{F}(x)} = \exp\left\{-\int_0^t \frac{a(x)}{a(x + v\,a(x))}\,dv\right\}. \tag{3.30}$$

We show that the integrand converges locally uniformly to 1. For given $\varepsilon > 0$ and $x \ge x_0(\varepsilon)$,

$$|a(x+va(x)) - a(x)| = \left| \int_x^{x+va(x)} a'(s)\,ds \right| \le \varepsilon |v| a(x) \le \varepsilon |t| a(x)\,,$$

where we used $a'(x) \to 0$ as $x \uparrow x_F$. This implies for $x \ge x_0(\varepsilon)$ that

$$\left| \frac{a(x+va(x))}{a(x)} - 1 \right| \le \varepsilon |t|\,.$$

The right–hand side can be made arbitrarily small, hence

$$\lim_{x \uparrow x_F} \frac{a(x)}{a(x + v\,a(x))} = 1\,, \tag{3.31}$$

uniformly on bounded v–intervals. This together with (3.30) yields

$$\lim_{x \uparrow x_F} \frac{\overline{F}(x + t\,a(x))}{\overline{F}(x)} = e^{-t} \tag{3.32}$$

uniformly on bounded t–intervals. Now choose the norming constants $d_n = (1/\overline{F})^{\leftarrow}(n)$ and $c_n = a(d_n)$. Then (3.32) implies

$$\lim_{n\to\infty} n\overline{F}\,(d_n + tc_n) = e^{-t} = -\ln \Lambda(t)\,, \quad t \in \mathbb{R}\,.$$

An application of Proposition 3.3.2 shows that $F \in \mathrm{MDA}(\Lambda)$. □

This result finishes our study of von Mises functions.

Characterisations of MDA(Λ)

Von Mises functions do not completely characterise the maximum domain of attraction of Λ. However, a slight modification of the defining relation (3.24) of a von Mises function yields a complete characterisation of MDA(Λ).

For a proof of the following result we refer to Resnick [530], Corollary 1.7 and Proposition 1.9.

Theorem 3.3.26 (Characterisation I of MDA(Λ))
The df F with right endpoint $x_F \le \infty$ belongs to the maximum domain of attraction of Λ if and only if there exists some $z < x_F$ such that F has representation

$$\overline{F}(x) = c(x) \exp\left\{ -\int_z^x \frac{g(t)}{a(t)}\,dt \right\}, \quad z < x < x_F\,, \tag{3.33}$$

where c and g are measurable functions satisfying $c(x) \to c > 0$, $g(x) \to 1$ as $x \uparrow x_F$, and $a(x)$ is a positive, absolutely continuous function (with respect to

Lebesgue measure) with density $a'(x)$ having $\lim_{x\uparrow x_F} a'(x) = 0$.

For F with representation (3.33) we can choose

$$d_n = F^{\leftarrow}(1 - n^{-1}) \quad \text{and} \quad c_n = a(d_n)$$

as norming constants.

A possible choice for the function a is

$$a(x) = \int_x^{x_F} \frac{\overline{F}(t)}{\overline{F}(x)}\, dt\,, \qquad x < x_F\,, \tag{3.34}$$

□

Motivated by von Mises functions, we call the function a in (3.33) an *auxiliary function* for F.

Remarks. 4) Representation (3.33) is not unique, there being some trade–off possible between the functions c and g. The following representation can be employed alternatively; see Resnick [530], Proposition 1.4:

$$\overline{F}(x) = c(x) \exp\left\{ - \int_z^x \frac{1}{a(t)}\, dt \right\}, \quad z < x < x_F\,, \tag{3.35}$$

for functions c and a with properties as in Theorem 3.3.26.

5) For a rv X the function $a(x)$ as defined in (3.34) is nothing but the *mean excess function*

$$a(x) = E(X - x \mid X > x)\,, \quad x < x_F\,;$$

see also Section 3.4 for a discussion on the use of this function. In Chapter 6 the mean excess function will turn out to be an important tool for statistical fitting of extremal event data. □

Another characterisation of MDA(Λ) was suggested in the proof of Proposition 3.3.25. There it was shown that every von Mises function satisfies (3.32), i.e. there exists a positive function $\widetilde{a}$ such that

$$\lim_{x\uparrow x_F} \frac{\overline{F}(x + t\widetilde{a}(x))}{\overline{F}(x)} = e^{-t}\,, \quad t \in \mathbb{R}\,. \tag{3.36}$$

Theorem 3.3.27 (Characterisation II of MDA(Λ))
The df F belongs to the maximum domain of attraction of Λ if and only if there exists some positive function $\widetilde{a}$ such that (3.36) holds. A possible choice is $\widetilde{a} = a$ as given in (3.34). □

The proof of this result is for instance to be found in de Haan [292], Theorem 2.5.1.

Now recall the notion of tail–equivalence (Definition 3.3.3). Similarly to the maximum domains of attraction of the Weibull and Fréchet distribution, tail–equivalence is an auxiliary tool to decide whether a particular distribution belongs to the maximum domain of attraction of Λ and to calculate the norming constants. In $\mathrm{MDA}(\Lambda)$ it is even more important because of the large variety of tails $\overline{F}$.

Proposition 3.3.28 (Closure property of $\mathrm{MDA}(\Lambda)$ under tail–equivalence) *Let F and G be dfs with the same right endpoint $x_F = x_G$ and assume that $F \in \mathrm{MDA}(\Lambda)$ with norming constants $c_n > 0$ and $d_n \in \mathbb{R}$; i.e.*

$$\lim_{n\to\infty} F^n\left(c_n x + d_n\right) = \Lambda(x)\,, \quad x \in \mathbb{R}. \tag{3.37}$$

Then

$$\lim_{n\to\infty} G^n\left(c_n x + d_n\right) = \Lambda(x+b)\,, \quad x \in \mathbb{R},$$

if and only if F and G are tail–equivalent with

$$\lim_{x\uparrow x_F} \overline{F}(x)/\overline{G}(x) = e^b\,.$$

Proof of the sufficiency. For a proof of the necessity see Resnick [530], Proposition 1.19. Suppose that $\overline{F}(x) \sim c\overline{G}(x)$ as $x \uparrow x_F$ for some $c > 0$. By Proposition 3.3.2 the limit relation (3.37) is equivalent to

$$\lim_{n\to\infty} n\overline{F}\left(c_n x + d_n\right) = e^{-x}\,, \quad x \in \mathbb{R}.$$

For such x, $c_n x + d_n \to x_F$ and hence, by tail–equivalence,

$$n\,\overline{G}\left(c_n x + d_n\right) \sim nc^{-1}\overline{F}\left(c_n x + d_n\right) \to c^{-1}e^{-x}\,, \quad x \in \mathbb{R}.$$

Therefore by Proposition 3.3.2,

$$\lim_{n\to\infty} G^n\left(c_n x + d_n\right) = \exp\left\{-e^{-(x+\ln c)}\right\} = \Lambda(x+\ln c)\,, \quad x \in \mathbb{R}.$$

Now set $\ln c = b$. □

The results of this section yield a further complete characterisation of $\mathrm{MDA}(\Lambda)$.

> $\mathrm{MDA}(\Lambda)$ consists of von Mises functions and their tail–equivalent dfs.

This statement and the examples discussed throughout this section show that MDA(Λ) consists of a large variety of distributions whose tails can be very different. Tails may range from moderately heavy (lognormal, heavy–tailed Weibull) to very light (exponential, dfs with support bounded to the right). Because of this, MDA(Λ) is perhaps the most interesting among all maximum domains of attraction. As a natural consequence of the variety of tails in MDA(Λ), the norming constants also vary considerably. Whereas in MDA(Φ_α) and MDA(Ψ_α) the norming constants are calculated by straightforward application of regular variation theory, more advanced results are needed for MDA(Λ). A complete theory has been developed by de Haan involving certain subclasses of $\mathcal{R}_{-\infty}$ and $\mathcal{R}_0$; see de Haan [292] or Bingham et al. [72], Chapter 3. Various examples below will illustrate the usefulness of results like Proposition 3.3.28.

Example 3.3.29 (Normal distribution)
See also Figure 3.3.30. Denote by Φ the df and by φ the density of the standard normal distribution. We first show that Φ is a von Mises function and check condition (3.25). An application of l'Hospital's rule to $\overline{\Phi}(x)/(x^{-1}\varphi(x))$ yields *Mill's ratio*, $\overline{\Phi}(x) \sim \varphi(x)/x$. Furthermore $\varphi'(x) = -x\varphi(x) < 0$ and

$$\lim_{x\to\infty} \frac{\overline{\Phi}(x)\,\varphi'(x)}{\varphi^2(x)} = -1\,.$$

Thus $\Phi \in$ MDA(Λ) by Example 3.3.23 and Proposition 3.3.25. We now calculate the norming constants. Use Mill's ratio again:

$$\overline{\Phi}(x) \sim \frac{\varphi(x)}{x} = \frac{1}{\sqrt{2\pi}\,x}\, e^{-x^2/2}\,, \quad x\to\infty\,, \tag{3.38}$$

and interpret the right–hand side as the tail of some df G. Then by Proposition 3.3.28, Φ and G have the same norming constants c_n and d_n. According to (3.29), $d_n = G^{\leftarrow}(1-n^{-1})$. Hence look for a solution of $-\ln\overline{G}(d_n) = \ln n$; i.e.

$$\frac{1}{2}\, d_n^2 + \ln d_n + \frac{1}{2}\ln 2\pi = \ln n\,. \tag{3.39}$$

Then a Taylor expansion in (3.39) yields

$$d_n = (2\ln n)^{1/2} - \frac{\ln\ln n + \ln 4\pi}{2(2\ln n)^{1/2}} + o\left((\ln n)^{-1/2}\right)$$

as a possible choice for d_n. Since we can take $a(x) = \overline{\Phi}(x)/\varphi(x)$ we have that $a(x) \sim x^{-1}$ and therefore

$$c_n = a\,(d_n) \sim (2\ln n)^{-1/2}\,.$$

As the c_n are unique up to asymptotic equivalence, we choose

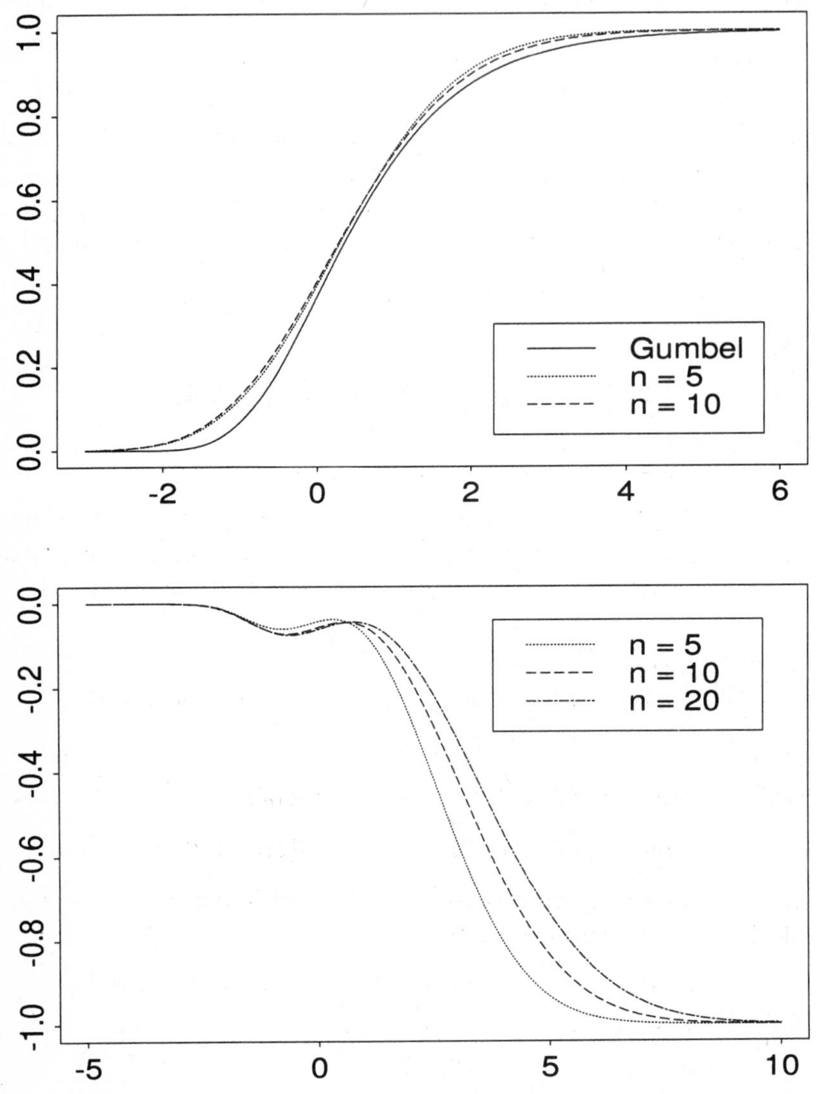

Figure 3.3.30 *Dfs of the normalised maxima of n standard normal rvs and the Gumbel df (top). In the bottom figure the relative error of this approximation for the tail is illustrated. The rate of convergence appears to be very slow.*

$$c_n = (2\ln n)^{-1/2}.$$

We conclude that

$$\sqrt{2\ln n}\left(M_n - \sqrt{2\ln n} + \frac{\ln\ln n + \ln 4\pi}{2(2\ln n)^{1/2}}\right) \xrightarrow{d} \Lambda. \qquad (3.40)$$

Note that $c_n \to 0$, i.e. the distribution of M_n becomes less spread around d_n as n increases. □

Similarly, it can be proved that the gamma distribution also belongs to MDA(Λ). The norming constants are given in Table 3.4.4.

Another useful trick to calculate the norming constants is via monotone transformations. If g is an increasing function and $\widetilde{x} = g(x)$, then obviously

$$\widetilde{M}_n = \max\left(\widetilde{X}_1, \ldots, \widetilde{X}_n\right) = g(M_n).$$

If $X \in \text{MDA}(\Lambda)$ with

$$\lim_{n\to\infty} P(M_n \le c_n x + d_n) = \Lambda(x), \quad x \in \mathbb{R},$$

then

$$\lim_{n\to\infty} P\left(\widetilde{M}_n \le g(c_n x + d_n)\right) = \Lambda(x), \quad x \in \mathbb{R}.$$

In some cases, g may be expanded in a Taylor series about d_n and just linear terms suffice to give the limit law for $\widetilde{M}_n$, with changed constants $\tilde{c}_n = c_n g'(d_n)$ and $\tilde{d}_n = g(d_n)$. We apply this method to the lognormal distribution.

Example 3.3.31 (Lognormal distribution)
Let X be a standard normal rv and $g(x) = e^{\mu+\sigma x}$, $\mu \in \mathbb{R}$, $\sigma > 0$. Then

$$\widetilde{X} = g(X) = e^{\mu+\sigma X}$$

defines a lognormal rv. Since $X \in \text{MDA}(\Lambda)$ we obtain

$$\lim_{n\to\infty} P\left(\widetilde{M}_n \le e^{\mu+\sigma(c_n x + d_n)}\right) = \Lambda(x), \quad x \in \mathbb{R},$$

where c_n, d_n are the norming constants of the standard normal distribution as calculated in Example 3.3.29. This implies

$$\lim_{n\to\infty} P\left(e^{-\mu-\sigma d_n}\,\widetilde{M}_n \le 1 + \sigma c_n x + o(c_n)\right) = \Lambda(x), \quad x \in \mathbb{R}.$$

Since $c_n \to 0$ it follows that

$$\frac{e^{-\mu-\sigma d_n}}{\sigma c_n}\left(\widetilde{M}_n - e^{\mu+\sigma d_n}\right) \xrightarrow{d} \Lambda,$$

so that $\widetilde{X} \in \text{MDA}(\Lambda)$ with norming constants

$$\widetilde{c}_n = \sigma c_n e^{\mu+\sigma d_n}, \qquad \widetilde{d}_n = e^{\mu+\sigma d_n}.$$

Explicit expressions for the norming constants of the lognormal distribution can be found in Table 3.4.4. □

Further Properties of Distributions in MDA(Λ)

In the remainder of this section we collect some further useful facts about distributions in MDA(Λ).

Corollary 3.3.32 (Existence of moments)
Assume that the rv X has df $F \in$ MDA(Λ) with infinite right endpoint. Then $\overline{F} \in \mathcal{R}_{-\infty}$. In particular, $E(X^+)^\alpha < \infty$ for every $\alpha > 0$, where $X^+ =$ $\max(0, X)$.

Proof. Every $F \in$ MDA(Λ) is tail–equivalent to a von Mises function. If $x_F = \infty$, the latter have rapidly varying tails; see Proposition 3.3.24(a), which also implies the statement about the moments; see Theorem A3.12(a). □

In Section 3.3.2 we showed that the maximum domains of attraction of Ψ_α and Φ_α are linked in a natural way. Now we show that MDA(Φ_α) can be embedded in MDA(Λ).

Example 3.3.33 (Embedding MDA(Φ_α) in MDA(Λ))
Let X have df $F \in$ MDA(Φ_α) with norming constants c_n. Define

$$X^* = \ln(1 \vee X)$$

with df F^*. By Proposition 3.3.2 and Theorem 3.3.7, $F \in$ MDA(Φ_α) if and only if

$$\lim_{n\to\infty} n\overline{F}(c_n x) = \lim_{n\to\infty} \frac{\overline{F}(c_n x)}{\overline{F}(c_n)} = x^{-\alpha}, \quad x > 0.$$

This implies that

$$\lim_{n\to\infty} \frac{\overline{F^*}\left(\alpha^{-1}x + \ln c_n\right)}{\overline{F^*}(\ln c_n)} = \lim_{n\to\infty} \frac{\overline{F}\left(c_n \exp\left\{\alpha^{-1}x\right\}\right)}{\overline{F}(c_n)} = e^{-x}, \quad x \in \mathbb{R}.$$

Hence $F^* \in$ MDA(Λ) with norming constants $c_n^* = \alpha^{-1}$ and $d_n^* = \ln c_n$. As auxiliary function one can take

$$a^*(x) = \int_x^\infty \frac{\overline{F^*}(y)}{\overline{F^*}(x)}\, dy$$ □

Example 3.3.34 (Closure of MDA(Λ) under logarithmic transformations)
Let X have df $F \in$ MDA(Λ) with $x_F = \infty$ and norming constants c_n, d_n, chosen according to Theorem 3.3.26. Define X^* and F^* as above. We intend to show that $F^* \in$ MDA(Λ) with norming constants $d_n^* = \ln d_n$ and $c_n^* = c_n/d_n$. Since $a'(x) \to 0$, (3.28) holds, and since $d_n = F^{\leftarrow}(1 - n^{-1}) \to \infty$, it follows that

$$\frac{c_n}{d_n} = \frac{a(d_n)}{d_n} \to 0 .$$

Moreover,

$$\begin{aligned}\overline{F^*}(c_n^* x + d_n^*) &= \overline{F}\left(\exp\left\{\frac{c_n}{d_n}x\right\} d_n\right)\\ &= \overline{F}\left(d_n\left(1+\frac{c_n}{d_n}x + o\left(\frac{c_n}{d_n}\right)\right)\right)\\ &= \overline{F}\left(c_n x + d_n + o\left(c_n\right)\right)\\ &\sim \overline{F}\left(c_n x + d_n\right) \sim n^{-1}e^{-x}, \quad n\to\infty .\end{aligned}$$

where we applied the uniformity of weak convergence to a continuous limit. The result follows from Proposition 3.3.2. □

Example 3.3.35 (Subexponential distributions and MDA(Λ))
Goldie and Resnick [276] characterise the dfs F that are both subexponential (we write $F \in \mathcal{S}$; see Definition 1.3.3) and in MDA(Λ). Starting from the representation (3.33) for $F \in \text{MDA}(\Lambda)$, they give necessary and sufficient conditions for $F \in \mathcal{S}$. In particular, $\lim_{x\to\infty} a(x) = \infty$ is necessary but not sufficient for $F \in \text{MDA}(\Lambda)\cap\mathcal{S}$. A simple sufficient condition for $F \in \text{MDA}(\Lambda) \cap \mathcal{S}$ is that a is eventually non–decreasing and that there exists some $t > 1$ such that

$$\liminf_{x\to\infty} \frac{a(tx)}{a(x)} > 1 . \tag{3.41}$$

This condition is easily checked for the following distributions which are all von Mises functions and hence in MDA(Λ):

– *Benktander–type–I*

$$\overline{F}(x) = (1+2(\beta/\alpha)\ln x)\exp\{-(\beta(\ln x)^2+(\alpha+1)\ln x)\}, \quad x \ge 1, \quad \alpha,\beta>0 .$$

Here one can choose

$$a(x) = \frac{x}{\alpha + 2\beta\ln x}, \quad x \ge 1 .$$

– *Benktander–type–II*

$$\overline{F}(x) = e^{\alpha/\beta}\, x^{-(1-\beta)} \exp\left\{-\frac{\alpha}{\beta}x^{\beta}\right\}, \quad x\ge 1, \quad \alpha>0, 0<\beta<1 ,$$

with auxiliary function

$$a(x) = \frac{x^{1-\beta}}{\alpha}, \quad x \ge 1 .$$

– *Weibull*

$$\overline{F}(x) = e^{-c\,x^\tau}\,, \quad x \ge 0\,, \quad 0 < \tau < 1\,, c > 0\,,$$

with auxiliary function

$$a(x) = c^{-1}\tau^{-1}x^{1-\tau}\,, \quad x \ge 0\,.$$

– *Lognormal*
with auxiliary function

$$a(x) = \frac{\overline{\Phi}(\sigma^{-1}(\ln x - \mu))\sigma x}{\varphi(\sigma^{-1}(\ln x - \mu))} \sim \frac{\sigma^2 x}{\ln x - \mu}\,, \quad x \to \infty\,.$$

The critical cases occur when F is in the tail close to an exponential distribution. For example, let

$$\overline{F}(x) \sim \exp\{-x(\ln x)^\alpha\}\,, \quad x \to \infty\,.$$

For $\alpha < 0$ we have $F \in \mathrm{MDA}(\Lambda) \cap \mathcal{S}$ in view of Theorem 2.7 in Goldie and Resnick [276], whereas for $\alpha \ge 0$, $F \in \mathrm{MDA}(\Lambda)$ but $F \notin \mathcal{S}$; see Example 1.4.3. □

Notes and Comments

There exist many results on the quality of convergence in extreme value limit theory. Topics include the convergence of moments, local limit theory and the convergence of densities, large deviations and uniform rates of convergence. We refer to Chapter 2 of Resnick [530] for a collection of such results.

Statistical methods based on extreme value theory are discussed in detail in Chapter 6. Various estimation methods will depend on an application of the Fisher–Tippett theorem and related results. The quality of those approximations will be crucial.

Figure 3.2.9 suggests a fast rate of convergence in the case of the exponential distribution: already for $n = 5$ the distribution of the normalised maximum is quite close to Λ, while for $n = 50$ they are almost indistinguishable. Indeed, it has been shown by Hall and Wellner [313] that for $F(x) = 1 - e^{-x}$, $x \ge 0$,

$$\sup_{x\in\mathbb{R}} \left|P\left(M_n - \ln n \le x\right) - \exp\left\{-e^{-x}\right\}\right| \le n^{-1}\left(2 + n^{-1}\right)\, e^{-2}\,.$$

In contrast to this rapid rate of convergence, the distribution of the normalised maximum of a sample of normal rvs converges extremely slowly to its limit distribution Λ; see Figure 3.3.30. This slow rate of convergence also depends on the particular choice of c_n and d_n. Hall [307] obtained an optimal rate by choosing c_n and d_n as solutions to

$$n c_n \varphi(c_n^{-1}) = 1 \quad \text{and} \quad d_n = c_n^{-1},$$

where φ denotes the standard normal density. Then there exist constants $0 < c < C \le 3$ such that

$$\frac{c}{\ln n} \le \sup_{x \in \mathbb{R}} \left| P(M_n \le c_n x + d_n) - \exp\{-e^{-x}\} \right| \le \frac{C}{\ln n}, \quad n \ge 2.$$

Leadbetter, Lindgren and Rootzén [418] and Resnick [530] derive various rates for $F \in \mathrm{MDA}(\Phi_\alpha)$ and $F \in \mathrm{MDA}(\Lambda)$. They also give numerical values for some explicit examples. See also Balkema and de Haan [38], Beirlant and Willekens [58], de Haan and Resnick [301], Goldie and Smith [278], Smith [587] and references therein.

In order to discuss the next point, we introduce a parametric family $(H_\xi)_{\xi \in \mathbb{R}}$ of dfs containing the standard extreme value distributions, namely

$$H_\xi = \begin{cases} \Phi_{1/\xi} & \text{if} \quad \xi > 0, \\ \Lambda & \text{if} \quad \xi = 0, \\ \Psi_{-1/\xi} & \text{if} \quad \xi < 0. \end{cases}$$

The df H_ξ above is referred to as the *generalised extreme value distribution* with parameter ξ; a detailed discussion is given in Section 3.4. The condition $F \in \mathrm{MDA}(H_\xi)$ then yields the so-called *ultimate approximation*

$$F^n\left(c_n x + d_n\right) \approx H_\xi(x)$$

for appropriate norming constants $c_n > 0$ and $d_n \in \mathbb{R}$. One method for improving the rate of convergence in the latter limit relation was already discussed in the classical Fisher–Tippett paper [240]. The basic idea is the following: the parameter ξ can be obtained as a limit. For instance in the Gumbel case $\xi = 0$, F has representation (3.33) with $a'(x) \to 0$ as $x \to \infty$. The *penultimate approximation* now consists of replacing ξ by $\xi_n = a'(d_n)$ leading to the relation

$$F^n\left(c_n x + d_n\right) \approx H_{\xi_n}(x).$$

Typically $\xi_n \neq 0$ so that in the Gumbel case, the penultimate approximation is based on a suitable Weibull ($\xi_n < 0$) or Fréchet ($\xi_n > 0$) approximation. The optimal rate of convergence $O\left((\ln n)^{-1}\right)$ for maxima of iid normal rvs in the ultimate approximation is improved to $O\left((\ln n)^{-2}\right)$ in the penultimate case, as shown in Cohen [128]. In special cases, further improvements can be given. For instance, using expansions of the df of normal maxima, Cohen [129] suggests

$$\Phi^n\left(\frac{x}{b_n}+b_n-\frac{1}{b_n}\right)\approx\Lambda(x)\left(1+e^{-x}\,\frac{x^2}{4\ln n}\right),$$

where $b_n^2\sim\ln n$. Further information is to be found in Joe [357] and Reiss [526].

3.4 The Generalised Extreme Value Distribution and the Generalised Pareto Distribution

In Section 3.2 we have shown that the standard extreme value distributions, together with their types, provide the only non-degenerate limit laws for affinely transformed maxima of iid rvs. As already mentioned in the Notes and Comments of the previous section, a one–parameter representation of the three standard cases in one family of dfs will turn out to be useful. They can be represented by introducing a parameter ξ so that

$\xi=\alpha^{-1}>0$	corresponds to the Fréchet distribution Φ_α ,
$\xi=0$	corresponds to the Gumbel distribution Λ ,
$\xi=-\alpha^{-1}<0$	corresponds to the Weibull distribution Ψ_α .

The following choice is by now widely accepted as *the* standard representation.

Definition 3.4.1 (Jenkinson–von Mises representation of the extreme value distributions: the generalised extreme value distribution (GEV))
Define the df H_ξ by

$$H_\xi(x)=\begin{cases}\exp\left\{-\left(1+\xi\,x\right)^{-1/\xi}\right\} & \text{if}\quad \xi\neq 0\,,\\ \exp\left\{-\exp\{-x\}\right\} & \text{if}\quad \xi=0\,,\end{cases}$$

where $1+\xi\,x>0$.

Table 3.4.2 *Maximum domain of attraction of the Fréchet distribution.*

Fréchet	$\Phi_\alpha(x) = \exp\{-x^{-\alpha}\}, \quad x > 0, \alpha > 0.$
MDA (Φ_α)	$x_F = \infty, \quad \overline{F}(x) = x^{-\alpha} L(x), \quad L \in \mathcal{R}_0.$
Norming constants	$c_n = F^{\leftarrow}(1 - n^{-1}) = n^{1/\alpha} L_1(n), \quad L_1 \in \mathcal{R}_0, \quad d_n = 0.$
Limit result	$c_n^{-1} M_n \xrightarrow{d} \Phi_\alpha$
	Examples
Cauchy	$f(x) = (\pi(1 + x^2))^{-1}, \quad x \in \mathbb{R}.$ $c_n = n/\pi$
Pareto Burr stable with index $\alpha < 2$	$\overline{F}(x) \sim K x^{-\alpha}, \quad K, \alpha > 0.$ $c_n = (Kn)^{1/\alpha}$
Loggamma	$f(x) = \dfrac{\alpha^\beta}{\Gamma(\beta)} (\ln x)^{\beta-1} x^{-\alpha-1}, \quad x > 1, \quad \alpha, \beta > 0.$ $c_n = \left((\Gamma(\beta))^{-1} (\ln n)^{\beta-1} n\right)^{1/\alpha}$

Table 3.4.3 *Maximum domain of attraction of the Weibull distribution.*

Weibull	$\Psi_\alpha(x) = \exp\{-(-x)^\alpha\}\,, \quad x < 0\,, \quad \alpha > 0\,.$
MDA (Ψ_α)	$x_F < \infty\,, \quad \overline{F}\left(x_F - x^{-1}\right) = x^{-\alpha} L(x)\,, \quad L \in \mathcal{R}_0\,.$
Norming constants	$c_n = x_F - F^{\leftarrow}(1 - n^{-1}) = n^{-1/\alpha} L_1(n)\,, \quad L_1 \in \mathcal{R}_0\,, \quad d_n = x_F\,.$
Limit result	$c_n^{-1}(M_n - x_F) \xrightarrow{d} \Psi_\alpha$
Examples	
Uniform	$f(x) = 1\,, \quad x \in (0,1)\,.$ $c_n = n^{-1}\,, \quad d_n = 1$
Power law behaviour at x_F	$\overline{F}(x) = K(x_F - x)^\alpha\,, \quad x_F - K^{-1/\alpha} \le x \le x_F\,, \quad K\,,\alpha > 0\,.$ $c_n = (K\,n)^{-1/\alpha}\,, \quad d_n = x_F$
Beta	$f(x) = \dfrac{\Gamma(a+b)}{\Gamma(a)\Gamma(b)}\, x^{a-1}\,(1-x)^{b-1}\,, \quad 0 < x < 1\,, \quad a\,,b > 0\,.$ $c_n = \left(n\,\dfrac{\Gamma(a+b)}{\Gamma(a)\,\Gamma(b+1)}\right)^{-1/b}\,, \quad d_n = 1$

Table 3.4.4 *Maximum domain of attraction of the Gumbel distribution.*

Gumbel	$\Lambda(x) = \exp\{-e^{-x}\}\,, \quad x \in \mathbb{R}\,.$
MDA(Λ)	$x_F \le \infty\,, \quad \overline{F}(x) = c(x)\exp\left\{-\int_{x_0}^{x} \frac{g(t)}{a(t)}\,dt\right\}, \quad x_0 < x < x_F\,,$ where $c(x) \to c > 0\,, \quad g(x) \to 1\,, \quad a'(x) \to 0 \quad$ as $x \uparrow x_F\,.$
Norming constants	$d_n = F^{\leftarrow}(1 - n^{-1})\,, \quad c_n = a(d_n)\,.$
Limit result	$c_n^{-1}(M_n - d_n) \xrightarrow{d} \Lambda$
	Examples
Exponential–like	$\overline{F}(x) \sim K e^{-\lambda x}\,, \quad K\,,\lambda > 0\,.$ $c_n = \lambda^{-1}$ $d_n = \lambda^{-1}\ln(Kn)$
Weibull–like	$\overline{F}(x) \sim K x^{\alpha}\exp\{-c x^{\tau}\}\,, \quad K\,,c\,,\tau > 0\,,\alpha \in \mathbb{R}\,.$ $c_n = (c\tau)^{-1}\,\left(c^{-1}\ln n\right)^{1/\tau - 1}$ $d_n = \left(c^{-1}\ln n\right)^{1/\tau} + \frac{1}{\tau}\left(c^{-1}\ln n\right)^{1/\tau-1}\left\{\frac{\alpha}{c\tau}\ln(c^{-1}\ln n) + \frac{\ln K}{c}\right\}$

Gamma	$f(x) = \frac{\beta^\alpha}{\Gamma(\alpha)}\, x^{\alpha-1}\, e^{-\beta x}\,, \quad x > 0\,, \quad \alpha\,,\beta > 0\,.$ $c_n = \beta^{-1}$ $d_n = \beta^{-1}(\ln n + (\alpha - 1)\ln\ln n - \ln \Gamma(\alpha))$
Normal	$\varphi(x) = \frac{1}{\sqrt{2\pi}}\, e^{-x^2/2}\,, \quad x \in \mathbb{R}\,.$ $c_n = (2\ln n)^{-1/2}$ $d_n = \sqrt{2\ln n} - \frac{\ln(4\pi) + \ln\ln n}{2(2\ln n)^{1/2}}$
Lognormal	$f(x) = \frac{1}{\sqrt{2\pi}\sigma x}\, e^{-(\ln x - \mu)^2/2\sigma^2}\,, \quad x > 0\,, \quad \mu \in \mathbb{R}\,, \sigma > 0\,.$ $c_n = \sigma(2\ln n)^{-1/2} d_n$ $d_n = \exp\left\{\mu + \sigma\left(\sqrt{2\ln n} - \frac{\ln(4\pi) + \ln\ln n}{2(2\ln n)^{1/2}}\right)\right\}$

Exponential behaviour at $x_F < \infty$	$\overline{F}(x) = K \exp\left\{-\frac{\alpha}{x_F - x}\right\}, \quad x < x_F, \quad \alpha, K > 0.$ $c_n = \frac{\alpha}{(\ln(Kn))^2}$ $d_n = x_F - \frac{\alpha}{\ln(Kn)}$
Benktander–type–I	$\overline{F}(x) = (1 + 2(\beta/\alpha)\ln x)\exp\{-(\beta(\ln x)^2 + (\alpha+1)\ln x)\}, x > 1, \alpha, \beta > 0$ $c_n = \frac{1}{2\sqrt{\beta \ln n}} \exp\left\{-\frac{\alpha+1}{2\beta} + \sqrt{\frac{\ln n}{\beta}}\right\}$ $d_n = \exp\left\{-\frac{\alpha+1}{2\beta} + \sqrt{\frac{\ln n}{\beta}} + \frac{\ln\ln n + \ln(4\beta/\alpha^2) + (\alpha+1)^2/(2\beta)}{4\sqrt{\beta \ln n}}\right\}$
Benktander–type–II	$\overline{F}(x) = x^{-(1-\beta)} \exp\{-\frac{\alpha}{\beta}(x^\beta - 1)\}, \quad x > 1, \quad \alpha > 0, 0 < \beta < 1$ $c_n = \frac{1}{\alpha}\left(\frac{\beta}{\alpha}\ln n\right)^{1/\beta - 1}$ $d_n = \left(\frac{\beta}{\alpha}\ln n\right)^{1/\beta} + \frac{1}{\alpha\beta}\left(\frac{\beta}{\alpha}\ln n\right)^{1/\beta-1}\left\{\alpha - (1-\beta)\left(\ln\ln n + \ln\frac{\beta}{\alpha}\right)\right\}$

Hence the support of H_ξ corresponds to

$$\begin{aligned} x &> -\xi^{-1} && \textit{for} \quad \xi > 0\,, \\ x &< -\xi^{-1} && \textit{for} \quad \xi < 0\,, \\ x &\in \mathbb{R} && \textit{for} \quad \xi = 0\,. \end{aligned}$$

H_ξ is called a standard generalised extreme value distribution (GEV). *One can introduce the related location–scale family $H_{\xi;\mu,\psi}$ by replacing the argument x above by $(x-\mu)/\psi$ for $\mu \in \mathbb{R}, \psi > 0$. The support has to be adjusted accordingly. We also refer to $H_{\xi;\mu,\psi}$ as GEV.* □

We consider the df H_0 as the limit of H_ξ for $\xi \to 0$. With this interpretation

$$H_\xi(x) = \exp\left\{-\left(1+\xi\, x\right)^{-1/\xi}\right\}\,, \quad 1+\xi\, x > 0\,,$$

serves as a representation for all $\xi \in \mathbb{R}$. The densities of the standard GEV for $\xi = -1, 0, 1$ are shown in Figure 3.2.4.

The GEV provides a convenient unifying representation of the three extreme value types Gumbel, Fréchet and Weibull. Its introduction is mainly motivated by statistical applications; we refer to Chapter 6 where this will become transparent. There GEV fitting will turn out to be one of the fundamental concepts.

The following theorem is one of the basic results in extreme value theory. In a concise analytical way, it gives the essential information collected in the previous section on maximum domains of attraction. Moreover, it constitutes the basis for numerous statistical techniques to be discussed in Chapter 6. First recall the notion of the quantile function $F^{\leftarrow}$ of a df F and define

$$U(t) = F^{\leftarrow}(1-t^{-1})\,, \quad t > 0\,.$$

Theorem 3.4.5 (Characterisation of MDA(H_ξ))
For $\xi \in \mathbb{R}$ the following assertions are equivalent:

(a) *$F \in \text{MDA}(H_\xi)$.*
(b) *There exists a positive, measurable function $a(\cdot)$ such that for $1+\xi x > 0$,*

$$\lim_{u \uparrow x_F} \frac{\overline{F}(u + x a(u))}{\overline{F}(u)} = \begin{cases} (1+\xi x)^{-1/\xi} & \textit{if} \quad \xi \neq 0\,, \\ e^{-x} & \textit{if} \quad \xi = 0\,. \end{cases} \tag{3.42}$$

(c) *For $x, y > 0$, $y \neq 1$,*

$$\lim_{s \to \infty} \frac{U(sx) - U(s)}{U(sy) - U(s)} = \begin{cases} \dfrac{x^\xi - 1}{y^\xi - 1} & \textit{if} \quad \xi \neq 0\,, \\[2ex] \dfrac{\ln x}{\ln y} & \textit{if} \quad \xi = 0\,. \end{cases} \tag{3.43}$$

Sketch of the proof. Below we give only the main ideas in order to show that the various conditions enter very naturally. Further details are to be found in the literature, for instance in de Haan [293].

(a)⇔(b) For $\xi = 0$ this is Theorem 3.3.27.

For $\xi > 0$ we have $H_\xi(x) = \Phi_\alpha(\alpha^{-1}(x+\alpha))$ for $\alpha = 1/\xi$. By Theorem 3.3.7, (a) is then equivalent to $\overline{F} \in \mathcal{R}_{-\alpha}$. By the representation theorem for regularly varying functions (Theorem A3.3), for some $z > 0$,

$$\overline{F}(x) = c(x) \exp\left\{ - \int_z^x \frac{1}{a(t)}\, dt \right\}, \quad z < x < \infty,$$

where $c(x) \to c > 0$ and $a(x)/x \to \alpha^{-1}$ as $x \to \infty$ locally uniformly. Hence

$$\lim_{u\to\infty} \frac{\overline{F}(u + xa(u))}{\overline{F}(u)} = \left(1 + \frac{x}{\alpha}\right)^{-\alpha},$$

which is (3.42). If (b) holds, choose $d_n = (1/\overline{F})^{\leftarrow}(n) = U(n)$, then

$$1/\overline{F}\,(d_n) \sim n\,,$$

and with $u = d_n$ in (3.42),

$$\left(1 + \frac{x}{\alpha}\right)^{-\alpha} = \lim_{n\to\infty} \frac{\overline{F}\,(d_n + xa\,(d_n))}{\overline{F}(d_n)} = \lim_{n\to\infty} n\overline{F}\,(d_n + x\,a(d_n))\,,$$

whence by Proposition 3.3.2, $F \in \mathrm{MDA}(H_\xi)$ for $\xi = \alpha^{-1}$.

The case $\xi < 0$ can be treated similarly.

(b)⇔(c) We restrict ourselves to the case $\xi \neq 0$, the proof for $\xi = 0$ being analogous. For simplicity, we assume that F is continuous and increasing on $(-\infty, x_F)$. Set $s = 1/\overline{F}(u)$, then (3.42) translates into

$$A_s(x) = \left(s\overline{F}(U(s) + xa(U(s)))\right)^{-1} \to (1 + \xi x)^{1/\xi}\,, \quad s \to \infty\,.$$

Now for every $s > 0$, $A_s(x)$ is decreasing and for $s \to \infty$ converges to a continuous function. Then because of Proposition A1.7, also $A_s^{\leftarrow}(t)$ converges pointwise to the inverse of the corresponding limit function, i.e.

$$\lim_{s\to\infty} \frac{U(st) - U(s)}{a(U(s))} = \frac{t^\xi - 1}{\xi}\,. \tag{3.44}$$

Now (3.43) is obtained by using (3.44) for $t = x$ and $t = y$ and taking the quotient. The proof of the converse can be given along the same lines. □

Remarks. 1) Condition (3.42) has an interesting probabilistic interpretation. Indeed, let X be a rv with df $F \in \mathrm{MDA}(H_\xi)$, then (3.42) reformulates as

$$\lim_{u \uparrow x_F} P\left(\frac{X-u}{a(u)} > x \,\middle|\, X > u\right) = \begin{cases} (1+\xi x)^{-1/\xi} & \text{if } \xi \neq 0, \\ e^{-x} & \text{if } \xi = 0. \end{cases} \tag{3.45}$$

Hence (3.45) gives a distributional approximation for the scaled excesses over the (high) threshold u. The appropriate scaling factor is $a(u)$. This interpretation will turn out to be crucial in many applications; see for instance Section 6.5.

2) In Section 6.4.4 we show how a reformulation of relation (3.43) immediately leads to an estimation procedure for quantiles outside the range of the data. A special case of (3.43) is also used to motivate the Pickands estimator of ξ; see Section 6.4.2.

3) In the proof of Theorem 3.4.5 there is implicitly given the result that $F \in \text{MDA}(H_\xi)$ for some $\xi \in \mathbb{R}$ if and only if there exists some positive function $a_1(\cdot)$ such that

$$\lim_{s\to\infty} \frac{U(st)-U(s)}{a_1(s)} = \frac{t^\xi - 1}{\xi}, \quad t>0. \tag{3.46}$$

If $\xi = 0$, the rhs of (3.46) has to be interpreted as $\ln t$. Moreover, for $\xi > 0$, (3.46) is equivalent to

$$\lim_{s\to\infty} \frac{U(st)}{U(s)} = t^\xi, \quad t>0,$$

i.e. $U \in \mathcal{R}_\xi$ and hence $a_1(s) \sim \xi U(s)$, $s \to \infty$. For $\xi < 0$, F has finite right endpoint x_F, hence $U(\infty) = x_F < \infty$, and (3.46) is equivalent to

$$U(\infty) - U \in \mathcal{R}_\xi .$$

In this case, $a_1(s) \sim -\xi(U(\infty) - U(s))$, $s \to \infty$. The above formulations are for instance to be found as Lemmas 2.1 and 2.2 in Dekkers and de Haan [174]; see de Haan [293] for proofs. The case $\xi = 0$ in (3.46) gives rise to the so–called class of *Π–varying functions*, a strict subclass of $\mathcal{R}_0$. The recognition of the importance of Π–variation for the description of $\text{MDA}(H_0)$ was one of the fundamental contributions to extreme value theory by de Haan [293]. □

In Remark 1 above we used the notion of excess. The following definition makes this precise.

Definition 3.4.6 (Excess distribution function, mean excess function)
Let X be a rv with df F and right endpoint x_F. For a fixed $u < x_F$,

$$F_u(x) = P(X - u \le x \mid X > u), \quad x \ge 0, \tag{3.47}$$

is the excess df of the rv X (of the df F) over the threshold u. *The function*

Pareto	$\dfrac{\kappa + u}{\alpha - 1}\,, \quad \alpha > 1$
Burr	$\dfrac{u}{\alpha\tau - 1}\,(1 + o(1))\,, \quad \alpha\tau > 1$
Loggamma	$\dfrac{u}{\alpha - 1}\,(1 + o(1))\,, \quad \alpha > 1$
Lognormal	$\dfrac{\sigma^2 u}{\ln u - \mu}\,(1 + o(1))$
Benktander–type–I	$\dfrac{u}{\alpha + 2\beta \ln u}$
Benktander–type–II	$\dfrac{u^{1-\beta}}{\alpha}$
Weibull	$\dfrac{u^{1-\tau}}{c\tau}\,(1 + o(1))$
Exponential	λ^{-1}
Gamma	$\beta^{-1}\left(1 + \dfrac{\alpha - 1}{\beta u} + o\left(\dfrac{1}{u}\right)\right)$
Truncated normal	$u^{-1}\,(1 + o(1))$

Table 3.4.7 *Mean excess functions for some standard distributions. The parametrisation is taken from Tables 1.2.5 and 1.2.6. The asymptotic relations are to be understood for $u \to \infty$.*

$$e(u) = E(X - u \mid X > u)$$

is called the mean excess function of X. □

Excesses over thresholds play a fundamental role in many fields. Different names arise from specific applications. For instance, F_u is known as the *excess–life* or *residual lifetime df* in reliability theory and medical statistics. In an insurance context, F_u is usually referred to as the *excess–of–loss df.* For a detailed discusssion of some basic properties and statistical applications of the mean excess function and the excess df, we refer to Sections 6.2.2 and 6.5.

Example 3.4.8 (Calculation of the mean excess function)
Using the definition of $e(u)$ and partial integration, the following formulae

are easily checked. They are useful for calculating the mean excess function in special cases. Suppose for ease of representation that X is a positive rv with df F and finite expectation; trivial changes allow for support (x_0, ∞) for some $x_0 > 0$. Then

$$
\begin{aligned}
e(u) &= \int_u^{x_F} (x-u)dF(x)/\overline{F}(u) \\
&= \frac{1}{\overline{F}(u)} \int_u^{x_F} \overline{F}(x)\,dx\,, \quad 0 < u < x_F\,.
\end{aligned} \tag{3.48}
$$

Whenever F is continuous,

$$
\overline{F}(x) = \frac{e(0)}{e(x)} \exp\left\{-\int_0^x \frac{1}{e(u)}\,du\right\}, \quad x > 0\,. \tag{3.49}
$$

It immediately follows from (3.49) that a continuous df is uniquely determined by its mean excess function. If, as in many cases of practical interest, $\overline{F} \in \mathcal{R}_{-\alpha}$ for some $\alpha > 1$, then an immediate application of Karamata's theorem (Theorem A3.6) yields $e(u) \sim u/(\alpha - 1)$ as $u \to \infty$. In Table 3.4.7 the mean excess functions of some standard distributions are summarised. □

The appearance of the rhs limit in (3.45) motivates the following definition.

Definition 3.4.9 (The generalised Pareto distribution (GPD))
Define the df G_ξ by

$$
G_\xi(x) = \begin{cases} 1 - (1+\xi x)^{-1/\xi} & \text{if } \xi \neq 0\,, \\ 1 - e^{-x} & \text{if } \xi = 0\,, \end{cases} \tag{3.50}
$$

where

$$
\begin{aligned}
x &\geq 0 & \text{if } \xi \geq 0\,, \\
0 \leq x &\leq -1/\xi & \text{if } \xi < 0\,.
\end{aligned}
$$

G_ξ *is called a* standard generalised Pareto distribution (GPD). *One can introduce the related location–scale family $G_{\xi;\nu,\beta}$ by replacing the argument x above by $(x-\nu)/\beta$ for $\nu \in \mathbb{R}$, $\beta > 0$. The support has to be adjusted accordingly. We also refer to $G_{\xi;\nu,\beta}$ as GPD.* □

As in the case of H_0, G_0 can also be interpreted as the limit of G_ξ as $\xi \to 0$. The df $G_{\xi;0,\beta}$ plays an important role in Section 6.5. To economise on notation, we will denote

$$
G_{\xi,\beta}(x) = 1 - \left(1 + \xi \frac{x}{\beta}\right)^{-1/\xi}, \quad x \in D(\xi, \beta)\,, \tag{3.51}
$$

where

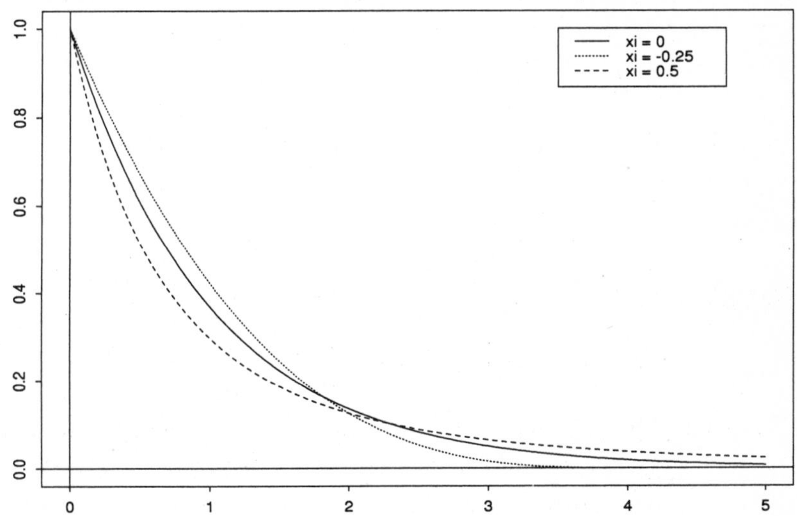

Figure 3.4.10 *Densities of the GPD for different parameters* ξ *and* $\beta = 1$.

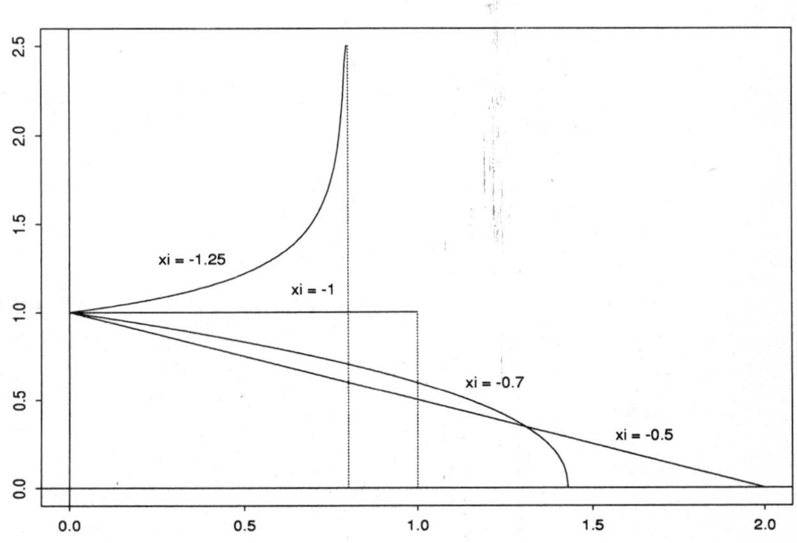

Figure 3.4.11 *Densities of the GPD for* $\xi < 0$ *and* $\beta = 1$. *Recall that they have compact support* $[0, -1/\xi]$.

$$x \in D(\xi,\beta) = \begin{cases} [0,\infty) & if \quad \xi \geq 0\,, \\ [0,-\beta/\xi] & if \quad \xi < 0\,. \end{cases}$$

Whenever we say that X has a GPD with parameters ξ and β, it is understood that X has df $G_{\xi,\beta}$.

Time to summarise:

> The GEV
>
> $$H_\xi\,, \quad \xi \in \mathbb{R}\,,$$
>
> describes the limit distributions of normalised maxima.
>
> The GPD
>
> $$G_{\xi,\beta}\,, \quad \xi \in \mathbb{R},\ \beta > 0\,,$$
>
> appears as the limit distribution of scaled excesses over high thresholds.

GPD fitting is one of the most useful concepts in the statistics of extremal events; see Section 6.5. Here we collect some basic probabilistic properties of the GPD.

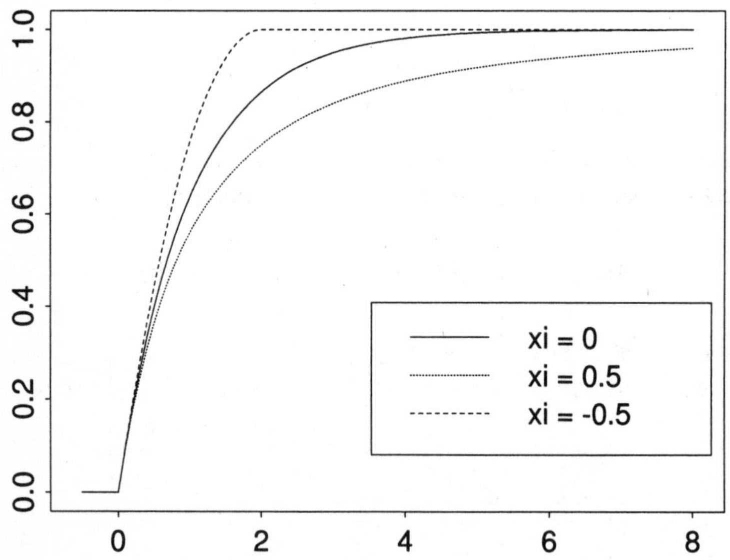

Figure 3.4.12 *GPD for different parameters ξ and $\beta = 1$.*

Theorem 3.4.13 (Properties of GPD)

(a) *Suppose X has a GPD with parameters ξ and β. Then $EX < \infty$ if and only if $\xi < 1$. In the latter case*

$$
\begin{aligned}
E\left(1+\frac{\xi}{\beta}X\right)^{-r} &= \frac{1}{1+\xi r}, \quad r > -1/\xi,\\
E\left(\ln\left(1+\frac{\xi}{\beta}X\right)\right)^{k} &= \xi^k\, k!, \quad k \in \mathbb{N},\\
EX\left(\overline{G}_{\xi,\beta}(X)\right)^r &= \frac{\beta}{(r+1-\xi)(r+1)}, \quad (r+1)/|\xi| > 0.
\end{aligned}
$$

If $\xi < 1/r$ with $r \in \mathbb{N}$, then

$$EX^r = \frac{\beta^r}{\xi^{r+1}} \frac{\Gamma(\xi^{-1}-r)}{\Gamma(1+\xi^{-1})}\, r!\,.$$

(b) *For every $\xi \in \mathbb{R}$, $F \in \mathrm{MDA}(H_\xi)$ if and only if*

$$\lim_{u\uparrow x_F} \sup_{0<x<x_F-u} |F_u(x) - G_{\xi,\beta(u)}(x)| = 0 \tag{3.52}$$

for some positive function β.

(c) *Suppose $x_i \in D(\xi,\beta), i = 1,2$, then*

$$\frac{\overline{G}_{\xi,\beta}(x_1+x_2)}{\overline{G}_{\xi,\beta}(x_1)} = \overline{G}_{\xi,\beta+\xi x_1}(x_2)\,. \tag{3.53}$$

(d) *Assume that N is Poi(λ), independent of the iid sequence (X_n) with a GPD with parameters ξ and β. Write $M_N = \max(X_1,\ldots,X_N)$. Then*

$$P(M_N \le x) = \exp\left\{-\lambda\left(1+\xi\frac{x}{\beta}\right)^{-1/\xi}\right\} = H_{\xi;\mu,\psi}(x)\,,$$

where $\mu = \beta\xi^{-1}(\lambda^\xi - 1)$ and $\psi = \beta\lambda^\xi$.

(e) *Suppose X has GPD with parameters $\xi < 1$ and β. Then for $u < x_F$,*

$$e(u) = E(X-u\,|\,X>u) = \frac{\beta+\xi u}{1-\xi}, \quad \beta + u\xi > 0\,.$$

Proof. (a) and (c) follow by direct verification.

(b) In Theorem 3.4.5 (see Remark 1) we have already proved that $F \in \mathrm{MDA}(H_\xi)$ if and only if

$$\lim_{u\uparrow x_F} |F_u(x) - G_{\xi,\beta(u)}(x)| = 0$$

where $\beta(u) = a(u)$. Because the GPD is continuous, the uniformity of the convergence follows; see Appendix A1.1.

(d) One immediately obtains

$$\begin{aligned} P(M_N \le x) &= \sum_{n=0}^{\infty} e^{-\lambda} \frac{\lambda^n}{n!} G^n_{\xi,\beta}(x) \\ &= \exp\{-\lambda \overline{G}_{\xi,\beta}(x)\} \\ &= \exp\left\{-\lambda \left(1 + \xi \frac{x}{\beta}\right)^{-1/\xi}\right\} \\ &= \exp\left\{-\left(1 + \xi \frac{x - \xi^{-1}\beta(\lambda^\xi - 1)}{\beta\lambda^\xi}\right)^{-1/\xi}\right\}, \quad \xi \neq 0 . \end{aligned}$$

The case $\xi = 0$ reduces to

$$P(M_N \le x) = \exp\left\{-e^{-(x - \beta \ln \lambda)/\beta}\right\} .$$

(e) This result immediately follows from the representation (3.48). □

Remarks. 4) Theorem 3.4.13 summarises various properties which are essential for the special role of the GPD in the statistical analysis of extremes. This will be made clear in Section 6.5.

5) The property (c) above is often reformulated as follows: the class of GPDs is closed with respect to changes of the threshold. Indeed the lhs in (3.53) is the conditional probability that, given our underlying rv exceeds x_1, it also exceeds the threshold $x_1 + x_2$. The rhs states that this probability is again of the generalised Pareto type. This closure property is important in reinsurance, where the GPDs are basic when treating excess–of–loss contracts; see for instance Conti [132]. In combination with property (d) it is also crucial for stop–loss treaties. For a discussion on different types of reinsurance treaties, see Section 8.7.

6) Property (b) above suggests a GPD as appropriate approximation of the excess df F_u for large u. This result goes back to Pickands [498] and is often formulated as follows. For some function β to be estimated from the data,

$$\overline{F}_u(x) = P(X - u > x \mid X > u) \approx \overline{G}_{\xi,\beta(u)}(x), \quad x > 0 .$$

Alternatively one considers for $x > u$,

$$P(X > x \mid X > u) \approx \overline{G}_{\xi;u,\beta(u)}(x) .$$

In both cases u has to be taken sufficiently large. See Section 6.5 for more details. Together (b) and (e) give us a nice graphical technique for choosing the threshold u so high that an approximation of the excess df F_u by a GPD is justified: given an iid sample $X_1, \ldots, X_n$, construct the empirical mean excess function $e_n(u)$ as sample version of the mean excess function $e(u)$. From (e) we have that the mean excess function of a GPD is linear, hence check for a u–region where the graph of $e_n(u)$ becomes roughly linear. For such u an approximation of F_u by a GPD seems reasonable. In Section 6.5 we will use this approach for fitting excesses over high thresholds.

7) From Proposition 3.1.1 (see also the succeeding Remark 1) we have learnt that the number of the exceedances of a high threshold is roughly Poisson. From Remark 6 we conclude that an approximation of the excess df F_u by a GPD may be justified. Moreover, it can be shown that the number of exceedances and the excesses are independent in an asymptotic sense; see Leadbetter [417].

8) Property (d) now says that in a model, where the number of exceedances is exactly Poisson and the excess df is an exact GDP, the maximum of these excesses has an exact GEV. □

The above remarks suggest the following approximate model for the exceedance times and the excesses of an iid sample:

- The number of exceedances of a high threshold follows a Poisson process.
- Excesses over high thresholds can be modelled by a GPD.
- An appropriate value of the high threshold can be found by plotting the empirical mean excess function.
- The distribution of the maximum of a Poisson number of iid excesses over a high threshold is a GEV.

In interpreting the above summary, do look at the precise formulation of the underlying theorems. If at this point you want to see some of the above in action; see for instance Smith [595], p. 461. Alternatively, consult the examples in Section 6.5

Notes and Comments

In this section we summarised some of the probabilistic properties of the GEV and the GPD. They are crucial for the statistical analysis of extremal events as provided in Chapter 6. The GEV will be used for statistical inference of data which occur as iid maxima of certain time series, for instance annual maxima of rainfall, windspeed etc. Theorem 3.4.5 opens the way to tail and high quantile estimation. Part (b) of this theorem leads immediately to the definition of the GPD, which goes back to Pickands [498]. An approximation of the excess df by the GPD has been suggested by hydrologists under the name *peaks over threshold* method; see Section 6.5 for a detailed discussion. Weak limit theory for excess dfs originates from Balkema and de Haan [37]. Richard Smith and his collaborators have further developed the theory and applications of the GPD in various fields. Basic properties of the GPD can for instance be found in Smith [591]. Detailed discussions on the use of the mean excess function in insurance are to be found in Beirlant et al. [57] and Hogg and Klugman [330].

3.5 Almost Sure Behaviour of Maxima

In this section we study the a.s. behaviour of the maxima

$$M_1 = X_1\,, \quad M_n = \max(X_1,\ldots,X_n)\,, \quad n \geq 2\,,$$

for an iid sequence $X, X_1, X_2, \ldots$ with common non–degenerate df F.

At the beginning we might ask:

What kind of results can we expect?
Is there, for example, a general theorem like the SLLN for iid sums?

The answer to the latter question is, unfortunately, negative. We have already found in the previous sections that the weak limit theory for (M_n) is very sensitive with respect to the tails $\overline{F}(x) = P(X > x)$. The same applies to the a.s. behaviour.

We first study the probabilities (i.o. stands for "infinitely often")

$$P(M_n > u_n \quad \text{i.o.}) \quad \text{and} \quad P(M_n \leq u_n \quad \text{i.o.})$$

for a non–decreasing sequence (u_n) of real numbers. We will fully characterise these probabilities in terms of the tails $\overline{F}(u_n)$.

We start with $P(M_n > u_n \text{ i.o.})$ which is not difficult to handle with the classical Borel–Cantelli lemmas. Recall from any textbook on probability theory that, for a sequence of events (A_n), $\{A_n \text{ i.o.}\}$ stands for

$$\limsup_{n\to\infty} A_n = \bigcap_{k=1}^{\infty} \bigcup_{n\geq k} A_n\,.$$

The standard Borel–Cantelli lemma states that

$$\sum_{n=1}^{\infty} P(A_n) < \infty \quad \text{implies} \quad P(A_n \quad \text{i.o.}) = 0\,.$$

Its partial converse for independent A_n tells us that

$$\sum_{n=1}^{\infty} P(A_n) = \infty \quad \text{implies} \quad P(A_n \quad \text{i.o.}) = 1\,.$$

A version of the following result for general independent rvs can be found in Galambos [249], Theorem 4.2.1.

Theorem 3.5.1 (Characterisation of the maximal a.s. growth of partial maxima)
Suppose that (u_n) is non-decreasing. Then

$$P(M_n > u_n \quad \text{i.o.}) = P(X_n > u_n \quad \text{i.o.})\,. \tag{3.54}$$

In particular,

$$P(M_n > u_n \quad \text{i.o.}) = 0 \quad \textit{or} \quad = 1$$

according as

$$\sum_{n=1}^{\infty} P(X > u_n) < \infty \quad \textit{or} \quad = \infty\,. \tag{3.55}$$

Notice that the second statement of the theorem is an immediate consequence of (3.54). Indeed, by the Borel–Cantelli lemma and its partial converse for independent events, $P(X_n > u_n \text{ i.o.}) = 0$ or $= 1$ according as (3.55) holds.

Proof. It suffices to show that (3.54) holds. Since $M_n \geq X_n$ for all n we need only to show that

$$P(M_n > u_n \quad \text{i.o.}) \leq P(X_n > u_n \quad \text{i.o.})\,. \tag{3.56}$$

Let x_F denote the right endpoint of the distribution F and suppose that $u_n \geq x_F$ for some n. Then

$$P(M_n > u_n) = P(X_n > u_n) = 0$$

for all large n, hence (3.56) is satisfied. Therefore assume that $u_n < x_F$ for all n. Then

$$F(u_n) < 1 \quad \text{for all} \quad n\,.$$

If $u_n \uparrow x_F$ and $M_n > u_n$ for infinitely many n then it is not difficult to see that there exist infinitely many n with the property $X_n > u_n$.
Now suppose that $u_n \uparrow b < x_F$. But then

$$\overline{F}(u_n) \geq \overline{F}(b) > 0\,.$$

By the converse Borel–Cantelli lemma, $1 = P(X_n > u_n \text{ i.o.})$, and then (3.54) necessarily holds. □

The determination of the probabilities $P(M_n \le u_n \text{ i.o.})$ is quite tricky. The following is a final improvement, due to Klass [383, 384], of a result of Barndorff–Nielsen [44, 45]. For the history of this problem see Galambos [249] and Klass [383].

Theorem 3.5.2 (Characterisation of the minimal a.s. growth of partial maxima)
Suppose that (u_n) is non–decreasing and that the following conditions hold:

$$\overline{F}(u_n) \rightarrow 0\,, \tag{3.57}$$

$$n\overline{F}(u_n) \rightarrow \infty\,. \tag{3.58}$$

Then

$$P(M_n \le u_n \quad \text{i.o.}) = 0 \quad \textit{or} \quad = 1$$

according as

$$\sum_{n=1}^{\infty} \overline{F}(u_n) \exp\left\{-n\overline{F}(u_n)\right\} < \infty \quad \textit{or} \quad = \infty\,. \tag{3.59}$$

Moreover, if

$$\overline{F}(u_n) \rightarrow c > 0\,, \quad \textit{then} \quad P(M_n \le u_n \quad \text{i.o.}) = 0\,,$$

while if

$$\liminf_{n\to\infty} n\overline{F}(u_n) < \infty\,, \qquad \textit{then} \quad P(M_n \le u_n \quad \text{i.o.}) = 1\,.$$

Remarks. 1) Conditions (3.57) and (3.58) are natural; (3.57) just means that one of the following conditions holds: $u_n \uparrow \infty$ or $u_n \uparrow x_F$ for a df F continuous at its right endpoint x_F or $u_n \uparrow b > x_F$. From Proposition 3.1.1 we know that (3.58) is equivalent to $P(M_n \le u_n) \rightarrow 0$.

2) Condition (3.59) is, at a first glance, a little bit mysterious, but from the proof below its meaning becomes more transparent. But already notice that $\overline{F}(u_n)\exp\{-n\overline{F}(u_n)\}$ is close to the probability

$$P(M_n \le u_n, X_{n+1} > u_n) = P(M_n \le u_n, M_{n+1} > u_n)\,.$$ □

Sketch of the proof. We first deal with the case that $\overline{F}(u_n) \rightarrow c > 0$. Then

$$\begin{aligned}
P(M_n \le u_n \quad \text{i.o.}) &= P\left(\bigcap_{i=1}^{\infty}\bigcup_{n\ge i}\{M_n \le u_n\}\right) \\
&= \lim_{i\to\infty} P\left(\bigcup_{n\ge i}\{M_n \le u_n\}\right) \\
&\le \lim_{i\to\infty}\sum_{n\ge i} P(M_n \le u_n) \\
&= \lim_{i\to\infty}\sum_{n\ge i} F^n(u_n) \\
&= 0,
\end{aligned}$$

since $F(u_n) < 1 - c + \epsilon < 1$ for a small ϵ and sufficiently large n .

Next suppose that $\liminf_{n\to\infty} n\overline{F}(u_n) < \infty$. Then

$$\begin{aligned}
P(M_n \le u_n \quad \text{i.o.}) &= \lim_{i\to\infty} P\left(\bigcup_{n\ge i}\{M_n \le u_n\}\right) \\
&\ge \limsup_{i\to\infty} P(M_i \le u_i) \\
&= \limsup_{i\to\infty} \exp\left\{i \ln\left(1 - \overline{F}(u_i)\right)\right\} \\
&= \limsup_{i\to\infty} \exp\left\{-i\overline{F}(u_i)(1 + o(1))\right\} \\
&= \exp\left\{-\liminf_{i\to\infty} i\overline{F}(u_i)\right\} \\
&> 0.
\end{aligned}$$

This and an application of the Hewitt–Savage 0–1 law prove that $P(M_n \le u_n \text{ i.o.}) = 1$.

Now we come to the main part of the theorem. We restrict ourselves to showing that the condition

$$\sum_{n=1}^{\infty} \overline{F}(u_n) \exp\left\{-n\overline{F}(u_n)\right\} < \infty \tag{3.60}$$

implies that

$$P(M_n \le u_n \quad \text{i.o.}) = 0.$$

Suppose that (3.60), (3.57) and (3.58) hold. We use a standard blocking technique for proving a.s. convergence results. Define the subsequence (n_k) as follows

$$n_1 = 1\,, \quad n_{k+1} = \min\left\{j > n_k : \; (j - n_k)\,\overline{F}\,(u_{n_k}) \geq 1\right\}\,.$$

This implies in particular that

$$(n_{k+1} - n_k)\,\overline{F}\,(u_{n_k}) \to 1\,. \tag{3.61}$$

Moreover, by (3.61), $n_{k+1}/n_k \to 1$. Hence there exists k_0 such that

$$n_j\overline{F}(u_{n_{j+1}}) \geq 1 \quad \text{for} \quad j \geq k_0\,.$$

Note also that the function $f(y) = y\exp\{-jy\}$ decreases for $y \geq j^{-1}$. Hence, for all $k \geq k_0$,

$$\begin{aligned}
&\sum_{n_k \leq j < n_{k+1}} \overline{F}\,(u_j)\exp\left\{-j\overline{F}\,(u_j)\right\} \\
&\geq \sum_{n_k \leq j < n_{k+1}} \overline{F}\,(u_{n_k})\exp\left\{-j\overline{F}\,(u_{n_k})\right\} \\
&\geq e^{-1}\,(n_{k+1} - n_k)\,\overline{F}\,(u_{n_k})\exp\left\{-n_k\overline{F}\,(u_{n_k})\right\} \\
&\geq e^{-1}\exp\left\{-n_k\overline{F}\,(u_{n_k})\right\}\,.
\end{aligned}$$

Thus (3.60) implies that

$$\sum_{k=1}^{\infty}\exp\left\{-n_k\overline{F}\,(u_{n_k})\right\} < \infty\,. \tag{3.62}$$

(It can as it happens be shown that (3.60) is equivalent to (3.62).) Notice that

$$\begin{aligned}
P\,(M_n \leq u_n \quad \text{i.o.}) &= P\left(\bigcap_{k=1}^{\infty}\bigcup_{n\geq k}\{M_n \leq u_n\}\right) \\
&= \lim_{k\to\infty} P\left(\bigcup_{n\geq k}\{M_n \leq u_n\}\right) \\
&= \lim_{l\to\infty} P\left(\bigcup_{k\geq l}\;\bigcup_{n_k\leq j<n_{k+1}}\{M_j \leq u_j\}\right)
\end{aligned}$$

$$\leq \lim_{l\to\infty}\sum_{k\geq l} P\left(\bigcup_{n_k\leq j<n_{k+1}} \{M_j \leq u_j\}\right)$$

$$\leq \lim_{l\to\infty}\sum_{k\geq l} P\left(M_{n_k} \leq u_{n_{k+1}}\right)$$

$$\leq \lim_{l\to\infty}\sum_{k\geq l} \frac{P\left(M_{n_{k+1}} \leq u_{n_{k+1}}\right)}{P\left(M_{n_{k+1}-n_k} \leq u_{n_{k+1}}\right)}. \qquad (3.63)$$

By construction of the n_k and by property (3.61),

$$\begin{aligned} P\left(M_{n_{k+1}-n_k} \leq u_{n_{k+1}}\right) &= \exp\left\{-(n_{k+1}-n_k)\overline{F}\left(u_{n_{k+1}}\right)(1+o(1))\right\} \\ &\geq \exp\left\{-(n_{k+1}-n_k)\overline{F}\left(u_{n_k}\right)(1+o(1))\right\} \\ &\to e^{-1}. \end{aligned}$$

This together with (3.63) and (3.62) yields $P(M_n \leq u_n \text{ i.o.}) = 0$.

The proof of the remaining part of the theorem is very technical. We refer to Klass [383, 384] for details. □

Recall (e.g. from Petrov [495, 496]) that the relation

$$\limsup_{n\to\infty} c_n^{-1}\left(M_n - d_n\right) = 1 \quad \text{a.s.}$$

for $c_n > 0$ and $d_n \in \mathbb{R}$ just means that

$$P\left(M_n > c_n(1+\epsilon) + d_n \quad \text{i.o.}\right) = 0 \quad \text{or} \quad = 1$$

according as $\epsilon > 0$ or $\epsilon < 0$ for small $|\epsilon|$. Similarly,

$$\liminf_{n\to\infty} c_n^{-1}\left(M_n - d_n\right) = 1 \quad \text{a.s.}$$

holds if and only if

$$P\left(M_n \leq c_n(1+\epsilon) + d_n \quad \text{i.o.}\right) = 0 \quad \text{or} \quad = 1$$

according as $\epsilon < 0$ or $\epsilon > 0$ for small $|\epsilon|$. Then the following is immediate from Theorems 3.5.1 and 3.5.2.

Corollary 3.5.3 (Characterisation of the upper and lower limits of maxima)

(a) *Assume that the sequences $u_n(\epsilon) = c_n(1+\epsilon) + d_n$, $n \in \mathbb{N}$, are non–decreasing for every $\epsilon \in (-\epsilon_0, \epsilon_0)$. Then the relation*

$$\sum_{n=1}^{\infty} \overline{F}\left(u_n(\epsilon)\right) < \infty \quad \textit{or} \quad = \infty$$

according as $\epsilon \in (0, \epsilon_0)$ *or* $\epsilon \in (-\epsilon_0, 0)$ *implies that*

$$\limsup_{n\to\infty} c_n^{-1}(M_n - d_n) = 1 \quad \text{a.s.}$$

(b) *Assume that the sequences* $u_n(\epsilon) = c_n(1+\epsilon) + d_n\,, n \in \mathbb{N}$, *are non-decreasing and satisfy* (3.57), (3.58) *for every* $\epsilon \in (-\epsilon_0, \epsilon_0)$. *Then the relation*

$$\sum_{n=1}^{\infty} \overline{F}(u_n(\epsilon)) \exp\left\{-n\overline{F}(u_n(\epsilon))\right\} < \infty \quad \textit{or} \quad = \infty$$

according as $\epsilon \in (-\epsilon_0, 0)$ *or* $\epsilon \in (0, \epsilon_0)$ *implies that*

$$\liminf_{n\to\infty} c_n^{-1}(M_n - d_n) = 1 \quad \text{a.s.}$$ □

We continue with several examples in order to illustrate the different options for the a.s. behaviour of the maxima M_n. Throughout we will use the following notation

$$\ln_0 x = x\,, \quad \ln_1 x = \max(0, \ln x)\,, \quad \ln_k x = \max\left(0, \ln_{k-1} x\right)\,, \quad k \geq 2\,,$$

i.e. $\ln_k x$ is the *kth iterated logarithm of* x.

Example 3.5.4 (Normal distribution, continuation of Example 3.3.29) Assume that $F = \Phi$ is the standard normal distribution. Then

$$\overline{\Phi}(x) \sim \frac{1}{\sqrt{2\pi x}} \exp\{-x^2/2\}\,. \tag{3.64}$$

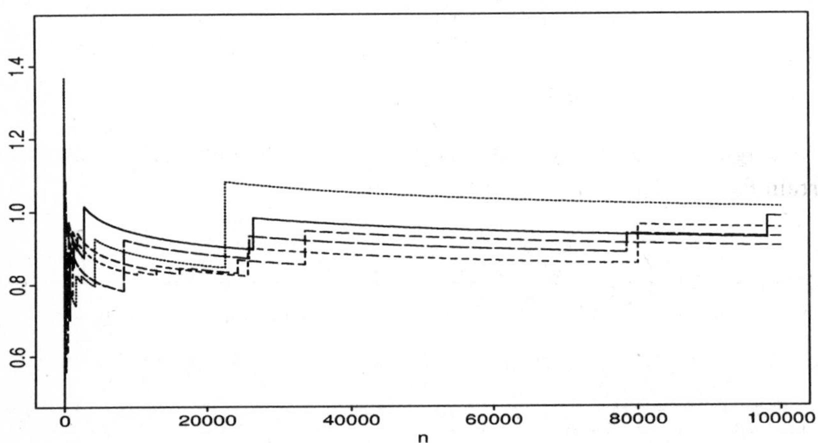

Figure 3.5.5 *Five sample paths of* $(M_n/\sqrt{2\ln n})$ *for* 100 000 *realisations of iid standard normal rvs. The rate of a.s. convergence to* 1 *appears to be very slow.*

From (3.40) we conclude

$$\frac{M_n}{\sqrt{2\ln n}} \stackrel{P}{\to} 1\,, \quad n \to \infty\,. \tag{3.65}$$

We are interested in a.s. refinements of this result.
Choose

$$u_n(\epsilon) = \sqrt{2\ln\left(\frac{(\ln_0 n \cdots \ln_r n)\ln_r^{\epsilon} n}{\sqrt{\ln n}}\right)}\,, \quad r \geq 0\,.$$

An application of Theorem 3.5.1 together with (3.64) yields

$$P\left(M_n > u_n(\epsilon) \quad \text{i.o.}\right) = 0 \quad \text{or} \quad = 1$$

according as $\epsilon > 0$ or $\epsilon < 0$ for small $|\epsilon|$ and hence, by Corollary 3.5.3,

$$\limsup_{n\to\infty} \frac{M_n}{\sqrt{2\ln n}} = 1 \quad \text{a.s.} \tag{3.66}$$

This result can further be refined. For example, notice that

$$\begin{aligned}
&P\left(M_n > \sqrt{2\ln\left(\frac{n\ln^{1+\epsilon} n}{\sqrt{\ln n}}\right)} \quad \text{i.o.}\right)\\
&= \quad P\left(M_n - \sqrt{2\ln\left(\frac{n}{\sqrt{\ln n}}\right)}\right.\\
&\qquad\qquad \left. > \sqrt{2\ln\left(\frac{n\ln^{1+\epsilon} n}{\sqrt{\ln n}}\right)} - \sqrt{2\ln\left(\frac{n}{\sqrt{\ln n}}\right)} \quad \text{i.o.}\right)\\
&= \quad 0 \quad \text{or} \quad = 1
\end{aligned}$$

according as $\epsilon > 0$ or $\epsilon \leq 0$. By the mean value theorem, for small $|\epsilon|$ and certain $\theta_n \in (0,1)$,

$$P\left(M_n - \sqrt{2\ln\left(\frac{n}{\sqrt{\ln n}}\right)} > \frac{1}{2}\,\frac{2(1+\epsilon)\ln_2 n}{\left(2\ln(n/\sqrt{\ln n}) + \theta_n 2(1+\epsilon)\ln_2 n\right)^{1/2}} \quad \text{i.o.}\right)$$

$$= 0 \quad \text{or} \quad = 1$$

according as $\epsilon > 0$ or $\epsilon \leq 0$. In view of Corollary 3.5.3 this just means that

$$\limsup_{n\to\infty} \frac{\sqrt{2\ln n}}{\ln_2 n}\left(M_n - \sqrt{2\ln\left(\frac{n}{\sqrt{\ln n}}\right)}\right) = 1 \quad \text{a.s.}$$

By the same arguments,

$$\limsup_{n\to\infty} \frac{\sqrt{2\ln n}}{\ln_{r+1} n}\left(M_n - \sqrt{2\ln\left(\frac{\ln_0 n \cdots \ln_{r-1} n}{\sqrt{\ln n}}\right)}\right) = 1 \quad \text{a.s.}, \quad r \geq 1.$$

Now choose

$$u_n'(\epsilon) = \sqrt{2\ln\left(\frac{n}{\sqrt{4\pi \ln n}\,\ln((\ln_1 n \cdots \ln_r n)\ln_r^{\epsilon} n)}\right)}, \quad r \geq 1.$$

An application of Theorem 3.5.2 yields that

$$P\left(M_n \leq u_n'(\epsilon) \quad \text{i.o.}\right) = 0 \quad \text{or} \quad = 1$$

according as $\epsilon > 0$ or $\epsilon < 0$ for small $|\epsilon|$. In particular, we may conclude that

$$\liminf_{n\to\infty} \frac{M_n}{\sqrt{2\ln n}} = 1 \quad \text{a.s.} \tag{3.67}$$

which, together with (3.66), yields the a.s. analogue to (3.65):

$$\lim_{n\to\infty} \frac{M_n}{\sqrt{2\ln n}} = 1 \quad \text{a.s.}$$

Refinements of relation (3.67) are possible in the same way as for lim sup. □

Example 3.5.6 (Exponential tails)
Let X be a rv with tail

$$\overline{F}(x) \sim K e^{-ax}, \quad x \to \infty,$$

for some $a, K > 0$. From Example 3.2.7 we conclude that

$$\frac{M_n}{\ln n} \xrightarrow{P} \frac{1}{a}, \quad n \to \infty. \tag{3.68}$$

We are interested in a.s. refinements of this result.
Choose

$$u_n(\epsilon) = \frac{1}{a}\ln\left(K\left(\ln_0 n \ln_1 n \cdots \ln_r n\right)\ln_r^{\epsilon}\right), \quad r \geq 0.$$

Then, for large n and small $|\epsilon|$,

$$\overline{F}\left(u_n(\epsilon)\right) \sim \frac{1}{(\ln_0 n \cdots \ln_r n)\ln_r^{\epsilon} n}.$$

An application of Theorem 3.5.1 yields that

$$P\left(M_n > u_n(\epsilon) \quad \text{i.o.}\right) = 0 \quad \text{or} \quad = 1$$

according as $\epsilon > 0$ or $\epsilon < 0$ for small $|\epsilon|$ and hence

$$\limsup_{n\to\infty} \frac{M_n}{\ln n} = \frac{1}{a} \quad \text{a.s.} \tag{3.69}$$

Now choose

$$u_n'(\epsilon) = \frac{1}{a} \ln \left(\frac{nK}{\ln(\ln_2 n (\ln_1 n \cdots \ln_r n) \ln_r^{\epsilon} n)} \right), \quad r \geq 1.$$

Then, by Theorem 3.5.2,

$$P(M_n \leq u_n'(\epsilon) \quad \text{i.o.}) = 0 \quad \text{or} \quad = 1 \tag{3.70}$$

according as $\epsilon > 0$ or $\epsilon < 0$ for small $|\epsilon|$. In particular, we may conclude that

$$\liminf_{n\to\infty} \frac{M_n}{\ln n} = \frac{1}{a} \quad \text{a.s.} \tag{3.71}$$

which, together with (3.69), yields an a.s. analogue of (3.68).

$$\lim_{n\to\infty} \frac{M_n}{\ln n} = \frac{1}{a} \quad \text{a.s.}$$

Refinements of (3.71) are possible. For example, for fixed $r \geq 1$,

$$\begin{aligned} & P(M_n \leq u_n'(\epsilon) \quad \text{i.o.}) \\ = \quad & P\bigg(M_n \leq \frac{1}{a}\bigg(\ln(nK) - \ln(\ln_3 n + (\ln_2 n + \ln_3 n + \cdots \end{aligned}$$

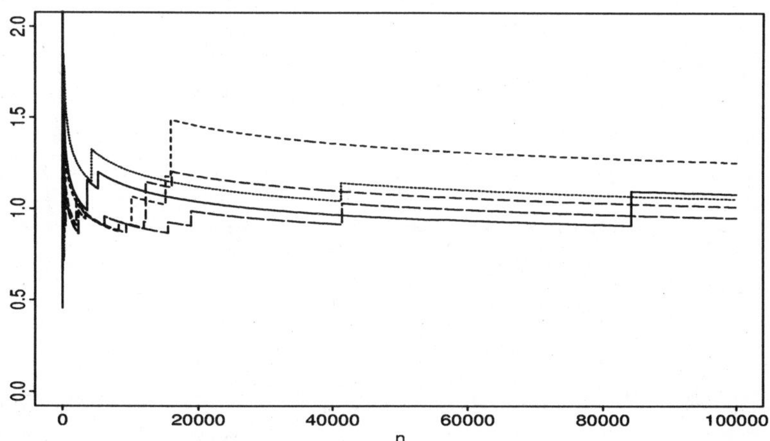

Figure 3.5.7 *Five sample paths of* $(M_n/\ln n)$ *for* 100 000 *realisations of iid standard exponential rvs. The rate of a.s. convergence to* 1 *appears to be very slow.*

$$
\begin{aligned}
& \qquad\qquad + \ln_{r+1} n + (1+\epsilon) \ln_{r+2} n)\Big)\Big) \quad \text{i.o.}\Big) \\
= \; & P\Big(M_n - \frac{1}{a}\Big(\ln(nK) - \ln(\ln_3 n + (\ln_2 n + \cdots + \ln_{r+1} n))\Big) \\
& \qquad \leq -\frac{1}{a} \ln\left(\frac{\ln_3 n + (\ln_2 n + \cdots + \ln_{r+1} n + (1+\epsilon)\ln_{r+2} n)}{\ln_3 n + (\ln_2 n + \cdots + \ln_{r+1} n)}\right) \text{ i.o.}\Big) \\
= \; & P\Big(M_n - \frac{1}{a}\Big(\ln(nK) - \ln\left(\ln_3 n + (\ln_2 n + \cdots + \ln_{r+1} n)\right)\Big) \\
& \qquad \leq -\frac{1}{a} \ln\left(1 + (1+\epsilon)\frac{\ln_{r+2} n}{\ln_3 n + (\ln_2 n + \cdots + \ln_{r+1} n)}\right) \quad \text{i.o.}\Big) .
\end{aligned}
$$

This together with (3.70), a mean value theorem argument and Corollary 3.5.3 imply that, for $r \geq 1$,

$$
\begin{aligned}
\liminf_{n\to\infty} \frac{\ln_2 n}{\ln_{r+2} n}\Big(M_n - \frac{1}{a}\Big(\ln(nK) - \ln\left(\ln_3 n + (\ln_2 n + \cdots + \ln_{r+1} n)\right)\Big)\Big) \\
= -\frac{1}{a} \quad \text{a.s.}
\end{aligned}
$$

□

Example 3.5.8 (Uniform distribution, continuation of Example 3.3.15) Let F be uniform on $(0,1)$. From Example 3.3.15 we know that

$$M_n \stackrel{\text{a.s.}}{\to} 1 .$$

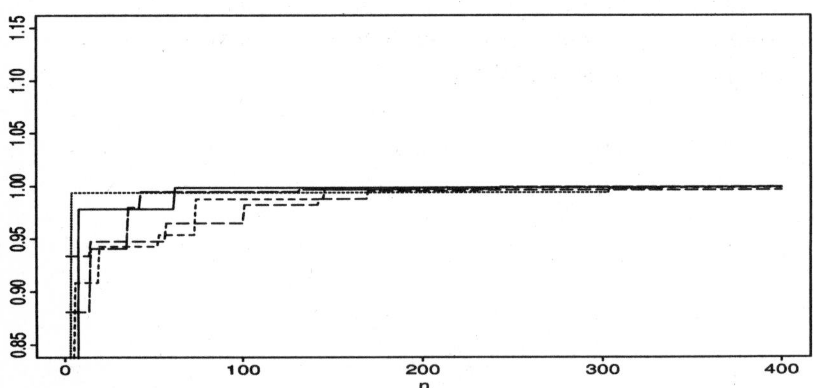

Figure 3.5.9 *Five sample paths of M_n for 400 realisations of iid $U(0,1)$ rvs. The rate of a.s. convergence to 1 is very fast.*

We derive some a.s. refinements of this limit result.
Choose

$$u_n(\epsilon) = 1 - \frac{1}{(\ln_0 n \ln_1 n \cdots \ln_r n) \ln_r^\epsilon n}, \quad r \geq 0 .$$

Then, by Theorem 3.5.1,

$$P(M_n > u_n(\epsilon) \quad \text{i.o.}) = 0 \quad \text{or} \quad = 1$$

according as $\epsilon > 0$ or $\epsilon < 0$ for small $|\epsilon|$.
Now choose

$$u'_n(\epsilon) = 1 - \frac{1}{n} \ln \left(\ln_2 n \, (\ln_1 n \cdots \ln_r n) \ln_r^\epsilon n\right), \quad r \geq 1 .$$

Then, by Theorem 3.5.2,

$$P(M_n \leq u'_n(\epsilon) \quad \text{i.o.}) = 0 \quad \text{or} \quad = 1$$

according as $\epsilon > 0$ or $\epsilon < 0$ for small $|\epsilon|$. □

Notes and Comments

There does not exist very much literature on the a.s. behaviour of maxima of iid rvs. An extensive account can be found in Galambos [249]. The treatment of results about the normalised lim inf of maxima started with work by Barndorff–Nielsen [44, 45] who proved necessary and sufficient conditions under certain restrictions. A result in the same spirit was obtained by Robbins and Siegmund [544]. Klass [383, 384] finally proved the criterion of Theorem 3.5.2 under minimal restrictions on the df F and on the behaviour of (u_n). Goldie and Maller [275] use point process techniques to derive a.s. convergence results for order statistics and records.

4

Fluctuations of Upper Order Statistics

After having investigated in Chapter 3 the behaviour of the maximum, i.e. the largest value of a sample, we now consider the joint behaviour of several upper order statistics. They provide information on the right tail of a df.

In Section 4.1, after some basic results on the ordered sample, we present various examples in connection with uniform and exponential order statistics and spacings. Just to mention two items, we touch on the subject of simulation of general upper order statistics (working from uniform random numbers) and prove the order statistics property of a homogeneous Poisson process. Here also Hill's estimator appears for the first time: we prove that it is a consistent estimator for the index of a regularly varying tail. Its importance will be made clear in Chapter 6.

In Section 4.2 we exploit the Poisson approximation, already used to derive limit laws of normalised maxima, in a more advanced way. This leads to the multivariate limit distribution of several upper order statistics. Such results provide the theoretical background when deriving limit properties for tail estimators, as we shall see in Chapter 6. Extensions to randomly indexed samples will be given in Section 4.3.

In Section 4.4 we show under what conditions the previous results can be extended to stationary sequences.

4.1 Order Statistics

Let $X, X_1, X_2, \ldots$ denote a sequence of iid non-degenerate rvs with common df F. In this section we summarise some important properties of the upper order statistics of a finite sample $X_1, \ldots, X_n$. To this end we define the *ordered sample*

$$X_{n,n} \le \cdots \le X_{1,n}\,.$$

Hence $X_{n,n} = \min(X_1, \ldots, X_n)$ and $X_{1,n} = M_n = \max(X_1, \ldots, X_n)$. The rv $X_{k,n}$ is called the *kth upper order statistic*. The notation for order statistics varies; some authors denote by $X_{1,n}$ the minimum and by $X_{n,n}$ the maximum of a sample. This leads to different representations of quantities involving order statistics.

The relationship between the order statistics and the empirical df of a sample is immediate: for $x \in \mathbb{R}$ we introduce the *empirical df* or *sample df*

$$F_n(x) = \frac{1}{n}\,\mathrm{card}\,\{i : 1 \le i \le n, X_i \le x\} = \frac{1}{n}\sum_{i=1}^{n} I_{\{X_i \le x\}}\,, \qquad x \in \mathbb{R},$$

where I_A denotes the indicator function of the set A. Now

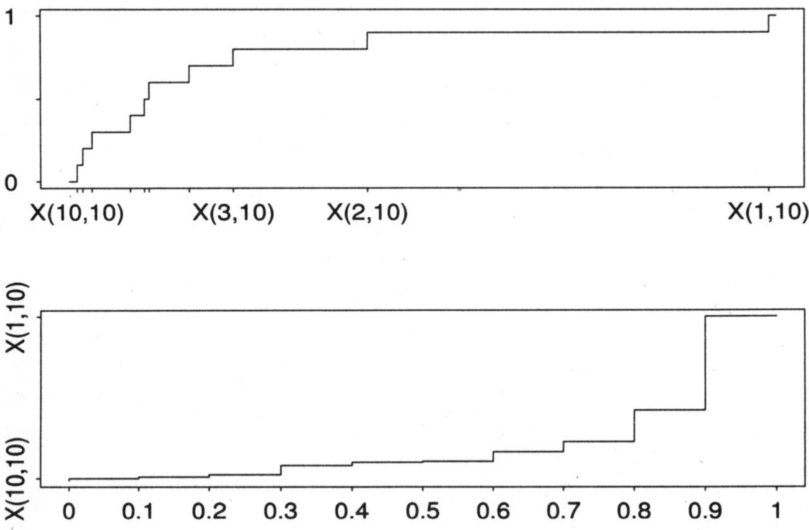

Figure 4.1.1 *Empirical df F_n (top) and empirical quantile function $F_n^{\leftarrow}$ (bottom) of a sample of* 10 *standard exponential rvs.*

$$X_{k,n} \le x \qquad \text{if and only if} \qquad \sum_{i=1}^{n} I_{\{X_i > x\}} < k\,, \tag{4.1}$$

which implies that

$$P\left(X_{k,n} \le x\right) = P\left(F_n(x) > 1 - \frac{k}{n}\right).$$

Upper order statistics estimate tails and quantiles, and also excess probabilities over certain thresholds. Recall the definition of the quantile function of the df F

$$F^{\leftarrow}(t) = \inf\{x \in \mathbb{R} : F(x) \ge t\}\,, \quad 0 < t < 1\,.$$

For a sample $X_1, \ldots, X_n$ we denote the empirical quantile function by $F_n^{\leftarrow}$. If F is a continuous function, then ties in the sample occur only with probability 0 and may thus be neglected, i.e. we may assume that $X_{n,n} < \cdots < X_{1,n}$. In this case $F_n^{\leftarrow}$ is a simple function of the order statistics, namely

$$F_n^{\leftarrow}(t) = X_{k,n} \quad \text{for } 1 - \frac{k}{n} < t \le 1 - \frac{k-1}{n}\,, \tag{4.2}$$

for $k = 1, \ldots, n$.

Next we calculate the df $F_{k,n}$ of the kth upper order statistic explicitly.

Proposition 4.1.2 (Distribution function of the kth upper order statistic) *For $k = 1, \ldots, n$ let $F_{k,n}$ denote the df of $X_{k,n}$. Then*

(a) $F_{k,n}(x) = \sum_{r=0}^{k-1} \binom{n}{r} \overline{F}^{r}(x)\, F^{n-r}(x)$.

(b) *If F is continuous, then*

$$F_{k,n}(x) = \int_{-\infty}^{x} f_{k,n}(z)\, dF(z)\,,$$

where

$$f_{k,n}(x) = \frac{n!}{(k-1)!\,(n-k)!}\, F^{n-k}(x)\, \overline{F}^{k-1}(x)\,;$$

i.e. $f_{k,n}$ is a density of $F_{k,n}$ with respect to F.

Proof. (a) For $n \in \mathbb{N}$ define

$$B_n = \sum_{i=1}^{n} I_{\{X_i > x\}}\,.$$

Then B_n is a sum of n iid Bernoulli variables with success probability

$$E I_{\{X > x\}} = P(X > x) = \overline{F}(x)\,.$$

Hence B_n is a binomial rv with parameters n and $\overline{F}(x)$. An application of (4.1) gives for $x \in \mathbb{R}$

$$\begin{aligned} F_{k,n}(x) &= P(B_n < k) \\ &= \sum_{r=0}^{k-1} P(B_n = r) \\ &= \sum_{r=0}^{k-1} \binom{n}{r} \overline{F}^r(x)\, F^{n-r}(x)\,. \end{aligned}$$

(b) Using the continuity of F, we calculate

$$\begin{aligned} &\frac{n!}{(k-1)!\,(n-k)!} \int_{-\infty}^{x} F^{n-k}(z)\, \overline{F}^{k-1}(z)\, dF(z) \\ &= \frac{n!}{(k-1)!\,(n-k)!} \int_{\overline{F}(x)}^{1} (1-t)^{n-k}\, t^{k-1}\, dt \\ &= \sum_{r=0}^{k-1} \binom{n}{r} \overline{F}^r(x)\, F^{n-r}(x) \;=\; F_{k,n}(x)\,. \end{aligned}$$

The latter follows from a representation of the incomplete beta function; it can be proved by multiple partial integration. See also Abramowitz and Stegun [3], formula 6.6.4. □

Similar arguments lead to the joint distribution of a finite number of different order statistics. If for instance F is absolutely continuous with density f, then the joint density of $(X_1, \ldots, X_n)$ is

$$f_{X_1,\ldots,X_n}(x_1,\ldots,x_n) = \prod_{i=1}^{n} f(x_i)\,, \qquad (x_1,\ldots,x_n) \in \mathbb{R}^n\,.$$

Since the n values of $(X_1, \ldots, X_n)$ can be rearranged in $n!$ ways (by absolute continuity there are a.s. no ties), every specific ordered collection $(X_{k,n})_{k=1,\ldots,n}$ could have come from $n!$ different samples. This heuristic argument can be made precise; see for instance Reiss [526], Theorem 1.4.1, or alternatively use the transformation theorem for densities. The joint density of the ordered sample becomes:

$$f_{X_{1,n},\ldots,X_{n,n}}(x_1,\ldots,x_n) = n! \prod_{i=1}^{n} f(x_i)\,, \qquad x_n < \cdots < x_1\,. \tag{4.3}$$

The following result on marginal densities is an immediate consequence of equation (4.3).

Theorem 4.1.3 (Joint density of k upper order statistics)
If F is absolutely continuous with density f, then

$$f_{X_{1,n},\ldots,X_{k,n}}(x_1,\ldots,x_k) = \frac{n!}{(n-k)!} F^{n-k}(x_k) \prod_{i=1}^{k} f(x_i)\,, \quad x_k < \cdots < x_1\,.$$

□

Further quantities which arise in a natural way are the *spacings*, i.e. the differences between successive order statistics. They are for instance the building blocks of Hill's estimator; see Example 4.1.12 and Section 6.4.2.

Definition 4.1.4 (Spacings of a sample)
For a sample $X_1,\ldots,X_n$ the spacings *are defined by*

$$X_{k,n} - X_{k+1,n}\,, \qquad k = 1,\ldots,n-1\,.$$

For rvs with finite left (right) endpoint $\widetilde{x}_F$ (x_F) we define the nth (0th) spacing as $X_{n,n} - X_{n+1,n} = X_{n,n} - \widetilde{x}_F$ $(X_{0,n} - X_{1,n} = x_F - X_{1,n})$. □

Example 4.1.5 (Order statistics and spacings of exponential rvs)
Let (E_n) denote a sequence of iid standard exponential rvs. An immediate consequence of (4.3) is the joint density of an ordered exponential sample $(E_{1,n},\ldots,E_{n,n})$:

$$f_{E_{1,n},\ldots,E_{n,n}}(x_1,\ldots,x_n) = n!\, \exp\left\{-\sum_{i=1}^{n} x_i\right\}, \qquad 0 < x_n < \cdots < x_1\,.$$

From this we derive the joint distribution of exponential spacings by an application of the transformation theorem for densities. Define the transformation

$$T(x_1,\ldots,x_n) = (x_1 - x_2, 2(x_2 - x_3),\ldots,n x_n)\,, \quad 0 < x_n < \cdots < x_1\,.$$

Then $\det(\partial T(\mathbf{x})/\partial \mathbf{x}) = n!$ and

$$T^{-1}(x_1,\ldots,x_n) = \left(\sum_{j=1}^{n}\frac{x_j}{j}, \sum_{j=2}^{n}\frac{x_j}{j},\ldots,\frac{x_n}{n}\right), \quad x_1,x_2,\ldots,x_n > 0\,.$$

Then the density g of $(E_{1,n} - E_{2,n}, 2(E_{2,n} - E_{3,n}),\ldots,nE_{n,n})$ is of the form

$$\begin{aligned} g(x_1,\ldots,x_n) &= \frac{1}{n!} f_{E_{1,n},\ldots,E_{n,n}}\left(\sum_{j=1}^{n}\frac{x_j}{j}, \sum_{j=2}^{n}\frac{x_j}{j},\ldots,\frac{x_n}{n}\right) \\ &= \exp\left\{-\sum_{i=1}^{n}\sum_{j=i}^{n}\frac{x_j}{j}\right\} \\ &= \exp\left\{-\sum_{i=1}^{n} x_i\right\}. \end{aligned}$$

This gives for $i = 1, \ldots, n$ that the rvs $i(E_{i,n} - E_{i+1,n})$ have joint density

$$g(x_1, \ldots, x_n) = \exp\left\{-\sum_{i=1}^{n} x_i\right\}, \qquad x_1, \ldots, x_n > 0.$$

This implies that the spacings

$$E_{1,n} - E_{2,n}, \; E_{2,n} - E_{3,n}, \ldots, E_{n,n}$$

are independent and exponentially distributed, and $E_{k,n} - E_{k+1,n}$ has mean $1/k$ for $k = 1, \ldots, n$, where we recall that $E_{n+1,n} = 0$. □

Example 4.1.6 (Markov property of order statistics)
When working with spacings from absolutely continuous dfs one can often make use of the fact that their order statistics form a Markov process, i.e.

$$\begin{aligned} & P(X_{k,n} \le y \mid X_{n,n} = x_n, \ldots, X_{k+1,n} = x_{k+1}) \\ = \;& P(X_{k,n} \le y \mid X_{k+1,n} = x_{k+1}). \end{aligned}$$

To be precise, $(X_{n,n}, \ldots, X_{1,n})$ is a non-homogeneous, discrete–time Markov process whose initial df is

$$P(X_{n,n} \le x) = 1 - \overline{F}^{\,n}(x),$$

and whose transition df $P(X_{k,n} \le y \mid X_{k+1,n} = x_{k+1})$ is the df of the minimum of k iid observations from the df F truncated at x_{k+1}. For $k = 1, \ldots, n-1$,

$$P(X_{k,n} > y \mid X_{k+1,n} = x_{k+1}) = \left(\frac{\overline{F}(y)}{\overline{F}(x_{k+1})}\right)^k, \quad y > x_{k+1}.$$

A proof of the Markov property is straightforward; see for instance Arnold, Balakrishnan and Nagaraja [20], Theorem 2.4.3. They also provide an example showing that the Markov property does not hold for general F; see their Section 3.4. □

Example 4.1.7 (Order statistics property of the Poisson process)
Let $N = (N(t))_{t \ge 0}$ be a homogeneous Poisson process with intensity $\lambda > 0$; for a definition see Example 2.5.2. Then the arrival times T_i of N in $(0, t]$, conditionally on $\{N(t) = n\}$, have the same distribution as the order statistics of a uniform sample on $(0, t)$ of size n; i.e.

$$P\big((T_1, T_2, \ldots, T_{N(t)}) \in A | N(t) = n\big) = P\big((U_{n,n}, \ldots, U_{1,n}) \in A\big)$$

for all Borel sets A in $\mathbb{R}_+$. This property is called the *order statistics property* of the Poisson process. It gives an intuitive description of the distribution of

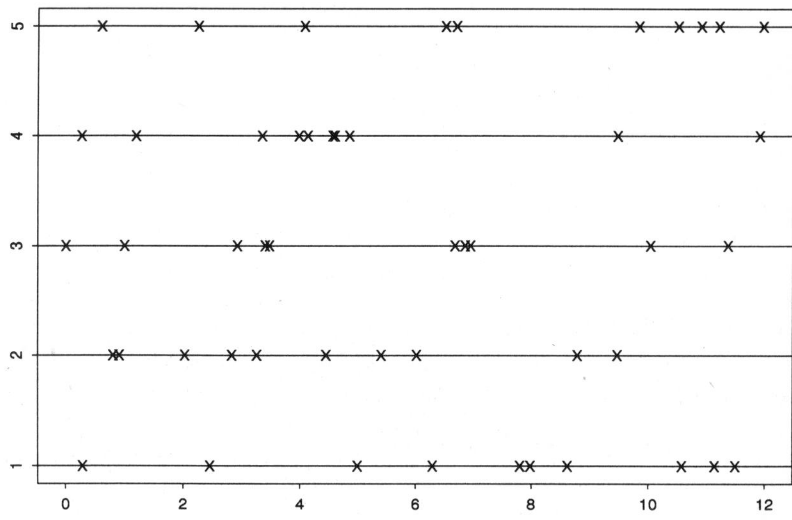

Figure 4.1.8 *Five realisations of the arrival–times of a Poisson process N with intensity* 1, *conditionally on* $\{N(12) = 10\}$. *They illustrate the order statistics property (Example 4.1.7).*

the arrival times of a Poisson process.

For a proof we assume that $0 < t_1 < \cdots < t_n < t$ and $h_1, \ldots, h_n$ are all positive, but small enough such that the intervals $J_i = (t_i, t_i + h_i]$, $i = 1, \ldots, n$, are disjoint. Then

$$\begin{aligned} & P(T_1 \in J_1, \ldots, T_n \in J_n \mid N(t) = n) \\ = \quad & P(T_1 \in J_1, \ldots, T_n \in J_n, N(t) = n) \,/\, P(N(t) = n)\,. \end{aligned}$$

Writing $N(J_i) = N(t_i + h_i) - N(t_i)$, $i = 1, \ldots, n$, and using the independence and stationarity of the increments of the Poisson process we obtain for the numerator that

$$\begin{aligned} & P(N(t_1) = 0, N(J_1) = 1, N(t_2) - N(t_1 + h_1) = 0, \\ & \qquad \ldots, N(t_n) - N(t_{n-1} + h_{n-1}) = 0, N(J_n) = 1, N(t) - N(t_n + h_n) = 0) \\ = \quad & P(N(t_1) = 0) P(N(J_1) = 1) P(N(t_2) - N(t_1 + h_1) = 0) \times \\ & \cdots \times P(N(J_n) = 1) P(N(t) - N(t_n + h_n) = 0) \end{aligned}$$

$$
\begin{aligned}
&= P(N(t_1)=0)P(N(h_1)=1)P(N(t_2-t_1-h_1)=0)\times \\
&\quad \cdots \times P(N(h_n)=1)P(N(t-(t_n+h_n))=0) \\
&= e^{-\lambda t_1}\times e^{-\lambda h_1}\lambda h_1 \times e^{-\lambda[t_2-(t_1+h_1)]}\times \\
&\quad \cdots \times e^{-\lambda[t_n-(t_{n-1}+h_{n-1})]}\times e^{-\lambda h_n}\lambda h_n \times e^{-\lambda(t-(t_n+h_n))} \\
&= \lambda^n e^{-\lambda t}\prod_{i=1}^n h_i\,.
\end{aligned}
$$

This implies

$$P(T_1\in J_1,\ldots,T_n\in J_n \mid N(t)=n)=\frac{n!}{t^n}\prod_{i=1}^n h_i\,.$$

The conditional densities are obtained by dividing both sides by $\prod_{i=1}^n h_i$ and taking the limit for $\max_{1\le i\le n} h_i\to 0$, yielding

$$f_{T_1,\ldots,T_n|N(t)}(t_1,\ldots,t_n|n)=\frac{n!}{t^n}\,,\quad 0<t_1<\cdots<t_n<t\,. \tag{4.4}$$

It follows from (4.3) that (4.4) is the density of the order statistics of n iid uniform rvs on $(0,t)$. □

The following concept is called *quantile transformation.* It is extremely useful since it often reduces a problem concerning order statistics to one concerning the corresponding order statistics from a uniform sample. The proof follows immediately from the definition of the uniform distribution.

Lemma 4.1.9 (Quantile transformation)
Let $X_1,\ldots,X_n$ be iid with df F. Furthermore, let $U_1,\ldots,U_n$ be iid rvs uniformly distributed on $(0,1)$ and denote by $U_{n,n}<\cdots<U_{1,n}$ the corresponding order statistics. Then the following results hold:

(a) $F^{\leftarrow}(U_1)\stackrel{d}{=}X_1$.

(b) *For every $n\in\mathbb{N}$,*

$$(X_{1,n},\ldots,X_{n,n})\stackrel{d}{=}\left(F^{\leftarrow}(U_{1,n}),\ldots,F^{\leftarrow}(U_{n,n})\right)\,.$$

(c) *The rv $F(X_1)$ has a uniform distribution on $(0,1)$ if and only if F is a continuous function.* □

Example 4.1.10 (Simulation of upper order statistics)
The quantile transformation above links the uniform distribution to some general distribution F. An immediate application of this result is random number generation. For instance, exponential random numbers can be obtained from uniform random numbers by the transformation $E_1 = -\ln(1-U_1)$. Simulation studies are widely used in an increasing number of applications. A simple algorithm for simulating order statistics of exponentials can be based on Example 4.1.5, which says that

$$(E_{i,n} - E_{i+1,n})_{i=1,\dots,n} \stackrel{d}{=} (i^{-1}E_i)_{i=1,\dots,n}\,,$$

with $E_{n+1,n} = 0$. This implies for the order statistics of an exponential sample that

$$(E_{i,n})_{i=1,\dots,n} \stackrel{d}{=} \left(\sum_{j=i}^{n} j^{-1}E_j\right)_{i=1,\dots,n}.$$

Order statistics and spacings of iid rvs U_i uniformly distributed on $(0,1)$ and standard exponential rvs E_i are linked by the following representations; see e.g. Reiss [526], Theorem 1.6.7 and Corollary 1.6.9. We write $\Gamma_n = E_1 + \cdots + E_n$, then

$$(U_{1,n}, U_{2,n}, \dots, U_{n,n}) \stackrel{d}{=} \left(\frac{\Gamma_n}{\Gamma_{n+1}}, \frac{\Gamma_{n-1}}{\Gamma_{n+1}}, \dots, \frac{\Gamma_1}{\Gamma_{n+1}}\right),$$

and

$$(1-U_{1,n}\,, U_{1,n}-U_{2,n}\,, \dots, U_{n,n}) \stackrel{d}{=} \left(\frac{E_{n+1}}{\Gamma_{n+1}}, \dots, \frac{E_1}{\Gamma_{n+1}}\right).$$

The four distributional identities above provide simple methods for generating upper order statistics or spacings of the exponential or uniform distribution. A statement for general F is given in (4.6) below. For more sophisticated methods based on related ideas we refer to Gerontidis and Smith [259] or Ripley [542], Section 4.1, and references therein. □

Example 4.1.11 (The limit of the ratio of two successive order statistics)
Consider $F \in \text{MDA}(\Phi_\alpha)$, equivalently $\overline{F} \in \mathcal{R}_{-\alpha}$, for some $\alpha > 0$. We want to show that

$$\frac{X_{k,n}}{X_{k+1,n}} \stackrel{P}{\to} 1\,, \quad k = k(n) \to \infty\,, \quad k/n \to 0\,. \tag{4.5}$$

The latter fact will frequently be used in Chapter 6.

For the proof we conclude from Lemma 4.1.9(b) and Example 4.1.10 that

$$\begin{aligned}(X_{1,n}, \dots, X_{n,n}) &\stackrel{d}{=} (F^{\leftarrow}(U_{1,n}), \dots, F^{\leftarrow}(U_{n,n}))\\ &\stackrel{d}{=} (F^{\leftarrow}(\Gamma_n/\Gamma_{n+1}), \dots, F^{\leftarrow}(\Gamma_1/\Gamma_{n+1}))\,, \end{aligned}\tag{4.6}$$

where $\Gamma_n = E_1 + \cdots + E_n$ and the E_i are iid standard exponential rvs. Notice that (4.6) holds only for every fixed n. However, we are interested in the weak convergence result (4.5), and therefore it suffices to show (4.5) for one special version of

$$((X_{k,n})_{k=1,\ldots,n})_{n\geq 1}\,.$$

In particular, we may choose this sequence by identifying the lhs and the rhs in (4.6) not only in distribution, but pathwise. Hence we get

$$\frac{X_{k,n}}{X_{k+1,n}} = \frac{F^{\leftarrow}(\Gamma_{n-k+1}/\Gamma_{n+1})}{F^{\leftarrow}(\Gamma_{n-k}/\Gamma_{n+1})}\,. \tag{4.7}$$

Since $\overline{F} \in \mathcal{R}_{-\alpha}$,

$$F^{\leftarrow}(1-t^{-1}) = t^{1/\alpha}L(t)\,, \quad t>0\,, \tag{4.8}$$

for some $L \in \mathcal{R}_0$; see Bingham, Goldie and Teugels [72], Corollary 2.3.4. By the SLLN and since $k/n \to 0$, $\Gamma_{n-k}/\Gamma_{n+1} \stackrel{\text{a.s.}}{\to} 1$. Hence, by (4.7) and (4.8) for sufficiently large n,

$$\frac{X_{k,n}}{X_{k+1,n}} = \left(\frac{\Gamma_{n+1}-\Gamma_{n-k}}{\Gamma_{n+1}-\Gamma_{n-k+1}}\right)^{1/\alpha} \frac{L(\Gamma_{n+1}/(\Gamma_{n+1}-\Gamma_{n-k+1}))}{L(\Gamma_{n+1}/(\Gamma_{n+1}-\Gamma_{n-k}))}\,. \tag{4.9}$$

Again using the SLLN, $k \to \infty$ and $k/n \to 0$,

$$\frac{\Gamma_{n+1}-\Gamma_{n-k}}{\Gamma_{n+1}-\Gamma_{n-k+1}} \stackrel{d}{=} \frac{\Gamma_{k+1}}{\Gamma_k} \stackrel{\text{a.s.}}{\to} 1\,, \tag{4.10}$$

$$\frac{\Gamma_{n+1}-\Gamma_{n-k+1}}{\Gamma_{n+1}} = \frac{\Gamma_{n+1}-\Gamma_{n-k}}{\Gamma_{n+1}}(1+o(1)) \stackrel{\text{a.s.}}{\to} 0 \tag{4.11}$$

Relations (4.9)–(4.11) and the uniform convergence theorem for $L \in \mathcal{R}_0$ (see Theorem A3.2) prove (4.5). □

Example 4.1.12 (Asymptotic properties of the Hill estimator)
Assume X is a positive rv with regularly varying tail $\overline{F}(x) = x^{-\alpha}L_0(x)$ for some $\alpha > 0$ and $L_0 \in \mathcal{R}_0$. For applications it is important to know α. In Section 6.4.2 several estimators of α are derived and their statistical properties are studied. The most popular estimator of α was proposed by Hill [326]. It is based on the k upper order statistics of an iid sample:

$$\widehat{\alpha}_n^{-1} = \frac{1}{k-1}\sum_{i=1}^{k-1} \ln\left(\frac{X_{i,n}}{X_{k,n}}\right) = \frac{1}{k-1}\sum_{i=1}^{k-1} \ln X_{i,n} - \ln X_{k,n}\,, \tag{4.12}$$

for $k \geq 2$. We suppress the dependence on k in the notation.

There exist many variations on the theme "Hill" with $k-1$ replaced by k

(and vice versa) at different places in (4.12). By (4.5) all these estimators have the same asymptotic properties provided $k = k(n) \to \infty$ and $k/n \to 0$. We are interested in the asymptotic properties of $\widehat{\alpha}_n^{-1}$ (consistency, asymptotic normality).

By Lemma 4.1.9(b) we may and do assume that $\widehat{\alpha}_n^{-1}$ has representation

$$\widehat{\alpha}_n^{-1} = \frac{1}{k-1}\sum_{i=1}^{k-1} \ln F^{\leftarrow}(U_{i,n}) - \ln F^{\leftarrow}(U_{k,n}) \tag{4.13}$$

for an ordered sample $U_{n,n} < \cdots < U_{i,n}$ from a uniform distribution on $(0,1)$. We are interested only in asymptotic distributional properties of $\widehat{\alpha}_n^{-1}$. For this it suffices to study the distribution of $\widehat{\alpha}_n^{-1}$ at every fixed n. If one wants to study a.s. convergence results one has to consider the distribution of the whole sequence $(\widehat{\alpha}_n)$. Then representation (4.13) is not useful. (Lemma 4.1.9(b) is applicable only for a finite vector of order statistics.) Regular variation of $\overline{F}$ implies that

$$F^{\leftarrow}(y) = (1-y)^{-1/\alpha} L\left((1-y)^{-1}\right) , \quad y \in (0,1) ,$$

for some $L \in \mathcal{R}_0$; see Bingham et al. [72], Corollary 2.3.4. Combining (4.13) with the representation of $(U_{k,n})$ via iid standard exponential rvs E_i (see Example 4.1.10) and writing

$$\Gamma_n = E_1 + \cdots + E_n , \quad n \geq 1 ,$$

we obtain the representation

$$\begin{aligned}
\widehat{\alpha}_n^{-1} &= \frac{1}{k-1}\sum_{i=1}^{k-1} \ln \left[\left(1-\frac{\Gamma_{n-i+1}}{\Gamma_{n+1}}\right)^{-1/\alpha} L\left(\left(1-\frac{\Gamma_{n-i+1}}{\Gamma_{n+1}}\right)^{-1}\right)\right] \\
&\qquad - \ln \left[\left(1-\frac{\Gamma_{n-k+1}}{\Gamma_{n+1}}\right)^{-1/\alpha} L\left(\left(1-\frac{\Gamma_{n-k+1}}{\Gamma_{n+1}}\right)^{-1}\right)\right] \\
&= \frac{1}{\alpha}\frac{1}{k-1}\sum_{i=1}^{k-1} \ln \frac{\Gamma_{n+1}-\Gamma_{n-k+1}}{\Gamma_{n+1}-\Gamma_{n-i+1}} \\
&\qquad + \frac{1}{k-1}\sum_{i=1}^{k-1} \ln \frac{L(\Gamma_{n+1}/(\Gamma_{n+1}-\Gamma_{n-i+1}))}{L(\Gamma_{n+1}/(\Gamma_{n+1}-\Gamma_{n-k+1}))} \\
&= \beta_n^{(1)} + \beta_n^{(2)} .
\end{aligned} \tag{4.14}$$

The leading term in this decomposition is $\beta_n^{(1)}$. It determines the asymptotic properties of the estimator $\widehat{\alpha}_n^{-1}$. Thus we first study $\beta_n^{(1)}$. Again applying Example 4.1.10 we see that for every $k \geq 2$,

$$\left(\frac{\Gamma_{n+1}-\Gamma_{n-i+1}}{\Gamma_{n+1}-\Gamma_{n-k+1}}\right)_{i=1,\ldots,k-1} \stackrel{d}{=} \left(\frac{\Gamma_i}{\Gamma_k}\right)_{i=1,\ldots,k-1} \stackrel{d}{=} (U_{k-i,k-1})_{i=1,\ldots,k-1}\,.$$

Hence, for iid (U_i) uniform on $(0,1)$,

$$\beta_n^{(1)} \stackrel{d}{=} -\frac{1}{\alpha}\frac{1}{k-1}\sum_{i=1}^{k-1}\ln U_i \stackrel{d}{=} \frac{1}{\alpha}\frac{1}{k-1}\sum_{i=1}^{k-1} E_i\,.$$

We immediately conclude from the SLLN and the CLT for iid rvs that

$$\beta_n^{(1)} \stackrel{P}{\to} \frac{1}{\alpha}\,,$$

$$\alpha\sqrt{k}\left(\beta_n^{(1)}-\frac{1}{\alpha}\right) \stackrel{d}{\to} \Phi\,,$$

where Φ is the standard normal distribution, provided that $k=k(n)\to\infty$. Notice that $\beta_n^{(2)}$ vanishes if the relation $\overline{F}(x)=cx^{-\alpha}$ holds for large x and constant $c>0$, and then the limit theory for $\beta_n^{(1)}$ and for $\widehat{\alpha}_n^{-1}$ is the same. However, for real data one can never assume that the tail $\overline{F}$ has exact power law behaviour. Therefore one also has to understand the limit behaviour of the second term $\beta_n^{(2)}$ in the decomposition (4.14). Recall from the representation theorem (Theorem A3.3) that the slowly varying function L can be written in the form

$$L(x)=c(x)\exp\left\{\int_z^x \frac{\delta(u)}{u}\,du\right\}\,,\quad x\ge z\,, \tag{4.15}$$

for some $z>0$, functions $c(x)\to c_0>0$ and $\delta(x)\to 0$ as $x\to\infty$. With this representation we obtain

$$\begin{aligned}\beta_n^{(2)} &= \frac{1}{k-1}\sum_{i=1}^{k-1}\left(\ln\frac{c\big(\Gamma_{n+1}/(\Gamma_{n+1}-\Gamma_{n-i+1})\big)}{c\big(\Gamma_{n+1}/(\Gamma_{n+1}-\Gamma_{n-k+1})\big)}\right.\\ &\qquad\left.+\int_{(1-\Gamma_{n-k+1}/\Gamma_{n+1})^{-1}}^{(1-\Gamma_{n-i+1}/\Gamma_{n+1})^{-1}}\frac{\delta(u)}{u}\,du\right)\\ &= \beta_n^{(3)}+\beta_n^{(4)}\,.\end{aligned}$$

If we assume that $k=k(n)\to\infty$ and $k/n\to 0$ then, by the SLLN, uniformly for $i\le k$,

$$\frac{\Gamma_{n-i+1}}{\Gamma_{n+1}} = \frac{n^{-1}\Gamma_{n-i+1}}{n^{-1}\Gamma_{n+1}} \stackrel{\text{a.s.}}{\to} 1\,.$$

This immediately implies that $\beta_n^{(3)}\stackrel{\text{a.s.}}{\to} 0$. Set

$$C_n=\sup\left\{|\delta(u)| : u\ge (1-\Gamma_{n-k+1}/\Gamma_{n+1})^{-1}\right\}$$

and notice that, by the remark above, $C_n \xrightarrow{\text{a.s.}} 0$. Thus

$$\begin{aligned}\beta_n^{(4)} &\le C_n \frac{1}{k-1}\sum_{i=1}^{k-1}\int_{(1-\Gamma_{n-k+1}/\Gamma_{n+1})^{-1}}^{(1-\Gamma_{n-i+1}/\Gamma_{n+1})^{-1}} \frac{1}{u}\,du \\ &= C_n\,\alpha\,\beta_n^{(1)}\,.\end{aligned}$$

This shows that $\beta_n^{(2)} \xrightarrow{P} 0$ provided $k = k(n) \to \infty$ and $k/n \to 0$. This, together with $\beta_n^{(1)} \xrightarrow{P} \alpha^{-1}$, proves the consistency of Hill's estimator $\widehat{\alpha}_n^{-1}$ whatever the slowly varying function L_0 in the tail $\overline{F}(x) = x^{-\alpha}L_0(x)$. Under conditions on the growth of $k(n)$ (e.g. $k(n) = [n^\gamma]$ for some $\gamma \in (0,1)$) it can even be shown that $\widehat{\alpha}_n^{-1} \xrightarrow{\text{a.s.}} \alpha$. We refer to Mason [445], whose arguments we followed closely in the discussion above, and to Deheuvels, Häusler and Mason [170].

From the course of the proof it is clear that, in order to show a CLT for $\widehat{\alpha}_n^{-1}$, one has to prove $\sqrt{k}\beta_n^{(2)} \xrightarrow{P} 0$. This means one has to impose some condition on the decay to zero of the function $\delta(\cdot)$ in the representation (4.15). Alternatively, one needs some regular variation condition with remainder term which has to be specified. We do not intend to go into detail here, but we refer to the discussion in Section 6.4.2 on the Hill estimator and related topics where sufficient conditions for the asymptotic normality (also under dependence of the X_n) are given.

Finally, we want to illustrate that the Hill estimator can perform very poorly if the slowly varying function in the tail is far away from a constant. For the sake of argument, assume that

$$F^{\leftarrow}(y) = (1-y)^{-1/\alpha}(-\ln(1-y))\,, \quad y \in (0,1)\,. \tag{4.16}$$

Observe that

$$\left(\frac{\Gamma_{n+1}-\Gamma_{n-i+1}}{\Gamma_{n+1}}\right)_{i=1,\ldots,k-1} \stackrel{d}{=} \left(\frac{\Gamma_i}{\Gamma_{n+1}}\right)_{i=1,\ldots,k-1}\,.$$

Then

$$\begin{aligned}\beta_n^{(2)} &= \frac{1}{k-1}\sum_{i=1}^{k-1}\ln\frac{\ln(1-\Gamma_{n-i+1}/\Gamma_{n+1})}{\ln(1-\Gamma_{n-k+1}/\Gamma_{n+1})} \\ &\stackrel{d}{=} \frac{1}{k-1}\sum_{i=1}^{k-1}\ln\frac{\ln(\Gamma_i/\Gamma_{n+1})}{\ln(\Gamma_k/\Gamma_{n+1})} \\ &= \frac{1}{k-1}\sum_{i=1}^{k-1}\ln\frac{1-\ln\Gamma_i/\ln\Gamma_{n+1}}{1-\ln\Gamma_k/\ln\Gamma_{n+1}}\,.\end{aligned}$$

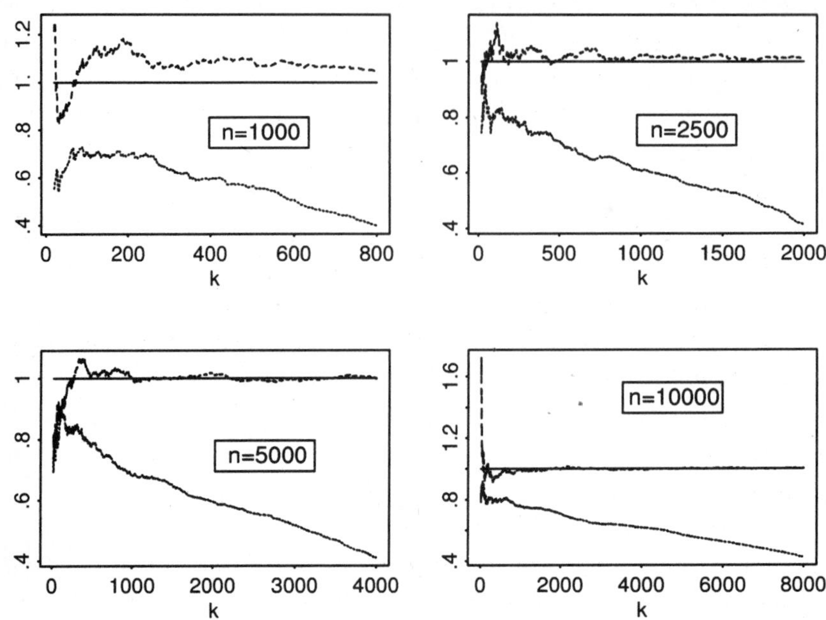

Figure 4.1.13 *A "Hill horror plot": the Hill estimator $\widehat{\alpha}_n^{-1}$ from n iid realisations with distribution tail $\overline{F}_1(x) = 1/x$ (top line) and $\overline{F}_2(x) = 1/(x \ln x)$ (bottom line). The solid line corresponds to $\alpha = 1$. The performance of the Hill estimate for F_2 is very poor. The value k is the number of upper order statistics used for the construction of the Hill estimator* (4.12).

The SLLN and a Taylor–expansion argument applied to the last relation show that, with probability 1, the rhs can be estimated as follows

$$
\begin{aligned}
&= -(1+o(1))\frac{1}{\ln \Gamma_{n+1}}\frac{1}{k-1}\sum_{i=1}^{k-1} \ln \frac{\Gamma_i}{\Gamma_k} \\
&\stackrel{d}{=} -(1+o(1))\frac{1}{\ln n}\frac{1}{k-1}\sum_{i=1}^{k-1} \ln U_i \\
&\stackrel{d}{=} (1+o(1))(\ln n)^{-1}(k-1)^{-1}\Gamma_{k-1} = O((\ln n)^{-1})\,.
\end{aligned}
$$

This means that $\beta_n^{(2)} \stackrel{P}{\to} 0$ at a logarithmic rate. Moreover, if we wanted to construct asymptotic confidence bands via a CLT for the Hill estimator, we would have to compensate for the (essentially $(1/\ln n)$–term) $\beta_n^{(2)}$ in the centring constants. Thus the centring constants in the CLT would depend on the (usually unknown) slowly varying function L. In other words, *there is no standard CLT for $\widehat{\alpha}_n^{-1}$ in the class of regularly varying tails.* These two facts concerning the quality of the Hill estimator should be a warning to everybody applying tail estimates. We also include a "Hill horror plot" (Figure 4.1.13) for the situation as in (4.16). For a further discussion of the Hill estimator we refer to Section 6.4.2. □

The asymptotic properties of the upper order statistic $X_{k,n}$ naturally enter when one studies tail and quantile estimators; see Chapter 6.

Proposition 4.1.14 (Almost sure convergence of order statistics)
Let F be a df with right (left) endpoint $x_F \le \infty$ ($\widetilde{x}_F \ge -\infty$) and $(k(n))$ a non-decreasing integer sequence such that

$$\lim_{n\to\infty} n^{-1}k(n) = c \in [0,1]\,.$$

(*a*) *Then $X_{k(n),n} \stackrel{\text{a.s.}}{\to} x_F$ ($\widetilde{x}_F$) according as $c = 0$ ($c = 1$).*

(*b*) *Assume that $c \in (0,1)$ is such that there is a unique solution $x(c)$ of the equation $\overline{F}(x) = c$. Then*

$$X_{k(n),n} \stackrel{\text{a.s.}}{\to} x(c).$$

Proof. We restrict ourselves to showing (b), the proof for (a) goes along the same lines. By (4.1) and the SLLN,

$$\begin{aligned} P(X_{k(n),n} \le x \quad \text{i.o.}) &= P\left(\frac{n}{k(n)}\frac{1}{n}\sum_{i=1}^{n} I_{\{X_i > x\}} < 1 \quad \text{i.o.}\right) \\ &= P\left(\overline{F}(x)(1+o(1)) < c \quad \text{i.o.}\right)\,. \end{aligned}$$

The latter probability is 0 or 1 according as $x < x(c)$ or $x > x(c)$. Hence $\liminf_{n\to\infty} X_{k(n),n} = x(c)$ a.s. In an analogous way one can show that the relation $\limsup_{n\to\infty} X_{k(n),n} = x(c)$ a.s. holds. This proves the proposition.□

Notes and Comments

A standard book on order statistics is David [156], while a more recent one is Arnold, Balakrishnan and Nagaraja [20]. Empirical distributions and processes are basic to all this material. Hence books such as Pollard [504], Shorack and Wellner [579], and van der Vaart and Wellner [628] provide the fundamentals for this section as well as others, and indeed go far beyond. Reiss [526] investigates in particular the link with statistical procedures based on extreme value theory in much greater detail. The latter reference also contains a wealth of interesting bibliographical notes. Two seminal papers on spacings were written by Pyke [518, 519]. They had a great impact on the field and are still worth reading.

4.2 The Limit Distribution of Upper Order Statistics

Let $X_1, \ldots, X_n$ be iid with df F. Recall from Proposition 3.1.1 that for a sequence (u_n) of thresholds and $0 \le \tau \le \infty$,

$$\lim_{n\to\infty} P(X_{1,n} \le u_n) = e^{-\tau} \quad \Leftrightarrow \quad \lim_{n\to\infty} n\overline{F}(u_n) = \tau\,. \tag{4.17}$$

In this section we ask:

Can we extend relation (4.17) *to any upper order statistic* $X_{k,n}$ *for a fixed* $k \in \mathbb{N}$*?*

Or even

Can we obtain joint limit probabilities for a fixed number k *of upper order statistics* $X_{k,n}, \ldots, X_{1,n}$*?*

Consider for $n \in \mathbb{N}$ the number of exceedances of the threshold u_n by $X_1, \ldots, X_n$:

$$B_n = \sum_{i=1}^{n} I_{\{X_i > u_n\}}\,.$$

Then B_n is a binomial rv with parameters n and $\overline{F}(u_n)$. In Proposition 4.1.2 we used this quantity for finite n to calculate the df of the kth upper order statistic. Basic to the following result is the fact that exceedances $\{X_i > u_n\}$ tend to become rarer when we raise the threshold. On the other hand, we raise the sample size. We balance two effects so that $EB_n = n\overline{F}(u_n) \to \tau$ as $n \to \infty$, and hence immediately the classical theorem of Poisson applies: $B_n \stackrel{d}{\to} Poi(\tau)$. The thresholds u_n are chosen such that the expected number of exceedances converges. The following result shows that $n\overline{F}(u_n) \to \tau$ is also necessary for the Poisson approximation to hold.

Theorem 4.2.1 (Limit law for the number of exceedances)
Suppose (u_n) *is a sequence in* $\mathbb{R}$ *such that* $n\overline{F}(u_n) \to \tau$ *for some* $\tau \in [0, \infty]$ *as* $n \to \infty$*. Then*

$$\lim_{n\to\infty} P(B_n \le k) = e^{-\tau} \sum_{r=0}^{k} \frac{\tau^r}{r!}\,, \qquad k \in \mathbb{N}_0\,. \tag{4.18}$$

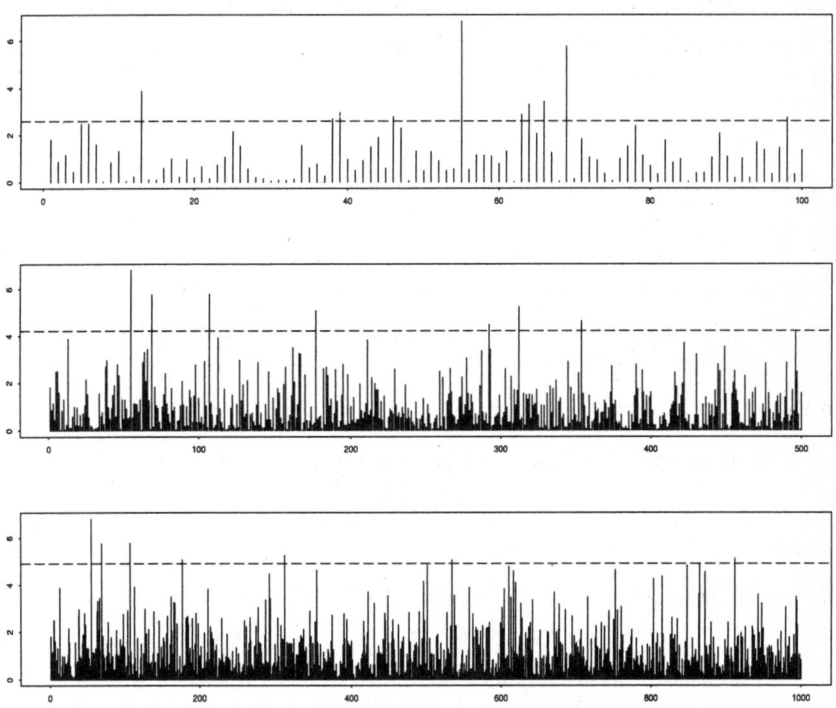

Figure 4.2.2 *Visualisation of the Poisson approximation for extremes of iid standard exponential rvs. The threshold increases with the sample size* $n = 100, 500, 1\,000$. *Notice that the first sample also appears at the beginning of the second and the second at the beginning of the third.*

For $\tau = 0$ *we interpret the rhs as* 1, *for* $\tau = \infty$ *as* 0.

If (4.18) *holds for some* $k \in \mathbb{N}_0$, *then* $n\overline{F}(u_n) \to \tau$ *as* $n \to \infty$, *and thus* (4.18) *holds for all* $k \in \mathbb{N}_0$.

Proof. For $\tau \in (0, \infty)$, sufficiency is simply the Poisson limit theorem as indicated above. For $\tau = 0$, we have

$$P(B_n \leq k) \geq P(B_n = 0) = \left(1 - \overline{F}(u_n)\right)^n = \left(1 + o\left(\frac{1}{n}\right)\right)^n \to 1\,.$$

For $\tau = \infty$ we have for arbitrary $\theta > 0$ that $n\overline{F}(u_n) \geq \theta$ for large n. Since the binomial df is decreasing in θ, we obtain

$$P(B_n \leq k) \leq \sum_{r=0}^{k} \binom{n}{r} \left(\frac{\theta}{n}\right)^r \left(1 - \frac{\theta}{n}\right)^{n-r}.$$

Thus for k fixed,

$$\limsup_{n\to\infty} P(B_n \le k) \le e^{-\theta} \sum_{r=0}^{k} \frac{\theta^r}{r!} \to 0, \qquad \theta \to \infty.$$

Hence $P(B_n \le k) \to 0$ as $n \to \infty$.

For the converse assume that (4.18) holds for some $k \in \mathbb{N}_0$, but $n\overline{F}(u_n) \not\to \tau$. Then there exists some $\tau' \ne \tau$ in $[0,\infty]$ and a subsequence (n_k) such that $n_k\overline{F}(u_{n_k}) \to \tau'$ as $k \to \infty$, and thus B_{n_k} converges weakly to a Poisson rv with parameter τ', contradicting (4.18). □

The Poisson approximation (4.18) allows us to derive asymptotics for the kth order statistic. The definition of B_n and (4.1) imply

$$P(B_n < k) = P(X_{k,n} \le u_n), \quad 1 \le k \le n, \tag{4.19}$$

which by (4.18) gives immediately the following result.

Theorem 4.2.3 (Limit probabilities for an upper order statistic)
Suppose (u_n) is a sequence in $\mathbb{R}$ such that $n\overline{F}(u_n) \to \tau \in [0,\infty]$ as $n \to \infty$. Then

$$\lim_{n\to\infty} P(X_{k,n} \le u_n) = e^{-\tau} \sum_{r=0}^{k-1} \frac{\tau^r}{r!}, \quad k \in \mathbb{N}. \tag{4.20}$$

For $\tau = 0$ we interpret the rhs as 1 and for $\tau = \infty$ as 0.

If (4.20) holds for some $k \in \mathbb{N}$, then $n\overline{F}(u_n) \to \tau$ as $n \to \infty$, and thus (4.20) holds for all $k \in \mathbb{N}$. □

For $u_n = c_n x + d_n$ and $\tau = \tau(x) = -\ln H(x)$ as in Proposition 3.3.2 we obtain the following corollary:

Corollary 4.2.4 (Limit distribution of an upper order statistic)
Suppose $F \in \mathrm{MDA}(H)$ with norming constants $c_n > 0$ and $d_n \in \mathbb{R}$. Define

$$H^{(k)}(x) = H(x) \sum_{r=0}^{k-1} \frac{(-\ln H(x))^r}{r!}, \quad x \in \mathbb{R}.$$

For x such that $H(x) = 0$ we interpret $H^{(k)}(x) = 0$. Then for each $k \in \mathbb{N}$,

$$\lim_{n\to\infty} P\left(c_n^{-1}(X_{k,n} - d_n) \le x\right) = H^{(k)}(x). \tag{4.21}$$

On the other hand, if for some $k \in \mathbb{N}$

$$\lim_{n\to\infty} P\left(c_n^{-1}(X_{k,n} - d_n) \le x\right) = G(x), \quad x \in \mathbb{R},$$

for a non-degenerate df G, then $G = H^{(k)}$ for some extreme value distribution H and (4.21) holds for all $k \in \mathbb{N}$. □

Example 4.2.5 (Upper order statistics of the Gumbel distribution)
By partial integration,

$$H^{(k)}(x) = \frac{1}{(k-1)!} \int_{-\ln H(x)}^{\infty} e^{-t}\, t^{k-1}\, dt = \Gamma_k\left(-\ln H(x)\right), \quad x \in \mathbb{R},$$

where Γ_k denotes the incomplete gamma function. In particular, if H is the Gumbel distribution Λ, then

$$\Lambda^{(k)}(x) = \frac{1}{(k-1)!} \int_{e^{-x}}^{\infty} e^{-t}\, t^{k-1}\, dt = P\left(\sum_{i=1}^{k} E_i > e^{-x}\right)$$

for $E_1, \ldots, E_k$ iid standard exponential rvs, where we used the well-known fact that $\sum_{i=1}^{k} E_i$ is $\Gamma(k,1)$-distributed. Hence, if $Y^{(k)}$ has df $\Lambda^{(k)}$, then $Y^{(k)} \stackrel{d}{=} -\ln \sum_{i=1}^{k} E_i$. □

The limit distribution of the kth upper order statistic was obtained by considering the number of exceedances of a level u_n by $X_1, \ldots, X_n$. Similar arguments can be adapted to prove convergence of the joint distribution of several upper order statistics.

To this end let for $k \in \mathbb{N}$ the levels $u_n^{(k)} \le \cdots \le u_n^{(1)}$ satisfy

$$\lim_{n\to\infty} n\overline{F}(u_n^{(i)}) = \tau_i, \quad i = 1, \ldots, k, \tag{4.22}$$

where $0 \le \tau_1 \le \tau_2 \le \cdots \le \tau_k \le \infty$, and define

$$B_n^{(j)} = \sum_{i=1}^{n} I_{\left\{X_i > u_n^{(j)}\right\}}, \quad j = 1, \ldots, k,$$

i.e. $B_n^{(j)}$ is the number of exceedances of $u_n^{(j)}$ by $X_1, \ldots, X_n$.

Theorem 4.2.6 (Multivariate limit law for the number of exceedances)
Suppose that the sequences $(u_n^{(j)})$ satisfy (4.22) for $j = 1, \ldots, k$. Then for $l_1, \ldots, l_k \in \mathbb{N}_0$,

$$\begin{aligned} &\lim_{n\to\infty} P\left(B_n^{(1)} = l_1, B_n^{(2)} = l_1 + l_2, \ldots, B_n^{(k)} = l_1 + \cdots + l_k\right) \\ &= \frac{\tau_1^{l_1}}{l_1!} \frac{(\tau_2 - \tau_1)^{l_2}}{l_2!} \cdots \frac{(\tau_k - \tau_{k-1})^{l_k}}{l_k!}\, e^{-\tau_k}. \end{aligned} \tag{4.23}$$

The rhs is interpreted as 0 if $\tau_k = \infty$.

Proof. We write $p_{n,j} = \overline{F}(u_n^{(j)})$. Using the defining properties of the multinomial distribution, we find that the lhs probability of (4.23) equals

$$\binom{n}{l_1} p_{n,1}^{l_1} \binom{n-l_1}{l_2} (p_{n,2} - p_{n,1})^{l_2} \cdots$$

$$\cdots \binom{n - l_1 - \cdots - l_{k-1}}{l_k} (p_{n,k} - p_{n,k-1})^{l_k} (1 - p_{n,k})^{n-l_1-\cdots-l_k} .$$

If $\tau_k < \infty$ then we obtain from (4.22) that

$$\binom{n}{l_1} p_{n,1}^{l_1} \sim \frac{(n\, p_{n,1})^{l_1}}{l_1!} \to \frac{\tau_1^{l_1}}{l_1!} ,$$

$$\binom{n - l_1 - \cdots - l_{i-1}}{l_i} (p_{n,1} - p_{n,i-1})^{l_i} \sim \frac{(np_{n,i} - np_{n,i-1})^{l_i}}{l_i!}$$

$$\to \frac{(\tau_i - \tau_{i-1})^{l_1}}{l_i!} , \quad \text{for } 2 \le i \le k ,$$

$$(1 - p_{n,k})^{n-l_1-\cdots-l_k} \sim \left(1 - \frac{np_{n,k}}{n}\right)^n \to e^{-\tau_k} ,$$

giving (4.23).
If $\tau_k = \infty$, the probability in (4.23) does not exceed $P(B_n^{(k)} = \sum_{i=1}^k l_i)$. By Theorem 4.2.1, the latter converges to 0. □

Clearly, as in (4.19),

$$\begin{aligned} & P\left(X_{1,n} \le u_n^{(1)}, \ldots, X_{k,n} \le u_n^{(k)}\right) \\ = \; & P\left(B_n^{(1)} = 0,\, B_n^{(2)} \le 1, \ldots, B_n^{(k)} \le k-1\right) , \end{aligned} \tag{4.24}$$

and thus the joint asymptotic distribution of the k upper order statistics can be obtained directly from Theorem 4.2.6. In particular, if $c_n^{-1}(X_{1,n} - d_n)$ converges weakly, then so does the vector

$$\left(c_n^{-1}\left(X_{1,n} - d_n\right), \ldots, c_n^{-1}\left(X_{k,n} - d_n\right)\right) .$$

Although for small k the joint limit distribution of the k upper order statistics can easily be derived from (4.24) and Theorem 4.2.6, the general case is rather complicated. If the df F is absolutely continuous with density f satisfying certain regularity conditions the following heuristic argument can be made precise (for details see Reiss [526], Theorem 5.3.4): suppose $F \in \mathrm{MDA}(H)$ with density f, then the df of the maximum $F^n(c_n x + d_n) = P(c_n^{-1}(X_{1,n} - d_n) \le x)$ has also a density such that for almost all $x \in \mathbb{R}$,

$$nc_n f(c_n x + d_n) F^{n-1}(c_n x + d_n) \sim nc_n f(c_n x + d_n) H(x) \to h(x),$$

where h is the density of the extreme value distribution H. Furthermore, for $k \in \mathbb{N}$ the weak limit of the random vector $(c_n^{-1}(X_{j,n} - d_n))_{j=1,\dots,k}$ has, by Theorem 4.1.3, the density

$$\lim_{n\to\infty} F^{n-k}(c_n x_k + d_n) \prod_{j=1}^{k} \Big((n-j+1)\, c_n\, f(c_n x_j + d_n) \Big)$$

$$= \quad H(x_k) \prod_{j=1}^{k} \frac{h(x_j)}{H(x_j)}, \qquad x_k < \cdots < x_1. \tag{4.25}$$

Definition 4.2.7 (k–dimensional H–extremal variate)
For any extreme value distribution H with density h define for $x_k < \cdots < x_1$ in the support of H

$$h^{(k)}(x_1,\dots,x_k) = H(x_k) \prod_{j=1}^{k} \frac{h(x_j)}{H(x_j)}.$$

A vector $\left(Y^{(1)},\dots,Y^{(k)}\right)$ of rvs with joint density $h^{(k)}$ is called a k–dimensional H–extremal variate. □

The heuristic argument (4.25) can be made precise and formulated as follows:

Theorem 4.2.8 (Joint limit distribution of k upper order statistics)
Assume that $F \in$ MDA(H) with norming constants $c_n > 0$ and $d_n \in \mathbb{R}$. Then, for every fixed $k \in \mathbb{N}$,

$$\left(c_n^{-1}(X_{i,n} - d_n)\right)_{i=1,\dots,k} \xrightarrow{d} (Y^{(i)})_{i=1,\dots,k}, \qquad n \to \infty,$$

where $(Y^{(1)},\dots,Y^{(k)})$ is a k–dimensional H–extremal variate. □

Example 4.2.9 (Density of a k–dimensional H–extremal variate)

$$H = \Phi_\alpha: \quad \varphi_\alpha(x_1,\cdots,x_k) \;=\; \alpha^k \exp\Big\{ -x_k^{-\alpha} - (\alpha+1)\sum_{j=1}^{k} \ln x_j \Big\},$$
$$0 < x_k < \cdots < x_1,$$

$$H = \Psi_\alpha: \quad \psi_\alpha(x_1,\cdots,x_k) \;=\; \alpha^k \exp\Big\{ -(-x_k)^\alpha + (\alpha-1)\sum_{j=1}^{k} \ln(-x_j) \Big\},$$
$$x_k < \cdots < x_1 < 0,$$

$$H = \Lambda: \quad \lambda(x_1,\dots,x_k) \;=\; \exp\Big\{ -e^{-x_k} - \sum_{j=1}^{k} x_j \Big\}, \quad x_k < \cdots < x_1.$$

□

In Example 4.1.5 we investigated the spacings of an exponential sample. Now we may ask

What is the joint limit df of the spacings of a sample of extremal rvs?

Example 4.2.10 (Spacings of Gumbel variables)
The exponential distribution is in MDA(Λ) (see Example 3.2.7) and hence for iid standard exponential rvs $E_1, \ldots, E_n$ we obtain

$$(E_{i,n} - \ln n)_{i=1,\ldots,k+1} \xrightarrow{d} (Y^{(i)})_{i=1,\ldots,k+1}\,, \qquad n \to \infty\,,$$

where $(Y^{(1)}, \ldots, Y^{(k+1)})$ is the $(k+1)$–dimensional Λ–extremal variate with density

$$h^{(k+1)}(x_1, \ldots, x_{k+1}) = \exp\left\{-e^{-x_{k+1}} - \sum_{i=1}^{k+1} x_i\right\}, \quad x_{k+1} < \cdots < x_1\,. \tag{4.26}$$

The continuous mapping theorem (Theorem A2.6) implies for exponential spacings that

$$\begin{aligned}(E_{i,n} - E_{i+1,n})_{i=1,\ldots,k} &= ((E_{i,n} - \ln n) - (E_{i+1,n} - \ln n))_{i=1,\ldots,k} \\ &\xrightarrow{d} \left(Y^{(i)} - Y^{(i+1)}\right)_{i=1,\ldots,k}\,, \quad n \to \infty\,.\end{aligned}$$

By Example 4.1.5 we obtain the representation

$$\left(Y^{(i)} - Y^{(i+1)}\right)_{i=1,\ldots,k} \stackrel{d}{=} \left(i^{-1}\,E_i\right)_{i=1,\ldots,k} \tag{4.27}$$

for iid standard exponential rvs $E_1, \ldots, E_k$. □

Corollary 4.2.11 (Joint limit distribution of upper spacings in MDA(Λ))
Suppose $F \in$ MDA(Λ) with norming constants $c_n > 0$, then

$$(a) \quad \left(c_n^{-1}\,(X_{i,n} - X_{i+1,n})\right)_{i=1,\ldots,k} \xrightarrow{d} \left(i^{-1}\,E_i\right)_{i=1,\ldots,k} \quad \text{for } k \geq 1\,,$$

$$(b) \quad c_n^{-1}\left(\sum_{i=1}^{k} X_{i,n} - k\,X_{k+1,n}\right) \xrightarrow{d} \sum_{i=1}^{k} E_i \qquad \text{for } k \geq 2\,,$$

where $E_1, \ldots, E_k$ are iid standard exponential rvs.

Proof. (a) This follows by the same argument as for the exponential rvs in Example 4.2.10.
(b) We apply the continuous mapping theorem (Theorem A2.6):

$$
\begin{aligned}
c_n^{-1}\left(\sum_{i=1}^k X_{i,n} - k\,X_{k+1,n}\right) &= c_n^{-1}\sum_{i=1}^k (X_{i,n} - X_{k+1,n}) \\
= c_n^{-1}\sum_{i=1}^k \sum_{j=i}^k (X_{j,n} - X_{j+1,n}) &= \sum_{i=1}^k c_n^{-1} i\,(X_{i,n} - X_{i+1,n}) \\
\xrightarrow{d} \sum_{i=1}^k i\left(Y^{(i)} - Y^{(i+1)}\right) &\stackrel{d}{=} \sum_{i=1}^k E_i\,.
\end{aligned}
$$

□

Example 4.2.12 (Spacings of Fréchet variables)
The joint density of the spacings of the $(k+1)$–dimensional Fréchet variate $(Y^{(1)},\dots,Y^{(k+1)})$ can also be calculated. We start with the joint density of $Y^{(1)}-Y^{(2)},\dots,Y^{(k)}-Y^{(k+1)},\ Y^{(k+1)}$. Define the transformation

$$
T\left(x_1,\dots,x_{k+1}\right) = \left(x_1-x_2, x_2-x_3,\dots,x_k-x_{k+1},x_{k+1}\right),
$$

for $x_{k+1}<\cdots<x_1$. Then $\det(\partial T(\mathbf{x})/\partial \mathbf{x}) = 1$ and for $x_1,x_2,\dots,x_{k+1}\in\mathbb{R}$ we obtain

$$
T^{-1}\left(x_1,\dots,x_{k+1}\right) = \left(\sum_{j=1}^{k+1} x_j, \sum_{j=2}^{k+1} x_j,\dots,x_k+x_{k+1},x_{k+1}\right).
$$

For the spacings of the $(k+1)$–dimensional Fréchet variate $(Y^{(1)},\dots,Y^{(k+1)})$ this yields the density

$$
\begin{aligned}
&g_{Y^{(1)}-Y^{(2)},\dots,Y^{(k)}-Y^{(k+1)},\,Y^{(k+1)}}\left(x_1,\dots,x_{k+1}\right) \\
&\quad= \alpha^{k+1}\exp\left\{-x_{k+1}^{-\alpha}\right\} x_{k+1}^{-\alpha-1}\left(x_{k+1}+x_k\right)^{-\alpha-1}\cdots\left(x_{k+1}+\cdots+x_1\right)^{-\alpha-1}
\end{aligned}
$$

for $x_1,\dots,x_{k+1}>0$.

From this density it is obvious that the spacings of the $(k+1)$–dimensional Fréchet variate are dependent. Hence such an elegant result as (4.27) cannot be expected for $F\in\mathrm{MDA}(\Phi_\alpha)$. □

By analogous calculations as above we find the joint limit density of the spacings of the upper order statistics of a sample from a df $F\in\mathrm{MDA}(\Phi_\alpha)$.

Corollary 4.2.13 (Joint limit distribution of upper spacings in $\mathrm{MDA}(\Phi_\alpha)$)
Suppose $F\in\mathrm{MDA}(\Phi_\alpha)$ with norming constants $c_n>0$. Let $(Y^{(1)},\dots,Y^{(k+1)})$ be the $(k+1)$–dimensional Fréchet variate. Then

$$(a)\quad \left(c_n^{-1}\left(X_{i,n}-X_{i+1,n}\right)\right)_{i=1,\ldots,k} \xrightarrow{d} \left(Y^{(i)}-Y^{(i+1)}\right)_{i=1,\ldots,k}\,,\quad k\geq 1\,,$$

$$(b)\quad c_n^{-1}\left(\sum_{i=1}^{k}X_{i,n}-k\,X_{k+1,n}\right) \xrightarrow{d} \sum_{i=1}^{k} i\left(Y^{(i)}-Y^{(i+1)}\right),\quad k\geq 2\,.$$

The limit variables in (a) and (b) are defined by the spacings $Y^{(1)}-Y^{(2)},\ldots,$ $Y^{(k)}-Y^{(k+1)}$ *which have joint density*

$$g_{Y^{(1)}-Y^{(2)},\ldots,Y^{(k)}-Y^{(k+1)}}(x_1,\ldots,x_k)$$

$$= \alpha^{k+1}\int_0^\infty \exp\left\{-y^{-\alpha}\right\}\left(y\,(y+x_k)\cdots(y+x_k+\cdots+x_1)\right)^{-\alpha-1}dy$$

for $x_1,\ldots,x_k>0$. □

Notes and Comments

The Poisson approximation which we applied in this section in order to prove weak limit laws for upper order statistics is a very powerful tool. Its importance in this field, particularly for the investigation of extremes of dependent sequences and stochastic processes, is uncontested. More generally, exceedances of a threshold can be modelled by a point process in the plane. This yields limit laws for maxima of stochastic sequences, allowing us to explain cluster effects in extremes of certain processes. The principle tool is weak convergence of point processes. In Chapter 5 an introduction to this important subject can be found. There also the extremal behaviour of special processes is treated.

There are many other applications of the Poisson approximation in various fields. Recent books are Aldous [7] and Barbour, Holst and Janson [43]; see also the review paper by Arratia, Goldstein and Gordon [22].

4.3 The Limit Distribution of Randomly Indexed Upper Order Statistics

In this section we compare the weak limit behaviour of a finite number of upper order statistics and of randomly indexed maxima for an iid sequence (X_n) of rvs with common df F. As usual, $(N(t))_{t\geq 0}$ is a process of integer–valued rvs which we also suppose to be independent of (X_n). We write

$$X_{n,n}\leq\cdots\leq X_{1,n}\quad\text{and}\quad X_{N(t),N(t)}\leq\cdots\leq X_{1,N(t)}$$

for the order statistics of the samples $X_1, \ldots, X_n$ and $X_1, \ldots, X_{N(t)}$, respectively, and we also use

$$M_n = X_{1,n} \quad \text{and} \quad M_{N(t)} = X_{1,N(t)}$$

for the corresponding sample maxima.

If F belongs to the maximum domain of attraction of the extreme value distribution H ($F \in \text{MDA}(H)$), there exist constants $c_n > 0$ and $d_n \in \mathbb{R}$ such that

$$c_n^{-1}\left(M_n - d_n\right) \xrightarrow{d} H\,. \tag{4.28}$$

It is a natural question to ask:

Does relation (4.28) *remain valid along the random index set* $(N(t))$*?*

From Lemma 2.5.6 we already know that (4.28) implies

$$c_{N(t)}^{-1}\left(M_{N(t)} - d_{N(t)}\right) \xrightarrow{d} H$$

provided $N(t) \xrightarrow{P} \infty$, but we want to keep the old norming sequences (c_n), (d_n) instead of the random processes $(c_{N(t)})$, $(d_{N(t)})$. This can be done under quite general conditions as we will soon see. However, the limit distribution will also then change.

We proceed as in Section 4.2. We introduce the variables

$$B_t^{(i)} = \sum_{j=1}^{N(t)} I_{\{X_j > u_t^{(i)}\}}\,, \quad i = 1, \ldots, k\,,$$

which count the number of exceedances of the (non–random) thresholds

$$u_t^{(k)} \le \cdots \le u_t^{(1)}\,, \quad t \ge 0\,, \tag{4.29}$$

by $X_1, \ldots, X_{N(t)}$. We also suppose that there exist numbers

$$0 \le \tau_1 \le \cdots \le \tau_k \le \infty$$

such that for $i = 1, \ldots, k$,

$$t\,p_{t,i} \;=\; t\,\overline{F}(u_t^{(i)}) \;\to\; \tau_i\,, \quad t \to \infty\,. \tag{4.30}$$

The following result is analogous to Theorem 4.2.6:

Theorem 4.3.1 (Multivariate limit law for the number of exceedances) *Suppose that* $(u_t^{(i)})_{t\geq 0}$, $i=1,\ldots,k$, *satisfy* (4.29) *and* (4.30). *Assume there exists a non–negative rv* Z *such that*

$$\frac{N(t)}{t} \xrightarrow{P} Z\,, \quad t\to\infty\,. \tag{4.31}$$

Then, for all integers $l_i \geq 0$, $i=1,\ldots,k$,

$$\begin{aligned}
&\lim_{t\to\infty} P\left(B_t^{(1)}=l_1, B_t^{(2)}=l_1+l_2,\ldots,B_t^{(k)}=l_1+\cdots+l_k\right)\\
&= E\left[\frac{(Z\tau_1)^{l_1}}{l_1!}\,\frac{(Z(\tau_2-\tau_1))^{l_2}}{l_2!}\cdots\frac{(Z(\tau_k-\tau_{k-1}))^{l_k}}{l_k!}\,e^{-Z\tau_k}\right].
\end{aligned}$$

The rhs is interpreted as 0 *if* $\tau_k=\infty$.

Proof. We proceed in a similar way as for the proof of Theorem 4.2.6. For the sake of simplicity we restrict ourselves to the case $k=2$. We condition on $N(t)$, use the independence of $(N(t))$ and (X_n) and apply (4.30) and (4.31):

$$\begin{aligned}
&P\left(B_t^{(1)}=l_1, B_t^{(2)}=l_1+l_2 \,\middle|\, N(t)\right) \qquad (4.32)\\
&= \binom{N(t)}{l_1} p_{t,1}^{l_1}\binom{N(t)-l_1}{l_2}(p_{t,2}-p_{t,1})^{l_2}(1-p_{t,2})^{N(t)-l_1-l_2}\\
&= (1+o_P(1))\,\frac{(N(t)p_{t,1})^{l_1}}{l_1!}\,\frac{(N(t)(p_{t,2}-p_{t,1}))^{l_2}}{l_2!}\,(1-p_{t,2})^{N(t)}\\
&= (1+o_P(1))\,\frac{\left(\frac{N(t)}{t}(tp_{t,1})\right)^{l_1}}{l_1!}\,\frac{\left(\frac{N(t)}{t}(t(p_{t,2}-p_{t,1}))\right)^{l_2}}{l_2!}\times\\
&\qquad\times\exp\left\{\frac{N(t)}{t}\left(t\ln(1-p_{t,2})\right)\right\}\\
&\xrightarrow{P} \frac{(Z\tau_1)^{l_1}}{l_1!}\,\frac{(Z(\tau_2-\tau_1))^{l_2}}{l_2!}\,e^{-Z\tau_2}\,, \quad t\to\infty\,. \qquad (4.33)
\end{aligned}$$

Notice that the expressions in (4.32) are uniformly integrable and that (4.33) is integrable. Hence we may conclude (see for instance Karr [373], Theorem 5.17) that

$$\begin{aligned}
P\left(B_t^{(1)}=l_1, B_t^{(2)}=l_1+l_2\right) &= E\left[P\left(B_t^{(1)}=l_1, B_t^{(2)}=l_1+l_2 \,\middle|\, N(t)\right)\right]\\
&\to E\left[\frac{(Z\tau_1)^{l_1}}{l_1!}\,\frac{(Z(\tau_2-\tau_1))^{l_2}}{l_2!}\,e^{-Z\tau_2}\right]
\end{aligned}$$

as $t \to \infty$, which concludes the proof. □

Now one could use the identity

$$P\left(X_{1,N(t)} \le u_t^{(1)}, \ldots, X_{k,N(t)} \le u_t^{(k)}\right)$$
$$= P\left(B_t^{(1)} = 0,\ B_t^{(2)} \le 1, \ldots, B_t^{(k)} \le k-1\right)$$

and Theorem 4.3.1 to derive the limit distribution of the vector of upper order statistics $(X_{1,N(t)}, \ldots, X_{k,N(t)})$. This, however, leads to quite complicated formulae, and so we restrict ourselves to some particular cases.

First we study the limit distribution of a single order statistic $X_{k,N(t)}$ for fixed $k \in \mathbb{N}$. For this reason we suppose that $F \in \mathrm{MDA}(H)$, i.e. (4.28) is satisfied for appropriate constants $c_n > 0$ and $d_n \in \mathbb{R}$. From Proposition 3.3.2 we know that (4.28) is equivalent to

$$\lim_{n\to\infty} n\overline{F}\,(c_n x + d_n) = -\ln H(x)\,, \quad x \in \mathbb{R}. \tag{4.34}$$

Under (4.34) it follows for every $k \in \mathbb{N}$ that the relation

$$\lim_{n\to\infty} P\left(c_n^{-1}\,(X_{k,n} - d_n) \le x\right) \;=\; \Gamma_k(-\ln H(x))\,, \qquad x \in \mathbb{R},$$

holds, where Γ_k denotes the incomplete gamma function; see Corollary 4.2.4 and Example 4.2.5. A similar statement is true for randomly indexed upper order statistics.

Theorem 4.3.2 (Limit distribution of the kth upper order statistic in a randomly indexed sample)
Suppose that $N(t)/t \xrightarrow{P} Z$ holds for a non–negative rv Z with df F_Z and that (4.34) is satisfied. Then

$$\begin{aligned}
&\lim_{n\to\infty} P\left(c_n^{-1}\left(X_{k,N(n)} - d_n\right) \le x\right) \\
&\quad= \int_0^\infty \Gamma_k(-z\ln H(x))\, dF_Z(z) \qquad (4.35)\\
&\quad= E\left[\Gamma_k\left(-\ln H^Z(x)\right)\right]\,, \qquad x \in \mathbb{R}.
\end{aligned}$$

Proof. We use the same ideas as in the proof of Theorem 4.3.1. Write

$$B_n = \sum_{j=1}^{N(n)} I_{\{X_j > c_n x + d_n\}}\,.$$

Conditioning on $N(n)$, we find that

$$
\begin{aligned}
& P\left(c_n^{-1}\left(X_{k,N(n)}-d_n\right) \le x \mid N(n)\right) \\
&= P\left(B_n \le k-1 \mid N(n)\right) \\
&= \sum_{i=0}^{k-1} \binom{N(n)}{i} \left(F\left(c_n x+d_n\right)\right)^{N(n)-i} \left(\overline{F}\left(c_n x+d_n\right)\right)^i \\
&= (1+o_P(1)) \sum_{i=0}^{k-1} \frac{1}{i!} \left(\frac{N(n)}{n}\left(n\overline{F}\left(c_n x+d_n\right)\right)\right)^i \times \\
&\qquad \times \exp\left\{\frac{N(n)}{n}\left(n \ln\left(1-\overline{F}\left(c_n x+d_n\right)\right)\right)\right\} \\
&\stackrel{P}{\rightarrow} \sum_{i=0}^{k-1} \frac{1}{i!}(-Z \ln H(x))^i e^{Z \ln H(x)} \\
&= H^Z(x) \sum_{i=0}^{k-1} \frac{\left(-\ln H^Z(x)\right)^i}{i!} .
\end{aligned}
$$

Taking expectations in the limit relation above, we arrive at (4.35). □

Example 4.3.3 Let $\tilde{N} = (\tilde{N}(t))_{t \ge 0}$ be a homogeneous Poisson process with intensity 1 and let Z be a positive rv independent of $\tilde{N}$. Then

$$
N(t) = \tilde{N}(Zt)\,, \qquad t \ge 0,
$$

defines a so–called *mixed Poisson process*. The latter class of processes has been recognized as important in insurance; see Section 1.3.3 and Grandell [284]. Notice that, conditionally upon Z, N is a homogeneous Poisson process with intensity Z. Hence, by the SLLN for renewal counting processes (Theorem 2.5.10),

$$
P\left(\frac{N(t)}{t} \rightarrow Z \,\middle|\, Z\right) = 1 \quad \text{a.s.}
$$

Thus taking expectations on both sides,

$$
P\left(\frac{N(t)}{t} \rightarrow Z\right) = 1.
$$

This shows that Theorems 4.3.1 and 4.3.2 are applicable to the order statistics of a sample indexed by a mixed Poisson process. □

For practical purposes, it often suffices to consider processes $(N(t))$ satisfying

$$
\frac{N(t)}{t} \stackrel{P}{\rightarrow} \lambda \tag{4.36}
$$

for some constant $\lambda > 0$. For example, the renewal counting processes, including the important homogeneous Poisson process, satisfy (4.36) under general conditions; see Section 2.5.2. Analogous arguments to those in Section 4.2 combined with the ones in the proof of Theorem 4.3.1 lead to the following:

Theorem 4.3.4 (Limit distribution of a vector of randomly indexed upper order statistics)
Assume that (4.36) *holds for a positive constant* λ *and that* $F \in \mathrm{MDA}(H)$ *for an extreme value distribution* H *such that* (4.28) *is satisfied. Then*

$$\left(c_n^{-1}\left(X_{i,N(n)} - d_n\right)\right)_{i=1,\ldots,k} \quad \xrightarrow{d} \quad (Y_\lambda^{(i)})_{i=1,\ldots,k}\,,$$

where $(Y_\lambda^{(1)},\ldots,Y_\lambda^{(k)})$ *denotes the* k*–dimensional extremal variate corresponding to the extreme value distribution* H^λ*. In particular,*

$$\lim_{n\to\infty} P\left(c_n^{-1}\left(X_{k,n} - d_n\right) \le x\right) = \Gamma_k\left(-\ln H^\lambda(x)\right)\,, \quad x \in \mathbb{R}. \qquad \square$$

Notes and Comments

The limit distribution of randomly indexed maxima and order statistics under general dependence assumptions between $(N(t))$ and (X_n) has been studied in Galambos [249] and in Barakat and El–Shandidy [42]. General randomly indexed sequences of rvs have been considered in Korolev [404]; see also the list of references therein.

Randomly indexed maxima and order statistics occur in a natural way when one is interested in the extreme value theory of the individual claims in an insurance portfolio up to time t. Randomly indexed order statistics are of particular interest for reinsurance where they occur explicitly as quantities in reinsurance treaties, as for instance when a reinsurer will cover the k largest claims of a company over a given period of time. This issue is discussed in more detail in Section 8.7.

4.4 Some Extreme Value Theory for Stationary Sequences

One of the natural generalisations of an iid sequence is a strictly stationary process: we say that the sequence of rvs (X_n) is *strictly stationary* if its finite–dimensional distributions are invariant under shifts of time, i.e.

$$(X_{t_1},\ldots,X_{t_m}) \stackrel{d}{=} (X_{t_1+h},\ldots,X_{t_m+h})$$

for any choice of indices $t_1 < \cdots < t_m$ and integers h; see also Appendix A2.1. It is common to define (X_n) with index set $\mathbb{Z}$. We can think of (X_n) as a time series of observations at discrete equidistant instants of time where the distribution of a block $(X_t, X_{t+1}, \ldots, X_{t+h})$ of length h is the same for all integers t.

For simplicity we use throughout the notion of a "stationary" sequence for a "strictly stationary" one. A strictly stationary sequence is naturally also *stationary in the wide sense* or *second order stationary* provided the second moment of $X = X_0$ is finite, i.e. $EX_n = EX$ for all n and $\operatorname{cov}(X_n, X_m) = \operatorname{cov}(X_0, X_{|n-m|})$ for all n and m.

It is impossible to build up a general extreme value theory for the class of all stationary sequences. Indeed, one has to specify the dependence structure of (X_n). For example, assume $X_n = X$ for all n. This relation defines a stationary sequence and

$$P(M_n \le x) = P(X \le x) = F(x)\,, \quad x \in \mathbb{R}.$$

Thus the distribution of the sample maxima can be *any* distribution F. This is not a reasonable basis for a general theory.

The other extreme of a stationary sequence occurs when the X_n are mutually independent, i.e. (X_n) is an iid sequence. In that case we studied the weak limit behaviour of the upper order statistics in Section 4.2. In particular, we know that there exist only three types of different limit laws: the *Fréchet distribution* Φ_α, the *Weibull distribution* Ψ_α and the *Gumbel distribution* Λ (Fisher–Tippett Theorem 3.2.7). The dfs of the type of $\Phi_\alpha, \Psi_\alpha, \Lambda$ are called extreme value distributions. In this section we give conditions on the stationary sequence (X_n) which ensure that its sample maxima (M_n) and the corresponding maxima $(\widetilde{M}_n)$ of an iid sequence $(\widetilde{X}_n)$ with common df $F(x) = P(\widetilde{X} \le x)$ exhibit a similar limit behaviour. We call $(\widetilde{X}_n)$ *an iid sequence associated with* (X_n) or simply *an associated iid sequence.* As before we write $F \in \operatorname{MDA}(H)$ for any of the extreme value distributions H if there exist constants $c_n > 0$ and $d_n \in \mathbb{R}$ such that $c_n^{-1}(\widetilde{M}_n - d_n) \xrightarrow{d} H$. For the derivation of the limit probability of $P(\widetilde{M}_n \le u_n)$ for a sequence of thresholds (u_n) we made heavy use of the following factorisation property:

$$\begin{aligned} P\left(\widetilde{M}_n \le u_n\right) &= P^n\left(\widetilde{X} \le u_n\right) \qquad (4.37) \\ &= \exp\left\{n \ln\left(1 - P\left(\widetilde{X} > u_n\right)\right)\right\} \\ &\approx \exp\left\{-n\overline{F}(u_n)\right\}. \end{aligned}$$

In particular, we concluded in Proposition 3.1.1 that, for any $\tau \in [0,\infty]$, $P(\widetilde{M}_n \le u_n) \to \exp\{-\tau\}$ if and only if $n\overline{F}(u_n) \to \tau \in [0,\infty]$. It is clear that we cannot directly apply (4.37) to maxima of a dependent stationary sequence. However, to overcome this problem we assume that there is *a specific type of asymptotic independence*:

Condition $D(u_n)$: *For any integers p, q and n*

$$1 \le i_1 < \cdots < i_p < j_1 < \cdots < j_q \le n$$

such that $j_1 - i_p \ge l$ we have

$$\left| P\left(\max_{i \in A_1 \cup A_2} X_i \le u_n \right) - P\left(\max_{i \in A_1} X_i \le u_n \right) P\left(\max_{i \in A_2} X_i \le u_n \right) \right| \le \alpha_{n,l}\,,$$

where $A_1 = \{i_1, \ldots, i_p\}$, $A_2 = \{j_1, \ldots, j_q\}$ and $\alpha_{n,l} \to 0$ as $n \to \infty$ for some sequence $l = l_n = o(n)$.

This condition as well as $D'(u_n)$ below and their modifications have been intensively applied to stationary sequences in the monograph by Leadbetter, Lindgren and Rootzén [418]. Condition $D(u_n)$ is a distributional mixing condition, weaker than most of the classical forms of dependence restrictions. A discussion of the role of $D(u_n)$ as a specific mixing condition can be found in Leadbetter et al. [418], Sections 3.1 and 3.2. Condition $D(u_n)$ implies, for example, that

$$P\left(M_n \le u_n\right) = P^k\left(M_{[n/k]} \le u_n\right) + o(1) \tag{4.38}$$

for constant or slowly increasing k. This relation already indicates that the limit behaviour of (M_n) and its associated sequence $(\widetilde{M}_n)$ must be closely related. The following result (Theorem 3.3.3 in Leadbetter et al. [418]) even shows that the classes of possible limit laws for the normalised and centred sequences (M_n) and $(\widetilde{M}_n)$ coincide.

Theorem 4.4.1 (Limit laws for maxima of a stationary sequence)
Suppose $c_n^{-1}(M_n - d_n) \xrightarrow{d} G$ for some distribution G and appropriate constants $c_n > 0$, $d_n \in \mathbb{R}$. If the condition $D(c_n x + d_n)$ holds for all real x, then G is an extreme value distribution.

Proof. Recall from Theorems 3.2.2, 3.2.3 and from Definition 3.2.6 that G is an extreme value distribution if and only if G is max–stable. By (4.38),

$$P\left(M_{nk} \le c_n x + d_n\right) \;=\; P^k\left(M_n \le c_n x + d_n\right) + o(1) \;\to\; G^k(x)$$

for every integer $k \ge 1$, and every continuity point x of G. On the other hand,

$$P\left(M_{nk} \le c_{nk} x + d_{nk}\right) \;\to\; G(x)\,.$$

Now we may proceed as in the proof of Theorem 3.2.2 to conclude that G is max–stable. □

Remark. 1) Theorem 4.4.1 does *not* mean that the relations $c_n^{-1}(M_n - d_n) \xrightarrow{d} G$ and $c_n^{-1}(\widetilde{M}_n - d_n) \xrightarrow{d} H$ hold with $G = H$. We will see later that G is often of the form H^θ for some $\theta \in [0,1]$ (see for instance Example 4.4.2 and Section 8.1); θ is then called *extremal index*. □

Thus max–stability of the limit distribution is *necessary* under the conditions $D(c_n x + d_n)$, $x \in \mathbb{R}$. Next we want to find *sufficient* conditions for convergence of the probabilities $P(M_n \le u_n)$ for a given threshold sequence (u_n) satisfying

$$n\overline{F}(u_n) \to \tau \tag{4.39}$$

for some $\tau \in [0, \infty)$. From Proposition 3.1.1 we know that (4.39) and $P(\widetilde{M}_n \le u_n) \to \exp\{-\tau\}$ are equivalent. But may we replace $(\widetilde{M}_n)$ by (M_n) under $D(u_n)$? The answer is, unfortunately, NO. All one can derive is

$$\liminf_{n\to\infty} P(M_n \le u_n) \ge e^{-\tau};$$

see the proof of Proposition 4.4.3 below.

Example 4.4.2 (See also Figure 4.4.5.) Assume that (Y_n) is a sequence of iid rvs with df $\sqrt{F}$ for some df F. Define the sequence (X_n) by

$$X_n = \max(Y_n, Y_{n+1}), \quad n \in \mathbb{N}.$$

Then (X_n) is a stationary sequence and X_n has df F for all $n \ge 1$. From this construction it is clear that maxima of (X_n) appear as pairs at consecutive indices.

Now assume that for $\tau \in (0, \infty)$ the sequence u_n satisfies $u_n \uparrow x_F$ (x_F is the right endpoint of F) and (4.39). Then $F(u_n) \to 1$ and

$$nP(Y_1 > u_n) = n\left(1 - \sqrt{F(u_n)}\right) = \frac{n\overline{F}(u_n)}{1 + \sqrt{F(u_n)}} \to \frac{\tau}{2}.$$

Hence, by Proposition 3.1.1,

$$\begin{aligned} P(M_n \le u_n) &= P(\max(Y_1, \ldots, Y_n, Y_{n+1}) \le u_n) \\ &= P(\max(Y_1, \ldots, Y_n) \le u_n)\, F(u_n) \\ &\to e^{-\tau/2}. \end{aligned}$$

Condition $D(u_n)$ is naturally satisfied: if A_1 and A_2 are chosen as in $D(u_n)$ and $l \ge 2$, then we can take $\alpha_{n,l} = 0$. □

This example supports the introduction of a second technical condition.

Condition $D'(u_n)$: *The relation*

$$\lim_{k\to\infty}\limsup_{n\to\infty} n\sum_{j=2}^{[n/k]} P(X_1>u_n, X_j>u_n)=0\,.$$

Remark. 2) $D'(u_n)$ is an "anti–clustering condition" on the stationary sequence (X_n). Indeed, notice that $D'(u_n)$ implies

$$E\sum_{1\le i<j\le [n/k]} I_{\{X_i>u_n,X_j>u_n\}} \le [n/k]\sum_{j=2}^{[n/k]} EI_{\{X_1>u_n,X_j>u_n\}}\to 0\,,$$

so that, in the mean, joint exceedances of u_n by pairs (X_i,X_j) become very unlikely for large n. □

Now we have introduced the conditions which are needed to formulate the following analogue of Proposition 3.1.1; see Theorem 3.4.1 in Leadbetter et al. [418]:

Proposition 4.4.3 (Limit probabilities for sample maxima)
Assume that the stationary sequence (X_n) *and the threshold sequence* (u_n) *satisfy* $D(u_n)$, $D'(u_n)$. *Suppose* $\tau\in[0,\infty)$. *Then condition* (4.39) *holds if and only if*

$$\lim_{n\to\infty} P(M_n\le u_n)=e^{-\tau}\,. \tag{4.40}$$

Proof. We restrict ourselves to the sufficiency part in order to illustrate the use of the conditions $D(u_n)$ and $D'(u_n)$. The necessity follows by similar arguments.
We have, for any $l\ge 1$,

$$\begin{aligned}\sum_{i=1}^{l} P(X_i>u_n) - \sum_{1\le i<j\le l} P(X_i>u_n,X_j>u_n)&\\ \le\ P(M_l>u_n)\ \le\ \sum_{i=1}^{l} P(X_i>u_n)\,.&\end{aligned} \tag{4.41}$$

Exploiting the stationarity of (X_n) we see that

$$\sum_{i=1}^{l} P(X_i>u_n) \;=\; l\,\overline{F}(u_n)\,,$$

$$\sum_{1\le i<j\le l} P(X_i>u_n,X_j>u_n)\;\le\; l\sum_{j=2}^{l} P(X_1>u_n,X_j>u_n)\,.$$

Combining this and (4.41) for $l = [n/k]$ ($[x]$ denotes the integer part of x) and for a fixed k, we derive upper and lower estimates for $P(M_{[n/k]} \le u_n)$:

$$\begin{aligned} 1 - [n/k]\overline{F}(u_n) &\le P\left(M_{[n/k]} \le u_n\right) \\ &\le 1 - [n/k]\overline{F}(u_n) + [n/k]\sum_{j=2}^{[n/k]} P(X_1 > u_n, X_j > u_n) . \end{aligned}$$

From (4.39) we immediately have

$$[n/k]\overline{F}(u_n) \to \tau/k , \quad n \to \infty ,$$

and, by condition $D'(u_n)$,

$$\limsup_{n\to\infty} [n/k] \sum_{j=2}^{[n/k]} P(X_1 > u_n, X_j > u_n) = o(1/k) , \quad k \to \infty .$$

Thus we get the bounds

$$1 - \frac{\tau}{k} \le \liminf_{n\to\infty} P\left(M_{[n/k]} \le u_n\right) \le \limsup_{n\to\infty} P\left(M_{[n/k]} \le u_n\right) \le 1 - \frac{\tau}{k} + o(1/k) .$$

This and relation (4.38) imply that

$$\begin{aligned} \left(1 - \frac{\tau}{k}\right)^k &\le \liminf_{n\to\infty} P(M_n \le u_n) \\ &\le \limsup_{n\to\infty} P(M_n \le u_n) \le \left(1 - \frac{\tau}{k} + o(1/k)\right)^k . \end{aligned}$$

Letting $k \to \infty$ we see that

$$\lim_{n\to\infty} P(M_n \le u_n) = e^{-\tau} .$$

This concludes the proof. □

Example 4.4.4 (Continuation of Example 4.4.2)
We observed in Example 4.4.2 that condition (4.39) implies $P(M_n \le u_n) \to \exp\{-\tau/2\}$. We have already checked that $D(u_n)$ is satisfied. Thus $D'(u_n)$ must go wrong. This can be easily seen: since X_1 and X_j are independent for $j \ge 2$ we conclude that

$$\begin{aligned} & n \sum_{j=2}^{[n/k]} P(X_1 > u_n, X_j > u_n) \\ &= nP(X_1 > u_n, X_2 > u_n) + n([n/k]-2)P^2(X_1 > u_n) \\ &= nP(\max(Y_1,Y_2) > u_n , \max(Y_2,Y_3) > u_n) + \tau^2/k + o(1) \\ &= n\left(P(Y_2 > u_n, Y_3 \le u_n) + P(Y_2 > u_n \text{ or } Y_1 > u_n , Y_3 > u_n)\right) \\ &\quad + \tau^2/k + o(1) , \quad n \to \infty . \end{aligned}$$

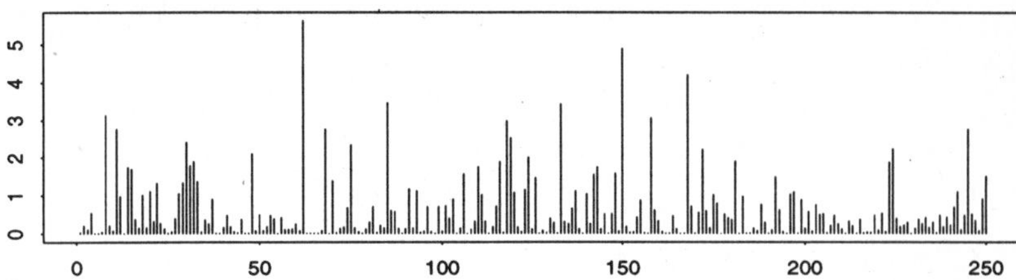

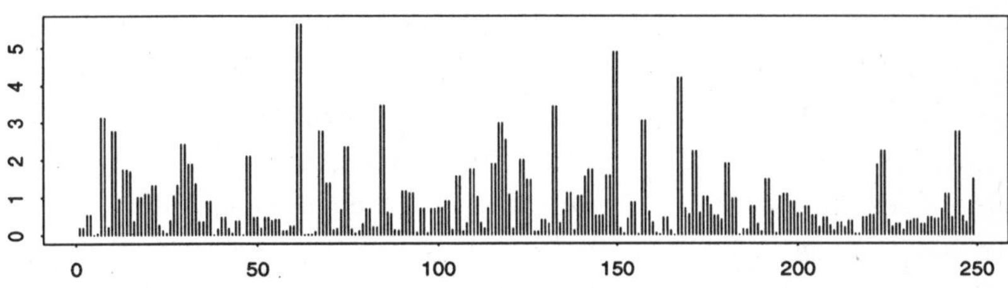

Figure 4.4.5 *A realisation of the sequences* (Y_n) *(top) and* (X_n) *(bottom) with* F *standard exponential as discussed in Examples 4.4.2 and 4.4.4. Extremes appear in clusters of size* 2.

We have

$$nP(Y_2 > u_n, Y_3 \le u_n) \sim nP(Y_1 > u_n) \to \tau/2\,.$$

Thus condition $D'(u_n)$ cannot be satisfied. The reason for this is that maxima in (X_n) appear in clusters of size 2. Notice that

$$E\left(\sum_{i=1}^{n} I_{\{X_i > u_n, X_{i+1} > u_n\}}\right) = nP(X_1 > u_n, X_2 > u_n) \to \tau/2 > 0\,,$$

so that in the long run the expected number of joint exceedances of u_n by the pairs (X_i, X_{i+1}) stabilises around a positive number. □

Proceeding precisely as in Section 3.3 we can now derive the limit distribution for the maxima M_n:

Theorem 4.4.6 (Limit distribution of maxima of a stationary sequence)
Let (X_n) *be a stationary sequence with common df* $F \in \mathrm{MDA}(H)$ *for some extreme value distribution* H*, i.e. there exist constants* $c_n > 0$, $d_n \in \mathbb{R}$ *such that*

$$\lim_{n\to\infty} n\overline{F}(c_n x + d_n) = -\ln H(x)\,, \quad x \in \mathbb{R}\,. \tag{4.42}$$

Assume that for $x \in \mathbb{R}$ the sequences $(u_n) = (c_n x + d_n)$ satisfy the conditions $D(u_n)$ and $D'(u_n)$. Then (4.42) is equivalent to each of the following relations:

$$c_n^{-1}(M_n - d_n) \xrightarrow{d} H, \tag{4.43}$$

$$c_n^{-1}(\widetilde{M}_n - d_n) \xrightarrow{d} H. \tag{4.44}$$

Proof. The equivalence of (4.42) and (4.44) is immediate from Proposition 3.3.2. The equivalence of (4.42) and (4.43) follows from Proposition 4.4.3. □

From the discussion above we are not surprised about the same limit behaviour of the maxima of a stationary sequence and its associated iid sequence; the conditions $D(c_n x + d_n)$ *and* $D'(c_n x + d_n)$ force the sequence (M_n) to behave very much like the maxima of an iid sequence. Notice that Theorem 4.4.6 also ensures that we can choose the sequences (c_n) and (d_n) in the same way as proposed in Section 3.3.

Thus the problem about the maxima of a stationary sequence has been reduced to a question about the extremes of iid rvs. However, now one has to verify the conditions $D(c_n x + d_n)$ and $D'(c_n x + d_n)$ which, in general, is tedious. Conditions $D(u_n)$ and $D'(u_n)$ have been discussed in detail in the monograph by Leadbetter et al. [418]. The case of a Gaussian stationary sequence is particularly nice: one can check $D(u_n)$ and $D'(u_n)$ via the asymptotic behaviour of the autocovariances

$$\gamma(h) = \operatorname{cov}(X_0, X_h), \quad h \geq 0.$$

The basic idea is that the distributions of two Gaussian vectors are "close" to each other if their covariance matrices are "close". Leadbetter et al. [418] make this concept precise by a so-called *normal comparison lemma* (their Theorem 4.2.1), a particular consequence of which is the estimate

$$\begin{aligned}&\left|P(X_{i_1} \leq u_n, \ldots, X_{i_k} \leq u_n) - \Phi^k(u_n)\right| \\ &\leq \; \text{const}\, n \sum_{h=1}^{n} |\gamma(h)| \exp\left\{\frac{-u_n^2}{1+|\gamma(h)|}\right\}\end{aligned}$$

for $1 \leq i_1 < \cdots < i_k \leq n$. Here (X_n) is stationary with marginal df the standard normal Φ, and it is assumed that $\sup_{h\geq 1}|\gamma(h)| < 1$. In particular,

$$|P(M_n \leq u_n) - \Phi^n(u_n)| \leq \text{const}\, n \sum_{h=1}^{n} |\gamma(h)| \exp\left\{\frac{-u_n^2}{1+|\gamma(h)|}\right\}. \tag{4.45}$$

Now it is not difficult to check conditions $D(u_n)$ and $D'(u_n)$. For details see Lemma 4.4.1 in Leadbetter et al. [418]:

Lemma 4.4.7 (Conditions for $D(u_n)$ and $D'(u_n)$ for a Gaussian stationary sequence)

Assume (X_n) *is stationary Gaussian and let* (u_n) *be a sequence of real numbers.*

(a) *Suppose the rhs in* (4.45) *tends to zero as* $n \to \infty$ *and* $\sup_{h\geq 1}|\gamma(h)| < 1$. *Then* $D(u_n)$ *holds.*
(b) *If in addition* $\limsup_{n\to\infty} n\overline{\Phi}(u_n) < \infty$ *then* $D'(u_n)$ *holds.*
(c) *If* $\gamma(n)\ln n \to 0$ *and* $\limsup_{n\to\infty} n\overline{\Phi}(u_n) < \infty$ *then both conditions* $D(u_n)$ *and* $D'(u_n)$ *are satisfied.* □

Now recall that the normal distribution Φ is in the maximum domain of attraction of the Gumbel law Λ; see Example 3.3.29. Then the following is a consequence of Lemma 4.4.7 and of Theorem 4.4.6. The constants c_n and d_n are chosen as in Example 3.3.29.

Theorem 4.4.8 (Limit distribution of the maxima of a Gaussian stationary sequence)
Let (X_n) *be a stationary sequence with common standard normal df* Φ. *Suppose that*

$$\lim_{n\to\infty} \gamma(n)\ln n = 0\,.$$

Then

$$\sqrt{2\ln n}\left(M_n - \sqrt{2\ln n} + \frac{\ln\ln n + \ln 4\pi}{2(2\ln n)^{1/2}}\right) \xrightarrow{d} \Lambda\,.$$

□

The assumption $\gamma(n)\ln n \to 0$ is called *Berman's condition* and is very weak. Thus Theorem 4.4.8 states that Gaussian stationary sequences have very much the same extremal behaviour as Gaussian iid sequences.

Example 4.4.9 (Gaussian linear processes)
An important class of stationary sequences is that of the linear processes (see Section 5.5 and Chapter 7), which have an infinite moving average representation

$$X_n = \sum_{j=-\infty}^{\infty} \psi_j Z_{n-j}\,, \quad n \in \mathbb{Z}\,, \tag{4.46}$$

where $(Z_n)_{n\in\mathbb{Z}}$ is an iid sequence and $\sum_j \psi_j^2 < \infty$. We also suppose that $EZ_1 = 0$ and $\sigma_Z^2 = \mathrm{var}(Z_1) < \infty$. If (Z_n) is Gaussian, so is (X_n). Conversely, most interesting Gaussian stationary processes have representation (4.46); see Brockwell and Davis [92], Theorem 5.7.1, in particular, the popular (causal) ARMA processes; see Example 7.1.1. In that case the coefficients ψ_j decrease to zero at an exponential rate. Hence the autocovariances of (X_n), i.e.

$$\gamma(h) = E(X_0 X_h) = \sigma_Z^2 \sum_{j=-\infty}^{\infty} \psi_j \psi_{j+h}\,, \quad h \ge 0\,,$$

decrease to zero exponentially as $h \to \infty$. Thus Theorem 4.4.8 is applicable to Gaussian ARMA processes.
Gaussian fractional ARIMA(p,d,q) processes with $p,q \ge 1, d \in (0,0.5)$, enjoy a (causal) representation (4.46) with $\psi_j = j^{d-1}L(j)$ for a slowly varying function L; see Brockwell and Davis [92], Section 13.2. It is not difficult to see that the assumptions of Theorem 4.4.8 also hold in this case. Fractional ARIMA processes with $d \in (0,0.5)$ are a standard example of long memory processes where the sequence $\gamma(h)$ is not supposed to be absolutely summable. This shows that the restriction $\gamma(n)\ln n \to 0$ is indeed very weak in the Gaussian case.
In Section 5.5 we will study the extreme value behaviour of linear processes with subexponential noise (Z_n) in MDA(Λ) or MDA(Φ_α). We will learn that the limit distributions of (M_n) are of the form H^θ for some $\theta \in (0,1]$ and an extreme value distribution H. This indicates the various forms of limit behaviour of maxima of linear processes, depending on their tail behaviour. □

Notes and Comments

Extreme value theory for stationary sequences has been treated in detail in Leadbetter et al. [418]. There one can also find some remarks on the history of the use of conditions $D(u_n)$ and $D'(u_n)$. A very recommendable review article is Leadbetter and Rootzén [419].

In summary, the conditions $D(u_n)$ and $D'(u_n)$ ensure that the extremes of the stationary sequence (X_n) have the same qualitative behaviour as the extremes of an associated iid sequence. The main problem is to verify conditions $D(u_n)$ and $D'(u_n)$. For Gaussian (X_n) this reduces to showing Berman's condition, namely that $\gamma(n) = \mathrm{cov}(X_0, X_n) = o(1/\ln n)$. It covers wide classes of Gaussian sequences, in particular ARMA and fractional ARIMA processes. We mention that Leadbetter et al. [418] also treated the cases $\gamma(n)\ln n \to c \in (0,\infty]$.

In Section 5.3.2 we will come back to stationary sequences satisfying the conditions $D(u_n)$ and $D'(u_n)$. There we will also study the behaviour of the upper order statistics.

5

An Approach to Extremes via Point Processes

Point process techniques give insight into the structure of limit variables and limit processes which occur in the theory of summation (see Chapter 2), in extreme value theory (see Chapters 3 and 4) and in time series analysis (see Chapter 7).

One can think of a point process N simply as a random distribution of points X_i in space. For a given configuration (X_i) and a set A, $N(A)$ counts the number of $X_i \in A$. It is convenient to imagine the distribution of N as the probabilities

$$P\left(N(A_1) = k_1, \ldots, N(A_m) = k_m\right)$$

for all possible choices of nice sets $A_1, \ldots, A_m$ and all non–negative integers $k_1, \ldots, k_m$.

The most important point processes are those for which $N(A)$ is Poisson distributed. This leads to the notion of a Poisson random measure N (see Definition 5.1.9) as a generalisation of the classical (homogeneous) Poisson process on $[0, \infty)$. Poisson random measures are basic for the understanding of links between extreme value theory and point processes; they occur in a natural way as weak limits of sample point processes N_n, say. This means, to over–simplify a little, that the relations

$$(N_n(A_1), \ldots, N_n(A_m)) \stackrel{d}{\to} (N(A_1), \ldots, N(A_m))$$

hold for any choice of sets A_i. Kallenberg's Theorem 5.2.2 gives surprisingly simple conditions for this convergence to hold.

These are the fundamental notions which we need throughout. They are made precise in Sections 5.1 and 5.2. The interrelationship between extremes, point processes and weak convergence is perhaps best illustrated by the *point process of exceedances* of a given threshold by a sequence of rvs; see Example 5.1.3 and Section 5.3. Then the reader is urged to go through the beautiful results on exceedances, limits of upper order statistics, joint convergence of maxima and minima, records etc. in order to get a general impression about the method; see Sections 5.4 and 5.5. Point process methods yield a unified and relatively easy approach to extreme value theory. In contrast to the classical techniques as used in Chapter 3 and 4, they do allow for the treatment of extremes of sequences more general than iid in a straightforward way.

In this chapter we need some tools from functional analysis and from measure theory as well as certain arguments from weak convergence in metric spaces; see Appendix A2. In our presentation we try to reduce these technicalities to a minimum, but we cannot avoid them completely.

Our discussion below closely follows Resnick [530].

5.1 Basic Facts About Point Processes

5.1.1 Definition and Examples

In this section we are concerned with the question

What is a point process, how can we describe its distribution, and what are simple examples?

For the moment, consider a sequence (X_n) of random vectors in the so-called *state space* E and define for $A \subset E$

$$N(A) = \text{card}\{i : X_i \in A\}\,,$$

i.e. $N(A)$ counts the number of X_i falling into A. Naturally, $N(A) = N(A, \omega)$ is random for a given set A and, under general conditions, $N(\cdot, \omega)$ defines a random counting measure with atoms X_n on a suitable σ-algebra $\mathcal{E}$ of subsets of E. This is the intuitive meaning of the *point process* N.

For our purposes, the state space E, where the points live, is a subset of a finite-dimensional Euclidean space possibly including points with an infinite coordinate, and E is equipped with the σ-algebra $\mathcal{E}$ of the Borel sets generated by the open sets. It is convenient to write a point process using *Dirac measure* ε_x for $x \in E$:

$$\varepsilon_x(A) = \begin{cases} 1 & \text{if} \quad x \in A\,, \\ 0 & \text{if} \quad x \notin A\,, \end{cases} \qquad A \in \mathcal{E}\,.$$

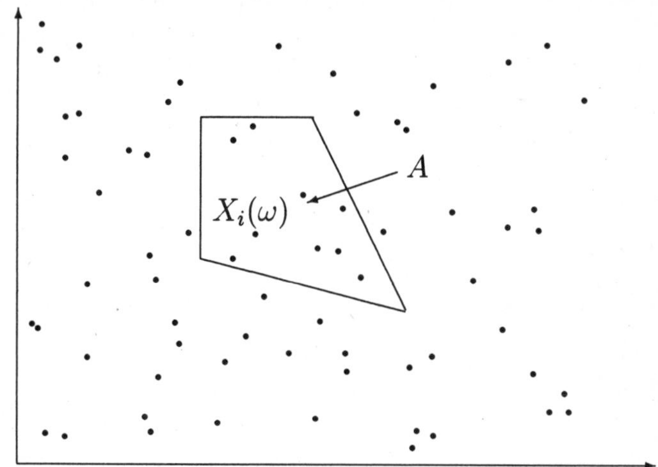

Figure 5.1.1 *A configuration of random points X_i in $\mathbb{R}_+ \times \mathbb{R}_+$. The number of points that fall into the set A constitute the counting variable $N(A)$; in this case $N(A,\omega) = 9$.*

For a given sequence $(x_i)_{i\geq 1}$ in E,

$$m(A) = \sum_{i=1}^{\infty} \varepsilon_{x_i}(A) = \sum_{i:x_i \in A} 1 = \operatorname{card}\{i : x_i \in A\}\,, \quad A \in \mathcal{E}\,,$$

defines a *counting measure* on $\mathcal{E}$ which is called a *point measure* if $m(K) < \infty$ for all compact sets $K \in E$. Let $M_p(E)$ be the space of *all* point measures on E equipped with an appropriate σ–algebra $\mathcal{M}_p(E)$.

Definition 5.1.2 (Definition of a point process)
A point process *on E is a measurable map*

$$N : [\Omega, \mathcal{F}, P] \to [M_p(E), \mathcal{M}_p(E)]\,. \qquad \square$$

Remarks. 1) The σ–algebra $\mathcal{M}_p(E)$ contains all sets of the form $\{m \in M_p(E) : m(A) \in B\}$ for $A \in \mathcal{E}$ and any Borel set $B \subset [0,\infty]$, i.e. it is the smallest σ–algebra making the maps $m \to m(A)$ measurable for all $A \in \mathcal{E}$.

2) A point process is a random element or a random function which assumes point measures as values. It is convenient to think of a point process as a collection $(N(A))_{A\in\mathcal{E}}$ of the extended rvs $N(A)$. (An extended rv can assume the value ∞ with positive probability.). Point processes are special *random measures*; see for instance Kallenberg [365].

3) The point processes we are interested in can often be written in the form

$$N = \sum_{i=1}^{\infty} \varepsilon_{X_i}$$

for a sequence (X_n) of d–dimensional random vectors. Then, for each $\omega \in \Omega$,

$$N(A, \omega) = \sum_{i=1}^{\infty} \varepsilon_{X_i(\omega)}(A)\,, \quad A \in \mathcal{E}\,,$$

defines a point measure on $\mathcal{E}$.

4) Assume that $m = \sum_{i=1}^{\infty} \varepsilon_{x_i}$ is a point measure on E. Let (y_i) be a subsequence of (x_i) containing all mutually distinct values (x_i) with no repeats. Define the *multiplicity of* y_i as

$$n_i = \operatorname{card}\{j : j \geq 1\,,\, y_i = x_j\}\,.$$

Then we may write

$$m = \sum_{i=1}^{\infty} n_i \varepsilon_{y_i}\,.$$

If $n_i = 1$ for all i, then m is called a *simple point measure*, and a *multiple* one, otherwise. Analogously, if the realisations of the point process N are only simple point measures, then N is a *simple point process*, and a *multiple* one, otherwise. Alternatively, a point process N is simple if

$$P(N(\{x\}) \leq 1\,,\, x \in E) = 1\,. \qquad \square$$

Example 5.1.3 (Point process of exceedances)
One of the point processes closely related to extreme value theory is the *point process of exceedances*: let u be a real number and (X_n) a sequence of rvs. Then the *point process of exceedances*

$$N_n(\cdot) = \sum_{i=1}^{n} \varepsilon_{n^{-1}i}(\cdot) I_{\{X_i > u\}}\,, \quad n = 1, 2, \ldots\,, \tag{5.1}$$

with state space $E = (0, 1]$ counts the number of exceedances of the threshold u by the sequence $X_1, \ldots, X_n$. For example, take the whole interval $(0, 1]$. Then

$$\begin{aligned} N_n(0,1] &= \operatorname{card}\left\{i : 0 < n^{-1} i \leq 1 \quad \text{and} \quad X_i > u\right\} \\ &= \operatorname{card}\left\{i \leq n : X_i > u\right\}\,. \end{aligned}$$

Here and in the sequel we write for a measure μ

$$\mu(a,b] = \mu((a,b])\,, \quad \mu[a,b] = \mu([a,b]) \quad \text{etc.}$$

We also see immediately the close link with extreme value theory. For example, let $X_{k,n}$ denote the kth largest order statistic of the sample $X_1,\dots,X_n$. Then

$$\begin{aligned}
\{N_n(0,1]=0\} &= \{\mathrm{card}\,\{i \le n : X_i > u\} = 0\} \\
&= \{\text{None of the } X_i,\ i \le n, \text{ exceeds } u\} \\
&= \{\max(X_1,\dots,X_n) \le u\} \\
\{N_n(0,1] < k\} &= \{\mathrm{card}\,\{i \le n : X_i > u\} < k\} \\
&= \{\text{Fewer than } k \text{ among the } X_i,\ i \le n, \text{ exceed } u\} \\
&= \{\text{The order statistic } X_{k,n} \text{ does not exceed } u\} \\
&= \{X_{k,n} \le u\}\,.
\end{aligned}$$

We notice that the point process of exceedances can be written in the (perhaps more intuitively appealing) form

$$N_n(\cdot) = \sum_{i=1}^{n} \varepsilon_{n^{-1}i,X_i}(\cdot)\,, \quad n = 1,2,\dots\,, \tag{5.2}$$

with two–dimensional state space $E = (0,1] \cap (u,\infty)$. In our presentation we prefer version (5.1) on $E = (0,1]$, with the exception of Section 5.5.1. The advantage of this approach is that weak convergence of (5.1) can be dealt with by simpler means than for (5.2); compare for instance the difficulty of the proofs in Sections 5.3 and 5.5.1. In Section 5.3 our interest will focus on the point process of exceedances for a sequence of non–decreasing thresholds $u = u_n$ which we will choose in such a way that (N_n) converges weakly, in the sense of Section 5.2, to a Poisson random measure; see Definition 5.1.9 below. □

Example 5.1.4 (Renewal counting process)
Let (Y_i) be a sequence of iid positive rvs, $T_n = Y_1 + \cdots + Y_n$, $n \ge 1$. Recall from Section 2.5.2 the *renewal counting process* generated by (Y_i):

$$N(t) = \mathrm{card}\{i : T_i \le t\}\,, \quad t \ge 0\,.$$

To this process we can relate the point process

$$N(A) = \sum_{i=1}^{\infty} \varepsilon_{T_i}(A)\,, \quad A \in \mathcal{E}\,,$$

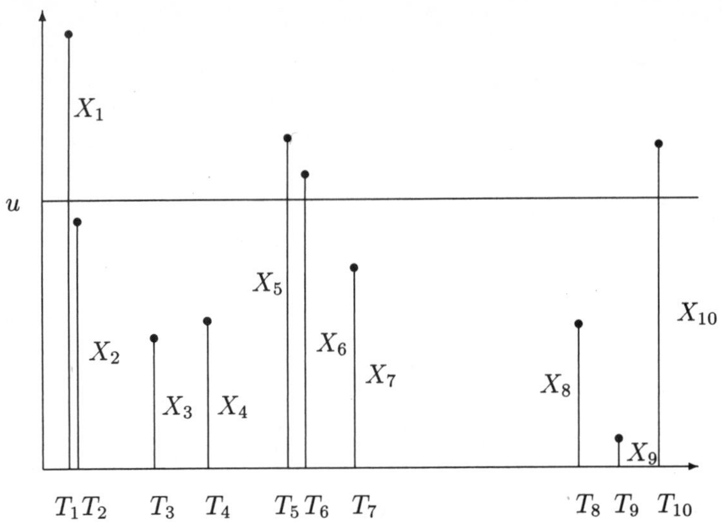

Figure 5.1.5 *Visualisation of the point process of exceedances corresponding to the random sums from Example 5.1.6.*

with state space $E = [0, \infty)$. Notice that for $A = [0, t]$ we obtain

$$N(t) = N[0, t], \quad t \geq 0.$$

In this sense, every renewal counting process corresponds to a point process. The point process defined in this way is simple since $0 < T_1 < T_2 < \cdots$ with probability 1. Recall that a homogeneous Poisson process (see Example 2.5.2) is a particular renewal counting process with exponential rvs Y_i. Hence a Poisson process defines a "Poisson point process". □

Example 5.1.6 (Random sums driven by a renewal counting process)
In Chapter 1 and Section 2.5.3 we considered random sums driven by a renewal counting process:

$$S(t) = \sum_{i=1}^{N(t)} X_i, \quad t \geq 0.$$

Here $(N(t))$ is a renewal counting process as defined in Example 5.1.4 and (X_i) is an iid sequence independent of $(N(t))$. Recall from Chapter 1 that random sums are closely related to the renewal risk model in which we can interpret the rv X_i as claim size arriving at time T_i. A point process related to $(S(t))$ is given by

$$\widetilde{N}(A) = \sum_{i=1}^{\infty} \varepsilon_{(T_i, X_i)}(A), \quad A \in \mathcal{E},$$

with state space $E = [0, \infty) \times \mathbb{R}$. For example, in the insurance context

$$\widetilde{N}((a,b] \times (u, \infty)) = \text{card}\{i : a < T_i \leq b,\, X_i > u\}$$

counts the number of claims arriving in the time interval $(a, b]$ and exceeding the threshold value u. Notice that $\widetilde{N}$ is very close in spirit to the point process of exceedances from Example 5.1.3. □

5.1.2 Distribution and Laplace Functional

The realisations of a point process N are point measures. Therefore the *distribution* or the *probability law* of N is defined on subsets of point measures:

$$P_N(A) = P(N \in A), \quad A \in \mathcal{M}_p(E).$$

This distribution is not easy to imagine. Fortunately, the distribution of N is uniquely determined by the family of the distributions of the finite–dimensional random vectors

$$(N(A_1), \ldots, N(A_m)) \tag{5.3}$$

for any choice of $A_1, \ldots, A_m \in \mathcal{E}$ and $m \geq 1$; see Daley and Vere–Jones [153], Proposition 6.2.III. The collection of all these distributions is called the *finite–dimensional distributions of the point process.* We can imagine the finite–dimensional distributions much more easily than the distribution P_N itself. Indeed, (5.3) is a random vector of integer–valued rvs which is completely given by the probabilities

$$P(N(A_1) = k_1, \ldots, N(A_m) = k_m), \quad k_i \geq 0, \quad i = 1, \ldots, m.$$

From a course on probability theory we know that it is often convenient to describe the distribution of a rv or of a random vector by some analytical means. For example, one uses a whole class of transforms: chfs, Laplace–Stieltjes transforms, generating functions etc. A similar tool exists for point processes:

Definition 5.1.7 (Laplace functional)
The Laplace functional *of the point process N is given by*

$$\begin{aligned} \Psi_N(g) &= E \exp\left\{-\int_E g\, dN\right\} \\ &= \int_{M_p(E)} \exp\left\{-\int_E g(x)\, dm(x)\right\} dP_N(m). \end{aligned} \tag{5.4}$$

It is defined for non–negative measurable functions g on the state space E. □

Remarks. 1) The Laplace functional Ψ_N determines the distribution of a point process completely; see Example 5.1.8 below.

2) Laplace functionals are an important tool for discovering the properties of point processes; they are particularly useful for studying the weak convergence of point processes; see Section 5.2.

3) The integral $\int_E g dN$ in (5.4) is well defined as a Lebesgue–Stieltjes integral. Write $N = \sum_{i=1}^{\infty} \varepsilon_{X_i}$ for random vectors with values in E; then

$$\int_E g dN = \sum_{i=1}^{\infty} g(X_i).$$

In particular, $\int_A dN = \int_E I_A \, dN = N(A)$. □

Example 5.1.8 (Laplace functional and Laplace transform)
To get an impression of the use of Laplace functionals we consider the special functions

$$g = z\, I_A\,, \quad z \geq 0\,, \quad A \in \mathcal{E}\,.$$

Then

$$\Psi_N(g) = E \exp\left\{-\int_E g\, dN\right\} = E \exp\left\{-z\, N(A)\right\}\,,$$

so that the notion of the ordinary Laplace transform of the rv $N(A)$ is embedded in the Laplace functional. Now suppose that $A_1, \ldots, A_m \in \mathcal{E}$. If we choose the functions

$$g = z_1\, I_{A_1} + \cdots + z_m\, I_{A_m}\,, \quad z_1 \geq 0\,, \ldots, z_m \geq 0\,,$$

then we obtain the joint Laplace transform of the finite–dimensional distributions, i.e. of the random vectors (5.3). From the remarks above we learnt that the finite–dimensional distributions determine the distribution of N. On the other hand, the finite–dimensional distributions are uniquely determined by their Laplace transforms, and hence the Laplace functional uniquely determines the distribution of N. □

5.1.3 Poisson Random Measures

Point processes are collections of counting variables. The simplest and perhaps most useful example of a counting variable is binomially distributed: $B_n = \sum_{i=1}^{n} I_{\{X_i \in A_n\}}$ for iid X_i counts the number of "successes" $\{X_i \in A_n\}$ among $X_1, \ldots, X_n$, and $p_n = P(X_1 \in A_n)$ is the "success probability". Then Poisson's theorem tells us that $B_n \stackrel{d}{\rightarrow} Poi(\lambda)$ provided $p_n \sim \lambda/n$. This simple limit result also suggests the following definition of a *Poisson random measure* which occurs as natural limit of many point processes; see for instance Section 5.3.

Let μ be a *Radon measure* on $\mathcal{E}$, i.e. $\mu(A) < \infty$ for compact sets $A \subset E$.

Definition 5.1.9 (Poisson random measure (PRM))
A point process N is called a Poisson process *or a* Poisson random measure with mean measure μ *(we write* PRM(μ)) *if the following two conditions are satisfied:*

(a) *For $A \in \mathcal{E}$,*

$$P(N(A) = k) = \begin{cases} e^{-\mu(A)} \dfrac{(\mu(A))^k}{k!} & \text{if } \mu(A) < \infty\,, \\ 0 & \text{if } \mu(A) = \infty\,, \end{cases} \qquad k \geq 0\,.$$

(b) *For any $m \geq 1$, if $A_1, \ldots, A_m$ are mutually disjoint sets in $\mathcal{E}$ then $N(A_1), \ldots, N(A_m)$ are independent rvs.* □

Remark. The name *mean measure* is justified by the fact that $EN(A) = \mu(A)$. Since a Poisson distribution is determined by its mean value, it follows from the above definition that PRM(μ) is determined by its mean measure μ. □

Example 5.1.10 (Homogeneous PRM)
Recall the notion of a homogeneous Poisson process $(N(t))_{t \geq 0}$ with intensity $\lambda > 0$ from Example 2.5.2. It is a process with stationary, independent increments such that $N(t)$ is $Poi(\lambda t)$ distributed. Hence

$$P(N(t) = k) = e^{-\lambda t} \frac{(\lambda t)^k}{k!}\,, \quad k = 0, 1, \ldots.$$

Since $(N(t))_{t \geq 0}$ is a non–decreasing process the construction

$$N(s,t] = N(t) - N(s)\,, \quad 0 \leq s < t < \infty\,,$$

and the extension theorem for measures define a point process N on the Borel sets of $E = [0, \infty)$. The stationary, independent increments of $(N(t))_{t \geq 0}$ imply that

$$\begin{aligned} & P\left(N\left(A_1\right) = k_1, \ldots, N\left(A_m\right) = k_m\right) \\ = \;& e^{-\lambda |A_1|} \frac{(\lambda |A_1|)^{k_1}}{k_1!} \cdots e^{-\lambda |A_m|} \frac{(\lambda |A_m|)^{k_m}}{k_m!} \end{aligned}$$

for any mutually disjoint A_i and integers $k_i \geq 0$. Here $|\cdot|$ denotes Lebesgue measure on $[0, \infty)$. This relation is immediate for disjoint intervals A_i, and in the general case one has to approximate the disjoint Borel sets A_i by intervals.

Alternatively, we saw from Example 5.1.4 that a homogeneous Poisson process with intensity λ can be defined as a simple point process $N = \sum_{i=1}^{\infty} \varepsilon_{T_i}$, where $T_i = Y_1 + \cdots + Y_i$ for iid exponential rvs Y_i with mean value $1/\lambda$.

Notice that N has mean measure

$$\mu(A) = \lambda|A| = \lambda \int_A dx\,, \quad A \in \mathcal{E}\,. \qquad \square$$

Now suppose that N is PRM$(\lambda|\cdot|)$ with state space $E \subset \overline{\mathbb{R}}^d$ ($\overline{\mathbb{R}} = \mathbb{R} \cup \{-\infty, \infty\}$), where $\lambda > 0$ and $|\cdot|$ denotes Lebesgue measure on E. As a generalisation of the homogeneous Poisson process on $[0,\infty)$ we call N a *homogeneous PRM* or *homogeneous Poisson process with intensity* λ. Moreover, if the mean measure μ of a PRM is absolutely continuous with respect to Lebesgue measure, i.e. there exists a non–negative function f such that

$$\mu(A) = \int_A f(x)\,dx\,, \quad A \in \mathcal{E}\,,$$

then f is the *intensity* or the *rate of the PRM.*

Alternatively, we can define a PRM(μ) by its Laplace functional:

Example 5.1.11 (Laplace functional of PRM(μ))

$$\Psi_N(g) = \exp\left\{-\int_E \left(1 - e^{-g(x)}\right) d\mu(x)\right\} \tag{5.5}$$

for any measurable $g \geq 0$. Formula (5.5) is a consequence of the more general Lemma 5.1.12 below. $\square$

Lemma 5.1.12 *Let N be* PRM(μ) *on $E \subset \overline{\mathbb{R}}^d$. Assume that the Lebesgue integral $\int_E (\exp\{f(x)\} - 1) d\mu(x)$ exists and is finite. Then $\int_E |f| dN < \infty$ a.s. and*

$$I_N(f) = E\exp\left\{\int_E f dN\right\} = \exp\left\{-\int_E \left(1 - e^{f(x)}\right) d\mu(x)\right\}.$$

Proof. For $A \in \mathcal{E}$ with $\mu(A) < \infty$ write $f = I_A$. Then

$$\begin{aligned} I_N(f) &= E\exp\left\{\int_E f\,dN\right\} = E\exp\{N(A)\} \\ &= \sum_{k=0}^{\infty} e^k \frac{(\mu(A))^k}{k!} e^{-\mu(A)} = e^{-\mu(A)(1-e)} \\ &= \exp\left\{-\int_E \left(1 - e^{f(x)}\right) d\mu(x)\right\}. \end{aligned}$$

For

$$f = \sum_{i=1}^{m} z_i \, I_{A_i} \,, \quad z_i \geq 0 \,, \quad i = 1, \ldots, m \,, \tag{5.6}$$

and disjoint $A_1, \ldots, A_m$ we can use the independence of $N(A_1), \ldots, N(A_m)$:

$$\begin{aligned} I_N(f) &= E\exp\left\{\sum_{i=1}^{m} z_i N(A_i)\right\} \\ &= \prod_{i=1}^{m} \exp\left\{-\int_E \left(1 - e^{z_i \, I_{A_i}}\right) d\mu(x)\right\} \\ &= \exp\left\{-\int_E \left(1 - e^{f(x)}\right) d\mu(x)\right\} . \end{aligned}$$

General non–negative f are the monotone limit of step functions (f_n) as in (5.6). Thus, applying the monotone convergence theorem, we obtain

$$\begin{aligned} I_N(f) &= \lim_{n\to\infty} E\exp\left\{\int_E f_n \, dN\right\} \\ &= \lim_{n\to\infty} \exp\left\{-\int_E \left(1 - e^{f_n(x)}\right) d\mu(x)\right\} \\ &= \exp\left\{-\int_E \left(1 - e^{f(x)}\right) d\mu(x)\right\} . \end{aligned}$$

Since the right–hand side is supposed to be finite, $E\exp\{\int_E f dN\} < \infty$, hence $\int_E f dN < \infty$ a.s.

For negative f one can proceed similarly. For general f, write $f = f^+ - f^-$. Notice that $\int_E f^+ dN$ and $\int_E f^- dN$ are independent since $E_+ = \{x \in E : f(x) > 0\}$ and $E_- = \{x \in E : f(x) < 0\}$ are disjoint. Hence

$$\begin{aligned} I_N(f) &= I_N(f^+) I_N(-f^-) \\ &= \exp\left\{-\int_{E_+} \left(1 - e^{f^+}\right) d\mu\right\} \exp\left\{-\int_{E_-} \left(1 - e^{-f^-}\right) d\mu\right\} \\ &= \exp\left\{-\int_E \left(1 - e^{f}\right) d\mu\right\} . \end{aligned}$$

This proves the lemma. □

PRM have an appealing property: they remain PRM under transformations of their points.

Proposition 5.1.13 (Transformed PRM are PRM)
Suppose N is PRM(μ) *with state space $E \subset \overline{\mathbb{R}}^d$. Assume that the points of N are transformed by a measurable map $\widetilde{T} : E \to E'$, where $E' \subset \overline{\mathbb{R}}^m$ for some $m \geq 1$. Then the resulting transformed point process is* PRM($\mu(\widetilde{T}^{-1})$) *on E', i.e. this PRM has mean measure $\mu(\widetilde{T}^{-1}(\cdot)) = \mu\{x \in E : \widetilde{T}(x) \in \cdot\}$.*

Proof. Assume that N has representation $N = \sum_{i=1}^{\infty} \varepsilon_{X_i}$. Then the transformed point process can be written as

$$\widetilde{N} = \sum_{i=1}^{\infty} \varepsilon_{\widetilde{T}(X_i)} \,.$$

We calculate the Laplace functional of $\widetilde{N}$:

$$\begin{aligned}
\Psi_{\widetilde{N}}(g) &= E \exp\left\{-\int_{E'} g d\widetilde{N}\right\} \\
&= E \exp\left\{-\sum_{i=1}^{\infty} g(\widetilde{T}(X_i))\right\} \\
&= E \exp\left\{-\int_{E} g(\widetilde{T}) dN\right\} \\
&= \exp\left\{-\int_{E} \left(1 - e^{-g(\widetilde{T}(x))}\right) d\mu(x)\right\} \\
&= \exp\left\{-\int_{E'} \left(1 - e^{-g(y)}\right) d\mu(\widetilde{T}^{-1}(y))\right\} .
\end{aligned}$$

This is the Laplace functional of PRM($\mu(\widetilde{T}^{-1})$) on E'; see Example 5.1.11. □

Example 5.1.14 Let Γ_k be the points of a homogeneous Poisson process on $[0, \infty)$ with intensity λ and $\widetilde{T}(x) = \exp\{x\}$. Then $\widetilde{N} = \sum_{i=1}^{\infty} \varepsilon_{\exp\{\Gamma_i\}}$ is PRM on $[1, \infty)$ with mean measure given by

$$\widetilde{\mu}(a, b] = \int_a^b d\mu(\widetilde{T}^{-1}(y)) = \lambda \int_{\ln a}^{\ln b} dx = \lambda \ln(b/a) \,, \quad 1 \leq a < b < \infty \,. \tag{5.7}$$

It is interesting to observe that the mean measure of the PRM $\widetilde{N}$ depends only on the fraction b/a, so that the mean measure is the same for all intervals $(ca, cb]$ for any $c > 0$.

Now assume that the PRM $\widetilde{N}$ is defined on the state space $\mathbb{R}_+$ where its mean measure is given by (5.7) for all $0 < a < b < \infty$. Since the distribution of a PRM is determined via its mean measure it follows that the PRM $\widetilde{N}(\cdot)$

and $\widetilde{N}(c\cdot)$ on $\mathbb{R}_+$ have the same distribution in $M_p(\mathbb{R}_+)$ for every positive constant c. □

Example 5.1.15 (Compound Poisson process)
Let (Γ_i) be the points of a homogeneous Poisson process N on $[0,\infty)$ with intensity $\lambda > 0$ and (ξ_i) be a sequence of iid non–negative integer–valued rvs, independent of N. Consider the multiple point process

$$\widetilde{N} = \sum_{i=1}^{\infty} \xi_i \varepsilon_{\Gamma_i}$$

and notice that

$$\widetilde{N}(0,t] = \sum_{i=1}^{\infty} \xi_i \varepsilon_{\Gamma_i}(0,t] = \sum_{i=1}^{N(t)} \xi_i \,, \quad t \geq 0 \,.$$

This is nothing but a particular (i.e. integer–valued) compound Poisson process as used for instance in Chapter 1 for the Cramér–Lundberg model. Therefore we call the point process $\widetilde{N}$ a *compound Poisson process with intensity* λ *and cluster sizes* ξ_i. The probabilities $\pi_k = P(\xi_1 = k)$, $k \geq 0$, are the *cluster probabilities* of $\widetilde{N}$.

The point process notion *compound Poisson process* as introduced above is perhaps not the most natural generalisation of the corresponding random sum process. One would like a *random measure* with property $\widetilde{N}(0,t] = \sum_{i=1}^{N(t)} \xi_i$ for iid ξ_i with any distribution. Since $\widetilde{N}(0,t]$ could then assume any real value this calls for the introduction of a *signed random measure*. For details we refer to Kallenberg [365].

Compound Poisson processes frequently occur as limits of the point processes of exceedances of a strictly stationary sequence; see for instance Sections 5.5 and 8.4. □

Notes and Comments

Point processes are special random measures; see Kallenberg [365]. Standard monographs on point processes and random measures are Cox and Isham [134], Daley and Vere–Jones [153], Kallenberg [365], Karr [372], Matthes, Kerstan and Mecke [447], Reiss [527]. Point processes are also treated in books on stochastic processes; see for instance Jacod and Shiryaev [352], Resnick [529, 531].

In our presentation we leave out certain details. This does not always leave the sufficient mathematical rigour. We are quite cavalier concerning measurability (for instance for point processes) and existence results (for instance for PRM), and we are not precise about compact sets in $E \subset \overline{\mathbb{R}}^d$. The disappointed reader is invited to read through Chapters 3 and 4 in Resnick [530] or to consult the first few chapters in Daley and Vere–Jones [153].

5.2 Weak Convergence of Point Processes

Weak convergence of point processes is a basic tool for dealing with the asymptotic theory of extreme values, linear time series and related fields. We give here a short introduction to the topic. First of all we have to clarify:

What does weak convergence of point processes actually mean?

This question cannot be answered at a completely elementary level. Consider point processes $N, N_1, N_2, \ldots$ on the same state space $E \subset \overline{\mathbb{R}}^d$. Then we know from Section 5.1.2 that the distribution of these point processes in $M_p(E)$, the space of all point measures on E, is determined by their finite–dimensional distributions. Thus a natural requirement for weak convergence of (N_n) towards N would be that, for any choice of "good" Borel sets $A_1, \ldots, A_m \in \mathcal{E}$ and for any integer $m \geq 1$,

$$P\left(N_n(A_1), \ldots, N_n(A_m)\right) \to P\left(N(A_1), \ldots, N(A_m)\right) . \tag{5.8}$$

On the other hand, every point process N can be considered as a stochastic process, i.e. as a collection of rvs $N(A)$ indexed by the sets $A \in \mathcal{E}$. Thus N is an infinite–dimensional object which must be treated in an appropriate way. A glance at Appendix A2 should convince us that we need something more than convergence of the finite–dimensional distributions, namely a condition which is called "tightness" meaning that the probability mass of the point processes N_n should not disappear from "good" (compact) sets in $M_p(E)$. This may sound fine, but such a condition is not easily verified. For example, we would have to make clear in what sense we understand compactness. This has been done in Appendix A2.6 by introducing an appropriate (so–called *vague*) metric in $M_p(E)$.

Perhaps unexpectedly, point processes are user–friendly in the sense that tightness follows from the convergence of their finite–dimensional distributions; see for instance Daley and Vere–Jones [153], Theorem 9.1.VI. Hence we obtain quite an intuitive notion of weak convergence:

Definition 5.2.1 (Weak convergence of point processes)
Let $N, N_1, N_2, \ldots$ be point processes on the state space $E \subset \overline{\mathbb{R}}^d$ equipped with the σ–algebra $\mathcal{E}$ of the Borel sets. We say that (N_n) converges weakly to N in $M_p(E)$ *(we write $N_n \xrightarrow{d} N$) if* (5.8) *is satisfied for all possible choices of*

sets $A_i \in \mathcal{E}$ satisfying $P(N(\partial A_i) = 0) = 1$, $i = 1, \ldots, m$, $m \geq 1$. (∂A denotes the boundary of A.) □

Assume for the moment that the state space E is an interval $(a, b] \subset \mathbb{R}$. Convergence of the finite–dimensional distributions can be checked by surprisingly simple means as the following result shows. Recall the notion of a *simple point process* from Remark 4 after Definition 5.1.2, i.e. it is a process whose points have multiplicity 0 or 1 with probability one.

Theorem 5.2.2 (Kallenberg's theorem for weak convergence to a simple point process on an interval)
Let (N_n) and N be point processes on $E = (a, b] \subset \mathbb{R}$ and let N be simple. Suppose the following two conditions hold:

$$EN_n(A) \to EN(A) \tag{5.9}$$

for all intervals $A = (c, d]$ with $a < c < d \leq b$ and

$$P(N_n(B) = 0) \to P(N(B) = 0) \tag{5.10}$$

for all unions $B = \cup_{i=1}^k (c_i, d_i]$ of mutually disjoint intervals $(c_i, d_i]$ such that

$$a < c_1 < d_1 < \cdots < c_k < d_k \leq b$$

and for every $k \geq 1$. Then $N_n \xrightarrow{d} N$ in $M_p(E)$. □

Remarks. 1) A result in the same spirit can also be formulated for point processes on intervals in $\mathbb{R}^d$.

2) In Section 5.3 we apply Kallenberg's theorem to point processes of exceedances (see also Example 5.1.3) which are closely related to extreme value theory. □

The Laplace functional (see Definition 5.1.7) is a useful theoretical tool for verifying the weak convergence of point processes. In much the same way as the weak convergence of a sequence of rvs is equivalent to the pointwise convergence of their chfs or Laplace–Stieltjes transforms, so the weak convergence of a sequence of point processes is equivalent to the convergence of their Laplace functionals over a suitable family of functions g. Specifically, recall that the real–valued function g has compact support if there exists a compact set $K \subset E$ such that $g(x) = 0$ on K^c, the complement of K. Then we define

$$\begin{aligned} C_K^+(E) \;=\; & \{g : g \text{ is a continuous, non–negative function on } E \\ & \text{with compact support}\}\,. \end{aligned}$$

Now we can formulate the following fundamental theorem; see Daley and Vere–Jones [153], Proposition 9.1.VII:

Theorem 5.2.3 (Criterion for weak convergence of point processes via convergence of Laplace functionals)
The point processes (N_n) converge weakly to the point process N in $M_p(E)$ if and only if the corresponding Laplace functionals converge for every $g \in C_K^+(E)$ as $n \to \infty$, i.e.

$$\Psi_{N_n}(g) = E \exp\left\{ - \int_E g \, dN_n \right\} \to \Psi_N(g) = E \exp\left\{ - \int_E g \, dN \right\} . \quad (5.11)$$

□

Remark. 3) We mention that (5.11) for every $g \in C_K^+(E)$ is equivalent to $\int_E g dN_n \xrightarrow{d} \int_E g dN$ for every $g \in C_K^+(E)$. Indeed, if $g \in C_K^+(E)$ then $zg \in C_K^+(E)$, $z > 0$. Thus (5.11) implies the convergence of the Laplace transforms of the rvs $\int_E g dN_n$ and vice versa. But convergence of the Laplace transforms of non–negative rvs is equivalent to their convergence in distribution. □

We consider another class of point processes which is important for applications. Assume

$$N_n = \sum_{i=1}^{\infty} \varepsilon_{(n^{-1}i, \xi_{n,i})} , \quad n = 1, 2, \ldots , \quad (5.12)$$

where the random vectors $\xi_{n,i}$ are iid for every n. It is convenient to interpret $n^{-1}i$ as a scaled (deterministic) time coordinate and $\xi_{n,i}$ as a scaled (random) space coordinate.

Theorem 5.2.4 (Weak convergence to a PRM)
Suppose (N_n) is a sequence of point processes (5.12) *with state space $\mathbb{R}_+ \times E$ and N is* PRM$(|\cdot| \times \mu)$, *where $|\cdot|$ is Lebesgue measure on $\mathbb{R}_+$. Then*

$$N_n \xrightarrow{d} N , \quad n \to \infty ,$$

in $M_p(\mathbb{R}_+ \times E)$ if and only if the relation

$$nP(\xi_{n,1} \in \cdot) \xrightarrow{v} \mu(\cdot) , \quad n \to \infty , \quad (5.13)$$

holds on $\mathcal{E}$.

Remark. 4) In (5.13), the relation $\mu_n \xrightarrow{v} \mu$ denotes *vague convergence* of the measures μ_n to the measure μ on E. For our purposes, E is a subset of $\overline{\mathbb{R}}^d$. Typically, $E = (0, \infty]$ or $E = [-\infty, \infty] \backslash \{0\}$ or $E = (-\infty, \infty]$. In this case, $\mu_n \xrightarrow{v} \mu$ amounts to showing that $\mu_n(a, b] \to \mu(a, b]$ for all $a < b$ ($b = \infty$ is possible). In the case $E = (-\infty, \infty] \backslash \{0\}$ the origin must not be included in $(a, b]$. A brief introduction to vague convergence and weak convergence is given in Appendix A2.6. For a general treatment we refer to Daley and Vere–Jones [153], Chapter 9, or Resnick [530], Chapter 3. □

Sketch of the proof. We restrict ourselves to the sufficiency part and give only the basic idea. For a full proof we refer to Resnick [530], Proposition 3.21. Let $g \in C_K^+(\mathbb{R}_+ \times E)$ and consider the Laplace functional

$$\begin{aligned}
\Psi_{N_n}(g) &= E\exp\left\{-\int_E g\, dN_n\right\} \\
&= E\exp\left\{-\sum_{i=1}^{\infty} g\left(n^{-1}i,\, \xi_{n,i}\right)\right\} \\
&= \prod_{i=1}^{\infty}\left(1-\int_E\left(1-e^{-g\left(n^{-1}i,\,\mathbf{x}\right)}\right) dP\left(\xi_{n,1}\le \mathbf{x}\right)\right).
\end{aligned}$$

Passing to logarithms, making use of a Taylor expansion for $\ln(1-x)$ and utilising the vague convergence in (5.13) one can show that

$$\begin{aligned}
-\ln \Psi_{N_n}(g) &= -\sum_i \ln\left(1-\int_E\left(1-e^{-g\left(n^{-1}i,\,\mathbf{x}\right)}\right) dP\left(\xi_{n,1}\le \mathbf{x}\right)\right) \\
&= n^{-1}\sum_i \int_E\left(1-e^{-g\left(n^{-1}i,\,\mathbf{x}\right)}\right) d\left[nP\left(\xi_{n,1}\le \mathbf{x}\right)\right] + o(1) \\
&\to \int_{\mathbb{R}_+}\int_E\left(1-e^{-g(s,\,\mathbf{x})}\right) ds\, d\mu(\mathbf{x}), \quad n\to\infty.
\end{aligned}$$

A glance at formula (5.5) convinces us that the last line in the above display is just $-\ln \Psi_N(g)$ where N is PRM($|\cdot| \times \mu$). Now an application of Theorem 5.2.3 yields the result. □

Notes and Comments

Weak convergence of point processes and random measures has been treated in all standard texts on the topic. We again refer here to Daley and Vere–Jones [153], Kallenberg [365], Matthes, Kerstan and Mecke [447], and Resnick [529, 530]. Resnick [530] gives an account of point process techniques particularly suited to extreme value theory. Leadbetter, Lindgren and Rootzén [418] use point process techniques for extremes of stationary sequences, and they provide the necessary background from point process theory in their Appendix.

As mentioned above, a rigorous treatment of weak convergence of point processes requires us to consider them as random elements in an appropriate metric space. A brief introduction to this topic is given in Appendix A2; the general theory can be found for instance in Billingsley [69] or Pollard

[504]. One way to metrize weak convergence of point processes is via vague convergence of measures; see Appendix A2.6. A rigorous treatment is given in Daley and Vere–Jones [153], Chapter 9, or Resnick [530], Chapter 3.

Weak convergence of point processes and vague convergence are closely related to regular variation in $\mathbb{R}_+^d$, see for instance de Haan and Resnick [299, 300] and Stam [605], also Bingham, Goldie and Teugels [72].

Theorems 5.2.2 and 5.2.4 are the basic tools in Sections 5.3–5.5. Theorem 5.2.2 is slightly more general in the sense that no vague convergence (or regular variation) assumption on the tails of the underlying dfs is required. Theorem 5.2.2 has been utilised in the monograph by Leadbetter et al. [418] on extremes of stationary sequences; see also Section 5.3.2. Theorem 5.2.4 will prove very effective in the case that the underlying sequence of random points has a special structure which can in some way be relaxed to an iid sequence, as is the case of linear processes (see Section 5.5) which are special stationary processes.

Resnick [529], pp. 134–135, gives a short resumé of advantages and disadvantages of point process techniques which we cite here in part:

Some Advantages:

(*a*) The methods are by and large dimensionless. Proofs work just as well in $\mathbb{R}^d$ as in $\mathbb{R}$.

(*b*) Computations are kept to a minimum and are often replaced by continuity or structural arguments. This makes proofs simpler and more instructive.

(*c*) The methods lend themselves naturally to proving weak convergence in a function–space setting. Functional limit theorems are more powerful and informative than the one–dimensional variety. Furthermore, they are often (despite common prejudices) simpler.

Some Disadvantages:

(*a*) The methods are not so effective for showing that regular variation is a necessary condition.

(*b*) The methods sail smoothly only when all random variables are non–negative.

(*c*) The methods rely heavily on continuity. Sometimes this can be seen as an advantage, as discussed above. But heavy reliance on continuity is also a limitation in that many questions which deal with quality of convergence (local limit theorems, rates of convergence, large deviations) are beyond the capabilities of continuity arguments.

(*d*) The point process technique cannot handle problems involving normality or Brownian motion.

(*e*) Those who prefer analytical methods may not find the approaches described here (in [529]) attractive.

(*f*) Effective use of weak convergence techniques depends on a detailed knowledge of the properties of the limit processes. Thus it is necessary to know something about stochastic processes.

Since the second list appears longer than the first, I am compelled to make some remarks about the disadvantages. Most serious in my view are (a) and (d). As for (a), there are notable exceptions to the remark that the methods are not suited to proving necessity. Regarding (d), it is sad that the point process technique fails miserably in the presence of normality, but other weak convergence methods often succeed admirably in this case. Disadvantage (d) is a nuisance, but one can usually avoid the obstacles created by two signs by using pruning techniques or random indices. As for (f) a method cannot handle problems for which it is inherently unsuited. The problem raised in (e) is simply one of taste. As for disadvantage (f), these weak convergence techniques frequently suggest interesting problems in stochastic processes. So if (f) were rephrased suitably, it could be moved to the plus column in the ledger.

5.3 Point Processes of Exceedances

In Example 5.1.3 we introduced the point process of exceedances of a threshold u_n by the rvs $X_1, \dots, X_n$:

$$N_n(\cdot) = \sum_{i=1}^{n} \varepsilon_{n^{-1}i}(\cdot) I_{\{X_i > u_n\}}\,, \quad n = 1, 2, \dots. \tag{5.14}$$

We also indicated the close link with extreme value theory: let $X_{n,n} \le \cdots \le X_{1,n}$ denote the order statistics of the sample $X_1, \dots, X_n$ and $M_n = X_{1,n}$. Then

$$\begin{aligned} \{N_n(0,1] = 0\} &= \{M_n \le u_n\}\,, \\ \{N_n(0,1] < k\} &= \{X_{k,n} \le u\}\,. \end{aligned} \tag{5.15}$$

In this section we show the weak convergence of a sequence (N_n) of such point processes to a homogeneous Poisson process N on the state space $E = (0, 1]$. The sequence (X_n) is supposed to be iid or strictly stationary satisfying the assumptions D and D' from Section 4.4. As a byproduct and for illustrative purposes we give alternative proofs of the limit results for maxima and upper order statistics provided in Chapters 3 and 4.

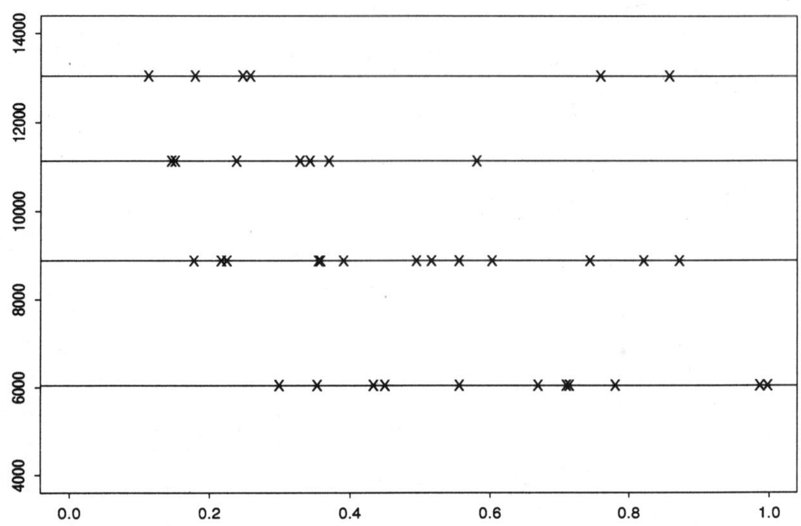

Figure 5.3.1 *Visualisation of the point processes of exceedances of insurance claim data caused by water, $n = 1\,762$ observations. For the threshold $u_1 = 6\,046$ we chose $n/4$ data points, correspondingly $u_2 = 8\,880$ and $n/2$, $u_3 = 11\,131$ and $3n/4$, $u_4 = 13\,051$ and n.*

5.3.1 The IID Case

Assume that the X_n are iid rvs and let (u_n) be a sequence of real thresholds. Recall from Proposition 3.1.1 that, for any $\tau \in [0,\infty]$, the relation $P(M_n \le u_n) \to \exp\{-\tau\}$ holds if and only if

$$n\overline{F}(u_n) = E\sum_{i=1}^{n} I_{\{X_i > u_n\}} \to \tau\,. \tag{5.16}$$

The latter condition ensures that there are on average roughly τ exceedances of the threshold u_n by $X_1,\ldots,X_n$. The Poisson approximation for extremes is visualised in Figure 4.2.2; see also Figure 5.3.1. Condition (5.16) also implies weak convergence of the point processes N_n:

Theorem 5.3.2 (Weak convergence of point processes of exceedances, iid case)
Suppose that (X_n) is a sequence of iid rvs with common df F. Let (u_n) be threshold values such that (5.16) *holds for some $\tau \in (0,\infty)$. Then the point processes of exceedances N_n, see* (5.14), *converge weakly in $M_p(E)$ to*

a homogeneous Poisson process N on $E = (0,1]$ with intensity τ, i.e. N is $\text{PRM}(\tau|\cdot|)$, *where $|\cdot|$ denotes Lebesgue measure on E.*

Proof. We may and do assume that the limit process N is embedded in a homogeneous Poisson process on $[0,\infty)$. In that case we argued that N must be simple; see Example 5.1.10. Hence we can apply Kallenberg's Theorem 5.2.2. Notice that for $A = (a,b] \subset (0,1]$ the rv

$$\begin{aligned} N_n(A) &= \sum_{i=1}^{n} \varepsilon_{n^{-1}i}(A)\, I_{\{X_i > u_n\}} \\ &= \sum_{a<n^{-1}i\le b} I_{\{X_i > u_n\}} \\ &= \sum_{i=[na]+1}^{[nb]} I_{\{X_i > u_n\}} \end{aligned}$$

is binomial with parameters $([nb]-[na], \overline{F}(u_n))$. Here $[x]$ denotes the integer part of x. Thus, by assumption (5.16),

$$EN_n(A) = ([nb]-[na])\overline{F}(u_n) \sim (n(b-a))\left(n^{-1}\tau\right) = \tau(b-a) = EN(A)\,,$$

which proves (5.9).

Thus it remains to show (5.10). Since $N_n(A)$ is binomial and in view of (5.16) we have

$$\begin{aligned} P(N_n(A)=0) &= F^{[nb]-[na]}(u_n) \\ &= \exp\left\{([nb]-[na])\ln\left(1-\overline{F}(u_n)\right)\right\} \\ &\to \exp\{-\tau(b-a)\}\,. \end{aligned} \tag{5.17}$$

Recalling the definition of the set B from (5.10) and taking the independence of the X_i into account we conclude from (5.17) that

$$\begin{aligned} P(N_n(B)=0) &= P\left(N_n(c_i,d_i]=0\,,\quad i=1,\ldots,k\right) \\ &= P\left(\max_{[nc_i]<j\le[nd_i]} X_j \le u_n\,,\quad i=1,\ldots,k\right) \\ &= \prod_{i=1}^{k} P\left(\max_{[nc_i]<j\le[nd_i]} X_j \le u_n\right) \\ &= \prod_{i=1}^{k} P\left(N_n(c_i,d_i]=0\right) \end{aligned}$$

$$\to \prod_{i=1}^{k} \exp\{-\tau(d_i - c_i)\} .$$

On the other hand, by the Poisson property of N,

$$P(N(B) = 0) = \exp\{-\tau|B|)\} = \exp\left\{-\tau \sum_{i=1}^{k} (d_i - c_i)\right\} .$$

This proves the theorem by virtue of Kallenberg's Theorem 5.2.2. □

The following example shows the close link between extreme value theory and the point processes of exceedances.

Example 5.3.3 (Continuation of Example 5.1.3)
An application of Theorem 5.3.2 together with (5.15) yields

$$P(X_{k,n} \le u_n) = P(N_n(0,1] < k) \to P(N(0,1] < k) = e^{-\tau} \sum_{i=0}^{k-1} \frac{\tau^i}{i!} .$$

This was the content of Theorem 4.2.3. Similar arguments as for Corollary 4.2.4 also allow us to derive the limit distribution of the kth order statistic for dfs F in the maximum domain of attraction of an extreme value distribution. □

In Example 5.1.6 we considered iid sum processes indexed by a renewal counting process and a corresponding point process: let (X_i) and (Y_i) be two independent sequences of iid rvs, suppose Y_1 is positive with probability 1 and set $T_i = Y_1 + \cdots + Y_i$. Then $N'(t) = \text{card}\{i : T_i \le t\}$ defines a renewal counting process and $S(t) = \sum_{i=1}^{N'(t)} X_i$ for $t > 0$ is the sum process. Here we consider the corresponding point process of exceedances

$$\widetilde{N}_n(\cdot) = \sum_{i=1}^{N'(n)} \varepsilon_{n^{-1}T_i}(\cdot) I_{\{X_i > u_n\}} \tag{5.18}$$

on the state space $E = (0,1]$. As before, (u_n) is a real–valued threshold sequence. The strong law of large numbers implies $n^{-1}T_{[nx]} \stackrel{\text{a.s.}}{\to} xEY_1 = x\lambda^{-1}$ for $x \in (0,1]$, and so we may hope that a result similar to Theorem 5.3.2 holds in this situation. That is indeed the case:

Theorem 5.3.4 (Weak convergence of point processes of exceedances, iid case and random index)
Let $(\widetilde{N}_n)$ be the point processes of exceedances (5.18) *of the threshold sequence (u_n). Assume that (u_n) satisfies* (5.16) *for some $\tau \in (0,\infty)$. Moreover, let $T_n = Y_1 + \cdots + Y_n$ be the points of a renewal counting process on $[0,\infty)$ with $EY_1 = \lambda^{-1} \in \mathbb{R}_+$. Then the relation $\widetilde{N}_n \stackrel{d}{\to} N$ holds in $M_p(E)$, where N is a homogeneous Poisson process on $E = (0,1]$ with intensity $\tau\lambda$.*

Proof. For an application of Kallenberg's Theorem 5.2.2 it remains to show the following two relations:

$$E\widetilde{N}_n(a,b] \;\to\; EN(a,b] = \tau\lambda(b-a)\,, \quad 0<a<b\le 1\,, \tag{5.19}$$

$$P(\widetilde{N}_n(B)=0) \;\to\; P(N(B)=0) \tag{5.20}$$

for all sets $B = \cup_{i=1}^m (c_i,d_i]$ with $0<c_1<d_1<\cdots<c_k<d_k\le 1$, $k\ge 1$. As above, we write $(N'(t))$ for the renewal counting process generated by (T_i). Then, by homogeneity of N' and in view of (5.16),

$$\begin{aligned}
E\widetilde{N}_n(a,b] &= E\sum_{i:a<n^{-1}T_i\le b} I_{\{X_i>u_n\}} \\
&= E\sum_{i=N'(na)+1}^{N'(nb)} I_{\{X_i>u_n\}} \\
&= E\sum_{i=1}^{N'(n(b-a))} I_{\{X_i>u_n\}} \\
&= \sum_{k=0}^{\infty} P\left(N'(n(b-a))=k\right) E\sum_{i=1}^{k} I_{\{X_i>u_n\}} \\
&= \sum_{k=0}^{\infty} P\left(N'(n(b-a))=k\right)\left(k\overline{F}(u_n)\right) \\
&= \left(n\overline{F}(u_n)\right)\left(n^{-1}EN'(n(b-a))\right) \\
&\to \tau\lambda(b-a) = EN(b-a)\,.
\end{aligned}$$

Here we also used that $n^{-1}EN'(n(b-a)) \sim \lambda(b-a)$; see Proposition 2.5.12. This proves (5.19). Next we turn to the proof of (5.20). For simplicity we restrict ourselves to the set $B=(c_1,d_1]\cup(c_2,d_2]$. Conditioning on N' and using the independence of (T_i) and (X_i) as well as the homogeneity of N', we obtain

$$\begin{aligned}
&P\left(\widetilde{N}_n(B)=0\right) \\
&= P\left(\max_{i:c_1<n^{-1}T_i\le d_1} X_i\le u_n\,,\ \max_{i:c_2<n^{-1}T_i\le d_2} X_i\le u_n\right) \\
&= P\left(\max_{N'(nc_1)<i\le N'(nd_1)} X_i\le u_n\,,\ \max_{N'(nc_2)<i\le N'(nd_2)} X_i\le u_n\right)
\end{aligned}$$

$$
\begin{aligned}
&= E\left((F(u_n))^{(N'(nd_1)-N'(nc_1))+(N'(nd_2)-N'(nc_2))}\right)\\
&= E\left((F(u_n))^{N'\left(n((d_1-c_1)+(d_2-c_2))\right)}\right)\\
&= E\exp\left\{\frac{N'(n((d_1-c_1)+(d_2-c_2)))}{n} n\ln(1-\overline{F}(u_n))\right\}\\
&\to \exp\left\{-\lambda\tau\Big((d_1-c_1)+(d_2-c_2)\Big)\right\}\\
&= P(N(B)=0)\,.
\end{aligned}
$$

In the last step we also used the SLLN for renewal counting processes (Theorem 2.5.10) and Lebesgue dominated convergence. This proves (5.20) and, by Kallenberg's theorem, also the assertion. □

Example 5.3.5 (Limit distribution for iid sequence with random index)
Let (X_i) be iid and independent of the renewal counting process $(N'(t))$ on $[0,\infty)$ with $EY_1 = \lambda^{-1}$. Denote by $X_{N'(t),N'(t)} \le \cdots \le X_{1,N'(t)}$ the order statistics of the random sample $X_1,\ldots,X_{N'(t)}$. Then we may conclude from Theorem 5.3.4 that

$$
\begin{aligned}
P\left(X_{k,N'(n)} \le u_n\right) &= P\left(N_n(0,1] < k\right)\\
&\to e^{-\tau\lambda}\sum_{i=0}^{k-1}\frac{(\tau\lambda)^i}{i!}\,,\qquad k=1,2,\ldots\,,
\end{aligned}
$$

provided $n\overline{F}(u_n) \to \tau \in (0,\infty)$. In particular, if $u_n = u_n(x) = c_n x + d_n$ and $n\overline{F}(u_n(x)) \to -\ln H(x)$, $x \in \mathbb{R}$, for some extreme value distribution H, then

$$
P\left(X_{k,N'(n)} \le c_n x + d_n\right) \to H^\lambda(x)\sum_{i=0}^{k-1}\frac{\left(-\ln H^\lambda(x)\right)^i}{i!}\,,\qquad k=1,2,\ldots\,.
$$

This result was given in Theorem 4.3.2 in a more general set–up. □

5.3.2 The Stationary Case

In this section we approach the problem of finding the limit distribution of the maxima M_n and of the upper order statistics of a sample from a strictly stationary sequence (X_n) via the point process of exceedances as introduced in (5.14). We assume that the conditions $D(u_n)$ and $D'(u_n)$ from Section 4.4 hold for a threshold sequence (u_n), and we cite them here for convenience:

Condition $D(u_n)$: *For any integers p, q and n*

$$1 \le i_1 < \cdots < i_p < j_1 < \cdots < j_q \le n$$

such that $j_1 - i_p \ge l$ *we have*

$$\left| P\left(\max_{i \in A_1 \cup A_2} X_i \le u_n\right) - P\left(\max_{i \in A_1} X_i \le u_n\right) P\left(\max_{i \in A_2} X_i \le u_n\right) \right| \le \alpha_{n,l}\,,$$

where $A_1 = \{i_1, \ldots, i_p\}$, $A_2 = \{j_1, \ldots, j_q\}$ *and* $\alpha_{n,l} \to 0$ *as* $n \to \infty$ *for some sequence* $l = l_n = o(n)$.

Condition $D'(u_n)$: *The relation*

$$\limsup_{n \to \infty} n \sum_{j=2}^{[n/k]} P\left(X_1 > u_n, X_j > u_n\right) \to 0$$

holds as $k \to \infty$.

Remark. For an interpretation of these conditions we refer to Section 4.4. We mention here that condition $D'(u_n)$ has an intuitive interpretation in the language of point processes: if (u_n) is chosen to satisfy $n\overline{F}(u_n) \to \tau \in (0, \infty)$ then there are on average approximately τ exceedances of u_n by $X_1, \ldots, X_n$, and hence τ/k among $X_1, \ldots, X_{[n/k]}$. Condition $D'(u_n)$ bounds the probability of more than one exceedance among $X_1, \ldots, X_{[n/k]}$. This will eventually ensure that there are no multiple points in the limiting Poisson process; i.e. this condition prevents clustering in the limit. In this context, Example 4.4.4 is quite instructive: condition $D'(u_n)$ is violated since maxima typically occur as pairs. □

Having in mind the results of Section 4.4 it is certainly not surprising that Theorem 5.3.2 remains valid for certain strictly stationary sequences:

Theorem 5.3.6 (Weak convergence of point processes of exceedances, stationary case)
Suppose (X_n) *is strictly stationary and* (u_n) *is a sequence of threshold values such that* (5.16), $D(u_n)$ *and* $D'(u_n)$ *hold. Let* (N_n) *be the processes* (5.14). *Then* $N_n \xrightarrow{d} N$ *in* $M_p(E)$, *where* N *is a homogeneous PRM on* $E = (0,1]$ *with intensity* τ.

Proof. We proceed as in the proof of Theorem 5.3.2 or 5.3.6, applying Kallenberg's Theorem 5.2.2. The proof of (5.9) is the same as in the iid case. Thus it remains to show (5.10) making use of $D(u_n)$ and $D'(u_n)$. For simplicity we restrict ourselves to sets $B = (c_1, d_1] \cup (c_2, d_2]$ with $0 < c_1 < d_1 < c_2 < d_2 \le 1$. The general case can be dealt with analogously.

Take $(a, b] \subset (0, 1]$. Using the stationarity of (X_n) and Proposition 4.4.3 we obtain

$$\begin{aligned} P\left(N_n(a,b]=0\right) &= P\left(\max_{i\le[nb]-[na]} X_i \le u_n\right) \\ &\to \exp\{-\tau(b-a)\} = P(N(a,b]=0)\,. \qquad (5.21) \end{aligned}$$

From condition $D(u_n)$ we conclude that

$$\begin{aligned} & P\left(N_n(B)=0\right) \\ &= P\left(N_n(c_1,d_1]=0\,, N_n(c_2,d_2]=0\right) \\ &= P\left(\max_{c_1<n^{-1}i\le d_1} X_i \le u_n\,,\ \max_{c_2<n^{-1}i\le d_2} X_i \le u_n\right) \\ &= P\left(\max_{c_1<n^{-1}i\le d_1} X_i \le u_n\right) P\left(\max_{c_2<n^{-1}i\le d_2} X_i \le u_n\right) + o(1)\,. \end{aligned}$$

Indeed, the distance between the two index sets

$$A_1 = \{[nc_1]+1,\ldots,[nd_1]\} \quad \text{and} \quad A_2 = \{[nc_2]+1,\ldots,[nd_2]\}$$

exceeds $(c_2-d_1)n > l_n = o(n)$ which implies that $\alpha_{n,l_n} \to 0$. Hence, by (5.21),

$$P\left(N_n(B)=0\right) \to \exp\left\{-\tau\Big((d_1-c_1)+(d_2-c_2)\Big)\right\} = P(N(B)=0)\,,$$

which concludes the proof of (5.10) and, by Kallenberg's theorem, proves the assertion. □

The following is analogous to Example 5.3.3:

Example 5.3.7 (Limit probabilities of upper order statistics)
As usual, let

$$X_{n,n} \le \cdots \le X_{1,n}$$

denote the order statistics of the sample $X_1,\ldots,X_n$. Suppose that the assumptions of Theorem 5.3.6 hold. Then

$$P\left(X_{k,n} \le u_n\right) = P\left(N_n(0,1] < k\right) \to P(N(0,1]<k) = e^{-\tau}\sum_{i=0}^{k-1}\frac{\tau^i}{i!}\,.$$

This extends Proposition 4.4.3 to the upper order statistics of a strictly stationary sequence. □

Now it is immediate that we can derive the limit distribution of an upper order statistic $X_{k,n}$ by the usual folklore. Let $(\widetilde{X}_n)$ be an associated iid sequence such that $X \stackrel{d}{=} \widetilde{X}$, and denote its order statistics in the natural way by $\widetilde{X}_{k,n}$.

Theorem 5.3.8 (Limit distribution of upper order statistics)
Let (X_n) be strictly stationary with common df $F \in \mathrm{MDA}(H)$ for an extreme value distribution H, i.e. there exist constants $c_n > 0$, $d_n \in \mathbb{R}$ such that

$$\lim_{n\to\infty} n\overline{F}\left(c_n x + d_n\right) = -\ln H(x)\,, \quad x \in \mathbb{R}\,.$$

Assume that the sequences $(u_n) = (c_n x + d_n)$, $x \in \mathbb{R}$, satisfy the conditions $D(u_n)$ and $D'(u_n)$. Then the relations

$$P\left(c_n^{-1}\left(X_{k,n} - d_n\right) \le x\right) \quad\to\quad H(x)\sum_{i=0}^{k-1}\frac{(-\ln H(x))^i}{i!}\,, \qquad x \in \mathbb{R}\,,$$

$$P(c_n^{-1}(\widetilde{X}_{k,n} - d_n) \le x) \quad\to\quad H(x)\sum_{i=0}^{k-1}\frac{(-\ln H(x))^i}{i!}\,, \qquad x \in \mathbb{R}\,,$$

hold for every $k \ge 1$. □

Theorem 5.3.8 shows the similarity between the asymptotic behaviour of the extremes of the stationary sequence (X_n) and of an associated iid sequence $(\widetilde{X}_n)$. This is again due to the conditions $D(u_n)$ and $D'(u_n)$.

In the following paragraphs we intend to generalise these results to a finite vector of order statistics. This means that we are interested in probabilities of the form

$$P\left(X_{1,n} \le u_n^{(1)}, \ldots, X_{k,n} \le u_n^{(k)}\right)$$

for k sequences of real numbers

$$u_n^{(k)} \le \cdots \le u_n^{(1)}\,. \tag{5.22}$$

Since we are dealing with k different sequences of thresholds $(u_n^{(i)})$, $i = 1, \ldots, k$, it seems appropriate to introduce a vector of k point processes of exceedances, one for each threshold sequence. However, the exceedances of the levels $u_n^{(i)}$ are very much related to each other. For example, an exceedance of $u_n^{(r)}$ is automatically an exceedance of $u_n^{(r+1)}$, and so it is possible by a geometric argument to reduce the problem of k exceedances to weak convergence of a point process on $(0,1] \times \mathbb{R}$. We refer to Leadbetter, Lindgren and Rootzén [418], Sections 5.5 and 5.6, for a complete description of their "thinning" procedure and omit details. We also omit the definition of the corresponding point processes and simply state the final result for the vector of exceedances. Before we can do this we have to introduce a k–dimensional analogue of condition $D(u_n)$ above. We suppose that the k sequences (5.22) are given.

Condition $D_k(\mathbf{u_n})$: *For any fixed* p, q *and for any integers*

$$1 \le i_1 < \cdots < i_p < j_1 < \cdots < j_q \le n$$

such that $j_1 - i_p \ge l$ *we have*

$$\left| P\left(X_{i_m} \le u_n^{(s_m)}\,,\, m = 1,\ldots,p, \quad X_{j_r} \le u_n^{(s'_r)}, r = 1,\ldots,q\right) \right.$$

$$\left. - \quad P\left(X_{i_m} \le u_n^{(s_m)}\,,\, m = 1,\ldots,p\right) P\left(X_{j_r} \le u_n^{(s'_r)}, r = 1,\ldots,q\right) \right| \le \alpha_{n,l}\,,$$

for any integers $1 \le s_l, s'_r \le k$, *and* $\alpha_{n,l} \to 0$ *as* $n \to \infty$ *for some sequence* $l = l_n = o(n)$.

It will not be necessary to define an extended $D'(\mathbf{u_n})$ condition, since we shall simply need to assume that $D'(u_n^{(i)})$ holds separately for each $i = 1,\ldots,k$.

As in Section 4.2 we write

$$B_n^{(i)} = \sum_{i=1}^{n} I_{\left\{X_i > u_n^{(i)}\right\}}\,, \quad n \ge 1\,, \quad i = 1,\ldots,k\,,$$

for the number of exceedances of $u_n^{(i)}$ by $X_1,\ldots,X_n$.

Theorem 5.3.9 (Joint weak convergence of the number of exceedances, stationary case)
Let (X_n) *be a strictly stationary sequence and suppose that the sequences* $(u_n^{(i)})$ *satisfy* (5.22) *and that* $n\overline{F}(u_n^{(i)}) \to \tau_i$ *for non-negative* τ_i, $i = 1,\ldots,k$. *Assume* $D_k(\mathbf{u_n})$ *and* $D'(u_n^{(i)})$ *for* $i = 1,\ldots,k$. *Then, for* $\ell_1,\ldots,\ell_k \ge 0$,

$$P\left(B_n^{(1)} = \ell_1, B_n^{(2)} = \ell_1 + \ell_2, \ldots, B_n^{(k)} = \ell_1 + \cdots + \ell_k\right)$$

$$\to \frac{\tau_1^{\ell_1}}{\ell_1!} \frac{(\tau_2 - \tau_1)^{\ell_2}}{\ell_2!} \cdots \frac{(\tau_k - \tau_{k-1})^{\ell_k}}{\ell_k!} e^{-\tau_k}\,, \qquad n \to \infty\,. \qquad \square$$

This theorem is completely analogous to the iid case; see Theorem 4.2.6. Moreover, as in the iid case, cf. Theorem 4.2.8, we obtain the joint limit law of the vector of upper order statistics:

Corollary 5.3.10 (Joint limit law of upper order statistics, stationary case)
Assume that $F \in \mathrm{MDA}(H)$ *with normalising constants* $c_n > 0$ *and centring constants* $d_n \in \mathbb{R}$. *Moreover, suppose that* $D_k(\mathbf{u_n})$ *and* $D'(u_n)$ *are satisfied for all sequences* $u_n = c_n x + d_n$, $x \in \mathbb{R}$. *Then the limit relation*

$$(c_n^{-1}(X_{i,n} - d_n))_{i=1,\ldots,k} \xrightarrow{d} (Y^{(i)})_{i=1,\ldots,k}\,, \quad k \ge 1\,, \quad n \to \infty\,,$$

holds, where $(Y^{(1)},\ldots,Y^{(k)})$ *is the* k*–dimensional extremal variate corresponding to the extreme value distribution* H. $\square$

Finally, we mention that all results for a vector of k upper order statistics which were given in Section 4.2 for the iid case remain valid for the strictly stationary case as well, provided that D and D' are satisfied.

Notes and Comments

The point process of exceeedances has been used extensively in the monograph by Leadbetter et al. [418] to build up an extreme value theory for iid and stationary sequences. There the theory presented above can be found in detail. In particular, they discuss the conditions $D(u_n)$ and $D'(u_n)$; see also Section 4.4. Further convergence results for the point process of exceedances are provided in Sections 5.5 and 8.4, where we consider linear and ARCH processes. In contrast to the present section the limiting point processes are not homogeneous Poisson but compound Poisson processes.

The point process techniques of this section could have been replaced by classical methods of extreme value theory. The latter were implicitly used for checking the assumptions of Kallenberg's theorem. Therefore the present section can be understood as an alternative approach to extreme value theory which is quite elegant in the case of stationary sequences. The real power of point process methods will become more transparent in Sections 5.4 and 5.5.

5.4 Applications of Point Process Methods to IID Sequences

In this section we apply point process techniques to the extremes of iid sequences (for some basic facts we refer to Chapters 3 and 4). We are mainly interested in records and record times. In Section 5.4.1 we give a short introduction to this topic. It is followed by some technical results (Section 5.4.2) which are used to embed the maxima of an iid sequence in an appropriate continuous–time process which in turn is a function of a PRM. This "coupling" construction is applied in Section 5.4.3 to derive limit results about the growth and the frequency of record times. In Section 5.4.4 we consider the weak convergence of maxima in a function space setting.

Throughout this section $X, X_1, X_2, \ldots$ is a sequence of iid rvs with common *continuous* df F. We also write

$$x_F^l = \inf\{x : F(x) > 0\} \quad \text{and} \quad x_F^r = \sup\{x : F(x) < 1\}$$

for the left and right endpoint of the distribution F. As usual, we denote the maximum of the first n rvs by

$$M_1 = X_1 \,, \quad M_n = \max_{i=1,\dots,n} (X_1, \dots, X_n) \,, \quad n \geq 2 \,.$$

Later on we will sometimes find it convenient to use $\wedge$, $\vee$ for min, max, respectively. The rvs Γ_i are always the points of a homogeneous Poisson process on $[0, \infty)$ with intensity 1. We can write them as

$$\Gamma_i = E_1 + \cdots + E_i \,, \quad i \geq 1 \,,$$

for an iid sequence of standard exponential rvs E_i.

5.4.1 Records and Record Times

In daily life we hear quite often about records; they are indeed omnipresent in sports, science, economy, environment etc. We hear about records of pollution, records of governmental debts, records in sports events, record insurance claims or record gains/losses in finance. Some clever people collect information about all sorts of records and write books about them.

What is a record in the context of extreme value theory?

If we consider observations X_n a record would be a temporary maximum (or minimum) in this sequence which will certainly change when time goes by. This is precisely the notion *record* which we intend to use in this chapter: a *record* X_n occurs if $X_n > M_{n-1}$. Clearly, the new maximum M_n coincides then with X_n. Notice that a record happens when there is a jump in the sequence (M_n). The times $L_1 < L_2 < \cdots$ when these jumps occur are random. For obvious reasons, they are called the *record times* of (X_n). In the insurance and financial context it is definitely an important issue to study both records and record times of sequences of rvs, dependent or independent. They give us some sort of prediction of the good or bad things which can happen in the future, in frequency and magnitude: big jumps in prices can lead to crashes of financial institutions; big claim sizes in an insurance portfolio can cause insolvency problems.

The following result describes the sequence of records (X_{L_n}) in terms of a PRM:

Theorem 5.4.1 (Point process description of records)
Let F be a continuous df with left endpoint x_F^l and right endpoint x_F^r. Then the records (X_{L_n}) of the iid sequence (X_n) are the points of a PRM(μ) *on (x_F^l, x_F^r) with mean measure μ given by*

$$\mu(a,b] = R(b) - R(a) \,, \quad x_F^l < a \leq b < x_F^r \,, \quad \textit{where} \quad R(x) = -\ln \overline{F}(x) \,.$$

In particular, if F is standard exponential then $R(t) = t$ and $(X_{L_n}) \stackrel{d}{=} (\Gamma_n)$ are the points of a homogeneous Poisson process on $\mathbb{R}_+$ with intensity 1.

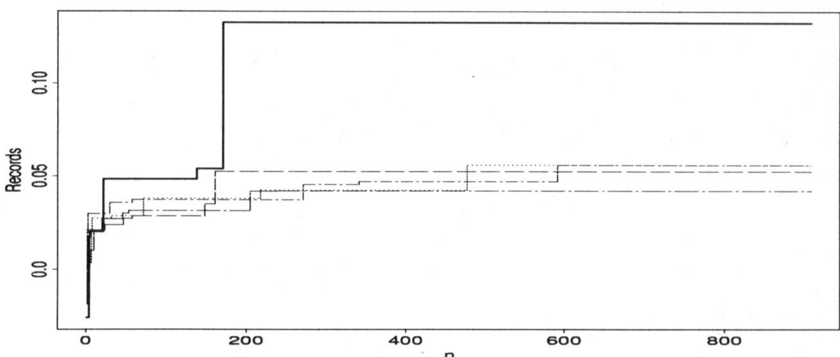

Figure 5.4.2 *Records (solid top line) of* 910 *daily log–returns of the Japanese stock index* NIKKEI *(February 22, 1990 – October 8, 1993) compared with four sample paths of records from* 910 *iid rvs. The rvs in the latter sequence are Gaussian with mean zero and the same variance as the* NIKKEI *data.*

Proof. Since F is continuous the function $R^{\leftarrow}$ is monotone increasing. Direct calculation yields

$$X_1 \stackrel{d}{=} R^{\leftarrow}(E_1)\,.$$

Indeed,

$$\begin{aligned} P\left(R^{\leftarrow}(E_1) \le x\right) &= P\left(E_1 \le R(x)\right) \\ &= 1 - e^{-R(x)} = F(x)\,. \end{aligned}$$

Hence the sequences (M_n) and

$$\left(\bigvee_{i=1}^{n} R^{\leftarrow}(E_i)\right) = \left(R^{\leftarrow}\left(\bigvee_{i=1}^{n} E_i\right)\right)$$

have the same distribution. Moreover, denoting by $(\widetilde{L}_n)(\stackrel{d}{=}(L_n))$ the record times of the sequence $(R^{\leftarrow}(E_i))$, we have for the sequences of records that

$$(X_{L_n}) \stackrel{d}{=} \left(R^{\leftarrow}\left(E_{\widetilde{L}_n}\right)\right)\,.$$

If F is standard exponential then the records $(R^{\leftarrow}(E_{\widetilde{L}_n})) = (E_{\widetilde{L}_n})$ are the points of a homogeneous Poisson process on $\mathbb{R}_+$ with intensity 1. This follows from the observation that $(E_{\widetilde{L}_n})$ is Markov with transition probabilities $\pi(x,(y,\infty)) = \exp\{-(y-x)\}$; see Resnick [530], Proposition 4.1. In view of Proposition 5.1.13, $(R^{\leftarrow}(E_{\widetilde{L}_n}))$ are then the points of a PRM with mean measure of $(a,b]$ given by

$$\left|(R^{\leftarrow})^{-1}(a,b]\right| = |\{s : a < R^{\leftarrow}(s) \le b\}| = R(b) - R(a)\,,$$

where $|\cdot|$ denotes Lebesgue measure. This concludes the proof. □

5.4.2 Embedding Maxima in Extremal Processes

The sequence $(M_n)_{n\ge1}$ defines a discrete–time stochastic process on the integers. We consider the finite–dimensional distributions of this process. We start with two dimensions: let $x_1 < x_2$ be real numbers and $t_1 < t_2$ be positive integers. Then

$$\begin{aligned} P(M_{t_1} \le x_1, M_{t_2} \le x_2) &= P\left(M_{t_1} \le x_1, \bigvee_{i=t_1+1}^{t_2} X_i \le x_2\right) \\ &= P(M_{t_1} \le x_1)\, P(M_{t_2-t_1} \le x_2) \\ &= F^{t_1}(x_1) F^{t_2-t_1}(x_2)\,. \end{aligned}$$

Moreover, if $x_1 > x_2$,

$$P(M_{t_1} \le x_1, M_{t_2} \le x_2) = F^{t_2}(x_2)\,.$$

Hence

$$P(M_{t_1} \le x_1, M_{t_2} \le x_2) = F^{t_1}(x_1 \wedge x_2)\; F^{t_2-t_1}(x_2)\,.$$

By induction we obtain

$$\begin{aligned} &P(M_{t_1} \le x_1, M_{t_2} \le x_2, \ldots, M_{t_m} \le x_m) \\ &= F^{t_1}\left(\bigwedge_{i=1}^{m} x_i\right) F^{t_2-t_1}\left(\bigwedge_{i=2}^{m} x_i\right) \cdots F^{t_m-t_{m-1}}(x_m) \end{aligned} \qquad (5.23)$$

for all positive integers $t_1 < t_2 < \cdots < t_m$, every $m \ge 1$ and real numbers x_i. From this representation it is not difficult to see that (M_n) is a Markov process; see Resnick [530], Section 4.1 or Breiman [90], Chapter 15.

We take (5.23) as the starting point for the definition of an *F–extremal process*: if we do not restrict ourselves to the non–negative integers, but if we allow for general real numbers $0 < t_1 < t_2 < \cdots < t_m$ then (5.23) defines a consistent family of distributions which, in view of Kolmogorov's consistency theorem, determines the distribution of a continuous–time process Y on $\mathbb{R}_+$.

Definition 5.4.3 (F–extremal process)
The process $Y = (Y(t))_{t>0}$ with finite-dimensional distributions (5.23) *is called an* extremal process generated by the df F *or an* F–extremal process. □

Thus the discrete–time process of the sample maxima (M_n) can be embedded in the continuous–time extremal process Y in the sense that

$$(M_n)_{n\geq 1} \stackrel{d}{=} (Y(n))_{n\geq 1} .$$

The latter relation is checked by a glance at the finite–dimensional distributions of (M_n) and Y at integer instants of time. The continuous–time process Y inherits the distributional properties of the sequence of maxima; it is a convenient tool for dealing with them.

An extremal process can be understood as a function of a PRM. Indeed, let

$$N = \sum_{k=1}^{\infty} \varepsilon_{(t_k, j_k)} \tag{5.24}$$

be PRM$(|\cdot| \times \mu)$ with state space $E = \mathbb{R}_+ \times \mathbb{R}$, where $|\cdot|$ denotes Lebesgue measure, and μ is given by the relation $\mu(a,b] = \ln F(b) - \ln F(a)$ for $a < b$. It is convenient to interpret (t_k, j_k) as coordinates of time (i.e. t_k) and space (i.e. j_k). Recall the definition of the Skorokhod space $\mathbb{D}$ of cadlag functions from Appendix A2.3 and define the mapping $\widetilde{T}_1 : M_p(E) \to \mathbb{D}\,(0,\infty)$ by

$$\widetilde{T}_1(N) = \widetilde{T}_1\left(\sum_{k=1}^{\infty} \varepsilon_{(t_k, j_k)}\right) = \sup\left\{j_k : t_k \leq \cdot\right\} . \tag{5.25}$$

Proposition 5.4.4 (Point process representation of F–extremal processes)
The F–extremal process $Y = (Y(t))_{t>0}$ has representation

$$Y(\cdot) \stackrel{d}{=} \sup\left\{j_k : t_k \leq \cdot\right\}$$

with respect to the PRM$(|\cdot| \times \mu)$ *defined in* (5.24).

Sketch of the proof. In view of the constructive definition of Y given above it suffices to show that the finite–dimensional distributions of Y and $\widetilde{T}_1(N)$ coincide. Fix $t > 0$. Notice that

$$\left\{\sup\left\{j_k : t_k \leq t\right\} \leq x\right\} \;=\; \left\{N\left((0,t] \times (x,\infty)\right) = 0\right\} .$$

Thus we have by definition of a PRM that

$$\begin{aligned} P\left(\sup\left\{j_k : t_k \leq t\right\} \leq x\right) &= P\left(N\left((0,t] \times (x,\infty)\right) = 0\right) \\ &= \exp\left\{-EN\left((0,t] \times (x,\infty)\right)\right\} \\ &= \exp\left\{-t\,\mu(x,\infty)\right\} \\ &= F^t(x) \\ &= P(Y(t) \leq x) . \end{aligned}$$

Similar arguments yield the finite–dimensional distributions in the general case. (The reader is urged to calculate them at least for two dimensions.) They can be shown to coincide with (5.23) which determine the whole distribution of Y. This concludes the proof. □

In the following we need another representation of an F–extremal process. It is a consequence of the following auxiliary result:

Lemma 5.4.5 *Assume F is continuous. Let N be the* PRM$(|\cdot|\times\mu)$ *on* $(0,t_0]\times(x_F^l,x_F^r)$, $t_0>0$, *as defined in* (5.24). *Then N has representation*

$$N'=\sum_{i=1}^{\infty}\varepsilon_{(U_i,Q^{\leftarrow}(\Gamma_i/t_0))}\,,$$

where (U_i) are iid uniform on $(0,t_0)$, independent of the points (Γ_i) of a homogeneous Poisson process on $[0,\infty)$ with intensity 1, *and*

$$Q^{\leftarrow}(y)=\inf\{s:Q(s)\le y\}\,,\quad Q(x)=-\ln F(x)\,.$$

Proof. In view of Remark 1 after Definition 5.1.7 it suffices to show that the Laplace functionals of N and N' coincide. Since N is PRM we know from Example 5.1.11 that

$$\Psi_N(g)=\exp\left\{-\int_{(0,t_0]}\int_{(x_F^l,x_F^r)}\left(1-e^{-g(t,x)}\right)d(\ln F(x))\,dt\right\}\,. \qquad (5.26)$$

Now, since F is continuous, $Q^{\leftarrow}$ is monotone decreasing. Conditioning on (Γ_i) and writing

$$g_1(x)=\frac{1}{t_0}\int_0^{t_0}\exp\{-g(t,x)\}\,dt=E\exp\{-g(U_1,x)\}\,,$$

we obtain

$$\begin{aligned}\Psi_{N'}(g)&=E\exp\left\{-\int_{(0,t_0]\times(x_F^l,x_F^r)}g\,dN'\right\}\\&=E\exp\left\{-\sum_{i=1}^{\infty}g\left(U_i,Q^{\leftarrow}(\Gamma_i/t_0)\right)\right\}\\&=E\prod_{i=1}^{\infty}g_1\left(Q^{\leftarrow}(\Gamma_i/t_0)\right)\\&=E\exp\left\{\sum_{i=1}^{\infty}\ln g_1\left(Q^{\leftarrow}(\Gamma_i/t_0)\right)\right\}\,.\end{aligned}$$

An application of Lemma 5.1.12 yields

$$\begin{aligned}\Psi_{N'}(g) &= \exp\left\{-\int_{\mathbb{R}_+}\left(1-g_1\left(Q^{\leftarrow}(z/t_0)\right)\right)dz\right\}\\ &= \exp\left\{-\int_{(0,t_0]}\int_{\mathbb{R}_+}\left(1-e^{-g(t,Q^{\leftarrow}(z))}\right)dz\,dt\right\}.\end{aligned}$$

Substituting x for $Q^{\leftarrow}(z)$ we arrive at the right–hand side of (5.26) which concludes the proof. □

An immediate consequence of Lemma 5.4.5 and Proposition 5.4.4 is the following

Corollary 5.4.6 *Let F be a continuous df, Y an F–extremal process. Then Y has representation*

$$Y(t) = \sup\left\{Q^{\leftarrow}(\Gamma_i/t_0) : U_i \le t\right\}, \quad t \in (0, t_0],$$

where (U_i) and (Γ_i) are defined in Lemma 5.4.5. □

The *jump times* τ_n of an F–extremal process are of particular interest since we may hope that jumps of (M_n) (the *records*) and of Y occur almost at the same time. This intuition will be made precise by a coupling argument in Section 5.4.3.

Theorem 5.4.7 (Point process of the jump times of an extremal process) *If F is continuous then*

$$N_\infty = \sum_{n=1}^{\infty} \varepsilon_{\tau_n} \tag{5.27}$$

is PRM(μ) *on $\mathbb{R}_+$ with intensity $f(t) = 1/t$, i.e.*

$$\mu(a,b] = \int_a^b f(t)\,dt = \ln b - \ln a \quad \text{for } a < b.$$

Proof. It suffices to show that N_∞ is PRM(μ) on $(0, t_0]$ for every fixed $t_0 > 0$. In view of Corollary 5.4.6 we may assume that the F–extremal process Y has representation

$$Y(t) = \sup\left\{Q^{\leftarrow}(\Gamma_i/t_0) : U_i \le t\right\}, \quad t \in (0, t_0].$$

Since F is continuous, $Q^{\leftarrow}$ is monotone decreasing, hence the jump times of the processes Y and $N_1(t) = \inf\{n \ge 1 : U_n \le t\}$ are identical. We may write

$$\begin{aligned} N_1(t) &= \inf\{n \geq 1 : U_n^{-1} \geq t^{-1}\} \\ &= \inf\left\{n \geq 1 : \bigvee_{i=1}^{n} U_i^{-1} \geq t^{-1}\right\}. \end{aligned}$$

Hence the jump times of N_1 in $(0, t_0]$ must be the records of $\max_{i=1,\ldots,n} U_i^{-1}$ in $[t_0^{-1}, \infty)$. By Theorem 5.4.1, the records of $\max_{i=1,\ldots,n} U_i^{-1}$ are the points of a PRM on $[t_0, \infty)$ with mean measure of $(a, b]$ given by

$$\begin{aligned} -\ln P\left(U_1^{-1} > b\right) - \left(-\ln P\left(U_1^{-1} > a\right)\right) &= -\ln(b^{-1}/t_0) + \ln(a^{-1}/t_0) \\ &= \ln(b/a). \end{aligned}$$

This concludes the proof. □

5.4.3 The Frequency of Records and the Growth of Record Times

In this section we use a special "coupling" construction of the jump times L_n of (M_n) and τ_n of the F–extremal process Y to derive information about the record times of the iid sequence (X_n). This will allow us to compare (L_n) and (τ_n) not only in distribution but also path by path.

By definition of Y (see Definition 5.4.3) $(M_n) \stackrel{d}{=} (Y(n))$. This relation allows us to assume that (L_n) and (τ_n) are defined on the same probability space in such a way that a jump of (M_n) (i.e. a *record*) at L_n (the *record time*) is also a jump of Y but the converse is not necessarily true. Indeed, Y is a continuous–time process, and so it may also have jumps in the open intervals $(L_n - 1, L_n)$. Recall the definition of the point process N_∞ of the jump times of Y from (5.27) and define the *point process of the record times* of (X_n) by

$$N = \sum_{i=1}^{\infty} \varepsilon_{L_i}. \qquad (5.28)$$

Then, given the above coupling construction of (L_n) and (τ_n),

$$\begin{aligned} \{N(n-1, n] = 1\} &= \{(X_i) \text{ has a record at time } n.\} \\ &= \{N_\infty(n-1, n] \geq 1\}. \end{aligned} \qquad (5.29)$$

The following question arises naturally:

How often does it actually happen that $N_\infty(n-1, n] > N(n-1, n]$?

The following result ensures that the sequences $(N_\infty(n-1, n])$ and $(N(n-1, n])$ are identical starting from a certain random index.

Proposition 5.4.8 (Coupling of N_∞ and N)
Assume the df F is continuous and that (L_n) and (τ_n) are constructed as above. Then there exists an integer–valued rv N_0 such that for almost every $\omega \in \Omega$,

$$N((n,n+1],\omega) = N_\infty((n,n+1],\omega)\,, \quad n \geq N_0(\omega)\,. \tag{5.30}$$

Proof. It suffices to show (see (5.29)) that the event $\{N_\infty(n,n+1] > 1\}$ occurs only finitely often with probability 1. By the Borel–Cantelli lemma, see Section 3.5, this is the case if

$$\sum_{n=1}^{\infty} P\left(N_\infty(n,n+1] > 1\right) < \infty\,. \tag{5.31}$$

Since N_∞ is PRM(μ) with $\mu(a,b] = \ln(b/a)$ (see Theorem 5.4.7), direct calculation shows that

$$\begin{aligned}
& P\left(N_\infty(n,n+1] > 1\right) \\
&= 1 - P\left(N_\infty(n,n+1] = 0\right) - P\left(N_\infty(n,n+1] = 1\right) \\
&= 1 - e^{-\ln(1+n^{-1})} - e^{-\ln(1+n^{-1})}\ln(1+n^{-1}) \\
&= 1 - (1+n^{-1})^{-1}\left(1+\ln(1+n^{-1})\right) \\
&\leq n^{-2}\,, \quad n \geq 1\,,
\end{aligned}$$

and (5.31) follows, which concludes the proof. □

Remark. The coupling relation (5.30) can be reformulated as follows: for almost every $\omega \in \Omega$ there exists an integer $j(\omega)$ such that

$$N_\infty((1,n],\omega) = j(\omega) + N((1,n],\omega)\,, \quad n \geq N_0(\omega)\,. \tag{5.32}$$

□

We use the coupling argument to answer the following question:

How often do records happen in a given period of time?

A first answer is supported by the following Poisson approximation to the point process N of the record times (see (5.28)):

Theorem 5.4.9 (Weak convergence of the point process of record times)
The limit relation

$$N_n(\cdot) = N(n\cdot) = \sum_{i=1}^{\infty} \varepsilon_{n^{-1}L_i}(\cdot) \xrightarrow{d} N_\infty(\cdot) = \sum_{i=1}^{\infty} \varepsilon_{\tau_i}(\cdot)$$

holds in $M_p(\mathbb{R}_+)$.

Proof. According to Theorem 5.2.3 and Remark 3 afterwards it suffices to show that

$$I_n = \int_{\mathbb{R}_+} g(x) dN_n(x) \xrightarrow{d} \int_{\mathbb{R}_+} g(x) dN_\infty(x) , \quad g \in C_K^+(\mathbb{R}_+) . \tag{5.33}$$

Since g has compact support, there exists an interval $[a,b] \subset \mathbb{R}_+$ such that $g(x) = 0$ for $x \notin [a,b]$. Hence

$$\begin{aligned} I_n &= \sum_{i=1}^{\infty} g(n^{-1} L_i) \\ &= \sum_{i: a \le n^{-1} i \le b} g(n^{-1} i) I_{\{N(i-1,i]=1\}} . \end{aligned}$$

Recalling the special construction (5.30) we obtain, for $na > N_0(\omega)$,

$$\begin{aligned} I_n &= \sum_{i: a \le n^{-1} i \le b} g(n^{-1} i) \; I_{\{N_\infty(i-1,i]=1\}} \\ &= \sum_{i: a \le n^{-1} i \le b} g(n^{-1} i) \; N_\infty(i-1, i] \\ &= \int_{\mathbb{R}_+} g_n(x) dN_\infty(x) = J_n , \end{aligned}$$

where $g_n(x) = \sum_{i=1}^{\infty} g(n^{-1} i) I_{(i-1,i]}(x)$. Thus we have shown that $I_n - J_n \xrightarrow{\text{a.s.}} 0$. By a Slutsky argument (see Appendix A2.5) it remains, for (5.33), to show that

$$J_n \xrightarrow{d} \int_{\mathbb{R}_+} g(x) dN_\infty(x) . \tag{5.34}$$

Recall from Example 5.1.14 that $N_\infty(\cdot)$ and $N_\infty(n \cdot)$ have the same distribution. Then

$$\begin{aligned} J_n &\stackrel{d}{=} \int_{\mathbb{R}_+} g_n(x) dN_\infty(nx) \\ &= \sum_{i=1}^{\infty} g(n^{-1} i) \; N_\infty \left(n^{-1}(i-1), n^{-1} i\right] \\ &\xrightarrow{\text{a.s.}} \int_{\mathbb{R}_+} g(x) dN_\infty(x) . \end{aligned}$$

In the last step we have used the defining properties of a Lebesgue integral, which here exists since g is continuous, has compact support and is bounded. This proves (5.34) and thus the theorem. □

It is an immediate consequence of this theorem that $N_n(a,b]$ is approximately $Poi(\ln(b/a))$ distributed:

$$\begin{aligned} N_n(a,b] &= \text{card}\left\{i : a < n^{-1}L_i \le b\right\} \\ &\xrightarrow{d} N_\infty(a,b] \stackrel{d}{=} Poi(\ln(b/a)) . \end{aligned}$$

Alternatively, the frequency of records in a given interval can be described by limit theorems for $N_\infty(1,t]$. *In the following we assume that* $\tau_1 > 1$. Otherwise we may consider only that part of the sequence (τ_n) for which $\tau_i > 1$. Since N_∞ is PRM(μ) on $\mathbb{R}_+$ with $\mu(a,b] = \ln(b/a)$ we may work with the representation (see Example 5.1.14)

$$N_\infty = \sum_{i=1}^{\infty} \varepsilon_{\exp\{\Gamma_i\}}, \tag{5.35}$$

where, as usual, (Γ_i) are the points of a homogeneous Poisson process on $[0,\infty)$ with intensity 1. Thus

$$N_\infty(1,t] \;=\; \text{card}\left\{i : 1 < e^{\Gamma_i} \le t\right\} = \text{card}\left\{i : 0 < \Gamma_i \le \ln t\right\} .$$

It is immediate that we can now apply the whole limit machinery for renewal counting processes from Section 2.5.2. For the time–changed renewal counting process $(N_\infty(1,t])$ we obtain the following: let Φ denote the standard normal distribution. Then

$$\left.\begin{aligned} &\text{SLLN} && \lim_{t\to\infty} (\ln t)^{-1} N_\infty(1,t] = 1 \text{ a.s.}, \\ &\text{LIL} && \limsup_{t\to\infty} \,(2\ln t \;\ln\ln\ln t)^{-1/2}\,(N_\infty(1,t] - \ln t) \\ &&& = -\liminf_{t\to\infty}\,(2\ln t\;\ln\ln\ln t)^{-1/2}\,(N_\infty(1,t] - \ln t) = 1 \text{ a.s.}, \\ &\text{CLT} && (\ln t)^{-1/2}\,(N_\infty(1,t] - \ln t) \xrightarrow{d} \Phi . \end{aligned}\right\} \tag{5.36}$$

The coupling construction (5.32) immediately implies that

$$c_n^{-1}\,(N_\infty(1,n] - N(1,n]) \xrightarrow{\text{a.s.}} 0$$

provided $c_n \to \infty$. This ensures that we may replace N_∞ by N and t by n, everywhere in (5.36):

Theorem 5.4.10 (Limit results for the frequency of records)
Suppose F has a continuous distribution, let (X_n) be an iid sequence with record times (L_n) and let N be the corresponding point process (5.28). *Then the following relations hold:*

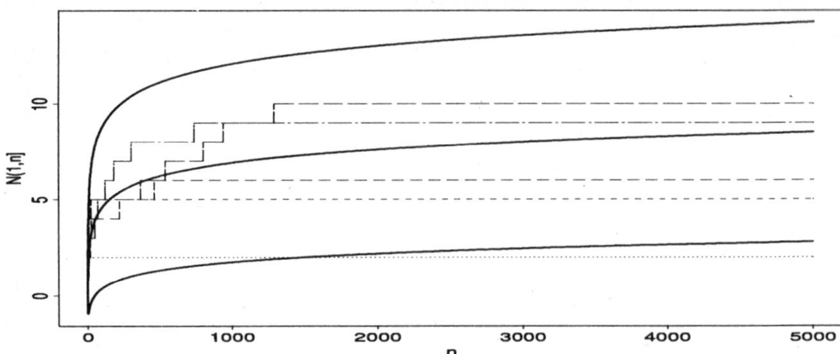

Figure 5.4.11 *The number of records $N(1,n]$, $n \le 5\,000$, from iid standard normal rvs. Five sample paths are given. The solid lines indicate the graphs of $\ln n$ (middle) and the 95% asymptotic confidence bands (top and bottom) based on Theorem 5.4.10.*

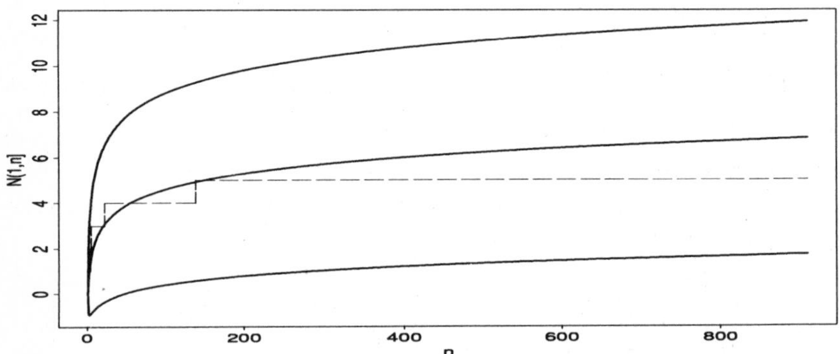

Figure 5.4.12 *The number of records $N(1,n]$, $n \le 910$, from 910 daily log-returns of the Japanese stock index* NIKKEI *(February 22, 1990 – October 8, 1993). The solid lines are the graphs of $\ln n$ (middle) and the 95% asymptotic confidence bands for the iid case; see Theorem 5.4.10.*

$$\begin{aligned}
&\text{SLLN} \quad \lim_{t\to\infty} (\ln t)^{-1} N(1,t] = 1 \quad \text{a.s.}\,,\\
&\text{LIL} \quad \limsup_{t\to\infty}\ (2\ln t\ \ln\ln\ln t)^{-1/2}\,(N(1,t]-\ln t)\\
&\qquad\quad = -\liminf_{t\to\infty}\ (2\ln t\ \ln\ln\ln t)^{-1/2}\,(N(1,t]-\ln t) = 1 \text{ a.s.}\,,\\
&\text{CLT} \quad (\ln t)^{-1/2}\,(N(1,t]-\ln t) \xrightarrow{d} \Phi\,,
\end{aligned}$$

where Φ denotes the standard normal distribution. □

Finally, we attack the following problem:

When do the records of (X_n) occur?

The coupling construction (5.32) again gives an answer: for $n \geq N_0(\omega)$ and almost every ω,

$$\left|L_n(\omega) - \tau_{n+j(\omega)}(\omega)\right| \leq 1\,.$$

Hence, by (5.35),

$$\begin{aligned}\ln L_n &= \ln\left(e^{\Gamma_{n+j}}\left(1 + O\left(e^{-\Gamma_{n+j}}\right)\right)\right)\\ &= \Gamma_{n+j} + o(1) \quad \text{a.s.}\end{aligned}$$

since $\Gamma_{n+j} = O(n)$ a.s. by the SLLN. We learnt in Example 3.5.6 that

$$\lim_{n\to\infty} (\ln n)^{-1} \max(E_1, \ldots, E_n) = 1 \quad \text{a.s.}$$

Hence

$$\ln L_n = \Gamma_n + (\Gamma_{n+j} - \Gamma_n) + o(1) = \Gamma_n + O(\ln n) \quad \text{a.s.}$$

This and the classical limit theory for sums of iid rvs (see Sections 2.1 and 2.2) yield the following:

Theorem 5.4.13 (Limit results for the growth of record times)
Assume F is continuous. Then the following relations hold for the record times L_n of an iid sequence (X_n) :

$$\begin{aligned}&\text{SLLN} \quad \lim_{n\to\infty} n^{-1} \ln L_n = 1 \quad \text{a.s.}\,,\\ &\text{LIL} \quad \limsup_{n\to\infty}\,(2n\,\ln\ln n)^{-1/2}\,(\ln L_n - n)\\ &\qquad = -\liminf_{n\to\infty}\,(2n\,\ln\ln n)^{-1/2}\,(\ln L_n - n) = 1 \text{ a.s.}\,,\\ &\text{CLT} \quad n^{-1/2}\,(\ln L_n - n) \stackrel{d}{\to} \Phi\,,\end{aligned}$$

where Φ denotes the standard normal distribution. □

In summary, the number of records in the interval $(1, t]$ is roughly of the order $\ln t$. Thus records become more and more unlikely for large t. Alternatively, the record times L_n grow roughly exponentially like $\exp\{\Gamma_n\}$ (or $\exp\{n\}$) and thus the period between two successive records becomes bigger and bigger.

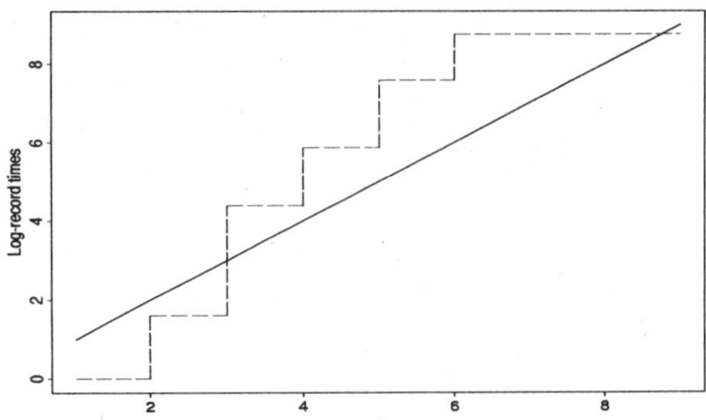

Figure 5.4.14 *The logarithmic record times of* 1 864 *daily log–returns of the* S&P *index. According to Theorem 5.4.13, the logarithmic record times should grow roughly linearly provided that they come from iid data.*

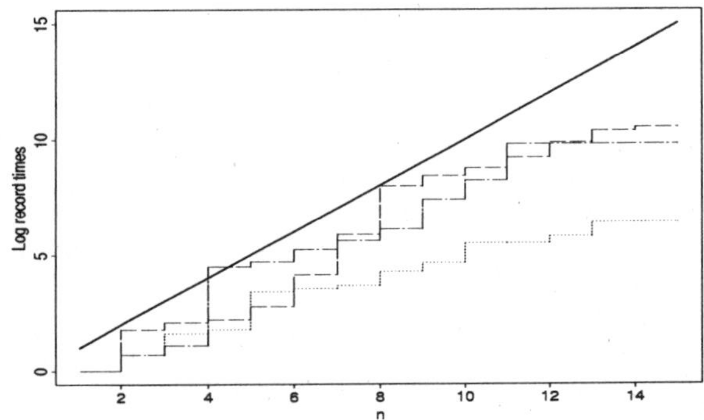

Figure 5.4.15 *The logarithmic record times of* 100 000 *iid standard normal rvs. Three sample paths are given. The straight line indicates the ideal asymptotic behaviour of these record times; see Theorem 5.4.13.*

5.4.4 Invariance Principle for Maxima

In Section 5.4.2 we embedded the sequence of the sample maxima (M_n) in a continuous–time F–extremal process Y. This was advantageous because we could make use of the hidden Poisson structure of Y to derive limit results about records and record times. In the sequel we are interested in the question:

How can we link the weak convergence of sample maxima with the weak convergence of point processes?

Intuitively, we try to translate the problem about the extremes of the sequence (X_n) for some particular df F into a question about the extremes of an iid sequence with common extreme value distribution H. Since there are only three standard extreme value distributions H, but infinitely many dfs F in the maximum domain of attraction of H ($F \in \mathrm{MDA}(H)$) this is quite a promising approach.

To make this idea precise suppose that F belongs to the maximum domain of attraction of H, i.e. there exist constants d_n and $c_n > 0$ such that

$$c_n^{-1}(M_n - d_n) \xrightarrow{d} H, \quad n \to \infty, \tag{5.37}$$

where H is one of the standard extreme value distributions (Weibull, Fréchet, Gumbel) as introduced in Definition 3.2.6. Set

$$Y_n(t) = \begin{cases} c_n^{-1}\left(M_{[nt]} - d_n\right) & \text{if} \quad t \geq n^{-1}, \\ c_n^{-1}(X_1 - d_n) & \text{if} \quad 0 < t < n^{-1}, \end{cases}$$

where $[x]$ denotes the integer part of x. Recall the notion of weak convergence in the Skorokhod space $\mathbb{D}(0,\infty)$ from Appendix A2.

The processes (Y_n) obey a result which parallels very much the Donsker invariance principle for sums of iid random variables; see Theorem 2.4.4.

Theorem 5.4.16 (Invariance principle for maxima)
Let H be one of the extreme value distributions and $Y = (Y(t))_{t>0}$ the corresponding H-extremal process. Then the relation

$$Y_n \xrightarrow{d} Y, \quad n \to \infty,$$

holds in $\mathbb{D}(0,\infty)$ if and only if (5.37) is satisfied.

Sketch of the proof. For a detailed proof see Resnick [530], Proposition 4.20. Take $t = 1$. Then $Y_n \xrightarrow{d} Y$ obviously implies (5.37), i.e.

$$Y_n(1) = c_n^{-1}(M_n - d_n) \xrightarrow{d} Y(1),$$

where $Y(1)$ has distribution H.
Now suppose that (5.37) holds. This is known to be equivalent to

$$n\overline{F}(c_n x + d_n) \to -\ln H(x) \tag{5.38}$$

on the support S of H; see Proposition 3.1.1. We define

$$\xi_{n,j} = \begin{cases} c_n^{-1}(X_j - d_n) & \text{if } c_n^{-1}(X_j - d_n) \in S, \\ \inf S & \text{otherwise}, \end{cases}$$

and

$$\mu(a,b] = \ln H(b) - \ln H(a)$$

for $(a,b] \subset S$. Topologising the state space E in the right way, (5.38) just means (see Proposition A2.12) that

$$nP(\xi_{n,1} \in \cdot) \stackrel{v}{\rightarrow} \mu(\cdot) \tag{5.39}$$

on the Borel sets of S. Now define

$$N_n = \sum_{k=1}^{\infty} \varepsilon_{(n^{-1}k, \xi_{n,k})}, \qquad N = \sum_{k=1}^{\infty} \varepsilon_{(t_k, j_k)},$$

where N is PRM$(|\cdot| \times \mu)$ on $\mathbb{R}_+ \times S$ and $|\cdot|$ denotes Lebesgue measure. Then Theorem 5.2.4 and (5.39) imply that $N_n \stackrel{d}{\rightarrow} N$. Recall the definition of the mapping $\widetilde{T}_1$ from (5.25). If we restrict ourselves to path spaces in which both N_n and N live then $\widetilde{T}_1$ can be shown to be a.s. continuous. An application of the continuous mapping theorem (see Theorem A2.6) yields that

$$\widetilde{T}_1(N_n) = \bigvee_{n^{-1}k \leq \cdot} \xi_{n,k} \quad \stackrel{d}{\rightarrow} \quad \widetilde{T}_1(N) = \bigvee_{t_k \leq \cdot} j_k$$

in $\mathbb{D}(0,\infty)$. Note that in view of Proposition 5.4.4

$$Y(\cdot) \stackrel{d}{=} \bigvee_{t_k \leq \cdot} j_k \,.$$

Moreover, one can show that in $\mathbb{D}(0,\infty)$ the relation

$$Y_n(\cdot) - \bigvee_{n^{-1}k \leq \cdot} \xi_{n,k} \stackrel{P}{\rightarrow} 0$$

is valid. This proves that $Y_n \stackrel{d}{\rightarrow} Y$. □

Remark. In the course of the proof above we left out all messy details. We also swept certain problems under the carpet which are related to the fact that the $\xi_{n,k}$ can be concentrated on the whole real line. This requires for instance a special treatment for $F \in \text{MDA}(\Phi_\alpha)$ (equivalently, $\overline{F} \in \mathcal{R}_{-\alpha}$) since a regular variation assumption on the right tail does naturally not influence the left tail of the distribution. Read Resnick [530], Section 4.4.2! □

This invariance principle encourages one to work with the H–extremal process Y instead of the process Y_n of sample maxima for F in the maximum domain of attraction of H. Thus, in an asymptotic sense, we are allowed to work with the distribution of Y instead of the one for Y_n. We stop here the discussion and refer to Section 5.5 where the weak convergence of extremal processes and of the underlying point processes is used to derive limit results about the upper extremes of dependent sequences.

Notes and Comments

We have seen in this section that point process techniques are very elegant tools for dealing with extremal properties of sequences of iid rvs. They allow us to derive deep results about the structure of extremal processes, of their jump times, about records, record times, exceedances etc. The basic idea is always to find the right point process, to show weak convergence to a PRM and possibly to apply the continuous mapping theorem in a suitable way.

The elegance of the method is one side of the coin. We have seen from the above outline of proofs that we have to be familiar with many tools from functional analysis, measure theory and stochastic processes. In particular, the proof of the a.s. continuity of the $\widetilde{T}$–mappings is never trivial and requires a deep understanding of stochastic processes. The a.s. continuity of the $\widetilde{T}$–mappings was treated for instance in Mori and Oodaira [468], Resnick [529], Serfozo [577].

Excellent references for extreme value theory in the context of point processes are Falk, Hüsler and Reiss [225], Leadbetter, Lindgren and Rootzén [418], Reiss [527] and Resnick [529, 530]. We followed closely the last source in our presentation.

In Section 6.2.5 we consider records as an exploratory statistical tool. There we also give some further references to literature on records.

5.5 Some Extreme Value Theory for Linear Processes

In Sections 4.4 and 5.3.2 we found conditions which ensured that the extremal behaviour of the strictly stationary sequence (X_n) is the same as that of an associated iid sequence $(\widetilde{X}_n)$, i.e. an iid sequence with the same common df F as $X = X_0$. Intuitively, those conditions $D(u_n)$ and $D'(u_n)$ guaranteed that high level exceedances by the sequence (X_n) were separated in time; i.e. clustering of extremes was avoided. This will change dramatically for the special class of strictly stationary sequences which we consider in this section. We suppose that (X_n) has representation as a *linear process*, i.e.

$$X_n = \sum_{j=-\infty}^{\infty} \psi_j Z_{n-j} \,, \quad n \in \mathbb{Z} \,,$$

where the *noise* sequence or the *innovations* (Z_n) are iid and the ψ_j are real numbers to be specified later. For simplicity we set $Z = Z_0$. Here we study linear processes from the point of view of extreme value theory. In Chapter 7 they are reconsidered from the point of view of time series analysis. Linear processes are basic in classical time series analysis. In particular, every ARMA process is linear, see Example 7.1.1, and most interesting Gaussian stationary sequences have a linear process representation.

Again we are interested in exceedances of a given deterministic sequence of thresholds (u_n) by the process (X_n), and in the joint distribution of a finite number of upper order statistics of a sample $X_1, \ldots, X_n$. We compare sequences of sample maxima for the noise (Z_n), the stationary sequence (X_n) and an associated iid sequence $(\widetilde{X}_n)$. As usual, (M_n) denotes the sequence of the sample maxima of (X_n).

5.5.1 Noise in the Maximum Domain of Attraction of the Fréchet Distribution Φ_α

We assume that Z satisfies the following condition:

$$\overline{F}_Z(x) = P(Z > x) = \frac{L(x)}{x^\alpha}, \quad x > 0, \tag{5.40}$$

for some $\alpha > 0$ and a slowly varying function L, i.e. $L(x)x^{-\alpha}$ is regularly varying with index $-\alpha$; see Appendix A3. By Theorem 3.3.7 this is equivalent to $Z \in \mathrm{MDA}(\Phi_\alpha)$ where

$$\Phi_\alpha(x) = e^{-x^{-\alpha}}, \quad x > 0,$$

denotes the standard Fréchet distribution which is one of the extreme value distributions; see Definition 3.2.6. Moreover, we assume that the tails are balanced in the sense that

$$\lim_{x\to\infty} \frac{P(Z > x)}{P(|Z| > x)} = p, \quad \lim_{x\to\infty} \frac{P(Z \le -x)}{P(|Z| > x)} = q, \tag{5.41}$$

for some $0 < p \le 1$ and such that $p + q = 1$. Thus we can combine (5.40) and (5.41):

$$\overline{F}_Z(x) = \frac{L(x)}{x^\alpha}, \quad x > 0, \quad F_Z(-x) \sim \frac{q}{p}\frac{L(x)}{x^\alpha}, \quad x \to \infty. \tag{5.42}$$

We also suppose that

$$\sum_{j=-\infty}^{\infty} |\psi_j|^\delta < \infty \quad \text{for some } 0 < \delta < \min(\alpha, 1). \tag{5.43}$$

This condition implies the absolute a.s. convergence of the linear process representation of X_n for every n; see also the discussion in Section 7.2. Note that the conditions here are very much like in Sections 7.2–7.5, but there we restrict ourselves to symmetric α–stable ($s\alpha s$) Z_n for some $\alpha < 2$. In that case,

$$\overline{F}_Z(x) \sim \frac{c}{x^\alpha}, \quad x \to \infty, \qquad \lim_{x\to\infty} \frac{P(Z > x)}{P(|Z| > x)} = \frac{1}{2},$$

hence (5.40) and (5.41) are naturally satisfied. We also mention that, if $\alpha < 2$, then the conditions (5.40) and (5.41) imply that Z has a distribution in the domain of attraction of an α–stable law; see Section 2.2.

We plan to reduce the study of the extremes of (X_n) to the study of the extremes of the iid sequence (Z_n). We choose the normalisation

$$c_n = (1/\overline{F}_Z)^{\leftarrow}(n), \tag{5.44}$$

where $f^{\leftarrow}$ denotes the generalised inverse of the function f. By (5.40) this implies that $\overline{F}_Z(c_n) \sim n^{-1}$. Then we also know that

$$c_n = n^{1/\alpha} L_1(n)$$

for a slowly varying function L_1. Moreover, from Theorem 3.3.7 we are confident of the limit behaviour

$$c_n^{-1} \max(Z_1, \ldots, Z_n) \xrightarrow{d} \Phi_\alpha .$$

So we may hope that c_n is also the right normalisation for the maxima M_n of the linear process (X_n).

We first embed $(c_n^{-1} X_k)_{k\ge1}$ in a point process and show its weak convergence to a function of a PRM. This is analogous to the proof of Theorem 5.4.16. Then we can proceed as in the iid case to derive information about the extremes of the sequence (X_n).

Theorem 5.5.1 (Weak convergence of the point processes of the embedded linear process)
Let $\sum_{k=1}^\infty \varepsilon_{(t_k, j_k)}$ be PRM$(|\cdot|\times\mu)$ *on $\mathbb{R}_+ \times E$, where $E = [-\infty,\infty]\backslash\{0\}$, $|\cdot|$ is Lebesgue measure and the measure μ on the Borel sets of E has density*

$$\alpha x^{-\alpha-1} I_{(0,\infty]}(x) + q p^{-1} \alpha (-x)^{-\alpha-1} I_{[-\infty,0)}(x), \quad x \in \mathbb{R}. \tag{5.45}$$

Suppose the conditions (5.42) *and* (5.43) *are satisfied. Then*

$$\sum_{k=1}^\infty \varepsilon_{\left(n^{-1}k, c_n^{-1} X_k\right)} \xrightarrow{d} \sum_{k=1}^\infty \sum_{i=-\infty}^\infty \varepsilon_{(t_k, \psi_i j_k)}, \quad n \to \infty,$$

in $M_p(\mathbb{R}_+ \times E)$.

Sketch of the proof. For a complete proof we refer to Davis and Resnick [160]; see also Resnick [530], Section 4.5.

We notice that condition (5.42) is equivalent to

$$nP\left(c_n^{-1} Z \in \cdot\right) \xrightarrow{v} \mu(\cdot)$$

on the Borel sets of E, where the measure μ on E is determined by (5.45). This holds by virtue of Proposition A2.12 and since, as $n \to \infty$,

$$nP\left(c_n^{-1}Z > x\right) \to x^{-\alpha} \quad \text{and} \quad nP\left(c_n^{-1}Z \le -x\right) \to qp^{-1}x^{-\alpha}\,, \quad x > 0\,.$$

It is then a consequence of Theorem 5.2.4 (see also the proof of Theorem 5.4.16) that

$$\sum_{k=1}^{\infty} \varepsilon_{\left(n^{-1}k,c_n^{-1}Z_k\right)} \stackrel{d}{\to} \sum_{k=1}^{\infty} \varepsilon_{(t_k,j_k)}\,, \quad n \to \infty\,, \tag{5.46}$$

in $M_p\left(\mathbb{R}_+ \times E\right)$, where the limit is PRM$(|\cdot| \times \mu)$.

The process $X_n = \sum_{j=-\infty}^{\infty} \psi_j Z_{n-j}$ is a (possibly infinite) moving average of the iid noise (Z_n). A naive argument suggests that we should first consider finite moving averages

$$X_n^{(m)} = \sum_{j=-m}^{m} \psi_j Z_{n-j}\,, \quad n \in \mathbb{Z}\,,$$

for a fixed integer m, then apply a Slutsky argument (see Appendix A2.5) and let $m \to \infty$.

For simplicity we restrict ourselves to the case $m = 1$ and we further assume that $(X_n^{(1)})$ is a moving average process of order 1 (MA(1)):

$$X_n^{(1)} = Z_n + \psi_1 Z_{n-1}\,, \quad n \in \mathbb{Z}\,.$$

We embed $(X_n^{(1)})$ in a point process which will be shown to converge weakly. We notice that $X_n^{(1)}$ is just a functional of the 2–dimensional vector

$$\mathbf{Z}_n = (Z_{n-1}, Z_n) = Z_{n-1}\mathbf{e}_1 + Z_n\mathbf{e}_2\,,$$

and so it is natural to consider the point process

$$\sum_{k=1}^{\infty} \varepsilon_{\left(n^{-1}k,c_n^{-1}\mathbf{Z}_k\right)}\,. \tag{5.47}$$

By some technical arguments it can be shown that (5.47) has the same weak limit behaviour as

$$\sum_{k=1}^{\infty} \left(\varepsilon_{\left(n^{-1}k,c_n^{-1}Z_k\mathbf{e}_1\right)} + \varepsilon_{\left(n^{-1}k,c_n^{-1}Z_k\mathbf{e}_2\right)}\right)\,.$$

Then an application of (5.46) and the continuous mapping theorem (see Theorem A2.6) yield that the point processes (5.47) converge weakly to

$$\sum_{k=1}^{\infty}\left(\varepsilon_{(t_k,j_k\mathbf{e}_1)}+\varepsilon_{(t_k,j_k\mathbf{e}_2)}\right).$$

Since we want to deal with the MA(1) process $(X_n^{(1)})$ we have to stick the coordinates of $\mathbf{Z}_n$ together, and this is again guaranteed by an a.s. continuous mapping $\widetilde{T}_2$, say, acting on the point processes:

$$\begin{aligned}
&\sum_{k=1}^{\infty}\varepsilon_{\left(n^{-1}k,c_n^{-1}(Z_k+\psi_1 Z_{k-1})\right)}\\
&= \widetilde{T}_2\left(\sum_{k=1}^{\infty}\varepsilon_{\left(n^{-1}k,c_n^{-1}\mathbf{Z}_k\right)}\right)\\
&\approx \widetilde{T}_2\left(\sum_{k=1}^{\infty}\left(\varepsilon_{\left(n^{-1}k,c_n^{-1}Z_k\mathbf{e}_1\right)}+\varepsilon_{\left(n^{-1}k,c_n^{-1}Z_k\mathbf{e}_2\right)}\right)\right)\\
&\xrightarrow{d} \widetilde{T}_2\left(\sum_{k=1}^{\infty}\left(\varepsilon_{(t_k,j_k\mathbf{e}_1)}+\varepsilon_{(t_k,j_k\mathbf{e}_2)}\right)\right)\\
&= \sum_{k=1}^{\infty}\left(\varepsilon_{(t_k,j_k)}+\varepsilon_{(t_k,\psi_1 j_k)}\right).
\end{aligned}$$

Similar arguments prove that

$$\sum_{k=1}^{\infty}\varepsilon_{\left(n^{-1}k,c_n^{-1}X_k^{(m)}\right)}\xrightarrow{d}\sum_{k=1}^{\infty}\sum_{i=-m}^{m}\varepsilon_{(t_k,\psi_i j_k)}$$

for every $m \geq 1$, and a Slutsky argument as $m \to \infty$ concludes the proof. $\square$

It is now our goal to consider some applications of this theorem. We suppose throughout that the assumptions of Theorem 5.5.1 are satisfied.

Extremal Processes and Limit Distributions of Maxima

Analogously to iid sample maxima we consider the continuous–time process

$$Y_n(t)=\begin{cases} c_n^{-1}M_{[nt]} & \text{if } \ t\geq n^{-1},\\ c_n^{-1}X_1 & \text{if } \ 0<t<n^{-1},\end{cases}$$

which is constructed from the sample maxima

$$M_n=\max(X_1,\ldots,X_n)\,,\quad n\geq 1\,.$$

Note that M_n is now the maximum of n dependent rvs. Define

$$\psi_+ = \max_j (\psi_j \vee 0) , \quad \psi_- = \max_j ((-\psi_j) \vee 0) . \tag{5.48}$$

Recall the definition of the mapping $\widetilde{T}_1$ from (5.25): for a point process $\sum_{k=1}^{\infty} \varepsilon_{(r_k,s_k)}$ set

$$\widetilde{T}_1\left(\sum_{k=1}^{\infty} \varepsilon_{(r_k,s_k)}\right) = \sup\{s_k : r_k \leq \cdot\} .$$

It is an a.s. continuous mapping from $M_p(\mathbb{R}_+ \times E)$ to $\mathbb{D}(0,\infty)$. This relation, Theorem 5.5.1 and the continuous mapping theorem yield that

$$\begin{aligned}
\widetilde{T}_1\left(\sum_{k=1}^{\infty} \varepsilon_{(n^{-1}k,c_n^{-1}X_k)}\right) &\stackrel{d}{=} Y_n(\cdot) \stackrel{d}{\to} \\
\widetilde{T}_1\left(\sum_{k=1}^{\infty}\sum_{i=-\infty}^{\infty} \varepsilon_{(t_k,\psi_i j_k)}\right) &= \bigvee_{t_k \leq \cdot}\left(\bigvee_{i=-\infty}^{\infty} \psi_i j_k\right) \\
&= \bigvee_{t_k \leq \cdot}\left((\psi_+ j_k) \vee (-\psi_- j_k)\right) = Y(\cdot) .
\end{aligned}$$

The process Y defined thus is indeed an extremal process (see Definition 5.4.3 and Proposition 5.4.4) since $Y = \widetilde{T}_1(\widetilde{N})$ where

$$\widetilde{N} = \sum_{k=1}^{\infty} \varepsilon_{(t_k,\psi_+ j_k)} + \sum_{k=1}^{\infty} \varepsilon_{(t_k,-\psi_- j_k)} ,$$

i.e. $\widetilde{N}$ is a PRM with mean measure of $(0,t] \times (x,\infty)$ equal to

$$t\left(\psi_+^{\alpha} + \psi_-^{\alpha} q p^{-1}\right) x^{-\alpha} \quad \text{for} \quad t > 0 , x > 0 .$$

By the definition of a PRM, for $t > 0$, $x > 0$,

$$\begin{aligned}
P(Y(t) \leq x) &= P\left(\widetilde{N}\left((0,t] \times (x,\infty)\right) = 0\right) \\
&= \exp\left\{-E\widetilde{N}\left((0,t] \times (x,\infty)\right)\right\} \\
&= \exp\left\{-t\left(\psi_+^{\alpha} + \psi_-^{\alpha} q p^{-1}\right) x^{-\alpha}\right\} .
\end{aligned}$$

Summarising the facts above we obtain an invariance principle for sample maxima which in the iid case is analogous to Theorem 5.4.16:

Theorem 5.5.2 (Invariance principle for the maxima of a linear process with noise in MDA(Φ_α))
Assume either $\psi_+ p > 0$ or $\psi_- q > 0$, that the conditions (5.42) *and* (5.43) *hold and let* (c_n) *be defined by* (5.44). *Then*

$$Y_n \xrightarrow{d} Y\,, \quad n \to \infty\,,$$

where Y is the extremal process generated by the extreme value distribution

$$\Phi_\alpha^{\psi_+^\alpha + \psi_-^\alpha q p^{-1}}(x) = \exp\left\{-\left(\psi_+^\alpha + \psi_-^\alpha q p^{-1}\right) x^{-\alpha}\right\}\,, \quad x > 0\,. \qquad \square$$

Remarks. 1) For (X_n) iid, $\psi_i = 0$ for $i \neq 0$ and $\psi_0 = 1$. Then Theorem 5.5.2 degenerates into the case of a Φ_α–extremal process Y.

2) The above method can be extended to get joint convergence of the processes generated by a finite number of upper extremes in the sample $X_1, \ldots, X_n$. $\square$

Corollary 5.5.3 (Limit laws for the maxima of a linear process with noise in MDA(Φ_α))
Assume that $Z \in$ MDA(Φ_α) for some $\alpha > 0$ and choose (c_n) according to (5.44). *Then*

$$c_n^{-1} \max(Z_1, \ldots, Z_n) \xrightarrow{d} \Phi_\alpha\,, \tag{5.49}$$

and, under the conditions of Theorem 5.5.2,

$$c_n^{-1} M_n \xrightarrow{d} \Phi_\alpha^{\psi_+^\alpha + \psi_-^\alpha q p^{-1}}\,. \tag{5.50}$$

Moreover, let $(\widetilde{X}_n)$ be an iid sequence associated with (X_n). Then

$$c_n^{-1} \widetilde{M}_n \xrightarrow{d} \Phi_\alpha^{\|\psi\|_\alpha^\alpha}\,, \tag{5.51}$$

where

$$\|\psi\|_\alpha^\alpha = \sum_{j=-\infty}^{\infty} |\psi_j|^\alpha \left(I_{\{\psi_j > 0\}} + q p^{-1} I_{\{\psi_j < 0\}}\right).$$

Proof. (5.49) and (5.50) follow from Theorem 5.5.2 and the fact that

$$Y_n(1) \stackrel{d}{=} c_n^{-1} M_n \xrightarrow{d} Y(1).$$

(5.51) is a consequence of (5.49), taking into consideration (see Lemma A3.26) that

$$P\left(\sum_{j=-\infty}^{\infty} \psi_j Z_j > x\right) \sim \|\psi\|_\alpha^\alpha P(|Z| > x)\,. \qquad \square$$

The latter relation suggests that classical estimators for the tail index α might also work for the tail of X_t. This is unfortunately not the case; see for instance Figure 5.5.4.

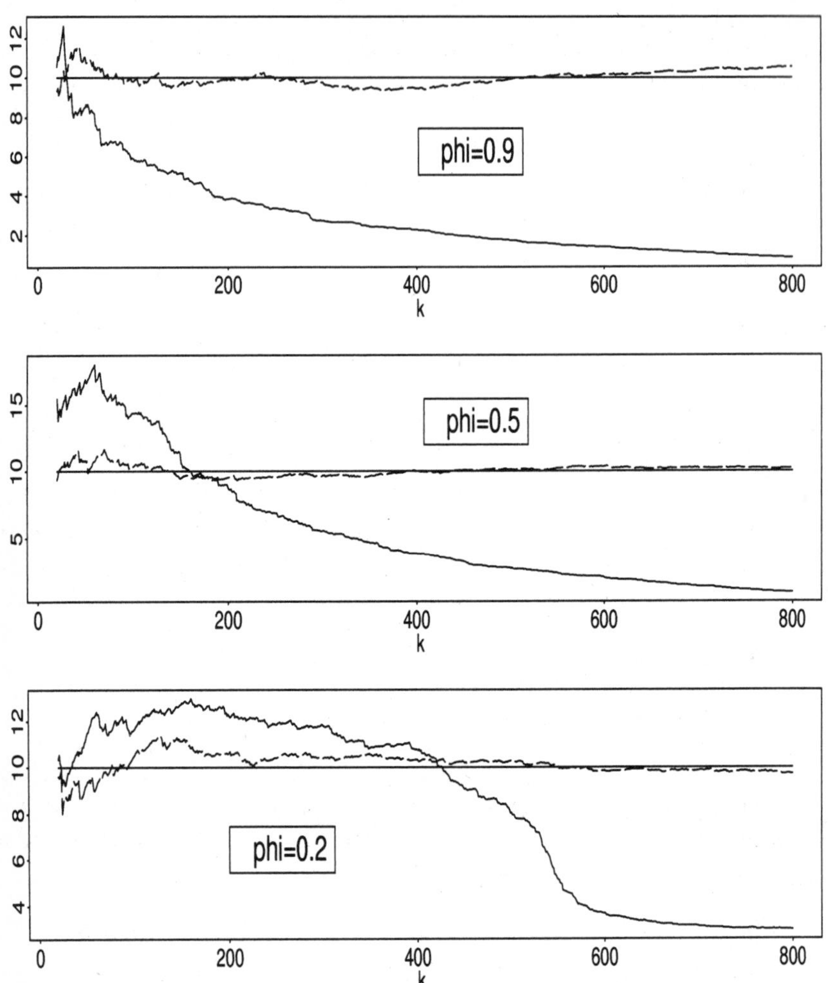

Figure 5.5.4 *A comparative study of the Hill–plots for* 1 000 *iid simulated data from an* AR(1) *process* $X_t = \phi X_{t-1} + Z_t$, $\phi \in \{0.9, 0.5, 0.2\}$. *The noise sequence* (Z_t) *comes from a symmetric distribution with exact Pareto tail* $P(Z > x) = 0.5x^{-10}$, $x \geq 1$. *According to Lemma A3.26,* $P(X > x) \sim c\,x^{-10}$. *The solid line corresponds to the Hill estimator of the* X_t *as a function of the* k *upper order statistics. The dotted line corresponds to the Hill estimator of the residuals* $\widehat{Z}_t = X_t - \widehat{\phi} X_{t-1}$, *where* $\widehat{\phi}$ *is the Yule–Walker estimator of* ϕ. *Obviously, the Hill estimator of the residuals yields much more accurate values. These figures indicate that the Hill estimator for correlated data has to be used with extreme care. Even for* $\phi = 0.2$ *the Hill estimator of the* X_t *cannot be considered as a satisfactory tool for estimating the index of regular variation. The corresponding theory for the Hill estimator of linear processes can be found in Resnick and Stărică* [535, 537].

Exceedances

Theorem 5.5.1 also allows us to derive results about the observations X_k/c_n exceeding a given threshold x, or equivalently about the linear process (X_k) exceeding the threshold $u_n = c_n x$. Without loss of generality we will assume that $|\psi_j| \le 1$ for all j.

Applying Theorem 5.5.1 and the continuous mapping theorem we find that the point process of points with ordinates bigger than $x > 0$ converges as $n \to \infty$. Thus let

$$E_x^+ = (x,\infty)\,,\quad E_x^- = (-\infty,-x)\,,\quad E_x = E_x^+ \cup E_x^-\,,\quad x>0\,;$$

then

$$\sum_{k=1}^{\infty} \varepsilon_{(n^{-1}k,c_n^{-1}X_k)}\left(\cdot \cap \mathbb{R}_+ \times E_x^+\right) \;\stackrel{d}{\to}\; \sum_{k=1}^{\infty}\sum_{i=-\infty}^{\infty} \varepsilon_{(t_k,\psi_i j_k)}\left(\cdot \cap \mathbb{R}_+ \times E_x^+\right) \tag{5.52}$$

in $M_p(\mathbb{R}_+ \times E_x^+)$. We can interpret this as weak convergence of the point processes of exceedances of xc_n by (X_k). We need the following auxiliary result.

Lemma 5.5.5 *The following relation holds in $M_p(\mathbb{R}_+ \times E_x)$:*

$$N_1 = \sum_{k=1}^{\infty} \varepsilon_{(t_k,j_k)} \stackrel{d}{=} N_2 = \sum_{k=1}^{\infty} \varepsilon_{(\Gamma_k,J_k)}\,,$$

where (Γ_k) is the sequence of points of a homogeneous Poisson process on $\mathbb{R}_+$ with intensity $\lambda = p^{-1}x^{-\alpha}$, independent of the iid sequence (J_k) with common density

$$\begin{aligned} g(y) &= \left(\alpha y^{-\alpha-1} I_{(x,\infty)}(y) + qp^{-1}\alpha(-y)^{-\alpha-1} I_{(-\infty,-x)}(y)\right)px^{\alpha} \\ &= f(y)\lambda^{-1}\,, \quad y \in \mathbb{R}\,. \end{aligned}$$

Proof. It suffices to show that the Laplace functionals of the point processes N_1 and N_2 coincide; see Example 5.1.8. Since N_1 is PRM$(|\cdot| \times \mu)$ on $\mathbb{R}_+ \times E_x$ we have by Example 5.1.11 that

$$\Psi_{N_1}(h) = \exp\left\{-\int_{\mathbb{R}_+}\int_{\mathbb{R}} \left(1 - e^{-h(t,z)}\right) f(z)dzdt\right\}.$$

On the other hand, conditioning on (Γ_k) and writing

$$h_1(t) = \int_{\mathbb{R}} \exp\{-h(t,z)\}\, g(z)dz = E\exp\{-h(t,J_1)\}\,,\quad t>0\,,$$

we obtain

$$
\begin{aligned}
\Psi_{N_2}(h) &= E\exp\left\{-\int_{\mathbb{R}_+\times\mathbb{R}} h dN_2\right\} \\
&= E\exp\left\{-\sum_{k=1}^{\infty} h(\Gamma_k, J_k)\right\} \\
&= E\prod_{k=1}^{\infty} h_1(\Gamma_k) \\
&= E\exp\left\{\sum_{k=1}^{\infty} \ln h_1(\Gamma_k)\right\}.
\end{aligned}
$$

The rvs Γ_k are the points of a homogeneous Poisson process with intensity $\lambda = p^{-1}x^{-\alpha}$. This and Lemma 5.1.12 yield

$$
\begin{aligned}
\Psi_{N_2}(h) &= E\exp\left\{-\lambda\int_{\mathbb{R}_+} (1-h_1(t))\, dt\right\} \\
&= \exp\left\{-\int_{\mathbb{R}_+}\int_{\mathbb{R}} \left(1-e^{-h(t,z)}\right) f(z)dzdt\right\}.
\end{aligned}
$$

This proves the lemma. □

Therefore the limit process in (5.52) has representation

$$
\sum_{k=1}^{\infty}\sum_{i=-\infty}^{\infty} \varepsilon_{(t_k,\psi_i j_k)} \stackrel{d}{=} \sum_{k=1}^{\infty}\sum_{i=-\infty}^{\infty} \varepsilon_{(\Gamma_k,\psi_i J_k)}
$$

in $M_p(\mathbb{R}_+ \times (x,\infty))$. Finally, we define the iid rvs $\xi_k = \operatorname{card}\{i : \psi_i J_k > x\}$. Now we can represent the limit process in (5.52) as a point process on $\mathbb{R}_+$:

$$
N_\infty = \sum_{k=1}^{\infty} \xi_k \varepsilon_{\Gamma_k}
$$

for independent (Γ_k) and (ξ_k). For any Borel set A in $\mathbb{R}_+$ this means that

$$
N_\infty(A) = \sum_{k=1}^{\infty} \xi_k \varepsilon_{\Gamma_k}(A) = \sum_{k:\Gamma_k\in A} \xi_k\,,
$$

i.e. N_∞ is a multiple point process with iid *multiplicities* or *cluster sizes* ξ_k. In particular, it is a compound Poisson process as defined in Example 5.1.15. This is completely different from the point processes of exceedances of iid or weakly dependent rvs (cf. Section 5.3) where the limit is homogeneous Poisson, hence simple. Thus the special dependence structure of linear processes yields clusters in the point process of exceedances.

Example 5.5.6 (AR(1) process)
We consider the AR(1) process $X_t = \varphi X_{t-1} + Z_t$, $t \in \mathbb{Z}$, for some $\varphi \in (0,1)$. It has a linear process representation

$$X_t = \sum_{j=0}^{\infty} \varphi^j Z_{t-j}\,, \quad t \in \mathbb{Z}.$$

The iid cluster sizes ξ_k have the following distribution:

$$\pi_0 = P(\xi_1 = 0) = P(J_1 \le x, \varphi J_1 \le x, \ldots) = P(J_1 \le x) = q\,, \quad x > 0\,,$$

and for $\ell \ge 1$,

$$\begin{aligned}
\pi_\ell &= P(\xi_1 = \ell) \\
&= P(J_1 > x, \ldots, \varphi^{\ell-1} J_1 > x, \varphi^\ell J_1 \le x) \\
&= P(\varphi^{\ell-1} J_1 > x, \varphi^\ell J_1 \le x) \\
&= p\varphi^{\alpha(\ell-1)}(1 - \varphi^\alpha).
\end{aligned}$$

□

Example 5.5.7 (MA(1) process)
We consider the MA(1) process $X_t = Z_t + \theta Z_{t-1}$, $t \in \mathbb{Z}$. Assume first $\theta > 0$. Direct calculation yields

$$\begin{aligned}
P(\xi_1 = 0) &= q \\
P(\xi_1 = 1) &= (1 - \theta^\alpha \wedge 1)\, p \\
P(\xi_1 = 2) &= (\theta^\alpha \wedge 1)\, p\,.
\end{aligned}$$

Thus the cluster sizes ξ_k may assume the values 0, 1 and 2 with positive probability for $\theta \in (0,1)$, whereas for $\theta > 1$ only the values 0 and 2 may occur.

Now assume $\theta < 0$, then

$$\begin{aligned}
P(\xi_1 = 0) &= (1 - (|\theta|^\alpha \wedge 1))\, q\,, \\
P(\xi_1 = 1) &= p + (|\theta|^\alpha \wedge 1)\, q\,.
\end{aligned}$$

Thus the cluster sizes ξ_k may assume only the values 0 and 1 for $\theta \le -1$, whereas for $\theta \in (-1,0)$, $\xi_k = 1$ a.s.

This means (in an asymptotic sense) that exceedances may only occur in clusters of 2 values if $\theta > 1$, whereas the cluster size may be 1 or 2 for $\theta \in (0,1)$. For $\theta < 0$ the point process of exceedances does not cluster. □

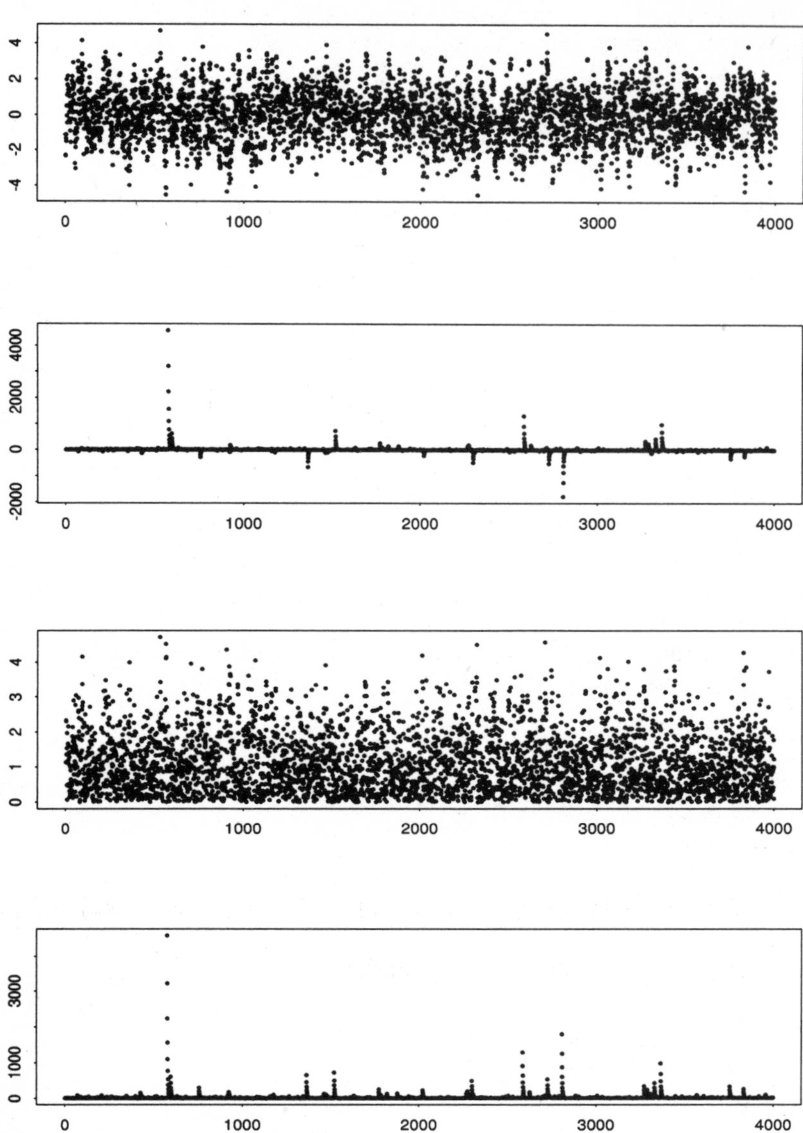

Figure 5.5.8 *Realisations of the* AR(1) *process* $X_t = 0.7X_{t-1} + Z_t$ *(top two) and of the corresponding absolute values (bottom two). In each pair of figures, the upper one corresponds to iid standard normal noise* (Z_t)*, the lower one to iid standard Cauchy noise. In the Cauchy case extremes tend to occur in clusters; see Example 5.5.6. In the Gaussian case clustering effects of extremal values are not present.*

Maxima and Minima

We consider the joint limit behaviour of the maxima (M_n) and of the minima

$$W_1 = X_1\,, \quad W_n = \min(X_1, \ldots, X_n)\,, \quad n \ge 2\,.$$

Choose $x > 0$ and $y < 0$ and write

$$A = (0,1] \times [(-\infty, y) \cup (x, \infty)]\,.$$

Then, by Theorem 5.5.1,

$$\begin{aligned} & P\left(c_n^{-1} M_n \le x\,, c_n^{-1} W_n > y\right) \\ = \quad & P\left(\sum_{k=1}^{\infty} \varepsilon_{(n^{-1}k, c_n^{-1} X_k)}(A) = 0\right) \\ \to \quad & P\left(\sum_{k=1}^{\infty} \sum_{i=-\infty}^{\infty} \varepsilon_{(t_k, \psi_i j_k)}(A) = 0\right)\,. \end{aligned} \tag{5.53}$$

We consider the event in (5.53) in detail. Notice that

$$\begin{aligned} & \Big\{\mathrm{card}\Big\{(k,i) : 0 < t_k \le 1 \text{ and } (\psi_i j_k < y \text{ or } \psi_i j_k > x)\Big\} = 0\Big\} \\ = \quad & \Big\{\mathrm{card}\Big\{k :\ 0 < t_k \le 1 \text{ and} \\ & (j_k < -x/\psi_- \text{ or } j_k > x/\psi_+ \text{ or } j_k < y/\psi_+ \text{ or } j_k > -y/\psi_-)\Big\} = 0\Big\}\,. \end{aligned}$$

Write

$$B = (0,1] \times \Big[\Big(-\infty, (-x/\psi_-) \vee (y/\psi_+)\Big) \cup \Big((x/\psi_+) \wedge (-y/\psi_-), \infty\Big)\Big]\,.$$

Then the right–hand side in (5.53) translates into

$$\begin{aligned} & P\left(\sum_{k=1}^{\infty} \varepsilon_{(t_k, j_k)}(B) = 0\right) \\ = \quad & \exp\Big\{-\mu\Big(\big(-\infty, (-x/\psi_-) \vee (y/\psi_+)\big) \cup \big((x/\psi_+) \wedge (-y/\psi_-), \infty\big)\Big)\Big\} \\ = \quad & \exp\left\{-\left(\left[\psi_+^\alpha x^{-\alpha} \vee \psi_-^\alpha (-y)^{-\alpha}\right] + q p^{-1}\left[\psi_-^\alpha x^{-\alpha} \vee \psi_+^\alpha (-y)^{-\alpha}\right]\right)\right\}, \end{aligned} \tag{5.54}$$

where ψ_+, ψ_- were defined in (5.48). Now introduce the two–dimensional df

$$G(x_1,x_2) = \begin{cases} \exp\left\{-\psi_+^\alpha x_1^{-\alpha}\right\} \wedge \exp\left\{-\psi_-^\alpha x_2^{-\alpha}\right\} & \text{for } x_1>0,\ x_2>0\,, \\ 0 & \text{otherwise}\,. \end{cases} \tag{5.55}$$

Thus the right–hand side of (5.54) can be written in the form

$$G(x,-y)G^{q/p}(-y,x)\,,$$

and using the relation

$$P\left(c_n^{-1}M_n \le x\,, c_n^{-1}W_n \le y\right) = P\left(c_n^{-1}M_n \le x\right) - P\left(c_n^{-1}M_n \le x\,, c_n^{-1}W_n > y\right)$$

we obtain the following:

Theorem 5.5.9 (Joint limit distribution of sample maxima and minima of linear process)
Assume that the conditions of Theorem 5.5.2 hold and let (c_n) be defined by (5.44). *Then, for all real x, y,*

$$P\left(c_n^{-1}M_n \le x\,, c_n^{-1}W_n \le y\right) \to G(x,\infty)G^{q/p}(\infty,x) - G(x,-y)G^{q/p}(-y,x)\,,$$

where $G(x,y)$ is defined by (5.55). □

Summary

Assume that $Z \in \mathrm{MDA}(\Phi_\alpha)$, i.e.

$$P(Z > x) = \frac{L(x)}{x^\alpha}\,, \quad x > 0\,,$$

for some $\alpha > 0$, and that

$$P(Z \le -x) \sim \frac{q}{p}\frac{L(x)}{x^\alpha}\,, \quad x \to \infty\,,$$

for non–negative p, q such that $p+q=1$ and $p>0$. Choose the constants c_n by

$$c_n = (1/\overline{F}_Z)^{\leftarrow}(n)\,.$$

Then

$$c_n^{-1}\max\left(Z_1,\ldots,Z_n\right) \xrightarrow{d} \Phi_\alpha$$

for the Fréchet distribution $\Phi_\alpha(x) = e^{-x^{-\alpha}}$, $x > 0$. Moreover, under the conditions of Theorem 5.5.2,

$$c_n^{-1}M_n \xrightarrow{d} \Phi_\alpha^{\psi_+^\alpha + \psi_-^\alpha qp^{-1}}\,, \quad x > 0\,,$$

where ψ_+, ψ_- are defined in (5.48). The point process of the exceedances of the threshold $c_n x$ by the linear process (X_k) converges weakly to a compound Poisson point process with iid cluster sizes which depend on the coefficients ψ_j.

5.5.2 Subexponential Noise in the Maximum Domain of Attraction of the Gumbel Distribution Λ

In this section we again consider the linear process $X_n = \sum_{j=-\infty}^{\infty} \psi_j Z_{n-j}$ driven by iid noise (Z_n) with common df F_Z. In contrast to Section 5.5.1 we assume that F_Z belongs to the maximum domain of attraction of the Gumbel distribution

$$\Lambda(x) = e^{-e^{-x}}, \quad x \in \mathbb{R}.$$

We know from Section 3.3.3 and Example 3.3.35 that MDA(Λ) contains a wide range of distributions with quite different tail behaviour. Indeed, F_Z may be subexponential (for instance the lognormal distribution), exponential or superexponential (for instance the normal distribution). We found in Example 4.4.9 that fairly general Gaussian linear processes (X_n) exhibit the same asymptotic extremal behaviour as their associated iid sequence $(\widetilde{X}_n)$. This changes dramatically for linear processes with subexponential noise as we have already learnt in Section 5.5.1 for regularly varying $\overline{F}_Z$. A similar statement holds when $F_Z \in \text{MDA}(\Lambda) \cap \mathcal{S}$, where $\mathcal{S}$ denotes the class of distributions F_Z with subexponential positive part Z^+; for the definition and properties of $\mathcal{S}$ see Section 1.3.2 and Appendix A3.2.

Before we state the main results for $F_Z \in \text{MDA}(\Lambda) \cap \mathcal{S}$ we introduce some conditions on the coefficients ψ_j and on the distribution F_Z. Throughout we suppose that the tail balance condition

$$\lim_{x\to\infty} \frac{P(Z > x)}{P(|Z| > x)} = p\,, \quad \lim_{x\to\infty} \frac{P(Z \le -x)}{P(|Z| > x)} = q \tag{5.56}$$

holds with $0 < p \le 1$, $p + q = 1$. We also assume

$$\sum_{j=-\infty}^{\infty} |\psi_j|^\delta < \infty \quad \text{for some } \delta \in (0,1)\,. \tag{5.57}$$

We have that $E|Z| < \infty$, which follows from the tail balance condition (5.56) and from the fact that $E(Z^+)^\beta < \infty$, $\beta > 0$, for $F_Z \in \text{MDA}(\Lambda)$; see Corollary 3.3.32. This and (5.57) guarantee the absolute a.s. convergence of the series X_n for every n. Without loss of generality we assume that

$$\max_j |\psi_j| = 1\,, \tag{5.58}$$

since otherwise we may consider the re–scaled process $X_n / \max_j |\psi_j|$. Then one or more of the ψ_j have absolute value one. The quantities

$$k^+ = \text{card}\,\{j : \psi_j = 1\}\,, \quad k^- = \text{card}\,\{j : \psi_j = -1\} \tag{5.59}$$

are crucial for the extremal behaviour of the sequence (X_n). The above conditions lead to the following result which is analogous to Theorem 5.5.1. Theorem 5.5.10 below is proved in Davis and Resnick [163], Theorem 3.3, in a more general situation.

Theorem 5.5.10 (Weak convergence of the point processes of the embedded linear process)
Suppose $F_Z \in \mathrm{MDA}(\Lambda) \cap \mathcal{S}$. Then there exist constants $c_n > 0$ and $d_n \in \mathbb{R}$ such that

$$n\overline{F}_Z\left(c_n x + d_n\right) \to -\ln \Lambda(x)\,, \quad x \in \mathbb{R}. \tag{5.60}$$

Furthermore, assume that conditions (5.56)–(5.58) *hold. Then*

$$\sum_{k=1}^{\infty} \varepsilon_{\left(n^{-1}k,\, c_n^{-1}(X_k - d_n)\right)} \quad \stackrel{d}{\to} \quad k^+ N_1 + k^- N_2$$

in $M_p\left(\mathbb{R}_+ \times E\right)$ with $E = (-\infty, \infty]$. Here

$$N_i = \sum_{k=1}^{\infty} \varepsilon_{(t_{ki}, j_{ki})}\,, \quad i = 1, 2\,,$$

are two independent $\mathrm{PRM}(|\cdot| \times \mu_i)$ *on $\mathbb{R}_+ \times E$, μ_1 has density $f_1(x) = e^{-x}$ and μ_2 has density $f_2(x) = (q/p)e^{-x}$, both with respect to Lebesgue measure.* □

Remarks. 1) If $k^+ > 1$ or $k^- > 1$, the limit point process $k^+ N_1 + k^- N_2$ is multiple with constant multiplicities k^+, k^-. The two independent processes $k^+ N_1$ and $k^- N_2$ are due to the contributions of those innovations Z_n for which $\psi_n = 1$ or $\psi_n = -1$.

2) A comparison of Theorems 5.5.10 and 5.5.1 shows that the limit point processes for $F_Z \in \mathrm{MDA}(\Lambda) \cap \mathcal{S}$ and $F_Z \in \mathrm{MDA}(\Phi_\alpha)$ are completely different although in both cases F_Z is subexponential. For $F_Z \in \mathrm{MDA}(\Phi_\alpha)$ the limit depends on *all* coefficients ψ_j whereas for $F_Z \in \mathrm{MDA}(\Lambda) \cap \mathcal{S}$ only the numbers k^+ and k^- defined by (5.59) are of interest. The differences are due to the completely different tail behaviour; $F_Z \in \mathrm{MDA}(\Phi_\alpha)$ implies regular variation of $\overline{F}_Z$, $F_Z \in \mathrm{MDA}(\Lambda) \cap \mathcal{S}$ rapid variation of $\overline{F}_Z$; see Corollary 3.3.32. This has immediate consequences for $P(X > x)$; see Appendix A3.3. □

In the sequel we again apply some standard arguments to derive information from Theorem 5.5.10 about the extremal behaviour of the linear process (X_n).

Extremal Processes and Limit Distribution of Maxima

Analogously to iid sample maxima we define the process

$$Y_n(t) = \begin{cases} c_n^{-1}\left(M_{[nt]} - d_n\right) & \text{if} \quad t \geq n^{-1}\,, \\ c_n^{-1}\left(X_1 - d_n\right) & \text{if} \quad 0 < t < n^{-1}\,. \end{cases}$$

Let

$$Y(t) = \bigvee_{t_{k1} \leq t} j_{k1} \vee \bigvee_{t_{k2} \leq t} j_{k2} = Y^+(t) \vee Y^-(t)\,, \quad t > 0\,.$$

We use the convention that $\max\emptyset = -\infty$. Then an application of the a.s. continuous mapping $\widetilde{T}_1$ from (5.25), the continuous mapping theorem and Theorem 5.5.10 yield that

$$\begin{aligned} Y_n &= \widetilde{T}_1\left(\sum_{k=1}^{\infty} \varepsilon_{\left(n^{-1}k,\, c_n^{-1}(X_k - d_n)\right)}\right) \\ &\xrightarrow{d} \widetilde{T}_1\left(k^+ \sum_{k=1}^{\infty} \varepsilon_{(t_{k1}, j_{k1})} + k^- \sum_{k=1}^{\infty} \varepsilon_{(t_{k2}, j_{k2})}\right) \\ &= Y \end{aligned}$$

in $\mathbb{D}(0,\infty)$. The cadlag processes Y^+ and Y^- are independent extremal processes and Y (being the maximum of them) is again an extremal process. Remember that $\sum_{k=1}^{\infty} \varepsilon_{(t_{k1}, j_{k1})}$ is PRM with the mean measure of $(0,t] \times (x,\infty)$ equal to te^{-x} and likewise $\sum_{k=1}^{\infty} \varepsilon_{(t_{k2}, j_{k2})}$ is PRM with the mean measure of $(0,t] \times (x,\infty)$ equal to $t(q/p)e^{-x}$. Thus Y^+ is Λ–extremal and Y^- is $\Lambda^{q/p}$–extremal. Hence $Y = Y^+ \vee Y^-$ is $\Lambda^{1+q/p}$–extremal. Then, for $t > 0$, $x \in \mathbb{R}$,

$$P(Y(t) \leq x) = \exp\left\{-t(1 + q/p)e^{-x}\right\} = \exp\left\{-tp^{-1}e^{-x}\right\}\,.$$

Theorem 5.5.11 (Invariance principle for the maxima of a linear process with noise in $\mathrm{MDA}(\Lambda) \cap \mathcal{S}$)
Assume that $F_Z \in \mathrm{MDA}(\Lambda) \cap \mathcal{S}$ *and that conditions* (5.56)–(5.58) *hold. Choose the constants* c_n, d_n *according to* (5.60). *Then*

$$Y_n \xrightarrow{d} Y\,, \qquad n \to \infty\,,$$

where Y *is the extremal process generated by the extreme value distribution*

$$\Lambda^{p^{-1}}(x) = \exp\left\{-p^{-1}e^{-x}\right\}\,, \quad x \in \mathbb{R}\,. \qquad \square$$

An immediate consequence is the following.

Corollary 5.5.12 (Limit laws for the maxima of a linear process with noise in MDA(Λ) $\cap\,\mathcal{S}$)
Under the conditions of Theorem 5.5.11 the following limit relations hold:

$$c_n^{-1}\left(\max(Z_1,\ldots,Z_n)-d_n\right) \xrightarrow{d} \Lambda\,, \tag{5.61}$$

$$c_n^{-1}\left(M_n-d_n\right) \xrightarrow{d} \Lambda^{p^{-1}}\,. \tag{5.62}$$

$$c_n^{-1}\left(\widetilde{M}_n-d_n\right) \xrightarrow{d} \Lambda^{k^+ + k^- q p^{-1}}\,. \tag{5.63}$$

Proof. (5.61) and (5.62) follow from Theorem 5.5.11, while (5.63) is a consequence of (5.61) taking into consideration (see Lemma A3.27) that

$$P\left(\sum_{j=-\infty}^{\infty}\psi_j Z_j > x\right) \sim (k^+p+k^-q)P\left(|Z|>x\right)\,. \qquad \square$$

Exceedances

For $x\in\mathbb{R}$ the point process of exceedances of $c_n x+d_n$ by the linear process (X_k) is given by

$$N_n(\cdot)=\sum_{k=1}^{\infty}\varepsilon_{n^{-1}k}(\cdot)I_{\{c_n^{-1}(X_k-d_n)>x\}}\,.$$

As a consequence of Theorem 5.5.10 and of the continuous mapping theorem we conclude that

$$N_n \xrightarrow{d} k^+\sum_{k=1}^{\infty}\varepsilon_{t_{k1}}I_{\{j_{k1}>x\}}+k^-\sum_{k=1}^{\infty}\varepsilon_{t_{k2}}I_{\{j_{k2}>x\}}=k^+N^+ + k^-N^- \tag{5.64}$$

in $M_p(\mathbb{R}_+)$. With a glance at the finite–dimensional distributions or at the Laplace functionals it is not difficult to check that N^+ and N^- are homogeneous Poisson processes on $\mathbb{R}_+$ with intensity e^{-x} and $(q/p)e^{-x}$, respectively. If (Γ_k^+) and (Γ_k^-) denote the sequences of the points of N^+ and N^- then we obtain the following result from (5.64):

Theorem 5.5.13 *Suppose that the assumptions of Theorem* 5.5.11 *hold. Then the point processes of exceedances of $c_n x+d_n$ by the linear process (X_k) converge weakly in $M_p(\mathbb{R}_+)$ as $n\to\infty$:*

$$\sum_{k=1}^{\infty}\varepsilon_{n^{-1}k}I_{\{c_n^{-1}(X_k-d_n)>x\}} \xrightarrow{d} \sum_{k=1}^{\infty}\left(k^+\varepsilon_{\Gamma_k^+}+k^-\varepsilon_{\Gamma_k^-}\right)\,.$$

Here (Γ_{k^+}) and (Γ_{k^-}) are the sequences of the points of two independent homogeneous Poisson processes on $\mathbb{R}_+$ with corresponding intensities e^{-x} and $(q/p)e^{-x}$. $\square$

We notice that the limit process of the point processes of exceedances is the sum of two independent compound Poisson processes where the cluster sizes are just constants k^+, k^-. This is in contrast to the iid or weakly dependent stationary case where the limit point process is a (simple) homogeneous Poisson process (see Section 5.3), but it is also different from the situation when $F_Z \in \text{MDA}(\Phi_\alpha)$. In the latter case the limit point process is compound Poisson with random cluster sizes (see Section 5.5.1).

Maxima and Minima

As in Section 5.5.1 point process methods can be used to derive the joint limit distribution of maxima and minima of linear processes with $F_Z \in \text{MDA}(\Lambda) \cap \mathcal{S}$. The approach is similar to the one in Section 5.5.1. We omit details and simply state a particular result. Let $W_n = \bigwedge_{i=1}^n X_i$ and suppose that $k^- = 0$, i.e. there is no index j with $\psi_j = -1$. Then

$$P\left(c_n^{-1}(M_n - d_n) \le x, \quad c_n^{-1}(W_n + d_n) \ge y\right) \to \Lambda(x)\Lambda^{q/p}(-y)$$

for $x, y > 0$. In general, the limit distribution depends on the fact whether $k^+ = 0$ or $k^- = 0$. For more details see Davis and Resnick [163].

Summary

Assume that $F_Z \in \text{MDA}(\Lambda) \cap \mathcal{S}$ with constants c_n and d_n chosen according to Theorem 3.3.26, i.e.

$$c_n^{-1}(\max(Z_1, \dots, Z_n) - d_n) \xrightarrow{d} \Lambda,$$

where Λ denotes the Gumbel distribution $\Lambda(x) = e^{-e^{-x}}$, $x \in \mathbb{R}$. Then, under the conditions of Theorem 5.5.11,

$$c_n^{-1}(M_n - d_n) \xrightarrow{d} \Lambda^{p^{-1}}.$$

Furthermore, the point processes of exceedances of the threshold $c_n x + d_n$ by the linear process (X_k) converge weakly to a multiple point process with constant multiplicities.

Notes and Comments

Asymptotic extreme value theory for linear processes with regularly varying tails is given in Resnick [530], Chapter 4.5. The latter is based on Davis and Resnick [160, 161, 162] who also treat more general aspects of time series analysis, see Chapter 7, and on Rootzén [549] and Leadbetter, Lindgren and Rootzén [418] who consider exceedances of linear processes.

Extremes of linear processes with exponential and subexponential noise variables were treated in Davis and Resnick [163]. Further interesting work in this context is due to Leadbetter and Rootzén [419] and Rootzén [549, 550].

Note that both the present section and Sections 4.4 and 5.3.2 deal with strictly stationary sequences. However, the assumptions and results are of different nature. The central conditions in the present section are regular variation of the tails $\overline{F}_Z$ or subexponentiality of the df F_Z. This allows one to embed the linear process in a point process and to derive elegant results which yield much information about the extremal behaviour of a linear process. The assumptions on the tails are much weaker in Sections 4.4 and 5.3.2. In particular, the df does not have to belong to any maximum domain of attraction. Thus more general classes of dfs can be covered. On the other hand, conditions of type $D(u_n)$ or $D'(u_n)$ ensure that we do not go too far away from the iid case. Linear processes seem to allow for "more dependence" in the sequence (X_n) although the kind of dependence is quite specific. We can also compare the different point processes of exceedances. In the case of linear processes we obtain multiple PRM in the limit. This is in contrast to Section 5.3.2, where the limit is a homogeneous Poisson process.

6

Statistical Methods for Extremal Events

6.1 Introduction

In the previous chapters we have introduced a multitude of probabilistic models in order to describe, in a mathematically sound way, extremal events in the one–dimensional case. The real world however often informs us about such events through *statistical data*: major insurance claims, flood levels of rivers, large decreases (or indeed increases) of stock market values over a certain period of time, extreme levels of environmental indicators such as ozone or carbon monoxide, wind–speed values at a certain site, wave heights during a storm or maximal and minimal performance values of a portfolio. All these, and indeed many more examples, have in common that they concern questions about extreme values of some underlying set of data. At this point it would be utterly foolish (and indeed very wrong) to say that all such problems can be cast into one or the other probabilistic model treated so far: this is definitely not the case! Applied mathematical (including statistical) modelling is all about trying to offer the applied researcher (the finance expert, the insurer, the environmentalist, the biologist, the hydrologist, the risk manager, ...) the necessary set of tools in order to deduce scientifically sound conclusions from data. It is however also very much about *reporting correctly:* the data have to be presented in a clear and objective way, precise questions have to be formulated, model–based answers given, always stressing the underlying assumptions. The whole process constitutes an art: statistical theory plays only a relatively small, though crucial role here.

The previous chapters have given us a whole battery of techniques with which to formulate in a mathematically precise way the basic questions underlying extreme value theory. This chapter aims at going one step further: based on data, we shall *present statistical tools allowing us to link questions asked in practice to a particular (though often non–unique) probabilistic model.* Our treatment as regards these statistical tools will definitely not be complete, though we hope it will be representative of current statistical methodology in this fast–expanding area. The reader will meet data, basic descriptive methods, and techniques from mathematical statistics concerning estimation and testing in extreme value models. We have tried to keep the technical level of the chapter down: the reader who has struggled through Chapter 5 on point processes may well be relieved! At the same time, chapters like the one on point processes are there to show how modern probability theory is capable of handling fairly complicated but realistic models. *The real expert on Extremal Event Modelling will definitely have to master both "extremes".*

After the mathematical theory of maxima, order statistics and heavy–tailed distributions presented in the previous chapters, we now turn to the crucial question:

How do extreme values manifest themselves in real data?

A full answer to this question would not only take most of the present chapter, one could write whole volumes on it. Let us start by seeing how in practice extremes in data manifest themselves. We do this through a series of partly hypothetical, partly real examples. At a later stage in the chapter, we will come back to some of the examples for a more detailed analysis.

Example 6.1.1 (River Nidd data)
A standard data–set in extreme value theory concerns flows of the river Nidd in Yorkshire, England; the source of the data is the Flood Studies Report NERC [477]. We are grateful to Richard Smith for having provided us with a copy of the data. The basic set contains 154 observations on flow data above 65 CUMECS over the 35–year period 1934–1970. A crude de–clustering technique was used by the hydrologists to prepare these data. Though the full set contains a series of values for each year, for a first analysis only the annual maxima are considered. In this way, intra–year dependencies are avoided and a valid assumption may be to suppose that the data $x_1, \ldots, x_{35}$ are realisations from a sequence $X_1, \ldots, X_{35}$ of iid rvs all with common *extreme value distribution* H say. Suppose we want to answer questions like:

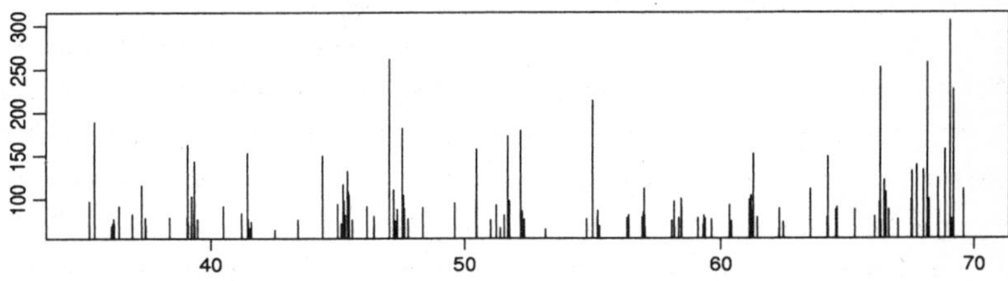

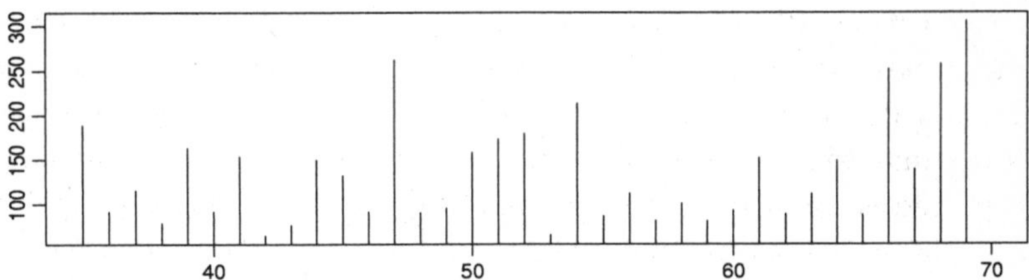

Figure 6.1.2 *The river Nidd data 1934–1970 (top) and the corresponding annual maxima (bottom). The data are measured in CUMECS.*

- What is the probability that the maximum flow for the next year will exceed a level x?
- What is the probability that the maximum flow for the next year exceeds all previous levels?
- What is the expected length of time (in years say) before the occurrence of a specific high quantity of flow?

Clearly, a crucial step forward in answering these questions would be our gaining knowledge of the df H. The theory of Chapter 3 gives us relevant parametric models for H; see the Fisher–Tippett theorem (Theorem 3.2.3) where the extreme value distributions enter. Standard statistical tools such as maximum likelihood estimation (MLE) are available. □

Example 6.1.3 (Insurance claims)
Suppose our data consist of fire insurance claims $x_1, \ldots, x_n$ over a specified period of time in a well–defined portfolio, as for instance presented in Figure 6.1.4. Depending on the type of fire causing the specific claims, a condition of the type "$x_1, \ldots, x_n$ come from an iid sample $X_1, \ldots, X_n$ with df F" may or may not be justified. Suppose for the sake of argument that the underlying

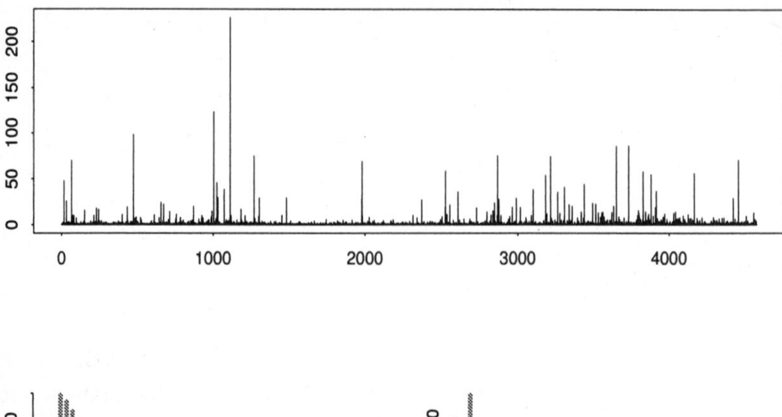

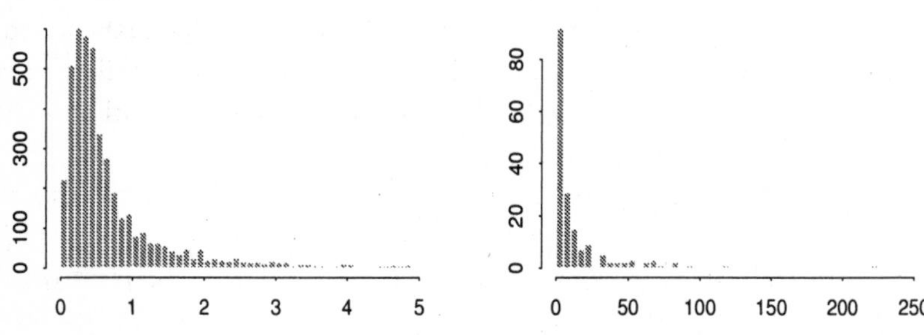

Figure 6.1.4 4 580 *claims from a fire insurance portfolio. The values are multiples of* 1 000 *SFr. The corresponding histogram of the claims* $\leq$ 5 000 *SFr (left) and of the remaining claims exceeding* 5 000 *SFr (right). The data are very skewed to the right. The x–axis of the histogram on the rhs reaches up to* 250 *due to a very large claim around* 225*; see also the top figure.*

portfolio is such that the above assumption can be made. Questions we want to answer (or tasks we want to perform) could be:

- Calculate next year's premium volume needed in order to cover, with sufficiently high probability, future losses in this portfolio.
- What is the *probable–maximum–loss* of this portfolio if the latter is defined as a high (for instance the 0.999–) quantile of the df F?
- Given that we want to write an excess–of–loss cover (see Example 8.7.4) with priority a_k (also referred to as *attachment point*) resulting in a one–in–k–year event, how do we calculate a_k? The latter means that we want to calculate a_k so that the probability of exceeding a_k equals $1/k$.

Again, as in the previous example, we are faced with a standard statistical fitting problem. The main difference is that in this case we do not immediately have a specific parametric model (such as the extreme value distributions in Example 6.1.1) in mind. We first have to learn about the data:

– Is F light– or heavy–tailed?
– What are its further shape properties: skewed, flat, unimodal,...?

In the heavy–tailed case fitting by a subexponential distribution (see Chapter 1 and Appendix A3.2) might be called for. The method of exceedances from Section 6.5 will be relevant. □

Example 6.1.5 (ECOMOR reinsurance)
The ECOMOR reinsurance contract stands for "Le Traité d'Excédent du Coût Moyen Relatif" and was introduced by the French actuary Thépaut [621] as a novel contract aiming to enlarge the reinsurer's flexibility in constructing customised products. A precise mathematical description is given in Example 8.7.7. Suppose over a period $[0,t]$, the claims $x_1,\dots,x_{n(t)}$ are received by the primary insurer. The ECOMOR contract binds the reinsurer to cover (for a specific premium) the excesses above the kth largest claim. This leads us to a model where $X_1,\dots,X_{N(t)}$ are (conditionally) iid with specific model assumptions on the underlying df F of X_i and on the counting process $(N(t))$; see Chapter 1. The relevant theory underlying the ECOMOR contracts, i.e. the distributional properties of the k largest order statistics $X_{1,N(t)},\dots,X_{k,N(t)}$ from a randomly indexed ordered sample, was given in Section 4.3. Standard models are hence at our disposal. It is perhaps worthwhile to stress that, though innovative in nature, ECOMOR never was a commercial success. □

Example 6.1.6 (Value–at-Risk)
Suppose a financial portfolio consists of a number of underlying assets (bonds, stocks, derivatives,...), all having individual (though correlated) values at any time t. Through the estimation of portfolio covariances, the portfolio manager then estimates the overall portfolio Profit–Loss (P&L) distribution. For details on this see for instance RiskMetrics [543]. Management and regulators may now be interested in setting "minimal requirements" or, for the sake of argument, a maximal limit on the potential losses. A possible quantity is the so–called Value–at-Risk (VaR) measure briefly treated in the discussion of Figure 4 of the Reader Guidelines. There the VaR is defined as the 5% quantile of the P&L distribution. The following questions are relevant.

– Estimate the VaR for a given portfolio.

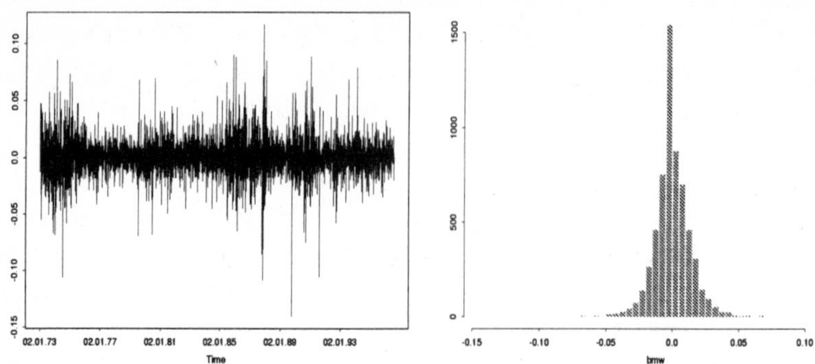

Figure 6.1.7 *Daily log–returns of BMW share prices for the period January 2, 1973 – July 23, 1996* ($n = 6\,146$)*, together with a histogram of the data.*

– Estimate the probability that, given we exceed the VaR, we exceed it by a certain amount. This corresponds to the calculation of the so–called *shortfall distribution.*

The first question concerns quantile estimation for an estimated df, in many cases outside the range of our data. The second question obviously concerns the estimation of the excess df as defined in Section 3.4 (modulo a change of sign: we are talking about losses!). The theory presented in the latter section advocates the use of the generalised Pareto distribution as a natural parametric model in this case. □

Example 6.1.8 (Fighting the arch–enemy with mathematics)
The above heading is the actual title of an interesting paper by de Haan [294] on the famous Dutch dyke project following the disastrous flooding of parts of the Dutch provinces of Holland and Zeeland on February 1, 1953, killing over 1 800 people. In it, de Haan gives an account of the theoretical and applied work done in connection with the problem of how to determine a safe height for the sea dykes in the Netherlands. More than with any other event, the resulting work by Dutch mathematicians under van Dantzig gave the statistical methodology of extremal events a decisive push. The statistical analyses also made a considerable contribution to the final decision making about the dyke heights. The problem faced was the following: given a small number p (in the range of 10^{-4} to 10^{-3}), determine the height of the sea dykes such that the probability that there is a flood in a given year equals p. Again, we are confronted with a quantile estimation problem. From the data available, it was clear that one needed estimates well outside the range of the data. The sea-

water level in the Netherlands is typically measured in (N.A.P. + x) meters (N.A.P. = Normaal Amsterdams Peil, the Dutch reference level corresponding to mean sea level). The 1953 flood was caused by a (N.A.P. + 3.85) m surge, whereas historical accounts estimate a (N.A.P. + 4) m for the 1570 flood, the worst recorded. The van Dantzig report estimated the $(1 - 10^{-4})$–quantile as (N.A.P. + 5.14) m for the annual maximum. That is, the one–in–ten–thousand–year surge height is estimated as (N.A.P. + 5.14) m. We urge all interested in extreme value statistics to read de Haan [294]. □

Many more examples with an increasing degree of complexity could have been given including:

- non–stationarity (seasonality, trends),
- sparse data,
- multivariate observations,
- infinite–dimensional data (for instance continuously monitored processes).

The literature cited throughout the book contains a multitude of examples. Besides the work mentioned already by Smith on the river Nidd and de Haan's paper on the dyke project, we call the following papers to the reader's attention:

- Rootzén and Tajvidi [553] where a careful analysis of Swedish wind storm losses (i.e. insurance data) is given. Besides the use of standard methodology (fitting of generalised extreme value and Pareto distributions), problems concerning trend analysis enter, together with a covariate analysis looking at the potential influence from numerous environmental factors.
- Resnick [532] considers heavy tail modelling in a huge data–set ($n \approx 50\,000$) in the field of the teletraffic industry. Besides giving a very readable and thought provoking review of some of the classical methods, extremes in time series models are specifically addressed. See also Sections 5.5 and 8.4.
- Smith [594] applies extreme value theory to the study of ozone in Houston, Texas. A key question concerns the detection of a possible trend in ground–level ozone. Such a study is particularly interesting as air–quality standards are often formulated in terms of the highest level of permitted emissions.

The above papers are not only written by *masters at their trade* (de Haan, Resnick, Rootzén, Smith), they also cover a variety of fields (hydrology, insurance, electrical engineering, environmental research).

Within the context of finance, numerous papers analysing specific data are being published; see Figure 6.1.7 for a typical example of financial return data. A paper which uses up–to–date statistical methodology on extremes is for instance Danielson and de Vries [154] where models for high frequency foreign exchange recordings are treated. See also Müller et al. [470] for more background on the data. Interesting case studies are also to be found in Barnett and Turkman [52], Falk, Hüsler and Reiss [225], and Longin [428]. The latter paper analyses US stock market data.

We hope that the examples above have singled out a series of problems. We now want to present their statistical solutions. There is no way in which we can achieve completeness concerning the statistical models now understood: the definitive book on this still awaits the writing. A formidable task indeed!

The following sections should offer the reader both hands–on experience of some basic methods, as well as a survival kit to get him/her safely through the "jungle of papers on extreme value statistics". The outcome should be a better understanding of those basic methods, together with a clear(er) overview of where the field is heading to. This chapter should also be a guide on where to look for further help on specific problems at hand.

Of the more modern textbooks containing a fair amount of statistical techniques we would like to single out Falk et al. [225] and Reiss [526]. The latter book also contains a large amount of historical notes. *It always pays to go back to the early papers and books written by the old masters*, and the annotated references in Reiss [526] could be your guide. *However, whatever you decide to read, don't miss out on Gumbel* [290]!

6.2 Exploratory Data Analysis for Extremes

One of the reasons why Gumbel's book [290] is such a feast to read is its inclusion of roughly 100 graphs and 50 tables. The author very much stresses the importance of *looking at data* before engaging in a detailed statistical analysis. In our age of nearly unlimited computing power this graphical data exploration is becoming increasingly important. The reader interested in some recent developments in this area may for instance consult Chambers et al. [109], Cleveland [122] or Tufts [627]. In the sections to follow we discuss some of the more useful graphical methods.

6.2.1 Probability and Quantile Plots

Given a set of data to be analysed, one usually starts with a histogram, one or more box–plots, a plot of the empirical df, in the multi–dimensional case a scatterplot or a so–called draughtsman's display which combines all 2×2 scatterplots in a graphical matrix form. Keeping to the main theme of the book, we restrict ourselves however to the one–dimensional case and start with a discussion of the problem:

Find a df F which is a good model for the iid data $X, X_1, \ldots, X_n$.

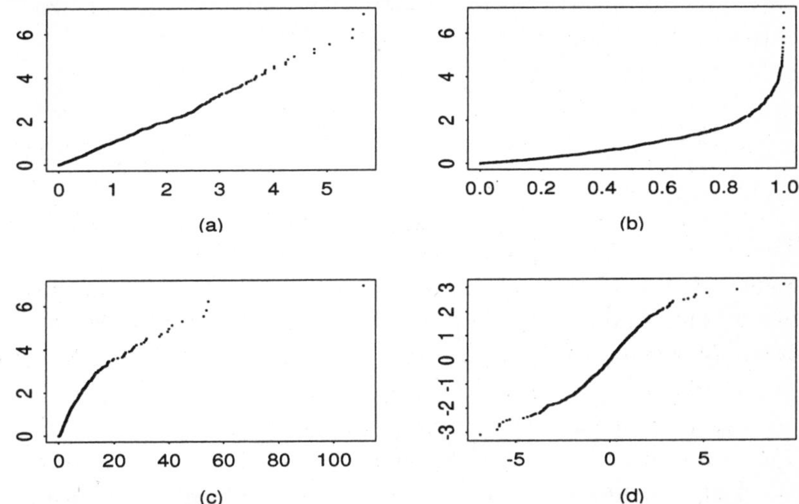

Figure 6.2.1 *QQ–plot of exponentially (a), uniformly (b), lognormally (c) distributed simulated data versus the exponential distribution. In (d) a QQ–plot of* t_4*–distributed data versus the standard normal distribution is given.*

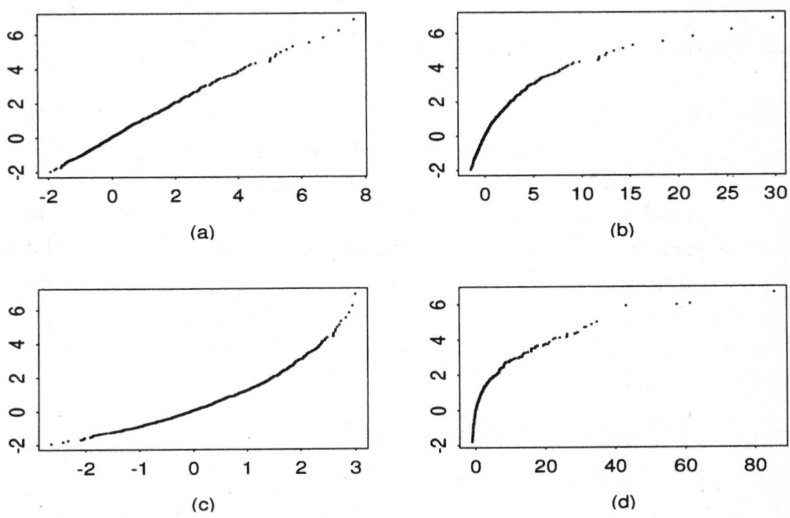

Figure 6.2.2 *QQ–plots: (a) Gumbel distributed simulated data versus Gumbel distribution. GEV distributed data with parameters (b):* $\xi = 0.3$*, (c):* $\xi = -0.3$*, (d):* $\xi = 0.7$*, versus Gumbel. The values* $\xi = 0.7$ *and* $\xi = 0.3$ *are chosen so that* $\alpha = 1/\xi$ *either belongs to the range* $(1, 2)$ *(typically encountered for insurance data) or* $(3, 4)$ *(corresponding to many examples in finance).*

Define the ordered sample $X_{n,n} \le \cdots \le X_{1,n}$. The theoretical basis that underlies probability plots is the quantile transformation of Lemma 4.1.9, which implies that for F continuous, the rvs $U_i = F(X_i)$, for $i = 1, \ldots, n$, are iid uniform on $(0,1)$. Moreover,

$$(F(X_{k,n}))_{k=1,\ldots,n} \stackrel{d}{=} (U_{k,n})_{k=1,\ldots,n} \,.$$

From this it follows that

$$EF(X_{k,n}) = \frac{n-k+1}{n+1}\,, \quad k = 1, \ldots, n\,.$$

Also note that $F_n(X_{k,n}) = (n-k+1)/n$, where F_n stands for the empirical df of F. The graph

$$\left\{ \left(F(X_{k,n})\,, \frac{n-k+1}{n+1} \right) : k = 1, \ldots, n \right\}$$

is called a *probability plot* (*PP–plot*). More common however is to plot the graph

$$\left\{ \left(X_{k,n}\,, F^{\leftarrow}\left(\frac{n-k+1}{n+1} \right) \right) : k = 1, \ldots, n \right\} \qquad (6.1)$$

typically referred to as the *quantile plot* (*QQ–plot*). In both cases, the *approximate linearity of the plot* is justified by the Glivenko–Cantelli theorem; see Example 2.1.4. The theory of weak convergence of empirical processes forms the basis for the construction of confidence bands around the graphs, leading to hypothesis testing. We refrain from entering into details here; see for instance Shorack and Wellner [579], p. 247.

There exist various variants of (6.1) of the type

$$\{ (X_{k,n}\,, F^{\leftarrow}(p_{k,n})) : k = 1, \ldots, n \}\,, \qquad (6.2)$$

where $p_{k,n}$ is a certain *plotting position*. Typical choices are

$$p_{k,n} = \frac{n-k+\delta_k}{n+\gamma_k}\,,$$

with (δ_k, γ_k) appropriately chosen allowing for some *continuity correction*. We shall mostly take (6.1) or (6.2) with

$$p_{k,n} = \frac{n-k+0.5}{n}\,.$$

For a Gumbel distribution

$$\Lambda(x) = \exp\left\{ -e^{-x} \right\}\,, \quad x \in \mathbb{R}\,,$$

the method is easily applied and leads to so-called *double logarithmic plotting*. Assume for instance that we want to test whether the sample $X_1, \ldots, X_n$ comes from Λ. To this end, we take the ordered sample and plot $X_{k,n}$ (more precisely the kth largest observation $x_{k,n}$) against $\Lambda^{\leftarrow}(p_{k,n}) = -\ln(-\ln p_{k,n})$, where $p_{k,n}$ is a plotting position as discussed above. If the Gumbel distribution provides a good fit to our data, then this QQ-plot should look roughly linear; see Figure 6.2.2(a).

Mostly, however, the data would be tested against a location-scale family $F((\cdot - \mu)/\psi)$ where in some cases (for instance when $F = \Phi$ standard normal) μ and ψ are the mean and standard deviation of X. A QQ-plot using F would still be linear, however with slope ψ and intercept μ. Using linear regression for instance, a quick estimate of both parameters can be deduced.

In summary, the main merits of QQ-plots stem from the following properties, taken from Chambers [108]; see also Barnett [50], Castillo [104], Section 6.2.1, David [156], Section 7.8, and Gnanadesikan [265].

(*a*) Comparison of distributions. *If the data were generated from a random sample of the reference distribution,* the plot should look roughly linear. *This remains true if the data come from a linear transformation of the distribution.*

(*b*) Outliers. *If one or a few of the data values are contaminated by gross error or for any reason are markedly different in value from the remaining values, the latter being more or less distributed like the reference distribution,* the outlying points may be easily identified on the plot.

(*c*) Location and scale. *Because a change of one of the distributions by a linear transformation* simply transforms the plot by the same transformation, *one may estimate graphically (through the intercept and slope) location and scale parameters for a sample of data, on the assumption that the data come from the reference distribution.*

(*d*) Shape. *Some difference in distributional shape may be deduced from the plot. For example if the reference distribution has heavier tails (tends to have more large values) the plot will curve down at the left and/or up at the right.*

For an illustration of (a) and (d) see Figure 6.2.1. For an illustration of (d) in a two-sided case see Figure 6.2.1(d).

So far we have considered only location-scale families. In the case of the generalised extreme value distribution (GEV), see Definition 3.4.1,

$$
\begin{aligned}
& H_{\xi;\mu,\psi}(x) \\
= & \exp\left\{-\left(1+\xi\frac{x-\mu}{\psi}\right)^{-1/\xi}\right\}, \quad 1+\xi(x-\mu)/\psi>0\,, \qquad (6.3) \\
= & \begin{cases} \Phi_\alpha(1+(x-\mu)/(\alpha\psi)) & \text{for } x>\mu-\psi\alpha, \quad \xi=1/\alpha>0\,, \\ \Psi_\alpha(-(1-(x-\mu)/(\alpha\psi))) & \text{for } x<\mu+\psi\alpha, \quad \xi=-1/\alpha<0\,, \\ \Lambda((x-\mu)/\psi) & \text{for } x\in\mathbb{R}, \qquad \xi=0\,, \end{cases}
\end{aligned}
$$

besides the location and scale parameters $\mu \in \mathbb{R}$, $\psi > 0$, a shape parameter $\xi \in \mathbb{R}$ enters, making immediate interpretation of a QQ–plot more delicate. Recall that Φ_α, Ψ_α and Λ denote the standard extreme value distributions Fréchet, Weibull and Gumbel; see Definition 3.2.6. A preferred method for testing graphically whether our sample comes from $H_{\xi;\mu,\psi}$ would be to first obtain an estimate $\widehat{\xi}$ for ξ either by guessing or by one of the methods given in Section 6.4.2, and consequently work out a QQ–plot using $H_{\widehat{\xi};0,1}$ where again μ and ψ may be estimated either by visual inspection or through linear regression. These preliminary estimates are often used as starting values in numerical iteration procedures.

6.2.2 The Mean Excess Function

Another useful graphical tool, in particular for discrimination in the tails, is the *mean excess function.* Note that we have already introduced this function in the context of the GEV; see Definition 3.4.6. We recall it here for convenience.

Definition 6.2.3 (Mean excess function)
Let X be a rv with right endpoint x_F; then

$$
e(u) = E(X-u \mid X>u)\,, \quad 0 \le u < x_F\,, \qquad (6.4)
$$

is called the mean excess function *of X.* □

The quantity $e(u)$ is often referred to as the *mean excess over the threshold value u.* This interpretation will be crucial in Section 6.5. In an insurance context, $e(u)$ can be interpreted as the expected claim size in the unlimited layer, over priority u. Here $e(u)$ is also called the *mean excess loss* function. In a reliability or medical context, $e(u)$ is referred to as the *mean residual life* function. In a financial risk management context, switching from the right

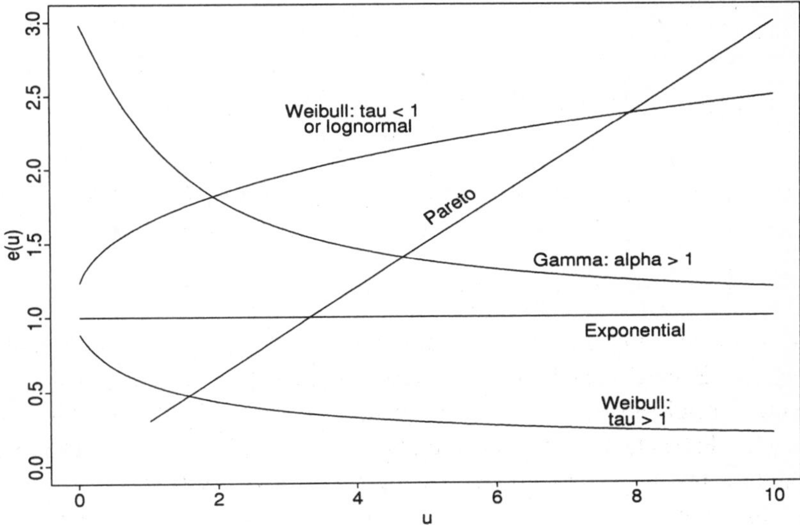

Figure 6.2.4 *Graphs of the mean excess function* $e(u)$ *of some standard distributions; see also Table 3.4.7. Note that heavy–tailed dfs typically have* $e(u)$ *tending to infinity.*

tail to the left tail, $e(u)$ is referred to as the *shortfall.* A summary of the most important mean excess functions is to be found in Table 3.4.7.

In Example 3.4.8 we already noted that any continuous df F is uniquely determined by its mean excess function; see (3.48) and (3.49) for the relevant formulae linking F to e and vice versa.

Example 6.2.5 (Some elementary properties of the mean excess function) If X is $Exp(\lambda)$ distributed, then $e(u) = \lambda^{-1}$ for all $u > 0$. Now assume that X is a rv with support unbounded to the right and df F. If for all $y \in \mathbb{R}$,

$$\lim_{x\to\infty} \frac{\overline{F}(x-y)}{\overline{F}(x)} = e^{\gamma y}\,, \tag{6.5}$$

for some $\gamma \in [0,\infty]$, then $\lim_{u\to\infty} e(u) = \gamma^{-1}$. For the proof use $e(u) = \int_u^\infty \overline{F}(y)\,dy/\overline{F}(u)$ and apply Karamata's theorem (Theorem A3.6) to $\overline{F}\circ\ln$. Notice that for $F \in \mathcal{S}$ (the class of subexponential distributions; see Definition 1.3.3), (6.5) is satisfied with $\gamma = 0$ so that in this heavy–tailed case, $e(u)$ tends to ∞ as $u \to \infty$. On the other hand, superexponential functions of the type $\overline{F}(x) \sim \exp\{-x^a\}$, $a > 1$, satisfy the limit relation (6.5) with $\gamma = \infty$ so that the mean excess function tends to 0. The intermediate cases are covered by the so-called $\mathcal{S}(\gamma)$-classes; see Definition 1.4.9, Embrechts and Goldie [205] and the references therein. □

Example 6.2.6 Recall that for X generalised Pareto the mean excess function is linear; see Theorem 3.4.13(e). The mean excess function of a heavy-tailed df, for large values of the argument, typically appears to be between a constant function (for $Exp(\lambda)$) and a straight line with positive slope (for the Pareto case). Consequently, interesting mean excess functions are of the form

$$e(u) = \begin{cases} u^{1-\beta}/\alpha\,, & \alpha > 0\,,\ 0 \le \beta < 1\,, \\ u/(\alpha + 2\beta \ln u)\,, & \alpha\,,\ \beta > 0\,. \end{cases}$$

Note that $e(u)$ increases but the rate of increase decreases with u. Benktander [60] introduced two families of distributions as claim size models with precisely such mean excess functions. Within the insurance world, they now bear his name. The Benktander-type-I and -type-II classes are defined in Table 1.2.6. □

A graphical test for tail behaviour can now be based on the *empirical mean excess function* $e_n(u)$. Suppose that $X_1, \ldots, X_n$ are iid with df F and let F_n denote the empirical df and $\Delta_n(u) = \{i : i = 1, \ldots, n, X_i > u\}$, then

$$e_n(u) = \frac{1}{\overline{F}_n(u)} \int_u^\infty \overline{F}_n(y)\, dy = \frac{1}{\mathrm{card}\Delta_n(u)} \sum_{i \in \Delta_n(u)} (X_i - u)\,, \quad u \ge 0\,, \tag{6.6}$$

with the convention that $0/0 = 0$. A *mean excess plot* (*ME-plot*) then consists of the graph

$$\{(X_{k,n}, e_n(X_{k,n})) : k = 1, \ldots, n\}\,.$$

The statistical properties of $e_n(u)$ can again be derived by using the relevant empirical process theory as explained in Shorack and Wellner [579], p. 778. For our purposes, the ME-plot is used *only* as a graphical method, mainly for distinguishing between light- and heavy-tailed models; see Figure 6.2.7 for some simulated examples. Indeed caution is called for when interpreting such plots. Due to the sparseness of the data available for calculating $e_n(u)$ for large u-values, the resulting plots are very sensitive to changes in the data towards the end of the range; see for instance Figure 6.2.8. For this reason, more robust versions like *median excess plots* and related procedures have been suggested; see for instance Beirlant, Teugels and Vynckier [57] or Rootzén and Tajvidi [553]. For a critical assessment concerning the use of mean excess functions in insurance see Rytgaard [562]. For a useful application of the ME-plot, see Section 6.5.1.

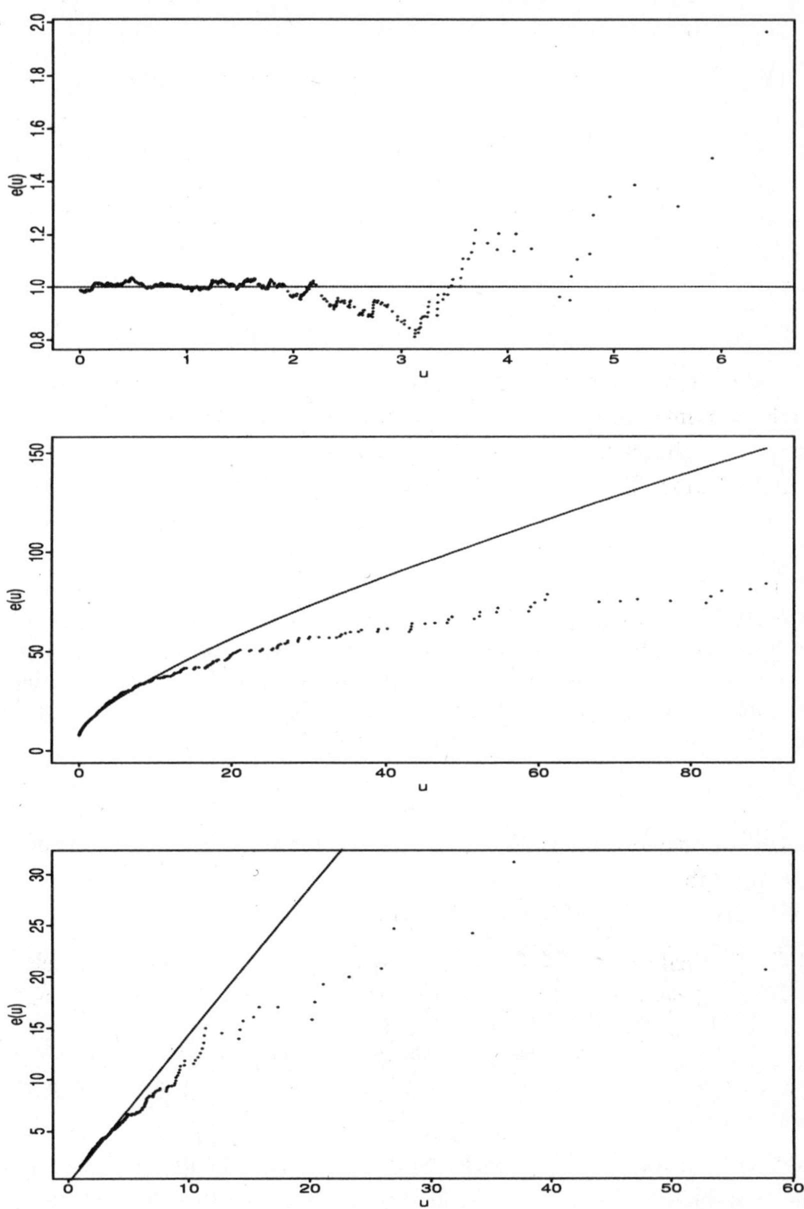

Figure 6.2.7 *The empirical mean excess function* $e_n(u)$ *of simulated data* ($n =$ 1 000) *compared with the corresponding theoretical mean excess function* $e(u)$ (*dashed line*)*: standard exponential* (*top*), *lognormal* (*middle*) *with* $\ln X$ $N(0,4)$, *Pareto* (*bottom*) *with tail index* 1.7.

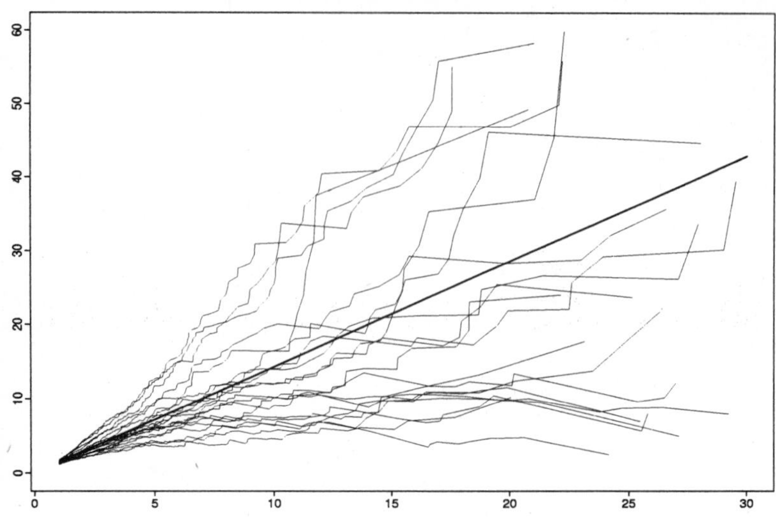

Figure 6.2.8 *The mean excess function of the Pareto distribution* $\overline{F}(x) = x^{-1.7}$, $x \geq 1$, *together with* 20 *empirical mean excess functions* $e_n(u)$ *each based on simulated data* ($n = 1\,000$) *from the above distribution. Note the very unstable behaviour, especially towards the higher values of* u. *This is typical and makes the precise interpretation of* $e_n(u)$ *difficult; see also Figure 6.2.7.*

Example 6.2.9 (Exploratory data analysis for some examples from insurance and finance)
In Figures 6.2.10–6.2.12 we have graphically summarised some properties of three real data–sets. Two come from insurance, one from finance. The data underlying Figure 6.2.11 correspond to Danish fire insurance claims in millions of Danish Kroner (1985 prices). The data were communicated to us by Mette Rytgaard and correspond to the period 1980–1993, inclusive. There is a total of $n = 2\,493$ observations. For a preliminary analysis of these data, see Rytgaard [562].

The second insurance data, presented in Figure 6.2.12, correspond to a portfolio of industrial fire data ($n = 8\,043$) reported over a two year period. This data–set is definitely considered by the portfolio manager as "dangerous", i.e. large claim considerations do enter substantially in the final premium calculation.

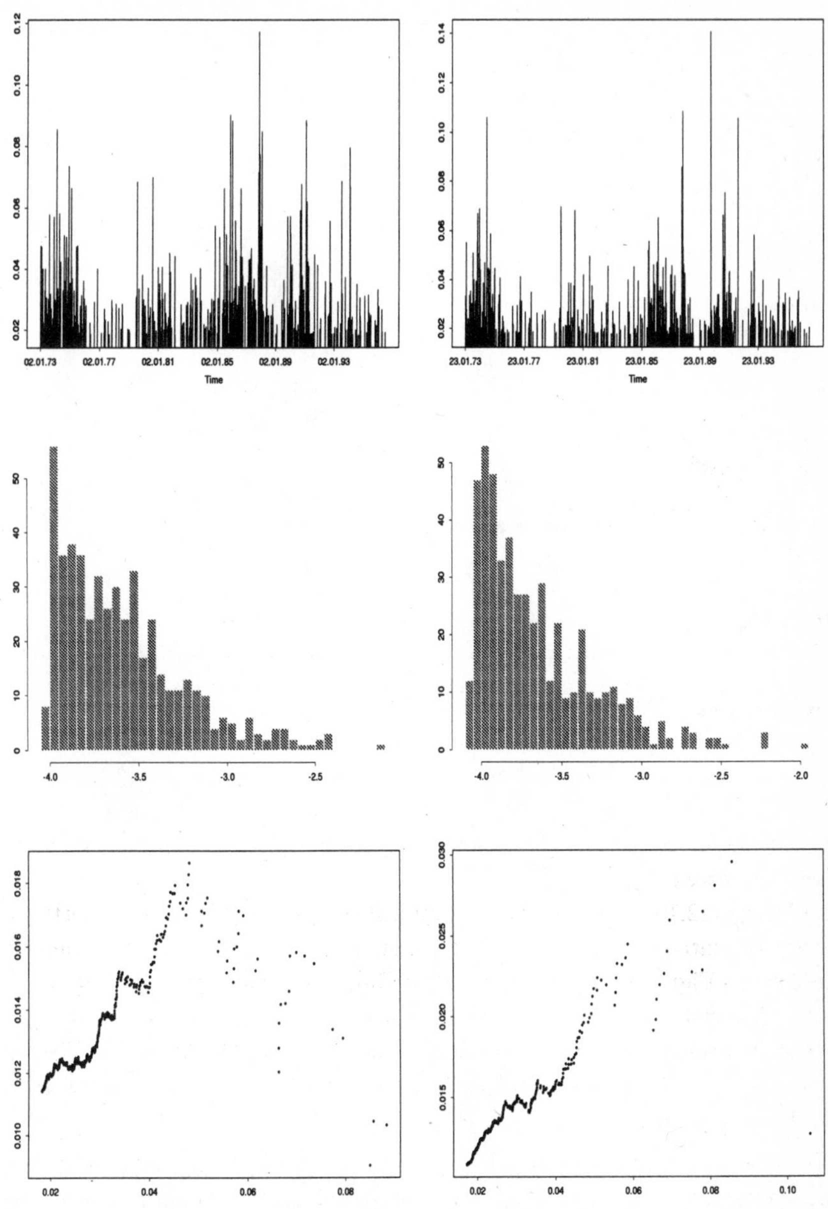

Figure 6.2.10 *Exploratory data analysis of BMW share prices. Top: the* 500 *largest values from the upper tail (positive returns) and lower tail (absolute negative returns). Middle: the corresponding log–histograms. Bottom: the ME–plots. See Example 6.2.9 for some comments.*

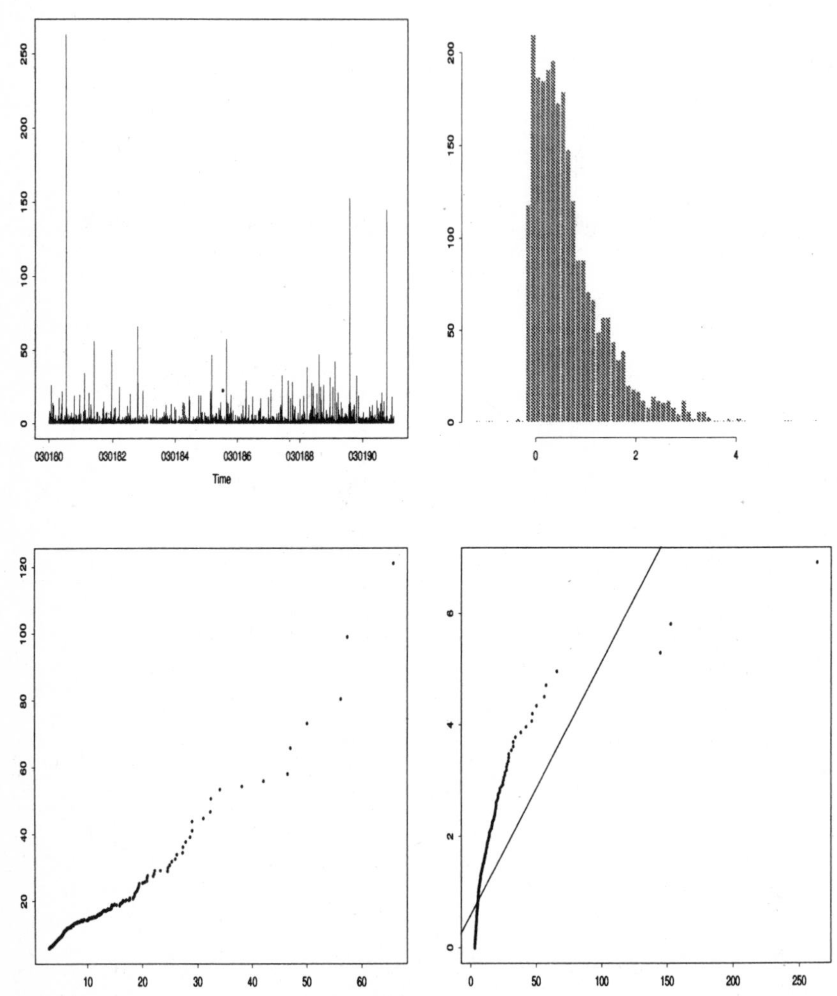

Figure 6.2.11 *Exploratory data analysis of Danish insurance claims caused by fire: the data (top left), the histogram of the log-transformed data (top right), the ME-plot (bottom left) and a QQ-plot against standard exponential quantiles (bottom right). See Example 6.2.9 for some comments.*

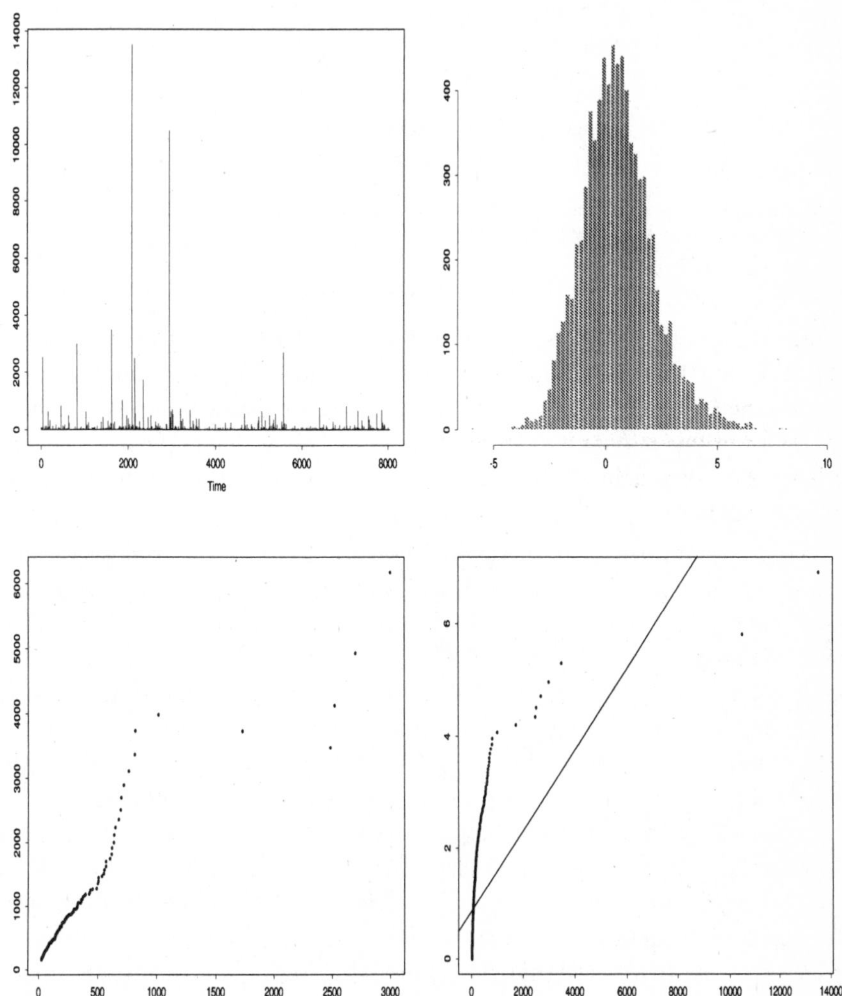

Figure 6.2.12 *Exploratory data analysis of insurance claims caused by industrial fire: the data (top left), the histogram of the log-transformed data (top right), the ME-plot (bottom left) and a QQ-plot against standard exponential quantiles (bottom right). See Example 6.2.9 for some comments.*

Data	Danish	Industrial
n	2 493	8 043
min	0.3134	0.003
1st quartile	1.157	0.587
median	1.634	1.526
mean	3.063	14.65
3rd quartile	2.645	4.488
max	263.3	13 520
$\widehat{x}_{0.99}$	24.61378	184.0009

Table 6.2.13 *Basic statistics for the Danish and the industrial fire data;* $\widehat{x}_{0.99}$ *stands for the empirical* 99%*-quantile.*

A first glance at the figures and Table 6.2.13 for both data–sets immediately reveals heavy–tailedness and skewness to the right. The corresponding mean excess functions are close to a straight line which indicates that the underlying distributions may be modelled by Pareto–like dfs. The QQ–plots against the standard exponential quantiles also clearly show tails much heavier than exponential ones.

Whereas often insurance data may be supposed to represent iid observations, this is typically not the case for finance data as the BMW daily log–return data underlying Figure 6.2.10. For the full data–set see Figure 6.1.7. The period covered is January 23, 1973 - July 12, 1996, resulting in $n = 6\ 146$ observations on the log–returns. Nevertheless, we may assume stationarity of the underlying times series so that many limit results (such as the SLLN) remain valid under general conditions. This would allow us to interpret the graphs of Figure 6.2.10 in a way similar to the iid case, i.e. we will assume that the empirical plots (histogram, empirical mean excess function, QQ–plot) are close to their theoretical counterparts. Note that we contrast these tools for the positive daily log–returns and the absolute values of the negative ones. The log–histograms again show skewedness to the right and heavy-tailedness. It is interesting to observe that the upper and lower tail of the distribution of the log–returns are different. Indeed, both the histograms and the ME-plots (mind the different slopes) indicate that the lower tail of the distribution is heavier than the upper one.

In Figure 6.2.10 we have singled out the 500 largest positive (left) and negative (right) log–returns over the above period. In Table 6.2.14 we have summarised some basic statistics for the three resulting data–sets: BMW—all, BMW–upper and BMW–lower. The nomenclature should be obvious.

We would like to stress that it is our aim to fit tail–probabilities (i.e. probabilities of extreme returns). Hence it is natural for such a fitting to disregard

Data	BMW–all	BMW–upper	BMW–lower
n	6 146	500	500
min	-0.14060	0.01818	0.01719
1st quartile	-0.006656	0.020710	0.019480
median	0.00000	0.02546	0.02331
mean	0.0003407	0.0295700	0.0279500
3rd quartile	0.007126	0.032920	0.031240
max	0.1172	0.1172	0.1406

Table 6.2.14 *Basic statistics for the BMW data.*

the "small" returns. The choice of 500 at this point is rather arbitrary; we will come back to this issue and indeed a more detailed analysis in Section 6.5.2. □

6.2.3 Gumbel's Method of Exceedances

There is a multitude of fairly easy analytic results concerning extremes which yield useful preliminary information on the data. The first method, *Gumbel's method of exceedances*, concerns the question:

How many values among future observations exceed past records?

Let $X_{n,n} < \cdots < X_{1,n}$ as usual be the order statistics of a sample $X_1, \ldots, X_n$ embedded in an infinite iid sequence (X_i) with continuous df F. Take the kth upper order statistic $X_{k,n}$ as a (random) threshold value and denote by $S_r^n(k)$, $r \geq 1$, the number of exceedances of $X_{k,n}$ among the next r observations $X_{n+1}, \ldots, X_{n+r}$, i.e.

$$S_r^n(k) = \sum_{i=1}^{r} I_{\{X_{n+i} > X_{k,n}\}} \,.$$

For ease of notation, we sometimes write S for $S_r^n(k)$ below.

Lemma 6.2.15 (Order statistics and the hypergeometric df)
The rv S defined above has a hypergeometric distribution, *i.e.*

$$P(S = j) = \frac{\binom{r+n-k-j}{n-k}\binom{j+k-1}{k-1}}{\binom{r+n}{n}}\,, \quad j = 0, 1, \ldots, r\,. \tag{6.7}$$

Proof. Conditioning yields

$$P(S = j) = \int_0^\infty P\left(S = j \mid X_{k,n} = u\right)\, dF_{k,n}(u)\,,$$

where $F_{k,n}$ denotes the df of $X_{k,n}$. Now use the fact that $(X_1,\ldots,X_n)$ and $(X_{n+1},\ldots,X_{n+r})$ are independent, that $\sum_{i=1}^r I_{\{X_i>u\}}$ has a binomial distribution with parameters r and $\overline{F}(u)$, and, from Proposition 4.1.2(b), that

$$dF_{k,n}(u) = \frac{n!}{(k-1)!(n-k)!}\,F^{n-k}(u)\,\overline{F}^{k-1}(u)\,dF(u)$$

to obtain (6.7). □

Remark. It readily follows from the definition of S and the argument given in the above proof that $ES = rk/(n+1)$ for the mean number of exceedances of the random threshold $X_{k,n}$. For a detailed discussion on the hypergeometric distribution see for instance Johnson and Kotz [361]. □

Example 6.2.16 Suppose $n = 100$, $r = 12$. We want to calculate the probabilities $p_k = P(S_{12}^{100}(k) = 0)$ that there are no exceedances of the level $X_{k,100}$, $k \geq 1$, in the next twelve observations. For $j = 0$, formula (6.7) reduces to

$$P\left(S_r^n(k) = 0\right) = \frac{n(n-1)\cdots(n-k+1)}{(r+n)(r+n-1)\cdots(r+n-k+1)}\,.$$

In tabulated form we obtain for $n = 100$ and $r = 12$,

k	1	2	3	4	5
p_k	0.893	0.796	0.709	0.631	0.561

So if we have, say, 100 monthly data points and set out to design a certain standard equal to the third largest observation, there is about a 70% chance that this level will not be exceeded during the next year. □

p	$k=1$	$k=2$	$k=3$	$k=4$	$k=5$
$j=0$	0.7778	0.6010	0.4612	0.3514	0.2657
$j=1$	0.1768	0.2795	0.3295	0.3428	0.3321
$j=2$	0.0370	0.0899	0.1446	0.1929	0.2299
$j=3$	0.0070	0.0234	0.0482	0.0791	0.1130
$j=4$	0.0012	0.0051	0.0130	0.0255	0.0427
$j=5$	0.0002	0.0009	0.0029	0.0066	0.0128
$j=6$	0.0000	0.0001	0.0005	0.0014	0.0031
$j=7$	0.0000	0.0000	0.0001	0.0002	0.0006
$j=8$	0.0000	0.0000	0.0000	0.0000	0.0001
$j=9$	0.0000	0.0000	0.0000	0.0000	0.0000

Table 6.2.17 *Exceedance probabilities of the river Nidd data. For given k (order statistic) and j (number of exceedances), $p = P(S_{10}^{35}(k) = j)$ as calculated in (6.7), is given; see Example 6.2.18.*

Example 6.2.18 (River Nidd data, continuation)
For the river Nidd annual data from Example 6.1.1 we have that $n = 35$. The exceedance probabilities (6.7) for the next $r = 10$ years are given in Table 6.2.17. For example, the probability of not exceeding during the next 10 years, the largest annual flow observed so far equals $P(S_{10}^{35}(1) = 0) = 0.7778$. The probability of exceeding at least once, during the next 10 years, the third highest level observed so far equals $1 - P(S_{10}^{35}(3) = 0) = 1 - 0.4612 = 0.5388$.

□

6.2.4 The Return Period

In this section we are interested in answering the question:

What is the mean waiting time between specific extremal events?

This question is usually made precise in the following way. Let (X_i) be a sequence of iid rvs with continuous df F and u a given threshold. We consider the sequence $(I_{\{X_i > u\}})$ of iid Bernoulli rvs with success probability $p = \overline{F}(u)$. Consequently, the time of the first success

$$L(u) = \min\{i \geq 1 : X_i > u\} ,$$

i.e. the time of the first exceedance of the threshold u, is a geometric rv with distribution

$$P(L(u) = k) = (1-p)^{k-1}p , \quad k = 1, 2, \ldots .$$

Notice that the iid rvs

$$L_1(u) = L(u) , \quad L_{n+1}(u) = \min\{i > L_n(u) : X_i > u\} , \quad n \geq 1 ,$$

describe the time periods between successive exceedances of u by (X_n). The *return period of the events* $\{X_i > u\}$ is then defined as $EL(u) = p^{-1} = (\overline{F}(u))^{-1}$, which increases to ∞ as $u \to \infty$. For ease of notation we take dfs with unbounded support above. All relevant questions concerning the return period can now be answered straightforwardly through the corresponding properties of the geometric distribution. Below we give some examples.

Define

$$r_k = P(L(u) \leq k) = p \sum_{i=1}^{k} (1-p)^{i-1} = 1 - (1-p)^k , \quad k \in \mathbb{N} .$$

Hence r_k is the probability that there will be at least one exceedance of u before time k (or within k observations). This gives a 1–1 relationship between r_k and the return period p^{-1}.

The probability that there will be an exceedance of u *before* the return period becomes

$$P(L(u) \le EL(u)) = P\left(L(u) \le [1/p]\right) = 1 - (1-p)^{[1/p]},$$

where $[x]$ denotes the integer part of x. For high thresholds u, i.e. for $u \uparrow \infty$ and consequently $p \downarrow 0$, we obtain

$$\begin{aligned} \lim_{u\uparrow\infty} P(L(u) \le EL(u)) &= \lim_{p\downarrow 0} \left(1 - (1-p)^{[1/p]}\right) \\ &= 1 - e^{-1} = 0.63212\,. \end{aligned}$$

This shows that for high thresholds the mean of $L(u)$ (the return period) is larger than its median.

Example 6.2.19 (Return period, t–year event)
Within an insurance context, a structure is to be insured on the basis that it will last at least 50 years with no more than 10% risk of failure. What does this information imply for the return period? Using the language above, the engineering requirement translates into

$$P(L(u) \le 50) \le 0.1\,.$$

Here we tacitly assumed that a structure failure for each year i can be modelled through the event $\{X_i > u\}$, where X_i is a structure–dependent critical component, say. We assume the iid property of the X_i. The above condition, solved for $P(L(u) \le 50) = 1-(1-p)^{50} = 0.1$, now immediately implies that $p = 0.002105$, i.e. $EL(u) = 475$. In insurance language one speaks in this case about a 475–*year event.*
The important next question concerns the implication of a t–*year event* requirement on the underlying threshold value. By definition this means that for the corresponding threshold u_t,

$$t = EL\left(u_t\right) = \frac{1}{\overline{F}\left(u_t\right)},$$

hence

$$u_t = F^{\leftarrow}\left(1 - t^{-1}\right).$$

In the present example, $u_{475} = F^{\leftarrow}(0.9979)$. This leads us once more to the crucial problem of high quantile estimation. □

Example 6.2.20 (Continuation of Example 6.1.8)
In the case of the Dutch dyke example, recall that, assuming stationarity among the annual maxima of sea levels, the last comparable flood before 1953 took place in November 1570, so that in the above language one would

speak about a 382–year event. The 1953 level hence corresponds roughly to the $(1-1/382)$–quantile of the distribution of the annual maximum. The subsequent government requirements demanded dykes to be built corresponding to a 1 000–to–10 000–year event! □

The above examples clearly stress the need for a solution to the following problems:

- Find reliable estimators for high quantiles from iid data.
- As most data in practice will exhibit dependence and/or non–stationarity find quantile estimation procedures for non–iid data.

6.2.5 Records as an Exploratory Tool

Suppose that the rvs X_i are iid with df F. Recall from Section 5.4 the definitions of records and record times: a *record* X_n occurs if $X_n > M_{n-1} = \max(X_1, \ldots, X_{n-1})$. By definition we take X_1 as a record. In Section 5.4 we used point process language in order to describe records and *record times* L_n. The latter are the random times at which the process (M_n) jumps. Define the *record counting process* as

$$N_1 = 1\,, \quad N_n = 1 + \sum_{k=2}^{n} I_{\{X_k > M_{k-1}\}}\,, \quad n \geq 2\,.$$

The following result (on the mean EN_n) may be surprising.

Lemma 6.2.21 (Moments of N_n)
Suppose (X_i) are iid with continuous df F and (N_n) defined as above. Then

$$EN_n = \sum_{k=1}^{n} \frac{1}{k} \quad \text{and} \quad \operatorname{var}(N_n) = \sum_{k=1}^{n} \left(\frac{1}{k} - \frac{1}{k^2} \right) .$$

Proof. From the definition of N_n we obtain

$$\begin{aligned} EN_n &= 1 + \sum_{k=2}^{n} P\left(X_k > M_{k-1}\right) \\ &= 1 + \sum_{k=2}^{n} \int_{-\infty}^{+\infty} P\left(X_k > u\right) dP\left(M_{k-1} \leq u\right) . \end{aligned}$$

Now use $P\left(M_{k-1} \leq u\right) = F^{k-1}(u)$ which immediately yields the result for EN_n. The same argument works for $\operatorname{var}(N_n)$. □

Notice that EN_n and $\operatorname{var}(N_n)$ are both of the order $\ln n$ as $n \to \infty$. More precisely, $EN_n - \ln n \to \gamma$, where $\gamma = 0.5772\ldots$ denotes Euler's constant. As a consequence:

the number of records of iid data grows very slowly!

Before reading further, guess the answer to the next question:

How many records do we expect in 100, 1 000 *or* 10 000 *iid observations?*

Table 6.2.22 contains the somewhat surprising answer; see also Figures 5.4.2 and 5.4.12.

$n = 10^k$, $k =$	EN_n	$\ln n$	$\ln n + \gamma$	D_n
1	2.9	2.3	2.9	1.2
2	5.2	4.6	5.2	1.9
3	7.5	7.0	7.5	2.4
4	9.8	9.2	9.8	2.8
5	12.1	11.5	12.1	3.2
6	14.4	13.8	14.4	3.6
7	16.7	16.1	16.7	3.9
8	19.0	18.4	19.0	4.2
9	21.3	20.7	21.3	4.4

Table 6.2.22 *Expected number of records EN_n in an iid sequence (X_n), together with the asymptotic approximations $\ln n$, $\ln n + \gamma$, and standard deviation $D_n = \sqrt{\operatorname{var}(N_n)}$, based on Lemma 6.2.21.*

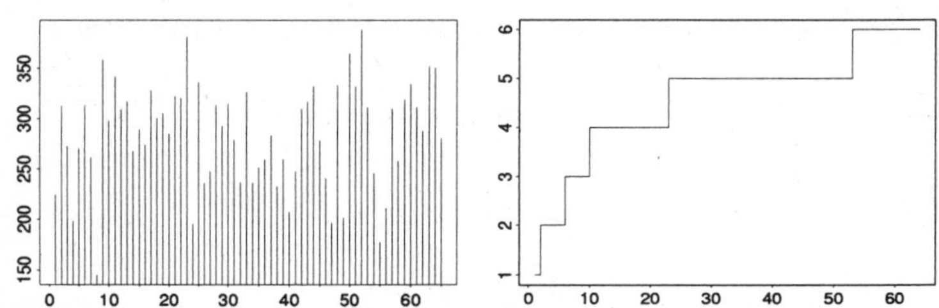

Figure 6.2.23 *Vancouver sunshine data and the corresponding numbers of records.*

Example 6.2.24 (Records in real data)
In Figure 6.2.23 the total amount of sunshine hours in Vancouver during the month of July from 1909 until 1973 is given. The data are taken from Glick [264]. There are 6 records in these $n = 64$ observations, namely for $i = 1, 2, 6, 10, 23, 53$. Clearly one would need a much larger n in order to test confidently the iid hypothesis for the underlying data $X_1, \ldots, X_{64}$ on the basis

of the record values. If the data were iid, then we would obtain $EN_{64} = 4.74$. The observed value of 6 agrees rather well. On the basis of these observations we have no reason to doubt the iid hypothesis. The picture however changes dramatically in Figure 3 of the Reader Guidelines, based on catastrophic insurance claims for the period 1970–1995. It is immediately clear that the number of records does not exhibit a logarithmic growth. □

6.2.6 The Ratio of Maximum and Sum

In this section we consider a further simple tool for detecting heavy tails of a distribution and for giving a rough estimate of the order of its finite moments. Suppose that the rvs $X, X_1, X_2, \ldots$ are iid and define for any positive p the quantities

$$S_n(p) = |X_1|^p + \cdots + |X_n|^p \,, \quad M_n(p) = \max(|X_1|^p, \ldots, |X_n|^p) \,, \quad n \geq 1 \,.$$

We also write $M_n = M_n(1)$ and $S_n = S_n(1)$ slightly abusing our usual notation. One way to study the underlying distribution is to look at the distributional or a.s. behaviour of functionals $f(S_n(p), M_n(p))$. For instance, in Section 8.2.4 we gained some information about the limit behaviour of the ratio M_n/S_n. In particular, we know the following facts (Y_1, Y_2 and $Y_2(p)$ are appropriate non–degenerate rvs):

$$\begin{aligned}
\frac{M_n}{S_n} &\stackrel{\text{a.s.}}{\to} 0 && \Leftrightarrow \quad E|X| < \infty \,, \\
\frac{M_n}{S_n} &\stackrel{P}{\to} 0 && \Leftrightarrow \quad E|X| I_{\{|X| \leq x\}} \in \mathcal{R}_0 \,, \\
\frac{S_n - nE|X|}{M_n} &\stackrel{d}{\to} Y_1 && \Leftrightarrow \quad P(|X| > x) \in \mathcal{R}_{-\alpha} \text{ for some } \alpha \in (1,2) \,, \\
\frac{M_n}{S_n} &\stackrel{d}{\to} Y_2 && \Leftrightarrow \quad P(|X| > x) \in \mathcal{R}_{-\alpha} \text{ for some } \alpha \in (0,1) \,, \\
\frac{M_n}{S_n} &\stackrel{P}{\to} 1 && \Leftrightarrow \quad P(|X| > x) \in \mathcal{R}_0 \,.
\end{aligned}$$

Writing

$$R_n(p) = \frac{M_n(p)}{S_n(p)} \,, \quad n \geq 1 \,, p > 0 \,, \tag{6.8}$$

we may conclude from the latter relations that the following equivalences hold:

$$R_n(p) \stackrel{\text{a.s.}}{\to} 0 \quad \Leftrightarrow \quad E|X|^p < \infty\,,$$

$$R_n(p) \stackrel{P}{\to} 0 \quad \Leftrightarrow \quad E|X|^p I_{\{|X|\le x\}} \in \mathcal{R}_0\,,$$

$$R_n(p) \stackrel{d}{\to} Y_2(p) \quad \Leftrightarrow \quad P(|X|>x) \in \mathcal{R}_{-\alpha p} \text{ for some } \alpha \in (0,1)\,,$$

$$R_n(p) \stackrel{P}{\to} 1 \quad \Leftrightarrow \quad P(|X|>x) \in \mathcal{R}_0\,.$$

Now it is immediate how one can use these limit results to obtain some preliminary information about $P(|X|>x)$: plot $R_n(p)$ against n for a variety of p–values. Then $R_n(p)$ should be small for large n provided that $E|X|^p < \infty$. On the other hand, if there are significant deviations of $R_n(p)$ from zero for large n, this is an indication for $E|X|^p$ being infinite; see Figures 6.2.25–6.2.29 for some examples of simulated and real data.

Clearly, what has been said about the absolute value of the X_i can be modified in the natural way to get information about the right or left distribution tail: replace everywhere $|X_i|^p$ by the pth power of the positive or negative part of the X_i. Moreover, the ratio of maximum over sum can be replaced by more complicated functionals of the upper order statistics of a sample; see for instance the definition of the *empirical large claim index* in Section 8.2.4. This allows to discriminate the distributions in a more subtle way.

Notes and Comments

The statistical properties of QQ–plots, with special emphasis on the heavy–tailed case, are studied for instance in Kratz and Resnick [407]. The importance of the mean excess function (or plot) as a diagnostic tool for insurance data is nicely demonstrated in Hogg and Klugman [330]; see also Beirlant et al. [57] and the references therein. Return periods and t–year events have a long history in hydrology; see for instance Castillo [104] and Rosbjerg [554]. For relevant statistical techniques coming more from a reliability context, see Crowder et al. [143]; methods more related to medical statistics are to be found in Andersen et al. [10].

Since the fundamental paper by Foster and Stuart [243], numerous papers have been published on records; see for instance Pfeifer [497], Kapitel 4, Resnick [530], Chapter 4, and the references cited therein, see also Goldie and Resnick [277], Nagaraja [475] and Nevsorov [478]. We find Glick [264] a very entertaining introduction. Smith [593] gives more information on statistical inference for records, especially in the non–iid case. In Section 5.4 we have discussed in more detail the relevant limit theorems for records and their connections with point process theory and extremal processes. Records in the

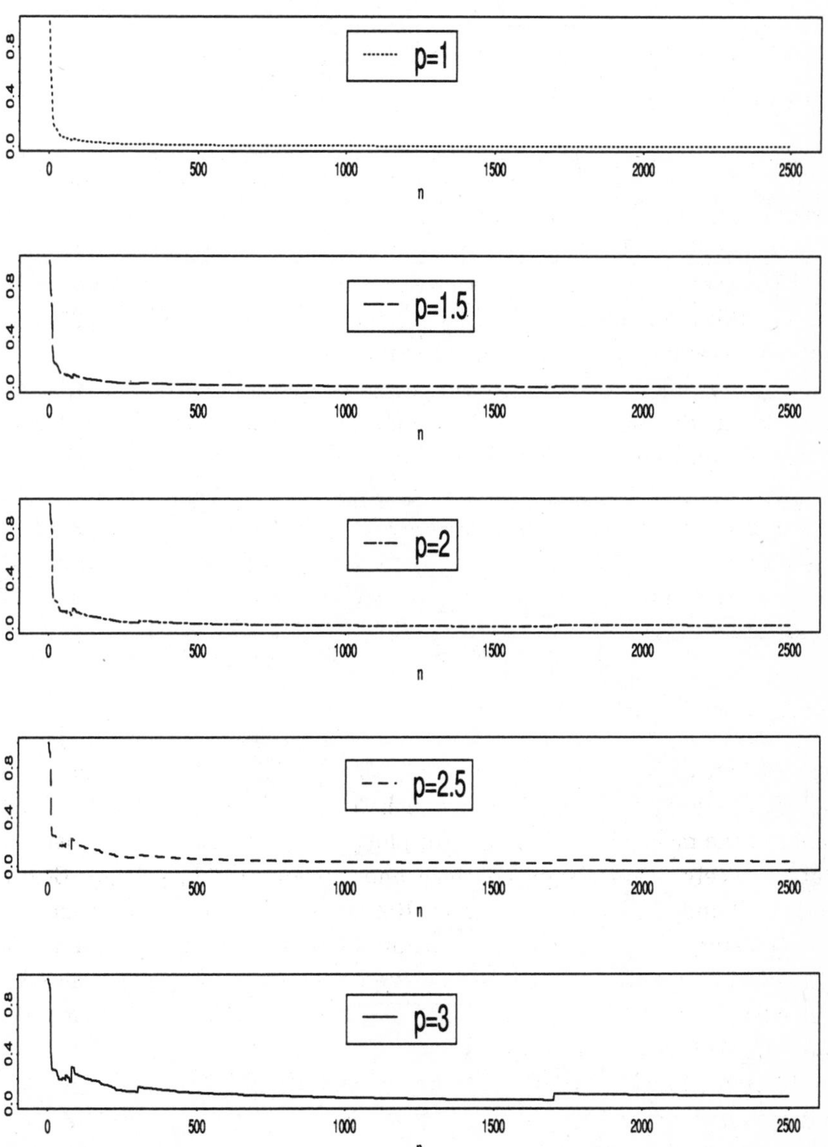

Figure 6.2.25 *The ratio $R_n(p)$ for different p. The X_i are 2 500 iid standard exponential data.*

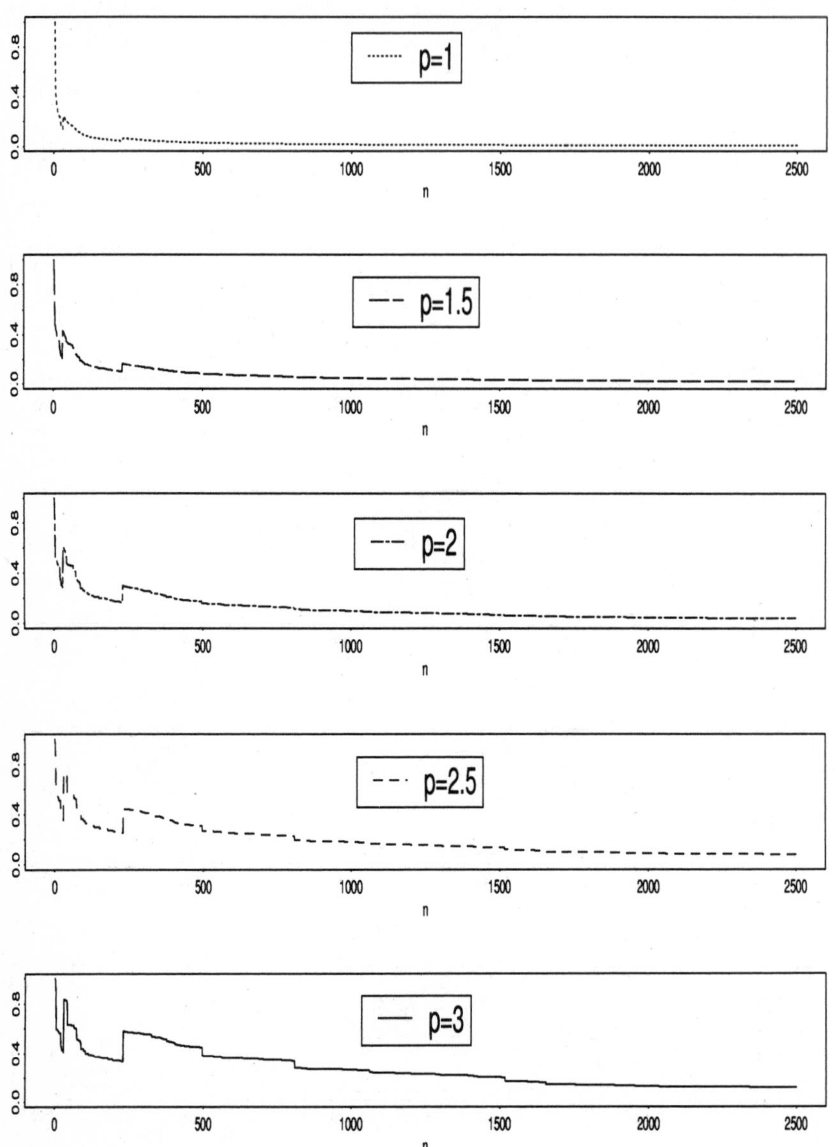

Figure 6.2.26 *The ratio $R_n(p)$ for different p. The X_i are 2 500 iid lognormal data.*

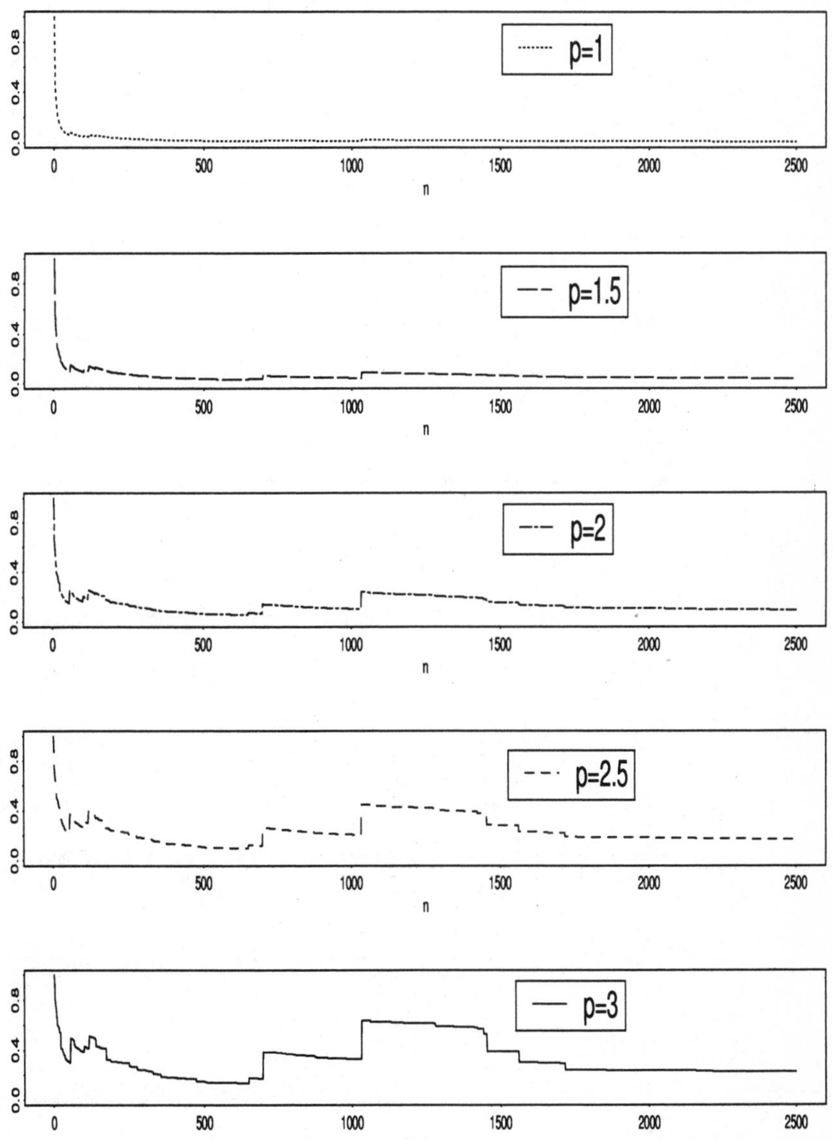

Figure 6.2.27 *The ratio $R_n(p)$ for different n and p. The X_i are 2 500 iid Pareto data with shape parameter 2.*

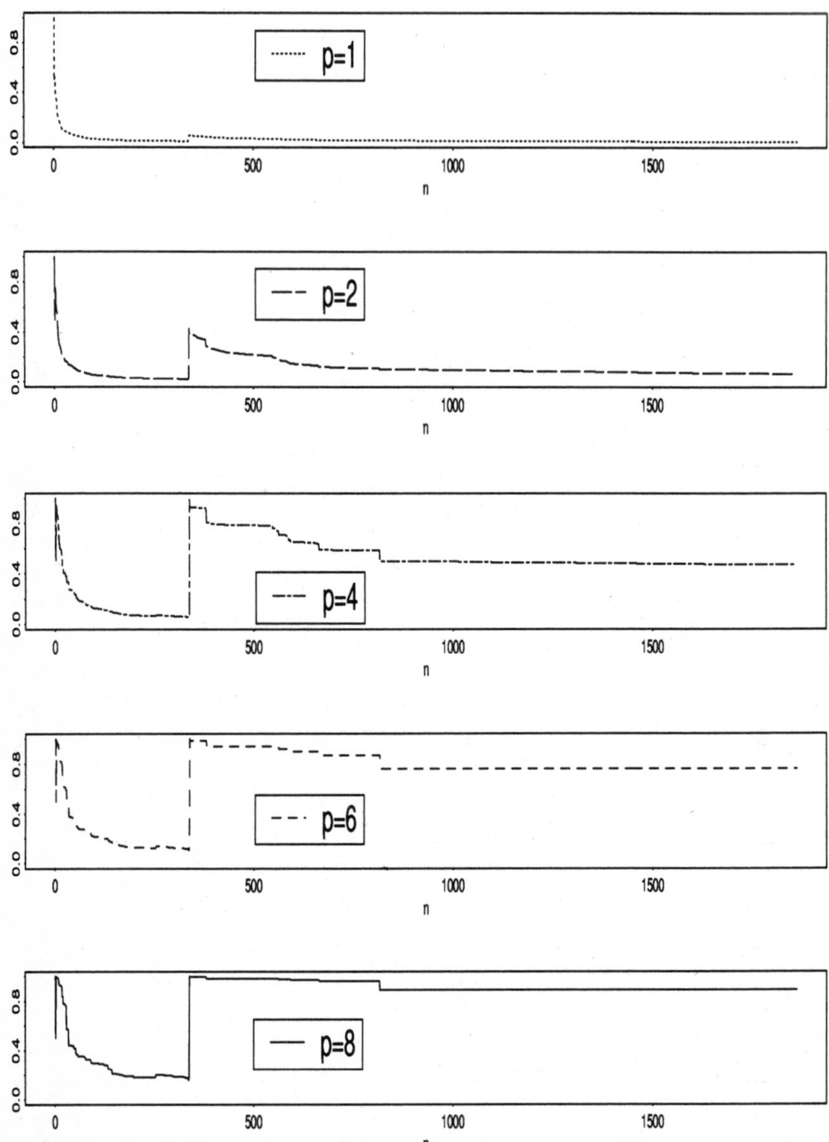

Figure 6.2.28 *The ratio $R_n(p)$ for different n and p. The X_i are* 1 864 *daily log–returns from the German stock index* DAX. *The behaviour of $R_n(p)$ indicates that these data come from a distribution with infinite* 4*th moment.*

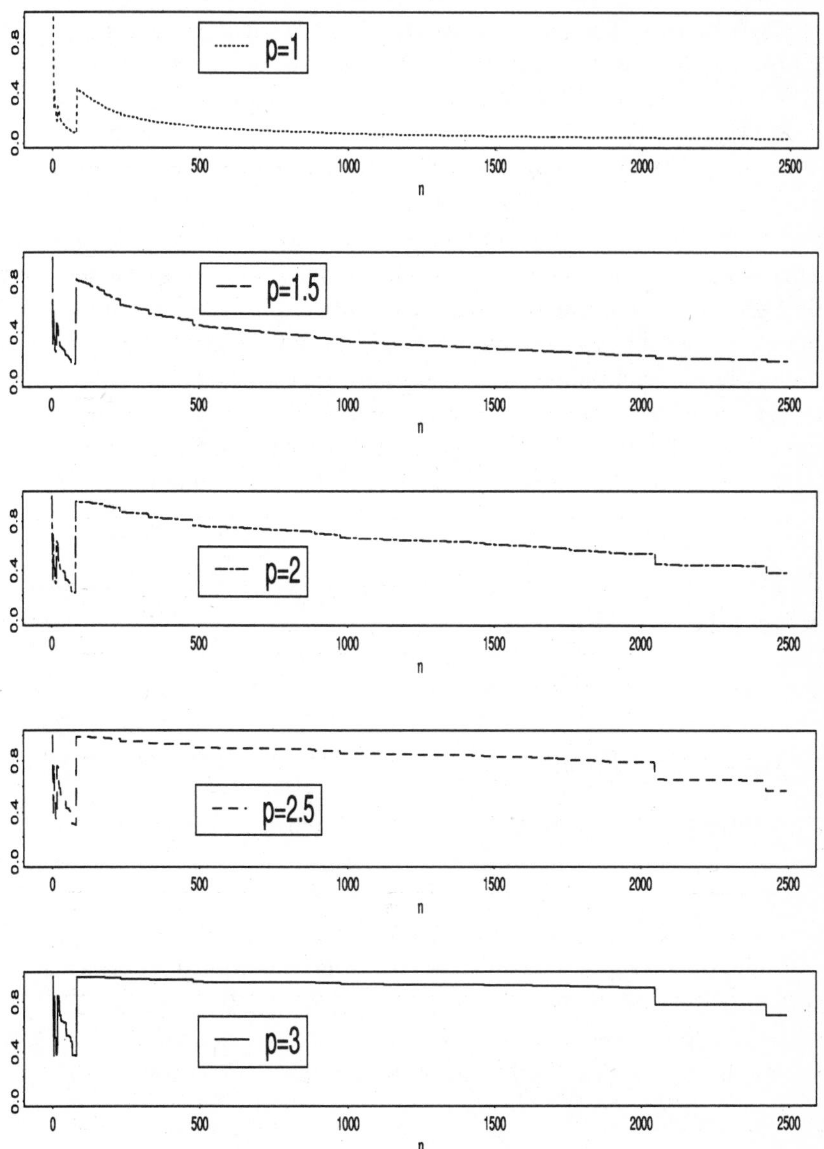

Figure 6.2.29 *The ratio $R_n(p)$ for different n and p. The X_i correspond to the Danish fire insurance data from Figure 6.2.11 ($n = 2\,493$). The behaviour of $R_n(p)$ indicates that these data come from a distribution with infinite 2nd moment. Also compare with Figures 6.2.25–6.2.27.*

presence of a trend have been investigated by several authors, in particular for sports data. A good place to start is Ballerini and Resnick [41] and the references therein. The behaviour of records in an increasing population is for instance described in Yang [644]. Smith [593] discusses the forecasting problem of records based on maximum likelihood methodology.

The exploratory techniques introduced so far all started from an iid assumption on the underlying data. Their interpretation becomes hazardous when applied in the non–iid case, as for instance to data exhibiting a trend. Various statistical detrending techniques exist within the realm of regression theory and time series analysis. These may range from fitting of a deterministic trend to the data, averaging, differencing,.... By one or more of these methods one would hope to filter out some iid residuals to which the previous methods again would apply; see for instance Brockwell and Davis [92], Section 1.4, Feigin and Resnick [231] or Kendall and Stuart [375], Chapter 46. It is perhaps worth stressing at this point that extremes in the detrended data do not necessarily correspond to extremes in the original data.

6.3 Parameter Estimation for the Generalised Extreme Value Distribution

Recall from (6.3) the *generalised extreme value distribution* (*GEV*)

$$H_{\xi;\mu,\psi}(x) = \exp\left\{-\left(1+\xi\frac{x-\mu}{\psi}\right)^{-1/\xi}\right\}, \quad 1+\xi\frac{x-\mu}{\psi} > 0\,. \tag{6.9}$$

As usual the case $\xi = 0$ corresponds to the Gumbel distribution

$$H_{0;\mu,\psi}(x) = \exp\left\{-e^{-(x-\mu)/\psi}\right\}, \quad x \in \mathbb{R}\,. \tag{6.10}$$

The parameter $\theta = (\xi, \mu, \psi) \in \mathbb{R} \times \mathbb{R} \times \mathbb{R}_+$ consists of a shape parameter ξ, location parameter μ and scale parameter ψ. For notational convenience, we shall either write H_ξ or H_θ depending on the case in hand. In Theorem 3.4.5 we saw that H_ξ arises as the limit distribution of normalised maxima of iid rvs. Standard statistical methodology from parametric estimation theory is available if our data consist of a sample

$$X_1, \ldots, X_n \quad \textit{iid from } H_\theta\,. \tag{6.11}$$

We mention here that the assumption of X_i having an *exact* extreme value distribution H_ξ is perhaps not the most realistic one. In the next section we turn to the more tenable assumption that the X_i are *approximately* H_ξ distributed. The "approximately" will be interpreted as "belonging to the maximum domain of attraction of".

Fitting of Annual Maxima

As already discussed in Example 6.1.1, data of the above type may become available when the X_i can be interpreted as maxima over disjoint time periods of length s say. In hydrology, which is the cradle of many of the ideas for statistics of extremal events, this period mostly consists of one year; see for instance the river Nidd data in Figure 6.1.2. The 1–year period is chosen in order to compensate for intra–year seasonalities. Therefore the original data may look like

$$\begin{aligned}\boldsymbol{X}^{(1)} &= \left(X_1^{(1)},\ldots,X_s^{(1)}\right)\\ \boldsymbol{X}^{(2)} &= \left(X_1^{(2)},\ldots,X_s^{(2)}\right)\\ \vdots & \qquad\quad \vdots\\ \boldsymbol{X}^{(n)} &= \left(X_1^{(n)},\ldots,X_s^{(n)}\right)\end{aligned}$$

where the vectors $(\boldsymbol{X}^{(i)})$ are assumed to be iid, but within each vector $\boldsymbol{X}^{(i)}$ the various components may (and mostly will) be dependent. The time length s is chosen so that the above conditions are likely to be satisfied. The basic iid sample from H_θ on which statistical inference is to be performed then consists of

$$X_i = \max(X_1^{(i)},\ldots,X_s^{(i)}), \quad i=1,\ldots,n\,. \tag{6.12}$$

For historical reasons and since s often corresponds to a 1–year period, statistical inference for H_θ based on data of the form (6.12) is referred to as

fitting of annual maxima.

Below we discuss some of the main techniques for estimating θ in the exact model (6.11).

6.3.1 Maximum Likelihood Estimation

The set–up (6.11) corresponds to the standard parametric case of statistical inference and hence *in principle* can be solved by *maximum likelihood methodology.* Suppose that H_θ has density h_θ. Then the *likelihood function* based on the data $\boldsymbol{X}=(X_1,\ldots,X_n)$ is given by

$$L(\theta;\boldsymbol{X}) = \prod_{i=1}^{n} h_\theta\left(X_i\right) I_{\{1+\xi(X_i-\mu)/\psi>0\}}\,.$$

Denote by $\ell(\theta;\boldsymbol{X}) = \ln L(\theta;\boldsymbol{X})$ the *log–likelihood function.* The *maximum likelihood estimator* (*MLE*) for θ then equals

$$\widehat{\theta}_n = \arg\max_{\theta\in\Theta} \ell(\theta; \boldsymbol{X}),$$

i.e. $\widehat{\theta}_n = \widehat{\theta}_n(X_1, \ldots, X_n)$ maximises $\ell(\theta; \boldsymbol{X})$ over an appropriate parameter space Θ. In the case of $H_{0;\mu,\psi}$ this gives us

$$\ell((0,\mu,\psi); \boldsymbol{X}) = -n\ln\psi - \sum_{i=1}^{n} \exp\left\{-\frac{X_i-\mu}{\psi}\right\} - \sum_{i=1}^{n} \frac{X_i-\mu}{\psi}.$$

Differentiating the latter function with respect to μ and ψ yields the likelihood equations in the Gumbel case:

$$\begin{aligned} 0 &= n - \sum_{i=1}^{n} \exp\left\{-\frac{X_i-\mu}{\psi}\right\}, \\ 0 &= n + \sum_{i=1}^{n} \frac{X_i-\mu}{\psi}\left(\exp\left\{-\frac{X_i-\mu}{\psi}\right\} - 1\right). \end{aligned}$$

Clearly no explicit solution exists to these equations. The situation for H_ξ when $\xi \neq 0$ is even more complicated, so that numerical procedures are called for. Jenkinson [355] and Prescott and Walden [510, 511] suggest variants of the Newton–Raphson scheme. With the existence of the Fortran algorithm published in Hosking [336] and its supplement in Macleod [432], the numerical calculation of the MLE $\widehat{\theta}_n$ for general H_θ poses no serious problem *in principle.*

Notice that we said *in principle.* Indeed in the so–called *regular cases* maximum likelihood estimation offers a technique yielding efficient, consistent and asymptotically normal estimators. See for instance Cox and Hinkley [133] and Lehmann [420] for a general discussion on maximum likelihood estimation. Relevant for applications in extreme value theory, typical *non–regular cases may occur whenever the support of the underlying df depends on the unknown parameters.* Therefore, although we have reliable numerical procedures for finding the MLE $\widehat{\theta}_n$, we are less certain about its properties, especially in the small sample case. For a discussion on this point see Smith [589]. In the latter paper it is shown that the classical (good) properties of the MLE hold whenever $\xi > -1/2$; this is not the case for $\xi \leq -1/2$.

As most distributions encountered in insurance and finance have support unbounded to the right (this is possible only for $\xi \geq 0$), the MLE technique offers a useful and reliable procedure in those fields.

At this point we would like to quantify a bit more the often encountered statement that *for applications in insurance (and finance for that matter) the case $\xi \geq 0$ is most important.* Clearly, all financial data *must* be bounded to the right; an obvious (though somewhat silly) bound is total wealth. The main point however is that in most data there does not seem to be clustering towards a well–defined upper limit but more a steady increase over time of the underlying maxima. The latter would then, for iid data, much more naturally be modelled within $\xi \geq 0$. A typical example is to be found in the Danish fire insurance data of Figure 6.2.11.

An example where a natural upper limit may exist is given in Figure 6.3.1. The data underlying this example correspond to a portfolio of water–damage insurance. In contrast to the industrial fire data of Figure 6.2.12, in this case the portfolio manager realises that large claims only play a minor role. Though the data again show an increasing ME–plot, for values above 5 000, the mean excess losses are growing much slower than to be expected from a really heavy–tailed model, unbounded to the right. The ME–plot for these data should be compared with those for the Danish fire data (Figure 6.2.11) and the industrial fire data (Figure 6.2.12). The Pickands estimator (to be introduced in Section 6.4.2) of the extreme value index in Figure 6.4.2 indicates that ξ could actually be negative. Compare also with the corresponding estimates of ξ for the fire data; see Figures 6.5.5 and 6.5.6.

An Extension to Upper Order Statistics

So far, our data have consisted of n iid observations of maxima which we have assumed to follow exactly a GEV H_θ. By appropriately defining the underlying time periods, we *design* independence into the model; see (6.12). Suppose now that, rather than just having the largest observation available, we possess the k largest of each period (year, say). In the notation of (6.12) this would amount to data

$$X_{k,s}^{(i)} \leq \cdots \leq X_{1,s}^{(i)} = X_i\,, \quad i, = 1, \ldots, n\,.$$

Maximum likelihood theory based on these $k \times n$ observations would use the joint density of the independent vectors $(X_{k,s}^{(i)}, \ldots, X_{1,s}^{(i)})$, $i = 1, \ldots, n$. Only rarely in practical cases could we assume that for each i the latter vectors are derived from iid data. If that were the case then maximum likelihood estimation should be based on the joint density of k upper order statistics from a GEV as discussed in Theorem 4.1.3:

$$\frac{s!}{(s-k)!}\, H_\theta^{s-k}(x_k) \prod_{\ell=1}^{k} h_\theta(x_\ell)\,, \quad x_k < \cdots < x_1\,,$$

where, depending on θ, the x–values satisfy the relevant domain restrictions. The standard error of the MLEs for μ and ψ can already be reduced considerably if $k = 2$, i.e. we take the two largest observations into account. For a brief discussion on this method see Smith [595], Section 4.18, and Smith

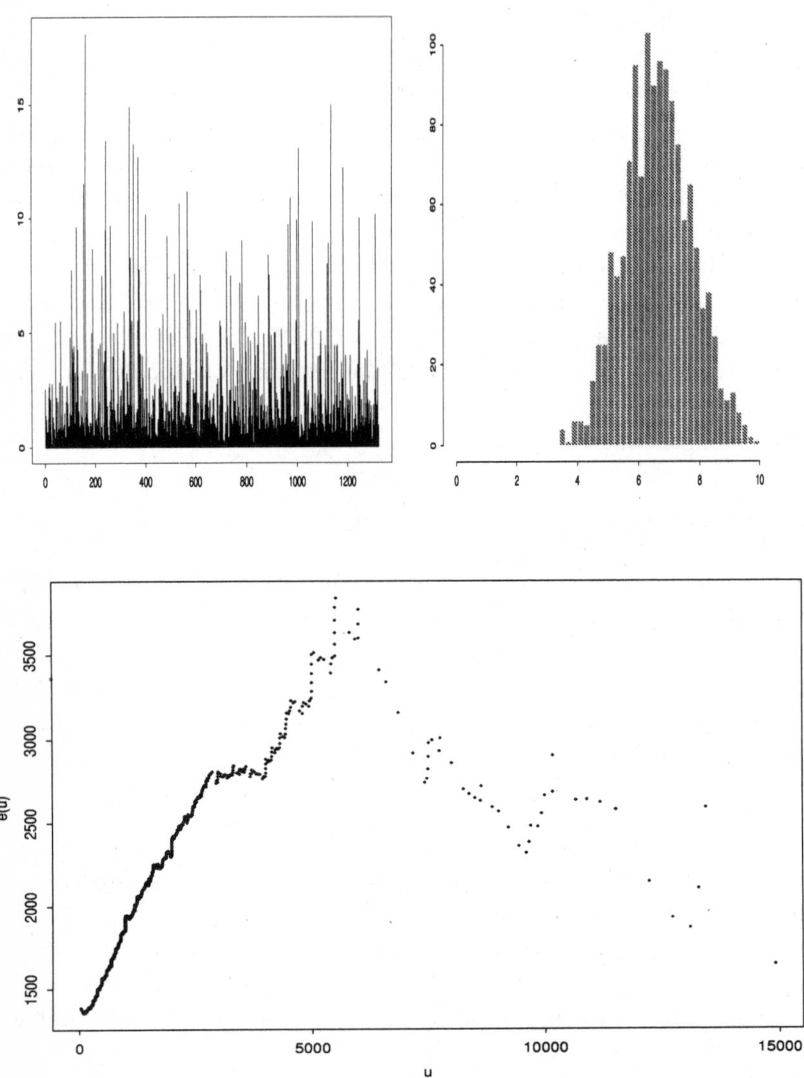

Figure 6.3.1 *Exploratory data analysis of insurance claims caused by water: the data (top, left), the histogram of the log-transformed data (top, right), the ME-plot (bottom). Notice the kink in the ME-plot in the range* (5 000, 6 000) *reflecting the fact that the data seem to cluster towards some specific upper value.*

[590], where also further references and examples are to be found. The case $n = 1$, i.e. only one year of observations say, and $k > 1$ was first discussed in Weissman [637].

A *final statement* concerning maximum likelihood methodology, again taken from Smith [595], is worth stressing:

> The big advantage of maximum likelihood procedures is that they can be generalised, with very little change in the basic methodology, to much more complicated models in which trends or other effects may be present.

If the above quote has made you curious, *do read* Smith [595]. We, however, would like to add that the MLE properties depend on where ξ falls in $(-\infty, \infty)$.

6.3.2 Method of Probability–Weighted Moments

Among all the ad–hoc methods used in parameter estimation, the method of moments has attracted a lot of interest. In full generality it consists of equating model–moments based on H_θ to the corresponding empirical moments based on the data. *Their general properties are notoriously unreliable on account of the poor sampling properties of second– and higher–order sample moments*, a statement taken from Smith [595], p. 447. The class of probability–weighted moment estimators stands out as more promising. This method goes back to Hosking, Wallis and Wood [338]. Define

$$w_r(\theta) = E\left(X H_\theta^r(X)\right), \quad r \in \mathbb{N}_0, \tag{6.13}$$

where H_θ is the GEV and X has df H_θ with parameter $\theta = (\xi, \mu, \psi)$. Recall that for $\xi \geq 1$, $\overline{H}_\theta$ is regularly varying with index $1/\xi$. Hence w_0 is infinite. Therefore we restrict ourselves to the case $\xi < 1$. Define the empirical analogue to (6.13),

$$\widehat{w}_r(\theta) = \int_{-\infty}^{+\infty} x\, H_\theta^r(x)\, dF_n(x), \quad r \in \mathbb{N}_0,$$

where F_n is the empirical df corresponding to the data $X_1, \ldots, X_n$. In order to estimate θ we solve the equations

$$w_r(\theta) = \widehat{w}_r(\theta), \quad r = 0, 1, 2.$$

We immediately obtain

$$\widehat{w}_r(\theta) = \frac{1}{n} \sum_{j=1}^{n} X_{j,n} H_\theta^r \left(X_{j,n}\right), \quad r = 0,1,2. \tag{6.14}$$

Recall the quantile transformation from Lemma 4.1.9(b):

$$\left(H_\theta(X_{n,n}), \ldots, H_\theta(X_{1,n})\right) \stackrel{d}{=} \left(U_{n,n}, \ldots, U_{1,n}\right),$$

where $U_{n,n} \leq \cdots \leq U_{1,n}$ are the order statistics of an iid sequence $U_1, \ldots, U_n$ uniformly distributed on $(0,1)$. With this interpretation, (6.14) can be written as

$$\widehat{w}_r(\theta) = \frac{1}{n} \sum_{j=1}^{n} X_{j,n} U_{j,n}^r, \quad r = 0,1,2. \tag{6.15}$$

Clearly, for $r = 0$, the rhs becomes $\overline{X}_n$, the sample mean. In order to calculate $w_r(\theta)$ for general r, observe that

$$w_r(\theta) = \int_{-\infty}^{+\infty} x\, H_\theta^r(x)\, dH_\theta(x) = \int_0^1 H_\theta^{\leftarrow}(y)\, y^r\, dy,$$

where for $0 < y < 1$,

$$H_\theta^{\leftarrow}(y) = \begin{cases} \mu - \dfrac{\psi}{\xi}\left(1 - (-\ln y)^{-\xi}\right) & \text{if } \xi \neq 0, \\ \mu - \psi \ln(-\ln y) & \text{if } \xi = 0. \end{cases}$$

This yields for $\xi < 1$ and $\xi \neq 0$, after some calculation,

$$w_r(\theta) = \frac{1}{r+1}\left\{\mu - \frac{\psi}{\xi}\left(1 - \Gamma(1-\xi)(1+r)^{\xi}\right)\right\}, \tag{6.16}$$

where Γ denotes the Gamma function $\Gamma(t) = \int_0^\infty e^{-u} u^{t-1}\, du$, $t > 0$. A combination of (6.15) and (6.16) gives us a probability–weighted moment estimator $\widehat{\theta}_n^{(1)}$. Further estimators can be obtained by replacing $U_{j,n}^r$ in (6.15) by some statistic. Examples are:

- $\widehat{\theta}_n^{(2)}$, where $U_{j,n}$ is replaced by any plotting position $p_{j,n}$ as defined in Section 6.2.1.

- $\widehat{\theta}_n^{(3)}$, where $U_{j,n}^r$ is replaced by

$$EU_{j,n}^r = \frac{(n-j)(n-j-1)\cdots(n-j-r+1)}{(n-1)(n-2)\cdots(n-r)}, \quad r = 1,2.$$

From (6.16), we immediately obtain

$$\begin{aligned} w_0(\theta) &= \mu - \frac{\psi}{\xi}(1 - \Gamma(1-\xi)), \\ 2w_1(\theta) - w_0(\theta) &= \frac{\psi}{\xi}\Gamma(1-\xi)\,(2^\xi - 1), \\ 3w_2(\theta) - w_0(\theta) &= \frac{\psi}{\xi}\Gamma(1-\xi)\,(3^\xi - 1), \end{aligned}$$

and hence

$$\frac{3w_2(\theta) - w_0(\theta)}{2w_1(\theta) - w_0(\theta)} = \frac{3^\xi - 1}{2^\xi - 1}.$$

Applying any of the estimators above to the last equation yields an estimator $\widehat{\xi}$ of ξ. Given $\widehat{\xi}$, the parameters μ and ψ are then estimated by

$$\begin{aligned} \widehat{\psi} &= \frac{(2\widehat{w}_1 - \widehat{w}_0)\,\widehat{\xi}}{\Gamma(1-\widehat{\xi})\,(2^{\widehat{\xi}} - 1)}, \\ \widehat{\mu} &= \widehat{w}_0 + \frac{\widehat{\psi}}{\widehat{\xi}}(1 - \Gamma(1-\widehat{\xi})), \end{aligned}$$

where $\widehat{w}_0$, $\widehat{w}_1$, $\widehat{w}_2$ are any of the empirical probability–weighted moments discussed above. The case $\xi = 0$ can of course also be covered by this method.

For a discussion on the behaviour of these estimators see Hosking et al. [338]. Smith [595] summarises as follows.

> The method is simple to apply and performs well in simulation studies. However, until there is some convincing theoretical explanation of its properties, it is unlikely to be universally accepted. There is also the disadvantage that, at present at least, it does not extend to more complicated situations such as regression models based on extreme value distributions.

6.3.3 Tail and Quantile Estimation, a First Go

Let us return to the basic set–up of (6.11) and (6.12), i.e. we have an iid sample $X_1, \ldots, X_n$ from H_θ. In this situation, a quantile estimator can be readily obtained. Indeed, by the methods discussed in the previous sections, we obtain an estimate $\widehat{\theta}$ of θ. Given any $p \in (0,1)$, the p–quantile x_p is defined via $x_p = H_\theta^{\leftarrow}(p)$; see Definition 3.3.5. A natural estimator for x_p, based on $X_1, \ldots, X_n$, then becomes

$$\widehat{x}_p = H_{\widehat{\theta}}^{\leftarrow}(p).$$

By the definition of H_θ this leads to

$$\widehat{x}_p = \widehat{\mu} - \frac{\widehat{\psi}}{\widehat{\xi}} \left(1 - (-\ln p)^{-\widehat{\xi}}\right) .$$

The corresponding tail estimate for $\overline{H}_\theta(x)$, for x in the appropriate domain, corresponds to

$$\overline{H}_{\widehat{\theta}}(x) = 1 - \exp\left\{ - \left(1 + \widehat{\xi}\, \frac{x - \widehat{\mu}}{\widehat{\psi}}\right)^{-1/\widehat{\xi}} \right\},$$

where $\widehat{\theta} = (\widehat{\xi}, \widehat{\mu}, \widehat{\psi})$ is either estimated by the MLE or by a probability–weighted moment estimator.

Notes and Comments

A recommendable account of estimation methods for the GEV, including a detailed discussion of the pros and cons of the different methods, is Buishand [102]. Hosking [335] discusses the problem of hypothesis testing within GEV.

If the extreme value distribution is known to be Fréchet, Gumbel or Weibull, the above methods can be adapted to the specific df under consideration. This may simplify the estimation problem in the case of $\xi \geq 0$ (Fréchet, Gumbel), but *not* for the Weibull distribution. The latter is due to non–regularity problems of the MLE as explained in Section 6.3.1. The vast amount of papers written on estimation for the three–parameter Weibull reflects this situation; see for instance Lawless [414, 415], Lockhart and Stephens [426], Mann [440], Smith and Naylor [596] and references therein. To indicate the sort of problems that may occur, we refer to Smith [589] who studies the Pareto–like probability densities

$$f(x; K, \alpha) \sim c\alpha(K - x)^{\alpha - 1}, \quad x \uparrow K,$$

where K and α are unknown parameters.

6.4 Estimating Under Maximum Domain of Attraction Conditions

6.4.1 Introduction

Relaxing condition (6.11), we assume in this section that for some $\xi \in \mathbb{R}$,

$$X_1, \ldots, X_n \quad \textit{are iid from } F \in \mathrm{MDA}\,(H_\xi)\,. \tag{6.17}$$

By Proposition 3.3.2, $F \in \mathrm{MDA}(H_\xi)$ is equivalent to

$$\lim_{n\to\infty} n\overline{F}\,(c_n x + d_n) = -\ln H_\xi(x) \tag{6.18}$$

for appropriate norming sequences (c_n) and (d_n), and x belongs to a suitable domain depending on the sign of ξ. Let us from the start be very clear about the fundamental difference between (6.11) and (6.17). Consider for illustrative purposes only the standard Fréchet case $\xi = 1/\alpha > 0$. Now (6.11) means that our sample $X_1, \ldots, X_n$ *exactly* follows a Fréchet distribution, i.e.

$$\overline{F}(x) = 1 - \exp\left\{-x^{-\alpha}\right\}\,, \quad x > 0\,.$$

On the other hand, by virtue of Theorem 3.3.7 assumption (6.17) reduces in the Fréchet case to

$$\overline{F}(x) = x^{-\alpha}\,L(x)\,, \quad x > 0\,,$$

for some slowly varying function L. Clearly, in this case the estimation of the tail $\overline{F}(x)$ is much more involved due to the non–parametric character of L. In various applications, one would mainly (in some cases, solely) be interested in α. So (6.11) amounts to full parametric assumptions, whereas (6.17) is essentially semi–parametric in nature: there is a parametric part α and a non–parametric part L. Because of this difference, (6.17) is much more generally considered as inference for *heavy–tailed distributions* as opposed to *inference for the GEV* in (6.11).

A handwaving consequence of (6.18) is that for large $u = c_n x + d_n$,

$$n\overline{F}(u) \approx \left(1 + \xi\,\frac{u - d_n}{c_n}\right)^{-1/\xi}\,,$$

so that a tail–estimator could take on the form

$$\widehat{\left(\overline{F}(u)\right)} = \frac{1}{n}\left(1 + \widehat{\xi}\,\frac{u - \widehat{d}_n}{\widehat{c}_n}\right)^{-1/\widehat{\xi}}\,, \tag{6.19}$$

for *appropriate estimators* $\widehat{\xi}, \widehat{c}_n$ and $\widehat{d}_n$. As (6.17) is essentially a *tail–property*, estimation of ξ may be based on k upper order statistics $X_{k,n} \le \cdots \le X_{1,n}$.

A whole battery of classical approaches has exploited this natural idea; see Section 6.4.2. The following mathematical conditions are usually imposed:

(a) $k(n) \to \infty$ use a sufficiently large number of order statistics, but

(b) $\dfrac{n}{k(n)} \to \infty$ as we are interested in a tail property, we should also make sure to concentrate only on the upper order statistics. *Let the tail speak for itself.*

(6.20)

When working out the details later, we will be able to see where exactly the properties on $(k(n))$ enter. Indeed it is precisely this degree of freedom k which will allow us to obtain the necessary statistical properties like consistency and asymptotic normality for our estimators.

From (6.19) we would *in principle* be in the position to estimate the quantile $x_p = F^{\leftarrow}(p)$, for fixed $p \in (0,1)$, as follows

$$\widehat{x}_p = \widehat{d}_n + \frac{\widehat{c}_n}{\widehat{\xi}}\left((n(1-p))^{-\widehat{\xi}} - 1\right) . \tag{6.21}$$

Typically, we will be interested in estimating high p–quantiles *outside the sample* $X_1, \ldots, X_n$. This means that $p = p_n$ is chosen in such a way that $p > 1 - 1/n$, hence the empirical df satisfies $\overline{F}_n(p) = 0$ and does not yield any information about such quantiles. In order to get good estimators for ξ, c_n and d_n in (6.21) a subsequence trick is needed. Assume for notational convenience that $n/k \in \mathbb{N}$. A standard approach now consists of passing to a subsequence (n/k) with $k = k(n)$ satisfying (6.20). The quantile x_p is then estimated by

$$\widehat{x}_p = \widehat{d}_{n/k} + \frac{\widehat{c}_{n/k}}{\widehat{\xi}}\left(\left(\frac{n}{k}(1-p_n)\right)^{-\widehat{\xi}} - 1\right) . \tag{6.22}$$

Why does this work? One reason behind this construction is that we need to estimate at two levels. First, we have to find a reliable estimate for ξ: this task will be worked out in Section 6.4.2. Condition (6.20) will appear very naturally. Second, we need to estimate the norming constants c_n and d_n which themselves are defined via quantiles of F. For instance, in the Fréchet case we know that $c_n = F^{\leftarrow}(1 - n^{-1})$; see Theorem 3.3.7. Hence estimating c_n is equivalent to the problem of estimating x_p at the boundary of our data range. By going to the subsequence (n/k), we move away from the critical boundary value $1 - n^{-1}$ to the safer $1 - (n/k)^{-1}$. Estimating $c_{n/k}$ is thus reduced to estimating quantiles *within* the range of our data. Similar arguments hold for $d_{n/k}$, and indeed for the Gumbel and Weibull case. We may therefore hope that the construction in (6.22) leads to a good estimator for x_p. The above discussion is only heuristic, a detailed statistical analysis shows that this approach can be made to work.

In the context of statistics of extremal events it may also be of interest to estimate the following quantity which is closely related to the quantiles x_p:

$$x_{p,r} = F^{\leftarrow}(p^{1/r}), \quad r \in \mathbb{N}.$$

Notice that $x_p = x_{p,1}$. The interpretation of $x_{p,r}$ is obvious from

$$p = F^r(x_{p,r}) = P(\max(X_{n+1}, \ldots, X_{n+r}) \le x_{p,r}),$$

so $x_{p,r}$ is that level which, with a given probability p, will not be exceeded by the *next* r observations $X_{n+1}, \ldots, X_{n+r}$. As an estimate we then obtain from (6.22)

$$\widehat{x}_{p,r} = \widehat{d}_{n/k} + \frac{\widehat{c}_{n/k}}{\widehat{\xi}} \left(\left(\frac{n}{k} \left(1 - p^{1/r} \right) \right)^{-\widehat{\xi}} - 1 \right).$$

In what follows we will concentrate only on estimation of x_p; from the definition of $x_{p,r}$ it is clear how one has to proceed for general r.

From the above heuristics we obtain a programme for the remainder of this section:

(*a*) Find appropriate estimators for the shape parameter ξ of the GEV.
(*b*) Find appropriate estimators for the norming constants c_n and d_n.
(*c*) Show that the estimators proposed above yield reasonable approximations to the distribution tail in its far end and to high quantiles.
(*d*) Determine the statistical properties of these estimators.

6.4.2 Estimating the Shape Parameter ξ

In this section we study different estimators of the shape parameter ξ for $F \in \mathrm{MDA}(H_\xi)$. We also give some of their statistical properties.

Method 1: Pickands's Estimator for $\xi \in \mathbb{R}$

The basic idea behind this estimator consists of finding a condition equivalent to $F \in \mathrm{MDA}(H_\xi)$ which involves the parameter ξ in an easy way. The key to Pickands's estimator and its various generalisations is Theorem 3.4.5, where it was shown that for $F \in \mathrm{MDA}(H_\xi)$, $U(t) = F^{\leftarrow}(1 - t^{-1})$ satisfies

$$\lim_{t\to\infty} \frac{U(2t) - U(t)}{U(t) - U(t/2)} = 2^{\xi}.$$

Furthermore, the following uniformity property holds: whenever $\lim_{t\to\infty} c(t) = 2$ for a positive function c,

$$\lim_{t\to\infty} \frac{U(c(t)t) - U(t)}{U(t) - U(t/c(t))} = 2^{\xi}. \tag{6.23}$$

The basic idea now consists of constructing an empirical estimator using (6.23). To that effect, let

$$V_{n,n} \le \cdots \le V_{1,n}$$

be the order statistics from an iid sample $V_1, \ldots, V_n$ with common Pareto df $F_V(x) = 1 - x^{-1}$, $x \ge 1$. It follows in the same way as for the quantile transformation, see Lemma 4.1.9(b), that

$$(X_{k,n})_{k=1,\ldots,n} \stackrel{d}{=} (U(V_{k,n}))_{k=1,\ldots,n},$$

where $X_1, \ldots, X_n$ are iid with df F. Notice that $V_{k,n}$ is the empirical $(1 - k/n)$–quantile of F_V. Using similar methods as in Examples 4.1.11 and 4.1.12, i.e. making use of the quantile transformation, it is not difficult to see that

$$\frac{k}{n} V_{k,n} \stackrel{P}{\to} 1, \quad n \to \infty,$$

whenever $k = k(n) \to \infty$ and $k/n \to 0$. In particular,

$$V_{k,n} \stackrel{P}{\to} \infty \quad \text{and} \quad \frac{V_{2k,n}}{V_{k,n}} \stackrel{P}{\to} \frac{1}{2}, \quad n \to \infty.$$

Combining this with (6.23) and using a subsequence argument, see Appendix A1.2, yields

$$\frac{U(V_{k,n}) - U(V_{2k,n})}{U(V_{2k,n}) - U(V_{4k,n})} \stackrel{P}{\to} 2^{\xi}, \quad n \to \infty.$$

Motivated by the discussion above and by (6.23), we now define the *Pickands estimator*

$$\widehat{\xi}^{(P)}_{k,n} = \frac{1}{\ln 2} \ln \frac{X_{k,n} - X_{2k,n}}{X_{2k,n} - X_{4k,n}}. \tag{6.24}$$

This estimator turns out to be weakly consistent provided $k \to \infty$, $k/n \to 0$:

$$\widehat{\xi}^{(P)}_{k,n} \stackrel{P}{\to} \xi, \quad n \to \infty.$$

This was already observed by Pickands [498]. A full analysis on $\widehat{\xi}^{(P)}_{k,n}$ is to be found in Dekkers and de Haan [173] from which the following result is taken.

Theorem 6.4.1 (Properties of the Pickands estimator)
Suppose (X_n) *is an iid sequence with df* $F \in \mathrm{MDA}(H_\xi)$, $\xi \in \mathbb{R}$. *Let* $\widehat{\xi}^{(P)} = \widehat{\xi}^{(P)}_{k,n}$ *be the Pickands estimator* (6.24).

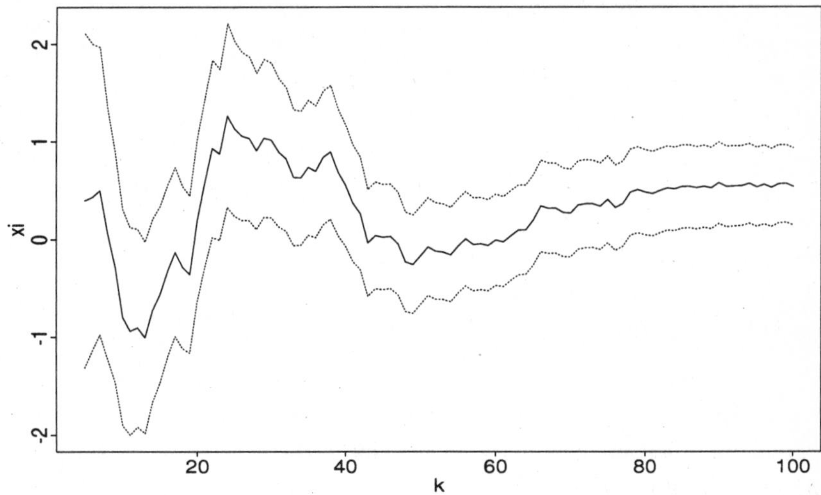

Figure 6.4.2 *Pickands–plot for the water–damage claim data; see Figure 6.3.1. The estimate of* ξ *appears to be close to* 0. *The upper and lower lines constitute asymptotic* 95% *confidence bands.*

(a) (Weak consistency) *If* $k \to \infty$, $k/n \to 0$ *for* $n \to \infty$, *then*

$$\widehat{\xi}^{(P)} \xrightarrow{P} \xi\,, \quad n \to \infty\,.$$

(b) (Strong consistency) *If* $k/n \to 0$, $k/\ln\ln n \to \infty$ *for* $n \to \infty$, *then*

$$\widehat{\xi}^{(P)} \xrightarrow{\text{a.s.}} \xi\,, \quad n \to \infty\,.$$

(c) (Asymptotic normality) *Under further conditions on* k *and* F (*see* Dekkers and de Haan [173], p. 1799),

$$\sqrt{k}\,(\widehat{\xi} - \xi) \xrightarrow{d} N(0, v(\xi))\,, \quad n \to \infty\,,$$

where

$$v(\xi) = \frac{\xi^2\left(2^{2\xi+1}+1\right)}{\left(2\left(2^{\xi}-1\right)\ln 2\right)^2}\,. \qquad \square$$

Remarks. 1) This theorem forms the core of a whole series of results obtained in Dekkers and de Haan [173]; on it one can base quantile and tail estimators and (asymptotic) confidence interval constructions. The quoted paper [173] also contains various simulated and real life examples in order *to see the theory in action.* We strongly advise the reader to go through it, perhaps avoiding upon first reading the (rather extensive) technical details. The main idea behind the above construction goes back to Pickands [498]. A nice summary is to be found in de Haan [294] from which the derivation above is taken.

2) In the spirit of Section 6.4.1 notice that the calculation of Pickands's estimator (6.24) involves a sequence of upper order statistics increasing with n. Consequently, one mostly includes a so-called *Pickands-plot* in the analysis, i.e.

$$\left\{(k, \widehat{\xi}^{(P)}_{k,n}) \,:\, k=1,\ldots,n\right\},$$

in order to allow for a choice depending on k. The interpretation of such plots, i.e. the optimal choice of k, is a delicate point for which no uniformly best solution exists. It is intuitively clear that one should choose $\widehat{\xi}^{(P)}_{k,n}$ from such a k-region where the plot is roughly horizontal. We shall come back to this point later; see the Summary at the end of this section. □

Method 2: Hill's Estimator for $\xi = \alpha^{-1} > 0$

Suppose $X_1,\ldots,X_n$ are iid with df $F \in \mathrm{MDA}(\Phi_\alpha)$, $\alpha > 0$, thus $\overline{F}(x) = x^{-\alpha}L(x)$, $x>0$, for a slowly varying function L; see Theorem 3.3.7. Distributions with such tails form the prime examples for modelling heavy-tailed phenomena; see for instance Section 1.3. For many applications the knowledge of the index α of regular variation is of major importance. If for instance $\alpha < 2$ then $EX_1^2 = \infty$. This case is often observed in the modelling of insurance data; see for instance Hogg and Klugman [330].

Empirical studies on the tails of daily log-returns in finance have indicated that one frequently encounters values α between 3 and 4; see for instance Guillaume et al. [289], Longin [428] and Loretan and Phillips [429]. Information of the latter type implies that, whereas covariances of such data would be well defined, the construction of confidence intervals for the sample autocovariances and autocorrelations on the basis of asymptotic (central limit) theory may be questionable as typically a finite fourth moment condition is asked for.

The Hill estimator of α takes on the following form:

$$\widehat{\alpha}^{(H)} = \widehat{\alpha}^{(H)}_{k,n} = \left(\frac{1}{k}\sum_{j=1}^{k} \ln X_{j,n} - \ln X_{k,n}\right)^{-1}, \tag{6.25}$$

where $k = k(n) \to \infty$ in an appropriate way, so that as in the case of Pickands's estimator, an increasing sequence of upper order statistics is used. One of the interesting facts concerning (6.25) is that various asymptotically equivalent versions of $\widehat{\alpha}^{(H)}$ can be derived through essentially different methods, showing that the Hill estimator is very natural. Below we discuss some derivations.

The MLE approach (Hill [326]). Assume for the moment that X is a rv with df F so that for $\alpha > 0$

$$P(X > x) = \overline{F}(x) = x^{-\alpha}\,, \quad x \geq 1\,.$$

Then it immediately follows that $Y = \ln X$ has df

$$P(Y > y) = e^{-\alpha y}\,, \quad y \geq 0\,,$$

i.e. Y is $Exp(\alpha)$ and hence the MLE of α is given by

$$\widehat{\alpha}_n = \overline{Y}_n^{\,-1} = \left(\frac{1}{n}\sum_{j=1}^{n} \ln X_j\right)^{-1} = \left(\frac{1}{n}\sum_{j=1}^{n} \ln X_{j,n}\right)^{-1}.$$

A trivial generalisation concerns

$$\overline{F}(x) = Cx^{-\alpha}\,, \quad x \geq u > 0\,, \tag{6.26}$$

with u known. If we interpret (6.26) as fully specified, i.e. $C = u^{\alpha}$, then we immediately obtain as MLE of α

$$\widehat{\alpha}_n = \left(\frac{1}{n}\sum_{j=1}^{n} \ln\left(\frac{X_{j,n}}{u}\right)\right)^{-1} = \left(\frac{1}{n}\sum_{j=1}^{n} \ln X_{j,n} - \ln u\right)^{-1}. \tag{6.27}$$

Now we often do not have the precise parametric information of these examples, but in the spirit of MDA(Φ_α) we assume that $\overline{F}$ behaves like a Pareto df above a certain known threshold u say. Let

$$K = \operatorname{card}\{i : X_{i,n} > u\,, i = 1, \ldots, n\}\,. \tag{6.28}$$

Conditionally on the event $\{K = k\}$, maximum likelihood estimation of α and C in (6.26) reduces to maximising the joint density of $(X_{k,n}, \ldots, X_{1,n})$. From Theorem 4.1.3 we deduce

$$\begin{aligned} & f_{X_{k,n},\ldots,X_{1,n}}(x_k, \ldots, x_1) \\ = \ & \frac{n!}{(n-k)!}\left(1 - Cx_k^{-\alpha}\right)^{n-k} C^k \alpha^k \prod_{i=1}^{k} x_i^{-(\alpha+1)}\,, \quad u < x_k < \cdots < x_1\,. \end{aligned}$$

A straightforward calculation yields the conditional MLEs

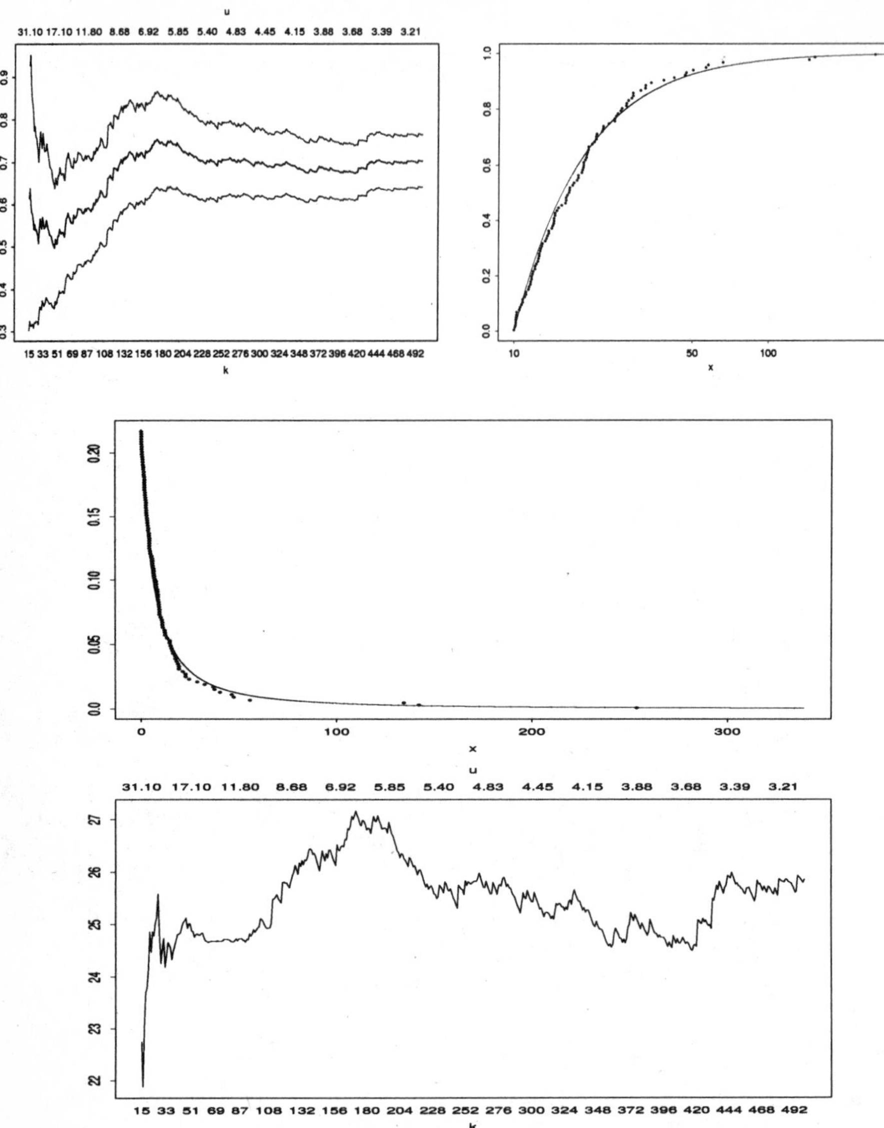

Figure 6.4.3 *Tail and quantile estimation based on a Hill–fit; see* (6.29) *and* (6.30). *The data are the Danish insurance claims from Example 6.2.9. Top, left: the Hill–plot for* $\xi = 1/\alpha$ *as a function of* k *upper order statistics (lower horizontal axis) and of the threshold* u *(upper horizontal axis), i.e. there are* k *exceedances of the threshold* u. *Top, right: the fit of the shifted excess df* $F_u(x-u)$, $x \geq u$, *on log–scale. Middle: tail–fit of* $\overline{F}(x+u)$, $x \geq 0$. *Bottom: estimation of the* 0.99*–quantile as a function of the* k *upper order statistics and of the corresponding threshold value* u. *The estimation of the tail is based on* $k = 109$ $(u = 10)$ *and* $\alpha = \xi^{-1} = 1.618$. *Compare also with the GPD–fit in Figure 6.5.5.*

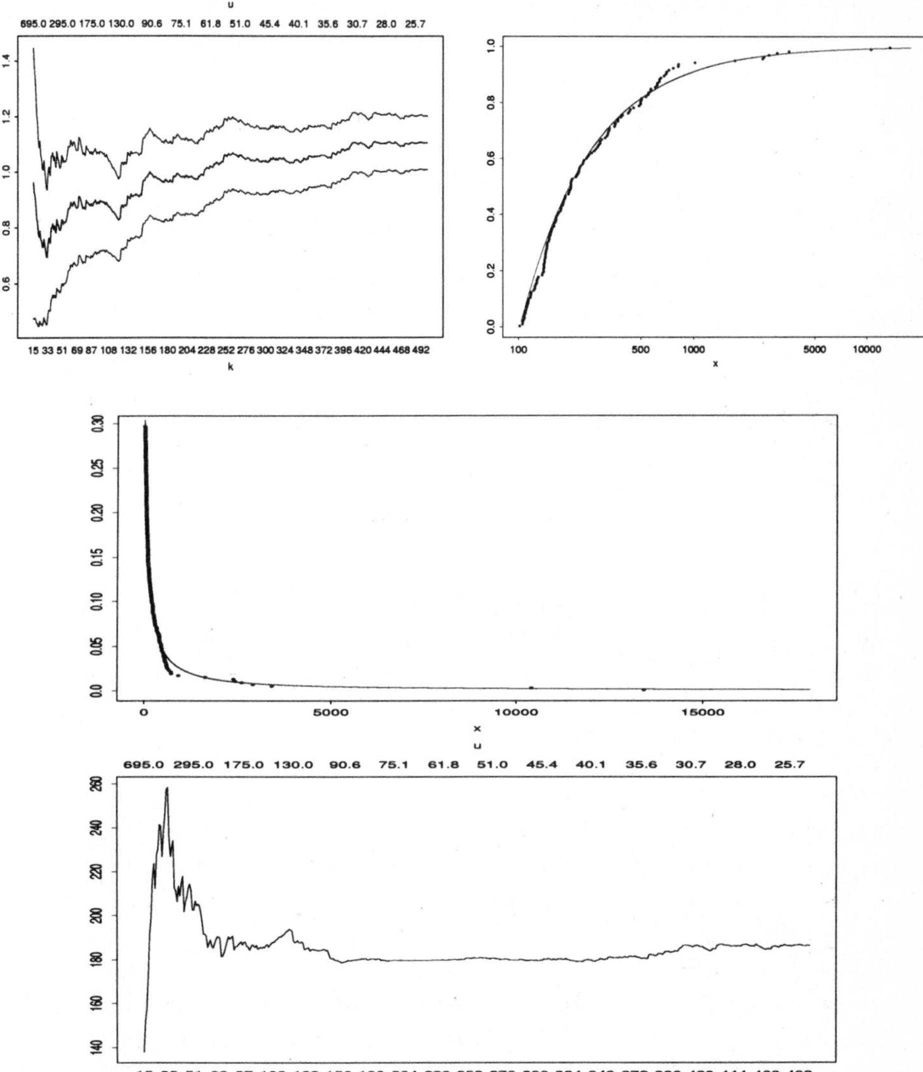

Figure 6.4.4 *Tail and quantile estimation based on a Hill–fit; see* (6.29) *and* (6.30). *The data are the industrial fire claims from Example 6.2.9. Top, left: the Hill–plot for* $\xi = 1/\alpha$ *as a function of* k *upper order statistics (lower horizontal axis) and of the threshold* u *(upper horizontal axis), i.e. there are* k *exceedances of the threshold* u. *Top, right: the fit of the shifted excess df* $F_u(x-u)$, $x \geq u$, *on log-scale. Middle: tail-fit of* $\overline{F}(x+u)$, $x \geq 0$. *Bottom: estimation of the* 0.99*–quantile as a function of the* k *upper order statistics and of the corresponding threshold value* u. *The estimation of the tail is based on* $k = 149$ ($u = 100$) *and* $\alpha = \xi^{-1} = 1.058$. *Compare also with the GPD–fit in Figure 6.5.5.*

$$\widehat{\alpha}_{k,n}^{(H)} = \left(\frac{1}{k}\sum_{j=1}^{k}\ln\left(\frac{X_{j,n}}{X_{k,n}}\right)\right)^{-1} = \left(\frac{1}{k}\sum_{j=1}^{k}\ln X_{j,n} - \ln X_{k,n}\right)^{-1}$$

$$\widehat{C}_{k,n} = \frac{k}{n}\,X_{k,n}^{\widehat{\alpha}_{k,n}^{(H)}}\,.$$

So Hill's estimator has the same form as the MLE in the exact model underlying (6.27) but now having *the deterministic u replaced by the random threshold* $X_{k,n}$, where k is defined through (6.28). We also immediately obtain an estimate for the tail $\overline{F}(x)$

$$\widehat{\left(\overline{F}(x)\right)} = \frac{k}{n}\left(\frac{x}{X_{k,n}}\right)^{-\widehat{\alpha}_{k,n}^{(H)}} \tag{6.29}$$

and for the p–quantile

$$\widehat{x}_p = \left(\frac{n}{k}(1-p)\right)^{-1/\widehat{\alpha}_{k,n}^{(H)}} X_{k,n}\,. \tag{6.30}$$

From (6.29) we obtain an estimator of the excess df $F_u(x-u)$, $x \ge u$, by using $F_u(x-u) = 1 - \overline{F}(x)/\overline{F}(u)$. Examples, based on these estimators are to be found in Figures 6.4.3 and 6.4.4 for the Danish, respectively industrial, fire insurance data. We will come back to these data more in detail in Section 6.5.2.

Example 6.4.5 (The Hill estimator at work)
In Figures 6.4.3 and 6.4.4 we have applied the above methods to the Danish fire insurance data (Figure 6.2.11) and the industrial fire insurance data (Figure 6.2.12); for a preliminary analysis see Example 6.2.9. For the Danish data we have chosen as an initial threshold $u = 10$ ($k = 109$). The corresponding Hill estimate has a value $\widehat{\xi} = 0.618$. When changed to $u = 18$ ($k = 47$), we obtain $\widehat{\xi} = 0.497$. The Hill–plot shows a fairly stable behaviour in the range $(0.5, 0.7)$. As in most applications, the quantities of main interest are the high quantiles. We therefore turn immediately to Figure 6.4.3 (bottom), where $\widehat{x}_{0.99}$ is plotted across all relevant u– (equivalently, k–) values. For k in the region $(45, 110)$ the quantile–Hill–plot shows a remarkably stable behaviour around the value 24.7. This agrees perfectly with the empirical estimate of 24.6 for $x_{0.99}$; see Table 6.2.13. This should be contrasted with the situation in Figure 6.4.4 for the industrial fire data. For the latter data, the estimate for ξ ranges from 0.945 for $u = 100$ ($k = 149$) over 0.745 for $u = 300$ ($k = 49$) to 0.799 for $u = 500$ ($k = 29$). All estimates clearly correspond to infinite variance models! An estimate for $x_{0.99}$ in the range $(180, 200)$ emerges, again in agreement with the empirical value of 184. We would like to stress

at this point that the above discussion represents only the beginning of a detailed analysis. The further discussions *have* to be conducted *together* with the actuary responsible for the underlying data. □

The regular variation approach (de Haan [295]). This approach is in the same spirit as the construction of Pickands's estimator, i.e. base the inference on a suitable reformulation of $F \in \mathrm{MDA}(\Phi_\alpha)$. Indeed $F \in \mathrm{MDA}(\Phi_\alpha)$ if and only if

$$\lim_{t\to\infty} \frac{\overline{F}(tx)}{\overline{F}(t)} = x^{-\alpha}\,, \quad x > 0\,.$$

Using partial integration, we obtain

$$\int_t^\infty (\ln x - \ln t)\, dF(x) = \int_t^\infty \frac{\overline{F}(x)}{x}\, dx\,,$$

so that by Karamata's theorem (Theorem A3.6)

$$\frac{1}{\overline{F}(t)} \int_t^\infty (\ln x - \ln t)\, dF(x) \to \frac{1}{\alpha}\,, \quad t \to \infty\,. \tag{6.31}$$

How do we find an estimator from this result? Two choices have to be made:

(*a*) replace F by an estimator, the obvious candidate here is the empirical distribution function

$$F_n(x) = \frac{1}{n}\sum_{i=1}^n I_{\{X_i \le x\}} = \frac{1}{n}\sum_{i=1}^n I_{\{X_{i,n} \le x\}}\,,$$

(*b*) replace t by an appropriate high, data dependent level (recall $t \to \infty$); we take $t = X_{k,n}$ for some $k = k(n)$.

The choice of t is motivated by the fact that $X_{k,n} \stackrel{\text{a.s.}}{\to} \infty$ provided $k = k(n) \to \infty$ and $k/n \to 0$; see Proposition 4.1.14. From (6.31) the following estimator results

$$\frac{1}{\overline{F}_n\left(X_{k,n}\right)} \int_{X_{k,n}}^\infty \left(\ln x - \ln X_{k,n}\right)\, dF_n(x) = \frac{1}{k-1}\sum_{j=1}^{k-1} \ln X_{j,n} - \ln X_{k,n}$$

which, modulo the factor $k-1$, is again of the form $(\widehat{\alpha}^{(H)})^{-1}$ in (6.25). Notice that the change from k to $k-1$ is asymptotically negligible; see Example 4.1.11.

The mean excess function approach. This is essentially a reformulation of the approach above; we prefer to list it separately because of its methodological merit. Suppose X is a rv with df $F \in \mathrm{MDA}(\Phi_\alpha)$, $\alpha > 0$, and for notational convenience assume that $X > 1$ a.s. One can now rewrite (6.31) as follows (see also Example 6.2.5)

$$E(\ln X - \ln t \mid \ln X > \ln t) \to \frac{1}{\alpha}, \quad t \to \infty.$$

So denoting $u = \ln t$ and $e^*(u)$ the mean excess function of $\ln X$ (see Definition 6.2.3) we obtain

$$e^*(u) \to \frac{1}{\alpha}, \quad u \to \infty.$$

Hill's estimator can then be interpreted as the empirical mean excess function of $\ln X$ calculated at the threshold $u = \ln X_{k,n}$, i.e. $e_n^*(\ln X_{k,n})$.

We summarise as follows.

Suppose $X_1, \ldots, X_n$ are iid with df $F \in \mathrm{MDA}(\Phi_\alpha)$, $\alpha > 0$, then a natural estimator for α is provided by the *Hill estimator*

$$\widehat{\alpha}_{k,n}^{(H)} = \left(\frac{1}{k}\sum_{j=1}^{k} \ln X_{j,n} - \ln X_{k,n}\right)^{-1}, \tag{6.32}$$

where $k = k(n)$ satisfies (6.20).

Below we summarise the main properties of the Hill estimator. Before looking at the theorem, you may want to refresh your memory on the meaning of linear processes (see Sections 5.5 and 7.1) and weakly dependent strictly stationary processes (see Section 4.4).

Theorem 6.4.6 (Properties of the Hill estimator)
Suppose (X_n) is strictly stationary with marginal distribution F satisfying for some $\alpha > 0$ and $L \in \mathcal{R}_0$,

$$\overline{F}(x) = P(X > x) = x^{-\alpha} L(x), \quad x > 0.$$

Let $\widehat{\alpha}^{(H)} = \widehat{\alpha}_{k,n}^{(H)}$ be the Hill estimator (6.32).

(a) (Weak consistency) *Assume that one of the following conditions is satisfied:*
- *(X_n) is iid* (Mason [445]),
- *(X_n) is weakly dependent* (Rootzén, Leadbetter and de Haan [552], Hsing [341]),

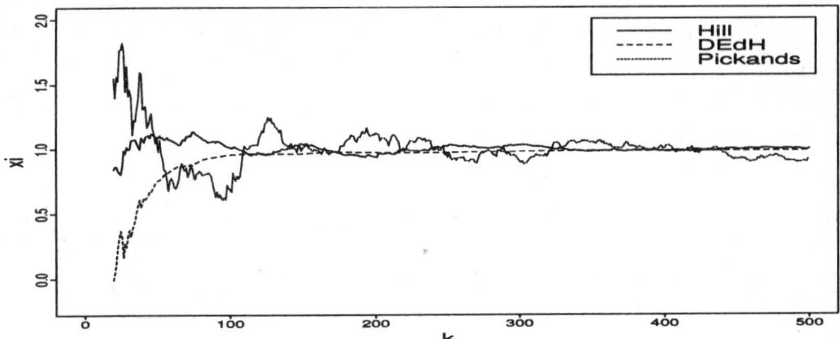

Figure 6.4.7 *Pickands–, Hill– and DEdH–plots for* 2 000 *simulated iid data with df given by* $\overline{F}(x) = x^{-1}$, $x \geq 1$. *For reasons of comparison, we choose the Hill estimator for* $\xi = 1$ *as* $\widehat{\xi}^{(H)} = (\widehat{\alpha}^{((H)})^{-1}$. *Various Hill– and related plots from simulated and real life data are for instance given in Figures 6.4.11, 6.4.12 and Section 6.5.2. See also Figure 4.1.13 for a "Hill horror plot" for the tail* $\overline{F}(x) = 1/(x \ln x)$ *and Figure 5.5.4 for the case of dependent data.*

– (X_n) *is a linear process* (Resnick and Stărică [535, 537]).

If $k \to \infty$, $k/n \to 0$ *for* $n \to \infty$, *then*

$$\widehat{\alpha}^{(H)} \xrightarrow{P} \alpha\,.$$

(*b*) (Strong consistency) (Deheuvels, Häusler and Mason [170]) *If* $k/n \to 0$, $k/\ln\ln n \to \infty$ *for* $n \to \infty$ *and* (X_n) *is an iid sequence, then*

$$\widehat{\alpha}^{(H)} \xrightarrow{\text{a.s.}} \alpha\,.$$

(*c*) (Asymptotic normality) *If further conditions on* k *and* F *are satisfied (see for instance the Notes and Comments below) and* (X_n) *is an iid sequence, then*

$$\sqrt{k}\left(\widehat{\alpha}^{(H)} - \alpha\right) \xrightarrow{d} N\left(0, \alpha^2\right)\,. \qquad \square$$

Remarks. 3) Theorem 6.4.6 should be viewed as a counterpart to Theorem 6.4.1 on the Pickands estimator. Because of the importance of $\widehat{\alpha}^{(H)}$, we prefer to formulate Theorem 6.4.6 in its present form for sequences (X_n) more general than iid.

4) Do not interpret this theorem as saying that *the Hill estimator is always fine.* The theorem *only* says that rather generally the standard statistical properties hold. One still needs a crucial set of conditions on $\overline{F}$ and $k(n)$. In particular, second–order regular variation assumptions on $\overline{F}$ have to be

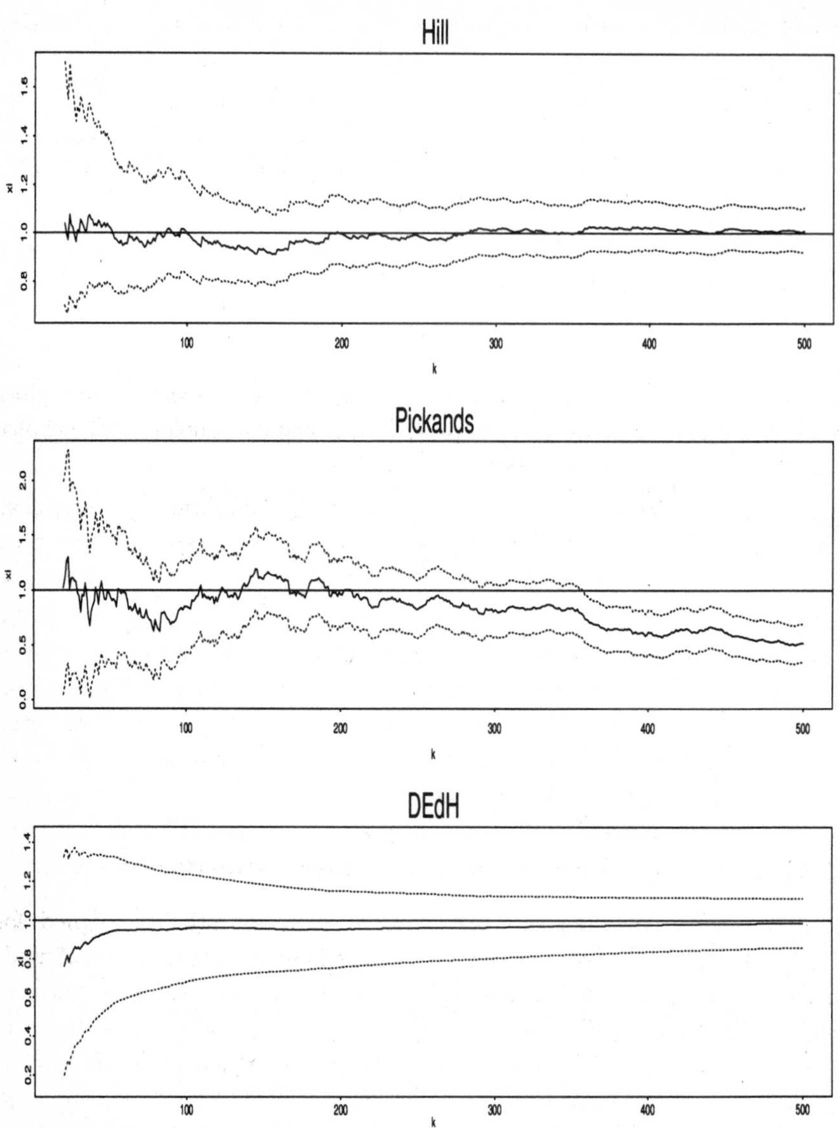

Figure 6.4.8 *Pickands–, Hill– and DEdH–plots with asymptotic 95% confidence bands for 2 000 absolute values of iid standard Cauchy rvs. The tail of the latter is Pareto–like with index $\xi = 1$. Recall that, for given k, the DEdH and the Hill estimator use the k upper order statistics of the sample, whereas the Pickands estimator uses $4k$ of them. In the case of the Pickands estimator one clearly sees the trade–off between variance and bias; see also the discussion in the Notes and Comments.*

imposed to derive the asymptotic normality of $\widehat{\alpha}^{(H)}$. The same applies to the case of the Pickands estimator. Notice that these conditions are *not* verifiable in practice.

5) In Example 4.1.12 we prove the weak consistency of Hill's estimator for the iid case and indicate how to prove its asymptotic normality. Moreover, we give an example showing that the rate of convergence of Hill's estimator can be arbitrarily slow.

6) As in the case of the Pickands estimator, an analysis based on the Hill estimator is usually summarised graphically. The *Hill–plot*

$$\left\{(k, \widehat{\alpha}_{k,n}^{(H)}) : k = 2, \dots, n\right\},$$

turns out to be instrumental in finding the *optimal* k. Smoothing Hill–plots over a specific range of k–values may defuse the critical problem of the choice of k; see for instance Resnick and Stărică [536].

7) The asymptotic variance of $\widehat{\alpha}^{(H)}$ depends on the unknown parameter α so that in order to calculate the asymptotic confidence intervals an appropriate estimator of α, typically $\widehat{\alpha}^{(H)}$, has to be inserted.

8) The Hill estimator is very sensitive with respect to dependence in the data; see for instance Figure 5.5.4 in the case of an autoregressive process. For ARMA and weakly dependent processes special techniques have been developed, for instance by first fitting an ARMA model to the data and then applying the Hill estimator to the residuals. See for instance the references mentioned under part (a) of Theorem 6.4.6. □

Method 3: The Deckers–Einmahl–de Haan Estimator for $\xi \in \mathbb{R}$

A disadvantage of Hill's estimator is that it is essentially designed for $F \in \mathrm{MDA}(H_\xi)$, $\xi > 0$. We have already stressed before that this class of models suffices for many applications in the realm of finance and insurance. In Dekkers, Einmahl und de Haan [175], Hill's estimator is extended to cover the whole class H_ξ, $\xi \in \mathbb{R}$. In Theorem 3.3.12 we saw that for $F \in \mathrm{MDA}(H_\xi)$, $\xi < 0$, the right endpoint x_F of F is finite. For simplicity we assume that $x_F > 0$. In Section 3.3 we found that the maximum domain of attraction conditions for H_ξ *all involve some kind of regular variation.* As for deriving the Pickands and Hill estimator, one can reformulate regular variation conditions to find estimators for any $\xi \in \mathbb{R}$. Dekkers et al. [175] come up with the following proposal:

$$\widehat{\xi} = 1 + H_n^{(1)} + \frac{1}{2}\left(\frac{(H_n^{(1)})^2}{H_n^{(2)}} - 1\right)^{-1}, \tag{6.33}$$

where

$$H_n^{(1)} = \frac{1}{k}\sum_{j=1}^{k}(\ln X_{j,n} - \ln X_{k+1,n})$$

is the reciprocal of Hill's estimator (modulo an unimportant change from k to $k+1$) and

$$H_n^{(2)} = \frac{1}{k}\sum_{j=1}^{k}(\ln X_{j,n} - \ln X_{k+1,n})^2 .$$

Because $H_n^{(1)}$ and $H_n^{(2)}$ can be interpreted as empirical moments, $\widehat{\xi}$ is also referred to as a *moment estimator of* ξ. To make sense, in all the estimators discussed so far we could (and actually should) replace $\ln x$ by $\ln(1 \vee x)$. In practice, this should not pose problems because we assume $k/n \to 0$. Hence the relation $X_{k,n} \stackrel{\text{a.s.}}{\to} x_F > 0$ holds; see Example 4.1.14.

At this point we *pause for a moment* and see where we are. First of all

Do we have all relevant approaches for estimating the shape parameter ξ*?*

Although various estimators have been presented, we have to answer this question by *no*! The above derivations were all motivated by analytical results on regular variation. In Chapter 5 however we have tried hard to convince you that point process methods are *the methodology* to use when one discusses extremes, and we possibly could use point process theory to find alternative estimation procedures. This can be made to work; one programme runs under the heading

Point process of exceedances,

or, as the hydrologists call it,

POT: the **P***eaks* **O***ver* **T***hreshold method.*

Because of its fundamental importance we decided to spend a whole section on this method; see Section 6.5.

Notes and Comments

In the previous sections we discussed some of the main issues underlying the statistical estimation of the shape parameter ξ. This general area is rapidly expanding so that an overview at any moment of time is immediately outdated. The references cited are therefore not exhaustive and reflect our personal interest. The fact that a particular paper does not appear in the list of references does not mean that it is considered less important.

The Hill Estimator: the Bias-Variance Trade–Off

Theorem 6.4.6 for iid data asserts that whenever $\overline{F}(x) = x^{-\alpha}L(x)$, $\alpha > 0$, then the Hill estimator $\widehat{\alpha}^{(H)} = \widehat{\alpha}_{k,n}^{(H)}$ satisfies

$$\sqrt{k}\left(\widehat{\alpha}^{(H)} - \alpha\right) \xrightarrow{d} N\left(0, \alpha^2\right),$$

where $k = k(n) \to \infty$ *at an appropriate rate.* However, in the formulation of Theorem 6.4.6 *we have not told you the whole story:* depending on the precise choice of k and on the slowly varying function L, there is an important trade–off between bias and variance possible. It all comes down to second–order behaviour of L, i.e. asymptotic behaviour *beyond* the defining property $L(tx) \sim L(x)$, $x \to \infty$. Typically, for increasing k the asymptotic variance α^2/k of $\widehat{\alpha}^{(H)}$ *decreases:* so let us take k as large as possible. Unfortunately,

when doing so, a bias may enter!

A fundamental paper introducing higher–order regular variation techniques for solving this problem is Goldie and Smith [278]. In our discussion below we follow de Haan and Peng [297]. Similar results are to be found in de Haan and Resnick [302], Hall [309], Häusler and Teugels [320] and Smith [587] for instance.

The second–order property needed beyond $\overline{F}(x) = x^{-\alpha}L(x)$ is that

$$\lim_{x\to\infty} \frac{\overline{F}(tx)/\overline{F}(x) - t^{-\alpha}}{a(x)} = t^{-\alpha}\,\frac{t^{\rho} - 1}{\rho}, \quad t > 0, \tag{6.34}$$

exists, where $a(x)$ is a measurable function of constant sign. The right–hand side of (6.34) is to be interpreted as $t^{-\alpha}\ln t$ if $\rho = 0$. The constant $\rho \le 0$ is the *second–order parameter* governing the *rate of convergence* of $\overline{F}(tx)/\overline{F}(x)$ to $t^{-\alpha}$. It necessarily follows that $|a(x)| \in \mathcal{R}_{\rho}$; see Geluk and de Haan [252], Theorem 1.9. In terms of $U(t) = F^{\leftarrow}(1 - t^{-1})$, (6.34) is equivalent to

$$\lim_{x\to\infty} \frac{U(tx)/U(x) - t^{1/\alpha}}{A(x)} = t^{1/\alpha}\,\frac{t^{\rho/\alpha} - 1}{\rho/\alpha}, \tag{6.35}$$

where $A(x) = \alpha^{-2}a(U(x))$.

The following result is proved as Theorem 1 in de Haan and Peng [297].

Theorem 6.4.9 (The bias–variance trade–off for the Hill estimator)
Suppose (6.35), *or equivalently* (6.34), *holds and* $k = k(n) \to \infty$, $k/n \to 0$ *as* $n \to \infty$. *If*

$$\lim_{n\to\infty} \sqrt{k}\,A\left(\frac{n}{k}\right) = \lambda \in \mathbb{R}, \tag{6.36}$$

then as $n \to \infty$

$$\sqrt{k}\left(\widehat{\alpha}^{(H)} - \alpha\right) \xrightarrow{d} N\left(\frac{\alpha^3\lambda}{\rho - \alpha}, \alpha^2\right). \qquad \square$$

Example 6.4.10 (The choice of the value k)
Consider the special case

$$\overline{F}(x) = cx^{-\alpha}(1 + x^{-\beta})$$

for positive constants c, α and β. We can choose

$$a(x) = \beta x^{-\beta},$$

giving $\rho = -\beta$ in (6.34). Since $U(t) = (ct)^{1/\alpha}(1 + o(1))$, we obtain

$$A(x) = \frac{\beta}{\alpha^2}(cx)^{-\beta/\alpha}(1 + o(1)).$$

Then (6.36) yields k such that

$$k \sim Cn^{(2\beta)/(2\beta+\alpha)}, \quad k \to \infty,$$

where C is a constant, depending on α, β, c and λ. Moreover, $\lambda = 0$ if and only if $C = 0$, hence $k = o(n^{(2\beta)/(2\beta+\alpha)})$. □

From (6.36) it follows that for k tending to infinity *sufficiently slowly*, i.e. taking only a moderate number of order statistics into account for the construction of the Hill estimator, $\lambda = 0$ will follow. In this case $\widehat{\alpha}^{(H)}$ is an asymptotically unbiased estimator for α, as announced in Theorem 6.4.6. The asymptotic mean squared error equals

$$\frac{1}{k}\left(\alpha^2 + \frac{\alpha^6\lambda^2}{(\rho - \alpha)^2}\right).$$

Theorem 6.4.9 also explains the typical behaviour of the Hill–plot showing large variations for small k versus small variations (leading to a biased estimate) for large k.

Results such as Theorem 6.4.9 are useful mainly from a methodological point of view. Condition (6.34) is *rarely verifiable in practice.* We shall come back to this point later; see the Summary at the end of this section.

A Comparison of Different Estimators of the Shape Parameter

The question as to which estimator for the shape parameter ξ one should use has no clear cut answer. It all depends on the possible values of ξ and, as we have seen, on the precise properties of the underlying df F. Some general statements can however be made. For $\xi = \alpha^{-1} > 0$ and dfs satisfying (6.34), de Haan and Peng [297] proved results similar to Theorem 6.4.9 for the Pickands estimator (6.24) and the Dekkers–Einmahl–de Haan (DEdH)

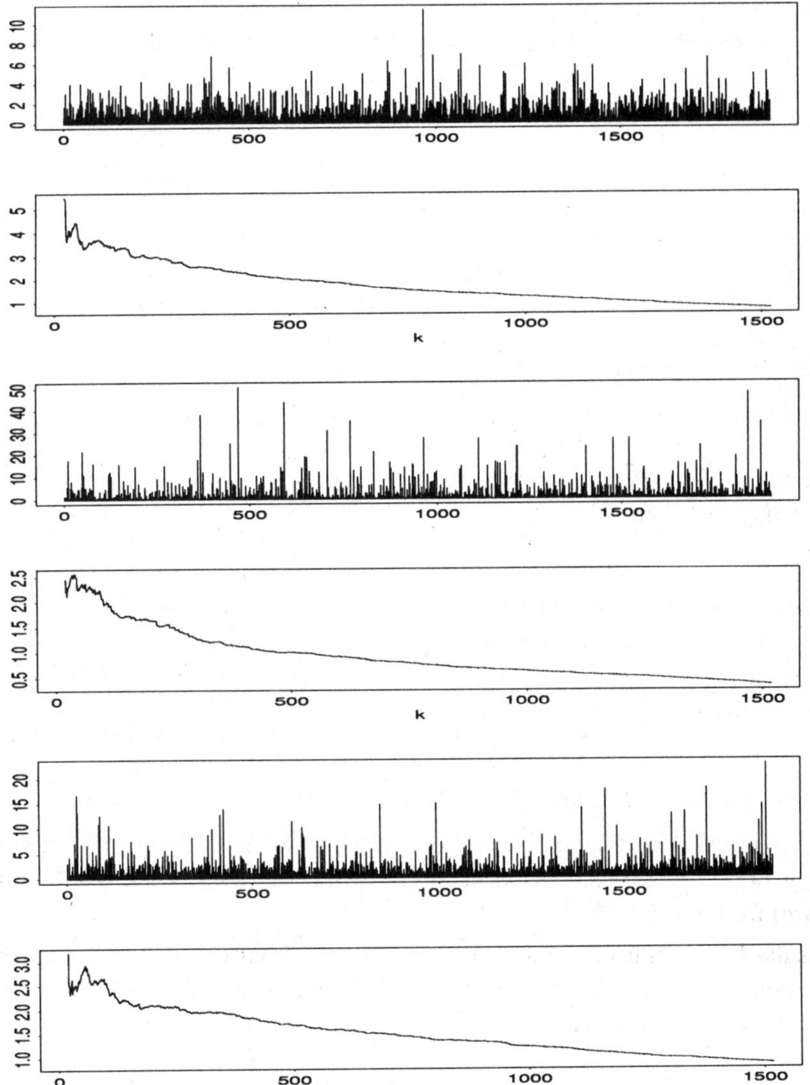

Figure 6.4.11 (Warning) *A comparative study of Hill–plots* $(20 \le k \le 1\,900)$ *for* $1\,900$ *iid simulated data from the distributions: standard exponential (top), heavy–tailed Weibull with shape parameter* $\alpha = 0.5$ *(middle), standard lognormal (bottom). The Hill estimator does* not *estimate anything reasonable in these cases. A (too) quick glance at these plots could give you an estimate of* 3 *for the exponential distribution. This should be a warning to everybody using Hill– and related plots! They must be treated with extreme care. One definitely has to contrast such estimates with the exploratory data analysis techniques from Section 6.2.*

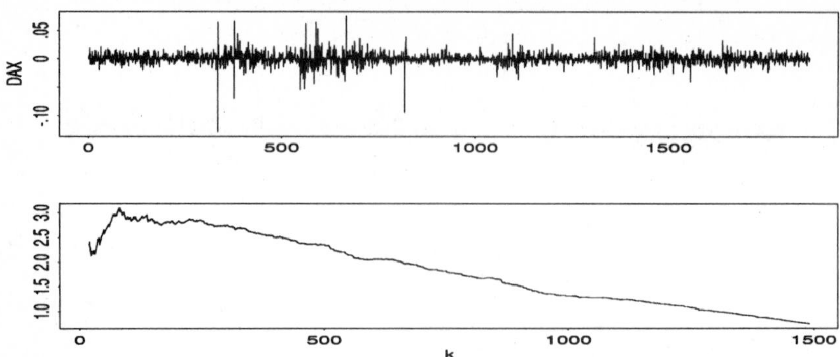

Figure 6.4.12 1 864 *daily log–returns (closing data) of the German stock index* DAX (*September 20, 1988 – August 24, 1995*) (*top*) *and the corresponding Hill–plot of the absolute values* (*bottom*). *It gives relatively stable estimates around the value* 2.8 *in the region* $100 \leq k \leq 300$. *This is much a wider region than in Figures 6.4.11. This Hill–plot is also qualitatively different from the exact Pareto case; see Figure 6.4.7. The deviations can be due to the complicated dependence structure of financial times series.*

estimator (6.33). It turns out that in the case $\rho = 0$ the Hill estimator has minimum mean squared error. The asymptotic relative efficiencies for these estimators critically depend on the interplay between ρ and α. Both the Pickands and the DEdH estimator work for general $\xi \in \mathbb{R}$. For $\xi > -2$ the DEdH estimator has lower variance than Pickands's. Moreover Pickands's estimator is difficult to use since it is rather unstable; see Figures 6.4.7 and 6.4.8. There exist various papers combining higher–order expansions of $\overline{F}$ together with resampling methods. The bootstrap for Hill's estimator has been studied for instance by Hall [310]. An application to the analysis of high frequency foreign exchange rate data is given by Danielson and de Vries [154]; see also Pictet, Dacorogna and Müller [499]. For applications to insurance data see Beirlant et al. [57].

Besides the many papers already referred to, we also would like to mention Anderson [13], Boos [79], Csörgő and Mason [148], Davis and Resnick [159], Drees [186], Falk [224], Häusler and Teugels [320], Lo [425] and Smith and Weissman [599].

More Estimators for the Index of Regular Variation

Hahn and Weiner [306] apply Karamata's theorem to derive a joint estimator of the index of regular variation and an asymmetry parameter for distribution tails. Their method essentially uses truncated moments. An alternative approach based on point process methods is discussed in Höpfner [332, 333] and Jacod and Höpfner [334]. Csörgő, Deheuvels and Mason [147] study kernel type estimates of α including the Hill estimator as a special case. Wei [635] proposes conditional maximum likelihood estimation under both, full and partial knowledge of the slowly varying function.

In the case of models like

$$\overline{F}(x) = \exp\left\{-x^{-\alpha}L(x)\right\}, \quad \alpha > 0, \quad L \in \mathcal{R}_0,$$

one should consult Beirlant and Teugels [56], Beirlant et al. [57], Chapter 4, Broniatowski [94], Keller and Klüppelberg [374] and Klüppelberg and Villaseñor [400].

Estimators for the Index of a Stable Law

Since the early work by Mandelbrot [436] numerous papers have been published concerning the hypothesis that logarithmic returns in financial data follow a stable process with parameter $0 < \alpha < 2$. Though the exact stability has been disputed, a growing consensus is formed around the heavy–tailedness of log–returns. Consequently, various authors focussed on parameter estimation in stable models. Examples include Koutrouvelis [406] using regression type estimators based on the empirical characteristic function; see also Feuerverger [237], Feuerverger and McDunnough [238] and references therein. McCulloch [450] suggests estimators based on functions of the sample quantiles; this paper also contains a good overall discussion. Though McCulloch's approach seems optimal in the exact stable case, the situation may dramatically change if only slight deviations from stability are present in the data. DuMouchel [194, 195] is a good place to start reading on this. For a detailed discussion on these problems together with an overview on the use of stable distributions and processes in finance see Mittnik and Rachev [465, 466] and the references therein.

6.4.3 Estimating the Norming Constants

In the previous section we obtained estimators for the shape parameter ξ given iid data $X_1, \ldots, X_n$ with df $F \in \mathrm{MDA}(H_\xi)$. Recall that the latter condition is equivalent to

$$c_n^{-1}(M_n - d_n) \xrightarrow{d} H_\xi$$

for appropriate norming constants $c_n > 0$ and $d_n \in \mathbb{R}$. We also know that this relation holds if and only if

$$n\overline{F}\left(c_n x + d_n\right) \to -\ln H_\xi(x)\,, \quad n \to \infty\,, \quad x \in \mathbb{R}\,.$$

As we have already seen in Section 6.4.1, norming constants enter in ***quantile and tail estimation***; see (6.19). Below we discuss one method how norming constants can be estimated. Towards the end of this section we give some further references to other methods. In Section 3.3 we gave analytic formulae linking the norming sequences (c_n) and (d_n) with the tail $\overline{F}$. For instance, in the Gumbel case $\xi = 0$ with right endpoint $x_F = \infty$ the following formulae were derived in Theorem 3.3.27

$$c_n = a\left(d_n\right)\,, \quad d_n = F^{\leftarrow}\left(1 - n^{-1}\right)\,, \tag{6.37}$$

where $a(\cdot)$ stands for the auxiliary function which can be taken in the form

$$a(x) = \int_x^\infty \frac{\overline{F}(y)}{\overline{F}(x)}\, dy\,.$$

Notice the ***problem***: on the one hand, we need the norming constants c_n and d_n in order to obtain quantile and tail estimates. On the other hand, (6.37) defines them as functions of just that tail, so it seems that

this surely is a race we cannot win!

Though this is partly true let us see how far we can get. We will try to convince you that the appropriate reformulation of the above sentence is:

this is a race which will be difficult to win!

Consider the more general set–up $F \in \mathrm{MDA}(H_\xi)$, $\xi \ge 0$, which includes for our purposes the most important cases of the Fréchet and the Gumbel distribution. In Examples 3.3.33 and 3.3.34 we showed how one can unify these two maximum domains of attraction by the logarithmic transformation

$$x^* = \ln(1 \vee x)\,, \quad x \in \mathbb{R}\,.$$

Together with Theorem 3.3.26 the following useful result can be obtained.

Lemma 6.4.13 (Embedding MDA (H_ξ), $\xi \ge 0$, in MDA(Λ))
Let $X_1, \ldots, X_n$ be iid with df $F \in \mathrm{MDA}(H_\xi)$, $\xi \ge 0$, with $x_F = \infty$ and norming constants $c_n > 0$ and $d_n \in \mathbb{R}$. Then $X_1^*, \ldots, X_n^*$ are iid with df $F^* \in \mathrm{MDA}(\Lambda)$ and auxiliary function

$$a^*(t) = \int_t^\infty \frac{\overline{F^*}(y)}{\overline{F^*}(t)}\, dy\,.$$

The norming constants can be chosen as

$$
\begin{aligned}
d_n^* &= (F^*)^{\leftarrow}(1-n^{-1}),\\
c_n^* &= a^*(d_n^*) = \int_{d_n^*}^{\infty} \frac{\overline{F^*}(y)}{\overline{F^*}(d_n^*)}\,dy \sim n \int_{d_n^*}^{\infty} \overline{F^*}(y)\,dy\,.
\end{aligned}
$$

□

In Section 6.4.1 we tried to convince you that our estimators have to be based on the k largest order statistics $X_{k,n},\ldots,X_{1,n}$, where $k=k(n)\to\infty$. From the above lemma we obtain estimators if we replace F^* by the empirical df F_n^*:

$$
\begin{aligned}
\widehat{d}^*_{n/k} &= X^*_{k+1,n} = \ln(1\vee X_{k+1,n})\\
\widehat{c}^*_{n/k} &= \frac{n}{k}\int_{\widehat{d}^*_{n/k}}^{\infty} \overline{F^*_n}(y)\,dy\\
&= \frac{n}{k}\int_{\ln X_{k+1,n}}^{\ln X_{1,n}} \overline{F^*_n}(y)\,dy\\
&= \frac{1}{k}\sum_{j=1}^{k}\ln X_{j,n} - \ln X_{k+1,n}\,. \qquad (6.38)
\end{aligned}
$$

The latter is a version of the Hill estimator. The change from k to $k+1$ in (6.38) is asymptotically unimportant; see Example 4.1.11.

Next we make the transformation back from F^* to F via

$$
\frac{n}{k}\,P\left(X^* > c^*_{n/k}\,x + d^*_{n/k}\right) = \frac{n}{k}\,P\left(X > \exp\left\{c^*_{n/k}\,x + d^*_{n/k}\right\}\right), \quad x>0\,.
$$

Finally we use that $F^*\in \mathrm{MDA}(\Lambda)$, hence the left–hand side converges to e^{-x} as $n\to\infty$, provided that $n/k\to\infty$. We thus obtain the tail estimator

$$
\begin{aligned}
\widehat{(\overline{F}(x))} &= \frac{k}{n}\left(\exp\left\{-\widehat{d}^*_{n/k}+\ln x\right\}\right)^{-1/\widehat{c}^*_{n/k}}\\
&= \frac{k}{n}\left(\frac{x}{X_{k+1,n}}\right)^{-1/\widehat{c}^*_{n/k}}\,.
\end{aligned}
$$

This tail estimator was already obtained by Hill [326] for the exact model (6.26); see (6.30). As a quantile estimator we obtain

$$
\widehat{x}_p = \left(\frac{n}{k}(1-p)\right)^{-\widehat{c}^*_{n,k}} X_{k+1,n}\,.
$$

Time to summarise all of this:

Let $\widehat{\xi}^{(H)}$ denotes the Hill estimator for ξ, i.e.

$$\widehat{\xi}^{(H)} = \frac{1}{k} \sum_{j=1}^{k} \ln X_{j,n} - \ln X_{k,n} \,.$$

Let $X_1, \ldots, X_n$ be a sample from $F \in \text{MDA}(H_\xi)$, $\xi \geq 0$, and $k = k(n) \to \infty$ such that $k/n \to 0$. Then for x large enough, a tail estimator for $\overline{F}(x)$ becomes

$$\widehat{\left(\overline{F}(x)\right)} = \frac{k}{n} \left(\frac{x}{X_{k+1,n}} \right)^{-1/\widehat{\xi}^{(H)}} .$$

The quantile x_p defined by $F(x_p) = p \in (0,1)$ can be estimated by

$$\widehat{x}_p = \left(\frac{n}{k} (1-p) \right)^{-\widehat{\xi}^{(H)}} X_{k+1,n} \,.$$

Notes and Comments

It should be clear from the above that similar quantile estimation methods can be worked out using alternative parameter estimators as discussed in the previous sections. For instance, both Hill [326] and Weissman [637] base their approach on maximum likelihood theory and the limit distribution of the k upper order statistics as in Theorem 4.2.8. They cover the whole range $\xi \in \mathbb{R}$. Other estimators of the norming constants were proposed by Dijk and de Haan [183] and Falk, Hüsler and Reiss [225].

6.4.4 Tail and Quantile Estimation

As before, assume that we consider a sample $X_1, \ldots, X_n$ of iid rvs with df $F \in \text{MDA}(H_\xi)$ for some $\xi \in \mathbb{R}$. Let $0 < p < 1$ and x_p denote the corresponding p–quantile.

The whole point behind the domain of attraction condition $F \in \text{MDA}(H_\xi)$ is to be able to estimate quantiles outside the range of the data, i.e. $p > 1 - 1/n$. The latter is of course equivalent to finding estimators for the tail $\overline{F}(x)$ with x large. In Sections 6.3.3 and 6.4.3 we have already discussed some possibilities. Indeed, whenever we have estimators for the shape parameter ξ and the norming constants c_n and d_n, natural estimators of x_p and $\overline{F}(x)$ can immediately be derived from the defining property of $F \in \text{MDA}(H_\xi)$. We want to discuss some of them more in detail and point at their most important properties and caveats. From the start we would like to stress that estimation outside the range of the data can be made only if *extra model assumptions* are imposed. There is no magical technique which yields reliable results for free. One could formulate as in finance:

There is no free lunch when it comes to high quantile estimation!

In our discussion below, we closely follow the paper by Dekkers and de Haan [173]. The main results are formulated in terms of conditions on $U(t) = F^{\leftarrow}(1-t^{-1})$ so that $x_p = U(1/(1-p))$. Denoting $U_n(t) = F_n^{\leftarrow}(1-t^{-1})$, where $F_n^{\leftarrow}$ is the empirical quantile function,

$$U_n\left(\frac{n}{k-1}\right) = F_n^{\leftarrow}\left(1-\frac{k-1}{n}\right) = X_{k,n}\,, \quad k=1,\ldots,n\,.$$

Hence $X_{k,n}$ appears as a natural estimator of the $(1-(k-1)/n)$–quantile. The range $[X_{n,n}, X_{1,n}]$ of the data allows one to make a within–sample estimation up to the $(1-n^{-1})$–quantile. Although in any practical application p is fixed, from a mathematical point of view the difference between high quantiles within and outside the sample can for instance be described as follows:

(a) *high quantiles within the sample:* $p = p_n \uparrow 1$, $n(1-p_n) \to c\,, c \in (1,\infty]\,,$
(b) *high quantiles outside the sample:* $p = p_n \uparrow 1$, $n(1-p_n) \to c$, $0 \le c < 1$.

Case (a) for $c = \infty$ is addressed by the following result which is Theorem 3.1 in Dekkers and de Haan [173]. It basically tells us that we can just use the empirical quantile function for estimating x_p.

Theorem 6.4.14 (Estimating high quantiles I)
Suppose $X_1,\ldots,X_n$ is an iid sample from $F \in \mathrm{MDA}(H_\xi)$, $\xi \in \mathbb{R}$, and F has a positive density f. Assume that the density U' is in $\mathcal{R}_{\xi-1}$. Write $p = p_n$ and $k = k(n) = [n(1-p_n)]$, where $[x]$ denotes the integer part of x. If the conditions

$$p_n \to 1 \quad \textit{and} \quad n\,(1-p_n) \to \infty$$

hold then

$$\sqrt{2k}\,\frac{X_{k,n}-x_p}{X_{k,n}-X_{2k,n}} \xrightarrow{d} N\left(0, 2^{2\xi+1}\xi^2/(2^\xi-1)^2\right)\,. \qquad \square$$

Remark. 1) The condition $U' \in \mathcal{R}_{\xi-1}$ can be reformulated in terms of F. For instance for $\xi > 0$, the condition becomes $f \in \mathcal{R}_{-1-1/\xi}$. $\square$

In Theorem 3.4.5 we characterised $F \in \mathrm{MDA}(H_\xi)$ through the asymptotic behaviour of U:

$$\lim_{t\to\infty} \frac{U(tx)-U(t)}{U(ty)-U(t)} = \frac{x^\xi-1}{y^\xi-1}\,, \quad x,y>0\,, \quad y \neq 1\,.$$

For $\xi = 0$ the latter limit has to be interpreted as $\ln x / \ln y$. We can rewrite the above as follows

$$U(tx) = \frac{x^\xi - 1}{1 - y^\xi} \left(U(t) - U(ty)\right)(1 + o(1)) + U(t) . \qquad (6.39)$$

Using this relation, a heuristic argument suggests an estimator for the quantiles x_p outside the range of the data. Indeed, replace U by U_n in (6.39) and put $y = 1/2$, $x = (k-1)/(n(1-p))$ and $t = n/(k-1)$. Substitute ξ by an appropriate estimator $\widehat{\xi}$. Doing so, and neglecting $o(1)$–terms one finds the following estimator of x_p:

$$\widehat{x}_p \quad = \quad \frac{(k/(n(1-p)))^{\widehat{\xi}} - 1}{1 - 2^{-\widehat{\xi}}} \left(X_{k,n} - X_{2k,n}\right) + X_{k,n} . \qquad (6.40)$$

The following result is Theorem 3.3 in Dekkers and de Haan [173].

Theorem 6.4.15 (Estimating high quantiles II)
Suppose $X_1, \ldots, X_n$ is an iid sample from $F \in \mathrm{MDA}(H_\xi)$, $\xi \in \mathbb{R}$, and assume that $\lim_{n\to\infty} n(1-p) = c$ for some $c > 0$. Let $\widehat{x}_p$ be defined by (6.40) with $\widehat{\xi}$ the Pickands estimator (6.24). Then for every fixed $k > c$,

$$\frac{\widehat{x}_p - x_p}{X_{k,n} - X_{2k,n}} \xrightarrow{d} Y ,$$

where

$$Y = \frac{(k/c)^\xi - 2^{-\xi}}{1 - 2^{-\xi}} + \frac{1 - (Q_k/c)^\xi}{\exp\{\xi H_k\} - 1} . \qquad (6.41)$$

The rvs H_k and Q_k are independent, Q_k has a gamma distribution with parameter $2k+1$ and

$$H_k = \sum_{j=k+1}^{2k} \frac{E_j}{j}$$

for iid standard exponential rvs $E_1, E_2, \ldots$. □

Remarks. 2) The case $0 < c < 1$ of Theorem 6.4.15 corresponds to extrapolation outside the range of the data. For the extreme case $c = 0$, a relevant result is to be found in de Haan [293], Theorem 5.1. Most of these results depend on highly technical conditions on the asymptotic behaviour of $\overline{F}$. There is a strong need for comparative numerical studies on these high quantile estimators.

3) Approximations to the df of Y in (6.41) can be worked out explicitly.

4) As for the situation of Theorem 6.4.14, no results seem to exist concerning the optimal choice of k. For the consistency of the Pickands estimator $\widehat{\xi}$, which is part of the estimator $\widehat{x}_p$, one actually needs $k = k(n) \to \infty$; see

Theorem 6.4.1.

5) In the case $\xi < 0$ similar results can be obtained for the estimation of the right endpoint x_F of F. We refer the interested reader to Dekkers and de Haan [173] for further details and some examples. □

Summary

Throughout Section 6.4, we have discovered various estimators for the important shape parameter ξ of dfs in the maximum domain of attraction of the GEV. From these and further estimators, either for the location and scale parameters and/or norming constants, estimators for the tail $\overline{F}$ and high quantiles resulted. The properties of these estimators crucially depend on the higher–order behaviour of the underlying distribution tail $\overline{F}$. The latter is unfortunately not verifiable in practice.

On various occasions we hinted at the fact that the determination of the number k of upper order statistics finally used remains a delicate point in the whole set–up. Various papers exist which offer a semi–automatic or automatic, so–called ”optimal”, choice of k. See for instance Beirlant et al. [57] for a regression based procedure with various examples to insurance data, and Danielson and de Vries [154] for an alternative method motivated by examples in finance. We personally prefer a rather pragmatic approach realising that, whatever method one chooses, the “Hill horror plot” (Figure 4.1.13) would fool most, if not all. It also serves to show how delicate a tail analysis in practice really is. On the other hand, in the ”nice case” of exact Pareto behaviour, all methods work well; see Figures 6.4.7.

Our experience in analysing data, especially in (re)insurance, shows that in practice one is often faced with data which are clearly heavy–tailed and for which ”exact” Pareto behaviour of $\overline{F}(x)$ sets in for relatively low values of x; see for instance Figure 6.2.12. The latter is not so obvious in the world of finance. This is mainly due to more complicated dependence structure in most of the finance data; compare for instance Figures 6.5.10 and 6.5.11. A ”nice” example from the realm of finance was discussed in Figure 6.4.12. The conclusion ”the data are heavy–tailed” invariably has to be backed up with information from the user who provided the data in the first place! Furthermore, any analysis performed has to be supported by exploratory data analysis techniques as outlined in Section 6.2. Otherwise, situations as explained in Figure 6.4.11 may occur.

It is our experience that in many cases one obtains a Hill– (or related) plot which tends to have a fairly noticeable horizontal stretch across different (often lower) k–values. A choice of k in such a region is to be preferred. Though the above may sound vague, we suggest the user of extremal event techniques to experiment on both simulated as well as real data in order to get a feeling for what is going on. A final piece of advice along this route: *never* go for one estimate only. Calculate and always plot estimates of the relevant quantities (a quantile say) across a wide range of k–values; for examples see Figures 6.4.3 and 6.4.4. In the next section we shall come back to this point, replacing k by a threshold u. We already warn the reader beforehand: the approach offered in Setion 6.5 suffers from the same problems as those discussed in this section.

Notes and Comments

So far, we only gave a rather brief discussion on the statistical estimation of parameters, tails and quantiles in the heavy–tailed case. This area is still under intensive investigation so that at present no complete picture can be given. Besides the availability of a whole series of mathematical results a lot of insight is obtained through simulation and real life examples. In the next section some further techniques and indeed practical examples will be discussed. The interested reader is strongly advised to consult the papers referred to so far. An interesting dicussion on the main issues is de Haan [295], where also applications to currency exchange rates, life span estimation and sea level data are given; see also Davis and Resnick [159]. In Einmahl [198] a critical discussion concerning the exact meaning of *extrapolating outside the data* is given. He stresses the usefulness of the empirical df as an estimator.

6.5 Fitting Excesses Over a Threshold

6.5.1 Fitting the GPD

Methodology introduced so far was obtained either on the assumption that the data come from a GEV (see Section 6.3) or belong to its maximum domain of attraction (see Section 6.4). We based statistical estimation of the relevant parameters on maximum likelihood, the method of probability–weighted moments or some appropriate condition of regular variation type. In Section 3.4 we laid the foundation to an alternative approach based on exceedances of high thresholds. The key idea of this approach is explained below.

Suppose

$$X, X_1, \ldots, X_n \quad \textit{are iid with df } F \in \mathrm{MDA}(H_\xi) \textit{ for some } \xi \in \mathbb{R}.$$

First, choose a high threshold u and denote by

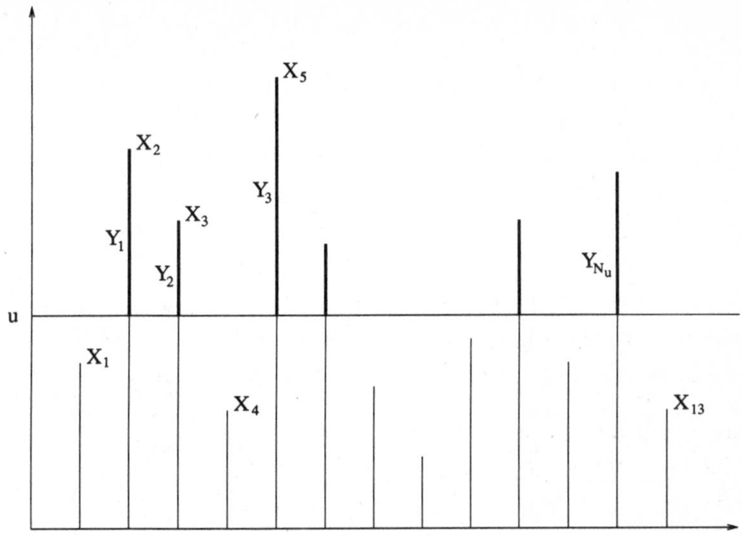

Figure 6.5.1 *Data* $X_1, \ldots, X_{13}$ *and the corresponding excesses* $Y_1, \ldots, Y_{N_u}$ *over the threshold* u.

$$N_u = \operatorname{card}\{i : i = 1, \ldots, n\,,\; X_i > u\}$$

the number of exceedances of u by $X_1, \ldots, X_n$. We denote the corresponding excesses by $Y_1, \ldots, Y_{N_u}$; see Figure 6.5.1. The *excess df* of X is given by

$$F_u(y) = P(X - u \le y \mid X > u) = P(Y \le y \mid X > u)\,, \quad y \ge 0\,;$$

see Definition 3.4.6. The latter relation can also be written as

$$\overline{F}(u+y) = \overline{F}(u)\,\overline{F}_u(y)\,. \tag{6.42}$$

Now recall the definition of the *generalised Pareto distribution* (*GPD*) from Definition 3.4.9: a GPD $G_{\xi,\beta}$ with parameters $\xi \in \mathbb{R}$ and $\beta > 0$ has distribution tail

$$\overline{G}_{\xi,\beta}(x) = \begin{cases} \left(1 + \xi \dfrac{x}{\beta}\right)^{-1/\xi} & \text{if } \xi \neq 0\,, \\ e^{-x/\beta} & \text{if } \xi = 0\,, \end{cases} \quad x \in D(\xi, \beta)\,,$$

where

$$D(\xi, \beta) = \begin{cases} [0, \infty) & \text{if } \xi \ge 0\,, \\ [0, -\beta/\xi] & \text{if } \xi < 0\,. \end{cases}$$

Theorem 3.4.13(b) gives a limit result for $\overline{F}_u(y)$, namely

$$\lim_{u \uparrow x_F} \sup_{0<x<x_F-u} |\overline{F}_u(x) - \overline{G}_{\xi,\beta(u)}(x)| = 0\,,$$

for an appropriate positive function β. Based on this result, for u large, the following approximation suggests itself:

$$\overline{F}_u(y) \approx \overline{G}_{\xi,\beta(u)}(y)\,. \tag{6.43}$$

It is important to note that β is a function of the threshold u. In practice, u will have to be taken sufficiently large. Given such a u, ξ and $\beta = \beta(u)$ are estimated from the excess data, so that the resulting estimates depend on u; see our discussion below.

Relation (6.42) then suggests a method for estimating the far end tail of F by estimating $\overline{F}_u(y)$ and $\overline{F}(u)$ separately. A natural estimator for $\overline{F}(u)$ is given by the empirical df

$$\widehat{(\overline{F}(u))} = \overline{F}_n(u) = \frac{1}{n}\sum_{i=1}^{n} I_{\{X_i>u\}} = \frac{N_u}{n}\,.$$

On the other hand, the generalised Pareto approximation (6.43) (remember that u is large!) motivates an estimator of the form

$$\widehat{(\overline{F}_u(y))} = \overline{G}_{\widehat{\xi},\widehat{\beta}}\,(y) \tag{6.44}$$

for appropriate $\widehat{\xi} = \widehat{\xi}_{N_u}$ and $\widehat{\beta} = \widehat{\beta}_{N_u}$.

A resulting estimator for the tail $\overline{F}(u+y)$ for $y > 0$ then takes on the form

$$\widehat{(\overline{F}(u+y))} = \frac{N_u}{n}\left(1 + \widehat{\xi}\,\frac{y}{\widehat{\beta}}\right)^{-1/\widehat{\xi}}\,. \tag{6.45}$$

In the Fréchet and Gumbel case ($\xi \ge 0$), the domain restriction in (6.45) is $y \ge 0$, clearly stressing that we estimate $\overline{F}$ in the upper tail. An estimator of the quantile x_p results immediately:

$$\widehat{x}_p = u + \frac{\widehat{\beta}}{\widehat{\xi}}\left(\left(\frac{n}{N_u}\,(1-p)\right)^{-\widehat{\xi}} - 1\right). \tag{6.46}$$

Furthermore, for $\widehat{\xi} < 0$ an estimator of the right endpoint x_F of F is given by

$$\widehat{x}_F = u - \frac{\widehat{\beta}}{\widehat{\xi}}\,.$$

The latter is obtained by putting $\widehat{x}_F = \widehat{x}_1$ (i.e. $p = 1$) in (6.46). In Section 6.4.1 we said that the method of exceedances belongs to the realm of point process theory. From (6.45) and (6.46) this is clear: statistical properties of the resulting estimators crucially depend on the distributional properties of the point process of exceedances (N_u); see for instance Example 5.1.3 and Theorem 5.3.2, and also the Notes and Comments below.

The above method is intuitively appealing. It goes back to hydrologists. Over the last 25 years they have developed this estimation procedure under the acronym of the **P***eaks* **O***ver* **T***hreshold* (*POT*) *method.* In order to work out the relevant estimators the following input is needed:

- reliable models for the point process of exceedances,
- a sufficiently high threshold u,
- estimators $\widehat{\xi}$ and $\widehat{\beta}$,
- and, if necessary, an estimator $\widehat{\nu}$ for location.

If one wants to choose an optimal threshold u one faces similar problems as for the choice of the number k of upper order statistics for the Hill estimator. A value of u *too high* results in too few exceedances and consequently high variance estimators. For u *too small* estimators become biased. *Theoretically*, it is possible to choose u asymptotically optimal by a quantification of a *bias versus variance trade–off*, very much in the same spirit as discussed in Theorem 6.4.9. In *reality* however, the same problems as already encountered for other tail estimators before do occur. We refer to the examples in Section 6.5.2 for illustrations on this.

One method which is of immediate use in practice is based on the linearity of the mean excess function $e(u)$ for the GPD. From Theorem 3.4.13(e) we know that for a rv X with df $G_{\xi,\beta}$,

$$e(u) = E(X - u \mid X > u) = \frac{\beta + \xi u}{1 - \xi}, \quad u \in D(\xi, \beta)\,,\ \xi < 1\,,$$

hence $e(u)$ is linear. Recall from (6.6) that the empirical mean excess function of a given sample $X_1, \ldots, X_n$ is defined by

$$e_n(u) = \frac{1}{N_u} \sum_{i \in \Delta_n(u)} (X_i - u)\,, \quad u > 0\,,$$

where as before $N_u = \operatorname{card}\{i : i = 1, \ldots, n, X_i > u\} = \operatorname{card}\Delta_n(u)$. The remark above now suggests a graphical approach for chosing u:

choose $u > 0$ *such that* $e_n(x)$ *is approximately linear for* $x \geq u$.

The key difficulty of course lies in the interpretation of *approximately.* Only practice can tell! One often observes a change in the slope of $e_n(u)$ for some value of u. Referring to some examples on sulphate and nitrate level in an acid rain study Smith [595], p. 460, says the following:

The general form of these mean excess plots is not atypical of real data, especially the change of slope near 100 *in both plots. Smith* [588] *observed similar behaviour in data on extreme insurance claims, and Davison and Smith* [166] *used a similar plot to identify a change in the distribution of the threshold form of the River Nidd data. Such plots therefore appear to be an extremely useful diagnostic in this form of analysis.*

The reader should never expect a unique choice of u to appear. We recommend using plots, to reinforce judgement and common sense and compare resulting estimates across a variety of u–values. In applications we often prefer plots indicating the threshold value u, as well as the number of exceedances used for the estimation, on the horizontal axes; the estimated value of the parameter or the quantile, say, is plotted on the vertical one. The latter is illustrated in Section 6.5.2. As can be seen from the examples, and indeed can be proved, all these plots exhibit the same behaviour as the Hill– and Pickands–plots before: high variability for u large (few observations) versus bias for u small (many observations, but at the same time the approximation (6.43) may not be applicable).

Concerning estimators for ξ and β, various methods similar to those discussed in Section 6.4.2 exist.

Maximum Likelihood Estimation

The following results are to be found in Smith [591].

Recall that our original data $\boldsymbol{X} = (X_1, \ldots, X_n)$ are iid with common df F. Assume F is GPD with parameters ξ and β, so that the density f is

$$f(x) = \frac{1}{\beta}\left(1 + \xi\,\frac{x}{\beta}\right)^{-\frac{1}{\xi}-1}, \quad x \in D(\xi, \beta)\,.$$

The log–likelihood function equals

$$\ell((\xi, \beta); \boldsymbol{X}) = -n \ln \beta - \left(\frac{1}{\xi} + 1\right) \sum_{i=1}^{n} \ln\left(1 + \frac{\xi}{\beta}\,X_i\right)\,.$$

Notice that the arguments of the above function have to satisfy the domain restriction $X_i \in D(\xi, \beta)$. For notational convenience, we have dropped that part from the likelihood function. Recall that $D(\xi, \beta) = [0, \infty)$ for $\xi \geq 0$. Now likelihood equations can be derived and solved numerically yielding the MLE $\widehat{\xi}_n$, $\widehat{\beta}_n$. This method works fine if $\xi > -1/2$, and in this case one can show that

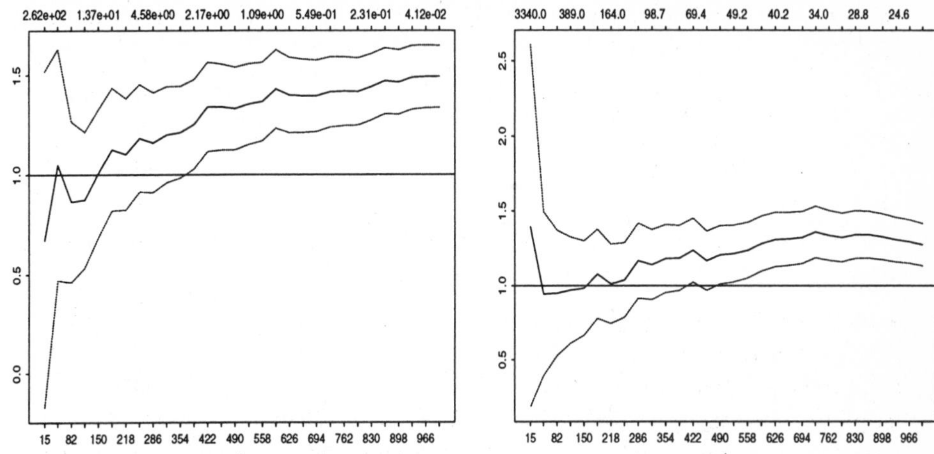

Figure 6.5.2 *A "horror plot" for the MLE of the shape parameter ξ of a GPD. The simulated data come from a df given by $\overline{F}(x) = 1/(x \ln x)$. Left: sample size* 1 000. *Right: sample size* 10 000. *The upper horizontal axis indicates the threshold value u, the lower one the corresponding number k of exceedances of u. As for Hill estimation (see Figure 4.1.13) MLE also becomes questionable for such perturbed Pareto tails.*

$$n^{1/2}\left(\widehat{\xi}_n - \xi\,,\, \frac{\widehat{\beta}_n}{\beta} - 1\right) \stackrel{d}{\to} N(0, M^{-1})\,, \quad n \to \infty\,,$$

where

$$M^{-1} = (1+\xi)\begin{pmatrix} 1+\xi & -1 \\ -1 & 2 \end{pmatrix}$$

and $N(\mu, \Sigma)$ stands for the bivariate normal distribution with mean vector μ and covariance matrix Σ. The usual MLE properties like consistency and asymptotic efficiency hold.

Because of (6.43), it is more realistic to assume a GPD for the excesses $Y_1, \ldots, Y_N$, where $N = N_u$ is independent of the Y_i. The resulting conditional likelihood equations can be solved best via a reparametrisation $(\xi, \beta) \to (\xi, \tau)$, where $\tau = -\xi/\beta$. This leads to the solution

$$\widehat{\xi} = \widehat{\xi}(\tau) = N^{-1}\sum_{i=1}^{N} \ln(1 - \tau Y_i)\,,$$

where τ satisfies

$$h(\tau) = \frac{1}{\tau} + \frac{1}{N}\left(\frac{1}{\widehat{\xi}(\tau)} + 1\right)\sum_{i=1}^{N} \frac{Y_i}{1 - \tau Y_i} = 0\,.$$

The function $h(\tau)$, defined for $\tau \in (-\infty, \max(Y_1, \dots, Y_N))$, is continuous at 0. Letting $u = u_n \to \infty$, Smith [591] derives various limit results for the distribution of $(\widehat{\xi}_N, \widehat{\beta}_N)$. As in the case of the Hill estimator (see Theorem 6.4.9), an asymptotic bias may enter. The latter again crucially depends on a second–order condition for $\overline{F}$.

Method of Probability–Weighted Moments

Similarly to our discussion in Section 6.3.2, Hosking and Wallis [337] also worked out a probability–weighted moment approach for the GPD. This is based on the quantitites (see Theorem 3.4.13(a))

$$w_r = EZ\left(\overline{G}_{\xi,\beta}(Z)\right)^r = \frac{\beta}{(r+1)(r+1-\xi)}\,, \qquad r = 0, 1\,,$$

where Z has GPD $G_{\xi,\beta}$. We immediately obtain

$$\beta = \frac{2w_0 w_1}{w_0 - 2w_1} \quad \text{and} \quad \xi = 2 - \frac{w_0}{w_0 - 2w_1}\,.$$

If we now replace w_0 and w_1 by empirical moment estimators, one obtains the probability–weighted moment estimators $\widehat{\beta}$ and $\widehat{\xi}$. Hosking and Wallis [337] give formulae for the approximate standard errors of these estimators. They compare their approach to the MLE approach and come to the conclusion that in the case $\xi \geq 0$ the method of probability–weighted moments offers a viable alternative. However, as we have already stressed in the case of a GEV, maximum likelihood methodology allows us to fit much more general models including time dependence of the parameters and the influence of explanatory variables.

6.5.2 An Application to Real Data

In the above discussion we have outlined the basic principles behind the GPD fitting programme. Turning to the practical applications, two main issues need to be addressed:

(*a*) fit the *conditional* df $F_u(x)$ for an appropriate range of x– (and indeed u–) values;

(*b*) fit the *unconditional* df $F(x)$, again for appropriate x–values.

Though formulae (6.44) and (6.45) *in principle* solve the problem, in practice care has to be taken about the precise range of the data available and/or the interval over which we want to fit. In our examples, we have used a set–up which is motivated mainly by insurance applications.

Take for instance the Danish fire insurance data. Looking at the ME–plot in Figure 6.2.11 we see that the data are clearly heavy–tailed. In order to estimate the shape parameter ξ a choice of the threshold u (equivalently, of the number k of exceedances) has to be made. In the light of the above discussion concerning the use of ME–plots at this stage, we suggest a first choice of $u = 10$ resulting in 109 exceedances. This means we choose u from a region above which the ME–plot is roughly linear. An alternative choice would perhaps be in the range $u \approx 18$. Figure 6.5.5 (top, left) gives the resulting estimates of ξ as a function of u (upper horizontal axis) and of the number k of exceedances of u (lower horizontal axis): the resulting plot is relatively stable with estimated values mainly in the range $(0.4, 0.6)$. Compare this plot with the Hill–plot in Figure 6.4.3. For $u = 10$ maximum likelihood estimates $\widehat{\xi} = 0.497$ (s.e.= 0.143) and $\widehat{\beta} = 6.98$ result. A change to $u = 18$ yields $\widehat{\xi} = 0.735$ (s.e.= 0.253) based on $k = 47$ exceedances.

From these estimates, using (6.44), an estimate for the (conditional) excess df $F_u(x)$ can be plotted. Following standard practice in reinsurance, in Figure 6.5.5 (top, right) we plot the shifted df $F_u(x-u)$, $x \geq u$. In the language of reinsurance the latter procedure estimates the probability that a claim lies in a given interval, given that the claim has indeed *pierced* the level $u = 10$.

Though the above estimation (once $u = 10$ is chosen) *only* uses the 109 largest claims, a crucial question still concerns where the layer ($u = 10$ and above) is to be positioned in the total portfolio; i.e. we also want to estimate the tail of the *unconditional* df F which yields information on the frequency with which a given high level u is pierced. At this point we need the full data–set and turn to formula (6.45). A straightforward calculation allows us to express $\widehat{F}(z)$ as a three–parameter GPD:

$$\widehat{F}(z) = 1 - \left(1 + \widehat{\xi}\,\frac{z - u - \widehat{\nu}}{\widehat{\beta}'}\right)^{-1/\widehat{\xi}}, \quad z \geq u, \tag{6.47}$$

where

$$\widehat{\nu} = \frac{\widehat{\beta}}{\widehat{\xi}}\left(\left(\frac{N_u}{n}\right)^{\widehat{\xi}} - 1\right) \quad \text{and} \quad \widehat{\beta}' = \widehat{\beta}\left(\frac{N_u}{n}\right)^{\widehat{\xi}}.$$

We would like to stress that the above df is designed only to fit the data well above the threshold u. Below u, where the data are typically abundant, various standard techniques can be used; for instance the empirical df. By combining both, GPD above u and empirical df below u, a good overall fit can be obtained. There are of course various possibilities to fine–tune such a construction.

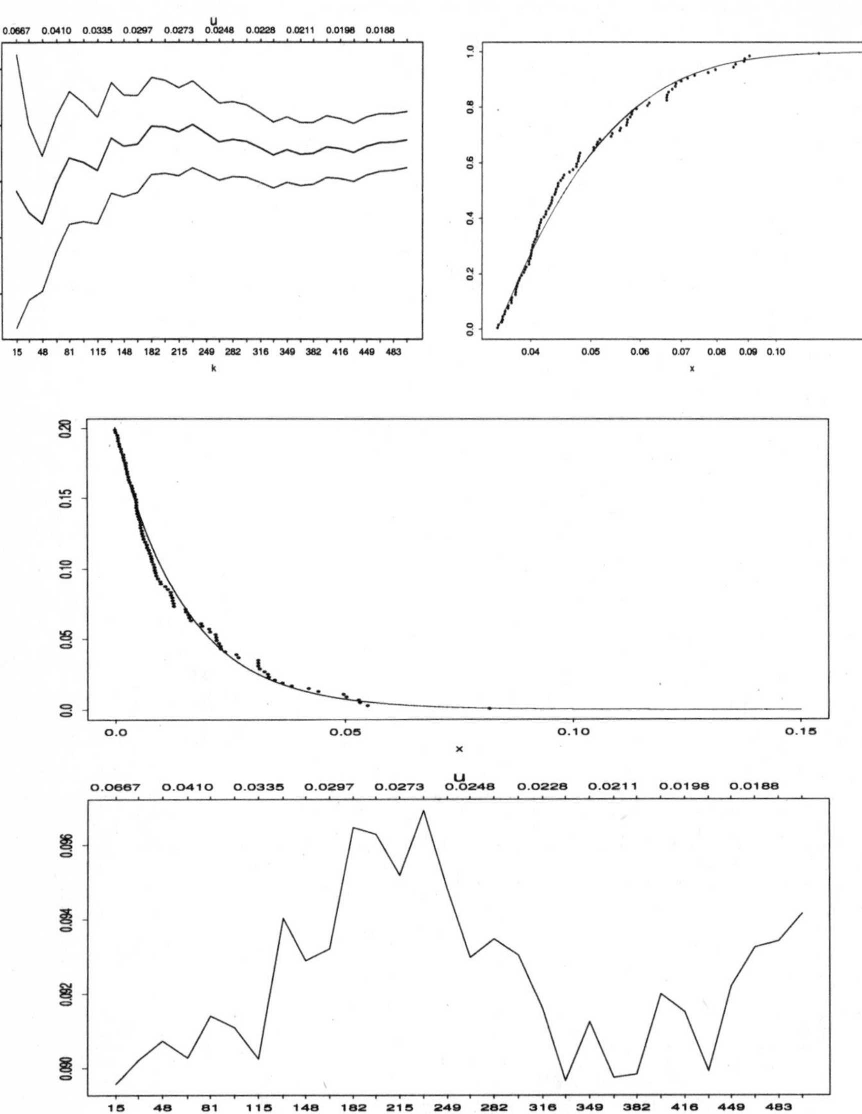

Figure 6.5.3 *The positive BMW log-returns from Figure 6.2.10. Top, left: MLE of ξ as a function of u and k with asymptotic 95% confidence band. Top, right: GPD-fit to $F_u(x-u)$, $x \geq u$, on log-scale. Middle: GPD tail-fit for $\overline{F}(x+u)$, $x \geq 0$. Bottom: estimates of the 99.9%-quantile for the positive returns as a function of the threshold u (upper horizontal axis) and of the corresponding number k of the upper order statistics (lower horizontal axis). A GPD with parameters $\xi = 0.0335$, $\beta = 0.0136$ is fitted, corresponding to $u = 0.0354$ and $k = 100$.*

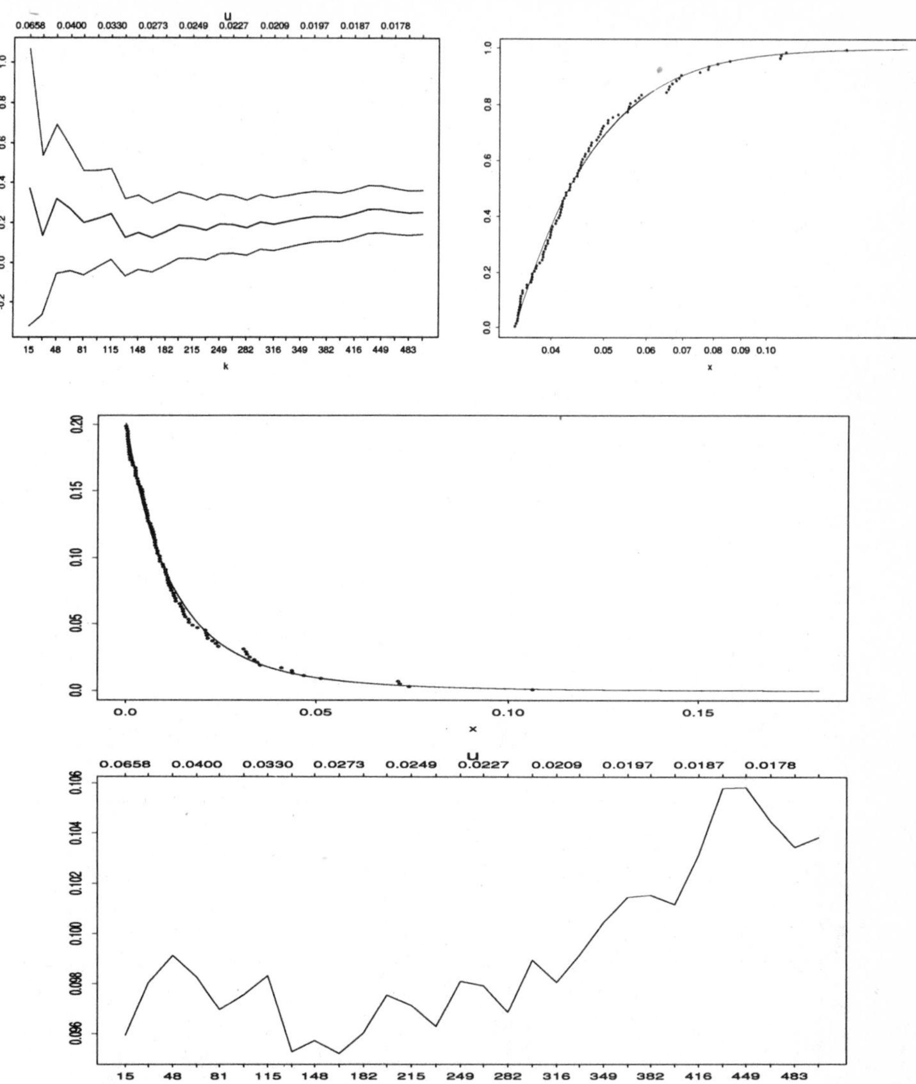

Figure 6.5.4 *The absolute values of the negative BMW log–returns from Figure 6.2.10. Top, left: MLE of* ξ *as a function of* u *and* k *with asymptotic* 95% *confidence band. Top, right: GPD–fit to* $F_u(x-u)$, $x \geq u$, *on log-scale. Middle: GPD tail–fit for* $\overline{F}(x+u)$, $x \geq 0$. *Bottom: estimates of the* 99.9%*–quantile for the absolute negative returns as a function of the threshold* u *(upper horizontal axis) and of the corresponding number* k *of the upper order statistics (lower horizontal axis). A GPD with parameters* $\xi = 0.207$, $\beta = 0.00849$ *is fitted, corresponding to* $u = 0.0343$ *and* $k = 100$*; i.e. the distribution has an infinite* 5*th moment. As mentioned in the discussion of Figure 6.2.10, the lower tail of the distribution appears to be heavier than the upper one.*

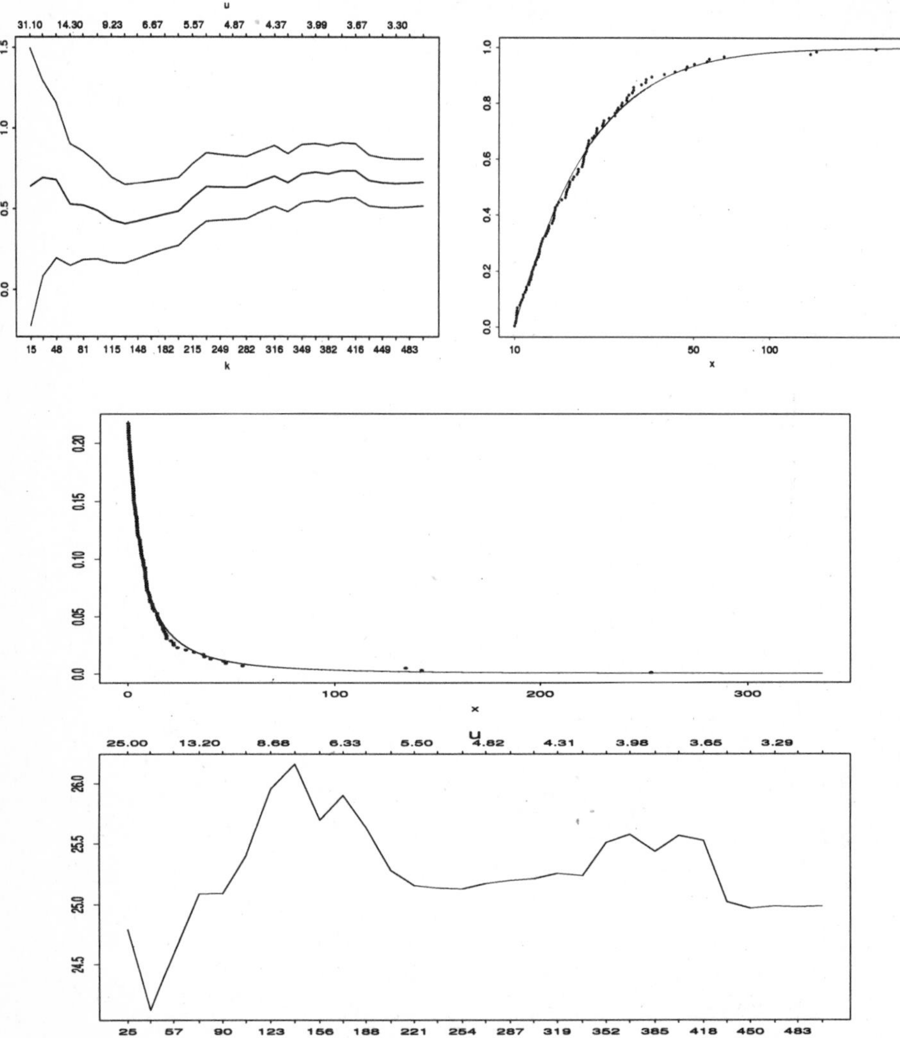

Figure 6.5.5 *The Danish fire insurance data; see Figure 6.2.11. Top, left: MLE for the shape parameter ξ of the GPD. The upper horizontal axis indicates the threshold u, the lower one the number k of exceedances/upper order statistics involved in the estimation. Top, right: fit of the shifted excess df $F_u(x-u)$, $x \geq u$, on log-scale. Middle: GPD tail-fit for $\overline{F}(x+u)$, $x \geq 0$. Bottom: estimates of the* 0.99*-quantile as a function of u (upper horizontal axis) and k (lower horizontal axis). A GPD with parameters $\xi = 0.497$ and $\beta = 6.98$ is fitted, corresponding to $k = 109$ exceedances of $u = 10$. Compare also with Figure 6.4.3 for the corresponding Hill-fit.*

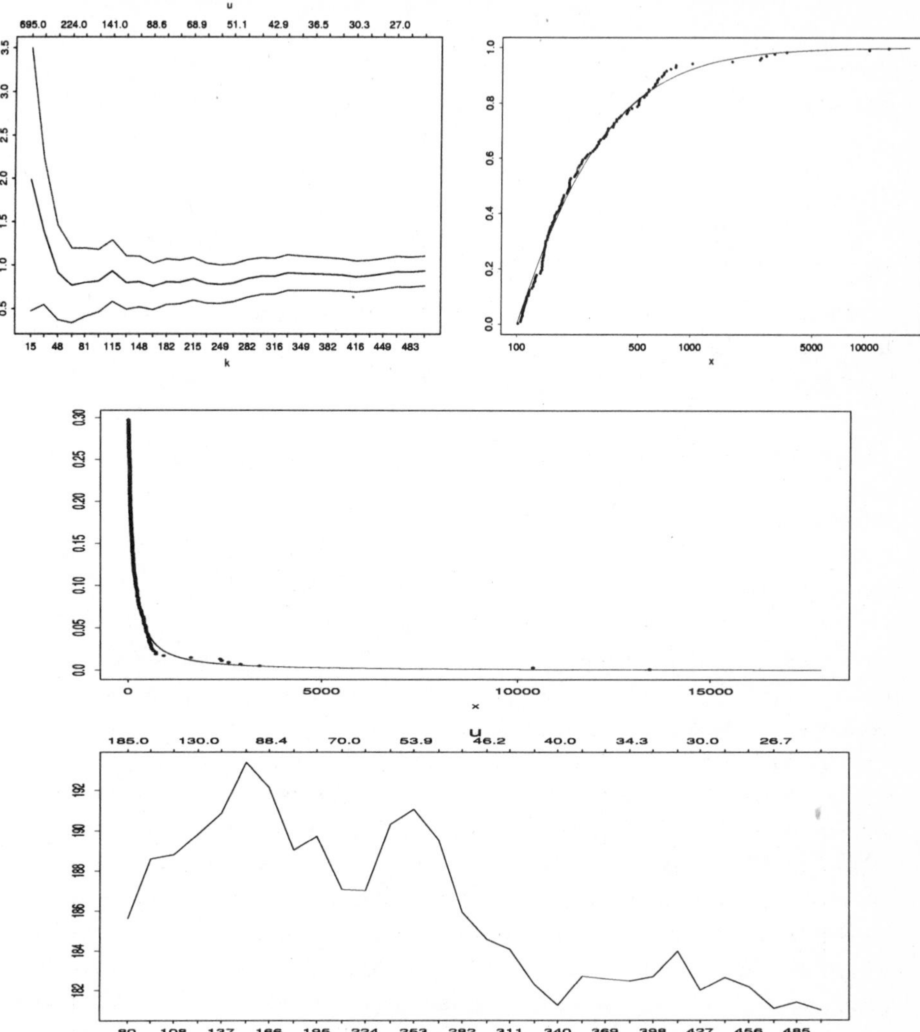

Figure 6.5.6 *The industrial fire insurance data; see Figure 6.2.12. Top, left: MLE for the shape parameter* ξ *of the GPD. The upper horizontal axis indicates the threshold* u, *the lower one the number* k *of exceedances/upper order statistics involved in the estimation. Top, right: fit of the shifted excess df* $F_u(x-u)$, $x \geq u$, *on log-scale. Middle: GPD tail-fit for* $\overline{F}(x+u)$, $x \geq 0$. *Bottom: estimates of the* 0.99*–quantile as a function of* u *(upper horizontal axis) and* k *(lower horizontal axis). A GPD with parameters* $\xi = 0.747$ *and* $\beta = 48.1$ *is fitted, corresponding to* $k = 149$ *exceedances of* $u = 100$. *Compare also with Figure 6.4.4 for the corresponding Hill–fit.*

Finally, using the above fit to $F(z)$, we can give estimates for the p–quantiles, $p \geq F(u)$. In Figure 6.5.5 (bottom) we have summarised the 0.99–quantile estimates obtained by the above method across a wide range of u–values (upper horizontal axis), i.e. for each u–value a new model was fitted and $x_{0.99}$ estimated. Alternatively, the number of exccedances of u is indicated on the lower horizontal axis. For these data a rather stable picture emerges. A value in the range (25,26) follows. Confidence intervals can be calculated. The software needed to do these, and further analyses are discussed in the Notes and Comments below.

Figure 6.5.6 for the industrial fire data (see Figure 6.2.12) and Figure 6.5.3 for the BMW share prices (see Figure 6.2.10) can be interpreted in a similar way.

Mission Improbable: How to Predict the Unpredictable

On studying the above data analyses, the reader may have wondered why we restricted our plots to $x_{0.99}$ for the Danish and industrial insurance data, and $x_{0.999}$ for the BMW data. In answering this question, we restrict attention to the insurance data. At various stages throughout the text we hinted at the fact that extreme value theory (EVT) offers methodology allowing for extrapolation outside the range of the available data. The reason why we are *very reluctant* to produce plots for high quantiles like 0.9999 or more, is that we feel that such estimates are to be treated *with extreme care*. Recall Richard Smith's statement from the Preface: "There is always going to be an element of doubt, as one is extrapolating into areas one doesn't know about. But what EVT is doing is making the best use of whatever data you have about extreme phenomena." Both fire insurance data–sets *have* information on extremes, and indeed EVT has produced models which make best use of whatever data we had at our disposal. Using these models, estimates for the p–quantiles x_p for every $p \in (0,1)$ can be given. The statistical reliability of these estimates becomes, as we have seen, very difficult to judge in general. Though we can work out approximate confidence intervals for these estimators, such constructions strongly rely on mathematical assumptions which are unverifiable in practice.

In Figures 6.5.7 and 6.5.8 we have reproduced the GPD estimates for $x_{0.999}$ and $x_{0.9999}$ for both the Danish and the industrial fire data. These plots should be interpreted with the above quote from Smith in mind. For instance, for the Danish fire insurance data we see that the estimate of about 25 for $x_{0.99}$ jumps at 90 for $x_{0.999}$ and at around 300 for $x_{0.9999}$. Likewise for the industrial fire, we get an increase from around 190 for $x_{0.99}$ to about 1 400 for $x_{0.999}$ and 10 000 for $x_{0.9999}$. These model–based estimates could form the

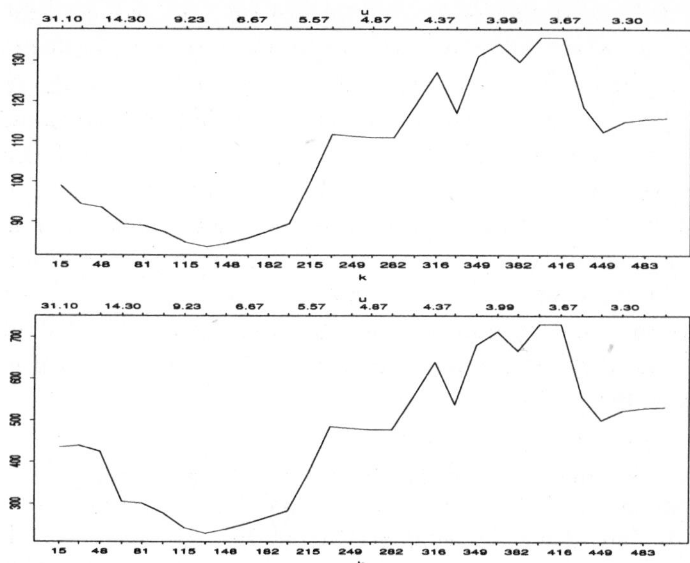

Figure 6.5.7 *GPD–model based estimates of the* 0.999*–quantile* (*top*) *and* 0.9999*–quantile* (*bottom*) *for the Danish fire insurance data.* **WARNING:** *for the interpretation of these plots,* read "Mission improbable" on p. 364.

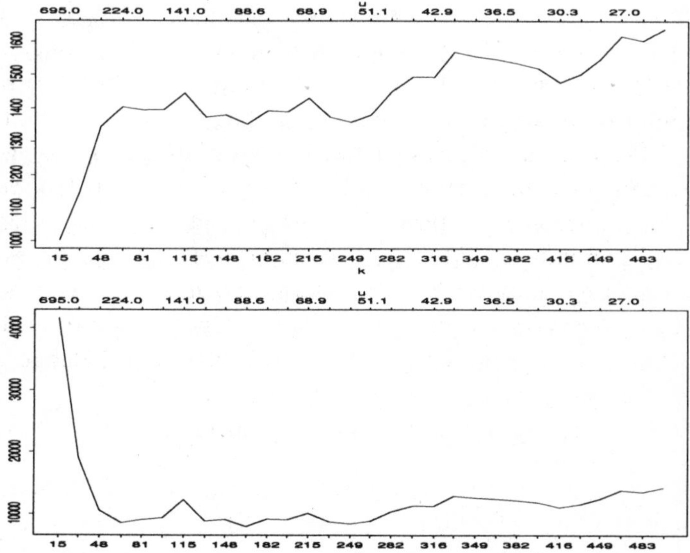

Figure 6.5.8 *GPD–model based estimates of the* 0.999*–quantile* (*top*) *and* 0.9999*–quantile* (*bottom*) *for the industrial fire data.* **WARNING:** *for the interpretation of these plots,* read "Mission improbable" on p. 364.

basis for a detailed discussion with the actuary/underwriter/broker/client responsible for these data. One can use them to calculate so-called *technical premiums*, which are to be interpreted as those premiums which we as statisticians believe to most honestly reflect the information available from the data. Clearly many other factors have to enter at this stage of the discussion. We already stressed before that in dealing with high layers/extremes one should *always consider total exposure* as an alternative. Economic considerations, management strategy, market forces will enter so that by using *all* these inputs we are able to come up with a premium acceptable both for the insurer as well as the insured. Finally, once the EVT model–machinery (GPD for instance) is put into place, it offers an ideal platform for simulation experiments and stress–scenarios. For instance, questions about the influence of single or few observations and model–robustness can be analysed in a straightforward way. Though we have restricted ourselves to a more detailed discussion for the examples from insurance, similar remarks apply to financial or indeed any other kind of data where extremal events play an important role.

Notes and Comments

The POT method has been used by hydrologists for more than 25 years. It has also been suggested for dealing with large claims in insurance; see for instance Kremer [409], Reiss [525] and Teugels [618, 619]. It may be viewed as an alternative approach to the more classical GEV fitting.

In the present section, we gave a brief *heuristic* introduction to the POT. The practical use of the GPD in extreme value modelling is best learnt from the fundamental papers by Smith [594], Davison [165], Davison and Smith [166], North [482] and the references therein. Falk [224] uses the POT method for estimating ξ. Its theoretical foundation was already laid by Pickands [498] and developed further for instance by Smith [591] and Leadbetter [417]. The statistical estimation of the parameters of the GPD is also studied in Tajvidi [611].

The POT model is usually formulated as follows:

(*a*) the excesses of an iid (or stationary) sequence over a high threshold u occur at the times of a Poisson process;
(*b*) the corresponding excesses over u are independent and have a GPD;
(*c*) excesses and exceedance times are independent of each other.

Here one basically looks at a space–time problem: excess sizes and exceedance times. Therefore it is natural to model this problem in a two–dimensional point process or in a marked point process setting; see Falk et al. [225], Section 2.3 and Section 10.3, and Leadbetter [417] for the necessary theoretical background. There also the stationary non–iid case is treated. Using these tools one can justify the above assumptions on the excesses and exceedance times *in an asymptotic sense.* A partial justification is to be found in Section 5.3.1 (weak convergence of the point process of exceedances to a Poisson limit) and in Theorem 3.4.13(b) (GPD approximation of the excess df).

The POT method allows for fitting GPD models with time–dependent parameters $\xi(t)$, $\beta(t)$ and $\nu(t)$, in particular one may include non–stationarity effects (trends, seasonality) into the model; see for instance Smith [594]. These are further attractive aspects of GPD fitting.

Example 6.5.9 (Diagnostic tools for checking the assumptions of the POT method)
In Figures 6.5.10 and 6.5.11 we consider some diagnostic tools (suggested by Smith and Shively [597]) for checking the Poisson process assumption for the exceedance times in the POT model. Figure 6.5.10 (top) shows the excesses over $u = 10$ by the Danish fire insurance claims; see Figure 6.2.11. The left middle figure is a plot of the first sample autocorrelations of the excesses. In interpreting the latter plot, the value of ξ is of course crucial. Indeed for $\xi > 1/2$, the theoretical autocorrelations do not exist and hence there is no straightforward interpretation of the sample autocorrelation plot. As we have seen, ξ takes values around $1/2$ for $u = 10$ and above $1/2$ for $u = 18$. Further analyses on this point gave however no reason to reject independence. In the right middle figure the corresponding inter–arrival times of the exceedances appear. If these times came from a homogeneous Poisson process they should be iid exponential; see Example 2.5.2. The (Lowess smoothed) curve in the figure can be used to indicate possible deviations from the stationary assumption; it is basically a smoothed mean value of the data and estimates the reciprocal of the intensity of the Poisson process. The curve is almost a straight line, parallel to the horizontal axis. In the left bottom figure a QQ–plot of the inter–arrival times versus exponential quantiles is given. The exponential fit is quite convincing. The sample autocorrelations of the inter–arrival times (bottom, right) yield no ground for rejecting the hypothesis of zero correlation. For a more detailed analysis of these data, see McNeil [452] and Resnick [534]. The last paper also discusses the problem of testing for independence when the underlying data possibly have infinite variance.

The picture changes for instance for the absolute values of the negative log–returns of the BMW share prices (see Figure 6.2.10). In Figure 6.5.11 the excesses over $u = 0.0343$ are given. Both autocorrelograms show a more intricate dependence structure often encountered in finance data. □

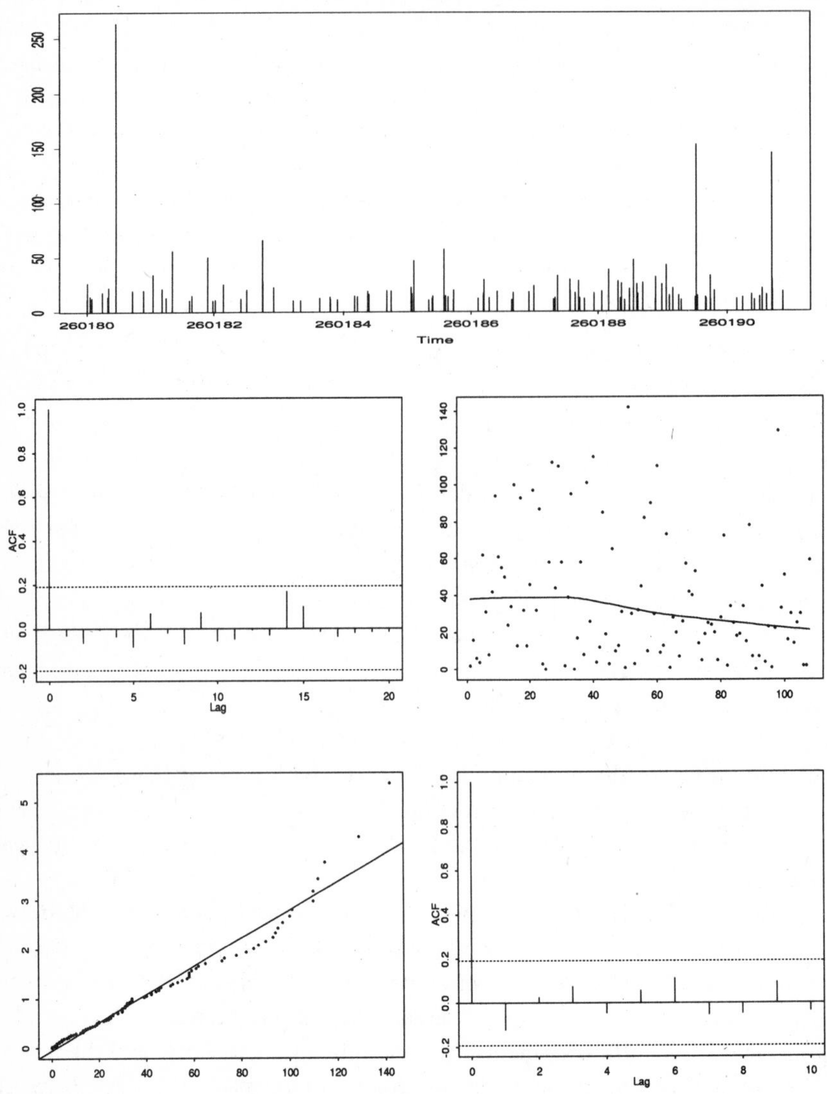

Figure 6.5.10 *Top: the excesses over* $u = 10$ *of the Danish fire insurance data; see Figure 6.2.11. Middle, left: the sample autocorrelations of the excesses. Middle, right: the inter–arrival times of the exceedances and smoothed mean values curve. Bottom, left: QQ–plot of the the inter–arrival times against exponential quantiles. Bottom, right: sample autocorrelations of these times. See Example 6.5.9 for further comments.*

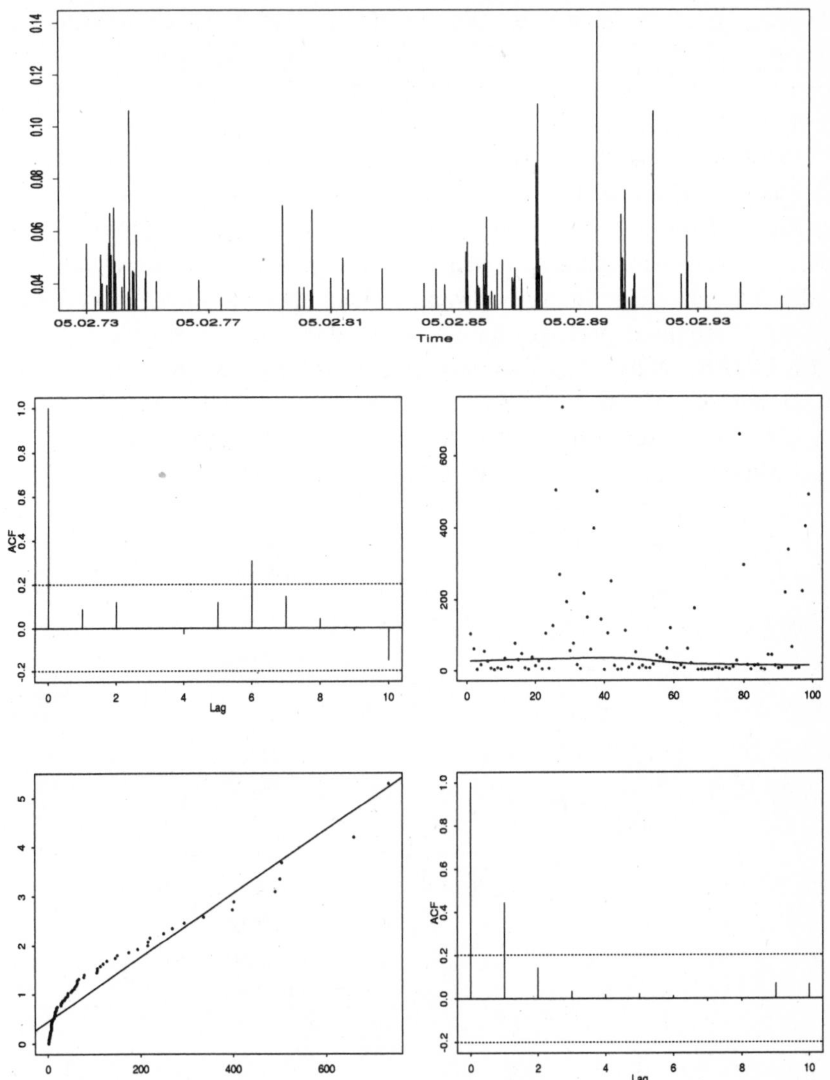

Figure 6.5.11 *Top: the excesses over* $u = 0.0343$ *of the absolute values of the negative BMW log-returns; see Figure 6.2.10. Middle, left: the sample autocorrelations of the excesses. Middle, right: the inter-arrival times of the exceedances and smoothed mean values curve. Bottom, left: QQ-plot of the inter-arrival times against exponential quantiles. Bottom, right: sample autocorrelations of these times. See Example 6.5.9 for further comments.*

Concerning **software**: most of the analyses done in this and other chapters may be performed in any statistical software environment; pick your favourite. We have mainly used **S–Plus**. An introduction to the latter package is for instance to be found in Spector [603]. Venables and Ripley [629] give a nice introduction to modern applied statistics with S–Plus. The programs used for the analyses of the present chapter were written by Alexander McNeil and can be obtained from http://www.math.ethz.ch/~mcneil/software.html. We thank Richard Smith for having made some code available forming the basis of the above programs. Further S-Plus programs providing confidence intervals for parameters in the GPD have been made available by Nader Tajvidi under http://www.math.chalmers.se/~nader/software.html. Various customised packages for extreme value fitting exist. Examples are **XTREMES**, which comes as part of Falk et al. [225], and **ANEX** [18]. In the context of risk management, **RiskMetrics** [543] forms an interesting software environment in which various of the techniques discussed so far, especially concerning quantile (VaR) estimation, are to be found.

7

Time Series Analysis for Heavy–Tailed Processes

In this chapter we present some recent research on time series with large fluctuations, relevant for many financial time series. We approach the problem starting from classical time series analysis presented in such a way that many standard results can also be used in the heavy–tailed case.

In Section 7.1 we give a short introduction to classical time series analysis stressing the basic definitions and properties. This summary clearly cannot replace a monograph on the topic, and so the interested reader who is not familiar with time series analysis should also consult a standard textbook. At the elementary level, Brockwell and Davis [93], and at the more advanced level Brockwell and Davis [92] provide the necessary background. In Section 7.2 linear processes with infinite variance are introduced. In Section 7.3 we concentrate on asymptotic properties of the sample correlations both in the finite and the infinite variance case. In Section 7.4 asymptotic properties of the periodogram under light or heavy tails of the observed time series are studied. Parameter estimation for ARMA processes is the topic of Section 7.5. We conclude with Section 7.6 in which notions such as "heteroscedasticity", "stochastic volatility" and their relationship to the previous sections are explained. We also give a short discussion about ARCH and related processes which are alternative models for time series exhibiting large fluctuations. A more profound analysis of ARCH processes is to be found in Section 8.4.

7.1 A Short Introduction to Classical Time Series Analysis

Classical time series analysis is mainly concerned with the statistical analysis of stationary processes and, in particular, of *linear processes*

$$X_t = \sum_{j=-\infty}^{\infty} \psi_j Z_{t-j}\,, \quad t \in \mathbb{Z}\,, \tag{7.1}$$

with iid real–valued *innovations* or *noise variables* $(Z_t)_{t\in\mathbb{Z}}$ which have mean zero and finite variance σ_Z^2. For reasons of standardisation we also require that $\psi_0 = 1$. In practical situations, one would of course consider so–called causal representations in (7.1), i.e. $\psi_j = 0$ for $j < 0$. For fixed t, the series in (7.1) converges a.s. provided the real–valued coefficients ψ_j satisfy the condition

$$\operatorname{var}(X_t) = \sigma_Z^2 \sum_{j=-\infty}^{\infty} \psi_j^2 < \infty\,. \tag{7.2}$$

The process $(X_t)_{t\in\mathbb{Z}}$ is strictly stationary, i.e. the finite–dimensional distributions of the process are invariant under shifts of the time index. Every strictly stationary finite variance process is also *stationary* (*in the wide sense*), i.e. there exists a constant μ such that $EX_t = \mu$ and EX_tX_{t+h} is only a function of h, not of t; see Appendix A2.1.

Example 7.1.1 (ARMA process)
The most popular linear processes are ARMA(p,q) *processes* (autoregressive–moving average processes of order (p,q)) which are given by the *difference equations*

$$X_t - \phi_1\, X_{t-1} - \cdots - \phi_p\, X_{t-p} \;=\; Z_t + \theta_1\, Z_{t-1} + \cdots + \theta_q\, Z_{t-q}\,, \quad t \in \mathbb{Z}\,. \tag{7.3}$$

The order (p,q) will typically be determined via an order selection criterion. The parameters ϕ_i and θ_i satisfy certain conditions in order to guarantee that equation (7.3) has a solution which can be expressed in the form (7.1). Special cases are the MA(q) or *moving average processes of order* q

$$X_t = Z_t + \theta_1\, Z_{t-1} + \cdots + \theta_q\, Z_{t-q}\,, \quad t \in \mathbb{Z}\,,$$

which only depend on the noise at the instants of time $t-q,\ldots,t$. Under additional assumptions on the parameters ϕ_i (see the discussion in Section 7.5), pure AR(p) or *autoregressive processes of order* p can be interpreted as genuine infinite moving average processes. For example, the AR(1) process can be written as

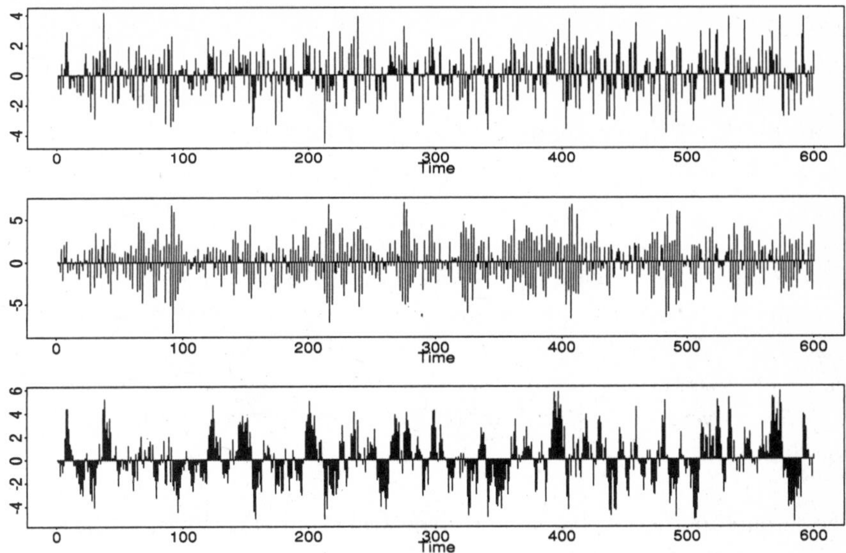

Figure 7.1.2 600 *realisations of iid* $N(0,2)$ *noise* (Z_t) *(top) and of the corresponding* AR(1) *processes* $X_t = -0.8X_{t-1} + Z_t$ *(middle) and* $X_t = 0.8X_{t-1} + Z_t$ *(bottom).*

$$
\begin{aligned}
X_t &= Z_t + \phi_1\, X_{t-1} \\
&= Z_t + \phi_1\, Z_{t-1} + \phi_1^2\, X_{t-2} \\
&= Z_t + \phi_1\, Z_{t-1} + \phi_1^2\, Z_{t-2} + \phi_1^3\, Z_{t-3} + \cdots \\
&= \sum_{j=0}^{\infty} \phi_1^j\, Z_{t-j}\,. \qquad (7.4)
\end{aligned}
$$

The condition $|\phi_1| < 1$ is obviously needed in order to justify the last equality; only in that case does the series (7.2) converge, and then (7.4) converges a.s. In the case of causal ARMA processes, see Section 7.5, the coefficients ψ_j in the representation (7.1) decrease to zero at an exponential rate and $\psi_j = 0$ for $j < 0$. In particular, for MA(q)–processes, $\psi_j = 0$ for $j > q$, and for an AR(1) process, $\psi_j = \phi_1^j$ for $j \geq 0$. □

In order to fit a model of type (7.1), the parameters ψ_j and σ_Z^2 have to be estimated. Roughly speaking, there exist two different approaches. In the *time domain* one studies the dependence structure in the series via the analysis of the *autocovariances* $\gamma(h) = \mathrm{cov}(X_t, X_{t+h})$ or *autocorrelations* $\rho(h) = \mathrm{corr}(X_t, X_{t+h})$:

$$\gamma(h) = \gamma(|h|) = E(X_t\, X_{t+h}) = E(X_0\, X_h)\,, \quad h \in \mathbb{Z}\,, \tag{7.5}$$

$$\rho(h) = \frac{\gamma(h)}{\gamma(0)}\,, \quad h \in \mathbb{Z}\,. \tag{7.6}$$

Example 7.1.3 (Continuation of Example 7.1.1)
Using the uncorrelatedness of the rvs Z_t it is easily seen that for the linear process (7.1) the relations

$$\gamma(h) = \sigma_Z^2 \sum_{j=-\infty}^{\infty} \psi_j \psi_{j+|h|}\,, \quad h \in \mathbb{Z}\,,$$

and

$$\rho(h) = \frac{\sum_{j=-\infty}^{\infty} \psi_j \psi_{j+|h|}}{\sum_{j=-\infty}^{\infty} \psi_j^2}\,, \quad h \in \mathbb{Z}\,,$$

hold. □

The second approach is based on the spectral (Fourier) analysis of the series, and is referred to as the *frequency domain* approach. The basic result for spectral analysis is the so–called *spectral representation* of a stationary (in the wide sense) process.

Theorem 7.1.4 (Spectral representation of stationary process)
Every (complex–valued) stationary mean–zero process $(X_t)_{t\in\mathbb{Z}}$ admits a stochastic integral representation

$$X_t = \int_{(-\pi,\pi]} e^{izt}\, dZ(z)\,, \quad t \in \mathbb{Z}\,, \tag{7.7}$$

with respect to a mean–zero complex–valued process $(Z(z))_{-\pi\le z\le\pi}$ with uncorrelated increments and such that

$$E\left|Z(z_2) - Z(z_1)\right|^2 = F(z_2) - F(z_1)\,, \quad -\pi \le z_1 < z_2 \le \pi\,,$$

for a right–continuous, non–decreasing, bounded function F on $[-\pi,\pi]$ with $F(-\pi) = 0$. □

Remark. The stochastic integrals in (7.7) are *not* (pathwise) Riemann or Riemann–Stieltjes integrals. They are defined as the mean square limit of Riemann–Stieltjes type integrals for step functions (similarly to the definition of an Itô integral). In particular, if (X_t) is a Gaussian process then $(Z(z))$ must also be a Gaussian process with independent increments.

Representation (7.7) of a *complex–valued* stationary process (X_t) is often preferred in the literature although in most cases of practical interest one deals with *real–valued* time series. In this case, (7.7) takes on the form

$$X_t = \int_{[-\pi,\pi]} \cos(\lambda t)\, dZ_1(\lambda) + \int_{[-\pi,\pi]} \sin(\lambda t)\, dZ_2(\lambda)\,, \qquad (7.8)$$

where Z_1 and $-Z_2$ are the real and the imaginary part of Z, respectively. Here we also assumed for simplicity that F is continuous at 0 and π. Moreover, for real–valued (X_t), F is a symmetric function about 0. This means that $F(\lambda) = F(\pi-) - F(-\lambda-)$, $\lambda \in (-\pi, \pi)$. In what follows we stick to the complex–valued representation of (X_t). This is for notational convenience; every statement can be interpreted in terms of real–valued processes by using (7.8). □

We can roughly think of the stochastic integral (7.7) as a sum

$$\sum_{k=1}^{n} \exp\{iz_{k-1}t\} \left(Z\left(z_k\right) - Z\left(z_{k-1}\right)\right) \qquad (7.9)$$

for a partition $((z_{k-1}, z_k])_{k=1,\dots,n}$ of the interval $(-\pi, \pi]$. Hence X_t is approximated by a linear combination of trigonometric functions $\exp\{iz_{k-1}t\}$ with random weights $Z(z_k) - Z(z_{k-1})$. It is clear that the trigonometric function $\exp\{iz_{k-1}t\}$ will have the more influence on the value of X_t the "larger" its random weight. A measure for the magnitude of this weight is the quantity $E|Z(z_k) - Z(z_{k-1})|^2 = F(z_k) - F(z_{k-1})$.

The function $F(x)$, $-\pi \le x \le \pi$, is called *the spectral distribution function* (*spectral df*) of the stationary process (X_t). Note that $F/F(\pi)$ is a probability df. The finite measure which is defined by the spectral df is called the *spectral distribution* of the stationary process (X_t). If F is absolutely continuous with respect to Lebesgue measure we can write it as an integral

$$F(x) = \int_{-\pi}^{x} f(z)\, dz\,, \quad x \in [-\pi, \pi]\,,$$

where the non–negative function $f(z)$ is called the *spectral density* of the process (X_t). Having the approximation (7.9) to the stationary process (7.7) in mind, we see that

$$E\left|Z\left(z_k\right) - Z\left(z_{k-1}\right)\right|^2 = \int_{z_{k-1}}^{z_k} f(z)dz\,.$$

In particular, if $f(z)$ has a large absolute maximum in the (sufficiently small) interval $[z_{k-1}, z_k]$ we may conclude that the summand $e^{iz_{k-1}t}(Z(z_k) - Z(z_{k-1}))$ makes a relevant contribution to the magnitude of the sum (7.9). Since the function $\exp\{iz\}$ has period 2π we have for $l \in \mathbb{Z}$, $\exp\{iz_{k-1}\} = \exp\{iz_{k-1}(1 + 2\pi l/z_{k-1})\}$. This means that, with high probability, the time series X_t assumes large values around the instants of time $1 + 2\pi l/z_{k-1}$.

Thus it exhibits some kind of cyclic behaviour with period $2\pi/z_{k-1}$ and random amplitude determined by $Z(z_k) - Z(z_{k-1})$. What has been said about the largest summand in (7.9) translates analogously to the other summands which clearly have less influence on X_t but they also create periodic subcycles with smaller amplitude.

Theorem 7.1.5 (Spectral density of linear process)
Each mean–zero linear process (7.1) *with finite variance* $\sigma_Z^2 = \mathrm{var}(Z_0)$ *admits a spectral representation* (7.7) *with spectral density*

$$f(z) = \frac{\sigma_Z^2}{2\pi} \left| \sum_{j=-\infty}^{\infty} \psi_j e^{-izj} \right|^2, \quad z \in [-\pi, \pi]. \tag{7.10}$$

□

From (7.10) it is clear that the spectral density of a linear process is basically determined by the function

$$\psi(z) = \sum_{j=-\infty}^{\infty} \psi_j e^{-izj}, \quad z \in [-\pi, \pi],$$

which is called the *transfer function of the linear filter* (ψ_j) or simply the *transfer function*. The function

$$|\psi(z)|^2, \quad z \in [-\pi, \pi], \tag{7.11}$$

is the *power transfer function*. Its estimation is of crucial importance for estimating the spectral density.

Example 7.1.6 (Continuation of Example 7.1.1)
The spectral density of an ARMA(p,q)–process has a particularly simple representation: consider the two polynomials

$$\phi(z) = 1 - \phi_1 z - \cdots - \phi_p z^p, \tag{7.12}$$

$$\theta(z) = 1 + \theta_1 z + \cdots + \theta_q z^q. \tag{7.13}$$

Then the spectral density of this ARMA process has representation

$$f(z) = \frac{\sigma_Z^2}{2\pi} \frac{|\theta(\exp\{-iz\})|^2}{|\phi(\exp\{-iz\})|^2}, \quad z \in [-\pi, \pi].$$

Notice that the constant spectral density $f(z) = \sigma_Z^2/(2\pi)$, $z \in [-\pi, \pi]$, corresponds to iid noise (Z_t). It is precisely for this reason (i.e. constant f) that the latter is often referred to as *white* noise: all frequencies contribute equally.

□

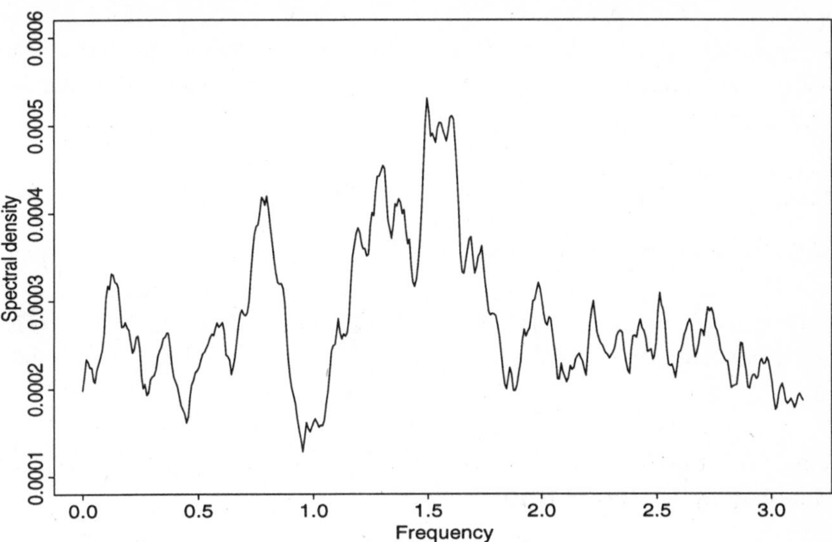

Figure 7.1.7 *Estimated spectral density of the closing log–returns of the Japanese stock index* NIKKEI *(February 22, 1990 – October 8, 1993). There is a peak around the frequency* $z' = 1.65$. *The spectral density is estimated from* $n = 910$ *daily data at the frequencies* $z_l = 2\pi l/n$. *Thus* z' *corresponds to* $l \approx 240$ *which is roughly the annual number of days at which the* NIKKEI *is evaluated. Thus the* NIKKEI *index has a cycle of* $n/l \approx 4$ *days, i.e. one business week, giving a natural explanation for the absolute maximum of the spectral density. On the other hand, there are plenty of submaxima in the spectral density whose interpretation is not immediate. Moreover, notice that the estimated spectral density is very flat, i.e. very close to the spectral density of an iid sequence. A detailed analysis would involve the construction of confidence intervals on the log–scale followed by appropriate testing for peaks.*

Time domain and frequency domain methods are equivalent analytic descriptions of a time series; they are actually two languages based on different tools. The following statement describes the close link between both domains.

Theorem 7.1.8 (Herglotz lemma)
A real–valued function $\gamma(h)$ *defined on the integers is the autocovariance function of a (real–valued) stationary process* $(X_t)_{t\in\mathbb{Z}}$ *if and only if there exists a spectral df* F *on* $[-\pi,\pi]$ *such that*

$$\gamma(h) = \int_{(-\pi,\pi]} \cos(hz)\, dF(z)\,, \quad h \in \mathbb{Z}\,. \tag{7.14}$$

□

As a consequence of (7.14), $(\gamma(h))$ uniquely determines F.

In the following sections we consider statistical methods which have been developed for linear processes in the classical case, i.e. when $EX_0^2 < \infty$. We will see that many methods of classical time series analysis in the time and in the frequency domain can be adapted to the heavy–tailed case, i.e. to infinite variance processes.

Notes and Comments

The basics of classical time series analysis as presented above can be found in any standard monograph on the topic; see for instance Anderson [16], Box and Jenkins [86], Brillinger [91], Brockwell and Davis [92, 93], Chatfield [111], Fuller [246], Hamilton [314], Hannan [315], Priestley [512] and Rosenblatt [555].

7.2 Heavy–Tailed Time Series

In this section we consider the (strictly stationary) linear process (7.1) with iid *innovations* or *noise* (Z_t), but we do not suppose that the variance σ_Z^2 is finite. To be more specific, we assume that $Z(= Z_0)$ has an $s\alpha s$ distribution, i.e. a symmetric α–stable distribution with chf

$$E\exp\{izZ\} = \exp\{-c|z|^\alpha\}\,, \quad z \in \mathbb{R}, \tag{7.15}$$

where $\alpha \in (0,2)$ and $c > 0$. We refer to Section 2.2 for more details on stable laws and their domains of attraction and we recall that Z has an infinite variance. In particular, by the properties of stable distributions we have the following identity in law for each t:

$$X_t \stackrel{d}{=} Z\left(\sum_{j=-\infty}^{\infty} |\psi_j|^\alpha\right)^{1/\alpha}\,. \tag{7.16}$$

This implies that X_t is $s\alpha s$, and one can even show that the finite–dimensional distributions of the process (X_t) are α–stable and therefore the process is stable ; see Section 8.8, in particular Example 8.8.12. We conclude from (7.16) that we need a specific condition on the coefficients ψ_j in order to guarantee the a.s. existence of the series in (7.1). By virtue of the 3–series theorem for a series of independent summands (for instance Petrov [495, 496], Billingsley [68], Theorem 22.8), X_t is well defined if and only if

$$\sum_{j=-\infty}^{\infty} |\psi_j|^\alpha < \infty\,. \tag{7.17}$$

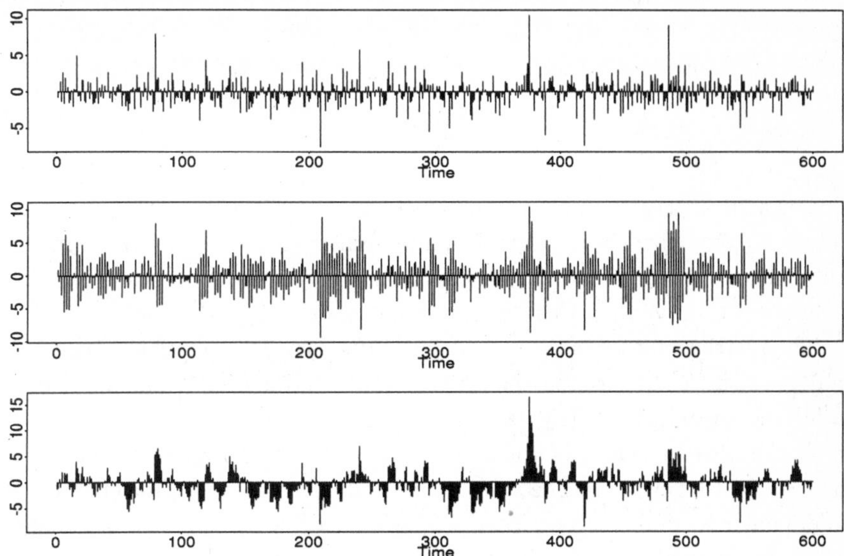

Figure 7.2.1 600 *realisations of iid symmetric* 1.8*–stable noise* (Z_t) *with chf* (7.15) (*top*) *and of the corresponding* AR(1) *processes* $X_t = -0.8X_{t-1} + Z_t$ (*middle*) *and* $X_t = 0.8X_{t-1} + Z_t$ (*bottom*). *Compare also with Figure 7.1.2 where the same time series models with Gaussian noise are considered.*

This condition fits nicely with (7.16). Note that condition (7.17) is satisfied if (X_t) is an ARMA(p, q) process which is given by the difference equation (7.3): as in the classical case one can show that (7.3) has a unique solution which can be expressed as a series (7.1). Moreover, the coefficients ψ_j are the same as in the classical case.

In our considerations below we will not only assume that (7.17) holds, but will require the more stringent condition

$$\sum_{j=-\infty}^{\infty} |\psi_j|^{\delta}\, j < \infty\,, \tag{7.18}$$

for some constant $\delta > 0$ such that

$$\begin{aligned} \delta &= 1 \quad \text{if} \quad \alpha > 1\,, \\ \delta &< \alpha \quad \text{if} \quad \alpha \le 1\,. \end{aligned}$$

The 3–series theorem and (7.18) ensure that the random series X_t converges absolutely with probability 1. Such a condition is necessary in order to guarantee that we can interchange limits, infinite sums and expectations.

Recall from Example 8.8.3 that the iid $s\alpha s$ rvs Z_t have representation $Z_t = A_t^{1/2} N_t$, where (A_t) and (N_t) are independent, A_t are iid $\alpha/2$–stable positive rvs and N_t are iid standard normal. Hence (X_t) has representation

$$X_t = \sum_{j=-\infty}^{\infty} \psi_j A_{t-j}^{1/2} N_{t-j} \,, \quad t \in \mathbb{Z} \,.$$

This can be interpreted as a linear process with Gaussian innovations N_t which are perturbed by potentially large multiplicative factors $A_t^{1/2}$. In this sense, the theory below may be considered as a modification of the classical (Gaussian) theory when there are large fluctuations in the noise.

The restriction to $s\alpha s$ rvs is not at all necessary. It is possible to consider analogous theorems for Z in the domain of attraction of a stable law or even under the assumption $E|Z|^p < \infty$ for some $p > 0$. The symmetry condition on Z can also be relaxed in a natural way, i.e. by appropriate centring of the rvs X_t. However, the theory then becomes even more technical since we would have to make several case studies according to different parameter values. In particular, we would have to introduce a possibly confusing variety of normalising and centring constants. Thus, for the sake of clarity and brevity of presentation, we restrict ourselves to this simple particular case.

Notes and Comments

There is a small but steadily increasing number of articles on heavy–tailed linear processes in the literature. An introduction to the topic can be found in Brockwell and Davis [92], Chapter 13.3, where a justification of the representation of an ARMA process as a linear process (7.1) is given; see also Samorodnitsky and Taqqu [565], Section 7.12. In the following sections we will cite more specific literature at the appropriate places. In a survey paper Klüppelberg and Mikosch [394] contrast results in the classical and in the heavy–tailed situation, both in the time and in the frequency domain.

In Section 7.1 we learnt about the spectral representation of a mean–zero stationary process. According to Theorem 7.1.4 every mean–zero stationary process has a representation via stochastic integrals (7.7) with respect to a process $(Z(z))$ with uncorrelated increments. For Gaussian (X_t) the process $(Z(z))$ is necessarily Gaussian with independent increments. In particular, every linear Gaussian process has a stochastic integral representation (7.7). We might ask whether we can obtain a similar integral representation for an $s\alpha s$ linear process. It is possible to define a stochastic integral (7.7) with respect to an $s\alpha s$ motion; see Section 8.8.3. One can even show that this process is strictly stationary and has some kind of a generalised spectral density. However, the *harmonisable* stable process (X_t) (see Example

8.8.17) does *not* have a representation as linear process (7.1) (see Cambanis and Soltani [103], Rosinski [556]), i.e. for $\alpha < 2$ the class of the *sαs* linear processes and the class of the harmonisable *sαs* processes are disjoint.

One of the main objectives in time series analysis is prediction. In the finite variance case, Hilbert space methods are used to make a linear prediction of a linear process; see for instance Chapter 5 in Brockwell and Davis [92]. In the infinite variance case, the L^2 distance cannot be used as a measure of error between the future value of the time series and its prediction. Another distance measure has to be introduced to build up a prediction theory analogous to the finite variance case. We refer to Cline and Brockwell [125] and to Brockwell and Davis [92], Section 13.3, who give some results in this spirit.

7.3 Estimation of the Autocorrelation Function

As pointed out in Section 7.1, in the time domain one studies the dependence structure of time series via autocovariances or autocorrelations. In this section we will consider some statistical estimation problems in the time domain for linear processes with or without finite variance.

In the classical case ($\sigma_Z^2 < \infty$) natural estimators for $\gamma(h)$, see (7.5), and $\rho(h)$, see (7.6), are given by the *sample autocovariance* $\widetilde{\gamma}_n(h)$ and the *sample autocorrelation* $\widetilde{\rho}_n(h)$:

$$\widetilde{\gamma}_n(h) = \frac{1}{n}\sum_{t=1}^{n-|h|} X_t\, X_{t+|h|}\,, \quad h \in \mathbb{Z}\,, \tag{7.19}$$

$$\widetilde{\rho}_n(h) = \frac{\widetilde{\gamma}_n(h)}{\widetilde{\gamma}_n(0)} = \frac{\sum_{t=1}^{n-|h|} X_t\, X_{t+|h|}}{\sum_{t=1}^{n} X_t^2}\,, \quad h \in \mathbb{Z}\,, \tag{7.20}$$

with the convention that $\widetilde{\gamma}_n(h) = \widetilde{\rho}_n(h) = 0$ for $|h| \ge n$. In the classical case, $\widetilde{\gamma}_n(h)$ and $\widetilde{\rho}_n(h)$ are consistent and asymptotically normal estimators of their deterministic counterparts. We restrict ourselves to autocorrelations.

Theorem 7.3.1 (Asymptotic normality of sample autocorrelations)
Let $(X_t)_{t\in\mathbb{Z}}$ be the mean–zero linear process (7.1). *Suppose that either*

$$\sum_{j=-\infty}^{\infty} |\psi_j| < \infty \quad \textit{and} \quad EZ^4 < \infty\,,$$

or

$$\sum_{j=-\infty}^{\infty} |\psi_j| < \infty\,, \quad \sum_{j=-\infty}^{\infty} \psi_j^2 |j| < \infty \quad \textit{and} \quad \sigma_Z^2 < \infty\,.$$

Then, for each $m \geq 1$,

$$\sqrt{n}\,(\widetilde{\rho}_n(h) - \rho(h))_{h=1,\dots,m} \quad \stackrel{d}{\to} \quad (Y_h)_{h=1,\dots,m}\,,$$

where

$$Y_h = \sum_{j=1}^{\infty} [\rho(h+j) + \rho(h-j) - 2\rho(j)\rho(h)]\; G_j\,, \quad h = 1,\dots,m\,, \qquad (7.21)$$

and $(G_j)_{j\geq 1}$ *are iid* $N(0,1)$ *rvs.* □

In particular, we obtain from Theorem 7.3.1 that, for each fixed $h \geq 1$,

$$\sqrt{n}\,(\widetilde{\rho}_n(h) - \rho(h)) \stackrel{d}{\to} \left(\sum_{j=1}^{\infty} |\rho(h+j) + \rho(h-j) - 2\rho(j)\rho(h)|^2\right)^{1/2} G_1\,.$$

Now suppose that $(Z_t)_{t\in\mathbb{Z}}$ is iid $s\alpha s$ noise with chf $E\exp\{izZ\} = \exp\{-c|z|^\alpha\}$ for some $\alpha < 2$. Since $\sigma_Z^2 = \mathrm{var}(Z) = \infty$ the notions of autocovariance and autocorrelation do not make sense. However, the corresponding sample analogues are obviously well defined finite rvs; see (7.19) and (7.20). Moreover, if the coefficients ψ_j satisfy (7.18) then the quantities

$$\rho(h) = \frac{\sum_{j=-\infty}^{\infty} \psi_j \psi_{j+|h|}}{\sum_{j=-\infty}^{\infty} \psi_j^2}\,, \quad h \in \mathbb{Z}\,,$$

are finite numbers although *they cannot be interpreted as autocorrelations of the process* (X_t). Nevertheless, we will use the same notation $\rho(h)$. Despite the fact that the autocorrelations are not defined, the sample autocorrelations are consistent estimators of the quantities $\rho(h)$ just as in the classical case.

Theorem 7.3.2 (Weak convergence of sample autocorrelations for $s\alpha s$ time series)
Let $(X_t)_{t\in\mathbb{Z}}$ *be the linear process* (7.1). *Suppose that* $(Z_t)_{t\in\mathbb{Z}}$ *is* $s\alpha s$ *noise with common chf* (7.15) *for some* $\alpha < 2$ *and that the coefficients* ψ_j *satisfy* (7.18). *Then, for each* $m \geq 1$,

$$(n/\ln n)^{1/\alpha}\,(\widetilde{\rho}_n(h) - \rho(h))_{h=1,\dots,m} \quad \stackrel{d}{\to} \quad (Y_h)_{h=1,\dots,m}\,,$$

where

$$Y_h = \sum_{j=1}^{\infty} [\rho(h+j) + \rho(h-j) - 2\rho(j)\rho(h)]\; \frac{G_j}{G_0}\,, \quad h = 1,\dots,m\,, \qquad (7.22)$$

and $(G_j)_{j\geq 0}$ *are independent stable rvs,* G_0 *is positive* $\alpha/2$*–stable with chf*

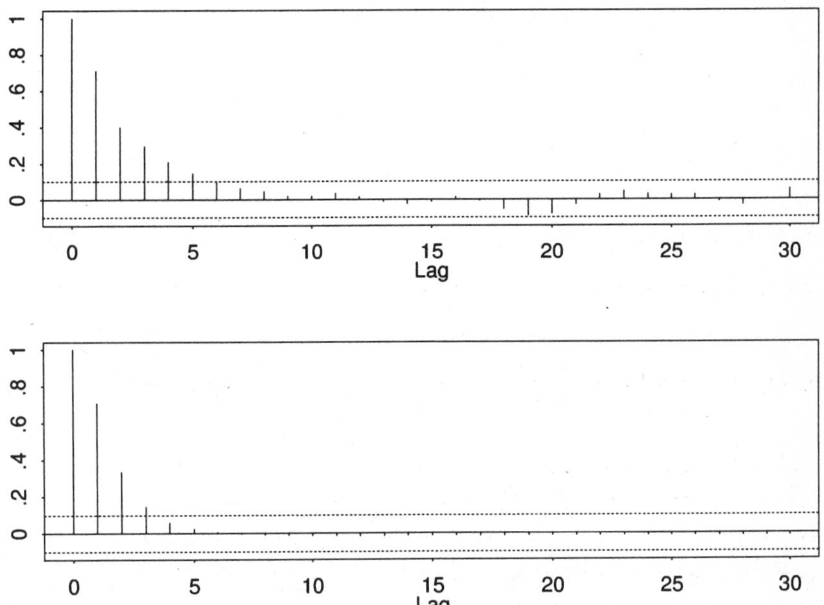

Figure 7.3.3 *The sample autocorrelations $\widetilde{\rho}_n(h)$ at lags $h \le 30$ from $n = 400$ values of the* ARMA(1,1) *process $X_t - 0.5X_{t-1} = Z_t + 0.5Z_t$ with $\rho(1) = 0.71$, $\rho(2) = 0.36$, $\rho(3) = 0.18$, $\rho(4) = 0.09$ etc. In the figure above, the noise is iid $N(0,1)$, in the figure below the noise is iid symmetric Cauchy. The dotted lines indicate the* 95% *asymptotic confidence band for the sample autocorrelations of iid $N(0,1)$ rvs.*

$$
\begin{aligned}
& E\exp\{izG_0\} && (7.23)\\
&= \exp\left\{-\Gamma(1-\alpha/2)\cos(\pi\alpha/4)|z|^{\alpha/2}\left(1 - i\,\mathrm{sign}(z)\tan(\pi\alpha/4)\right)\right\},
\end{aligned}
$$

and $(G_j)_{j\ge 1}$ are iid sαs rvs with chf

$$
E\exp\{izG_1\} = \begin{cases} \exp\{-\Gamma(1-\alpha)\cos(\pi\alpha/2)|z|^{\alpha}\} & \text{if } \alpha \neq 1\,,\\ \exp\{-\pi|z|/2\} & \text{if } \alpha = 1\,. \end{cases} \tag{7.24}
$$

□

Remarks. 1) Theorem 7.3.2 can be compared with Theorem 7.3.1: if we specialise Theorem 7.3.2 to one component then we obtain, for each fixed $h \ge 1$,

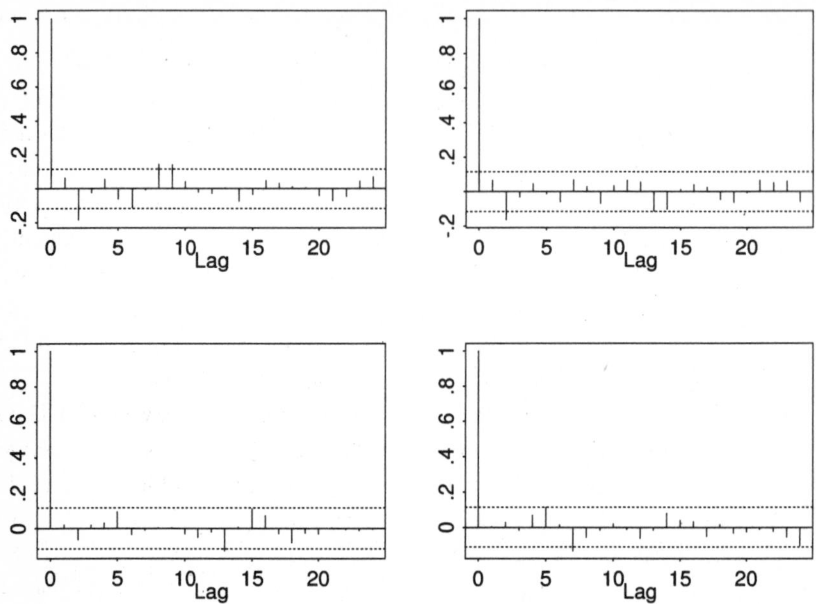

Figure 7.3.4 *The sample autocorrelations $\widetilde{\rho}_n(h)$ at lags $h \le 25$ of the closing log–returns of the* NIKKEI *index (February 22, 1990 – October 8, 1993). The autocorrelations of $X_1, \ldots, X_{300}$ (top left), of $X_{200}, \ldots, X_{500}$ (top right), of $X_{400}, \ldots, X_{700}$ (bottom left) and of $X_{600}, \ldots, X_{910}$ (bottom right) are given. The dotted lines indicate the* 95% *asymptotic confidence band for the sample autocorrelations of iid $N(0,1)$ rvs. Notice that only the sample autocorrelation at lag 2 is significantly different from zero in the top figures. This significant value may, however, be spurious.*

$$\begin{aligned}(n/\ln n)^{1/\alpha}\left(\widetilde{\rho}_n(h)-\rho(h)\right) \\ \xrightarrow{d}\ \left(\sum_{j=1}^{\infty}\left|\rho(h+j)+\rho(h-j)-2\rho(j)\rho(h)\right|^{\alpha}\right)^{1/\alpha}\frac{G_1}{G_0}\,. \qquad (7.25)\end{aligned}$$

In particular,

$$\widetilde{\rho}_n(h)=\rho(h)+O_P\left((n/\ln n)^{-1/\alpha}\right),$$

which compares favourably with the slower rate $O_P(n^{-1/2})$ in the classical case. We can interpret this faster rate of convergence in the sense that *large fluctuations in the innovations stabilise the estimation of $\rho(h)$*. The sample autocorrelation $\widetilde{\rho}_n(h)$ is a *studentised* or *self–normalised* version of the sample autocovariance. Self–normalisation has the effect that we replace the original observations $X_t, t = 1, \ldots, n$, by $X_t/(\sum_{s=1}^n X_s^2)^{1/2}, t = 1, \ldots, n$. In contrast to the former rvs the latter are bounded quantities and have a finite variance.

2) A close look at the proof of this result shows that the rv G_0 in the denominator is (up to a multiplicative constant) nothing but the limit of $n^{-2/\alpha} \sum_{t=1}^n X_t^2$. In the classical case the latter corresponds to the quantity $n^{-1} \sum_{t=1}^n X_t^2$ which converges in distribution to a constant and therefore self–normalisation is not necessary to guarantee consistent estimation of the sample autocovariances. However, it is common practice to estimate the autocorrelations; they are standardised autocovariances and therefore the dependence structure of different time series is made comparable from one series to another. Theorem 7.3.2 explains to some extent what happens to the estimation of $\rho(h)$ when there are large fluctuations in the noise.
A glance at the proofs in the finite variance case (see Brockwell and Davis [92], Chapter 7), and in the infinite variance case (Davis and Resnick [160, 161, 162]) explains why the structure of the limit variables Y_h in (7.21) and (7.22) is so similar. These limits are "built up" by the rvs G_h which are the limits of the normalised sample autocovariances $n^{-1} \sum_{t=1}^{n-|h|} Z_t Z_{t+|h|}$ of the iid noise (Z_t). Notice that, for $m > 1$, the normalised sample autocovariances $(n^{-1/2} \sum_{t=1}^{n-h} Z_t Z_{t+h})_{h=1,\ldots,m}$ converge in distribution to iid Gaussian $(G_h)_{h=1,\ldots,m}$, provided $\sigma_Z^2 < \infty$. If Z is $s\alpha s$, the random vector $((n \ln n)^{-1/\alpha} \sum_{t=1}^{n-h} Z_t Z_{t+h})_{h=1,\ldots,m}$ converges in distribution to iid $s\alpha s$ $(G_h)_{h=1,\ldots,m}$.
3) The distribution of G_1/G_0 in (7.25) is quite unfamiliar. It is given in Brockwell and Davis [92], formula (13.3.17), and can be expressed via some special functions; see Klüppelberg and Mikosch [397]. In particular, $E|G_1/G_0|^\delta = \infty$ or $< \infty$ according as $\delta > \alpha$ or $\delta < \alpha$. Quantiles of this distribution can be found by Monte–Carlo simulation of G_1/G_0. The limit distribution depends on α which has to be estimated; see Section 6.4.2. □

Notes and Comments

Theorem 7.3.1 and related results can be found in Brockwell and Davis [92], Chapter 7. Theorem 7.3.2 and more asymptotic theory for the sample autocovariance and the sample autocorrelation in the heavy–tailed situation are given in Davis and Resnick [160, 161, 162]; see also Brockwell and Davis [92], Chapter 13.3. We note that Davis and Resnick also treat the joint asymptotic behaviour of the (properly centred and normalised) sample autocovariances $(\widetilde{\gamma}_n(0), \widetilde{\gamma}_n(1), \ldots, \widetilde{\gamma}_n(m))$. In that case, various subcases must be considered:

a) $EZ^4 < \infty$. Then the limit vector is jointly Gaussian.

b) $EZ^4 = \infty$, $\sigma_Z^2 < \infty$ and Z^2 has a regularly varying tail. If (X_t) is a genuine linear process (i.e. not an iid sequence) then the limit vector consists of a stable rv multiplied by a constant vector; see Theorem 2.2 in Davis and Resnick [162]. This result can be refined for iid noise (Z_t) in which case the limit vector consists of a stable rv in the first component and of iid Gaussian variables in the remaining components.

c) Z is in the domain of attraction of an α–stable law, $\alpha < 2$. The limit vector is positive $\alpha/2$–stable in the first component and α–stable in the remaining components.

Case b) has recently attracted some attention in the econometrics literature, where the results by Davis and Resnick were partly reproved. The interest is based on the empirical evidence that some financial time series have regularly varying tails with index between 2 and 4; see for instance Longin [428] or Loretan and Philips [429]. The paper by Müller, Dacorogna and Pictet [471] contains a detailed analysis confirming fat–tailedness in the foreign exchange market and the inter–bank market of cash interest rates.

The similarity of the results in Theorems 7.3.1 and 7.3.2 is a consequence of the linearity of the process (X_t). Recent results by Davis and Resnick [164] and Resnick [533] for a bilinear process $X_t = cX_{t-1}Z_{t-1} + Z_t$ with infinite variance innovations Z_t show that the sample autocorrelations converge to non–degenerate limit laws.

7.4 Estimation of the Power Transfer Function

In this section we concentrate on the estimation of the power transfer function $|\psi(z)|^2$, see (7.11), of the linear process (7.1). We again commence with the classical finite variance case. As we know from Theorem 7.1.5, the spectral density of a linear process is given by

$$f(z) = \frac{\sigma_Z^2}{2\pi}\left|\sum_{j=-\infty}^{\infty}\psi_j e^{-izj}\right|^2 = \frac{\sigma_Z^2}{2\pi}\,|\psi(z)|^2\,, \quad z \in [-\pi,\pi]\,.$$

Therefore it is important to estimate the power transfer function $|\psi(z)|^2$. Its classical estimator is the *periodogram*

$$I_{n,X}(z) = \frac{1}{n}\left|\sum_{t=1}^{n} X_t e^{-izt}\right|^2\,, \quad z \in [-\pi,\pi]\,. \tag{7.26}$$

It is a natural estimator because it is constructed as empirical analogue to the power transfer function. Indeed, notice that

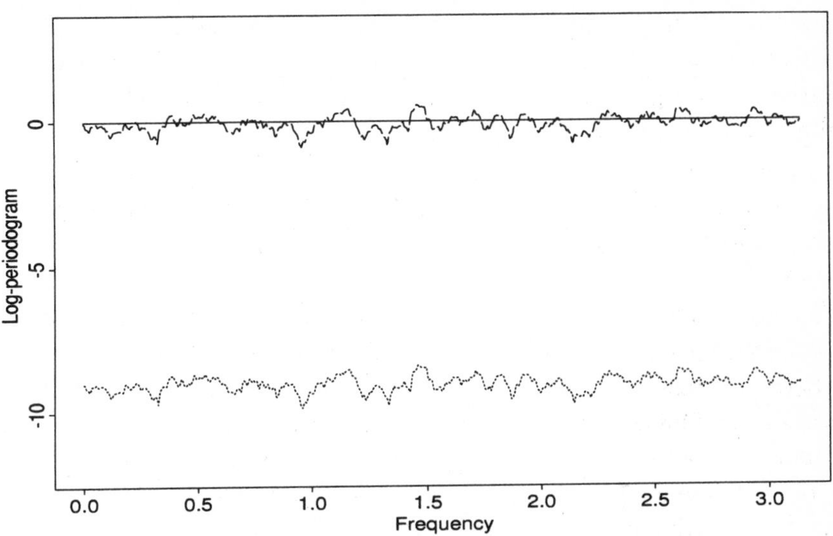

Figure 7.4.1 *The log–transformed smoothed periodogram* (7.26) *for* $n = 1\,864$ *daily log–returns (closing data) of the German stock index* DAX (*September 20, 1988 – August 24, 1995*) (*bottom*) *and the corresponding log–transformed smoothed self–normalised version of the periodogram* (*top*). *Both lines are almost constant indicating that the* DAX *log–returns perform very much like an iid noise sequence.*

$$\sigma_Z^2 \left|\psi\left(z\right)\right|^2 = \sum_{h=-\infty}^{\infty} \gamma(h)e^{-izh}$$

and that

$$I_{n,X}(z) = \sum_{|h|<n} \widetilde{\gamma}_n(h)e^{-izh}\,.$$

The periodogram is *not* a consistent estimator of the power transfer function, but under mild conditions we are not far away from consistency.

Theorem 7.4.2 (Limit distribution of the periodogram, classical case) *Let* $(X_t)_{t\in\mathbb{Z}}$ *be the linear process* (7.1). *Suppose that* $EZ = 0$, $\sigma_Z^2 < \infty$,

$$\sum_{j=-\infty}^{\infty} |\psi_j| < \infty \quad and \quad |\psi(z)|^2 > 0\,, \quad z \in [-\pi, \pi]\,.$$

Then, for any frequencies $0 < z_1 < \cdots < z_m < \pi$,

$$\left(I_{n,X}\left(z_i\right)\right)_{i=1,\ldots,m} \quad \stackrel{d}{\rightarrow} \quad \frac{\sigma_Z^2}{2}\left(\left|\psi(z_i)\right|^2\left(\alpha_i^2+\beta_i^2\right)\right)_{i=1,\ldots,m}\,,$$

where $\alpha_1, \beta_1, \ldots, \alpha_m, \beta_m$ *are iid* $N(0,1)$ *rvs.*

Note that $(\alpha_i^2 + \beta_i^2)/2$, as the sum of two independent χ^2 rvs, is a standard exponential rv. This is the form in which this classical result is usually formulated.

Sketch of the proof. As a first step one can show that

$$I_{n,X}(z) = |\psi(z)|^2 \, I_{n,Z}(z) + o_P(1)\,, \quad n \to \infty\,, \tag{7.27}$$

where $I_{n,Z}(z)$ denotes the periodogram of the iid sequence (Z_t). Hence the asymptotic behaviour of $I_{n,X}(z)$ depends only on the periodogram of an iid sequence. We write

$$\begin{aligned} I_{n,Z}(z) & \\ &= \frac{1}{2}\,\alpha_n^2(z) + \frac{1}{2}\,\beta_n^2(z) \\ &= \frac{1}{2}\left(\left(\frac{2}{n}\right)^{1/2}\sum_{t=1}^{n} Z_t\cos(zt)\right)^2 + \frac{1}{2}\left(\left(\frac{2}{n}\right)^{1/2}\sum_{t=1}^{n} Z_t\sin(zt)\right)^2 . \end{aligned}$$

Thus we have to study the weak convergence of the vector $(\alpha_n(z), \beta_n(z))$ which is not too difficult. The normalisation $\sqrt{n}$ in both terms $\alpha_n(z)$ and $\beta_n(z)$ at once suggests applying a two–dimensional CLT for non–iid summands. This argument indeed works (also for a finite number of periodogram ordinates) and proves that

$$(\alpha_n(z), \beta_n(z)) \quad \xrightarrow{d} \quad \sigma_Z\,(\alpha_1, \beta_1)\,,$$

for iid $N(0,1)$ rvs α_1 and β_1. An application of the continuous mapping theorem concludes the proof. □

Now suppose that (Z_t) is a sequence of iid *sαs* rvs with chf (7.15) for some $\alpha < 2$ and $c > 0$. We learnt from Theorem 7.1.4 that the notions of a spectral distribution and of a spectral density are very much related to stationary processes in the wide sense, i.e. processes with finite covariance function. Thus, for linear processes (X_t) with iid *sαs* noise, the notion of a spectral density does not make sense. However, the power transfer function of such a process is well defined under quite general conditions. And one can even show that a result similar to Theorem 7.4.2 is valid: we browse through the arguments of the proof of Theorem 7.4.2. First, (7.27) remains true if we redefine the periodogram in a suitable way:

$$I_{n,X}(z) = n^{-2/\alpha}\left|\sum_{t=1}^{n} X_t e^{-izt}\right|^2, \quad z \in [-\pi, \pi]\,. \tag{7.28}$$

Whenever we work with *sαs* time series we will assume that we deal with this modification of the periodogram, i.e. we suppress its dependence on α in the notation. Similar arguments as above show that we have to study the joint convergence of

$$(\alpha_n(z)\,,\beta_n(z)) = \left(n^{-1/\alpha}\sum_{t=1}^{n} Z_t\cos(zt)\,,\, n^{-1/\alpha}\sum_{t=1}^{n} Z_t\sin(zt)\right),$$

and the normalisation $n^{1/\alpha}$ already suggests that we have to apply a CLT for rvs in the domain of normal attraction of an α–stable law; see Section 2.2. However, $\alpha_n(z)$ and $\beta_n(z)$ are sums of non–iid rvs, so one has to modify the theory for weighted sums of iid *sαs* rvs. Finally, one arrives at the following result:

Theorem 7.4.3 (Limit distribution of the periodogram, *sαs* case)
Let $(X_t)_{t\in\mathbb{Z}}$ *be the linear process* (7.1) *with iid sαs noise* $(Z_t)_{t\in\mathbb{Z}}$ *and common chf* (7.15) *for some* $\alpha < 2$. *Suppose* (7.18) *and* $|\psi(z)|^2 > 0\,, z\in[-\pi,\pi]$. *Then, for any frequencies* $0 < z_1 < \cdots < z_m < \pi$,

$$\left(I_{n,X}\left(z_i\right)\right)_{i=1,\ldots,m} \stackrel{d}{\to} \left(|\psi(z_i)|^2\left(\alpha^2\left(z_i\right)+\beta^2\left(z_i\right)\right)\right)_{i=1,\ldots,m},$$

where $(\alpha(z_1),\beta(z_1),\ldots,\alpha(z_m),\beta(z_m))$ *is an sαs random vector in* $\mathbb{R}^{2m}$. *Moreover, there do not exist any two components in this vector which are independent.* □

Remark. For a precise formulation of this result we would need rather sophisticated arguments. The definition of an *sαs* random vector via its chf is given in Section 8.8.1. If we compare Theorems 7.4.3 and 7.4.2 there are some similarities in structure, but we also see significant differences. Concerning the latter, the components in the limit vector are dependent and their distribution depends on the frequencies z_k. A more detailed analysis of Theorem 7.4.3 shows that the limit distribution of $I_{n,Z}(z)$ is identical for all frequencies z which are irrational multiples of π. In many situations the components $(\alpha^2(z_k)+\beta^2(z_k))_{k=1,\ldots,m}$ of the limit vector are exchangeable in the sense that they can be embedded in a sequence of conditionally independent rvs. On the other hand, it seems difficult to apply such a result since it depends very much on the form of the frequencies and creates a non–tractable form of dependence in the limit. □

An application of Theorem 7.4.3 requires the knowledge of α which appears in the normalisation of (7.28). One way to overcome this problem is by *self–normalisation* or *studentisation.* But notice that α still appears in the limit distribution. This technique was already mentioned in the context of sample autocovariances and sample autocorrelations. Results similar to Theorem 7.4.3 hold true for the *self–normalised periodogram*

$$\widetilde{I}_{n,X}(z) = \frac{\left|\sum_{t=1}^{n} X_t \exp\{-izt\}\right|^2}{\sum_{t=1}^{n} X_t^2}, \quad z \in [-\pi, \pi] .$$

The motivation for such an estimator is given by the fact that

$$n^{-2/\alpha} \sum_{t=1}^{n} X_t^2 = n^{-2/\alpha} \Psi^2 \sum_{t=1}^{n} Z_t^2 \left(1 + o_P(1)\right) \stackrel{d}{\to} \Psi^2 G_0 ,$$

where G_0 is a positive $\alpha/2$–stable rv and $\Psi^2 = \sum_{j=-\infty}^{\infty} \psi_j^2$. This means that $\sum_{t=1}^{n} X_t^2$ and $n^{2/\alpha}$ are roughly of the same order. It can be shown both in the classical and in the *sαs* case that the following holds true for any frequency $z \in (0, \pi)$:

$$\widetilde{I}_{n,X}(z) \stackrel{d}{\to} \frac{|\psi(z)|^2}{\Psi^2} \frac{\alpha^2(z) + \beta^2(z)}{G_0} = \frac{|\psi(z)|^2}{\Psi^2} \left(1 + T(z)\right) .$$

In the classical case, $G_0 = 2$ and $(\alpha^2(z) + \beta^2(z))/2$ is a standard exponential rv, thus independent of the frequency z. In the *sαs* situation, $\alpha(z)$, $\beta(z)$ and G_0 are dependent and $(\alpha(z), \beta(z))$ has an *sαs* distribution in $\mathbb{R}^2$ depending on z. In spite of these differences, in both cases

$$P(1 + T(z) > x) \leq \exp\{-cx\}, \quad x > 0, \tag{7.29}$$

for a constant $c > 0$ independent of the distribution of Z and

$$ET(z) = 0, \quad \operatorname{cov}\left(T(z), T(z')\right) = 0, \quad 0 < z \neq z' < \pi . \tag{7.30}$$

These statements are trivial in the classical situation. They show how close the classical and the self–normalised *sαs* case actually are. Self–normalisation under the condition of a finite variance is clearly not necessary for convergence because $n^{-1} \sum_{t=1}^{n} X_t^2$ satisfies a fairly general LLN.

In the classical situation, the properties (7.29) and (7.30) suggest estimating the power transfer function via a smoothed version of the periodogram. The same methods also work in the *sαs* situation although it seems difficult to derive good confidence intervals in that case. In order to illustrate the method we restrict ourselves to a simple *discrete weighted average estimator;* similar results can be obtained for kernel type smoothers.

We introduce the following class of weights: let $(w_n(k))_{|k| \leq m}$ be non–negative real numbers such that

$$w_n(k) = w_n(-k), \quad \sum_{|k| \leq m} w_n(k) = 1, \quad \sum_{|k| \leq m} w_n^2(k) \to 0,$$

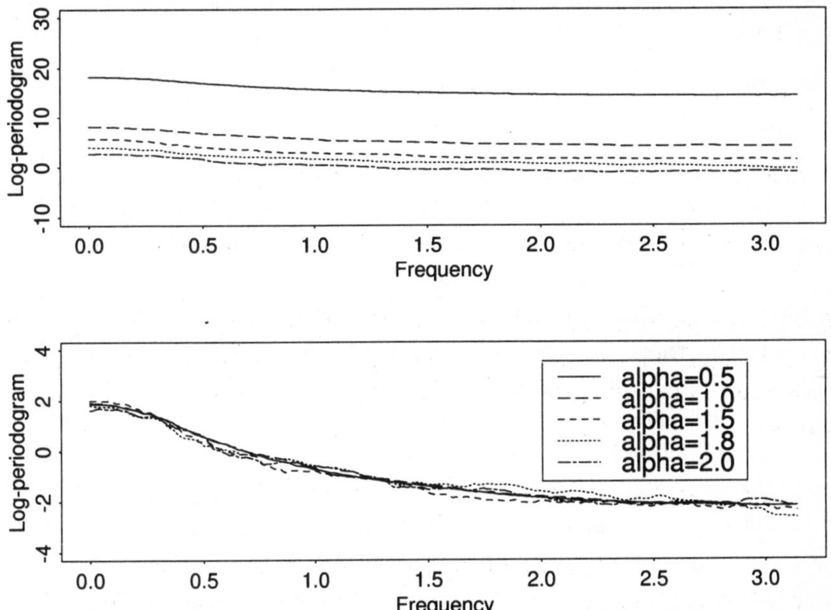

Figure 7.4.4 *The figure above shows the logarithm of the smoothed classical periodogram* (7.26) *for* $n = 500$ *realisations of the* AR(1) *process* $X_t = 0.8X_{t-1} + Z_t$. *The iid noise* (Z_t) *has either a common standard normal distribution* $(\alpha = 2)$ *or a common sαs distribution* $(\alpha \in \{1.8, 1.5, 1.0, 0.5\})$. *The estimated curves are almost parallel indicating that the normalisation in the cases* $\alpha \neq 2$ *is not correctly chosen. The figure below shows the logarithm of the smoothed, self-normalised periodogram as used in Proposition 7.4.5 for the same realisations of the* AR(1) *process with weights* (7.31), $m = 15$.

where we also require that $m = m(n) \to \infty$ and $m/n \to 0$ as $n \to \infty$. The simplest example of such weights is given by

$$w_n(k) = \frac{1}{2m+1}, \quad m = [n^\gamma], \quad \text{for some } \gamma \in (0,1), \tag{7.31}$$

where $[\cdot]$ denotes the integer part.

Proposition 7.4.5 *Suppose that the iid noise rvs* Z_t *are either sαs for some* $\alpha < 2$ *or that* $\operatorname{var}(Z) = \sigma_Z^2 < \infty$ *and* $EZ = 0$. *Then, for* $0 < z < \pi$,

$$\sum_{|k| \le m} w_n(k)\, \widetilde{I}_{n,X}(z_k) \quad \stackrel{P}{\to} \quad \frac{|\psi(z)|^2}{\Psi^2}, \quad n \to \infty.$$

Here $z_0 = 2\pi l/n$ for some $l \in \mathbb{N}$ is the Fourier frequency closest to z from the left, and $z_k = 2\pi(l+k)/n$ for the same l and $|k| \le m$. □

Notes and Comments

The classical asymptotic theory for the periodogram (Theorem 7.4.2) can for instance be found in Brockwell and Davis [92], Section 10.3, or in Priestley [512]. The asymptotic independence of the periodogram ordinates at different frequencies gives an intuitive explanation for the fact that one cannot improve the pointwise weak convergence (i.e. convergence at fixed frequencies) towards a FCLT. However, the asymptotic independence suggests considering integrated versions of the periodogram as analogues to the empirical df. A rule of thumb is that *any asymptotic result which holds for the empirical df of iid rvs has some analogue in the language of the integrated periodogram.* There exist many results which show the close relationship between the theory for the integrated periodogram and for the empirical df; see for instance Dzhaparidze [196], Grenander and Rosenblatt [285] or Priestley [512]. For this reason the integrated periodogram is sometimes also called *the empirical spectral distribution function.* FCLTs for the integrated periodogram can be used for constructing goodness–of–fit tests or for detecting a changepoint of the spectral distribution function via Kolmogorov–Smirnov or Cramér–von Mises type statistics. A recent account of the asymptotic theory in the classical case has been given by Anderson [17]; see also Bartlett [44, 45] and Grenander and Rosenblatt [285] as classical references, Klüppelberg and Mikosch [396] (for changepoint detection via the integrated periodogram with a limiting Kiefer–Müller process, see also the literature cited therein), Dahlhaus [151] and Mikosch and Norvaiša [462] (for uniform CLTs and LLNs of the integrated periodogram indexed by classes of square–integrable functions). In the α–stable case Klüppelberg and Mikosch [395] show FCLTs for the integrated periodogram with limiting processes which can be considered as α–stable analogues of the Brownian bridge; see Example 8.8.14. For long memory processes, Kokoszka and Mikosch [402] show analogous results both in the classical situation and in the infinite variance case. We also refer to the survey papers by Klüppelberg and Mikosch [397] and Mikosch [457] .

The theory for the estimation of the power transfer function in the *sαs* and more general cases has been developed in Klüppelberg and Mikosch [390, 391]. The exact formulation of Theorem 7.4.3 is given in [390], and related results for the self–normalised version of the periodogram are contained in [391]; see also Bhansali [66]. It should be mentioned that the effect of "robustification" of periodogram estimators via self–normalisation has been observed in the time series context for a long time; see for instance Anderson [17] or Priestley [512], Chapter 6.

Results for the smoothed periodogram (such as Proposition 7.4.5) in the classical setting can be found for instance in Brockwell and Davis [92], Chapter 10, in Grenander and Rosenblatt [285] or in Priestley [512], Chapter 6. In the $s\alpha s$ situation, Proposition 7.4.5 is formulated in Klüppelberg and Mikosch [391].

The estimation of the pseudo–spectral density of harmonisable stable processes (see the Notes and Comments at the end of Section 7.2) has been considered for instance in Masry and Cambanis [446] and Hsing [343].

7.5 Parameter Estimation for ARMA Processes

From Section 7.1 we recall the notion of an ARMA(p,q) process, which is a linear process given by the difference equations

$$X_t - \phi_1 X_{t-1} - \cdots - \phi_p X_{t-p} = Z_t + \theta_1 Z_{t-1} + \cdots + \theta_q Z_{t-q} , \quad t \in \mathbb{Z} , \tag{7.32}$$

for a fixed order (p,q). We write

$$\beta = (\phi_1, \ldots, \phi_p, \theta_1, \ldots, \theta_q)^T$$

and use the symbol β_0 for the *true, but unknown* parameter vector. The observed time series $X_1, \ldots, X_n$ is supposed to come from the model (7.32) with $\beta = \beta_0$.

In the classical setting, there exist three basic techniques for estimating β_0: *Gaussian maximum likelihood, least squares* and *Whittle estimation.* The latter two methods provide approximations to the Gaussian maximum likelihood estimator, the least squares estimator in the time domain and the Whittle estimator in the frequency domain. They can be shown to be asymptotically equivalent in the sense that they yield strongly consistent and asymptotically normal estimators of β_0. Moreover, the Whittle estimator, when restricted to pure AR processes, is the celebrated *Yule–Walker estimator.* The Yule–Walker estimator is introduced as moment estimator: the parameter vector is chosen such that the theoretical and empirical autorcorrelations of an AR process match. It has been extended to ARMA processes and is commonly used as preliminary estimator for more advanced optimisation procedures; see Brockwell and Davis [92], Sections 8.1–8.5. In the following we restrict ourselves to Whittle estimation which has been important within asymptotic estimation theory since its discovery; see Whittle [641]. It works

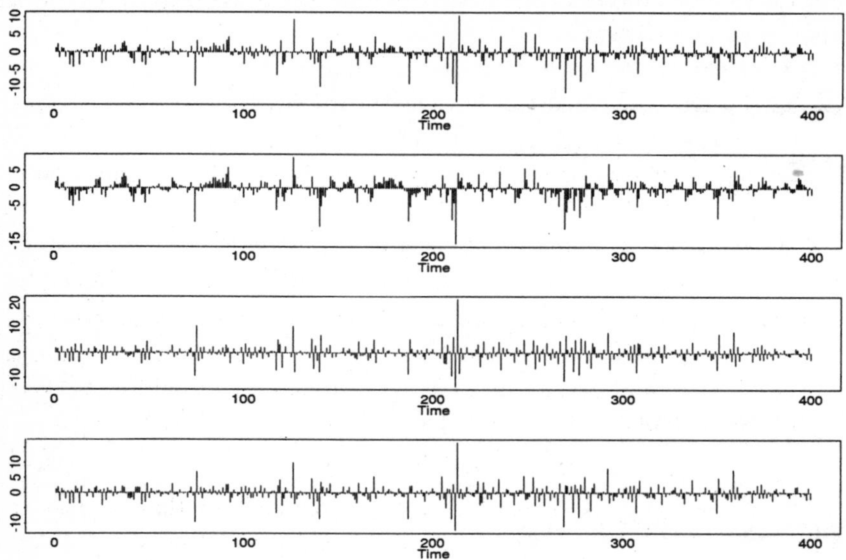

Figure 7.5.1 400 *realisations of iid* s1.5s *noise (top). They are used to generate the* ARMA *processes* $X_t - 0.4X_{t-1} = Z_t$, $X_t = Z_t + 0.8Z_{t-1}$ *and* $X_t - 0.4\ X_{t-1} = Z_t + 0.8\ Z_{t-1}$ *(from top to bottom).*

for processes under very different conditions such as long– or short–range dependence, heavy or light tails. It can easily be calculated by means of the fast Fourier transform algorithm; see Brockwell and Davis [92], Section 10.7. First we formulate the classical finite variance result and then turn to the $s\alpha s$ case.

For a given parameter vector β the coefficients $\psi_j = \psi_j(\beta)$, $j \geq 0$, of the linear process define the corresponding power transfer function

$$|\psi(z,\beta)|^2 = \left|\sum_{j=0}^{\infty} \psi_j(\beta) e^{-izj}\right|^2, \quad z \in [-\pi,\pi].$$

We suppose that β_0 belongs to the natural parameter space

$$\begin{aligned} \mathcal{C} \quad = \quad & \{\beta \in \mathbb{R}^{p+q} : \phi_p \neq 0, \theta_q \neq 0, \\ & \quad \phi(z) \text{ and } \theta(z) \text{ have no common zeros,} \\ & \quad \phi(z)\,\theta(z) \neq 0 \text{ for } |z| \leq 1\}, \end{aligned}$$

where the polynomials $\phi(z)$ and $\theta(z)$ are defined in (7.12) and (7.13), respectively. In that case the difference equations (7.32) have a unique solution which is a linear process (7.1) whose coefficients $\psi_j(\beta)$ decrease to zero at an exponential rate. Define

$$\widetilde{\sigma}_n^2(\beta) = \int_{-\pi}^{\pi} \frac{I_{n,X}(z)}{|\psi(z,\beta)|^2}\, dz\,.$$

The Whittle estimator is motivated by the observation that the function

$$g(\beta) = \int_{-\pi}^{\pi} \frac{|\psi(z,\beta_0)|^2}{|\psi(z,\beta)|^2}\, dz$$

assumes its absolute minimum on the closure of C precisely at the point $\beta = \beta_0$; see Brockwell and Davis [92], Proposition 10.8.1. Thus, if we replaced $|\psi(z,\beta_0)|^2$ by an appropriate periodogram estimator we would expect that the *Whittle estimator*

$$\widetilde{\beta}_n = \arg\min_{\beta\in C} \widetilde{\sigma}_n^2(\beta) \tag{7.33}$$

should be close to β_0. Indeed, this approach works. Note that we may replace the ordinary periodogram $I_{n,X}$, see (7.26), in $\widetilde{\sigma}_n^2(\beta)$ by any other renormalised version of it: the value of $\widetilde{\beta}_n$ is not influenced. However, if we are interested in the limiting behaviour of $\widetilde{\sigma}_n^2(\widetilde{\beta}_n)$ then the normalisation of $I_{n,X}$ is important.

For practical purposes, the following version of the Whittle estimator (7.33) is more appropriate: define the *discretised version* of $\widetilde{\sigma}_n^2(\beta)$ as

$$\widehat{\sigma}_n^2(\beta) = \frac{2\pi}{n} \sum_{z_t\in(-\pi,\pi]} \frac{I_{n,X}(z_t)}{|\psi(z_t,\beta)|^2}\,,$$

where $z_t = 2\pi t/n$ denote the Fourier frequencies and

$$\widehat{\beta}_n = \arg\min_{\beta\in C} \widehat{\sigma}_n^2(\beta)\,. \tag{7.34}$$

Both versions of Whittle's estimator have the same asymptotic properties:

Theorem 7.5.2 (Asymptotic normality of Whittle estimator, classical case) *Suppose $(X_t)_{t\in\mathbb{Z}}$ is the* ARMA(p,q) *process given by* (7.32) *with $EZ=0$ and $\sigma_Z^2 = \mathrm{var}(Z) < \infty$. Then*

$$\sqrt{n}\,(\widetilde{\beta}_n - \beta_0) \quad \xrightarrow{d} \quad 8\pi^2 W^{-1}(\beta_0) \sum_{j=0}^{\infty} b_j G_j \tag{7.35}$$

in $\mathbb{R}^{p+q}$, where (G_j) are iid $N(0,1)$ rvs, $W^{-1}(\beta_0)$ is the inverse of the matrix

$$W(\beta_0) \quad = \quad \int_{-\pi}^{\pi} \left[\frac{\partial \ln|\psi(z,\beta_0)|^2}{\partial\beta}\right] \left[\frac{\partial \ln|\psi(z,\beta_0)|^2}{\partial\beta}\right]^T dz\,, \tag{7.36}$$

and

$$b_j = \frac{1}{2\pi} \int_{-\pi}^{\pi} e^{-ijz} \left|\psi(z, \beta_0)\right|^2 \frac{\partial \left|\psi(z, \beta_0)\right|^{-2}}{\partial \beta} \, dz . \tag{7.37}$$

Relation (7.35) *remains valid with* $\widetilde{\beta}_n$ *replaced by* $\widehat{\beta}_n$. *Moreover,*

$$\widetilde{\sigma}_n^2(\widetilde{\beta}_n) \overset{\text{a.s.}}{\to} 2\pi\sigma_Z^2 , \qquad \widehat{\sigma}_n^2(\widehat{\beta}_n) \overset{\text{a.s.}}{\to} 2\pi\sigma_Z^2 . \qquad \square$$

Remarks. 1) The limit vector in (7.35) is mean–zero Gaussian with covariance matrix $4\pi W^{-1}(\beta_0)$; see Brockwell and Davis [92], Theorem 10.8.2.

2) The estimators which are based on Gaussian maximum likelihood or least squares yield exactly the same asymptotic results; see Brockwell and Davis [92], Theorem 10.8.2. Moreover, if we restrict ourselves to AR processes then the estimator $\widetilde{\beta}_n$ coincides with the commonly used Yule–Walker estimator; for a definition see Brockwell and Davis [92], Section 8.1. □

The basic idea for the proof of Theorem 7.5.2 is a Taylor expansion of $\partial\widetilde{\sigma}_n^2(\beta_0)/\partial\beta$ about $\beta = \widetilde{\beta}_n$ and then an approximation of the expansion via linear combinations of a finite number of sample autocovariances of the iid noise which are jointly asymptotically normal; see Theorem 7.3.1. The same idea also works in the heavy–tailed case since we have joint weak convergence of a vector of sample autocorrelations of the iid noise; see Theorem 7.3.2.

The following is analogous to the classical result of Theorem 7.5.2. *Recall that we define the periodogram for* $s\alpha s$ (X_t) *by* (7.28).

Theorem 7.5.3 (Limit distribution of Whittle estimator, $s\alpha s$ case)
Suppose $(X_t)_{t\in\mathbb{Z}}$ *is the* ARMA(p,q) *process given by* (7.32) *for iid* $s\alpha s$ *noise* $(Z_t)_{t\in\mathbb{Z}}$ *with common chf* (7.15) *for some* $\alpha < 2$. *Then*

$$(n/\ln n)^{1/\alpha} \left(\widetilde{\beta}_n - \beta_0\right) \overset{d}{\to} 8\pi^2 W^{-1}(\beta_0) \sum_{j=1}^{\infty} b_j \frac{G_j}{G_0} \tag{7.38}$$

in $\mathbb{R}^{p+q}$, *where* G_0 *and* $(G_j)_{j\ge 1}$ *are independent rvs with chfs* (7.23) *and* (7.24), *respectively,* $W^{-1}(\beta_0)$ *is the inverse of the matrix* (7.36) *and, for* $j \ge 1$, b_j *is the vector* (7.37). *Relation* (7.35) *remains valid with* $\widetilde{\beta}_n$ *replaced by* $\widehat{\beta}_n$. *Moreover,*

$$\widetilde{\sigma}_n^2(\widetilde{\beta}_n) \overset{d}{\to} 2\pi G_0 ,$$

$$\widehat{\sigma}_n^2(\widehat{\beta}_n) \overset{d}{\to} 2\pi G_0 ,$$

$$n^{2/\alpha}\left(\sum_{t=1}^{n} X_t^2\right)^{-1} \widetilde{\sigma}_n^2(\widetilde{\beta}_n) \overset{P}{\to} 2\pi/\Psi^2 ,$$

$$n^{2/\alpha}\Big(\sum_{t=1}^{n} X_t^2\Big)^{-1}\widehat{\sigma}_n^2(\widehat{\beta}_n) \ \xrightarrow{P}\ 2\pi/\Psi^2 . \qquad \square$$

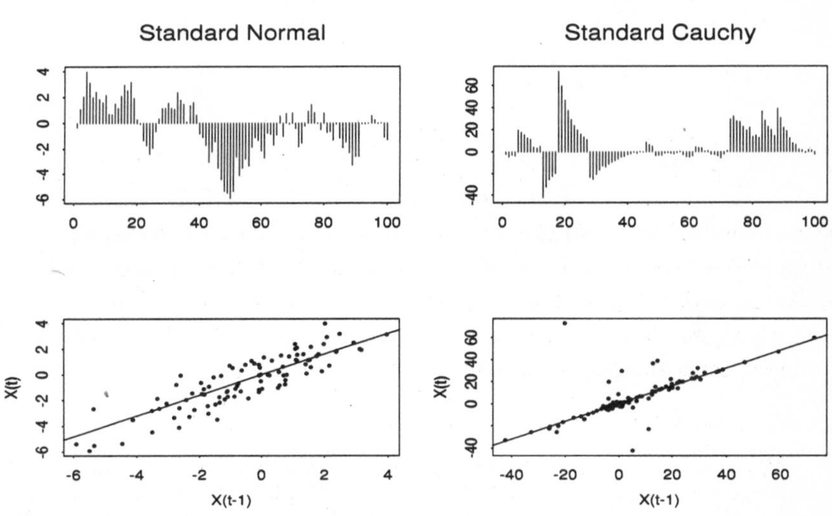

Figure 7.5.4 100 *realisations of the* AR(1) *process* $X_t = 0.8X_{t-1} + Z_t$ *with standard normal noise (top left) and standard Cauchy noise (top right). In the two bottom figures the respective scatterplots and a straight line with slope* 0.8 *are shown. The latter two graphs indicate that a regression estimator of the value* 0.8 *would yield more precise results in the Cauchy case. These graphs indicate to some extent why parameter estimators usually work better in the infinite variance case: large values of the noise show the dependence structure of the observed time series much better than for small values* Z_t.

Remarks. 3) Theorem 7.5.3 shows that one of the important classical estimation procedures also works for heavy–tailed processes. The rate of convergence $O_P((n/\ln n)^{-1/\alpha})$ in the *sαs* case is faster than $O_P(n^{-1/2})$ in the classical case.

4) Remark 3 in Section 7.3 concerning the calculation of the distribution of G_1/G_0 also applies to the situation of Theorem 7.5.3. $\square$

To get some idea of how the Whittle estimator behaves in the heavy–tailed situation, we ran a small simulation study using the estimator $\widehat{\beta}_n$ in (7.34) based on the summed periodogram $\widehat{\sigma}_n^2$. It should be emphasized that the estimation requires knowledge of neither the stability parameter α nor the scale parameter c of the data; see the chf (7.15).

The following table summarizes some of our simulation results. We generated 100 observations from each of the models

1. $X_t - 0.4\, X_{t-1} = Z_t,$
2. $X_t = Z_t + 0.8\, Z_{t-1},$
3. $X_t - 0.4\, X_{t-1} = Z_t + 0.8\, Z_{t-1},$

where the innovations sequence (Z_t) was either iid α-stable with $\alpha = 1.5$ and scale parameter $c = 1$, or, for comparison purposes, $N(0,2)$. In the stable case we relied on the algorithm given by Chambers, Mallows and Stuck [110] for the generation of the innovation process. We ran 1 000 such simulations for each model. In the stable example we estimated the ARMA parameters via the estimator $\widehat{\beta}_n$, and in the Gaussian case via the usual ML estimator (MLE). The results were as follows:

Model	True	Whittle estimate		Maximum likelihood	
No.	values	mean	st. dev.	mean	st. dev.
1	$\phi_1 = 0.4$	0.384	0.093	0.394	0.102
2	$\theta_1 = 0.8$	0.782	0.097	0.831	0.099
3	$\phi_1 = 0.4$	0.397	0.100	0.385	0.106
	$\theta_1 = 0.8$	0.736	0.124	0.815	0.082

Table 7.5.5 *Estimating the parameters of stable and normal* ARMA *processes via Whittle and MLE estimates.*

We point out that the accuracy of the Whittle estimator in the stable case seems to be comparable to that of the MLE in the Gaussian case. See also the comparative empirical study of different parameter estimators given in Figures 7.5.6 and 7.5.7.

Notes and Comments

The classical estimation theory for ARMA processes can be found in any standard textbook on times series analysis; see for instance Box and Jenkins [86] as a classical monograph or Brockwell and Davis [92]. The asymptotic theory for the Yule–Walker, the Gaussian maximum likelihood, the least squares and the Whittle estimator (Theorem 7.5.2) is given in Brockwell and Davis [92],

Chapters 8 and 10. There the estimator $\widehat{\beta}_n$, see (7.34), is treated, but similar arguments prove the results for $\widetilde{\beta}_n$.

The general methodology for the Whittle estimator (see Whittle [641]), with many subsequent contributions, such as from Hannan [315, 316], Fox and Taqqu [244], Dahlhaus [152], or Giraitis and Surgailis [263], has evolved steadily towards a unified theory. Heyde and Gay [324] give an overview of the existing literature in the univariate and multivariate case for Gaussian/non–Gaussian processes/fields with or without long–range dependence.

Theorem 7.5.3 and related results for the heavy–tailed case can be found in Mikosch et al. [458]. There a discussion of other parameter estimators for heavy–tailed processes is also given. We provide here an outline:

"There is a small, but interesting and rapidly growing, literature on parametric estimation for ARMA processes with infinite variance innovations. The difficulties in developing a maximum likelihood estimator have led to a number of essentially *ad hoc* procedures, each of which generalises some aspect of the Gaussian case. Nevertheless, a relatively consistent picture, at least as far as rates of convergence are concerned, has developed. Not surprisingly, the first estimator studied was a *Yule–Walker* (YW) *type estimator* for the parameters of an AR(p) process.

The YW–estimates $\widehat{\phi}_{YW}$ of the true values ϕ_0 of an AR(p) process are defined as the solution of

$$\widetilde{\Xi}\phi_{YW} = \widetilde{\rho}$$

where $\widetilde{\Xi} = [\widetilde{\rho}_n(i-j)]_{i,j=1}^p$, $\widetilde{\rho} = (\widetilde{\rho}_n(1), \dots, \widetilde{\rho}_n(p))^T$, and $\widetilde{\rho}_n(h)$ is the sample autocorrelation function. In the autoregressive case it is not difficult to see that the YW–estimate coincides with the Whittle estimate based on $\widetilde{\sigma}_n^2(\beta)$.

Hannan and Kanter [317] showed that if $0 < \alpha < 2$, and $\delta > \alpha$, then

$$n^{1/\delta}\left(\widehat{\phi}_{YW} - \phi_0\right) \stackrel{\text{a.s.}}{\to} 0\,, \quad n \to \infty\,.$$

More recently, Davis and Resnick [162] showed that there exists a slowly varying function L_0 such that

$$n^{1/\alpha}\,L_0(n)\left(\widehat{\phi}_{YW} - \phi_0\right) \quad \stackrel{d}{\to} \quad \mathbf{Y}\,, \quad n \to \infty$$

where the structure of $\mathbf{Y}$ is closely related to the rhs of (7.38).

A somewhat different approach to parameter estimation, still in the purely autoregressive case, is based on a *least absolute deviation* (LAD) *estimator*, which we denote by $\widehat{\phi}_{LAD}$. The LAD–estimate of ϕ_0 is defined as the minimiser of

$$\sum_{t=1}^{n} |X_t - \phi_1\,X_{t-1} - \cdots - \phi_p\,X_{t-p}|$$

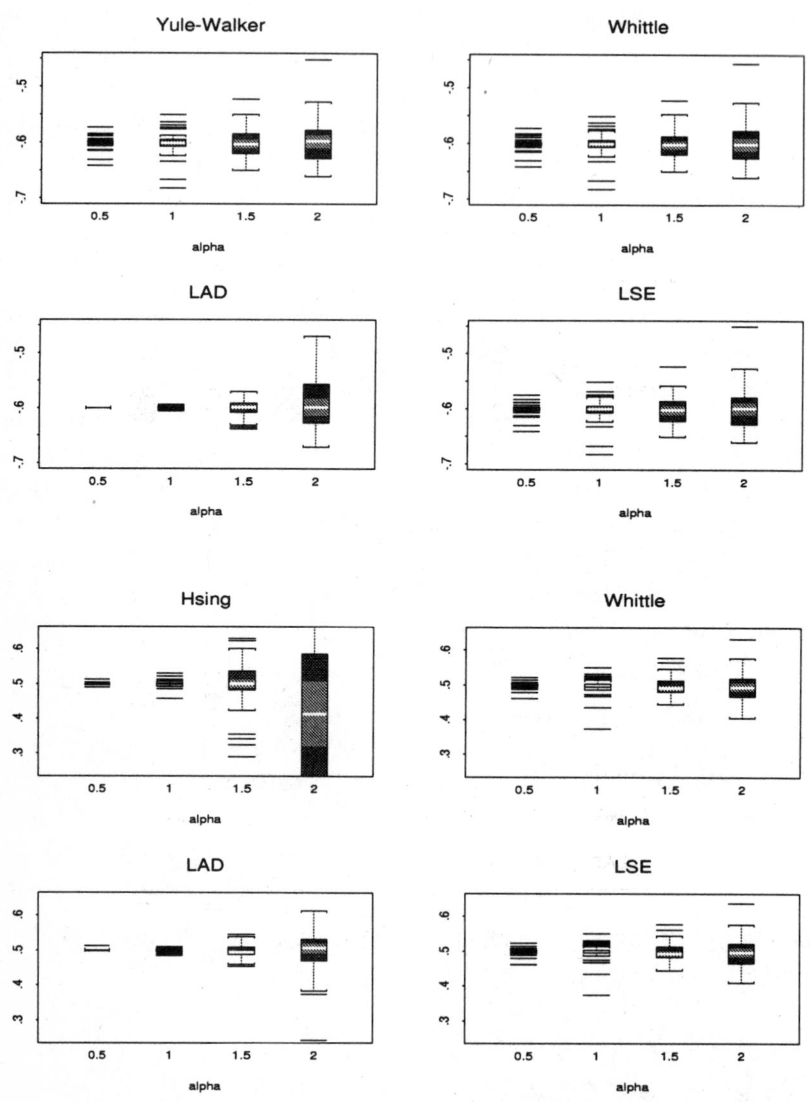

Figure 7.5.6 *Boxplots of four parameter estimators of* $\phi_1 = -0.6$ *in the* AR(1) *model* $X_t + 0.6X_{t-1} = Z_t$ *(top) and of* $\theta_1 = 0.5$ *in the* MA(1) *model* $X_t = Z_t + 0.5Z_{t-1}$ *(bottom). They are based on* 50 *simulations of a time series of length* 500, Z *is* $s\alpha s$ *with scale parameter* $c = 1$, $\alpha \in \{0.5, 1, 1.5, 2\}$.

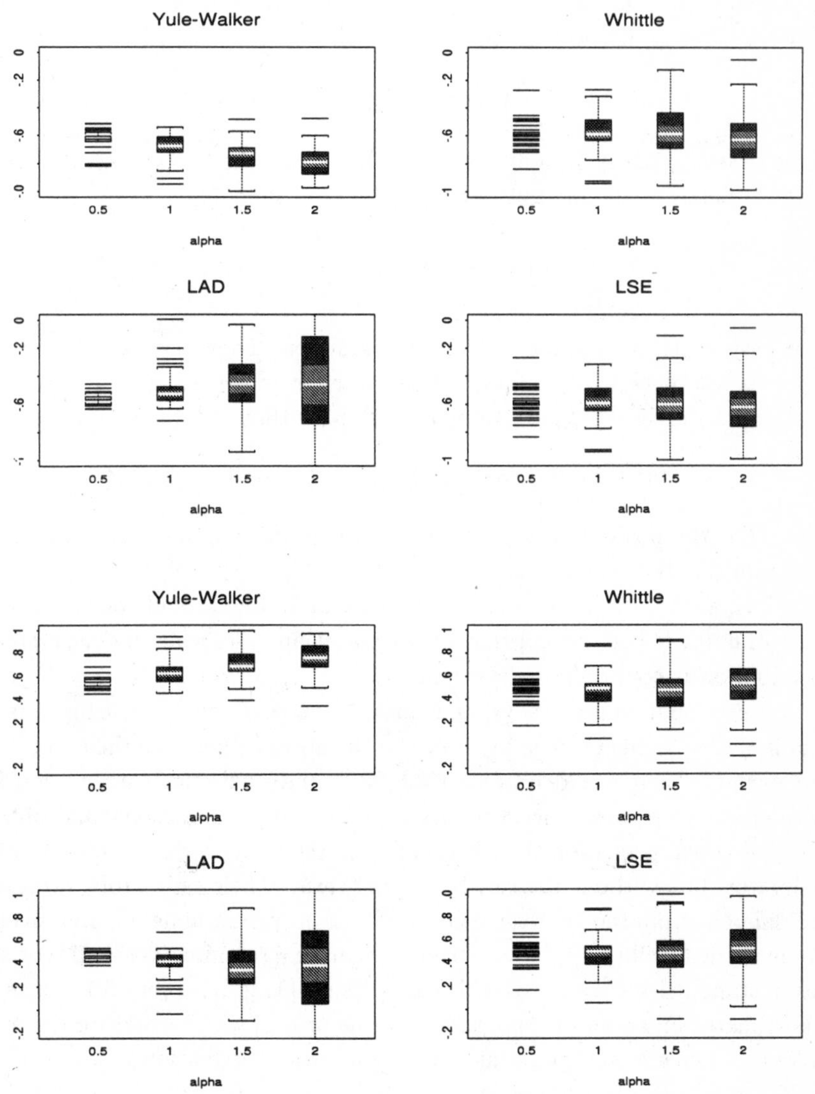

Figure 7.5.7 *Boxplots of four parameter estimators for* $\phi_1 = -0.6$ *(top) and* $\theta_1 = 0.5$ *(bottom) in the* ARMA(1, 1) *model* $X_t + 0.6X_{t-1} = Z_t + 0.5Z_{t-1}$. *They are based on* 50 *simulations of a time series of length* 500, Z *is s*α*s with scale parameter* $c = 1$, $\alpha \in \{0.5, 1, 1.5, 2\}$.

with respect to $\phi = (\phi_1, \ldots, \phi_p)^T$.

An and Chen [9] showed that if Z has a unique median at zero and Z is in the domain of attraction of a stable distribution with index $\alpha \in (1,2)$, or Z has a Cauchy distribution centred at zero, then, for $\delta > \alpha$,

$$n^{1/\delta}\left(\widehat{\phi}_{LAD} - \phi_0\right) \xrightarrow{P} 0\,, \quad n \to \infty\,.$$

More recently, Davis, Knight and Liu [158] defined the M–*estimate* $\widehat{\phi}_M$ *of an* AR(p) *process* as the minimiser of the objective function

$$\sum_{t=p+1}^{n} r\left(X_t - \phi_1 X_{t-1} - \cdots - \phi_p X_{t-p}\right)$$

with respect to ϕ, where r is some loss function. They also established the weak convergence of $\widehat{\phi}_M$, for the case when r is convex with a Lipschitz continuous derivative. Specifically, they showed that

$$n^{1/\alpha} L_1(n)\left(\widehat{\phi}_M - \phi_0\right) \quad \xrightarrow{d} \quad \xi\,, \qquad n \to \infty\,,$$

where ξ is the position of the minimum of a certain random field, and L_1 is a certain slowly varying function.

Thus, as is the case for the Whittle estimator, the rate of convergence of the estimator is better than that in the Gaussian case, while the asymptotic distribution is considerably less familiar.

We note that "more rapid than Gaussian" rates of convergence for estimators in heavy–tailed problems seem to be the norm rather than the exception. For example, Feigin and Resnick [229, 230] study parameter estimation for autoregressive processes with *positive*, heavy–tailed innovations, and obtain rates of convergence for their estimator of the same order as the Whittle estimator, but without the slowly varying term. Their estimators, however, are different from the Whittle estimator both in spirit and detail, and involve the numerical solution of a non–trivial linear programming problem. For the latter standard software exists. Finally, Hsing [342], Theorem 3.1, suggests an estimator based on *extreme value considerations*, which works for the pure MA case. Once again, he obtains an asymptotic distribution reminiscent of (7.38), with a similar rate of convergence."

There are some more recent contributions to parameter estimation of heavy–tailed ARMA and related processes. Kokoszka and Taqqu [403] prove that Theorem 7.5.3 remains valid for fractional ARIMA processes with noise distribution in the domain of normal attraction of an α–stable law, $1 < \alpha < 2$. In that case, the coefficients ψ_j are not absolutely summable, but of order $\psi_j \sim c j^{d-1}$ for some $d < 1 - 1/\alpha$. In analogy to the finite variance case, such processes are said to have *long memory* or *long–range dependence*. Davis [157] proves results on M, LAD and Gauss–Newton estimates of infinite variance ARMA processes. He shows that LAD estimates outperform Gauss–Newton and Whittle estimation in the sense that $(n^{1/\alpha}(\widehat{\beta}_{LAD} - \beta_0))$ converges in distribution. Thus one can avoid the logarithmic term in the normalisation of (7.38). The performance of these estimators is illustrated in Figures 7.5.6 and 7.5.7.

We mention that some of the classical *procedures for determining the order* (p, q) of an ARMA process, for instance the AIC for AR processes (see Brockwell and Davis [92], Chapter 9.2), also work in the heavy–tailed case (see for instance Bhansali [64, 65], Knight [401]).

7.6 Some Remarks About Non–Linear Heavy–Tailed Models

In the literature on financial time series and finance one often finds claims like *"real financial data come from non–linear, non–stationary processes with heavy–tailed, leptokurtic marginal distributions"* or *"financial data are heteroscedastic"* or *"the price process is highly volatile"*. It is the aim of this section to give a brief explanation of some of these catchy words which are used to describe irregular behaviour of financial time series and processes.

Though price or exchange rate processes (X_t) themselves can rarily be described as a *stationary* time series, in most cases a straightforward transformation brings them back (or closer) to a stationary model. For instance, share prices, exchange rates, stock indexes etc. are believed to grow roughly exponentially when time goes by. Therefore time series analysts and econometricians mostly agree on the fact that *daily logarithmic differences* or *log–returns*

$$R_t = \ln\left(\frac{X_t}{X_{t-1}}\right) = \ln X_t - \ln X_{t-1}$$

constitute a stationary process (strictly or in the wide sense); see for instance Taylor [616]. Clearly, "daily" can be changed into any relevant period of time. Note that, by Taylor's formula,

$$\ln\left(\frac{X_t}{X_{t-1}}\right) = \ln\left(1 + \frac{X_t - X_{t-1}}{X_{t-1}}\right) \approx \frac{X_t - X_{t-1}}{X_{t-1}}.$$

Hence (R_t) can be considered as the sequence of *daily relative returns*. Notice that log–differencing of financial data makes them comparable; only the relative change over time is then of interest. In particular, they become independent of the monetary unit.

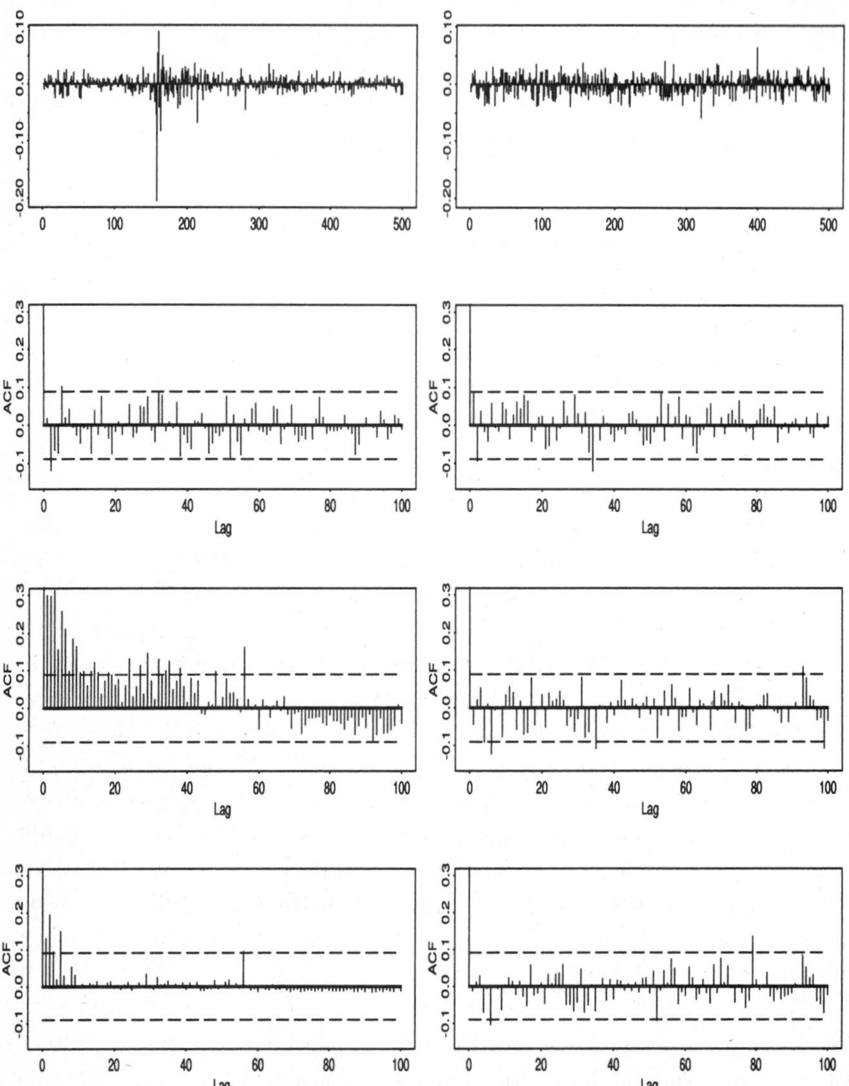

Figure 7.6.1 *Top:* 500 *daily log–returns of the* S&P *index* (*left*) *and* 500 *realisations of iid Gaussian noise* (*right*) *with the same mean and variance. The corresponding sample autocorrelations of the data* (*second row*), *of their absolute values* (*third row*) *and of their squares* (*bottom*). *A comparion shows that the* S&P *data have a difficult dependence structure different from an iid sequence. The dashed lines indicate* 95% *asymptotic confidence bands for the sample autocorrelations of iid Gaussian rvs.*

In what follows we always consider log–returns or log–differences which are supposed to come from a stationary sequence. Stationarity of log–returns may hence be accepted as a working hypothesis. Clearly, this assumption becomes questionable if one considers times series over too long periods of time, or in the case of so–called high frequency or tick–by–tick data, where the sampling interval may go down to second or minute level.

A glance at any series of log–returns shows very convincingly that there are values which are significantly larger than the others; see for instance the S&P series in Figure 7.6.1. Therefore the one–dimensional marginal distribution is certainly not light–tailed, in particular not Gaussian. We discussed an example of this in the analysis of the BMW data; see Figures 6.2.10 and 6.5.3. The notion of "heavy–tailedness" is obviously not defined in a unique way; in this book we tend to use it as "not having exponential moments". The class of subexponential distributions, see Section 1.3.2, satisfies this assumption. There exist several approaches for modelling heavy–tailed log–returns, for instance infinite variance processes and the ARCH family. Below we give a short description of both classes.

In the previous sections we considered infinite variance stable processes, in particular ARMA and linear processes, and related statistical problems. Our conclusion was that many classical techniques work for such processes, and the rate of convergence in the estimation of parameters and functions is usually better than in the finite variance case, although the limit distributions are in general not easy to handle. In the financial literature there has been interest in infinite variance stable distributions and processes for a long time. Articles on infinite variance processes in finance usually refer to the two, by now classical, sources Mandelbrot [436] and Fama [226]. These authors propagated the use of stable and Pareto distributions in finance. However, articles supporting the hypothesis of α–stable distributions in finance do not always mention the discussion which started afterwards and has never been finished; see for instance Groenendijk, Lucas and de Vries [286] and Ghose and Kroner [260] for some recent contributions. We cite here Taylor [616], pp. 46–47, which is a standard monograph on financial time series:

"Fama and Roll [228] describe a practical method for estimating α. Estimates are always between the special cases $\alpha = 1$ for Cauchy distributions and $\alpha = 2$ for normal distributions. Many researchers find the conclusion of infinite variance, when $\alpha < 2$, unacceptable. Detailed studies of stock returns have conclusively rejected the stable distributions (Blattberg and Gonedes [76]; Hagerman [304]; Perry [492]). Hagerman, for example, shows that estimates of α steadily increase from about 1.5 for daily returns to about 1.9 for returns measured over 35 days. Returns over a month or more have distributions much closer to the normal shape than daily returns. A decade after his 1965 paper, Fama prefers to use normal distributions for monthly returns and so discard stable distributions for daily returns (Fama [227], Chapter 1)."

These remarks should be read in the spirit of Chapter 6 where, among other topics, we considered the hypothesis of Pareto tails and certain infinite moments for financial and insurance data. One of our main conclusions was that it is very difficult to make a final decision about the value of a tail index α or the finiteness of a certain power moment. However, the examples considered there show convincingly that log–returns have certain infinite power moments. This is already seen by using simple diagnostic tools such as QQ– or mean excess plots, and later reinforced by more subtle means such as tail index estimators.

Log–returns exhibit a complicated dependence structure; see for instance the sample autocorrelations of the transformed S&P data in Figure 7.6.1. Therefore the direct use of standard tail index estimates (Hill, Pickands) may become questionable. These estimators are very sensitive with respect to dependence in the data; see for instance Figure 5.5.4 and Resnick and Stărică [535, 537]. The dependence structure gets even more complicated when data are aggregated over weeks or months. Moreover, we have seen in Chapter 6 that tail index estimation requires large sample sizes. Thus some of the problems mentioned in the above citation could be due to the dependence in the data and/or too small sample sizes. In our opinion, tail estimation methods do not allow for a precise conclusion concerning patterns in estimates for α based on data at different time scales.

Clearly, much of the finance literature is based on the notions of volatility and correlation, i.e. finite second moments are necessarily required in such models, and therefore the infinite variance case has gained only marginal popularity. The emergence of quantitative techniques for risk management has changed this attitude considerably. It is a fact (or a so–called *stylized fact*, as finance experts like to call it) that most *financial data are heavy–tailed*! The infinite variance linear processes discussed in the previous sections offer only one possible model for such data. A more careful look at financial data quickly reveals the need for much more versatile models. For instance, the detailed analysis of high–frequency foreign exchange data and data on the inter–bank market of cash interest rates as summarised in Müller et al. [471] shows that though variances are finite, third or fourth moments may be infinite. See also Longin [428], Loretan and Phillips [429] and various examples in Chapter 6 for some empirical evidence. If one accepts an infinite fourth moment for a stationary process, standard time series procedures may not work. For instance, asymptotic confidence bands for sample autocorrelations are based on a CLT for which a finite fourth moment is required. Therefore, the interpretation of sample autocorrelations may become problematic.

In the context of heavy-tailed models the notion of *leptokurtosis* often occurs. As noted for instance in Eberlein and Keller [197] upon studying data consisting of daily prices of the 30 DAX shares over a three-year period: "...there is considerably more mass around the origin and in the tails than the standard normal distribution can provide." The *kurtosis* of a rv X with df F is defined either as

$$k_m = \frac{E(X-\mu)^4}{(E(X-\mu)^2)^2}, \tag{7.39}$$

or

$$k_p = \frac{\frac{1}{2}(x_{0.75} - x_{0.25})}{x_{0.9} - x_{0.1}}, \tag{7.40}$$

where for $0 < p < 1$, the quantiles x_p are defined via $x_p = F^{\leftarrow}(p)$. The advantage of k_p over k_m is that k_p is always well defined. For the normal distribution, $k_m = 3$ and $k_p = 0.263$. A df F is now called *leptokurtic* if either $k_m > 3$ or $k_p < 0.263$; see Medhi [453], Section 4.10.2, for a further discussion. As stated in Mood, Graybill and Boes [467], p. 76, "these measures do not always measure what they suppose to", and indeed practice in the finance literature has evolved to using the notion "leptokurtic" for indicating "excess peakedness *and* heavy tails".

The notion *stochastic volatility* is used for describing random changes of the variance as a function of time, the latter mainly in the context of solutions to stochastic differential equations (SDE). To set the scene, first consider the linear SDE

$$dX_t = cX_t\,dt + \sigma_0 X_t\,dB_t\,, \quad t \in [0,T]\,, \tag{7.41}$$

where for the moment $c \in \mathbb{R}$ and $\sigma_0 > 0$ are constants. The driving process (B_t) is Brownian motion, see Section 2.4, and the differential dB_t has to be interpreted in the sense of Itô calculus. It is well known that (7.41) has a unique strong solution

$$X_t = X_0 \exp\left\{\left(c - \frac{1}{2}\,\sigma_0^2\right) t + \sigma_0 B_t\right\}, \quad t \in [0,T]\,. \tag{7.42}$$

The stochastic process (X_t) in (7.42) is called *geometric Brownian motion* and is *the* standard, so-called Black-Scholes, model for financial price processes. The value σ_0 is called *volatility*; it constitutes the main parameter to be estimated from the data. If (X_t) were geometric Brownian motion we would obtain for the log-returns

$$R_t = \left(c - \frac{1}{2}\sigma_0^2\right) + \sigma_0\,(B_t - B_{t-1})\,.$$

Hence log–returns should look like iid normal data with variance σ_0^2. If we compare this with real data we see that this rarely is the case; see for instance Figure 7.6.1. Hence the SDE (7.41) is clearly an idealisation of the real world. The main, obvious, reason in favour of (7.42) is its tractability with respect to actual calculations. Also micro–economic considerations may be given leading to processes of the type (7.42). Though as a first approximation, the statistical fit of (7.42) to data may be reasonable, there are various arguments going *against* it. These include:

- Real processes are not continuous in time; in the real world prices do not change in intervals of microsecond length. On the other hand, the self–similarity property of Brownian motion suggests that in any time interval the driving process generates some noise.
- The linear SDE (7.41) with constant coefficients does not explain the fluctuations and jumps in real data. Nor does it explain the observed leptokurtic behaviour of price distributions.
- The SDE (7.41) suggests that the noise dB_t of the market is independent for disjoint time intervals. This is certainly not the case, at least not in small intervals of time.

"Over the last twenty years, the Black–Scholes option pricing model has proved to be a valuable tool for the pricing of options and, more generally, for the management of hedged positions in the derivative markets. However, a number of systematic biases in the model prices suggests that *the underlying security volatility may be stochastic.* This observation is further reinforced by empirical evidence from the underlying asset prices." This statement is taken from the review paper by Ball [39] on stochastic volatility. What does the latter mean? The word "stochastic" actually refers to the fact that the volatility σ_0 in the SDE (7.41) is a random function of time: $\sigma_0 = \sigma_0(t, \omega)$. Stochastic volatility is one of the current main research topics in mathematical finance. By allowing for a random and time–dependent volatility one can quite flexibly describe the change of the variance of log–returns; for some more references see the Notes and Comments.

Discrete–time versions of stochastic volatility models are also relevant. They are usually referred to as "conditional heteroscedasticity" models in the econometrics and time series literature. In their simplest form these models can often be written in multiplicative form

$$R_t = \mu + V_t Z_t , \quad t \in \mathbb{Z} , \tag{7.43}$$

where R_t is again interpreted as a daily log–return, say, and μ is the expectation of R_t. With few exceptions, V_t and Z_t are supposed to be independent,

Z_t are iid standard normal rvs and V_t are non-negative rvs such that V_t is a function of $R_{t-1}, R_{t-2}, \ldots$. Then

$$ER_t = \mu\,, \quad \operatorname{var}(R_t | \, R_{t-1}, R_{t-2}, \ldots) = V_t^2\,,$$

i.e. V_t^2 is the conditional variance of R_t given the past $R_{t-1}, R_{t-2}, \ldots$. As a consequence, R_t is mixed Gaussian.

The argument which is usually given in favour of the model (7.43) is the following: many real financial time series have negligible sample autocorrelations at all lags with possible exceptions at lags 1, 2 or 3, although these values are fairly small. This fits well with the fact that

$$\operatorname{cov}(R_t, R_{t+h}) = E\,(V_t\, Z_t V_{t+h} Z_{t+h}) = 0\,, \quad h \geq 1\,.$$

On the other hand, the sample autocorrelations of the absolute value and of the squares of real returns are usually greater than 0.2 at the first few lags which is supported in part by the theoretical model (7.43). Figure 7.6.1 shows a financial time series with typical behaviour of the sample autocorrelations. If one fits an ARMA model to these data and simulates an ARMA process of the same length with the same parameters, one can see that the autocorrelations of the real and of the simulated data almost coincide, but the absolute values and the squares of the ARMA process do not have autocorrelations similar to the real data.

The processes which are most popular in econometrics and which belong to the class (7.43) are the *ARCH* (*autoregressive-conditionally-heteroscedastic*) *models* and their variants *GARCH* (*generalised ARCH*), *ARMACH* (*autoregressive-moving-average-conditionally-heteroscedastic*) etc. For example, an ARCH(p)-process (ARCH of order p) with mean $\mu = 0$ is given by the equation

$$R_t = \left(\phi_0 + \sum_{i=1}^{p} \phi_i R_{t-i}^2\right)^{1/2} Z_t\,, \quad t \in \mathbb{Z}\,,$$

for iid $N(0,1)$ rvs Z_t, and non-negative parameters ϕ_i. Then

$$V_t^2 = \operatorname{var}(R_t | \, R_{t-1}, R_{t-2}, \ldots) = \phi_0 + \sum_{i=1}^{p} \phi_i\, R_{t-i}^2\,,$$

i.e. the conditional variance of the return R_t, given the returns in the past, is just a function of the last p returns. Thus, if V_t^2 is large, the order of magnitude of R_t and future returns is very much influenced by this conditional variance. This may lead to clusters of large values of R_t; see Figure 7.6.2. One can show that a stationary version of (R_t) exists if the coefficients ϕ_i belong to the parameter set C (see Section 7.5), i.e. if the corresponding AR difference

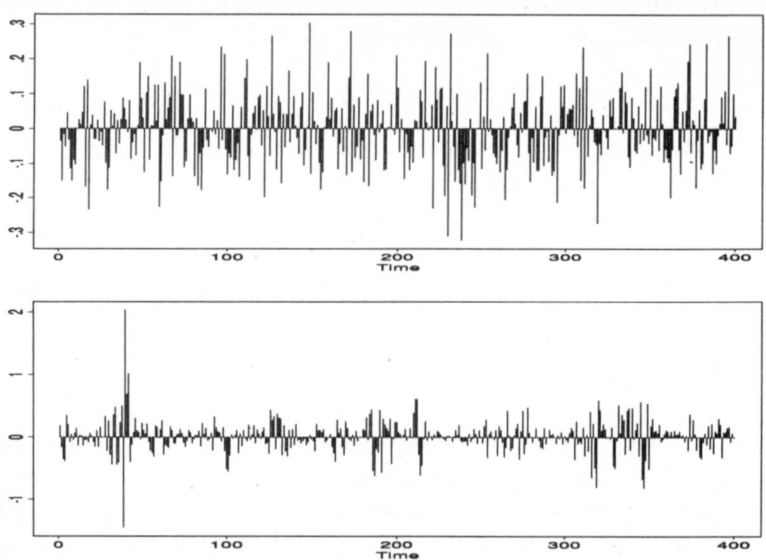

Figure 7.6.2 400 *realisations of the ARCH processes* $X_t = (0.01 + 0.1X_{t-1}^2 + 0.01X_{t-2}^2)^{1/2}Z_t$ *(top) and* $X_t = (0.01 + 0.9X_{t-1}^2 + 0.1X_{t-2}^2)^{1/2}Z_t$ *(bottom) for iid* $N(0,1)$ *rvs* Z_t.

equations have a unique a.s. convergent solution. We refer to Section 8.4 for a more detailed analysis of ARCH(1) processes. There we show in particular that such processes have Pareto-like tails.

Notes and Comments

Standard references for SDE are the monographs by Chung and Williams [121] (it contains a derivation of the Black-Scholes formula), Karatzas and Shreve [368] (it also has a chapter about stochastic finance, the recent monograph [369] of these authors is solely devoted to mathematical finance), Protter [514] (this is an excellent reference to stochastic integration with respect to semimartingales). The modern theory of stochastic calculus is masterfully presented in Rogers and Williams [547]. Chapter II of Volume One of the latter is "a highly systematic account, with detailed proofs, of what every young probabilist *must* know". We always found Revuz and Yor [541] a most reliable source on martingales and Brownian motion when studying new continuous-time models in finance. Kloeden and Platen [385] is a compendium on numerical solutions and applications of SDE; see also the companion book Kloeden, Platen and Schurz [386] which is understood as an introduction to SDE and their numerical solution aimed at the non–mathematician. The extremal behaviour of financial models given by SDE has been investigated in Borkovec and Klüppelberg [81].

Parameter estimation in financial models is an important subject. In the by now classical case of (7.41), the volatility σ_0 appears as the main parameter of interest. It is traditionally either estimated using historical data (*historical volatility*) or by inverting certain option pricing formulae containing the volatility parameter in a 1-to-1 way (*implied volatility*). Most standard textbooks on mathematical finance contain a discussion on this topic. The interested reader may also want to consult for instance Garman and Klass [251] or Rogers and Satchell [546] for some basic ideas underlying volatility estimation. More generally, however, the topic of parameter estimation for stochastic processes relevant in finance is becoming an important area of research. A key question concerns estimation of parameters in continuous–time models from discrete–time observations. The statistical literature on this topic is huge! We personally found the accumulated contributions from the "Aarhus School" particularly useful. Relevant references from the latter are Bibby and Sørensen [67], Pedersen [490] and the recent review paper by Sørensen [602]; see also Florens-Zmirou [241], Genon-Catalot and Jacod [254], Kessler [377]. Göing [269] summarises the various alternative approaches; see also the references in that paper. Especially the problem of discretisation and estimation in stochastic volatility models receives attention. Melino [455] is a further recent review on estimation of continuous–time models in finance.

From the growing literature on stochastic volatility in SDE and option pricing we give a short list: Ball and Roma [40], Eisenberg and Jarrow [199], Frey [245], Ghysels, Harvey and Renault [261], Hull and White [347, 348], Rydberg [560, 561], Scott [571], Sin [583], Stein and Stein [606]. We especially found Frey [245] a most informative and readable introduction.

An alternative approach to modelling changing volatility in financial data is by assuming that the driving process in the underlying SDE is *not* Brownian motion, but a process with marginal distribution tails heavier than Gaussian ones. Candidates for such processes may come from the class of semimartingales, including for instance Lévy and infinite variance stable processes. In this context, Barndorff–Nielsen [46, 47] discusses the use of generalised inverse Gaussian distributions for modelling financial data; see also Eberlein and Keller [197] and Küchler et al. [410].

The literature on links between time series and SDE should be viewed more generally within the context of embedding discrete–time (and possibly space–time) models in continuous–time processes. A paper to start with is de Haan and Karandikar [296]; see also Nelson [476] and references therein.

The models (7.43) can be understood as solutions to stochastic recurrence equations. Work by Kesten [379] and Goldie [272] shows for some particular cases, including the ARCH(1) process, that $P(X_t > x) \sim cx^{-p}$ for certain positive constants c and p. This will be shown explicitly in Section 8.4.

Samorodnitsky [564] explains the effect of changes in the estimation of the tail index α, when different time aggregations are used, by a shot noise process with heavy tails. The reader interested in some of the more recent analysis on this topic may consult Müller at al. [471] and the references therein.

A useful model of the type (7.43) is the so-called heterogeneous ARCH (or HARCH) model introduced in Müller et al. [469]. The HARCH(k) process with $k \geq 1$ satisfies

$$R_t = \left(\phi_0 + \sum_{i=1}^{k} \phi_i \left(\sum_{j=1}^{i} R_{t-j} \right)^2 \right)^{1/2} Z_t \,, \quad t \in \mathbb{Z} \,, \tag{7.44}$$

for iid rvs Z_t and non-negative parameters ϕ_i. Note that HARCH(1) processes are ARCH(1). They allow for instance to model the empirically observed fact that for foreign exchange intra-day data, volatility over a coarse time grid significantly predicts volatility defined over a fine grid. The conditions for the existence of stationary solutions of (7.44) together with necessary and/or sufficient conditions for the existence of moments are given in Embrechts et al. [217].

Links between GARCH and stable processes were considered by Diebold [181] and by de Vries [634]. Mittnik and Rachev [465, 466] model asset returns with "alternative" stable distributions, i.e. with distributions which are stable with respect to certain operations; for ordinary summation the α-stable laws appear, for maxima the max-stable limit distributions.

A good introduction to the problem of modelling financial time series is given by Taylor [616]. He discusses different models of multiplicative type, see (7.43), and compares the performance of these models with the behaviour of a large amount of real financial data sets. The literature on ARCH models, their ramifications and related models is rapidly increasing. There exist more than 50 different such models which fact does not make it easy to distinguish between them. We refer here to Bollerslev, Chou and Kroner [78], Shephard [578] and references therein.

8

Special Topics

8.1 The Extremal Index

8.1.1 Definition and Elementary Properties

In Chapters 3–6 we presented a wealth of material on extremes. In most cases we restricted ourselves to iid observations. However, in reality extremal events often tend to occur in clusters caused by local dependence in the data. For instance, large claims in insurance are mainly due to hurricanes, storms, floods, earthquakes etc. Claims are then linked with these events and do not occur independently. The same can be observed with financial data such as exchange rates and asset prices. If one large value in such a time series occurs we can usually observe a cluster of large values over a short period afterwards.

The *extremal index* is a quantity which, in an intuitive way, allows one to characterise the relationship between the dependence structure of the data and their extremal behaviour. To understand this notion we first recall some of the examples of extremal behaviour for a strictly stationary sequence (X_n) with marginal df F. In this section we consider only this kind of model. As usual, M_n stands for the maximum of the sample $X_1, \ldots, X_n$, $(\widetilde{X}_n)$ is an associated iid sequence (i.e. with common df F) and $(\widetilde{M}_n)$ denotes the corresponding sequence of maxima.

Example 8.1.1 Assume that the condition

$$n\overline{F}(u_n) \to \tau \in (0, \infty) \tag{8.1}$$

holds for some non–decreasing sequence (u_n).

(a) For an iid sequence (X_n) we know from Proposition 3.1.1 that (8.1) is equivalent to the relation

$$\lim_{n\to\infty} P(M_n \le u_n) = e^{-\tau}\,. \tag{8.2}$$

Moreover, from Theorem 5.3.2 we conclude that the point processes of exceedances

$$N_n(\cdot) = \sum_{i=1}^{n} \varepsilon_{n^{-1}i}(\cdot) I_{\{X_i > u_n\}}$$

of u_n by $X_1, \ldots, X_n$ converge weakly to a homogeneous Poisson process N with intensity τ.

(b) Recall the conditions $D(u_n)$ and $D'(u_n)$ from Section 4.4. They ensure that the strictly stationary sequence (X_n) has the same asymptotic extremal behaviour as an associated iid sequence. In particular, (8.1) implies (8.2) (see Proposition 4.4.3) and $N_n \xrightarrow{d} N$ where N is again a homogeneous Poisson process with intensity τ; see Theorem 5.3.6. Conditions $D(u_n)$ and $D'(u_n)$ are satisfied for large classes of Gaussian stationary sequences, including many Gaussian linear processes (for instance Gaussian ARMA and fractional ARIMA processes).

(c) Recall the situation of Example 4.4.2: starting with an iid sequence (Y_n) with df $\sqrt{F}$, the strictly stationary sequence

$$X_n = \max(Y_n, Y_{n+1})\,, \quad n \in \mathbb{N},$$

has df F and

$$M_n = \max(Y_1, \ldots, Y_{n+1})\,, \quad n \in \mathbb{N}.$$

If (8.1) is satisfied then

$$\lim_{n\to\infty} P(M_n \le u_n) = e^{-\tau/2}\,.$$

We know from Example 4.4.2 that condition $D(u_n)$ is satisfied in this case, but $D'(u_n)$ is not; see Example 4.4.4.

(d) Let (X_n) be a linear process

$$X_n = \sum_{j=-\infty}^{\infty} \psi_j Z_{n-j}\,, \quad n \in \mathbb{Z}\,,$$

driven by iid noise (Z_t). Assume $P(Z_1 > x) = x^{-\alpha}L(x)$ as $x \to \infty$ and that the tail balance condition

$$P(Z_1 > x) \sim pP(|Z_1| > x) \quad \text{and} \quad P(Z_1 \le -x) \sim qP(|Z_1| > x) \tag{8.3}$$

holds for some $p \in (0,1]$, $q = 1-p$. This implies that $F \in \mathrm{MDA}(\Phi_\alpha)$ for some $\alpha > 0$. Then we obtain from Corollary 5.5.3 that there exist constants $c_n > 0$ such that $u_n = u_n(x) = c_n x$ satisfies (8.1) for $\tau = \tau(x) = x^{-\alpha}$, $x > 0$, and

$$\begin{aligned}
\lim_{n\to\infty} P(\widetilde{M}_n \le u_n(x)) &= \Phi_\alpha(x)\,, \quad x \in \mathbb{R}_+\,,\\
\lim_{n\to\infty} P(M_n \le u_n(x)) &= \Phi_\alpha^\theta(x)\,, \quad x \in \mathbb{R}_+\,.
\end{aligned}$$

Here Φ_α denotes the standard Fréchet distribution and

$$\theta = \left(\psi_+^\alpha\, p + \psi_-^\alpha\, q\right) / \|\psi\|_\alpha^\alpha\,,$$

where

$$\psi_+ = \max_j(\psi_j \vee 0) \quad \text{and} \quad \psi_- = \max_j((-\psi_j) \vee 0)\,,$$

$$\|\psi\|_\alpha^\alpha = \sum_{j=-\infty}^{\infty} |\psi_j|^\alpha (p\, I_{\{\psi_j>0\}} + q\, I_{\{\psi_j<0\}})\,.$$

The point processes N_n of the exceedances of $u_n(x)$ by $X_1, \ldots, X_n$ converge weakly to a compound Poisson process

$$N_\infty = \sum_{k=1}^{\infty} \xi_k \varepsilon_{\Gamma_k}\,;$$

see Section 5.5.1. The Γ_k are the points of a homogeneous Poisson process, and the ξ_k are iid cluster sizes.

(e) Let (X_n) be a linear process driven by iid subexponential noise (Z_t) with $F \in \mathrm{MDA}(\Lambda)$, where Λ denotes the standard Gumbel distribution. We also assume the tail balance condition (8.3) and $\max_j |\psi_j| = 1$. Then we know from Corollary 5.5.12 that there exist constants $c_n > 0$ and $d_n \in \mathbb{R}$ such that $u_n = u_n(x) = c_n x + d_n$ satisfies (8.1) for $\tau = \tau(x) = \exp\{-x\}$, $x \in \mathbb{R}$, and

$$\begin{aligned}
\lim_{n\to\infty} P(\widetilde{M}_n \le u_n(x)) &= \Lambda(x)\,, \quad x \in \mathbb{R}\,,\\
\lim_{n\to\infty} P(M_n \le u_n(x)) &= \Lambda^\theta(x)\,, \quad x \in \mathbb{R}\,,
\end{aligned}$$

where

$$\theta = (k^+ p + k^- q)^{-1}\,,$$

and

$$k^+ = \mathrm{card}\{j : \psi_j = 1\} \quad \text{and} \quad k^- = \mathrm{card}\{j : \psi_j = -1\}\,.$$

The point processes N_n of exceedances of $u_n(x)$ converge weakly to a compound Poisson process N (see Theorem 5.5.13) such that

$$N = \sum_{k=1}^{\infty} \left(k^+ \varepsilon_{\Gamma_k^+} + k^- \varepsilon_{\Gamma_k^-} \right) ,$$

where the sequences of the points (Γ_k^+) and (Γ_k^-) are independent of each other, each of them representing the points of a homogeneous Poisson process.

(f) Recall the definition of an ARCH(1) process from Section 8.4.2:

$$X_n = \sqrt{\beta + \lambda X_{n-1}^2}\, Z_n \,, \quad n \in \mathbb{N} ,$$

for $\beta \geq 0$, $\lambda \in (0, 2e^{\gamma})$ and iid standard normal rvs Z_n. We also know that (X_n) is a strictly stationary sequence provided X_0 is appropriately chosen. Moreover, by Theorem 8.4.20, there exist constants $c_n > 0$ such that $u_n = u_n(x) = c_n x$ satisfies condition (8.1) for certain $\tau = \tau(x)$, and

$$P\left(M_n \leq u_n(x)\right) = \Phi_\kappa^\theta(x) \,, \quad x \in \mathbb{R}_+ ,$$

for some $\kappa = \kappa(\lambda)$ and $\theta \in (0,1)$, whereas an associated iid sequence has limit distribution Φ_κ provided the same norming constants c_n are used. Moreover, the point processes N_n of exceedances of the threshold $u_n(x)$ by $X_1, \ldots, X_n$ converge weakly to a compound Poisson process whose structure is described in Theorem 8.4.20. □

The examples above follow similar patterns. Indeed, it is typical for stationary (X_n) and (u_n) satisfying (8.1) that $P(M_n \leq u_n) \to \exp\{-\theta\tau\}$ for some $\theta \in (0,1]$, whereas $P(\widetilde{M}_n \leq u_n) \to \exp\{-\tau\}$. Moreover, in the iid and weakly dependent cases, the limit of the point processes of exceedances is homogeneous Poisson, whereas it is compound Poisson for the case of "stronger" dependence. The latter fact indicates that exceedances of high threshold values u_n tend to occur in clusters for dependent data. This is something we might have expected when looking at real data–sets.

The above examples suggest the following definition which allows us to distinguish between the extremal behaviour of different dependence structures.

Definition 8.1.2 (Extremal index)
Let (X_n) be a strictly stationary sequence and θ a non–negative number. Assume that for every $\tau > 0$ there exists a sequence (u_n) such that

$$\lim_{n\to\infty} n\overline{F}(u_n) = \tau , \tag{8.4}$$

$$\lim_{n\to\infty} P\left(M_n \leq u_n\right) = e^{-\theta\tau} . \tag{8.5}$$

Then θ is called the extremal index *of the sequence (X_n).* □

Remarks. 1) The definition of the extremal index can be shown to be independent of the particular sequence (u_n). More precisely, if (X_n) has extremal index $\theta > 0$ then, for any sequence of real numbers (u_n) and $\tau \in [0, \infty]$, the relations (8.4), (8.5) and $P(\widetilde{M}_n \le u_n) \to \exp\{-\tau\}$ are equivalent; see Leadbetter [416]. A particular consequence is the following: if $F \in \text{MDA}(H)$ for some extreme value distribution H then

$$c_n^{-1}(\widetilde{M}_n - d_n) \xrightarrow{d} H \quad \Leftrightarrow \quad c_n^{-1}(M_n - d_n) \xrightarrow{d} H^{\theta} \tag{8.6}$$

for appropriate norming constants $c_n > 0$ and $d_n \in \mathbb{R}$.

2) Since an extreme value distribution H is max–stable (see Definition 3.2.1), H^{θ} is of the same type as H, i.e. there exist constants $c > 0, d \in \mathbb{R}$ such that $H^{\theta}(x) = H(cx + d)$. This also implies that the limits in (8.6) can be chosen to be identical after a simple change of the norming constants. □

Example 8.1.3 (Continuation of Example 8.1.1)
From the discussion in Example 8.1.1 it is immediate that the cases (a) and (b) (iid and weakly dependent stationary sequences) yield the extremal index $\theta = 1$. In the case (c), $\theta = 0.5$ (this type of example can naturally be extended for constructing stationary sequences with extremal index $\theta = 1/k$ for any integer $k \ge 1$). The examples (d)–(f) (linear and ARCH(1) processes) show that we can get any number $\theta \in (0, 1]$ as extremal index. □

From these examples two natural questions arise:

How can we interpret the extremal index θ?

and

What is the range of the extremal index?

Section 8.1.2 is devoted to the first problem. The second one has a simple solution:

θ always belongs to the interval $[0, 1]$.

From Example 8.1.3 we already know that any number $\theta \in (0, 1]$ can be an extremal index. The case $\theta = 0$ is somewhat pathological. We refer to Leadbetter, Lindgren and Rootzén [418] and Leadbetter [417] for some examples. The cases $\theta > 0$ are of particular practical interest. It remains to show that $\theta > 1$ is impossible, but this follows from the following easy argument:

$$P(M_n \le u_n) = 1 - P\left(\bigcup_{i=1}^{n} \{X_i > u_n\}\right) \ge 1 - n\overline{F}(u_n) \,.$$

By definition of the extremal index, the lhs converges to $e^{-\theta\tau}$ whereas the rhs has limit $1-\tau$. Hence $e^{-\theta\tau} \geq 1-\tau$ for all $\tau > 0$ which is possible only if $\theta \leq 1$.

Next we ask:

Does every strictly stationary sequence have an extremal index?

Life would be easy if this were true. Indeed, extreme value theory for stationary sequences could then be derived from the corresponding results for iid sequences. The answer to the above question is (unfortunately) *no.*

Example 8.1.4 Assume (X_n) is iid with $F \in \text{MDA}(\Phi_\alpha)$ and norming constants $c_n > 0$. Assume A is a positive rv independent of (X_n). Then

$$\begin{aligned} & P\left(c_n^{-1} \max(AX_1, \ldots, AX_n) \leq x\right) \\ = \quad & P\left(c_n^{-1} M_n \leq A^{-1} x\right) \\ = \quad & EP\left(c_n^{-1} M_n \leq A^{-1} x \,\middle|\, A\right) \\ \to \quad & E \exp\left\{-x^{-\alpha} A^{\alpha}\right\}, \quad x > 0. \end{aligned}$$

□

It is worthwhile mentioning that, for large classes of stationary sequences (X_n), there exist real numbers $0 \leq \theta' \leq \theta'' \leq 1$ such that

$$e^{-\theta''\tau} \leq \liminf_{n\to\infty} P\left(M_n \leq u_n\right) \leq \limsup_{n\to\infty} P\left(M_n \leq u_n\right) \leq e^{-\theta'\tau}, \quad \tau > 0,$$

for every sequence (u_n) satisfying (8.4). A proof of this result under condition $D(u_n)$ is to be found in Leadbetter, Lindgren and Rootzén [418], Theorem 3.7.1.

8.1.2 Interpretation and Estimation of the Extremal Index

We start with a somewhat simplistic example (taken from Weissman [638]) showing the relevance of the notion of extremal index.

Example 8.1.5 Assume a dyke has to be built at the seashore to protect against floods with 95% certainty for the next 100 years. Suppose it has been established that the 99.9 and 99.95 percentiles of the annual wave–height are 10 m and 11 m, respectively. If the annual maxima are believed to be iid, then the dyke should be 11 m high $\left(0.9995^{100} \approx 0.95\right)$. But if the annual maxima are stationary with extremal index $\theta = 0.5$, then a height of 10 m is sufficient $\left(0.999^{50} \approx 0.95\right)$. □

This example brings out already that estimation of the extremal index θ must be a central issue in extreme value statistics for dependent data. Estimation of θ will be based on a number of different probabilistic interpretations of the extremal index, leading to the construction of different estimators. Throughout we exclude the degenerate case $\theta = 0$.

A First (Naive) Approach to the Estimation of θ: the Blocks Method

Starting from the definition of the extremal index θ, we have

$$P(M_n \le u_n) \approx P^\theta(\widetilde{M}_n \le u_n) = F^{\theta n}(u_n)\,,$$

provided $n\overline{F}(u_n) \to \tau > 0$. Hence

$$\lim_{n\to\infty} \frac{\ln P(M_n \le u_n)}{n \ln F(u_n)} = \theta\,. \tag{8.7}$$

This simple limit relation suggests constructing naive estimators of θ. Since we do not know $F(u_n)$ and $P(M_n \le u_n)$, these quantities have to be replaced by estimators. An obvious candidate for estimating the tail $\overline{F}(u_n)$ is its empirical version

$$\frac{N}{n} = \frac{1}{n}\sum_{i=1}^{n} I_{\{X_i > u_n\}}\,.$$

This choice is motivated by the Glivenko–Cantelli theorem for stationary ergodic sequences (X_n); see Example 2.1.4. To find an empirical estimator for $P(M_n \le u_n)$ is not straightforward. Recall from Section 4.4 that condition $D(u_n)$ implies

$$P(M_n \le u_n) \approx P^k(M_{[n/k]} \le u_n) \tag{8.8}$$

for constant k or slowly increasing $k = k(n)$. The approximation (8.8) forms the basis for the *blocks method.* For the sake of argument assume that $n = rk$ for integers $r = r(n) \to \infty$ and $k = k(n) \to \infty$. Otherwise, let $r = [n/k]$. This divides the sample $X_1, \ldots, X_n$ into k blocks of size r:

$$X_1, \ldots, X_r; \ldots; X_{(k-1)r+1}, \ldots, X_{kr}\,. \tag{8.9}$$

For each block we calculate the maximum

$$M_r^{(i)} = \max\left(X_{(i-1)r+1}, \ldots, X_{ir}\right)\,, \quad i = 1, \ldots, k\,.$$

Relation (8.8) then suggests the approximation

$$\begin{aligned} P(M_n \le u_n) &= P\left(\max_{1\le i\le k} M_r^{(i)} \le u_n\right) \approx P^k(M_r \le u_n) \\ &\approx \left(\frac{1}{k}\sum_{i=1}^{k} I_{\{M_r^{(i)}\le u_n\}}\right)^k = \left(1-\frac{K}{k}\right)^k . \end{aligned}$$

A combination of these heuristic arguments with (8.7) leads to the following estimator of θ:

$$\widehat{\theta}_n^{(1)} = \frac{k}{n}\frac{\ln(1-K/k)}{\ln(1-N/n)} = \frac{1}{r}\frac{\ln(1-K/k)}{\ln(1-N/n)} . \tag{8.10}$$

Here N is the number of exceedances of u_n by $X_1,\dots,X_n$ and K is the number of blocks with one or more exceedances. A Taylor expansion argument yields a second estimator

$$\widehat{\theta}_n^{(2)} = \frac{K}{N} = \frac{1}{r}\frac{K/k}{N/n} \approx \widehat{\theta}_n^{(1)} . \tag{8.11}$$

The blocks method accounts for *clustering* in the data. If the event

$$\left\{M_r^{(i)} > u_n\right\} = \bigcup_{j=1}^{r}\{X_{(i-1)r+j} > u_n\}$$

happens one says that *a cluster occurred in the ith block.* These events characterise the extremal behaviour of (X_n) if we assume that the size $r(n)$ of the blocks increases slowly with n. This gives us some feeling for the dependence structure in the sequence (X_n). In this sense, the extremal index is a measure of the clustering tendency of high–threshold exceedances in a stationary sequence.

There has been plenty of hand–waving in the course of the derivation of the estimators $\widehat{\theta}_n^{(i)}$. Therefore the following questions naturally arise:

What are the statistical properties of $\widehat{\theta}_n^{(1)}$ and $\widehat{\theta}_n^{(2)}$ as estimators of θ?

and

Given that we have n observations, how do we choose the values of r (or k) and u_n?

These are important questions. Partial answers are to be found in the literature; see the Notes and Comments. A flavour of what one can expect as an answer is summarised in the following remark from Weissman and Cohen [639]:

It turns out that it is not easy to obtain accurate estimates of θ.

It seems that we are in a similar situation to that in Chapter 6, where we tried to estimate the index ξ of an extreme value distribution. This should not discourage us from considering some more estimators of θ, especially as we will meet alternative interpretations of the extremal index along the way.

The Extremal Index as Reciprocal of the Mean Cluster Size

This approach is based on results by Hsing, Hüsler and Leadbetter [345], Theorems 4.1 and 4.2. They show that, under a mixing condition which is slightly stronger than $D(u_n)$, the point processes of exceedances

$$N_n = \sum_{i=1}^{n} \varepsilon_{n^{-1}i} I_{\{X_i > u_n\}}$$

converge weakly to a compound Poisson process (see Example 5.1.15)

$$N(\cdot) = \sum_{i=1}^{\infty} \xi_i \varepsilon_{\Gamma_i}(\cdot)$$

provided $n\overline{F}(u_n) \to \tau > 0$. Note that this is in accordance with the results for linear and ARCH processes; see Example 8.1.1(d)–(f). The homogeneous Poisson process underlying $N(\cdot)$ has intensity $\theta\tau$, and the iid cluster sizes ξ_j of $N(\cdot)$ have distribution (π_j) on $\mathbb{N}$. Also notice that

$$\begin{aligned} EN(0,1] &= E\sum_{i=1}^{\infty} \xi_i \varepsilon_{\Gamma_i}(0,1] \\ &= E\sum_{i=1}^{\infty} \varepsilon_{\Gamma_i}(0,1]\, E\xi_1 \\ &= \theta\tau\, E\xi_1 \,. \end{aligned} \qquad (8.12)$$

Under general assumptions the following relation holds:

$$\begin{aligned} \pi_j &= \lim_{n\to\infty} \pi_j(n) \\ &= \lim_{n\to\infty} P\left(\sum_{i=1}^{r} I_{\{X_i > u_n\}} = j \,\middle|\, \sum_{i=1}^{r} I_{\{X_i > u_n\}} > 0 \right), \quad j \in \mathbb{N}. \end{aligned}$$

Here again we have used the blocks as defined by (8.9), and $r = r(n)$ is the size of such a block. The integer sequence $(r(n))$ has to satisfy $r(n) \to \infty$ and $r(n)/n \to 0$ and some more specific growth conditions. Moreover, under an additional summability condition on $(\pi_j(n))$,

$$\lim_{n\to\infty}\sum_{j=1}^{\infty} j\pi_j(n) = \sum_{j=1}^{\infty} j\pi_j = E\xi_1 \,. \tag{8.13}$$

(Some summability condition is indeed needed; see Smith [592].) Recalling (8.12), we see that (8.13) has the following interesting interpretation:

$$\begin{aligned}\tau = \lim_{n\to\infty} n\overline{F}(u_n) &= \lim_{n\to\infty} EN_n(0,1] = \lim_{n\to\infty} E\sum_{i=1}^{n} I_{\{X_i>u_n\}} \\ &= EN(0,1] = \theta\tau\ E\xi_1 .\end{aligned}$$

This means that $\theta = (E\xi_1)^{-1}$ can be interpreted as the *reciprocal of the mean cluster size* of the limiting compound Poisson process N.

This interpretation of θ suggests an estimator based on the blocks method:

$$\widehat{\theta}_n^{(2)} = \frac{\sum_{i=1}^{k} I_{\{M_r^{(i)}>u_n\}}}{\sum_{i=1}^{n} I_{\{X_i>u_n\}}} = \frac{K}{N}\,,$$

i.e. number K of clusters of exceedances divided by the total number N of exceedances. The same estimator has already been suggested as an approximation to $\widehat{\theta}_n^{(1)}$.

The Extremal Index as Conditional Probability: the Runs Method

O'Brien [485] proved, under a weak mixing condition, that the following limit relation holds:

$$\begin{aligned}P(M_n \le u_n) &= (F(u_n))^{nP(M_{2,s}\le u_n|X_1>u_n)} + o(1) \\ &= \exp\{-nP(X_1>u_n, M_{2,s}\le u_n)\} + o(1)\end{aligned}$$

provided $n\overline{F}(u_n) \to \tau$. Here

$$M_{2,s} = \max(X_2,\ldots,X_s)\,,$$

and $s = s(n)$ satisfies $s/n \to 0$, $s \to \infty$ and some more specific growth conditions. On the other hand, by definition of the extremal index,

$$P(M_n \le u_n) = \exp\{-\theta\tau\} + o(1)\,.$$

Hence, under the conditions above,

$$\lim_{n\to\infty} \theta_n(s(n),u_n) = \lim_{n\to\infty} P(M_{2,s}\le u_n \mid X_1>u_n) = \theta\,.$$

Thus θ can be interpreted as a limiting conditional probability. The conditional probability $\theta_n(s(n),u_n)$ is some measure of the clustering tendency of high threshold exceedances: $M_{2,s}$ can be less than u_n only if X_1 is the last element in a cluster of values which exceed u_n. If large values appear in clusters, there must be longer intervals between different clusters. As a consequence, $P(M_n \le u_n)$ will typically be larger than for independent observations.

O'Brien's result has been used to construct an estimator of θ based on *runs*:

$$\widehat{\theta}_n^{(3)} = \frac{\sum_{i=1}^{n-r} I_{A_{i,n}}}{\sum_{i=1}^{n} I_{\{X_i > u_n\}}} = \frac{\sum_{i=1}^{n-r} I_{A_{i,n}}}{N} , \tag{8.14}$$

where

$$A_{i,n} = \{X_i > u_n, X_{i+1} \le u_n, \ldots, X_{i+r} \le u_n\} . \tag{8.15}$$

This means we take any sequence of $r = r(n)$ consecutive observations below the threshold as separating two clusters.

Example 8.1.6 We re–consider Example 8.1.1(c); see also Example 4.4.2. We show that $\theta = 1/2$ can be calculated explicitly in the three different ways explained above. Assume that $n\overline{F}(u_n) \to \tau > 0$.

(a) The following is immediate from the definition of X_n:

$$\frac{\ln P\left(M_n \le u_n\right)}{n \ln F\left(u_n\right)} \to \frac{1}{2} .$$

(b) High threshold exceedances of (X_n) typically appear in pairs. Hence $\pi_2 = 1$ and $E\xi_1 = \sum_{j=1}^{\infty} j\pi_j = 2$. Since θ is the reciprocal of the mean cluster size $E\xi_1$, $\theta = 2^{-1}$.

(c) Finally, consider the conditional probability

$$\begin{aligned} & P\left(X_2 \le u_n, \ldots, X_s \le u_n \mid X_1 > u_n\right) \\ = \; & \frac{P\left(X_1 > u_n, X_2 \le u_n, \ldots, X_s \le u_n\right)}{P\left(X_1 > u_n\right)} \\ = \; & \frac{F^{s/2}(u_n) - F^{(s+1)/2}(u_n)}{\overline{F}(u_n)} = \frac{F^{s/2}(u_n)}{1 + F^{1/2}(u_n)} . \end{aligned}$$

The latter expression converges to $1/2$ provided

$$F^{s/2}(u_n) = \exp\{2^{-1} s \ln(1 - \overline{F}(u_n)\} \to 1 .$$

This is clearly satisfied if $s = s(n) = o(n)$. □

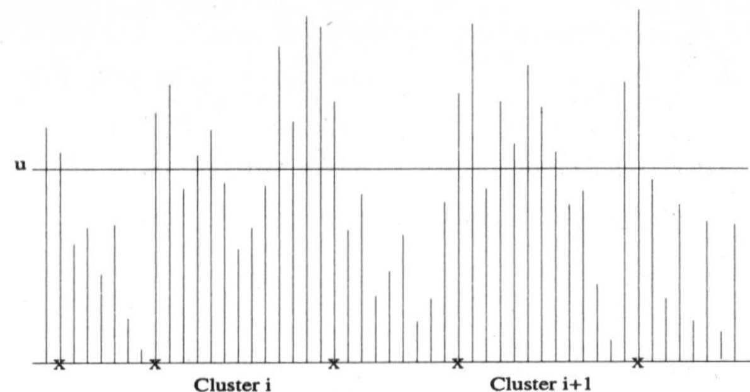

Figure 8.1.7 *Clusters defined by the runs method. We chose* $r = 5$*; the cluster size is equal to* 9 *for both clusters.*

8.1.3 Estimating the Extremal Index from Data

In this section we compare the performance of the estimators $\widehat{\theta}_n^{(i)}$ of the extremal index θ both, for real and simulated data. Table 8.1.8 summarises the results for the exchange rate data presented in Figure 8.4.4. For data of this type one often claims that ARCH or GARCH models yield a reasonable fit. An ARCH(1) fit, based on maximum likelihood estimation, yields

$$X_n = \sqrt{1.9 \cdot 10^{-5} + 0.5\, X_{n-1}^2}\; Z_n\,, \quad n \in \mathbb{N}, \tag{8.16}$$

for iid $N(0,1)$ noise (Z_n); see Section 8.4.2. For the above model an ARCH(1) time series with the same length as for the exchange rate data was simulated. The estimators $\widehat{\theta}^{(i)}$ are given in Table 8.1.9. From Table 8.4.23 we may read off the corresponding theoretical value $\theta = 0.835$. This shows that θ is clearly underestimated. Also notice that the estimates strongly depend on the chosen threshold value u and the size r.

In Figures 8.1.10–8.1.13 the number of exceedances of a given threshold u in a *cluster of observations* is visualised for the above data. Both, the blocks and the runs method, are illustrated. For the former, r denotes the block size as defined in (8.9). Every block is regarded as a cluster. For the same r we define a cluster in the runs method as follows: it is a set of successive observations separated from the neighbouring sets by a least r values below u. See Figure 8.1.7 for an illustration. The *cluster size* is then the number of exceedances of u in the cluster.

Every figure consists of three pairs of graphs. For each pair the upper (lower) graph illustrates the blocks (runs) method for the same u and r.

u	r	$\widehat{\theta}^{(1)}$	$\widehat{\theta}^{(2)}$	$\widehat{\theta}^{(3)}$	r	$\widehat{\theta}^{(1)}$	$\widehat{\theta}^{(2)}$	$\widehat{\theta}^{(3)}$
0.015	100	0.524	0.347	0.147	50	0.613	0.480	0.213
0.020	100	0.653	0.536	0.321	50	0.715	0.643	0.464
0.025	100	0.689	0.636	0.545	50	0.758	0.727	0.636

Table 8.1.8 *Estimators of θ for the exchange rate data of Figure 8.4.4* ($n = 4\,274$) *for different thresholds u and sizes r.*

u	r	$\widehat{\theta}^{(1)}$	$\widehat{\theta}^{(2)}$	$\widehat{\theta}^{(3)}$	r	$\widehat{\theta}^{(1)}$	$\widehat{\theta}^{(2)}$	$\widehat{\theta}^{(3)}$
0.015	100	0.625	0.403	0.149	50	0.820	0.612	0.403
0.020	100	0.708	0.571	0.393	50	0.714	0.643	0.536
0.025	100	0.632	0.583	0.583	50	0.694	0.667	0.583

Table 8.1.9 *Estimators of θ for $n = 4\,274$ simulated data from the* ARCH(1) *model given in (8.16) for different thresholds u and sizes r.*

Notes and Comments

The concept of extremal index originates from Newell [479], Loynes [430] and O'Brien [483]. A firm definition was given by Leadbetter [416]. An overview of results concerning the extremal index is given in Smith and Weissman [600] and Weissman [638].

Weissman and Cohen [639] present various models where the extremal index can be calculated explicitly. Special methods have been developed for ARMA processes (see Section 5.5) and Markov processes (see Leadbetter and Rootzén [419], Perfekt [491] and Rootzén [551]).

The presence of exceedance clustering also affects the asymptotic distribution of the upper order statistics. The following result is a consequence of Theorem 6.1 in Hsing et al. [345] (which holds under certain mixing conditions): whenever $(c_n^{-1}(M_n - d_n))$ converges weakly, the limit distribution is equal to H^θ for an extreme value distribution H and

$$\lim_{n\to\infty} P\left(c_n^{-1}\left(X_{k,n} - d_n\right) \le x\right) = H^\theta(x) \sum_{j=0}^{k-1} \frac{\left(-\ln H^\theta(x)\right)^j}{j!} \sum_{i=j}^{k-1} \Pi^{j*}(i)\,,$$

where $\Pi^{0*} = 1$ and Π^{j*} is the j–fold convolution of $\Pi = (\pi_j)$; see also Cohen [127], Hsing [339] and Leadbetter and Rootzén [419].

Hsing [339, 340, 341, 342] investigates the asymptotic properties of the estimators of θ. Aspects of bias, variance and the optimal choice of $r(n)$ and u_n for the $\widehat{\theta}_n^{(i)}$ are discussed in Smith and Weissman [600]. Real–life data analyses involving extremal index estimation have been carried out by Buishand [102], Coles [131], Davison and Smith [166], Smith, Tawn and Coles [598], and Tawn [615].

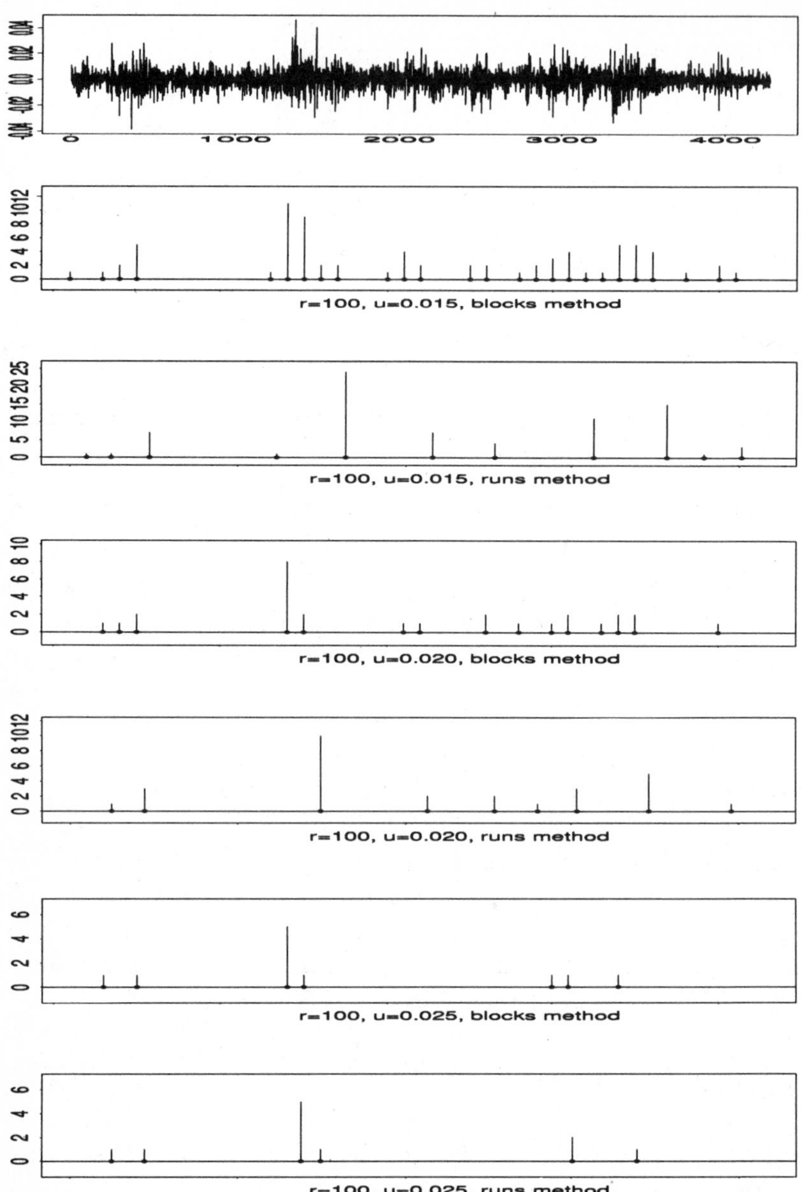

Figure 8.1.10 *Clusters of exceedances by the blocks method (top figures) in comparison with the runs method (bottom figures) for the exchange rate data from Figure 8.4.4 (top). The chosen values are* $r = 100$ *and* $u = 0.015$ *(top two),* $u = 0.020$ *(middle two) and* $u = 0.025$ *(bottom two). These figures clearly indicate the dependence in the data.*

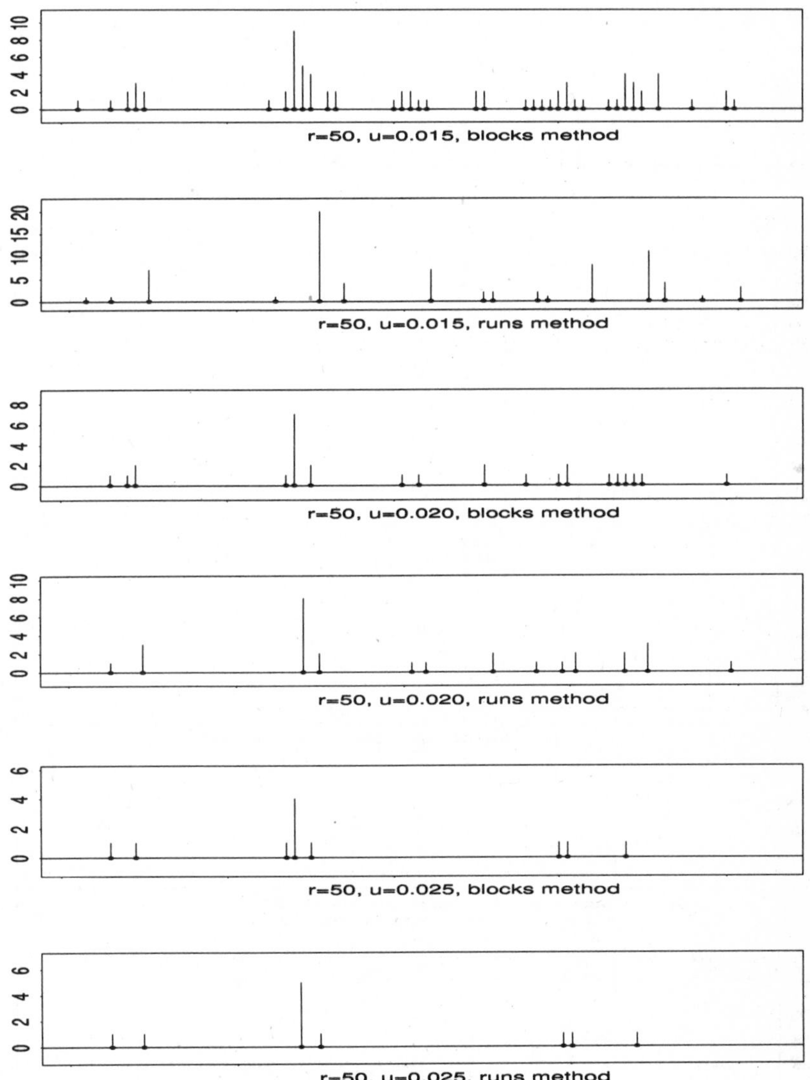

Figure 8.1.11 *Continuation of Figure 8.1.10. The chosen values are* $r = 50$ *and* $u = 0.015$ *(top two),* $u = 0.020$ *(middle two) and* $u = 0.025$ *(bottom two).*

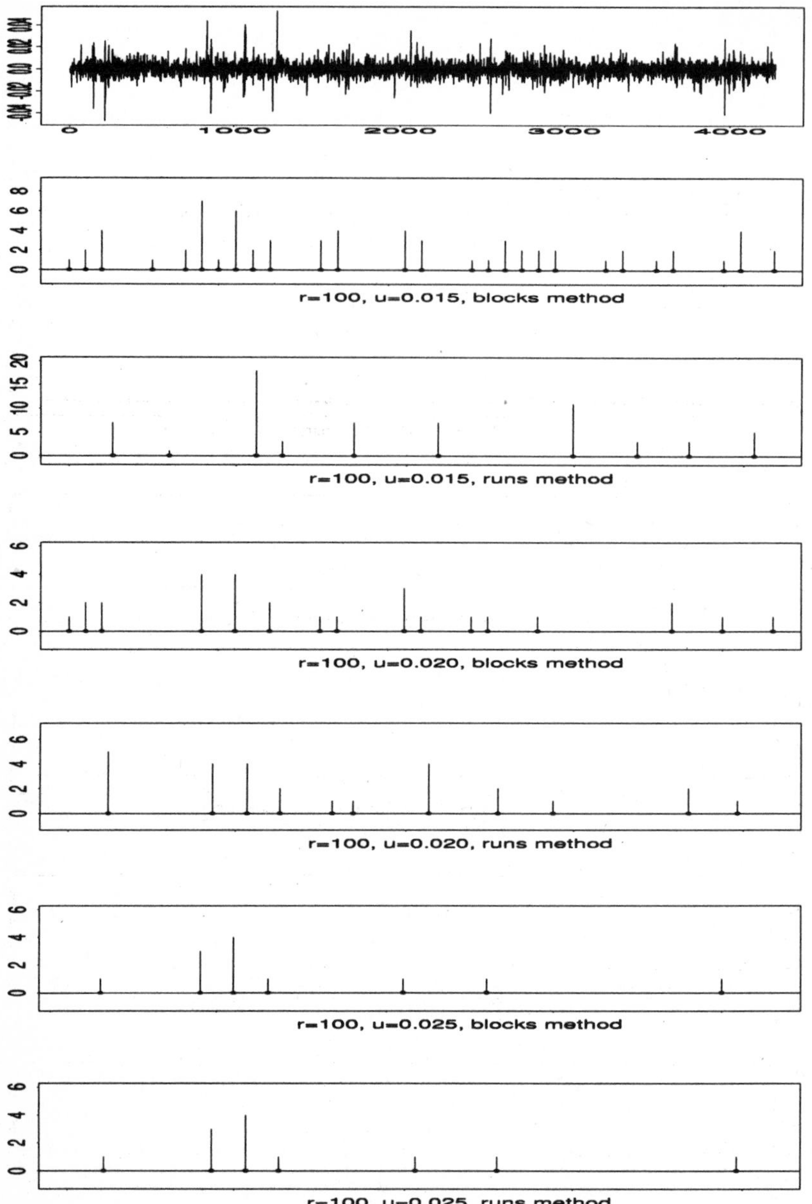

Figure 8.1.12 *Clusters of exceedances by the blocks method (top figures) in comparison with the runs method (bottom figures) for simulated* ARCH(1) *data (top) with parameters* $\widehat{\lambda} = 0.5$ *and* $\widehat{\beta} = 1.9 \cdot 10^{-5}$. *The chosen values are* $r = 100$ *and* $u = 0.015$ *(top two),* $u = 0.020$ *(middle two) and* $u = 0.025$ *(bottom two),*

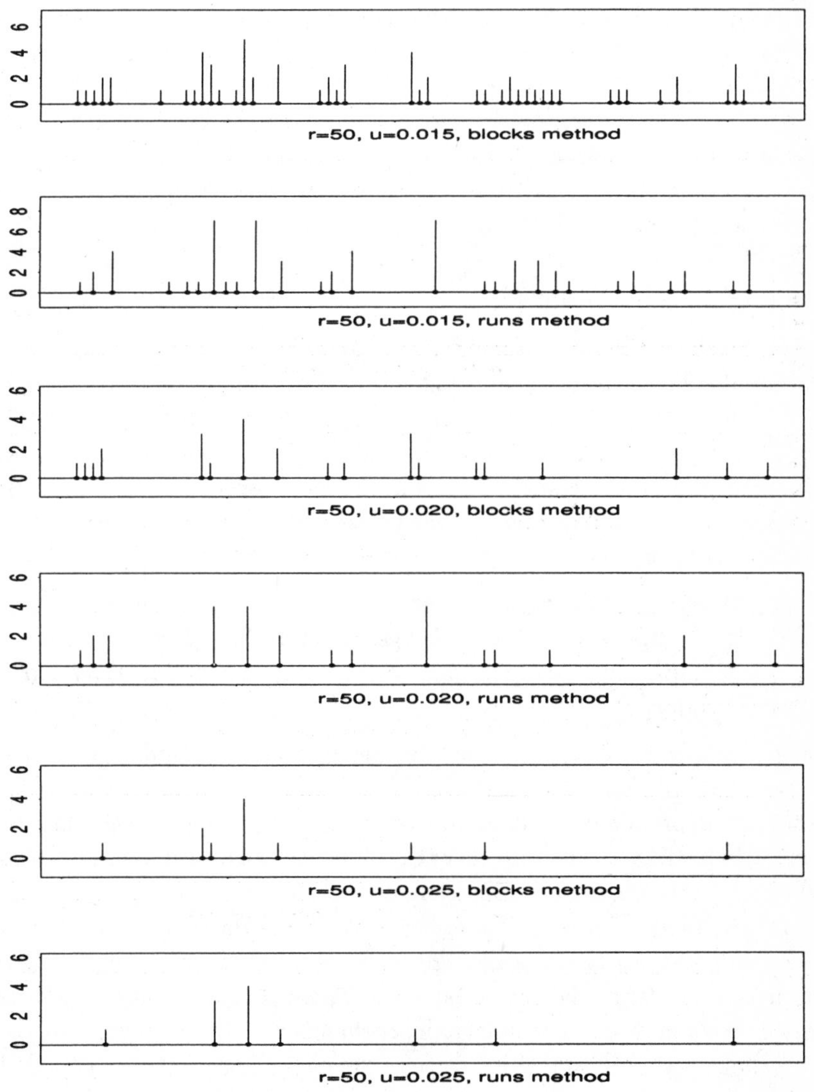

Figure 8.1.13 *Continuation of Figure 8.1.12. The chosen values are* $r = 50$ *and* $u = 0.015$ *(top two),* $u = 0.020$ *(middle two) and* $u = 0.025$ *(bottom two).*

8.2 A Large Claim Index

8.2.1 The Problem

Throughout the book, we have seen various examples where a comparison between the behaviour of the partial sum S_n and the partial maximum M_n (or indeed more general order statistics) of an iid sequence $X, X_1, \ldots, X_n$ with df F was instrumental in deriving estimates on extreme values. One very prominent example for non-negative rvs was the introduction of the family $\mathcal{S}$ of subexponential distributions as *the* kind of dfs fit for modelling heavy-tailed phenomena. Recall from Definition 1.3.3 that for $F \in \mathcal{S}$

$$\lim_{x\to\infty} \frac{P(S_n > x)}{P(M_n > x)} = 1\,, \quad n \geq 1\,.$$

Hence the tail of the partial maximum *essentially* determines the tail of the partial sum. The idea that heavy-tailedness corresponds to a statement of the type: "The behaviour of S_n is mainly determined by few upper order statistics", has been discussed in many publications. For instance, in Rootzén and Tajvidi [553] on the accumulated loss in the most severe storms encountered by a Swedish insurance group over a 12-year period 1982–1993, the following summary is to be found.

> *It can be seen that the most costly storm contributes about* 25% *of the total amount for the period, that it is* 2.7 *times bigger than the second worst storm, and that four storms together make up about half of the claims.*

Some of the results of this type will be treated in this section.

We would like to start our discussion however by a story. In a consulting discussion on premium calculations of one of us with two non-mathematicians working in insurance, the problem of the influence of extreme values on rating came up. Early on in the discussion it was realised that we were talking about two different kinds of Pareto law. Indeed, the Pareto law they referred to corresponded to the so-called 20–80 rule-of-thumb used by practicing actuaries when large claims are involved. This rule states that 20% of the individual claims are responsible for more than 80% of the total claim amount in a well defined portfolio. Using some of the methods developed so far in the book, we would like to answer the following question:

Can one characterise those portfolios where the 20–80 rule applies?

Or more precisely,

Classify those claim size dfs for which the 20–80 rule holds.

The next section will be devoted to the answer of the latter question.

8.2.2 The Index

Let $X_1, \ldots, X_n$ denote the first n claims in a portfolio. The X_i are assumed to be iid with continuous df F and finite mean $\mu > 0$. As before, we denote by $X_{n,n} \le \cdots \le X_{1,n}$ the corresponding order statistics. The total claim amount of the first n claims is denoted by $S_n = \sum_{k=1}^n X_k$ and $F_n^{\leftarrow}$ stands for the empirical quantile function. Consequently,

$$F_n^{\leftarrow}(y) = X_{k,n} \quad \text{for} \quad 1 - \frac{k}{n} < y \le 1 - \frac{k-1}{n}\,.$$

In particular, for $i = 1, \ldots, n$, $F_n^{\leftarrow}(i/n) = X_{n-i+1,n}$. The rvs needed in the analysis of the problem posed in the introduction are

$$T_n(p) = \frac{X_{1,n} + X_{2,n} + \cdots + X_{[pn],n}}{S_n}\,, \quad \frac{1}{n} < p \le 1\,.$$

Hence $T_n(p)$ is the proportion of the $[np]$ largest claims to the aggregate claim amount S_n. The 20–80 rule now says that $T_n(0.2)$ accounts for 80% of S_n. The following result gives us the behaviour of $T_n(p)$ for n large. For its formulation we introduce the function

$$D_G(p) = \frac{1}{\mu_G} \int_{1-p}^{1} G^{\leftarrow}(y)dy\,, \quad p \in [0,1]\,,$$

where G is the continuous df of a positive rv Y and

$$\mu_G = EY = \int_0^\infty y dG(y) = \int_0^1 G^{\leftarrow}(y)dy\,.$$

Theorem 8.2.1 (Order statistics versus sums, asymptotic behaviour)
Suppose $X_1, \ldots, X_n$ are iid positive rvs with continuous df F and finite mean μ. Then as $n \to \infty$,

$$\sup_{p\in[0,1]} |T_n(p) - D_F(p)| \to 0 \quad \text{a.s.}$$

Proof. Observe that

$$\begin{aligned}
\frac{1}{n}\left(X_{1,n} + X_{2,n} + \cdots + X_{[np],n}\right) &= \frac{1}{n} \sum_{i=0}^{[np]-1} F_n^{\leftarrow}\left(\frac{n-i}{n}\right) \\
&= \frac{1}{n} \int_0^{[np]-1} F_n^{\leftarrow}\left(\frac{n-y}{n}\right) dy \\
&= \int_{1-\frac{[np]}{n}+\frac{1}{n}}^{1} F_n^{\leftarrow}(v)\, dv \\
&= \mu_{F_n} D_{F_n}\left(\frac{[np]}{n} - \frac{1}{n}\right),
\end{aligned}$$

where

$$\mu_{F_n} = \overline{X}_n = n^{-1}S_n = \int_0^\infty y dF_n(y) .$$

We have by the SLLN that $\overline{X}_n \stackrel{\text{a.s.}}{\to} \mu$. Moreover, a Borel–Cantelli argument yields

$$\begin{aligned} \mu < \infty \quad &\Leftrightarrow \quad \sum_{n=1}^{\infty} P(X_n > \varepsilon n) < \infty\,, \quad \forall \varepsilon > 0 \\ &\Leftrightarrow \quad P(X_n > \varepsilon n \quad \text{i.o.}) = 0\,, \quad \forall \varepsilon > 0 \\ &\Leftrightarrow \quad \lim_{n\to\infty} n^{-1} X_n = 0 \quad \text{a.s.} \\ &\Leftrightarrow \quad \lim_{n\to\infty} n^{-1} M_n = 0 \quad \text{a.s.} \end{aligned} \tag{8.17}$$

Hence

$$\left| D_{F_n}\left(\frac{[np]}{n} - \frac{1}{n}\right) - D_{F_n}(p) \right| \;\leq\; \frac{2n^{-1}M_n}{\overline{X}_n} \stackrel{\text{a.s.}}{\to} 0 .$$

We thus have $T_n(p) = (1 + o(1))D_{F_n}(p)$ a.s. uniformly for p and hence it suffices to show that

$$\sup_{p\in[0,1]} |D_F(p) - D_{F_n}(p)| \to 0 \quad \text{a.s.}$$

Notice that

$$\begin{aligned} &\left| \int_{1-p}^{1} F^{\leftarrow}(y)dy - \int_{1-p}^{1} F_n^{\leftarrow}(y)dy \right| \\ &\leq \int_{1-p}^{1} |F^{\leftarrow}(y) - F_n^{\leftarrow}(y)|\, dy \\ &\leq \int_{0}^{1} |F^{\leftarrow}(y) - F_n^{\leftarrow}(y)|\, dy . \end{aligned} \tag{8.18}$$

From the Glivenko–Cantelli theorem (Example 2.1.4) we know that

$$\sup_x |F(x) - F_n(x)| \stackrel{\text{a.s.}}{\to} 0 .$$

Hence by Proposition A1.7, $F_n^{\leftarrow}(y) \to F^{\leftarrow}(y)$ a.s. for every continuity point y of $F^{\leftarrow}$. Moreover, the function $|F^{\leftarrow}(y) - F_n^{\leftarrow}(y)|$ is dominated by $F^{\leftarrow}(y) + F_n^{\leftarrow}(y)$ and

$$\begin{aligned} \int_0^1 |F^{\leftarrow}(y) - F_n^{\leftarrow}(y)|\, dy \;&\leq\; \int_0^1 F^{\leftarrow}(y) + \int_0^1 F_n^{\leftarrow}(y)dy \\ &=\; \mu + \mu_{F_n} \;=\; 2\mu + o(1) \quad \text{a.s.}, \end{aligned}$$

where we used the SLLN. Combining the argument above and Pratt's lemma (see Pratt [509]), we may conclude that the right–hand side in (8.18) converges to zero with probability 1. This concludes the proof. □

Motivated by the previous result, we have the following definition.

Definition 8.2.2 (A large claim index)
Let F be a continuous df on $(0,\infty)$ with finite mean μ. For $0 \le p \le 1$, we define the large claim index of F at p *by*

$$D_F(p) = \frac{1}{\mu} \int_{1-p}^{1} F^{\leftarrow}(y)\, dy\,. \qquad \square$$

Remarks. 1) The value $D_F(p)$ measures the extent to which the $100p\,\%$ largest claims in a portfolio contribute to the total claim amount. Theorem 8.2.1 suggests to call $T_n(p) = D_{F_n}([np]/n - 1/n)$ the *empirical large claim index.*

2) If one defines $L_F(1-p) = 1 - D_F(p)$, then L_F becomes the so–called *Lorenz curve.* See for instance Csörgő, Csörgő and Horváth [144], Goldie [270] and references therein for a detailed discussion on properties of L_F.

3) The condition $\mu < \infty$ in Theorem 8.2.1 can be dropped, yielding $T_n(p) \to 1$ in probability, whenever $\mu = \infty$ and the boundedness condition $\overline{F}(x) \le (1+x)^{-\alpha}$ for some $0 < \alpha \le 1$ and x large holds. For a proof see Aebi, Embrechts and Mikosch [5], Theorem 2.

4) In Hipp [327] the asymptotic distribution of $\sqrt{n}(T_n(p) - D_F(p))$ is studied by using the delta–method and the CLT for empirical processes; see Pollard [504]. This allows one to construct asymptotic confidence bands for $T_n(p)$. See also Csörgő et al. [144].

8.2.3 Some Examples

The basic question from the introduction can now be formulated as follows:

Give examples of dfs F for which $D_F(0.2)$ is approximately 0.8.

Below we have summarised D_F–values for the most important classes of claim size dfs. The parametrisations used are (see Tables 1.2.5 and 1.2.6 for details):

- Pareto (α) : $\overline{F}(x) = (1+x)^{-\alpha}$, $x \ge 0$,
- loggamma (α, β),
- lognormal (μ, σ),
- gamma (α, β),
- exponential.

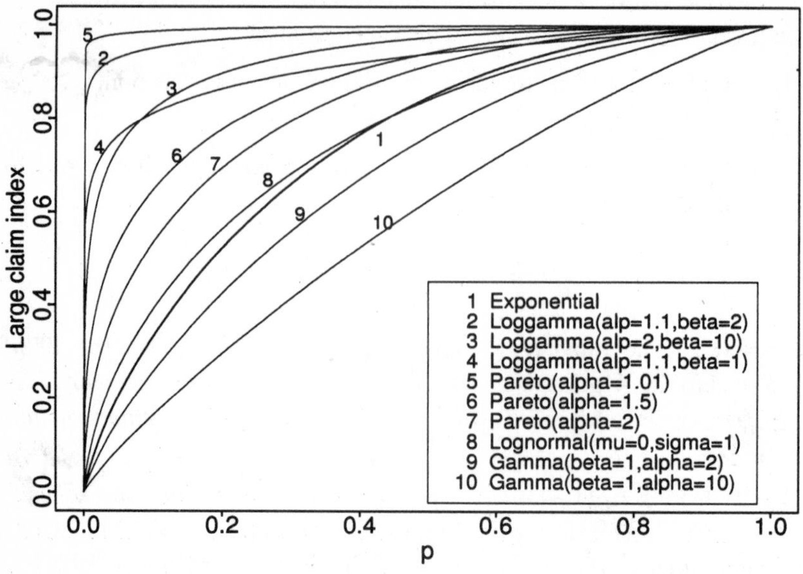

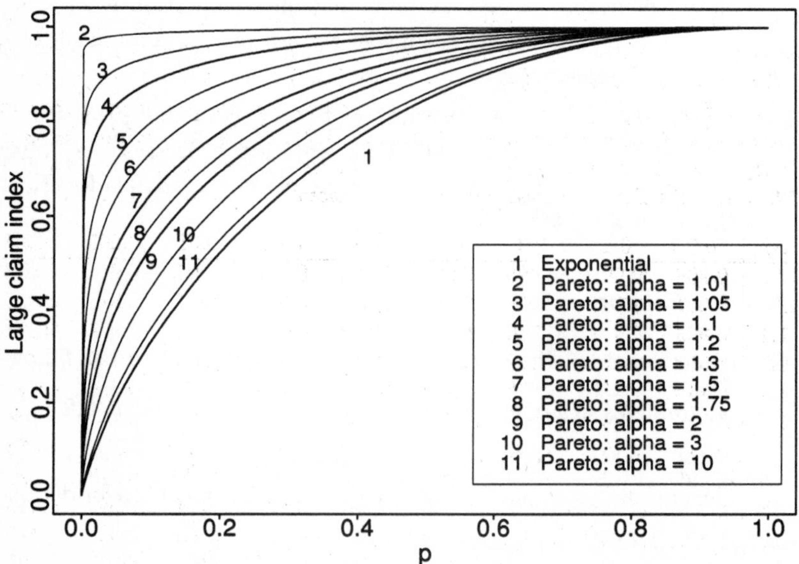

Figure 8.2.3 *The large claim index $D_F(p)$ across a wide class of potential claim size distributions (top) and for the family of the Pareto distributions with the exponential as a light–tailed limiting case.*

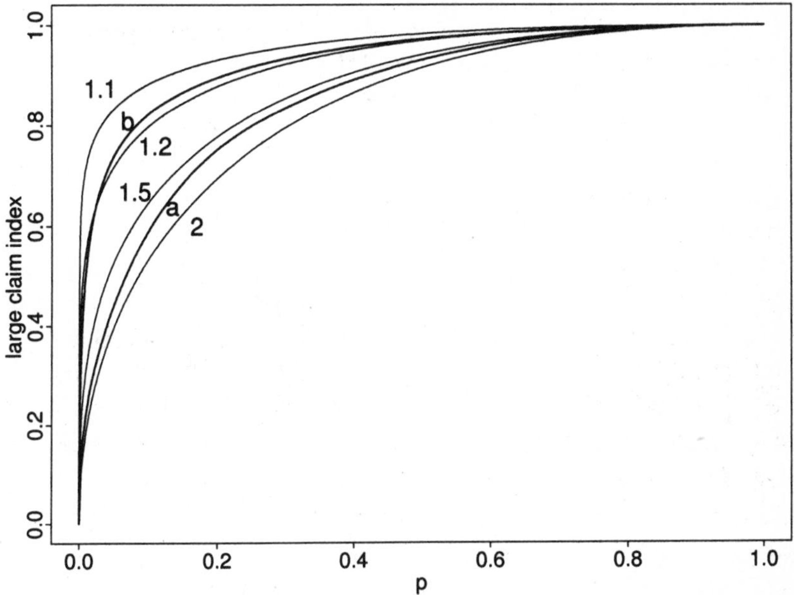

Figure 8.2.4 *The empirical large claim indices $T_n(p)$ for 647 excesses of 2.5 millions Danish Kroner corresponding to 2493 Danish fire insurance losses from 1980 to 1990 (curve (a)) and for 8043 industrial fire losses from 1992 – 1994 (curve (b)). For comparison the theoretical large claim index $D_F(p)$ is plotted for the Pareto distribution with shape parameters $\alpha \in \{1.1, 1.2, 1.5, 2\}$. The industrial fire losses are very heavy-tailed; their index curve appears between the Pareto index for $\alpha = 1.1$ and $\alpha = 1.2$; the Danish excess data appear between $\alpha = 1.5$ and $\alpha = 2$.*

$\alpha\backslash p$	0.5	0.4	0.3	0.2	0.1	0.05	0.01	0.005	0.001
1.01	0.998	0.997	0.995	0.992	0.986	0.980	0.965	0.958	0.943
1.05	0.991	0.985	0.976	0.963	0.936	0.908	0.843	0.816	0.756
1.1	0.983	0.972	0.956	0.930	0.882	0.833	0.723	0.679	0.587
1.2	0.969	0.950	0.922	0.878	0.798	0.718	0.555	0.495	0.379
1.4	0.948	0.918	0.873	0.804	0.685	0.575	0.372	0.306	0.194
1.7	0.928	0.886	0.825	0.736	0.589	0.460	0.248	0.188	0.098
2	0.914	0.865	0.795	0.694	0.532	0.397	0.190	0.136	0.062
3	0.890	0.829	0.744	0.626	0.446	0.307	0.119	0.078	0.028
5	0.872	0.802	0.708	0.580	0.392	0.255	0.086	0.052	0.016
10	0.859	0.784	0.684	0.549	0.359	0.225	0.068	0.040	0.011
1000	0.847	0.767	0.661	0.522	0.331	0.200	0.056	0.032	0.008
Exp	0.847	0.767	0.661	0.522	0.330	0.200	0.056	0.031	0.008

Table 8.2.5 *Large claim index $D_F(p)$ for different Pareto laws with index α and the exponential distribution.*

In Figure 8.2.3, D_F is plotted for a wide family of potential claim size distributions. The 20–80 rule seems to apply for Pareto dfs with parameter α in the range 1.3 to 1.5. Table 8.2.5 contains the calculated values of D_F for specific Pareto dfs and p–values. It is for instance seen that for F Pareto with $\alpha = 1.4$, $D_F(0.2) = 0.804$ exactly explaining the 20–80 rule. The information coming out of Figure 8.2.4 once more confirms the heavy–tailed behaviour often encountered in non–life insurance data. Both the Danish fire insurance data, as well as the industrial fire insurance data, though having finite mean, correspond to Pareto models with infinite variance. The Danish data also appear less heavy–tailed than the industrial fire data, a conclusion we already reached in Sections 6.2.2 and 6.5.2. At this point we would like to stress that the large claim index introduced in this section should be viewed *only* as a quick diagnostic tool; it therefore could have been included in the set of exploratory data analysis techniques in Section 6.2. We decided to spend a separate section on it because of the importance of 20–80 type rules often used by applied actuaries.

8.2.4 On Sums and Extremes

Suppose $X, X_1, \ldots, X_n$ are iid with df F concentrated on $(0, \infty)$. It immediately follows from (8.17) and the SLLN that $\mu = EX < \infty$ implies $M_n/S_n \stackrel{\text{a.s.}}{\to} 0$. O'Brien [484] also showed that the converse implication holds. Thus

$$\lim_{n\to\infty} \frac{M_n}{S_n} = 0 \quad \text{a.s.} \quad \text{if and only if} \quad EX < \infty\,, \tag{8.19}$$

so that for rvs with finite mean, in a strong (a.s.) sense,

the contribution of the maximum to the sum is asymptotically neglible.

Statement (8.19) has been further refined by O'Brien [484]. indeed, by weakening a.s. convergence to convergence in probability, he obtained

$$M_n/S_n \stackrel{P}{\to} 0 \quad \text{if and only if} \quad \int_0^x y\,dF(y) \in \mathcal{R}_0\,. \tag{8.20}$$

The above results (8.19) and (8.20) give conditions so that the maximum (and a–fortiori every order statistic) is asymptotically negligible with respect to the sum. A natural question concerns the other extreme case, namely

Under what conditions does $M_n/S_n \to 1$ in a specified way?

The following result was proved in Arov and Bobrov [21] (sufficiency) and Maller and Resnick [435] (necessity):

$$M_n/S_n \stackrel{P}{\to} 1 \quad \text{if and only if} \quad \overline{F} \in \mathcal{R}_0\,. \tag{8.21}$$

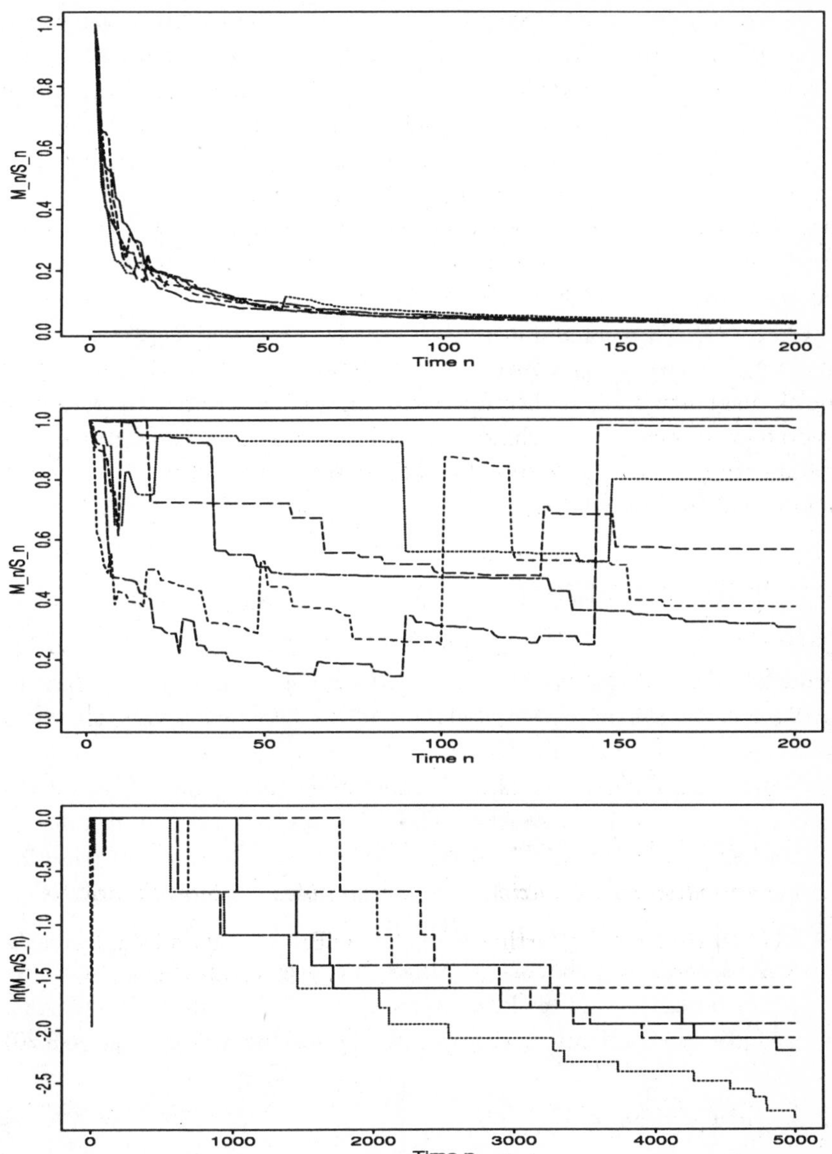

Figure 8.2.6 *Five realisations of* (M_n/S_n) *for iid standard exponential rvs (top) and 0.5–stable positive rvs (middle). In the first case* $M_n/S_n \stackrel{\text{a.s.}}{\rightarrow} 0$, *in the second one* (M_n/S_n) *converges in distribution. The bottom graph shows realisations of* $(\ln(M_n/S_n))$ *for a df with tail* $\overline{F}(x) = 1/\ln(x)$, $x \geq e$. *In this case,* $M_n/S_n \stackrel{P}{\rightarrow} 1$, *hence* $\ln(M_n/S_n) \stackrel{P}{\rightarrow} 0$. *The convergence appears to be very slow.*

Pruitt [516] gives necessary and sufficient conditions for the relation $M_n/S_n \to 1$ a.s.; this clearly means that one has to consider a subclass of distributions with $\overline{F} \in \mathcal{R}_0$. The latter condition is a very strong one indeed, it implies that $EX^\varepsilon = \infty$ for all $\varepsilon > 0$. A natural question therefore concerns situations in between (8.20) and (8.21). Twice the class $\mathcal{R}_0$ of slowly varying functions entered as *the* characterising class. The following result ((a1) $\Leftrightarrow$ (a2)) may therefore not come as a surprise.

(*a*) Equivalent are:

(*a1*) $M_n/S_n \stackrel{d}{\to} Y_1$ for some non–degenerate Y_1,

(*a2*) $\overline{F} \in \mathcal{R}_{-\alpha}$ for some $\alpha \in (0,1)$,

(*a3*) $\lim_{n\to\infty} E(S_n/M_n) = c_1 \in (1,\infty)$.

(*b*) If $\mu < \infty$, then equivalent are:

(*b1*) $(S_n - n\mu)/M_n \stackrel{d}{\to} Y_2$ for some non–degenerate Y_2,

(*b2*) $\overline{F} \in \mathcal{R}_{-\alpha}$ for some $\alpha \in (1,2)$,

(*b3*) $\lim_{n\to\infty} E((S_n - n\mu)/M_n) = c_2 \in (1,\infty)$.

The implication (a1)$\Rightarrow$(a2) is proved in Breiman [89], (a2)$\Rightarrow$(a1) in Chow and Teugels [117], for the rest see Bingham and Teugels [73].

Remark. Reconsidering the results of this section, we have learnt that the probabilistic behaviour of the ratio M_n/S_n for iid positive rvs (claims in the insurance context) characterises the underlying df only for $\overline{F} \in \mathcal{R}_{-\alpha}$, $0 \le \alpha < 2$. Although this class of distributions is not unimportant for the purposes of insurance it is nevertheless a relatively small class. For instance, it does not help to discriminate data from a lognormal or a Pareto distribution with finite variance. Therefore the large claim index introduced above offers some more flexibility for discriminating dfs F with heavy tails.

Nevertheless, the limit behaviour of M_n/S_n and the corresponding quantities for X_i^p can be used as an exploratory statistical tool for detecting whether EX^p is finite; see Section 6.2.6 where several data–sets were considered. □

Notes and Comments

In our discussion in Section 8.2.2 we closely followed Aebi et al. [5]. Some of the earlier work concerning the behaviour of M_n/S_n in the iid case is to be found in Arov and Bobrov [21] and Darling [155]. An early paper linking the asymptotic behaviour of partial sums with that of order statistics is Smirnov [586]. Since then many publications on the relation between maxima and sums appeared. In order to get up–to–date concerning sums, trimmed sums (i.e. sums minus some order statistics) and maxima for iid rvs with general df F, read Kesten and Maller [380] and consult references therein. See also Hahn, Mason and Weiner [305] on this topic.

The questions discussed above for the iid case are more naturally studied in a two–dimensional set–up, i.e. analyse the asymptotic behaviour of the vectors (S_n, M_n) properly normalised. Assume that there exist norming sequences (a_n), (b_n), (c_n) and (d_n) such that

$$\widetilde{S}_n = a_n^{-1}\,(S_n - b_n) \quad \text{and} \quad \widetilde{M}_n = c_n^{-1}\,(M_n - d_n)$$

converge weakly to $\widetilde{S}$, respectively $\widetilde{M}$. This means that $F \in \mathrm{DA}(\alpha) \cap \mathrm{MDA}(H)$ for some $\alpha \in (0,2]$ and extreme value distribution H. Chow and Teugels [117] show that the latter conditions holds if and only if the normalised *joint* weak limit of $(\widetilde{S}_n, \widetilde{M}_n)$ exists. Resnick [529] gives a detailed discussion of the properties of the limiting random vector, using point process techniques. Chow and Teugels [117] show that the limiting variables are independent if $\alpha = 2$. This result has been generalised by Anderson and Turkman [14, 15] and Hsing [344] to certain stationary sequences.

8.3 When and How Ruin Occurs

8.3.1 Introduction

In this section we return to the insurance risk process introduced in Chapter 1. There we studied, mainly by analytical methods, the asymptotic behaviour of the ruin probability when the initial capital increases to infinity. In this section we first review these results from a probabilistic point of view. We also continue the analysis of the risk process. In particular, we are interested in the question:

What does a sample path of the risk process leading to ruin look like?

This will be answered in Sections 8.3.2 and 8.3.3, both for the light– and the heavy–tailed case. An important issue in our analysis concerns information on the claim(s) causing ruin.

Throughout we consider the classical Cramér–Lundberg model as introduced in Definition 1.1.1:

(*a*) The *claim sizes* $X, X_1, X_2, \ldots$ are iid positive rvs with common non–lattice df F and finite mean $\mu = EX_1$.

(*b*) The claim X_k arrives at time $T_k = Y_1 + \cdots + Y_k$, where $Y, Y_1, Y_2, \ldots$ are iid $Exp(\lambda)$ rvs for some $\lambda > 0$. The corresponding *claim numbers* $N(t) = \sup\{k \ge 1 : T_k \le t\}$, $t \ge 0$, constitute a homogeneous Poisson process with intensity $\lambda > 0$.

(c) The processes $(N(t))$ and (X_k) are independent.

The corresponding *risk process* is then defined by

$$U(t) = u + ct - S(t)\,, \quad t \geq 0\,,$$

where u is the *initial capital*, $S(t) = \sum_{n=1}^{N(t)} X_n$ the *total claim amount* until time t and $c > 0$ is the *premium income rate*. We also assume the *net profit condition* $\lambda\mu - c < 0$.

In Chapter 1 we mainly concentrated on estimating the *ruin probability in infinite time*:

$$\psi(u) \;=\; P\left(\inf_{0\leq t<\infty} U(t) < 0\right).$$

For our purposes, it turns out to be convenient to express $\psi(u)$ in terms of the Lévy process R, where

$$R(t) = S(t) - ct = u - U(t)\,, \quad t \geq 0\,.$$

Therefore R can be considered as a continuous–time analogue to a random walk with negative drift. Consequently, we expect various results from random walk theory to be useful in this context; see also Chapter 1, equation (1.9) and the related discussion. In some cases the translation is a straightforward application of the so–called *method of the discrete skeleton* which is described below.

Since $c > 0$ ruin can occur only at the claim arrival times T_k when R jumps upwards; see Figure 8.3.1 (top). By virtue of the net profit condition the discrete–time process

$$R_n = \sum_{k=1}^{n}(X_k - cY_k) = \sum_{k=1}^{n} Z_k\,, \quad n \in \mathbb{N},$$

constitutes a random walk with negative drift which is generated by the iid sequence $Z, Z_1, Z_2, \ldots$. Moreover,

$$\psi(u) = P\left(\sup_{t\geq 0} R(t) > u\right) = P\left(\sup_{n\geq 1} R_n > u\right). \tag{8.22}$$

It is an immediate consequence of the SLLN that

$$M \;=\; \sup_{t\geq 0} R(t) \;=\; \sup_{n\geq 1} R_n \;<\; \infty \quad \text{a.s.} \tag{8.23}$$

The random walk (R_n) is referred to as a *discrete skeleton* embedded in the continuous–time Lévy process R. Such a construction allows us to use renewal theoretic arguments on the skeleton and to translate the results to

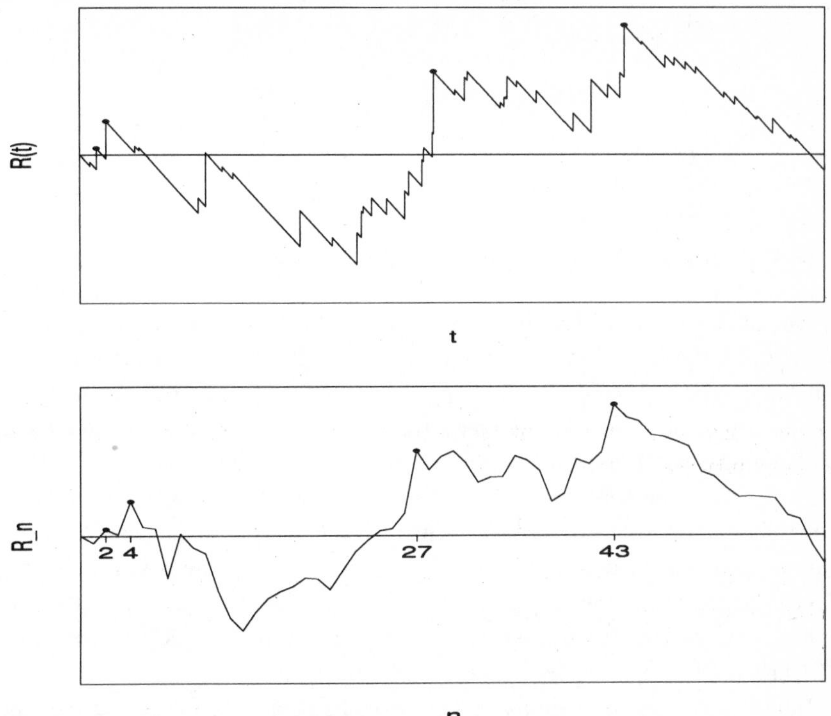

Figure 8.3.1 *A sample path of $(R(t))$ (top) and its discrete skeleton random walk (R_n) (bottom). The ladder points are indicated by $\bullet$. They appear for R_n at the indices (i.e. claim numbers) $n = 2, 4, 27, 43$. The ladder heights of (R_n) and $(R(t))$ coincide; they are also the records of (R_n).*

the process R. Thus representation (8.22) suggests using standard theory for the maximum of a random walk with negative drift. Below we present some of the main ideas and refer to the monographs Asmussen [27, 28], Feller [235] or Resnick [530] for more details.

Recall from equation (1.11) in Chapter 1 that

$$1 - \psi(u) = (1 - \alpha) \sum_{n=0}^{\infty} \alpha^n F_I^{n*}(u), \quad u \geq 0, \tag{8.24}$$

where

$$\alpha = \psi(0) = \frac{\lambda\mu}{c} \in (0, 1). \tag{8.25}$$

Here F_I is the *integrated tail df* $F_I(u) = \mu^{-1} \int_0^u \overline{F}(y)dy$.

In what follows we give a probabilistic interpretation of (8.24) by exploiting the relation $1 - \psi(u) = P(M \le u)$. We start by introducing another discrete skeleton for R. The quantities which suggest themselves in a natural way are the *ladder indices*

$$\begin{aligned} \tau_+(0) &= 0\,, \\ \tau_+(1) &= \inf\{t > 0 : R(t) > 0\}\,, \\ \tau_+(k+1) &= \inf\{t > \tau_+(k) : R(t) > R(\tau_+(k))\}\,, \quad k \in \mathbb{N}, \end{aligned}$$

and the *ladder heights* $R(\tau_+(k))$; see Figure 8.3.1 (top). Here, as usual, $\inf \emptyset = \infty$. In the language of Section 5.4 one could call $R(\tau_+(k))$ a record of the continuous–time process R and $\tau_+(k)$ the corresponding record time.

The process between two consecutive ladder indices is called a *ladder segment*. Due to the independent and stationary increments property of R, it is intuitively clear that at each *ladder point* $(\tau_+(k), R(\tau_+(k)))$ the process starts anew, and that the ladder segments constitute a sequence of iid stochastic processes. A detailed proof of these results uses the so–called *strong Markov property*; we refrain from going into details here. The resulting regenerative nature of the process is nicely discussed in Resnick [531], Sections 3.7.1 and 3.12.

Before we return to formula (8.24), we collect some useful facts about the first ladder segment. Writing

$$V = R(\tau_+(1)) \quad \text{and} \quad \widehat{Z} = -R(\tau_+(1)-)\,,$$

it follows that $A = V + \widehat{Z}$ is the size of the claim leading to ruin, given the initial capital $u = 0$. See Figure 8.3.8 for an illustration with $V = V_1$ and $\widehat{Z} = \widehat{Z}_1$. The following statement about V is classical; see for instance Cramér [138] or Feller [235], Section XI.4. The results for $\widehat{Z}$ and A are to be found in Dufresne and Gerber [193]. Here and in what follows we write

$$P^{(u)}(\cdot) = P(\cdot \,|\, \tau(u) < \infty)\,, \quad u \ge 0\,, \tag{8.26}$$

where

$$\tau(u) = \inf\{t > 0 : R(t) > u\}$$

is the *ruin time*, given the initial capital u.

Proposition 8.3.2 (Ruin with initial capital $u = 0$)
The following statements hold:

(a) $P^{(0)}(V \le x) = P^{(0)}(\widehat{Z} \le x) = F_I(x)$, (*b*) $P^{(0)}(A \le x) = \mu^{-1} \int_0^x y\, dF(y)$,

(*c*) *Let U be uniform on* $(0,1)$, *independent of A. Then the vectors* $(V, \widehat{Z})$ *and* $(UA, (1-U)A)$ *have the same distribution with respect to* $P^{(0)}$. □

Remark. Statement (c) can be translated into

$$\begin{aligned} P^{(0)}(V > v, \widehat{Z} > z) &= \frac{1}{\mu} \int_{v+z}^{\infty} P^{(0)}\left(\frac{v}{y} \le U \le 1 - \frac{z}{y}\right) y\, dF(y) \\ &= \frac{1}{\mu} \int_{v+z}^{\infty} (y - z - v) dF(y) \\ &= \overline{F}_I(v+z)\,, \quad v, z \ge 0\,, \end{aligned} \tag{8.27}$$

where we applied partial integration for the last equality. In the heavy–tailed case, for initial capital u tending to ∞, we shall derive a formula similar to (8.27); see (8.40). □

Now we return to formula (8.24). From (8.23) it follows that the ladder heights $R(\tau_+(k))$ determine the distribution of the maximum M. A precise formulation of the regenerative property of R at its ladder points implies that $R(\tau_+(k)) - R(\tau_+(k-1))$ are iid positive rvs. By Proposition 8.3.2(a), they have distribution tail

$$P\left(R(\tau_+(1)) > x\right) = P(\tau(0) < \infty) P^{(0)}(V > x) = \alpha \overline{F}_I(x)\,, \quad x > 0\,,$$

where $\alpha \in (0,1)$ is defined in (8.25). Here we used that, if $\tau(0) = \infty$, then $R(t) \le 0$ for all $t > 0$. The ladder indices constitute a renewal process which is transient; see Remark 3 in Appendix A4. This means that the total number K of renewals has a geometric distribution with parameter $1 - \alpha$, where $\alpha = \psi(0) = P(\tau_+(1) < \infty)$. Indeed, using the iid property of the increments $\tau_+(k) - \tau_+(k-1)$, we obtain

$$\begin{aligned} P(K = n) &= P\left(\tau_+(n) < \infty\,, \tau_+(n+1) = \infty\right) \\ &= P\left(\max_{k=1,\ldots,n} (\tau_+(k) - \tau_+(k-1)) < \infty\,, \tau_+(n+1) = \infty\right) \\ &= \alpha^n (1-\alpha)\,, \quad n \ge 0\,. \end{aligned}$$

Since

$$M = \sum_{k=1}^{K} \left(R(\tau_+(k)) - R(\tau_+(k-1))\right),$$

we have that

$$\begin{aligned}\psi(u) &= P(M > u) \\ &= \sum_{n=1}^{\infty} P(M > u\,,\, K = n) \\ &= \sum_{n=1}^{\infty} P\left(\sum_{k=1}^{n}\Big(R(\tau_+(k)) - R(\tau_+(k-1))\Big) > u\,,\, K = n\right) \\ &= (1-\alpha)\sum_{n=1}^{\infty} \alpha^n \left(1 - F_I^{n*}(u)\right)\,, \quad u \geq 0\,.\end{aligned}$$

This yields (8.24).

In the following sections we further exploit the underlying random walk structure of R. There we study a sample path of the risk process leading to ruin. The problem and its solution will be formulated as *conditional limit theorems* in terms of the conditional probability measures $P^{(u)}$ as defined in (8.26). We give a short review of the sample path description obtainable and contrast ruin under a small and large claim regime. These results will also give us asymptotic expressions for the *ruin probability in finite time*, i.e.

$$\psi(u,T) = P\left(\sup_{0\leq t\leq T} R(t) > u\right) = P\left(\tau(u) \leq T\right)\,, \quad 0 < T < \infty\,.$$

In our presentation we follow Asmussen [24] in the Cramér–Lundberg case and Asmussen and Klüppelberg [33] in the subexponential case. Those readers who want to study the mathematical methods in detail should consult these papers or the monograph by Asmussen [28].

8.3.2 The Cramér–Lundberg Case

Throughout this section we assume that the assumptions of the Cramér–Lundberg Theorem 1.2.2 hold. This means in particular that X has a moment generating function which is finite in some neighbourhood of the origin and that the *Lundberg exponent*, i.e. the solution of the equation $\int_0^\infty e^{\nu x}\overline{F}(x)dx = c/\lambda$, exists. Moreover, we assume $\int_0^\infty x e^{\nu x}\overline{F}(x)dx < \infty$. Theorem 1.2.2 then gives us the approximation

$$\psi(u) \quad \sim \quad Ce^{-\nu u}\,, \quad u \to \infty\,, \tag{8.28}$$

where C is a positive constant depending on the parameters of the risk process.

Consider the moment generating function of $Z = X - cY$:

$$\kappa(s) \quad = \quad Ee^{sZ}\,,$$

for appropriate s–values. The Lundberg exponent ν is the unique positive solution of the equation $\kappa(s) = 1$ and, by the net profit condition, $\kappa'(0) = (\lambda\mu - c)/\lambda < 0$. Notice that $\kappa'(s) < 0$ in a neighbourhood of the origin, $\kappa(s)$ is strictly convex and hence $\kappa'(\nu) > 0$. For an illustration of κ see Figure 8.3.3.

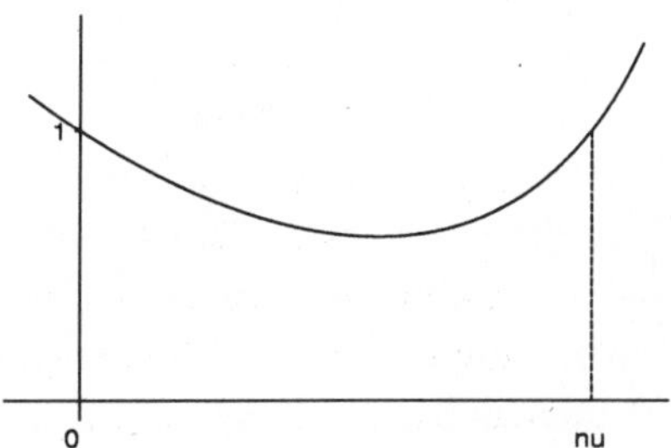

Figure 8.3.3 *A typical example of* $\kappa(s)$ *with the Lundberg coefficient* ν.

Let H_Z denote the df of $Z = X - cY$. The corresponding *Esscher transformed* or *exponentially tilted df* is given by

$$H_\nu(x) = \int_{-\infty}^{x} e^{\nu y} dH_Z(y) \,, \quad x \in \mathbb{R} \,.$$

Since $Ee^{\nu Z} = 1$, the df H_ν is proper with positive finite mean

$$\int_{-\infty}^{\infty} x dH_\nu(x) = \int_{-\infty}^{\infty} x e^{\nu x} dH_Z(x) = \kappa'(\nu) > 0 \,. \tag{8.29}$$

Following Feller [235], p. 406, we call a random walk with increment df H_ν *associated.* The main idea for dealing with the sample paths of R (with negative drift) under the condition $\tau(u) < \infty$ now consists of switching to an associated random walk (with positive drift). Write $(\widetilde{Z}_k)$ for an iid sequence with df H_ν. Observe that

$$\begin{aligned} & P^{(u)}(Z_1 \le x_1, \ldots, Z_n \le x_n) \\ = \; & P(Z_1 \le x_1, \ldots, Z_n \le x_n, M > u)/P(M > u) \\ = \; & \frac{1}{P(M > u)} \int_{-\infty}^{x_1} \cdots \int_{-\infty}^{x_n} P(M > u - y_1 - \cdots - y_n) \\ & \quad dH_Z(y_1) \cdots dH_Z(y_n) \end{aligned}$$

$$
\begin{aligned}
&= \frac{1}{\psi(u)} \int_{-\infty}^{x_1} \cdots \int_{-\infty}^{x_n} \psi(u - y_1 - \cdots - y_n) dH_Z(y_1) \cdots dH_Z(y_n) \\
&\sim \int_{-\infty}^{x_1} \cdots \int_{-\infty}^{x_n} e^{\nu(y_1+\cdots+y_n)} dH_Z(y_1) \cdots dH_Z(y_n) \qquad (8.30) \\
&= H_\nu(x_1) \cdots H_\nu(x_n) \\
&= P(\widetilde{Z}_1 \le x_1, \ldots, \widetilde{Z}_n \le x_n) .
\end{aligned}
$$

In (8.30) we used the ruin estimate (8.28), as $u \to \infty$, combined with dominated convergence. The latter is justified by a two–sided bound on $\psi(u)$ using both (8.28) and (1.14). This calculation shows that the distribution of the random walk (R_n) and of R, given that ruin occurs in finite time, is closely related to the distribution of the associated random walk.

This intuitive argument is further supported by the following facts: let H_n be the empirical df of the sample $Z_1, \ldots, Z_n$, i.e.

$$
H_n(x) = \frac{1}{n} \sum_{k=1}^{n} I_{\{Z_k \le x\}}, \quad x \in \mathbb{R}. \qquad (8.31)
$$

An application of the Glivenko–Cantelli theorem (Example 2.1.4) yields that

$$
\sup_{x\in\mathbb{R}} |H_n(x) - H_Z(x)| \to 0 \quad \text{a.s.}
$$

This changes completely if ruin occurs and n is replaced by the ruin times $\tau(u)$; the following result indicates that the increment df H_Z of the random walk (R_n), conditioned on the event that ruin occurs, is close to H_ν, the increment df of the asscociated random walk.

Proposition 8.3.4 *The following relation holds as $u \to \infty$:*

$$
\sup_{x\in\mathbb{R}} \left| H_{\tau(u)}(x) - H_\nu(x) \right| \to 0 \quad \textit{in } P^{(u)}\textit{–probability.} \qquad \square
$$

Remark. 1) Let A and A_u be rvs. Here and in what follows we write $A_u \to A$ in $P^{(u)}$–probability if

$$
\lim_{u\to\infty} P^{(u)}(|A_u - A| > \varepsilon) = 0, \quad \varepsilon > 0.
$$

Analogously convergence in $P^{(u)}$–distribution of a sequence of random vectors $A_u \to A$ as $u \to \infty$, as used in Theorem 8.3.9 below, is defined as $E^{(u)} f(A_u) \to Ef(A)$ for every bounded, continuous functional f; see also Appendix A2. $\square$

The following quantities describe the ruin event: the level $R(\tau(u)-)$ of the process R just before ruin, the level $R(\tau(u))$ at the ruin time, the excess

$R(\tau(u)) - u$ over u by R at the ruin time and the size $R(\tau(u)) - R(\tau(u)-)$ of the claim causing ruin. For $u \to \infty$ their asymptotic distributional behaviour is described below.

Theorem 8.3.5 (Ruin in the Cramér–Lundberg case)
The following relations hold for $u \to \infty$:

(a) $\sup_{t\in[0,1]} \left| \dfrac{R(t\tau(u))}{\tau(u)} - \kappa'(\nu)\, t \right| \to 0$ *in $P^{(u)}$–probability.*

(b) $\dfrac{\tau(u) - u/\kappa'(\nu)}{\sqrt{u}s(\nu)} \to N$ *in $P^{(u)}$–distribution,*

where N is a standard normal rv and $s(\nu)$ is a quantity involving the Lundberg exponent and certain moments of X.

(c) *The quantities $R(\tau(u)) - u$, $u - R(\tau(u)-)$ and $R(\tau(u)) - R(\tau(u)-)$ converge jointly in $P^{(u)}$–distribution to a non-degenerate limit distribution. Moreover, $\tau(u)$ and $R(\tau(u)) - u$ are asymptotically independent.* □

Remarks. 2) The above results (a) and (b) indicate that a sample path of R leading to ruin has *locally* a linear drift with slope $\kappa'(\nu) > 0$ just before ruin happens. Notice that this is in contrast to the *global* picture where the drift of R is negative. This is due to the close relationship between the sequences (Z_k) and $(\widetilde{Z}_k)$ for which $E\widetilde{Z}_k = \kappa'(\nu) > 0$; see (8.29). The link of these two sequences has been indicated above. The precise description of this relationship would lead us too far; we refer to Asmussen [24], where also Proposition 8.3.4 is taken from.

3) Part (c) implies in particular that all these quantities converge to finite limits. This is in contrast to the behaviour of the claim leading to ruin in the heavy–tailed case; see Theorem 8.3.9. □

For completeness, we conclude this section with some results on the ruin probability in finite time. Recall that the ruin probability for the interval $[0,T]$ is given by

$$\psi(u,T) = P\left(\inf_{0\le t\le T} U(t) < 0\right). \tag{8.32}$$

Approximations to $\psi(u,T)$ may for instance be derived from FCLTs. In Example 2.5.18 a diffusion approximation to $\psi(u,T)$ was presented. It is based on a FCLT for the total claim amount process $(S(t))$. Alternatively, Proposition 8.3.5 can be exploited.

Corollary 8.3.6 (Ruin in finite time in the Cramér–Lundberg case)

$$\lim_{u\to\infty} \sup_{0\le T<\infty} \left| e^{\nu u}\psi(u,T) - C\Phi\left(\frac{T - u/\kappa'(\nu)}{s(\nu)\sqrt{u}}\right)\right| = 0\,. \tag{8.33}$$

Here C is the same constant as in (8.28), Φ denotes the standard normal df, and $s(\nu)$ is a deterministic scaling factor involving the Lundberg exponent and certain moments of the claim size distribution.

Proof. First note that by the definition of $P^{(u)}$

$$P^{(u)}(\tau(u) \le T) = \psi(u,T)/\psi(u)\,.$$

Hence

$$\begin{aligned}&\sup_{0\le T<\infty}\left|e^{\nu u}\psi(u,T) - CP^{(u)}(\tau(u)\le T)\right|\\ &= \sup_{0\le T<\infty} e^{\nu u}\psi(u,T)\left|1-\frac{C}{e^{\nu u}\psi(u)}\right|\,.\end{aligned}$$

The right-hand side tends to 0 as $u\to\infty$ since $\psi(u,T)\le\psi(u)$ and $\exp\{\nu u\}\psi(u)\to C$ by (8.28). Since weak convergence to a continuous limit implies uniform convergence of the dfs the result follows from Theorem 8.3.5(b). □

Remark. 4) The limit relation (8.33) suggests as approximation of the ruin probability in finite time

$$\psi(u,T)\approx Ce^{-\nu u}\Phi\left(\frac{T-u/\kappa'(\nu)}{s(\nu)\sqrt{u}}\right)\,. \tag{8.34}$$

This approximation has to be treated with care: since no rate of convergence is given the remainder term in this limit relation may be larger than the term to be approximated. Further refinements and higher order approximations are to be found in Asmussen [25]. □

We may summarise the situation under the Cramér–Lundberg regime as follows:

> The behaviour of the sample path of R just before ruin occurs is as if the increment distribution changed from H_Z to H_ν, and the main dramatic feature we see in the sample path is a change of drift causing ruin. The intuitive picture is that rare events leading to ruin occur as a consequence of a build–up of claims which locally force the underlying random walk to behave like a random walk with positive drift. The increment distribution of such a random walk is given by the Esscher transformed df H_ν.

8.3.3 The Large Claim Case

In contrast to the Cramér–Lundberg case, under a large claim regime rare events causing ruin happen out of the blue. The process evolves in its "typical" way up to the ruin time. Then ruin occurs as a consequence of one single large claim.

Recall the definition of a subexponential df F ($F \in \mathcal{S}$):

$$P(X_1 + \cdots + X_n > x) \sim P(\max(X_1, \ldots, X_n) > x), \quad x \to \infty, \qquad (8.35)$$

for $n \geq 2$; see (1.26). In Chapter 1 various estimates of the ruin probability $\psi(u)$ for heavy-tailed F were presented. In particular, we may conclude from Theorem 1.3.6 that $F_I \in \mathcal{S}$ implies

$$\psi(u) \sim \frac{\alpha}{1-\alpha}\overline{F}_I(u) = \frac{\lambda}{c - \lambda\mu}\int_u^\infty \overline{F}(y)dy, \quad u \to \infty. \qquad (8.36)$$

Below we want to give answers to the following questions: *given that ruin occurs in finite time,*

(a) *How big is the claim leading to ruin?*
(b) *What is the asymptotic distribution of the ruin time?*
(c) *What does* "the process evolves in its typical way up to the ruin time" *actually mean?*

A first indicator of the fact that the risk process evolves typically up to the ruin time is provided by the behaviour of the empirical df H_n of the increments Z_n of the embedded discrete skeleton random walk R_n; see (8.31).

Proposition 8.3.7 *Under the conditions of Theorem 8.3.9,*

$$\sup_{x\in\mathbb{R}} \left|H_{\tau(u)}(x) - H_Z(x)\right| \to 0 \quad \textit{in } P^{(u)}\textit{-probability.} \qquad \square$$

Compare this result with Proposition 8.3.4 in the Cramér–Lundberg case; there the limit of $H_{\tau(u)}$ turns out to be the Esscher transformed df H_ν.

For a precise description of the ruin event itself we consider the $P^{(u)}$–distribution of the following quantities (see Figure 8.3.8 for an illustration):

(a) $-Z(u) = R(\tau(u)-)$, the level of R *just before* ruin,
(b) $V(u) = R(\tau(u))$, the level of R *at* the ruin time, and
(c) $V(u) - u$, the excess over u by R at the ruin time.

Again we may use the regenerative structure of the process R. The negative drift of R ensures that there are with probability one only finitely many ladder points. Let

$$K(u) = \inf\{k \in \mathrm{N} : R(\tau_+(k)) > u\},$$

with $\inf\emptyset = \infty$, denote the *ladder index causing ruin.* The increments $V_n = R(\tau_+(n)) - R(\tau_+(n-1))$ are iid and $P^{(0)}(V_1 \le x) = F_I(x)$ by Proposition 8.3.2(a). Then

$$\begin{aligned} &P(K(u)=n)\\ &= P(K(u)=n\,,\,\tau(u)<\infty)\\ &= P(\tau_+(n)<\infty)\,P(V_1+\cdots+V_{n-1}\le u\,,V_1+\cdots+V_n>u\,|\tau_+(n)<\infty)\\ &= P(\tau_+(n)<\infty)\,p_n(u)\,. \end{aligned}$$

A justification of the following arguments is given in the proof of Lemma 2.6 in Asmussen and Klüppelberg [33]; the crucial assumption used is subexponentiality of F_I implying that (8.35) holds for the V_n with respect to $P^{(0)}$:

$$\begin{aligned} p_n(u) &\sim P\Big(\max(V_1,\ldots,V_{n-1})\le u\,,\,\max(V_1,\ldots,V_n)>u\,\Big|\tau_+(n)<\infty\Big)\\ &\sim P(V_n>u\,|\,\tau_+(n)-\tau_+(n-1)<\infty)\\ &= P^{(0)}(V_1>u)=\overline{F}_I(u)\,,\quad u\to\infty\,. \end{aligned}$$

Also notice that for α as in (8.25),

$$\begin{aligned} P(\tau_+(n)<\infty) &= P\left(\max_{k=1,\ldots,n}\,(\tau_+(k)-\tau_+(k-1))<\infty\right)\\ &= P^n(\tau_+(1)<\infty)=\alpha^n\,. \end{aligned} \tag{8.37}$$

Combining (8.36)–(8.37), we conclude that

$$\begin{aligned} P^{(u)}(K(u)=n) &= \frac{P(K(u)=n,\tau(u)<\infty)}{\psi(u)}\\ &\sim \frac{\alpha^n\overline{F}_I(u)}{\psi(u)}\\ &\to (1-\alpha)\alpha^{n-1}\,, \end{aligned} \tag{8.38}$$

i.e. the number of ladder segments until ruin has asymptotically a geometric distribution with parameter $1-\alpha$.

The path $(S(t))_{t\in[0,\tau(u))}$ can be decomposed into $K(u)$ ladder segments, where $K(u)$ is asymptotically geometric as in (8.38), the first $K(u)-1$ segments are all "typical" as described in Proposition 8.3.2. However, the last segment leading to ruin behaves differently. Ruin is caused by one single large claim. Therefore one may expect that classical extreme value theory enters for describing the ruin event.

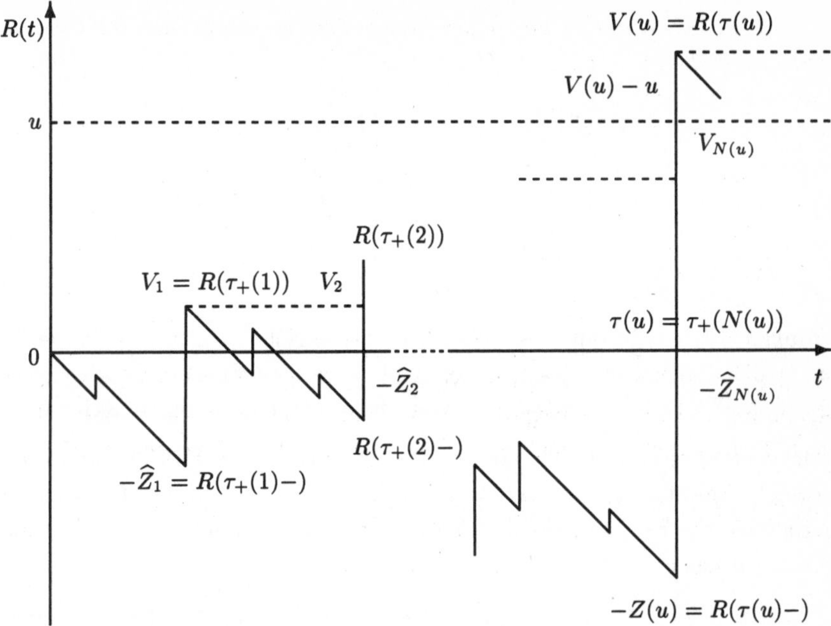

Figure 8.3.8 *Idealised sample path leading to ruin.*

Subexponential distributions are heavy–tailed in the sense that their tails decrease to 0 more slowly than any exponential tail; see Lemma 1.3.5. Their tails can be regularly varying, but also the lognormal and heavy–tailed Weibull distributions are subexponential. This implies that subexponential distributions may belong to the maximum domain of attraction of the Fréchet distribution Φ_α (see Section 3.3.1) or of the Gumbel distribution Λ (see Example 3.3.35). We distinguish between these two cases.

Let $(Z(u), V(u))$ be a random vector having the same $P^{(u)}$–distribution as $(-R(\tau(u)-), R(\tau(u)))$. The following result describes the sample path up to ruin and the ruin event itself.

Theorem 8.3.9 (Ruin in the subexponential case)
Assume that either $\overline{F}_I \in \mathcal{R}_{-1/\xi}$ for $\xi \in (0,\infty)$ or $F_I \in \mathrm{MDA}(\Lambda) \cap \mathcal{S}$ (this corresponds to $\xi = 0$ below), and let $a(u) = \int_u^\infty \overline{F}(x)dx/\overline{F}(u)$. Then

$$\sup_{t\in[0,1]} \left| \frac{R(t\tau(u))}{\tau(u)} + c(1-\alpha)t \right| \to 0 \tag{8.39}$$

and

$$(a(u))^{-1}\,(c(1-\alpha)\tau(u)\,,\,Z(u)\,,\,V(u)-u) \to (Z_\xi\,,\,Z_\xi\,,\,V_\xi)$$

as $u \to \infty$ in $P^{(u)}$–distribution. The rvs V_ξ and Z_ξ are both generalised Pareto distributed with

$$P(V_\xi > v\,, Z_\xi > z) = \overline{G}_\xi(v+z)\,, \quad v,z \geq 0\,, \tag{8.40}$$

where

$$\overline{G}_\xi(x) = \begin{cases} (1+\xi\, x)^{-1/\xi}\,, & \xi \in (0,\infty)\,, \\ e^{-x}\,, & \xi = 0\,. \end{cases} \tag{8.41}$$

□

Remarks. 1) Recall from Remark 1 after Proposition 8.3.5, that convergence in $P^{(u)}$–distribution of a sequence of random vectors $A_u \to A$ as $u \to \infty$ is defined as $E^{(u)} f(A_u) \to Ef(A)$ for every bounded, continuous functional f.

2) The generalised Pareto distribution appears as limit law for the normalised excesses of an iid sequence over high thresholds; see Section 3.4. This is similar to Theorem 8.3.9, where the excess $V(u) - u$ of the process R over the threshold u has a similar limit behaviour.

3) Relation (8.39) intuitively supports the statement that the process evolves "typically" up to time $\tau(u)$ because $R(t)/t \overset{\text{a.s.}}{\to} -c(1-\alpha) = \lambda\mu - c < 0$ by the SLLN. Also notice that $Z(u)/\tau(u) \to c(1-\alpha)$ in $P^{(u)}$–probability. This again indicates "typical" behaviour until ruin occurs.

4) The above theorem should be compared with Theorem 8.3.5. Notice in particular that, in the Cramér–Lundberg case, the excesses $V(u) - u$ converge weakly to a non–degenerate limit, while in the subexponential case the excesses tend to ∞. Notice that the normalising function $a(u)$ is the mean excess function which tends to infinity for subexponential F_I; see Example 3.3.34. The claim causing ruin starts at a "typical" level, then shoots all the way up, crosses the high level u and even shoots over this high level by a very large amount.

5) Equation (8.40) should be compared with (8.27). Although the last ladder segment has completely different probabilistic properties than the other ones, the excess $V(u) - u$ of R at the time of ruin and the level $Z(u)$ of R immediately before ruin occurs show a similar probabilistic relation for $u = 0$ and in the limit for $u \to \infty$. □

From Theorem 8.3.9 we immediately obtain for all $\xi \in [0,\infty)$

$$\frac{Z(u) + V(u) - u}{a(u)} \to Z_\xi + V_\xi \quad \text{in } P^{(u)}\text{–distribution}\,.$$

This yields information about the size $R(\tau(u)) - R(\tau(u)-) = V(u) + Z(u)$ of the claim causing ruin.

Corollary 8.3.10 (Size of the claim causing ruin in the subexponential case)

(a) *Assume that* $\overline{F}_I \in \mathcal{R}_{-1/\xi}$ *for* $\xi \in (0,\infty)$. *Then* $a(u) \sim \xi u$ *and*

$$\lim_{u\to\infty} P^{(u)}\left(\frac{V(u)+Z(u)}{u} > x\right) = \left(1+\frac{1}{\xi}\left(1-\frac{1}{x}\right)\right)x^{-1/\xi}, \quad x \geq 1.$$

(b) *Assume that* $F_I \in \mathrm{MDA}(\Lambda) \cap \mathcal{S}$. *Then* $a(u) = \int_u^\infty \overline{F}(x)dx/\overline{F}(u)$ *and*

$$\lim_{u\to\infty} P^{(u)}\left(\frac{V(u)+Z(u)-u}{a(u)} > x\right) = (1+x)e^{-x}, \quad x \geq 0.$$

□

We conclude with some results on the ruin probability in finite time, given by (8.32) for the interval $[0,T]$. Since

$$\psi(u,T)/\psi(u) = P(\tau(u) \leq T \mid \tau(u) < \infty) = P^{(u)}(\tau(u) \leq T),$$

the following can be derived from Theorem 8.3.9.

Corollary 8.3.11 (Ruin in finite time in the subexponential case)

(a) *Assume that* $\overline{F}_I \in \mathcal{R}_{-1/\xi}$ *for* $\xi \in (0,\infty)$. *Then* $a(u) \sim \xi u$ *and*

$$\lim_{u\to\infty} \frac{\psi(u,uT)}{\psi(u)} = 1-(1+c(1-\alpha)T)^{-1/\xi}.$$

(b) *Assume that* $F_I \in \mathrm{MDA}(\Lambda) \cap \mathcal{S}$. *Then* $a(u) = \int_u^\infty \overline{F}(x)dx/\overline{F}(u)$ *and*

$$\lim_{u\to\infty} \frac{\psi(u,a(u)T)}{\psi(u)} = 1-e^{-c(1-\alpha)T}.$$

□

We can summarise the situation under the subexponential regime as follows:

> The behaviour of the sample path of R just before ruin happens appears completely normal: it looks exactly as any sample path for which ruin never occurs. Ruin then happens out of the blue, caused by a single large claim. It is so large that, in order to obtain a finite non–degenerate limit of the excess of R over the threshold u, we have to normalise by a function which tends to ∞.

Notes and Comments

Notice that the problems discussed above have similarities with the estimation of VaR (ruin say) and the shortfall (the excess of the claim causing ruin) as discussed in Example 6.1.6.

Some early exposition of approximations to the ruin probability in finite time is to be found in Cramér [140] and Segerdahl [574], who first derived approximation (8.34). Later results exploit the diffusion approximation of the total claim amount process as a useful tool; see Iglehart [351] and Grandell [281] and also Example 2.5.18. A systematic approach with respect to FCLTs for R and $(\tau(u))$ is provided by Asmussen [24], see in particular his Corollary 3.1. Because of the small claim regime, the underlying limit processes are Gaussian. New ideas by Siegmund [580] from sequential analysis led to refinements and new variants.

The above results are closely related to large deviations. In the context of insurance risk models, they have been studied by Martin–Löf [444],[443], Djehiche [184], Slud and Hoesman [585] and Barndorff–Nielsen and Schmidli [49]. All this refers to the small claim regime. See also Section 8.6 for a review of large deviation results in the heavy–tailed case.

The results of Section 8.3.2 were derived in Asmussen [24] , the heavy–tailed case is treated in Asmussen and Klüppelberg [33]. Extending results of Asmussen, Fløe–Henriksen and Klüppelberg [31] and Asmussen and Klüppelberg [33], an extremal event analysis for the Markov–modulated risk model was carried out by Asmussen and Højgaard [32].

The conditional limit theorems above have been applied to the efficient simulation of ruin probabilities: in the light–tailed case by Asmussen [26] and in the heavy–tailed case by Asmussen and Binswanger [30].

8.4 Perpetuities and ARCH Processes

Random recurrence equations have been used in various fields of applied probability (to name a few references: Kesten [379] and Vervaat [631] were very stimulating papers in this field; the monograph by Brandt, Franken and Lisek [87]; the overview paper by Embrechts and Goldie [206]). In particular, ARCH and GARCH processes (see also Section 7.6 for some remarks on the topic) are given by stochastic recurrence equations. They serve as special exchange rate or log–return models with stochastic volatility and are very popular in econometrics. A flood of papers has been published on ARCH and related models, mainly in the context of statistics. See for instance the recent review paper by Shephard [578] and references therein.

Our interest in these models arose from the fact that ARCH processes with light–tailed input (i.e. Gaussian innovations) are indeed heavy–tailed (i.e. Pareto–like) time series. This was first observed in 1973 by Kesten [379]. It is the aim of this section to explain where the heavy tails come from and to study the behaviour of the maxima of an ARCH(1) sequence. In our presentation we follow Goldie [272] and de Haan et al. [303] who treat solutions of general recurrence equations and their extremal behaviour, respectively.

8.4.1 Stochastic Recurrence Equations and Perpetuities

In finance and insurance applications we are often confronted with two contrary phenomena: accumulating and discounting. It is the aim of this section to shed some light on discrete time accumulation and discounting techniques which are closely linked to stochastic recurrence equations. In what follows we introduce two basic concepts in those equations.

Example 8.4.1 (Accumulation and perpetuities)
Suppose you invest at times $0, 1, 2, \ldots$ one unit (say, \$1) in a bond with interest rate $\delta \in (0, 1)$. What is the accumulated value Y_t at time t of the interest payments made at times $0, \ldots, t$ for $t \geq 1$, assuming $Y_0 = 1$? Simple calculation shows that

$$Y_0 = 1\,, \quad Y_1 = 1 + (1+\delta)\,, \quad Y_2 = 1 + (1+\delta) + (1+\delta)^2\,, \ldots\,.$$

In particular, we observe that Y_t and Y_{t-1} are linked by the recursion

$$Y_t = 1 + (1+\delta) Y_{t-1}\,, \quad t \in \mathbb{N}\,. \tag{8.42}$$

Now assume that the interest rate δ depends on time t. The recursion (8.42) can immediately be modified:

$$Y_t = 1 + (1+\delta_t) Y_{t-1}\,, \quad t \in \mathbb{N}\,,$$

yielding

$$Y_0 = 1\,, \quad Y_1 = 1 + (1+\delta_1)\,, \quad Y_2 = 1 + (1+\delta_2) + (1+\delta_2)(1+\delta_1)\,, \ldots\,.$$

We can also imagine a more complicated situation: assume that the interest rate δ and the invested amount A are time dependent. Setting $B_t = 1 + \delta_t$, this leads to the recursion

$$Y_t = A_t + B_t Y_{t-1}\,, \quad t \in \mathbb{N}\,, \tag{8.43}$$

or explicitly

$$\begin{aligned} Y_1 &= Y_0B_1 + A_1 \, , \\ Y_2 &= Y_0B_1B_2 + A_1B_2 + A_2 \, , \\ Y_3 &= Y_0B_1B_2B_3 + A_1B_2B_3 + A_2B_3 + A_3 \, , \ldots \, , \end{aligned}$$

and in general (by convention, $\prod_{j=l}^{l-1} a_j = 1$ and $\sum_{j=l}^{l-1} a_j = 0$)

$$Y_t = Y_0 \prod_{j=1}^{t} B_j + \sum_{m=1}^{t} A_m \prod_{j=m+1}^{t} B_j \, , \quad t \in \mathbb{N} \, . \tag{8.44}$$

If Y_0, A_t, B_t are rvs, (8.43) is called a *(forward) stochastic recurrence equation* or a *(forward) stochastic difference equation.* The word "forward" is related to the fact that, starting from an initial value Y_0, we successively apply the *random affine mappings* $\Psi_t(x) = A_t + B_t x$ such that $Y_t = \Psi_t(Y_{t-1})$. The latter relation is also called an "outer iteration"; see for instance Embrechts and Goldie [206].

In the insurance context, (Y_t) as given by (8.44) can be interpreted as the value of a *perpetuity*: the payments A_t are made at the beginning of each period and the accumulated payments Y_{t-1} are subject to interest. The name "perpetuity" comes from "perpetual payment streams" and recently gained some popularity in the literature on stochastic recurrence equations. In the form (8.44), (Y_t) is referred to as a *perpetuity–due.* Gerber [257], Section 2.6, gives a brief introduction to perpetuities from a life insurance point of view. See also Dufresne [191, 192] and the references therein for applications in insurance, mainly to pension funding. An introduction within the realm of finance is for instance to be found in Brealey and Myers [88], p. 33; see also Geman and Yor [253]. More background is to be found in the recent paper by Goldie and Grübel [273]. □

Example 8.4.2 (Discounting)
The reverse problem to accumulation is discounting. Suppose payments of one unit are made at times $0, 1, 2, \ldots$. Given the interest rate $\delta \in (0,1)$, the discounted value at time 0 of those payments, made till time t, is

$$U_t = 1 + (1+\delta)^{-1} + (1+\delta)^{-2} + \cdots + (1+\delta)^{-t} \, , \quad t \in \mathbb{N} \, .$$

Allowing for time–dependent interest rates δ and payments A' and setting $C_t = (1+\delta_t)^{-1}$, we obtain in a similar fashion

$$\begin{aligned} U_t &= A_1' + A_2'C_1 + A_3'C_1C_2 + \cdots + A_t'C_1 \cdots C_{t-1} + U_0 C_1 \cdots C_t \\ &= \sum_{m=1}^{t} A_m' \prod_{j=1}^{m-1} C_j + U_0 \prod_{j=1}^{t} C_j \, , \quad t \in \mathbb{N} \, . \end{aligned} \tag{8.45}$$

(The value U_0, which may be viewed as the final (time t) down–payment, is unimportant when we are interested in the behaviour of U_t for large t. Under weak assumptions, the last term in (8.45) can be shown to converge to zero a.s.; see the proof of Proposition 8.4.3(b) below.) If we assume that U_0, (A'_t), (C_t) are sequences of rvs, (8.45) can be written using a so–called (*backward*) *stochastic recurrence equation* or a (*backward*) *stochastic difference equation.* For the interpretation of (8.45) as an "inner iteration" of random affine maps see Embrechts and Goldie [206].

A glance at (8.45) and (8.44) convinces us that the structure of the discounted and accumulated sequences (U_t) and (Y_t) is very similar. This also concerns the distribution of these rvs: assume Y_0 is independent of the iid sequence $((A_t, B_t))_{t\geq 1}$ and U_0 is independent of the iid sequence $((A'_t, C_t))_{t\geq 1}$. Observe that for every $t \in \mathbb{N}$

$$\left(Y_0, ((A_k, B_k)_{1\leq k\leq t}\right) \stackrel{d}{=} \left(Y_0, ((A_{t-k+1}, B_{t-k+1}))_{1\leq k\leq t}\right) ,$$

implying that

$$\begin{aligned} Y_t &= Y_0 \prod_{j=1}^{t} B_j + \sum_{m=1}^{t} A_m \prod_{j=m+1}^{t} Bj \\ &\stackrel{d}{=} Y_0 \prod_{j=1}^{t} B_j + \sum_{m=1}^{t} A_m \prod_{j=1}^{m-1} Bj \,. \end{aligned}$$

Immediately, if $Y_0 \stackrel{d}{=} U_0$ and $(A_1, B_1) \stackrel{d}{=} (A'_1, C_1)$, then $U_t \stackrel{d}{=} Y_t$. □

Throughout this section we assume that Y_0 *is independent of the iid sequence* $((A_t, B_t))_{t\geq 1}$. *We also write for convenience* $(A, B) = (A_1, B_1)$. In what follows we are concerned with properties of the perpetuity sequence (Y_t) defined by (8.44). It follows from the discussion in Example 8.4.2 that every statement about the distribution of Y_t is also about the distribution of U_t defined by (8.45). The question we want to answer is

Which assumptions on (A, B) *and* Y_0 *guarantee convergence of* (Y_t) *in distribution, and what are the properties of the limit distribution?*

The answer can be found in Vervaat [631]; see also Kesten [379].

Proposition 8.4.3 (Distribution and moments of perpetuities)
Let (Y_t) *be the stochastic process defined by* (8.44) *and assume that*

$$E \ln^+ |A| < \infty \quad \textit{and} \quad -\infty \leq E \ln |B| < 0 \,. \tag{8.46}$$

(a) $Y_t \xrightarrow{d} Y$ *for some rv* Y *and* Y *satisfies the identity in law*

$$Y \stackrel{d}{=} A + BY\,, \tag{8.47}$$

where Y *and* (A,B) *are independent.*

(b) *Equation* (8.47) *has a solution, unique in distribution, which is given by*

$$Y \stackrel{d}{=} \sum_{m=1}^{\infty} A_m \prod_{j=1}^{m-1} B_j\,. \tag{8.48}$$

The rhs of (8.48) *converges absolutely with probability 1.*

(c) *If we choose* $Y_0 \stackrel{d}{=} Y$ *as in* (8.48), *then the process* $(Y_t)_{t\ge 0}$ *is strictly stationary.*

Now assume the moment conditions

$$E|A|^p < \infty \quad \text{and} \quad E|B|^p < 1 \quad \text{for some } p \in [1,\infty)\quad .$$

(d) *Then* $E|Y|^p < \infty$, *and the series in* (8.48) *converges in pth mean.*

(e) *If* $E|Y_0|^p < \infty$, *then* (Y_t) *converges to* Y *in pth mean, and in particular* $E|Y_t|^p \to E|Y|^p$ *as* $t \to \infty$.

(f) *The moments* EY^m *are uniquely determined by the equations*

$$EY^m = \sum_{k=0}^{m} \binom{m}{k} E\left(B^k A^{m-k}\right) EY^k\,, \quad m = 1,\ldots,[p]\,, \tag{8.49}$$

where $[p]$ *denotes the integer part of* p.

Proof. (a) The existence of the weak limit of (Y_t) is shown in part (b) below. Then it is immediate that

$$(A_t, B_t, Y_{t-1}) \xrightarrow{d} (A, B, Y)$$

with (A,B) and Y independent. This and the continuous mapping theorem prove (a).

(b) Iterating equation (8.43) yields

$$Y_t = Y_0 \prod_{j=1}^{t} B_j + \sum_{m=1}^{t} A_m \prod_{j=m+1}^{t} B_j\,, \quad t \ge 0\,. \tag{8.50}$$

We write $Y_t(Y_0) = Y_t$ when we want to emphasize the dependence on the initial value Y_0. Starting with different Y_0' and Y_0'', we obtain

$$Y_t(Y_0') - Y_t(Y_0'') = (Y_0' - Y_0'') \prod_{j=1}^{t} B_j\,, \quad t \in \mathbb{N}. \tag{8.51}$$

The SLLN and (8.46) imply

$$\frac{1}{t}\sum_{j=1}^{t}\ln|B_j| \stackrel{\text{a.s.}}{\to} E\ln|B| < 0\,. \tag{8.52}$$

Hence

$$\left|\prod_{j=1}^{t} B_j\right| = \exp\left\{\sum_{j=1}^{t}\ln|B_j|\right\} \stackrel{\text{a.s.}}{\to} 0\,.$$

From this and (8.51) we conclude, that if $Y_t(Y_0) \stackrel{d}{\to} Y$ for some Y_0, then the latter relation holds for any initial value Y_0. In particular, if there is a Y such that $Y_t \stackrel{d}{\to} Y$ and $Y_0 \stackrel{d}{=} Y$ is independent of $(A_t, B_t)_{t\geq 1}$, then $Y_t(Y_0) \stackrel{d}{=} Y$, and therefore Y in (8.47) is unique in distribution. Thus it remains to show that (Y_t) converges in distribution to Y given by (8.48).

Set $Y_0^* = 0$ and

$$Y_t^* = \sum_{m=1}^{t} A_m \prod_{j=1}^{m-1} B_j\,, \quad t\in\mathbb{N}. \tag{8.53}$$

Then

$$Y_t \stackrel{d}{=} Y_t^* + Y_0\prod_{j=1}^{t} B_j\,.$$

which immediately follows from the discussion in Example 8.4.2. Note that the Y_t^* are the partial sums of the infinite series in (8.48). So a sufficient condition for Y_t^* to converge in distribution is a.s. convergence of the series (8.48). Now, by the SLLN (8.52), and since $m^{-1}\ln^+|A_m| \stackrel{\text{a.s.}}{\to} 0$ (this is also a consequence of the SLLN and of (8.46)),

$$\left|A_m\prod_{j=1}^{m-1} B_j\right| \leq \exp\left\{m\left(\frac{1}{m}\ln^+|A_m| + \frac{1}{m}\sum_{j=1}^{m-1}\ln|B_j|\right)\right\} < e^{-am}$$

for some $a\in(0, |E\ln|B||)$ and sufficiently large m, with probability 1. Hence the rhs in (8.48) converges almost surely. This proves (b).

The proof of (c) follows from the special structure of the process (Y_t) for any vector $(Y_t,\ldots,Y_{t+h})$ for $t,h\in\mathbb{N}$, by a straightforward generalisation of the following argument: by (8.47) we obtain for any $t\in\mathbb{N}$

$$\begin{aligned}(Y_t, Y_{t+1}) &= (Y_t, A_{t+1} + B_{t+1}Y_t) \\ &\stackrel{d}{=} (Y_0, A_1 + B_1 Y_0) = (Y_0, Y_1)\,.\end{aligned} \tag{8.54}$$

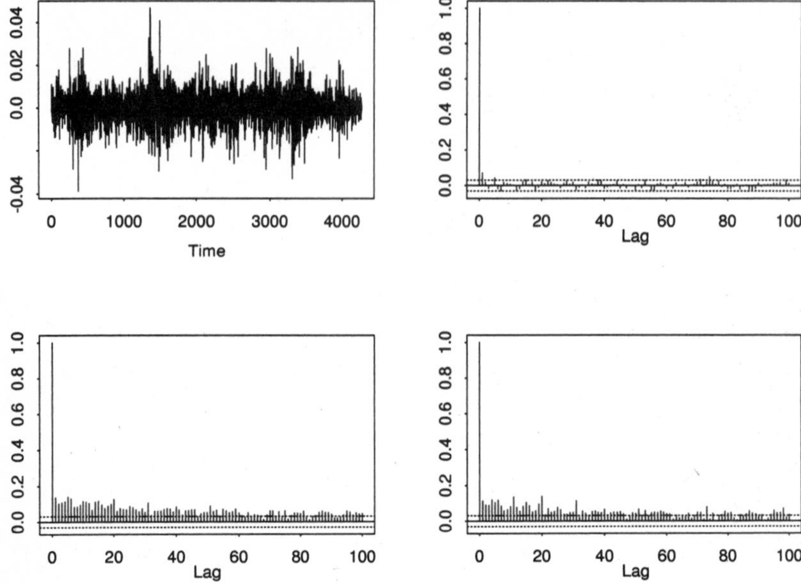

Figure 8.4.4 *Log-returns of the exchange rate \$US/£UK, January 2, 1980–May 21, 1996 (top, left) and the corresponding sample autocorrelations of this time series (top, right), of its absolute values (bottom, left) and of its squares (bottom, right). The dotted lines indicate the* 95% *asymptotic confidence band for the sample autocorrelations of iid Gaussian rvs.*

(d) For any rv Z set $\|Z\|_p = (E|Z|^p)^{1/p}$. Since $\|A\|_p < \infty$ and $\|B\|_p < 1$, an application of Jensen's inequality ensures that (8.46), and hence Proposition 8.4.3(a)–(c) hold. Moreover,

$$E\|Y\|_p \le \sum_{t=1}^{\infty} \left\| A_t \prod_{j=1}^{t-1} B_j \right\|_p = \|A\|_p \sum_{t=1}^{\infty} \|B\|_p^{t-1} < \infty .$$

Hence $E|Y|^p < \infty$ and the series in (8.48) converges in pth mean.

(e) Y_t^* as defined in (8.53) converges a.s. to the rhs of (8.48). By (a)–(c) and dominated convergence, $E|Y_t^*|^p \to E|Y|^p$. Thus, for $Y_0 = 0$ a.s., $E|Y_t|^p \to E|Y|^p$, using $Y_t(0) \stackrel{d}{=} Y_t^*$. For general Y_0 it follows from (8.51) that

$$E|Y_t(Y_0) - Y_t(0)|^p \le (E|B|^p)^t E|Y_0|^p \to 0, \quad t \to \infty .$$

(f) From (8.47) we conclude that (8.49) holds, whenever the occurring expectations exist, in particular for $m = 1, \ldots, [p]$. The equations determine

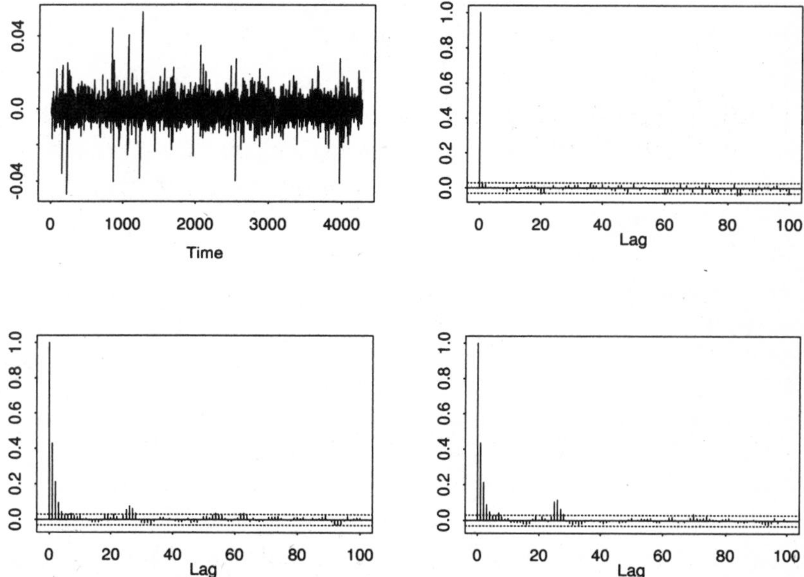

Figure 8.4.5 *Simulated sample path of an* ARCH(1) *process with parameters* $\lambda = 0.50001$ *and* $\beta = 1.9 \cdot 10^{-5}$*, the estimated parameters for the data of Figure 8.4.4, and the corresponding sample autocorrelations of this time series (top, right), of its absolute values (bottom, left) and of its squares (bottom, right). The dotted lines indicate the* 95% *asymptotic confidence band for the sample autocorrelations of iid Gaussian rvs.*

EY^m successively for $m = 1, \ldots, [p]$: the coefficient EB^m of EY^m on the rhs satisfies

$$|EB^m| \leq E|B|^m < 1\,,$$

since $q^{-1} \ln E|B|^q$ is a convex function in q on $(0, p]$ with non–positive values at the endpoints. □

8.4.2 Basic Properties of ARCH Processes

In Chapter 7 we suggested linear processes for modelling financial data. They may be appropriate as a first approximation, but often do not capture the more detailed structure of financial data. Such data often exhibit the following features:

(*a*) Almost no correlation in the data.
(*b*) Volatility changes in time.

(*c*) Data are heavy–tailed.
(*d*) High correlation of the squares and absolute values of the data.
(*e*) High threshold exceedances appear in clusters.

Some of these features are not captured by linear processes. Various models have been introduced aiming at properties (a)–(e). This section is devoted to one particular class of such models. In 1982, Engle [219] introduced the **A***uto* **R***egressive* **C***onditionally* **H***eteroscedastic process of order* p (ARCH(p)) for $p \in \mathbb{N}$. This class was extended by Bollerslev [77] who suggested an alternative and more flexible dependence structure for describing log–returns, the *generalised* ARCH or GARCH(p,q) model ($p,q \in \mathbb{N}$). It is defined by the equations

$$X_t = \sigma_t Z_t \,, \quad t \in \mathbb{N}, \tag{8.55}$$

where (Z_t) is a sequence of iid standard normal rvs and σ_t obeys the relation

$$\sigma_t^2 = \beta + \sum_{i=1}^{p} \lambda_i X_{t-i}^2 + \sum_{j=1}^{q} \delta_j \sigma_{t-j}^2 \,, \quad t \in \mathbb{N}, \tag{8.56}$$

with fixed non–negative constants β, λ_i and δ_j. Notice that (X_t) is a Gaussian mixture model. In contrast to linear processes such as ARMA models, where the noise is additive, here the noise (Z_t) appears multiplicatively. The variance σ_t^2 of X_t, conditionally on the past observations, is given by the GARCH(p,q) equation (8.56). Thus the conditional variance σ_t^2 depends linearly on the past via the earlier squared log–returns X_{t-i}^2 for $i = 1, \ldots, p$ and the conditional variances σ_{t-j}^2 for $j = 1, \ldots, q$. It means that high volatility may result from large absolute log–returns $|X_{t-i}|$ or from large volatility σ_{t-j} in preceding time periods.

Fitting these models to financial data has been a major issue of econometrics during recent years. However, this is not the topic of this section and we refer to the book by Harvey [319], the review papers by Bollerslev [77], Shephard [578], and the vast amount of references therein; see also Section 7.6. Another book in this context, also containing interesting case studies, is Taylor [616]. A textbook treatment on the statistical analysis of financial time series, including various sections on (G)ARCH processes is Mills [464].

We fitted an ARCH(1) process to the exchange rates presented in Figure 8.4.4. We obtained parameter estimators $\widehat{\lambda} = 0.50001$ and $\widehat{\beta} = 1.9 \cdot 10^{-5}$ by standard maximum likelihood estimation. Figure 8.4.5 shows a simulated sample path of an ARCH(1) process with the estimated parameter values.

We concentrate, however, rather on probabilistic properties implicated by (a)-(e) above, and in particular we study the upper tail of the one–dimensional marginal distribution. The extremes of certain stationary sequences are particularly tractable by the methods presented in Sections 4.4,

5.3 and 5.5. The behaviour of the ARCH(1) process in its extremes is also well understood and might serve as an indicator of the features of the more general class of GARCH(p, q) models; see Section 8.4.3.

For $q = 0$ and $p = 1$, (8.55) and (8.56) reduce to the ARCH(1) *process*, which is defined by the equations

$$X_t = \sqrt{\beta + \lambda X_{t-1}^2}\, Z_t\,, \quad t \in \mathbb{N}, \tag{8.57}$$

for some initial rv X_0 independent of (Z_t), parameters $\beta > 0$ and $\lambda > 0$. It is a Markov process, given by the explicit autoregressive structure (8.57). By construction, X_{t-1} and Z_t are independent for every $t \in \mathbb{N}$, and X_t are mean–zero, uncorrelated rvs provided $EX_0^2 < \infty$.

For deriving probabilistic properties of the ARCH(1) process we will make extensive use of the fact that the *squared* ARCH(1) *process* (X_t^2) satisfies the stochastic recurrence equations

$$X_t^2 = (\beta + \lambda X_{t-1}^2)\, Z_t^2 = A_t + B_t X_{t-1}^2\,, \quad t \in \mathbb{N}, \tag{8.58}$$

where

$$(A_t, B_t) = (\beta Z_t^2, \lambda Z_t^2)\,, \quad t \in \mathbb{N}\,. \tag{8.59}$$

Assuming that X_0 is independent of (Z_t) and setting $Y_t = X_t^2$, we are immediately in the framework of Section 8.4.1. We intend to apply Proposition 8.4.3 to the sequence (X_t^2). This requires a check of the assumptions of that result. The following elementary lemma serves that purpose. In combination with Proposition 8.4.3 it will become the *key* to the results in Sections 8.4.2 and 8.4.3.

Lemma 8.4.6 *For a standard normal rv Z and $\lambda \in (0, 2e^{\gamma})$, where $\gamma \approx 0.5772$ is Euler's constant, define*

$$h(u) = E(\lambda Z^2)^u\,, \quad u \geq 0\,.$$

Then

$$h(u) = \frac{(2\lambda)^u}{\sqrt{\pi}}\,\Gamma\left(u + \frac{1}{2}\right)\,, \quad u \geq 0\,. \tag{8.60}$$

The function h is strictly convex in u, and there exists a unique solution $\kappa = \kappa(\lambda) > 0$ to the equation $h(u) = 1$. Moreover,

$$\kappa(\lambda) \begin{cases} > 1\,, & \lambda \in (0,1)\,, \\ = 1\,, & \lambda = 1\,, \\ < 1\,, & \lambda \in (1, 2e^{\gamma})\,, \end{cases} \tag{8.61}$$

and

$$E\left[(\lambda Z^2)^{\kappa} \ln(\lambda Z^2)\right] > 0 \quad \textit{and} \quad E\left[\ln(\lambda Z^2)\right] < 0\,. \tag{8.62}$$

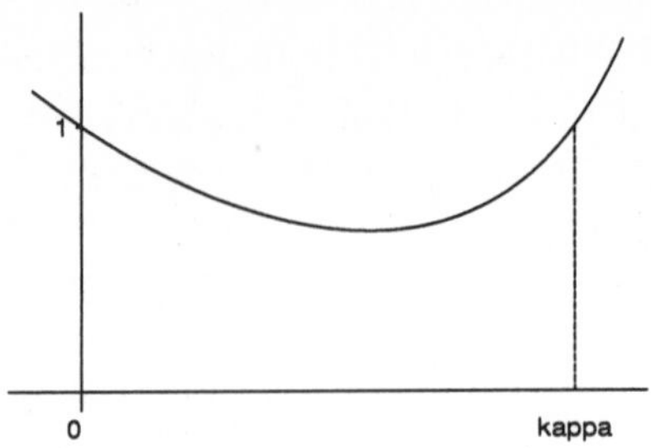

Figure 8.4.7 *One possible situation for $h(u)$ as described in Lemma 8.4.6.*

Proof. Notice first that $h(0) = 1$ for all λ. Furthermore, h has derivatives of all orders. In particular,

$$h'(u) = E\left[(\lambda Z^2)^u \ln(\lambda Z^2)\right] \tag{8.63}$$

$$h''(u) = E\left[(\lambda Z^2)^u (\ln(\lambda Z^2))^2\right] > 0\,. \tag{8.64}$$

(8.63) implies that

$$\begin{aligned} h'(0) &= E[\ln(\lambda Z^2)] &&= \ln\lambda + E\ln Z^2 \\ &= \ln(2\lambda) + \Gamma'(\tfrac{1}{2})/\Gamma(\tfrac{1}{2}) &&= \ln(2\lambda) - \gamma - 2\ln 2 \\ &= \ln\lambda - \ln 2 - \gamma &&< 0 \end{aligned} \tag{8.65}$$

for $0 < \lambda < 2e^{\gamma}$, where γ is Euler's constant. (8.64) implies that h is strictly convex on $\mathbb{R}_+$. By symmetry of the normal density and partial integration, we obtain

$$\begin{aligned} h(u) = E\left(\lambda Z^2\right)^u &= \frac{\lambda^u}{\sqrt{2\pi}} \int_{-\infty}^{\infty} x^{2u} e^{-x^2/2} dx \\ &= \frac{\lambda^u}{\sqrt{2\pi}} 2 \int_0^{\infty} (2y)^{u-1/2} e^{-y} dy \\ &= \frac{(2\lambda)^u}{\sqrt{\pi}} \Gamma\left(u + \frac{1}{2}\right), \end{aligned}$$

giving (8.60). Furthermore,

$$h(u) \geq E[(\lambda Z^2)^u I_{\{\lambda Z^2 > 2\}}] \geq 2^u P(\lambda Z^2 > 2) \to \infty\,, \quad u \to \infty\,.$$

The latter fact, together with $h(0) = 1$ and convexity of h implies that there exists a unique $\kappa > 0$ such that $h(\kappa) = 1$. Furthermore, $h'(\kappa) > 0$, giving together with (8.63) and (8.65) the inequalities (8.62). Since $h(1) = \lambda$, (8.61) follows by a monotonicity argument. □

The value $\kappa = \kappa(\lambda)$ is crucial for the tail behaviour of the marginal distribution, the existence of moments and the extremal behaviour of the ARCH(1) process. The equation $h(u) = 1$ cannot be solved explicitly, but numerical solutions can be found in Table 8.4.8.

λ	0.1	0.3	0.5	0.7	0.9	1.0	1.5	2.0	2.5	3.0	3.5
κ	13.24	4.18	2.37	1.59	1.15	1.0	.54	.31	.17	.075	.007

Table 8.4.8 *Values of* $\kappa = \kappa(\lambda)$ *for* $\lambda \in (0, 2e^{\gamma})$.

Now we are well prepared to apply Proposition 8.4.3 to (X_t^2).

Theorem 8.4.9 (Properties of the squared ARCH(1) process)
Let (X_t) *be an* ARCH(1) *process given by* (8.57) *for fixed* $\beta > 0$ *and* $\lambda \in (0, 2e^{\gamma})$, *where* γ *is Euler's constant, and assume that* X_0 *is independent of* (Z_t).

(a) *The process* (X_t) *is strictly stationary if*

$$X_0^2 \stackrel{d}{=} \beta \sum_{m=1}^{\infty} Z_m^2 \prod_{j=1}^{m-1} (\lambda Z_j^2) \,. \tag{8.66}$$

Moreover, every strictly stationary ARCH(1) *process* (X_t) *has marginal distribution*

$$X_t \stackrel{d}{=} |X_0| \, r_0 \,,$$

with X_0^2 *satisfying* (8.66), *and* r_0 *is a Bernoulli rv with* $P(r_0 = \pm 1) = 0.5$, *independent of* $|X_0|$.

(b) *Assume that* (X_t) *is strictly stationary and write* $X = X_0$, $Z = Z_1$. *Let* κ *be the unique positive solution of the equation* $h(u) = E(\lambda Z^2)^u = 1$. *Then* $E(X^2)^u < \infty$ *for* $0 \le u < \kappa$. *Denote by* p *the largest integer strictly less than* κ. *Then for* $m = 1, \ldots, p$,

$$EX^{2m} = (1 - E(\lambda Z^2)^m)^{-1} EZ^{2m} \sum_{k=0}^{m-1} \binom{m}{k} \lambda^k \beta^{m-k} EX^{2k} < \infty \,. \tag{8.67}$$

Proof. (a) Set $Y_t = X_t^2$, $A_t = \beta Z_t^2$ and $B_t = \lambda Z_t^2$. By (8.62), $E \ln^+ |A| < \infty$ and $E \ln |B| < 0$. An application of Proposition 8.4.3 yields that (Y_t) is strictly stationary with unique marginal distribution (8.66). An argument similar to (8.54) proves that (X_t) is also strictly stationary.

Now assume that (X_t) is strictly stationary. Then, by Proposition 8.4.3, X_0^2 necessarily satisfies (8.66). Since X_t is symmetric, we have the identity in law

$$X_t \stackrel{d}{=} |X_t|\, r_0$$

for a symmetric Bernoulli rv r_0 independent of X_t. This concludes the proof of (a).

(b) By Lemma 8.4.6, the function $h(u)$ is strictly convex and satisfies $h(0) = h(\kappa) = 1$. Hence $h(u) < 1$ for $0 < u < \kappa$. According to Proposition 8.4.3(d,f), $E(X^2)^u < \infty$ for $u < \kappa$ and $X^2 \stackrel{d}{=} (\beta + \lambda X^2)Z^2$. Then (8.67) follows from (8.49). □

Corollary 8.4.10 *Let (X_t) be a stationary* ARCH(1) *process with parameters $\beta > 0$ and $\lambda \in (0,1)$. Then the following relations hold:*

(a) $EX^2 = \beta/(1-\lambda)$.
(b) If $\lambda^2 < 1/3$, then $EX^4 < \infty$ and $\operatorname{corr}\left(X_t^2, X_0^2\right) = \lambda^t$ *for all $t \in \mathbb{N}$.*

Proof. (a) Choose $m = 1$ in (8.67). Then from $h(1) = \lambda < 1$,

$$EX^2 = \beta(1 - E(\lambda Z^2))^{-1} = \beta(1 - h(1))^{-1} = \beta/(1-\lambda)\,.$$

(b) By Theorem 8.4.9, $EX^4 < \infty$ for $0 < \lambda^2 < 1/3$. We obtain as in (a)

$$EX^4 = (1 - \lambda^2 EZ^4)^{-1} EZ^4 (\beta^2 + 2\lambda\beta EX^2) = \frac{3\beta^2}{1-3\lambda^2}\,\frac{1+\lambda}{1-\lambda}\,.$$

Iterating the ARCH(1) equation yields

$$\begin{aligned} EX_t^2 X_0^2 &= \frac{\beta^2}{1-\lambda}\sum_{k=0}^{t-1} \lambda^k + \lambda^t EX^4 \\ &= (EX^2)^2 + \lambda^t \operatorname{var}(X^2)\,. \end{aligned}$$

This concludes the proof. □

Remarks. 1) By now we have realised that λ is a crucial parameter of the ARCH(1) process (8.57):

- for $\lambda = 0$, (X_t) is normal noise,
- for $\lambda \in (0,1)$, (X_t) is stationary with finite variance,
- for $1 \le \lambda < 2e^{\gamma} \approx 3.56856$, (X_t) is stationary with infinite variance; see Theorem 8.4.12.

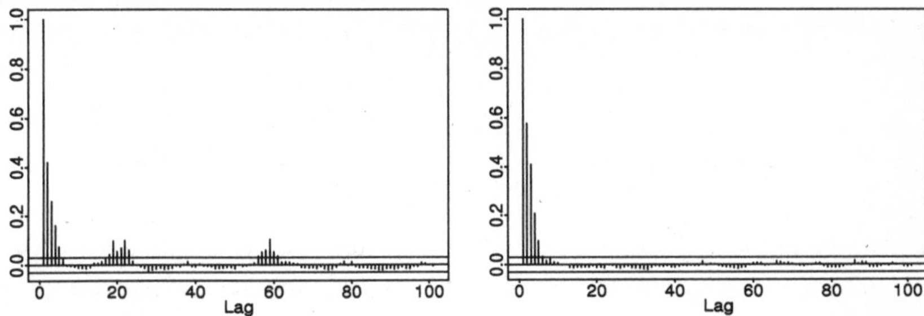

Figure 8.4.11 *Sample autocorrelation functions of two different simulated sample paths of an* ARCH(1) *process with* $\lambda = 0.7$. *The contrast between the two graphs indicates that the sample autocorrelations do not converge in probability to constants.*

2) We anticipate at this point that $EX^4 = \infty$ for $\lambda^2 \geq 1/3$. This follows from $h(2) = 1$ for $\lambda = 1/\sqrt{3} \approx 0.577$ together with (8.68) below. In this case the notion of autocorrelation does not make sense for the squared ARCH(1) process. Nevertheless, the sample autocorrelations $\widetilde{\rho}_{X^2}(t)$ are well defined for $X_1^2, \ldots, X_n^2$. In contrast to linear processes (see Theorems 7.3.1 and 7.3.2) we are not aware of a consistency result for $\widetilde{\rho}_{X^2}(t)$ which explains its behaviour for large n.

3) For $\lambda^2 < 1/3$ it is interesting to observe that the squared ARCH(1) process has the same autocorrelation structure as an AR(1) process $X_t = \lambda X_{t-1} + Z_t$, $t \in \mathbb{Z}$; see Figure 8.4.15. For $\lambda^2 \geq 1/3$ this is no longer true; see Figures 8.4.11 and 8.4.16. □

In the following result we describe the tail of the marginal distribution of an ARCH(1) process. It gives a precise meaning to the statement "light–tailed input causes heavy–tailed output", a fact already observed by Kesten [379] in the general context of stochastic recurrence equations. The renewal argument given below is due to Goldie [272] who also calculated the precise constant (8.69). Theorem 8.4.12 opens the door to extreme value theory for ARCH(1) processes; see Section 8.4.3. Goldie proved the following result for general perpetuity models as introduced in Section 8.4.1. For simplicity we restrict ourselves to the ARCH(1) case. A special property of the normal law (spread–out) allows us to shorten the renewal reasoning in Goldie [272] for the specific case we deal with.

Theorem 8.4.12 (The tail behaviour of an ARCH(1) process)
Let (X_t) *be a stationary* ARCH(1) *process with parameters* $\beta > 0$ *and* $\lambda \in (0, 2e^{\gamma})$, *where* γ *is Euler's constant. Let* $\kappa > 0$ *be the unique positive solution of the equation* $h(u) = 1$. *Then*

$$P(X > x) \sim \frac{c}{2} x^{-2\kappa}, \quad x \to \infty, \tag{8.68}$$

where

$$c = \frac{E\left[((\beta + \lambda X^2)^\kappa - (\lambda X^2)^\kappa)(Z^2)^\kappa\right]}{\kappa E\left[(\lambda Z^2)^\kappa \ln(\lambda Z^2)\right]} \in (0, \infty), \tag{8.69}$$

for a standard normal rv Z, independent of $X = X_0$.

Before we can prove this result we have to ensure that the constant (8.69) is well defined. Indeed, its numerator is the expected difference of two quantities each of which having infinite expectation.

Lemma 8.4.13 *Under the assumptions of Theorem 8.4.12,*

$$0 < E\left[(\beta + \lambda X^2)^\kappa - (\lambda X^2)^\kappa\right] < \infty. \tag{8.70}$$

Proof. For $\kappa \leq 1$ the function $(\beta + \lambda x^2)^\kappa - (\lambda x^2)^\kappa$ is positive and bounded. Thus (8.70) follows. For $\kappa > 1$, the function $f(z) = z^\kappa$ is strictly convex on $\mathbb{R}$ and hence

$$\beta\kappa(\lambda x^2)^{\kappa-1} \leq (\beta + \lambda x^2)^\kappa - (\lambda x^2)^\kappa \leq \beta\kappa(\beta + \lambda x^2)^{\kappa-1}.$$

This implies that

$$\begin{aligned} 0 < \beta\kappa E[(\lambda X^2)^{\kappa-1}] &\leq E\left[(\beta + \lambda X^2)^\kappa - (\lambda X^2)^\kappa\right]\right] \\ &\leq \beta\kappa E[(\beta + \lambda X^2)^{\kappa-1}], \end{aligned}$$

the last expression being finite by Theorem 8.4.9(b). □

We also need the following elementary tool. Notice that this statement is trivial if $EY^\delta < \infty$.

Lemma 8.4.14 *Let $X \geq Y$* a.s. *be non-negative rvs. Then*

$$\int_0^\infty (P(X > t) - P(Y > t))\, t^{\delta-1} dt = \delta^{-1} E\left[X^\delta - Y^\delta\right], \quad \delta \neq 0,$$

for the rhs finite or infinite.

Proof.

$$\begin{aligned} \delta \int_0^\infty (P(X > t) - P(Y > t))\, t^{\delta-1} dt &= \delta \int_0^\infty P(Y \leq t < X) t^{\delta-1} dt \\ &= \delta E \int_Y^X t^{\delta-1} dt \\ &= E\left[X^\delta - Y^\delta\right]. \end{aligned}$$

□

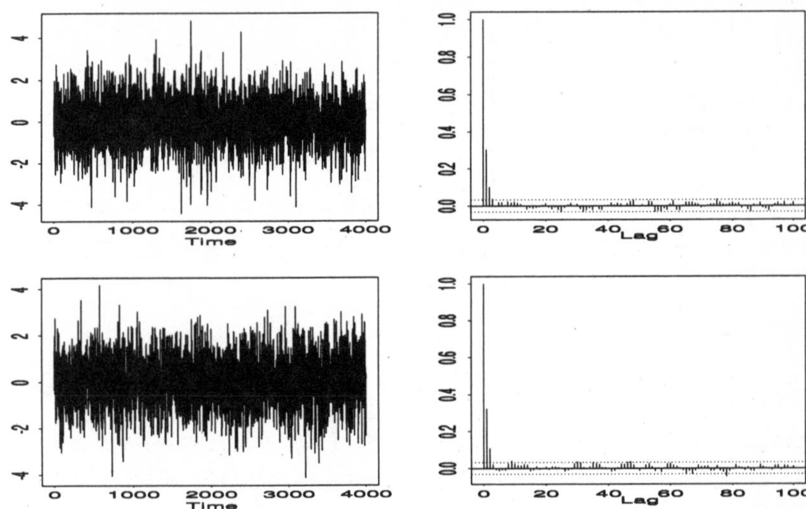

Figure 8.4.15 *Sample path of the* ARCH(1) *process* $X_t = \sqrt{0.3X_{t-1}^2 + 0.75}\, Z_t$ *and the sample autocorrelations of the corresponding squared* ARCH *process (top) and a sample path of the* AR(1) *process* $X_t = 0.3X_{t-1} + Z_t$ *and its sample autocorrelations (bottom).*

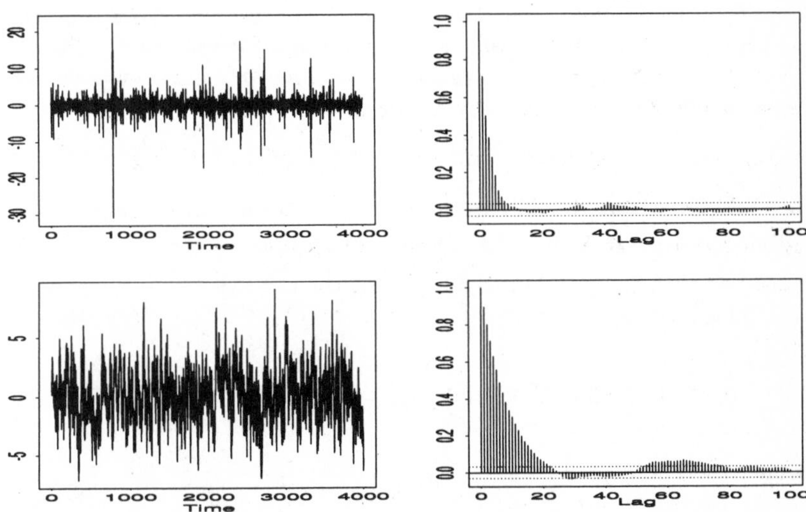

Figure 8.4.16 *Sample path of the* ARCH(1) *process* $X_t = \sqrt{0.9X_{t-1}^2 + 0.75}\, Z_t$ *and the sample autocorrelations of the corresponding squared* ARCH *process (top) and a sample path of the* AR(1) *process* $X_t = 0.9X_{t-1} + Z_t$ *and its sample autocorrelations (bottom).*

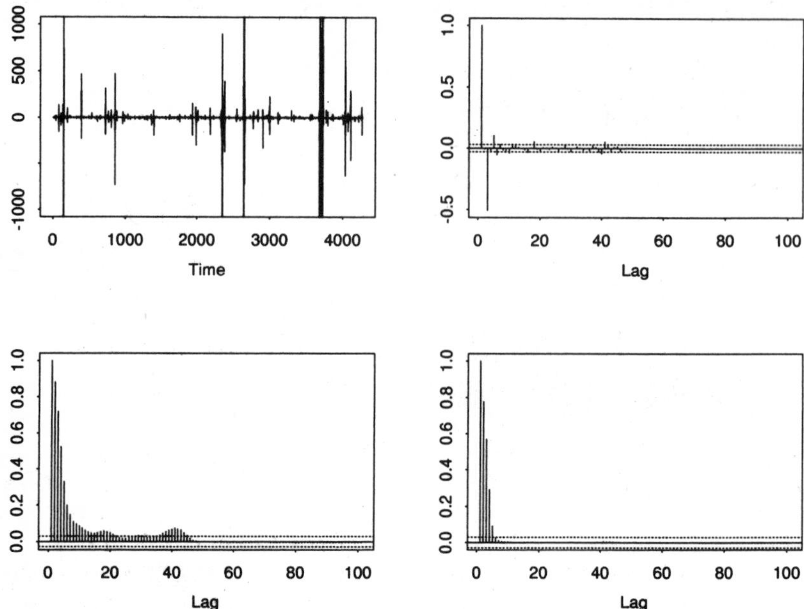

Figure 8.4.17 *Simulated sample path of an* ARCH(1) *process with parameter* $\lambda = 2.0$ *and the corresponding sample autocorrelations of this time series (top, right), of its absolute values (bottom, left) and of its squares (bottom, right). The dotted lines indicate the* 95% *asymptotic confidence band for the sample autocorrelations of iid Gaussian rvs.*

Proof of Theorem 8.4.12. Both the denominator and the numerator in (8.69) are positive in view of (8.70) and (8.62). Hence $c \in (0, \infty)$.

We proceed with the proof of (8.68). Define the random walk (S_n) generated by the iid sequence $(\ln(\lambda Z_t^2))$:

$$S_0 = 0 \,, \; S_n = \sum_{t=1}^{n} \ln(\lambda Z_t^2) \quad \text{and} \quad \Pi_0 = 1 \,, \; \Pi_n = e^{S_n} = \prod_{t=1}^{n} (\lambda Z_t^2) \,. \quad (8.71)$$

Consider the telescoping sum

$$\begin{aligned} & P(X^2 > e^x) - P\left(X^2\, \Pi_n > e^x\right) \\ = & \sum_{k=0}^{n-1} \left(P\left(X^2\, \Pi_k > e^x\right) - P\left(X^2\, \Pi_{k+1} > e^x\right)\right) \end{aligned}$$

$$= \sum_{k=0}^{n-1} \left(E\left[P\left(X^2\,\Pi_k > e^x \mid \Pi_k\right)\right] - E\left[P\left(\lambda Z_{k+1}^2\, X^2\,\Pi_k > e^x \mid \Pi_k\right)\right]\right)$$

$$= \sum_{k=0}^{n-1} \int_{\mathbb{R}} \left(P\left(X^2 > e^{x-y}\right) - P\left(\lambda Z^2 X^2 > e^{x-y}\right)\right) dP\left(S_k \le y\right) .$$

Introduce the measures ν_n by

$$d\nu_n(x) = \sum_{k=0}^{n} e^{\kappa x} dP\left(S_k \le x\right) , \quad n \in \mathbb{N}, \tag{8.72}$$

and define

$$g(x) = e^{\kappa x}\left(P\left(X^2 > e^x\right) - P\left(\lambda Z^2\, X^2 > e^x\right)\right) , \quad x \in \mathbb{R},$$

$$\delta_n(x) = e^{\kappa x} P\left(X^2\,\Pi_n > e^x\right) , \quad x \in \mathbb{R}, \quad n \in \mathbb{N},$$

and

$$r(x) = e^{\kappa x} P(X^2 > e^x) .$$

Notice that $X_t^2 \ge \lambda Z_t^2 X_{t-1}^2$ a.s., and hence g is non–negative. Now we obtain

$$\begin{aligned} r(x) &= e^{\kappa x} P(X^2 > e^x) \\ &= e^{\kappa x} \sum_{k=0}^{n-1} \int_{\mathbb{R}} \left(P\left(X^2 > e^{x-y}\right) - P\left(\lambda Z^2 X^2 > e^{x-y}\right)\right) \\ &\qquad\qquad dP\left(S_k \le y\right) + \delta_n(x) \\ &= \int_{\mathbb{R}} e^{\kappa(x-y)} \left(P\left(X^2 > e^{x-y}\right) - P\left(\lambda Z^2 X^2 > e^{x-y}\right)\right) \\ &\qquad\qquad e^{\kappa y} \sum_{k=0}^{n-1} dP\left(S_k \le y\right) + \delta_n(x) \\ &= \int_{\mathbb{R}} g(x-y) d\nu_{n-1}(y) + \delta_n(x) \\ &= g * \nu_{n-1}(x) + \delta_n(x) . \end{aligned} \tag{8.73}$$

For the proof of (8.68) one needs to study the behaviour of $r(x)$ as $x \to \infty$. For this reason we will apply the key renewal theorem in a similar fashion to that in the proof of the Cramér–Lundberg theorem (Theorem 1.2.2).
Define the measure η by

$$d\eta(x) = e^{\kappa x} dP\left(\ln(\lambda Z^2) \le x\right) .$$

Notice that, by definition of κ,

$$\int_{\mathbb{R}} d\eta(x) = \int_{\mathbb{R}} e^{\kappa x} dP\left(\ln(\lambda Z^2) \leq x\right) = E(\lambda Z^2)^\kappa = 1\,.$$

Hence η defines a probability measure on $\mathbb{R}$ (obviously not lattice) with mean (see (8.62))

$$\mu(\eta) = E\left[(\lambda Z^2)^\kappa \ln(\lambda Z^2)\right] > 0\,.$$

Let ν be the renewal measure corresponding to η, i.e.

$$d\nu(x) = \sum_{k=0}^{\infty} d\eta^{k*}(x) = \sum_{k=0}^{\infty} e^{\kappa x} dP\left(S_k \leq x\right)\,.$$

Notice that $\nu_n = \sum_{k=0}^{n} \eta^{k*}$ for $n \in \mathrm{N}$, where ν_n is defined in (8.72). From Lemmas 8.4.13 and 8.4.14 we conclude that

$$g \text{ is integrable, it is bounded and } \lim_{x\to\infty} g(x) = 0. \tag{8.74}$$

Since $\mu(\eta) > 0$, the renewal measure ν has the property

$$g * \nu(x) < \infty\,, \quad x \in \mathbb{R}\,.$$

Moreover,

$$g * \nu_{n-1}(x) \uparrow g * \nu(x)\,, \quad x \in \mathbb{R}\,. \tag{8.75}$$

Furthermore, by (8.62), $E\ln(\lambda Z^2) < 0$. Thus the random walk (S_n) has a negative drift and

$$\lim_{n\to\infty} \delta_n(x) = \lim_{n\to\infty} e^{\kappa x} P(X^2 e^{S_n} > e^x) = 0$$

for every $x \in \mathbb{R}$. This together with (8.75) and (8.73) implies that r satisfies

$$r(x) = \lim_{n\to\infty} \left(g * \nu_{n-1}(x) + \delta_n(x)\right) = g * \nu(x)\,, \quad x \in \mathbb{R}\,.$$

The function g satisfies (8.74) and η is absolutely continuous. The latter two facts allow us to apply a version of the key renewal theorem (Satz 3.1.7 in Alsmeyer [8]), which in turn, together with Lemma 8.4.14, leads to the relation

$$\lim_{x\to\infty} r(x) = \frac{1}{\mu(\eta)} \int_{\mathbb{R}} g(y)dy = \frac{1}{\mu(\eta)\kappa} E\left[((\beta + \lambda X^2)^\kappa - (\lambda X^2)^\kappa)(Z^2)^\kappa\right] = c$$

with c as in (8.69). Hence by the definition of r we conclude that $P(X^2 > x) \sim cx^{-\kappa}$, and by symmetry of X,

$$P(X > x) = \frac{1}{2} P(X^2 > x^2) \sim \frac{c}{2} x^{-2\kappa}\,, \quad x \to \infty\,. \qquad \square$$

Remark. 4) The literature on ARCH and its various generalisations abounds

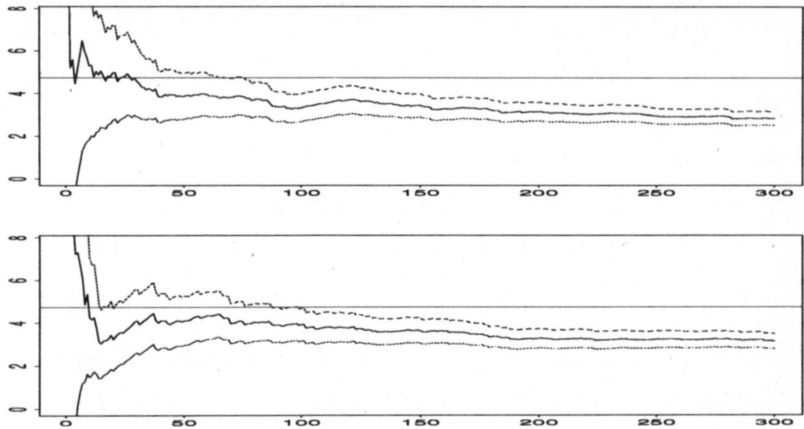

Figure 8.4.18 *The Hill estimator, see Section 6.4.2, for the data from Figure 8.4.4 (top) and for the* ARCH(1) *sequence of Figure 8.4.5 (bottom). For the latter we obtain from Table 8.4.8 the value* $\kappa = 2.365$*, hence the tail index* $2\kappa = 4.73$*, which is indicated by a straight line. See also Theorem 8.4.12.*

with often vague statements on stationarity, moment conditions and tail behaviour. We therefore consider it important to prove Theorem 8.4.12 in detail, especially as the main ideas for the Pareto asymptotics were made clear some time ago in Kesten [379]. In the latter paper it was stressed that (see (8.66))

$$X_0^2 \stackrel{d}{=} \sum_{m=1}^{\infty} (\beta Z_m^2) \prod_{j=1}^{m-1} (\lambda Z_j^2)$$

has a tail comparable to that of

$$\bigvee_{m=1}^{\infty} (\beta Z_m^2) \prod_{j=1}^{m-1} (\lambda Z_j^2) = \bigvee_{m=1}^{\infty} (\beta Z_m^2)\, e^{S_{m-1}} ,$$

and the tail of the latter quantity is determined in part by $\max_{n\geq 0} S_n$. Since S_n has a negative expectation, results for the distribution of a random walk with negative drift by means of defective renewal theory apply (cf. Feller [235], Section XII.5, Example c), giving the Pareto–like tails of X. □

8.4.3 Extremes of ARCH Processes

In this section we investigate the extremal behaviour of an ARCH(1) process (X_t) with parameters $\beta > 0$ and $\lambda \in (0, 2e^{\gamma})$, where γ is Euler's constant. Recall from (8.68) that $P(X > x) \sim cx^{-2\kappa}/2$. In the case that (X_t) is an iid sequence such that $X_1 \stackrel{d}{=} X$, we conclude from Theorem 3.3.7 that $X \in \text{MDA}(\Phi_{2\kappa})$. In particular,

$$\lim_{n\to\infty} P\left(n^{-1/(2\kappa)}\, M_n \le x\right) = \exp\left\{-\frac{c}{2}x^{-2\kappa}\right\}, \tag{8.76}$$

where, as usual, $M_n = \max(X_1, \ldots, X_n)$. It is now natural to ask:

Does the strictly stationary ARCH(1) *process have similar extremal behaviour as an iid sequence with the same marginal distribution?*

A full answer to this question was given in de Haan et al. [303], where also the case of general perpetuities was discussed.

As in the previous sections, it is convenient first to study the extremes of the squared ARCH(1) process (X_t^2). The following result is based on a fundamental theorem by Rootzén [551] on maxima of strictly stationary sequences, an application of which depends on the verification of the condition $D(u_n)$ for $u_n = xn^{-1/\kappa}$; cf. Section 4.4. The ARCH(1) process is strong mixing (see Diebolt and Guegan [182]) which entails $D(u_n)$.

Recall from Chapter 5 the basic notions of point process theory, and from Section 8.1 the definition and interpretation of the extremal index of a stationary sequence. As in the previous sections we write $X = X_0$ and $Z = Z_1$.

Theorem 8.4.19 (The extremes of a squared ARCH(1) process)
Let (X_t^2) *be a stationary squared* ARCH(1) *process and* $M_n^{(2)} = \max(X_1^2, \ldots, X_n^2)$. *Then*

$$\lim_{n\to\infty} P(n^{-1/\kappa} M_n^{(2)} \le x) = \exp\{-c\,\theta^{(2)} x^{-\kappa}\}, \quad x > 0, \tag{8.77}$$

where κ *is the positive solution of the equation* $E(\lambda Z^2)^u = 1$, c *is defined by* (8.69) *and*

$$\theta^{(2)} = \kappa \int_1^\infty P\left(\max_{n\ge 1} \prod_{t=1}^n (\lambda Z_t^2) \le y^{-1}\right) y^{-\kappa-1} dy\,. \tag{8.78}$$

For $x > 0$, *let*

$$N_n^{(2)}(\cdot) = \sum_{i=1}^n \varepsilon_{n^{-1}i}(\cdot) I_{\{X_i^2 > xn^{1/\kappa}\}}$$

be the point process of exceedances of the threshold $xn^{1/\kappa}$ *by* $X_1^2, \ldots, X_n^2$. *Then*

$$N_n^{(2)} \stackrel{d}{\to} N^{(2)}, \quad n \to \infty,$$

in $M_p((0,1])$, *where* $N^{(2)}$ *is a compound Poisson process with intensity* $c\theta^{(2)} x^{-\kappa}$ *and cluster probabilities*

$$\pi_k^{(2)} = \frac{\theta_k^{(2)} - \theta_{k+1}^{(2)}}{\theta^{(2)}}, \quad k \in \mathbb{N}, \tag{8.79}$$

where

$$\theta_k^{(2)} = \kappa \int_1^\infty P\left(\operatorname{card}\left\{n \in \mathbb{N} : \prod_{t=1}^n (\lambda Z_t^2) > y^{-1}\right\} = k-1\right) y^{-\kappa-1} dy\,.$$

In particular, $\theta_1^{(2)} = \theta^{(2)}$. □

Remarks. 1) Recall from Section 8.1 the definition of the extremal index of a strictly stationary sequence. A comparison of (8.76) and (8.77) shows that $\theta^{(2)}$ is nothing but the extremal index of the squared ARCH(1) process.

2) An alternative expression for $\theta_k^{(2)}$ can be obtained in terms of the random walk

$$S_0 = 0 \quad , \quad S_n = \sum_{t=1}^n \ln(\lambda Z_t^2) \tag{8.80}$$

which played a crucial role in proving the tail estimate of Theorem 8.4.12. Write

$$\infty = T_0 \geq T_1 \geq T_2 \geq \cdots$$

for the ordered values of the sequence (S_n). Naturally, $T_1 = \sup_{n \geq 0} S_n < \infty$ a.s. since (S_n) is a random walk with negative drift; see (8.62). We observe that

$$\operatorname{card}\left\{n \in \mathbb{N} : \prod_{t=1}^n (\lambda Z_t^2) > y^{-1}\right\} = \sum_{n=1}^\infty I_{\{S_n > -\ln y\}} = k-1$$

if and only if

$$T_{k-1} > -\ln y \quad \text{and} \quad T_k \leq -\ln y\,.$$

This implies that

$$\begin{aligned}
\theta_k^{(2)} &= \kappa \int_1^\infty \left(P(T_{k-1} > -\ln y) - P(T_k > -\ln y)\right) y^{-\kappa-1} dy \\
&= \int_0^1 \left(P(\exp\{\kappa T_{k-1}\} > y) - P(\exp\{\kappa T_k\} > y)\right) dy \qquad (8.81) \\
&= E\left[\exp\{\kappa \min(T_{k-1}, 0)\} - \exp\{\kappa \min(T_k, 0)\}\right].
\end{aligned}$$

In particular, $\theta^{(2)} = \theta_1^{(2)} = 1 - E\exp\{\kappa \min(T_1, 0)\}$. Plugging the values $\theta_k^{(2)}$ into formula (8.79), one can then calculate the cluster probabilities $\pi_k^{(2)}$. □

With Theorem 8.4.19 we have the extremes of (X_t^2), hence of $(|X_t|)$ under control, but how can we use this knowledge for the extremes of the ARCH process itself? We observe by the symmetry and iid properties of (Z_t) that $(|X_t|)$ and $(r_t) = (\text{sign}(X_t)) = (\text{sign}(Z_t))$ are mutually independent and

$$(X_t) = (r_t|X_t|)\,, \tag{8.82}$$

where $P(r_1 = \pm 1) = 0.5$. Hence the partial maxima of (X_t) and of $(r_t|X_t|)$ have the same distribution. Thus, in order to study exceedances of the high level threshold $u > 0$ by (X_t), we may try to use the known result for $(|X_t|)$, but we would have to rule out exceedances of the form $\{X_t < -u\}$. This is the basic idea of the derivation given below. For the formulation of the main result on the extremes of an ARCH(1) process we also need the probability generating function of the cluster probabilities $(\pi_k^{(2)})$ defined in (8.79):

$$\Pi^{(2)}(u) = \sum_{k=1}^{\infty} \pi_k^{(2)} u^k\,.$$

Theorem 8.4.20 (The extremes of an ARCH(1) process)
Let (X_t) be a stationary ARCH(1) process. Then

$$\lim_{n\to\infty} P\left(n^{-1/(2\kappa)} M_n \le x\right) = \exp\left\{-c\theta^{(2)}\left(1 - \Pi^{(2)}(0.5)\right) x^{-2\kappa}\right\}. \tag{8.83}$$

Further, for $x > 0$ let

$$N_n(\cdot) = \sum_{i=1}^{n} \varepsilon_{n^{-1}i}(\cdot) I_{\{X_i > x n^{1/(2\kappa)}\}}$$

be the point process of exceedances of the threshold $x n^{1/(2\kappa)}$ by $X_1, \dots, X_n$. Then

$$N_n \xrightarrow{d} N\,, \quad n \to \infty\,,$$

in $M_p((0,1])$, where N is a compound Poisson process with intensity $c\theta x^{-2\kappa}$ and cluster probabilities

$$\pi_k = \left(1 - \Pi^{(2)}(0.5)\right)^{-1} \sum_{m=k}^{\infty} \binom{m}{k} \pi_m^{(2)} 2^{-m}\,, \quad k \in \mathbb{N}_0\,. \tag{8.84}$$

Remark. 3) A comparison with (8.76) shows that $\theta = 2\theta^{(2)}(1 - \Pi^{(2)}(0.5))$ is the extremal index of the process (X_t). □

Sketch of the proof. We restrict ourselves to explain how the moment generating function occurs in the limit distribution of the maxima. For more details see de Haan et al. [303].
Set $u_n = x n^{1/(2\kappa)}$ for $x > 0$, let $N_n^{(2)}$ be the point process of the exceedances of u_n^2 by $X_1^2, \dots, X_n^2$ and let $1 \le \tau_1 < \tau_2 < \cdots \le n$ be the times when the

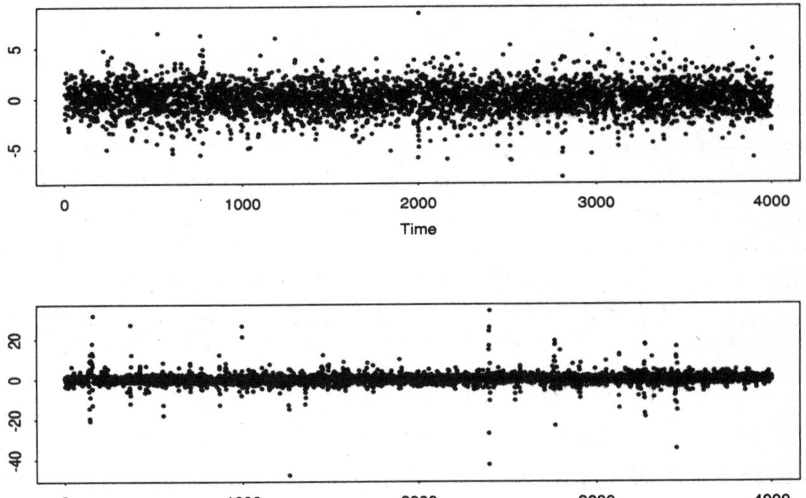

Figure 8.4.21 *Simulated sample path of an* ARCH(1) *process with parameters* $\lambda = 0.5$ *(top) and* $\lambda = 0.9$ *(bottom). In both cases* $\beta = 1$. *We conclude from the two graphs that clusters of large values become more prominent for larger values of* λ. *This is in accordance with Table 8.4.23: the extremal index* $\theta = \theta(\lambda)$ *becomes the smaller the larger* λ *is. Small values of* θ *indicate that there is more dependence in the data; see Section 8.1.*

exceedances occur. Now we use the ideas of the discussion before Theorem 8.4.20, in particular (8.82):

$$
\begin{aligned}
& P(M_n \le u_n) \\
& = P\left(\bigcup_{k=0}^{n} \left\{N_n^{(2)}(0,1] = k\,,\ X_{\tau_1} \le -u_n, \ldots, X_{\tau_k} \le -u_n\right\}\right) \\
& = \sum_{k=0}^{n} P\left(N_n^{(2)}(0,1] = k\,,\ r_{\tau_1} = -1, \ldots, r_{\tau_k} = -1\right) \\
& = \sum_{k=0}^{n} P\left(N_n^{(2)}(0,1] = k\right) 2^{-k} \\
& \to \sum_{k=0}^{\infty} P\left(N^{(2)}(0,1] = k\right) 2^{-k}\,, \qquad (8.85)
\end{aligned}
$$

where $N^{(2)}$ is the limit compound Poisson process given in Theorem 8.4.19 with x replaced by x^2. Hence we may write

$$N^{(2)} = \sum_{k=1}^{\infty} \xi_k \varepsilon_{\Gamma_k} ,$$

where the iid cluster sizes ξ_k are independent of the points Γ_k of the underlying homogeneous Poisson process $\widetilde{N}^{(2)}$ with intensity $\widetilde{\lambda} = c\theta^{(2)} x^{-2\kappa}$. We also write

$$\pi^{m*}(k) = P(\xi_1 + \cdots + \xi_m = k) , \quad m \in \mathbb{N} , k \in \mathbb{N}_0 .$$

Conditioning on the Poisson process $\widetilde{N}^{(2)}$, the rhs in (8.85) can be transformed as follows:

$$\begin{aligned}
&\sum_{k=0}^{\infty} P\left(N^{(2)}(0,1] = k\right) 2^{-k} \\
&= \sum_{k=0}^{\infty} \sum_{m=0}^{k} P\left(\widetilde{N}^{(2)}(0,1] = m\right) P\left(N^{(2)}(0,1] = k \middle| \widetilde{N}^{(2)}(0,1] = m\right) 2^{-k} \\
&= \sum_{k=0}^{\infty} \left(\sum_{m=0}^{k} e^{-\widetilde{\lambda}} \frac{\widetilde{\lambda}^m}{m!} \pi^{m*}(k) \right) 2^{-k} \\
&= \sum_{m=0}^{\infty} \left(\sum_{k=m}^{\infty} \pi^{m*}(k) 2^{-k} \right) e^{-\widetilde{\lambda}} \frac{\widetilde{\lambda}^m}{m!} \\
&= \sum_{m=0}^{\infty} (\Pi(0.5))^m \, e^{-\widetilde{\lambda}} \frac{\widetilde{\lambda}^m}{m!} \\
&= \exp\left\{ -\widetilde{\lambda}(1 - \Pi(0.5)) \right\} .
\end{aligned}$$

Inserting $\widetilde{\lambda} = c\theta^{(2)} x^{-2\kappa}$ gives (8.83). □

In the rest of this section we try to answer the question:

Given that β and λ are known, is there an easy way of calculating, at least approximately, the quantities $\theta^{(2)}$ and $\theta_k^{(2)}$?

An answer is offered by the following lemma.

Lemma 8.4.22 *Let E_κ be an exponential rv with parameter κ, independent of the random walk (S_n) defined by (8.80). Then*

$$\theta^{(2)} = \theta_k^{(1)} = P\left(\max_{n\geq 1} S_n \leq -E_\kappa\right), \tag{8.86}$$

$$\theta_k^{(2)} = P\left(\sum_{n=1}^{\infty} I_{\{S_n > -E_\kappa\}} = k-1\right), \quad k \geq 2. \tag{8.87}$$

Proof. We have

$$\begin{aligned}
& E\left[P\left(\max_{n\geq 1} S_n > -E_\kappa \middle| E_\kappa\right)\right] \\
&= \kappa \int_0^\infty P\left(\max_{n\geq 1}\sum_{t=1}^{n} \ln(\lambda Z_t^2) > -z\right) e^{-\kappa z} dz \\
&= \kappa \int_1^\infty P\left(\max_{n\geq 1}\sum_{t=1}^{n} \ln(\lambda Z_t^2) > -\ln y\right) y^{-\kappa-1} dy \\
&= \kappa \int_1^\infty P\left(\max_{n\geq 1}\prod_{t=1}^{n} (\lambda Z_t^2) > y^{-1}\right) y^{-\kappa-1} dy \\
&= 1-\theta^{(2)}.
\end{aligned}$$

This proves (8.86). Using the argument in Remark 2 above, the proof of (8.87) is similar. □

This lemma suggests simulating independent replications of (S_n) and, independently, exponential rvs E_κ, and then counting the events $\{S_n > -E_\kappa\}$ for each replication separately. Since the random walk has negative drift, the number of such exceedances is finite with probability 1. Practical problems (for instance how long one has to run the random walk and how many replications are necessary) are discussed in de Haan et al. [303].

Based on these considerations, natural estimators for $\theta^{(2)}$ and $\theta_k^{(2)}$ are

$$\widehat{\theta}^{(2)} = 1 - \frac{1}{N}\sum_{i=1}^{N} I_{\left\{\max_{n\geq 1} S_n^{(i)} > -E_\kappa^{(i)}\right\}},$$

$$\widehat{\theta}_k^{(2)} = \frac{1}{N}\sum_{i=1}^{N} I_{A_{ik}},$$

where

$$A_{ik} = \left\{\sum_{n=1}^{\infty} I_{G_i} = k-1\right\}, \quad G_i = \left\{S_n^{(i)} > -E_\kappa^{(i)}\right\},$$

and $((S_n^{(i)}), E_\kappa^{(i)})$ are N iid replications of $((S_n), E_\kappa)$. The corresponding θ– and π–quantities for the ARCH(1) process can now be calculated from (8.79) and (8.84). This idea has been used to generate Table 8.4.23 taken from de Haan et al. [303]. It is based on $N = 1000$ Monte–Carlo simulations of $((S_n), E_\kappa)$, each with maximal length $n = 1000$ of the random walk S_n.

λ	$\widehat{\theta}$	$\widehat{\pi}_1$	$\widehat{\pi}_2$	$\widehat{\pi}_3$	$\widehat{\pi}_4$	$\widehat{\pi}_5$
0.1	0.999	0.998	0.002	0.000	0.000	0.000
0.3	0.939	0.941	0.054	0.004	0.001	0.000
0.5	0.835	0.844	0.124	0.025	0.006	0.001
0.7	0.721	0.742	0.176	0.054	0.018	0.007
0.9	0.612	0.651	0.203	0.079	0.034	0.016
0.95	0.589	0.631	0.203	0.088	0.040	0.019
0.99	0.571	0.621	0.202	0.088	0.042	0.021

Table 8.4.23 *The estimated extremal index θ and the cluster probabilities π_k of the* ARCH(1) *process depending on the parameter λ.*

For the data of Figure 8.4.4 we estimated $\widehat{\lambda} = 0.5$. Hence we can read off the estimators $\widehat{\theta} = 0.835$ and $(\widehat{\pi}_k)_{k\in\mathbb{N}}$ immediately from Table 8.4.23. In Section 8.1.3 we present other methods for estimating θ. They are also applied to the data used for Figure 8.4.4.

Notes and Comments

A recent paper on perpetuities mainly looking at light–tailed behaviour is Goldie and Grübel [273]; their work is partly motivated by probabilistic selection algorithms in the style of quicksort and by shot–noise processes with exponentially decaying after–effect. A nice summary of the basic extreme value theory for ARCH processes is to be found in Borkovec [80]. The literature on ARCH–type processes is huge and the most important survey papers and relevant textbooks were already mentioned at the beginning of Section 8.4.2. Recently, within the finance community the search for alternative models from the large ARCH family has been spurred by the increasing availability of high–frequency data. New names have appeared on the ARCH–firmament; one example is the HARCH process as a model for heterogeneous volatilities. For a description of the latter see Müller et al. [469]. The proceedings [513] contain a wealth of material on the econometric modelling of high–frequency data. A further extension of the GARCH(p,q) model is the so–called ARCH–in–mean or ARCH–M model; see Kallsen and Taqqu [366], where also an option pricing formula in ARCH–type models can be found.

The work on extremes of Markov chains with applications to random recurrence equations has been extended by Perfekt [491].

In order to calculate (or approximate) $\theta^{(2)}$ from equation (8.86) the distribution of the maximum of a random walk is required. This is a problem well–studied in risk theory for estimating the Lundberg exponent; see Chapter 1. Alternatives to the simulation method leading to Table 8.4.23 exist. A numerical method based on the fast Fourier transform has been suggested by Grübel [287] and applied to the estimation of $\theta^{(2)}$ by Hooghiemstra and Meester [331].

8.5 On the Longest Success–Run

The following story is told in Révész [539]. It concerns a teaching experiment of T. Varga related to success–runs in a coin tossing sequence.

A class of school children is divided into two sections. In one of the sections each child is given a coin which they throw two hundred times, recording the resulting head and tail sequence on a piece of paper. In the other section the children do not receive coins, but are told instead that they should try to write down a "random" head and tail sequence of length two hundred. Collecting these slips of paper, he then tries to subdivide them into their original groups. Most of the times he succeeds quite well. His secret is that he had observed that in a randomly produced sequence of length two hundred, there are, say, head runs of length seven. On the other hand, he had also observed that most of those children who were to write down an imaginary random sequence are usually afraid of writing down runs of longer than four. Hence, in order to find the slips coming from the coin tossing group, he simply selects the ones which contain runs longer than five.

The experiment led T. Varga to ask:

What is the length of the longest run of pure heads in n Bernoulli trials?

The second story along the same lines comes from a discussion with an amateur gambler on the topic of the game of roulette. He claimed that together with friends (gambling colleagues) they had recorded the outcome of all roulette games for a particular casino over a period of three years. Obviously, they were after a pattern they could then use to beat the system. He told us that one evening they recorded 16 times red in a row! Clearly this was at the odds with the assumption of randomness in roulette! Or is it not?

Both examples above concern the question of the longest sequence of consecutive successes in a dichotomous experiment. Below we give a mathematical analysis of the problem.

We consider a very simple model: the rvs $X, X_1, X_2, \ldots$ are iid Bernoulli with success probability p, i.e.

$$P(X=1)=p\,, \quad P(X=0)=q=1-p\,,$$

for some $p \in (0,1)$. A *run of 1s of length j in $X_1, \ldots, X_n$* is defined as a subsequence $(X_{i+1}, \ldots, X_{i+j})$ of $(X_1, \ldots, X_n)$ such that

$$X_i = 0\,, \quad X_{i+1} = \cdots = X_{i+j} = 1\,, \quad X_{i+j+1} = 0\,,$$

where we formally set $X_0 = X_{n+1} = 0$. We try to answer the following question:

How long is the longest run of 1s *in $X_1, \ldots, X_n$?*

An alternative formulation is given via the *random walk (S_n) generated by (X_n)*:

$$S_0 = 0\,, \quad S_n = X_1 + \cdots + X_n\,.$$

Let

$$I_n(j) = \max_{0 \le i \le n-j} (S_{i+j} - S_i)\,, \quad 1 \le j \le n\,,$$

and Z_n be the largest integer such that $I_n(Z_n) = Z_n$. Then Z_n is the *length of the longest run of* 1s *in $X_1, \ldots, X_n$*.

Example 8.5.1 (Longest run in insurance)
Let Y_i be iid rvs denoting the claim sizes in a specific portfolio, and $u \ge 0$ a given threshold. Introduce the Bernoulli rvs

$$X_i = I_{\{Y_i > u\}}\,, \quad i \ge 1\,,$$

with success probability $p = P(Y_1 > u)$. The longest run of 1s in the Bernoulli sequence (X_i) corresponds to the longest consecutive sequence of exceedances of the threshold u. Instead of these particular X_i we could consider any sequence of Bernoulli rvs of the form $(I_{\{X_i \in A\}})$ for any Borel set A. A typical example would be to take the layer $A = (D_1, D_2]$ as in the reinsurance Example 8.7.6. □

Example 8.5.2 (Longest run in finance)
The standard Cox–Ross–Rubinstein model in finance assumes that risky assets either go up with probability p or down with probability $1-p$. The resulting binomial tree model serves as a skeleton for many of the more advanced models including the Black–Scholes model. A run of 1s in this set–up would correspond to consecutive increases in the price of the risky asset. For a description of these standard models in finance see Baxter and Rennie [53], Cox and Rubinstein [135], Duffie [190], Hull [346] or Karatzas and Shreve [369]. □

In the following we collect some useful facts about the length of runs of 1s in a sequence of iid rvs. We start with a precise distributional result.

Example 8.5.3 (The precise distribution of Z_n for a symmetric random walk)
A symmetric random walk corresponds to $p = 0.5$. Székely and Tusnády [610], see Révész [540], p. 18, gave the precise distribution for the largest integer Z_n such that $Z_n = I_n(Z_n)$ for a symmetric random walk, by combinatorial methods. For each $j = 1, \ldots, n$,

$$P(Z_n < j) = \frac{1}{2^n} \sum_{i=1}^{n+1} \sum_{k=0}^{i} (-1)^k \binom{i}{k} \binom{n-kj}{i-1} . \tag{8.88}$$

Formula (8.88) is of restricted value since only for small n can it be applied in a reasonable way. The computer time to evaluate formula (8.88) increases dramatically with n. Also, numerical approximations of the binomial terms via Stirling's formula or other methods do not give satisfactory answers. □

In the rest of this section we apply asymptotic methods to describe the growth of Z_n as $n \to \infty$.

8.5.1 The Total Variation Distance to a Poisson Distribution

Since the random walk (S_n) consists of binomial rvs a Poisson approximation argument seems appropriate. The following can be found in Barbour, Holst and Janson [43], pp. 244–249. These authors apply the celebrated Stein–Chen method to derive bounds on the *total variation distance* between the law F_{W_j} of

$$W_j = \sum_{i=1}^{n} I_{\{X_i=0, X_{i+1}=\cdots=X_{i+j}=1, X_{i+j+1}=0\}}$$

and the Poisson distribution with parameter

$$\lambda = nq^2p^j = EW_j .$$

Notice that W_j counts all runs of 1s with length j. The *total variation distance* between two distributions F_1 and F_2 on the non–negative integers is given by

$$d_{TV}(F_1, F_2) = \sup_{A \subset \mathbb{N}_0} \left| F_1(A) - F_2(A) \right| = \frac{1}{2} \sum_{k=0}^{\infty} \left| F_1(\{k\}) - F_2(\{k\}) \right| .$$

In particular, an upper bound for $d_{TV}(F_{W_j}, Poi\,(\lambda))$ provides also an estimate of the individual distances

$$\left|P(W_j = k) - e^{-\lambda}\frac{\lambda^k}{k!}\right|, \quad k \geq 0.$$

The following is formula (5.1) on p. 244 in Barbour et al. [43]:

$$d_{TV}\left(F_{W_j}, Poi(nq^2p^j)\right) \leq ((2j-1)q+2)qp^j, \quad n > 2j+2.$$

Thus, in particular,

$$\begin{aligned}
&\sup_{k\geq 0}\left|P(W_j = k) - e^{-\lambda}\frac{\lambda^k}{k!}\right| \\
&= \sup_{k\geq 0}\left|P(\text{There are precisely } k \text{ runs of length } j) - e^{-\lambda}\frac{\lambda^k}{k!}\right| \\
&\leq ((2j-1)q+2)qp^j. \qquad (8.89)
\end{aligned}$$

This estimate provides useful information when the rhs $((2j-1)q+2)qp^j$ is small compared with the Poisson probabilities on the lhs.

Example 8.5.4 (Continuation of Example 8.5.1)
In Example 8.5.1 we considered the largest number of consecutive exceedances of a threshold u by iid rvs Y_i. If we increase the threshold $u = u_n$ with n in such a way that

$$nP(Y_1 > u_n) = np_n \to \tau \in (0,\infty),$$

then the Poisson approximation of Proposition 3.1.1 yields

$$P(\max(Y_1,\ldots,Y_n) \leq u_n) \to e^{-\tau}. \qquad (8.90)$$

Set

$$X_i = I_{\{Y_i > u_n\}}, \quad i = 1,2,\ldots,$$

where we suppress the dependence of the X_i on n. From (8.89) we see that for $k \geq 1$

$$\begin{aligned}
&P(W_j = k) \\
&= P(\text{There are precisely } k \text{ consecutive exceedances of } u_n \text{ of length } j \\
&\qquad \text{by } Y_1,\ldots,Y_{n+j+1}) \\
&= \exp\left\{-\tau(1+o(1))p^{j-1}\right\}\frac{\left(\tau p^{j-1}\right)^k(1+o(1))}{k!} + O\left(p^j\right) \\
&= \exp\left\{-\tau^j(1+o(1))n^{-j+1}\right\}\frac{\left(\tau^j n^{-j+1}\right)^k(1+o(1))}{k!} + O\left(n^{-j}\right) \\
&= O\left(n^{-j+1}\right), \quad j \geq 2, \quad n \to \infty.
\end{aligned}$$

This result might be surprising at the first sight. However, it tells us only that the probability of a fixed number of runs of 1s in $X_1, \ldots, X_{n+j+1}$ with the same length $j \geq 2$ is negligible for large n. Notice that we can also evaluate the probability $P(W_j = 0)$, i.e. the probability that there are no consecutive exceedances of u_n of length j by $Y_1, \ldots, Y_{n+j+1}$. It is not difficult to see that

$$P(W_j = 0) = 1 + O\left(n^{-j}\right), \, j \geq 2, \quad P(W_1 = 0) = e^{-\tau} + o(1),$$

which is an interesting complement to (8.90).

If we assume that $\lambda = np_n^j$ converges to a positive constant then (8.89) gives a reasonable approximation by a $Poi(\lambda)$ law, with rate of convergence n^{-1}.

A Poisson approximation to the distribution of W_j for $p_n \to 0$ can also be obtained by point process methods; see Chapter 5. Introduce the point processes

$$N_n(\cdot) = \sum_{i=1}^{n} \varepsilon_{n^{-1}i}(\cdot) I_{\{Y_i \leq u_n, Y_{i+1} > u_n, \ldots, Y_{i+j} > u_n, Y_{i+j+1} \leq u_n\}}, \quad n \geq 1,$$

on the state space $E = (0, 1]$. We suppress the dependence of N_n on j. Notice that N_n is very close in spirit to the point process of exceedances used in extreme value theory; see Section 5.3. Similar methods also apply here to show that N_n converges to a Poisson random measure. We have

$$\{N_n(0,1] = k\} = \{W_j = k\}, \quad k \geq 0,$$

which links N_n with W_j. Assume that $np_n^j \to \tau \in \mathbb{R}_+$. Then it is not difficult to verify that

$$EN_n(a,b] \to \tau\,(b-a), \quad P(N_n(B) = 0) \to e^{-\tau}, \quad n \to \infty,$$

for any $0 < a < b \leq 1$, any finite union B of disjoint intervals $(c,d] \subset (0,1]$. Hence, by Kallenberg's theorem (Theorem 5.2.2),

$$N_n \xrightarrow{d} N \quad \text{in} \quad M_p((0,1]),$$

where N is a homogeneous Poisson process on $(0,1]$ with intensity τ. Hence

$$P(N_n(0,1] = k) = P(W_j = k) \to e^{-\tau}\frac{\tau^k}{k!}, \quad k \geq 0,$$

in particular $d_{TV}(F_{W_j}, Poi(\tau)) \to 0$. The latter follows from Scheffé's lemma; see Williams [643], Section 5.10. □

8.5.2 The Almost Sure Behaviour

Erdös and Rényi [220], see also Rényi [528], proved a result on the a.s. growth of the length of the longest run of 1s in a random walk:

Theorem 8.5.5 (A.s. growth of the length Z_n of the longest run of 1s)
For every fixed $p \in (0,1)$,

$$\lim_{n\to\infty} \frac{Z_n}{\ln n} = \frac{1}{-\ln p} \quad \text{a.s.} \tag{8.91}$$

□

Below we indicate how this result can be proved by classical limit theory.

Thus the longest run of 1s is roughly of the order $-\ln n/\ln p$, so it increases very slowly with n; see Table 8.5.11. It is natural to ask where the logarithmic normalisation in the SLLN (8.91) comes from. The basic idea is the following: write

$$\begin{aligned} L_1 &= \min\{n \geq 1 : X_n = 0\}\,, \quad \text{and for } k \geq 2, \\ L_k &= \min\{n : n > L_1 + \cdots + L_{k-1}\,, X_n = 0\} - (L_1 + \cdots + L_{k-1})\,. \end{aligned}$$

Since the X_n are iid we conclude from the Markov property of (S_n) that the L_n are iid positive rvs. It is not difficult to see that

$$P(L_1 = k) = qp^{k-1}\,, \quad k \geq 1\,.$$

Thus L_1 has a geometric distribution. By construction, the values $L_i - 1 \geq 1$ are the lengths of the runs of 1s and the rv

$$N(n) = \operatorname{card}\left\{m : \sum_{i=1}^{m} L_i \leq n\right\}$$

counts the number of zeros among $X_1, \ldots, X_n$. Hence it is binomial with parameters (n,q). Thus, in order to determine the length of the longest run of 1s in $X_1, \ldots, X_n$, we have to study maxima of iid geometric rvs along a randomly indexed sequence. Indeed,

$$\max_{i \leq N(n)} L_i - 1 \leq Z_n \leq \max_{i \leq N(n)+1} L_i - 1\,. \tag{8.92}$$

For the proof of Theorem 8.5.5 we need the following auxiliary result.

Proposition 8.5.6 (Characterisation of minimal and maximal a.s. growth of maxima of geometric rvs)
Let d_n *be positive integers such that* $d_n \uparrow \infty$.

(*a*) *The relation*

$$P\left(\max_{i\le n} L_i > d_n \quad \text{i.o.}\right) = 0 \quad or \quad 1$$

holds according as

$$\sum_{n=1}^{\infty} p^{d_n} < \infty \quad or \quad = \infty\,.$$

(*b*) *Suppose in addition that* $np^{d_n} \to \infty$. *Then*

$$P\left(\max_{i\le n} L_i \le d_n \quad \text{i.o.}\right) = 0 \quad or \quad 1$$

according as

$$\sum_{n=1}^{\infty} p^{d_n} \exp\left\{-np^{d_n}\right\} < \infty \quad or \quad = \infty\,.$$

Moreover, if $\liminf_{n\to\infty} np^{d_n} < \infty$ *then*

$$P\left(\max_{i\le n} L_i \le d_n \quad \text{i.o.}\right) = 1\,.$$

(*c*) (*a*) *and* (*b*) *remain valid if* n *is everywhere replaced by* $[nc]$, *where* $[x]$ *denotes the integer part of* x *and* c *is a positive constant.*

Proof. The first part immediately follows from Theorem 3.5.1 since $P(L_1 > d_n) = p^{d_n}$. The second part follows from Theorem 3.5.2; the assumptions

$$P(L_1 > d_n) \to 0 \quad \text{and} \quad nP(L_1 > d_n) = np^{d_n} \to \infty$$

are satisfied. For the third part, one has to modify the proofs of Theorems 3.5.1 and 3.5.2 step by step along the subsequence $([nc])$. This goes through without major difficulties. □

Remarks. 1) The restriction to integer sequences (d_n) is natural since we are dealing with integer–valued rvs and events of the form

$$\left\{\max_{i\le n} L_i \le x\right\} = \left\{\max_{i\le n} L_i \le [x]\right\}\,.$$

Here, and in the rest of this discussion, $[x]$ denotes as usual the integer part of x, and we shall also use $\{x\}$ for the fractional part $x - [x]$ of x. *The latter notation does not mean fractional part when delimiting the argument of a function!* Thus $\exp\{x\}$ is just e^x as usual.

2) In Example 3.5.6 we studied the a.s. behaviour of maxima of rvs with exponential tails of the form $P(X > x) \sim Ke^{-ax}$. Unfortunately, that theory is not directly applicable to (L_n), but has to be modified. The reason is that

1 1 1 0 1 1 1 0 1 1 0 0 1 0 0 0 1 0 1 0 0 0 1 1 1 1 0 0 0 0
0 0 0 1 0 1 0 1 1 1 0 1 1 1 0 1 1 0 1 0 1 1 0 0 1 0 1 1 1 1
1 0 1 1 1 1 1 0 1 0 1 1 1 0 0 1 1 1 1 0 0 1 1 1 0 0 0 1 1 1
0 1 0 1 1 0 1 1 0 1 0 0 0 1 1 0 1 1 0 0 0 0 1 1 0 0 1 0 0 0
1 1 0 0 1 0 1 1 1 0 0 1 0 0 1 0 0 1 1 0 1 0 0 1 0 1 0 0 1 0
1 1 1 0 0 1 1 0 0 0 1 1 1 0 1 1 1 0 0 1 0 0 0 1 0 0 1 0 0 0
0 1 0 1 1 1 1 0 1 1 1 0 0 1 1 0 0 1 0 1 0 1 0 0 0 1 1 0 1 1
1 0 1 1 0 1 1 1 0 1 0 1 0 0 0 1 1 0 1 1 0 1 0 1 0 1 0 1 0 1
0 0 1 0 0 0 0 1 0 0 1 0 0 0 0 0 0 0 1 0 1 0 0 1 1 1 1 0 0 0
0 1 1 1 0 1 1 1 1 1 1 0 0 0 0 1 1 1 0 1 0 1 1 0 0 0 0 1 1 0
0 1 1 1 0 0 1 0 1 1 1 1 0 0 1 0 1 1 1 0 1 1 1 1 0 0 1 0 1 0
1 0 1 1 1 0 1 0 1 1 0 0 0 0 1 0 1 0 1 0 0 1 0 1 1 1 0 0 1 0
0 0 1 1 1 0 1 1 1 1 0 1 1 1 1 1 0 0 0 0 0 0 0 0 1 0 0 0 1 1
1 0 0 1 1 1 1 0 1 1 1 0 0 0 0 0 0 0 1 0 0 0 1 1 1 1 1 0 1 1
1 1 0 0 0 0 0 0 0 1 0 0 0 1 0 1 0 1 0 1 0 1 1 1 1 0 0 1 1 0
0 0 1 1 1 1 1 1 0 0 0 1 0 0 1 0 1 1 1 0 0 0 0 1 1 1 0 1 1 0
1 0 1 1 0 1 0 0 1 1 0 0 0 1 1 1 1 0 0 1 1 0 0 1 1 0 1 0 1 0
0 0 1 0 1 1 1 0 1 0 0 0 1 0 0 1 1 1 1 1 1 1 0 1 0 0 1 1 1 0
0 0 0 0 0 0 1 1 0 0 0 0 1 0 0 1 1 0 0 0 0 1 1 0 0 1 0 1 1 1
0 1 0 0 1 1 0 0 0 1 0 1 1 1 0 0 0 1 1 0 0 0 0 0 0 1 0 0 1 0
1 0 1 1 0 1 0 1 1 1 0 0 1 1 0 | 1 1 1 1 1 1 1 1 1 1 1 1 | 0 1 0
0 0 1 0 1 0 1 0 1 1 1 1 0 0 0 0 1 1 0 0 0 1 0 0 1 0 1 0 0 0
1 1 1 1 0 0 0 0 1 0 1 1 1 1 1 0 1 1 0 1 0 1 0 0 0 1 0 1 0 0
0 1 1 0 1 1 0 1 0 1 1 0 1 0 1 1 1 0 1 1 1 0 0 1 0 1 0 0 0 0
1 0 0 0 0 1 1 0 1 0 0 1 0 1 0 0 1 1 0 1 0 1 0 1 0 1 0 0 0 1
1 1 0 0 1 0 0 1 0 1 1 1 0 0 1 0 1 0 1 0 1 0 1 1 0 0 1 1 0 1
0 1 0 0 0 1 0 1 0 0 0 0 0 0 1 0 0 1 0 0 1 1 1 1 0 0 1 0 1 0
1 1 0 0 1 1 1 0 0 0 0 1 0 0 0 1 0 1 1 1 0 1 1 0 1 0 0 1 0 0
0 1 0 0 1 1 0 1 1 0 1 0 0 1 1 0 1 1 1 1 1 0 1 1 0 0 1 0 0 0
1 0 1 0 1 0 1 1 1 0 0 1 1 0 0 0 0 1 0 0 0 0 0 0 0 0 1 0 1 1
1 1 1 0 1 1 0 1 0 1 0 1 0 1 1 0 0 0 1 0 1 0 0 0 1 0 1 1 0 0
0 0 0 1 1 1 1 0 1 1 0 1 1 1 1 0 1 0 1 1 1 0 0 0 1 0 1 1 1 1
0 1 0 0 0 1 0 0 1 1 0 1 1 0 0 0 1 0 1 0 0 1 0 1 0 0 1 0 0 0

Figure 8.5.7 *A random sequence of* 990 *values of* 0*s and* 1*s. Both values occur with the same chance. The longest run of* 1*s has length* 12. *This is in agreement with Table 8.5.11.*

$$P(L_1 > x) = P(L_1 > [x]) = p^{[x]} = p^{x-\{x\}}. \tag{8.93}$$

Thus the relation $P(L_1 > x) \sim p^x$ does not hold as $x \to \infty$. □

The proof of Theorem 8.5.5 is now a consequence of Proposition 8.5.6 and of relation (8.92). However, there is still a minor problem: from Proposition 8.5.6 we get only

$$\max_{i \le n} L_i / \ln n \stackrel{\text{a.s.}}{\to} -1/\ln p. \tag{8.94}$$

Thus we have to replace n by the random index $N(n)$ or by $N(n)+1$, but this does not provide any difficulties by virtue of Lemma 2.5.3 and the fact that $(N(n))$ can be interpreted as a renewal counting process observed at integer instants of time, hence $N(n)/n \stackrel{\text{a.s.}}{\to} q$; see Theorem 2.5.10.

Applying similar techniques one can prove results more precise than Theorem 8.5.5. Results of the following type can be found in Erdös and Révész [221], Guibas and Odlyzko [288], Deheuvels [168] (who proved more subtle theorems on the a.s. behaviour of Z_n as well as on the a.s. behaviour of the kth longest run of 1s in $X_1, \ldots, X_n$) or in Gordon et al. [280]. The latter also showed results about the length of runs of 1s that are interrupted by a given number of 0s. As in Section 3.5 we use the notation

$$\ln_0 x = x\,, \ln_1 x = \max(0, \ln x)\,, \ln_k x = \max(0, \ln_{k-1} x)\,, \quad x > 0\,, k \ge 2\,.$$

Let furthermore $[x]$ denote the integer part of x.

Theorem 8.5.8 (Almost sure behaviour of the length of the longest run of 1s)

(a) *For each $r \in \mathbb{N}$ the following relations hold:*

$$P\left(Z_n > \left[\frac{\ln_1(nq) + \cdots + \ln_r(nq) + \varepsilon \ln_r(nq)}{-\ln p}\right] \quad \text{i.o.}\right)$$

$$= \begin{cases} 0 & \text{if} \quad \varepsilon > 0\,, \\ 1 & \text{if} \quad \varepsilon < 0\,. \end{cases}$$

(b) *For each $\varepsilon > 0$,*

$$P\left(Z_n \le \left[\frac{\ln(nq) - \ln_3(nq) - \varepsilon}{-\ln p}\right] - 1 \quad \text{i.o.}\right) = 0$$

and

$$P\left(Z_n \le \left[\frac{\ln(nq) - \ln_3(nq)}{-\ln p}\right] + 1 \quad \text{i.o.}\right) = 1\,.$$

Proof. For fixed $r \geq 1$, $c > 0$ and small ε write

$$b_n(\varepsilon, c) = \left[\frac{\ln_1(nc) + \ln_2(nc) + \cdots + \ln_r(nc) + \varepsilon \ln_r(nc)}{-\ln p}\right].$$

Then it is easily checked that

$$\sum_{n=1}^{\infty} p^{b_n(\varepsilon,c)} < \infty \quad \text{or} \quad = \infty$$

according as $\varepsilon > 0$ or $\varepsilon \leq 0$. Hence, by Proposition 8.5.6,

$$P\left(\max_{i \leq [nc]} L_i - 1 > b_n(\varepsilon, c) \quad \text{i.o.}\right) = 0 \quad \text{or} \quad = 1 \tag{8.95}$$

according as $\varepsilon > 0$ or $\varepsilon \leq 0$. Since $N(n)/n \stackrel{\text{a.s.}}{\to} q$ it follows that, for each small fixed $\delta > 0$ and for large n, with probability 1,

$$n(1-\delta)q \leq N(n) \leq n(1+\delta)q - 1. \tag{8.96}$$

Also notice that, for large n,

$$b_n(\varepsilon/2, q(1+\delta)) \leq b_n(\varepsilon, q). \tag{8.97}$$

Having (8.96), (8.97) and (8.92) in mind we obtain

$$P\left(\max_{i \leq n(1+\delta)q} L_i - 1 > b_n(\varepsilon/2, (1+\delta)q) \quad \text{i.o.}\right) \geq P\left(Z_n > b_n(\varepsilon, q) \quad \text{i.o.}\right).$$

This together with (8.95) proves the first part of the theorem for $\varepsilon > 0$. For $\varepsilon < 0$ one may proceed in a similar way.

We proceed similarly in the second part. For $\varepsilon > 0$ set $\varepsilon' = \ln(1+\varepsilon)$. Write, for $\varepsilon > 0$, $c > 0$ and large n,

$$b'_n(\varepsilon, c) = \left[\frac{\ln(nc) - \ln_3(nc) - \varepsilon}{-\ln p}\right] = \left[\frac{\ln(nc) - \ln((1+\varepsilon')\ln_2(nc))}{-\ln p}\right]$$

and (fractional part)

$$z_n = \left\{\frac{\ln(nc) - \ln_3(nc) - \varepsilon}{-\ln p}\right\}.$$

Then by (8.93), for large n,

$$\begin{aligned}
& p^{b'_n(\varepsilon,c)} \exp\left\{-[nc]\, p^{b'_n(\varepsilon,c)}\right\} \\
&= \frac{\ln_2(nc)}{nc}(1+\varepsilon')\, p^{-z_n} \exp\left\{-\frac{[nc]}{nc} \ln_2(nc)(1+\varepsilon')\, p^{-z_n}\right\} \\
&\leq \text{const} \frac{\ln_2 n}{n(\ln n)^{(1+\varepsilon'/2)}}.
\end{aligned}$$

But the latter sequence is summable, and it follows from Proposition 8.5.6 that

$$P\left(\max_{i\le nc} L_i \le b_n'(\varepsilon, c) \quad \text{i.o.}\right) = 0\,. \tag{8.98}$$

Similarly, let

$$b_n''(c) = \left[\frac{\ln(nc) - \ln_3(nc)}{-\ln p}\right] + 1$$

and

$$z_n' = \left\{\frac{\ln(nc) - \ln_3(nc)}{-\ln p}\right\}.$$

Then

$$\begin{aligned} & p^{b_n''(c)} \exp\left\{-[nc]\, p^{b_n''(c)}\right\} \\ = \quad & \frac{\ln_2(nc)}{nc} p^{1-z_n'} \exp\left\{-\frac{[nc]}{nc} \ln_2(nc)\, p^{1-z_n'}\right\} \\ \ge \quad & \text{const}\, \frac{\ln_2 n}{n \ln n}\,. \end{aligned}$$

The latter sequence is not summable and hence, because of Proposition 8.5.6,

$$P\left(\max_{i\le nc} L_i \le b_n''(c) \quad \text{i.o.}\right) = 1\,.$$

It remains to switch to the random index sequence $(N(n))$. We proceed as in the first part of the proof: choose $\delta > 0$ small and $\varepsilon'' > 0$ such that for large n,

$$\frac{\ln(nq) - \ln_3(nq) - \varepsilon}{-\ln p} \le \frac{\ln(nq(1-\delta)) - \ln_3(nq(1-\delta)) - \varepsilon''}{-\ln p}\,.$$

Then, by (8.98),

$$P\left(Z_n \le b_n'(\varepsilon, q) - 1 \quad \text{i.o.}\right) \le P\left(\max_{i\le n(1-\delta)q} L_i \le b_n'(\varepsilon'', (1-\delta)q) \quad \text{i.o.}\right) = 0\,.$$

One can similarly proceed with the sequences $(b_n''(c))$, but we omit details. This proves the second statement of the theorem . □

These results show the very subtle a.s. behaviour of the length of the longest run of 1s. We can deduce the following statement:

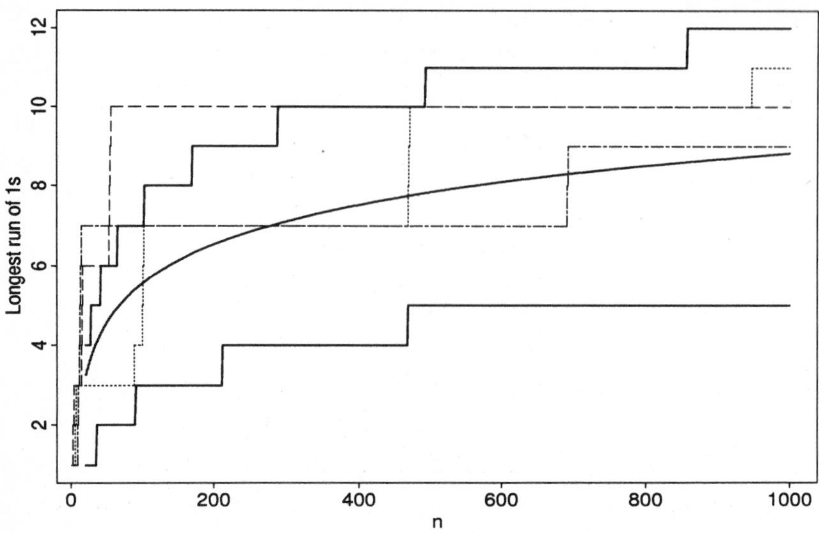

Figure 8.5.9 *Three simulated sample paths of the length Z_n of the longest run of* 1*s in a random walk. For comparison the curves (solid lines) of* $\ln n/\ln 2$ *(middle),* α_n *(bottom) and* β_n *(top) are drawn; see* (8.100).

Corollary 8.5.10 *For every fixed* $\varepsilon > 0$ *and* $r \in \mathbb{N}$, *with probability* 1 *the length of the longest run of* 1*s in* $X_1, \ldots, X_n$ *falls for large* n *into the interval* $[\alpha_n, \beta_n]$, *where*

$$\begin{aligned} \alpha_n &= \left[\frac{\ln(nq) - \ln_3(nq) - \varepsilon}{-\ln p}\right] - 1, \\ \beta_n &= \left[\frac{\ln(nq) + \cdots + \ln_r(nq) + \varepsilon \ln_r(nq)}{-\ln p}\right]. \end{aligned} \tag{8.99}$$

□

In Table 8.5.11 we compare the a.s. growth rate for Z_n, i.e. $-\ln n/\ln p$, with the lower and upper bounds in (8.99). We choose $p = 1/2$ and $r = 3$, $\varepsilon = 0.001$ and the particular bounds

$$\begin{aligned} \alpha_n &= \left[\frac{\ln(n/2) - \ln_3(n/2) - 0.001}{\ln 2}\right] - 1, \\ \beta_n &= \left[\frac{\ln(n/2) + \ln_2(n/2) + 1.001 \ln_3(n/2)}{\ln 2}\right]. \end{aligned} \tag{8.100}$$

n	$\ln n/\ln 2$	α_n	β_n
20	3.32	2	4
50	5.64	3	6
100	6.64	4	8
150	7.23	4	8
200	7.64	5	9
250	7.96	5	9
500	8.96	6	11
750	9.55	6	11
1000	9.96	7	12
1500	10.55	7	13
2000	10.96	8	13
5000	12.28	9	15
10000	13.28	10	16
50000	15.61	12	19
100000	16.61	13	20
1000000	19.93	16	24

Table 8.5.11 *Almost sure bounds α_n and β_n with $\varepsilon = 0.001$ and $r = 3$, see* (8.100), *for the longest run Z_n of 1s in a random walk.*

8.5.3 The Distributional Behaviour

From Example 3.1.5 we learnt that the maxima of geometric rvs do not have a limit distribution whatever the centring and normalising constants. However, the tail of the geometric distribution is very close to the tail of the exponential law. To be precise,

$$L_1 - 1 \stackrel{d}{=} [E_1/(-\ln p)] \tag{8.101}$$

for a standard exponential rv E_1, $[x]$ again stands for the integer part of x. This is easily seen:

$$\begin{aligned} qp^k &= P(L_1 - 1 = k) = p^k - p^{k+1} \\ &= P(E_1 \in [k(-\ln p), (k+1)(-\ln p)]) \\ &= P([E_1/(-\ln p)] = k) \,. \end{aligned}$$

Let now (E_n) be iid standard exponential rvs. We know from Example 3.2.7 that

$$\max(E_1, \ldots, E_n) - \ln n \xrightarrow{d} \Lambda\,, \tag{8.102}$$

where $\Lambda(x) = e^{-e^{-x}}$, $x \in \mathbb{R}$, stands for the Gumbel distribution. Having (8.92) in mind we may also hope that the distribution of Z_n is not too far away from the Gumbel distribution.

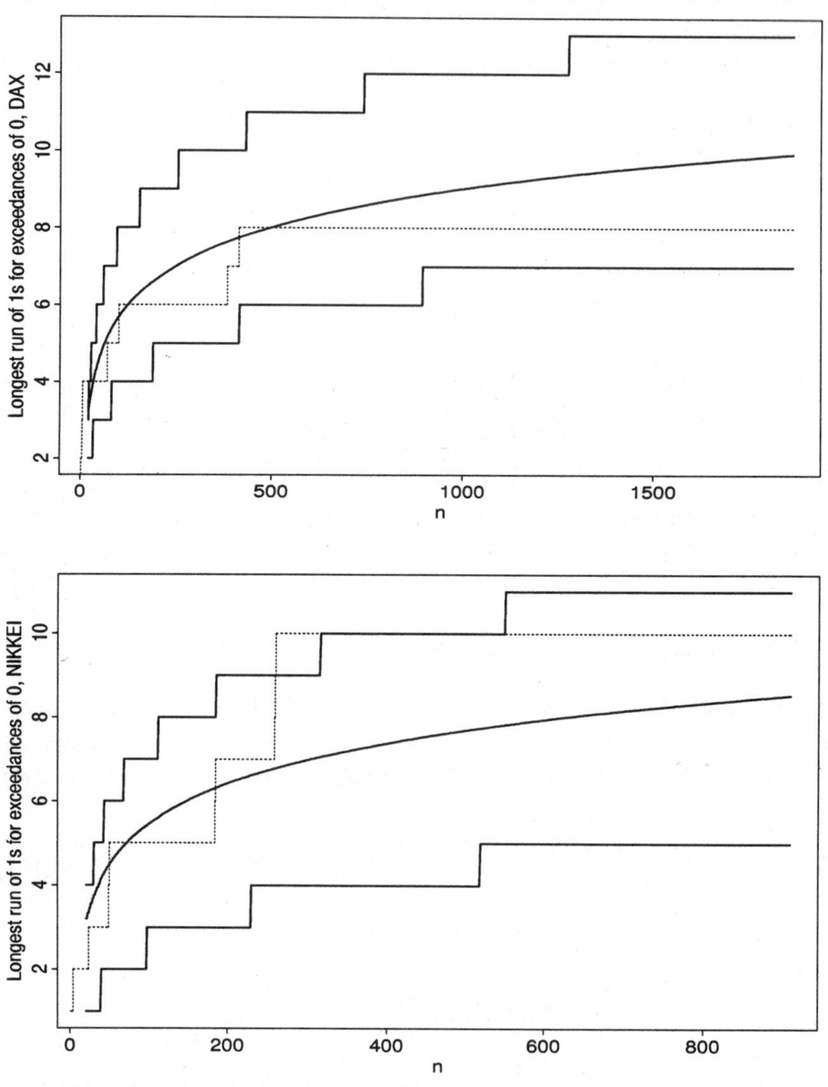

Figure 8.5.12 *The longest run of* 1*s (dotted line) generated by the indicators* $I_{\{Y_n>0\}}$ *for financial data.*
Top*: the underlying time series* (Y_n) *consists of* 1 864 *daily log–returns (closing data) of the German stock index* DAX*, July 1, 1988 – August 24, 1995.*
Bottom*: the underlying time series* (Y_n) *consists of* 910 *daily log–returns (closing data) of the Japanese stock index* NIKKEI*, February 22, 1990 – August 8, 1993. The longest runs* Z_n *exhibit a behaviour similar to the longest runs in a random walk with* $p = 0.5$. *For comparison, in each figure the solid curves of* $\ln n/\ln 2$ *(middle),* α_n *(bottom) and* β_n *(top) are drawn, see* (8.100)*, using an estimated* $p = 0.5$.

Theorem 8.5.13 (Asymptotic distributional behaviour of the length of the longest run of 1s)
Let Y be a rv with the Gumbel distribution Λ. Then

$$\sup_{k\in\mathbb{Z}}\left|P\left(Z_n-\left[\frac{\ln(nq)}{-\ln p}\right]\le k\right)-P\left(\left[\frac{Y}{-\ln p}+\left\{\frac{\ln(nq)}{-\ln p}\right\}\right]\le k\right)\right|\to 0\,.$$

Here $\{x\}$ denotes the fractional part of x and $[x]$ its integer part.

Proof. From (8.102) and Lemma 2.5.6 we may conclude that

$$\frac{\max_{i\le N(n)} E_i}{-\ln p}-\frac{\ln N(n)}{-\ln p}\xrightarrow{d}\frac{Y}{-\ln p}\,,\tag{8.103}$$

given that an Anscombe condition holds and Y has distribution Λ. But this can be seen by the following arguments:

$$\begin{aligned}
&P\left(\max_{n(1-\delta)q<m\le n(1+\delta)q}\left|\left(\max_{i\le m}E_i-\ln m\right)-\left(\max_{i\le nq}E_i-\ln([nq])\right)\right|>\varepsilon\right)\\
\le\;& P\left(\max_{i\le n(1+\delta)q}E_i-\max_{i\le n(1-\delta)q}E_i>\varepsilon/2\right)+I_{(\varepsilon/2,\infty)}\left(\ln\left(\frac{[n(1+\delta)q]}{[n(1-\delta)q]}\right)\right)\\
=\;& p_1+p_2\,,
\end{aligned}$$

say. The quantity p_2 is equal to 0 for δ small and n large. Moreover,

$$\begin{aligned}
p_1 &= P\left(\max_{n(1-\delta)q<i\le n(1+\delta)q}E_i-\max_{i\le n(1-\delta)q}E_i>\varepsilon/2\right)\\
&= P\left(\max_{i\le[n(1+\delta)q]-[n(1-\delta)q]}E_i'-\ln([n(1+\delta)q]-[n(1-\delta)q])\right.\\
&\qquad-\max_{i\le[n(1-\delta)q]}E_i-\ln([n(1-\delta)q])\\
&\qquad\left.>\varepsilon/2-\ln\left(\frac{[n(1+\delta)q]-[n(1-\delta)q]}{[n(1-\delta)q]}\right)\right),
\end{aligned}$$

where (E_i') is an independent copy of (E_i). In view of (8.102), we see that p_1 converges to

$$P\left(Y_1-Y_2>\varepsilon/2-\ln(2\delta/(1-\delta))\right)$$

for iid Y_i with a common Gumbel distribution. Now, the rhs in the latter limit probability can be made arbitrarily small by choosing δ sufficiently small. Hence an Anscombe condition holds. Since $N(n)/n\xrightarrow{\text{a.s.}}q$ we may now replace the expression $\ln N(n)/(-\ln p)$ in relation (8.103) by $\ln(nq)/(-\ln p)$. Because Λ is continuous we immediately obtain from (8.103) that

$$\sup_t \left| P\left(\frac{\max_{i\le N(n)} E_i}{-\ln p} - \frac{\ln(nq)}{-\ln p} \le t\right) - P\left(\frac{Y}{-\ln p} \le t\right)\right| \to 0 .$$

This implies that

$$\begin{aligned} &\sup_t \left| P\left(\left[\frac{\max_{i\le N(n)} E_i}{-\ln p}\right] - \left[\frac{\ln(nq)}{-\ln p}\right] \le t\right)\right. \\ &\qquad\qquad \left. -P\left(\left[\frac{Y}{-\ln p} + \left\{\frac{\ln(nq)}{-\ln p}\right\}\right] \le t\right)\right| \\ = &\sup_t \left| P\left(\max_{i\le N(n)} \left[\frac{E_i}{-\ln p}\right] - \left[\frac{\ln(nq)}{-\ln p}\right] \le t\right)\right. \\ &\qquad\qquad \left. -P\left(\left[\frac{Y}{-\ln p} + \left\{\frac{\ln(nq)}{-\ln p}\right\}\right] \le t\right)\right| \quad \to \quad 0 . \end{aligned}$$

Since the quantities involved in the latter limit relation are integer–valued the supremum over all real t reduces to the supremum over the integers. This allows us to complete the proof of the theorem in view of representation (8.101) and inequality (8.92). □

A version of this result was proved in Gordon et al. [280] who also derived bounds on the expectation and the variance of Z_n.

From Theorem 8.5.13 we may conclude the following about the asymptotic distributional behaviour of Z_n:

$$\begin{aligned} &P\left(Z_n \le k + [\ln(nq)/(-\ln p)]\right) \\ = &\; P\left(\left[\frac{Y}{-\ln p} + \left\{\frac{\ln(nq)}{-\ln p}\right\}\right] \le k\right) + o(1) \\ = &\; P\left(\frac{Y}{-\ln p} + \left\{\frac{\ln(nq)}{-\ln p}\right\} < k+1\right) + o(1) \\ = &\; P\left(Y < (k+1)(-\ln p) - \left\{\frac{\ln(nq)}{-\ln p}\right\}(-\ln p)\right) + o(1) \\ = &\; \exp\left\{-p^{k+1-\{\ln(nq)/(-\ln p)\}}\right\} + o(1) \end{aligned}$$

for all integers $k \ge 1 - [\ln(nq)/(-\ln p)]$. In particular, for k positive,

$$P\left(Z_n = k + \left[\frac{\ln(nq)}{-\ln p}\right]\right) \approx qp^{k-\{\ln(nq)/(-\ln p)\}} .$$

Notes and Comments

The question about the longest run of 1s and related problems have attracted much attention in the literature. Applications lie not only in extreme value theory, finance and insurance but also in molecular biology (longest matching sequences in two DNA strings), pattern recognition (longest repetitive patterns in random sequences) and many other fields. A few relevant references are Arratia and Waterman [23], Gordon, Schilling and Waterman [280], Guibas and Odlyzko [288], Karlin and Ost [370]. A paper on the longest success–run in the context of insurance and finance is Binswanger and Embrechts [75].

The problem of the longest run of 1s is closely related to the question about the order of magnitude of the increments of a general random walk (S_n). We noted that a convenient tool to describe the order of the increments is given by

$$I_n(j) = \max_{0 \le i \le n-j} (S_{i+j} - S_i)\,, \quad 1 \le j \le n\,.$$

For the particular case of iid Bernoulli rvs with success probability $p \in (0,1)$ we may conclude from Section 8.5.2 that

$$\lim_{n\to\infty} \frac{I_n\left(\left[\frac{\ln n}{-\ln p}\right]\right)}{\left[\frac{\ln n}{-\ln p}\right]} = 1 \quad \text{a.s.}$$

This result already shows the typical order of the increments for a general random walk. Now assume that $EX = 0$ and that X has moment generating function

$$M(h) = Ee^{hX}\,.$$

Let

$$h_0 = \sup\{h : M(h) < \infty\} \ge 0$$

and define the number $c = c(\alpha)$ by

$$e^{-1/c} = \inf_h e^{-h\alpha} M(h)\,, \quad \alpha > 0\,. \tag{8.104}$$

It is easy to see that if $h_0 > 0$ then the infimum lies striclty between 0 and 1, so that c is positive.

The following is a classical result due to Erdös and Rényi [220]:

Theorem 8.5.14 (Erdös–Rényi SLLN for the increments of a random walk)
The relation

$$\lim_{n\to\infty} \frac{I_n([c\ln n])}{[c\ln n]} = \alpha \quad \text{a.s.}$$

holds for each

$$\alpha \in \left\{ \frac{M'(h)}{M(h)} : 0 < h < h_0 \right\} .$$

Here $c = c(\alpha)$ is given by equation (8.104). □

Numerous generalisations and extensions of the Erdös–Rényi SLLN have been proved. They depend very much on large deviation techniques (therefore the existence of the moment generating function; see Sections 2.3 and 8.6) and on generalisations of classical renewal theory. Results of iterated logarithm type, see Section 2.1, have been shown for the largest increments $I_n(b_n)$ for various sequences $b_n \uparrow \infty$. We refer to work by Csörgő and Steinebach [146], Deheuvels [168], Deheuvels and Devroye [169], Deheuvels and Steinebach [171, 172], Steinebach [607] and the references therein.

8.6 Some Results on Large Deviations

In Section 2.3 we touched on the question of *large deviation probabilities* for sums $S_n = X_1 + \cdots + X_n$ of iid rvs X_n. In the present section we intend to give some results in this direction as a preliminary step towards dealing with reinsurance treaties in Section 8.7.

A *large deviation probability* is an asymptotic evaluation of $P(S_n > x_n)$ for a given sequence $x_n \to \infty$, where (x_n) is such that $P(S_n > x_n) = o(1)$. Or, alternatively, it can be understood as an asymptotic evaluation of $P(S_n > x)$ uniformly over some x–region depending on n, where now the region is such that $P(S_n > x_n) = o(1)$ uniformly over it. Thus large deviation probabilities tell us about the probability that S_n exceeds a large threshold value x or x_n. When dealing with extremal events it is of particular interest to get analytic expressions or estimates for those probabilities.

In this section we consider *precise large deviations.* This means we evaluate the probability $P(S_n > x)$ to at least the accuracy $P(S_n > x) \sim a_n(x)$ for an explicit sequence of positive functions or numbers (a_n). This is in contrast to *rough large deviations* which are evaluations to the accuracy $\ln P(S_n > x) \sim b_n$ for some explicit sequence (b_n).

In Section 2.3 we learnt about *two types of precise large deviation results.* The first, Cramér's theorem (Theorem 2.3.3), tells us about the validity of the normal approximation to the df of the sums S_n under the very strong condition that the moment generating function

$$M(h) = Ee^{hX} \tag{8.105}$$

exists in a neighbourhood of the origin. In particular, if $\text{var}(X) = 1$ then

$$\begin{aligned} P\left(S_n - n\mu > x\right) &= \overline{\Phi}\left(x/\sqrt{n}\right)(1+o(1)) \,, \\ P\left(S_n - n\mu \le -x\right) &= \overline{\Phi}\left(x/\sqrt{n}\right)(1+o(1)) \,, \end{aligned} \tag{8.106}$$

uniformly for $x = o(n^{1/6})$. This result is of restricted use for the purposes of insurance and finance in view of the condition on the exponential moments of X. However, precise large deviation results under the existence of the moment generating function are more the rule than the exception. Under that condition several theorems about the order of $P(S_n - n\mu > x)$ have been proved in the critical region where x is of the same order as $ES_n = n\mu$. Final results are due to Bahadur and Rao [36] and Petrov [493, 494]; see Petrov [495], Bucklew [96]. For completeness and in order to get an impression of the difficulty of the problem we state here Petrov's theorem. Recall the notion of a *lattice–distributed rv* X: there exist $d > 0$ and $a \in \mathbb{R}$ such that

$$\sum_{k=-\infty}^{\infty} P(X = kd + a) = 1 \,. \tag{8.107}$$

We call the largest possible d in (8.107) the *maximal step* of the df F.

Theorem 8.6.1 (Petrov's theorem on precise large deviations under an exponential moment condition)
Suppose that the moment generating function M, see (8.105), exists in a neighbourhood of the origin. Let

$$b = \sup\left\{h : \int_0^\infty e^{hx}\, dF(x) < \infty\right\},$$

and $h = h(x)$ be the unique solution of the equation

$$m(h) = \frac{M'(h)}{M(h)} = x \,.$$

Set

$$\sigma^2(h) = m'(h) \,, \quad a_0 = \lim_{h \uparrow b} m(h) \,,$$

and assume a_0 finite.

(a) Suppose F is non–lattice. Then

$$P\left(S_n - n\mu > x\right) = \frac{\exp\{n(\ln M(h) - hx)\}}{h\sigma(h)\sqrt{2\pi n}}(1 + o(1))$$

uniformly for $x \in [\varepsilon n, (a_0 - \varepsilon)n]$.

(*b*) *Suppose F is lattice with maximal step d; see* (8.107). *Then*

$$P\left(S_n - n\mu > x\right) = \frac{d\exp\left\{n(\ln M(h) - hx)\right\}}{\sigma(h)\sqrt{2\pi n}\left(1-e^{-dh}\right)}\left(1 + O\left(\frac{1}{n}\right)\right),$$

uniformly for $x \in [\varepsilon n, (a_0 - \varepsilon)n]$. □

The very formulation of these results shows that it is not an easy matter to apply them in a given situation. In particular, solving the equation $m(h) = x$ is in general troublesome. Basically, the same problems occur as for determining the Lundberg exponent in risk theory; see Definition 1.2.3.

Since we emphasise problems related to heavy tails we also want to give some idea about precise large deviations in that case. We gained a first impression from Heyde's theorem for a symmetric F in the domain of attraction of an α–stable law with $\alpha < 2$, i.e. $\overline{F} \in \mathcal{R}_{-\alpha}$, as discussed in Theorem 2.3.5. There we found that the condition $n\overline{F}(x_n) \to 0$ implies the relation

$$P(S_n > x_n) = n\overline{F}(x_n)(1 + o(1)) = P(M_n > x_n), \tag{8.108}$$

where, as usual, M_n denotes the maximum of the first n values of the X_i. This is the typical relation that we can expect in the general case of regularly varying tails. Notice that the F with regularly varying tails form a subclass of the subexponential distributions, see Section 1.3, which are defined by the relation

$$P\left(S_n > x\right) = P\left(M_n > x\right)\left(1 + o(1)\right)$$

for every fixed $n \geq 1$, as $x \to \infty$. Thus (8.108) is just an extension of the latter limit relation to the case when *both x and n tend to infinity.* It again shows the dominating role of the maximum term over the sum of iid rvs.

Relation (8.108) remains valid for a wider class of distributions. This is exemplified by the following theorem which is due to A. Nagaev [472, 473]; an independent probabilistic proof of the relation (8.109) below has been given by S. Nagaev [474].

Theorem 8.6.2 (Precise large deviations with regularly varying tails, I) *Suppose that $\overline{F} \in \mathcal{R}_{-\alpha}$ for some $\alpha > 2$, $E|X|^{2+\delta} < \infty$ for some $\delta > 0$ and that* $\operatorname{var}(X) = 1$. *Then*

$$P\left(S_n - n\mu > x\right) = \overline{\Phi}\left(\frac{x}{\sqrt{n}}\right)(1 + o(1)) + n\overline{F}(x)(1 + o(1))$$

uniformly for $x \geq \sqrt{n}$. In particular,

$$P\left(S_n - n\mu > x\right) = \overline{\Phi}\left(\frac{x}{\sqrt{n}}\right)(1 + o(1))$$

for $\sqrt{n} \le x \le a(n \ln n)^{1/2}$ *and* $a < \sqrt{\alpha - 2}$, *and*

$$P(S_n - n\mu > x) = n\overline{F}(x)(1 + o(1)) = P(M_n > x) \qquad (8.109)$$

for $x > a(n \ln n)^{1/2}$ *and* $a > \sqrt{\alpha - 2}$. □

Finally, we give here a unified result for $\overline{F} \in \mathcal{R}_{-\alpha}$ for any $\alpha > 1$.

Theorem 8.6.3 (Precise large deviations with regularly varying tails, II) *Suppose that* $\overline{F} \in \mathcal{R}_{-\alpha}$ *for some* $\alpha > 1$. *Then for every fixed* $\gamma > 0$,

$$P(S_n - n\mu > x) = n\overline{F}(x)(1 + o(1))$$

uniformly for $x \ge \gamma n$. □

For $\alpha < 2$ this result is due to Heyde [321, 322, 323], for $\alpha > 2$ it was proved by A. Nagaev, as already mentioned. A unified approach for regularly varying $\overline{F}$ (and more general classes of tails) has been given by Cline and Hsing [126].

Cline and Hsing [126] proved a result of type

$$P(S_n > x) \sim P(M_n > x)$$

uniformly for certain x–regions, for $\overline{F}$ of extended regular variation, i.e.

$$c^{-\beta} \le \liminf_{x\to\infty} \frac{\overline{F}(cx)}{\overline{F}(x)} \le \limsup_{x\to\infty} \frac{\overline{F}(cx)}{\overline{F}(x)} \le c^{-\alpha}$$

for some $0 < \alpha \le \beta < \infty$ and every $c \ge 1$. They also extend their results to certain randomly indexed sums and maxima. Precise large deviation problems for other subexponential distributions, for example of the type $\overline{F}(x) = \exp\{-L(x)x^{\alpha}\}$ for a slowly varying L and $\alpha \in [0,1)$, have also been treated. They depend very much on the x–region and on the particular form of $\overline{F}$. They are not so easily formulated as Theorems 8.6.2 and 8.6.3. The most complete picture about precise large deviations for subexponential distributions can be found in Pinelis [500], in the more recent survey paper by Rozovski [559] and in the monograph Vinogradov [632].

Precise large deviations for random sums with heavy tails are our next goal. We will point out in Section 8.7 how probabilities of precise large deviations occur in a natural way in problems related to reinsurance treaties. We restrict ourselves to the compound process

$$S(t) = \sum_{i=1}^{N(t)} X_i\,, \quad t \ge 0\,,$$

where X_n are iid non–negative, non–degenerate rvs independent of the counting process $(N(t))_{t\ge 0}$. We extend the standard compound Poisson process in the sense that, for every t, the rv $N(t)$ is Poisson, but its mean value is not necessarily of the form λt for a constant intensity $\lambda > 0$. This makes sense in the context of insurance futures (see Section 8.7) where $N(t)$ can be large due to "high density arrival times", i.e. even in small intervals $[0, t]$ the mean value $EN(t)$ is huge. In this sense, $(S(t))$ may be considered as a process indexed by the "operational time" $EN(t)$ which increases to infinity when t increases.

Recall that

$$\mu(t) = ES(t) = \mu \, EN(t) \,.$$

The following is analogous to Theorem 8.6.3:

Theorem 8.6.4 (Precise large deviations for random sums with regularly varying tails)
Suppose that $(S(t))_{t\geq 0}$ is a compound process where $(N(t))_{t\geq 0}$ are Poisson rvs such that $EN(t) \to \infty$ as $t \to t_0$ for some $t_0 \in (0, \infty]$, and X is a.s. non-negative and non-degenerate. Moreover, $\overline{F} \in \mathcal{R}_{-\alpha}$ for some $\alpha > 1$. Then

$$P\left(S(t) - \mu(t) > x\right) = EN(t)\overline{F}(x)(1 + o(1)) \,, \tag{8.110}$$

uniformly for $x \geq \gamma EN(t)$, for every fixed $\gamma > 0$ as $t \to t_0$. □

A proof of this result is given in Klüppelberg and Mikosch [398]. Note that the rhs in (8.110) is of the same order as $P(M_{N(t)} > x)$ as $t \to t_0$.

Notes and Comments

Rough large deviations are widely applied in physics and mathematics; see for instance the books by Dembo and Zeitouni [177] and Deuschel and Strook [178], and for an introductory level Bucklew [96]. In that context, it is mostly supposed that the random objects considered (not necessarily sums) have a moment generating function finite in some neighbourhood of the origin. This is motivated by Cramér's result (see Theorem 2.3.3) and by its various generalisations and extensions. There does not exist so much literature for subexponential distributions. A survey of precise large deviation results was provided by S. Nagaev [474] with many useful precise large deviation estimates and an extensive reference list. More recent accounts are the papers by Doney [185], Pinelis [500] and Rozovski [559] mentioned previously; see also the monograph Vinogradov [632]. Gantert [250] considers rough large deviations for sums of rvs which are subexponential, constitute a stationary ergodic sequence and satisfy certain mixing conditions.

Large deviation techniques in the context of insurance are not uncommon; see for instance Asmussen and Klüppelberg [33] and Klüppelberg and

Mikosch [398] in the heavy–tailed case, and Djehiche [184], Martin–Löf [444], Slud and Hoesman [585] under exponential moment conditions. These papers mainly emphasise relations between estimates of ruin probabilities and large deviation results.

8.7 Reinsurance Treaties

8.7.1 Introduction

Extreme value theory has an important role to play in the pricing of reinsurance contracts, especially in the area of contracts for single events or few events, involving high layers. The prime example is the *CatXL reinsurance treaty* which corresponds in financial option theory to a bull spread with the market loss ratio assuming the role of the underlying. The discussion on CatXL below is taken from *Sigma* [581], p. 6.

In catastrophe reinsurance, the dominant type of reinsurance treaty is the "Catastrophe Excess–of–Loss Cover per Event" - or CatXL for short. In contrast to the proportional reinsurance treaty, in which the reinsurer shares in equal parts in the premiums written and the claims incurred by the primary insurers, with the non–proportional treaty the reinsurer pays only from a contractually agreed amount of loss (deductible) up to a defined maximum (exit point). The losses considered here are those which are attributable to specific occurrences (mainly natural events like windstorm, earthquake etc. ... but in certain cases also conflagration or strike and riot) and occur within a contractually agreed period of time. Both loss amounts which do not reach the lower limit and those exceeding the upper limit must be borne by the primary insurer. The span between the deductible and the exit point is called the "cover" or "line".

The need for extreme value theory modelling becomes clear when discussing the so–called *reference loss*; the definition below is again taken from *Sigma* [581], p. 10–11.

The reference loss is a value which corresponds to a major loss which insurance companies with average capitalisation should take as a basis for deciding on the level of CatXL cover they require. The reference losses chosen are such that they are rare but nevertheless possible. For European markets, a realistic windstorm loss scenario was chosen for each market, while for markets at risk from hurricanes a loss event with a return period of 100 *years was chosen. For the earthquake reference losses, return periods of* 100 *years (in countries prone to high seismic activity),* 500 *years (in countries prone to moderate*

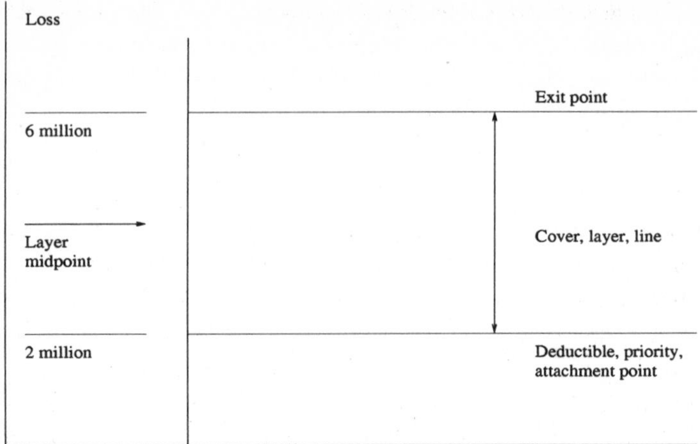

Figure 8.7.1 *Example of an XL cover.*

seismic activity) and 1 000 *years (in countries with low seismic activity) were assumed.*

The reference loss with a predetermined return period can be read off from the so–called loss frequency curve of the relevant portfolio. In the language of Section 6.2.4, estimation of the reference loss corresponding to a t–year event comes down to

$$\widehat{u}_t = \widehat{F}^{\leftarrow}(1 - t^{-1}) .$$

Here $\widehat{F}$ stands for the estimated claim size distribution (in the *Sigma*–language above: the loss–frequency curve). The methodology discussed in Section 6.5 turns out to be particularly useful for estimating these high quantiles in CatXL contracts. Also the contract–specific calculation of the deductible, the exit point and the resulting premium can be based *in part* on extreme value theory. We emphasise "in part" here: indeed, in the determination of premiums many more market factors and alternative techniques enter. Important examples of the latter are simulation methodology and the analysis of stress scenarios. Also for single event large claims it is paramount to estimate (mostly by non–stochastic means) the *total exposure*, i.e. the cost of the total loss of a system to be insured.

An ever recurring theme, or indeed question, is:

In the event of disastrous natural events,
have primary insurers bought sufficient cover?

Or equivalently,

How do the estimated reference losses compare with a contract's exit point?

And indeed,

If more cover is needed,
where does the necessary risk capital come from?

The new word appearing on the market is *securitisation of risk*. To stay in line with the above discussion, we briefly discuss below just one of the finance industry's answers to the last question: CAT futures and PCS options. In 1992, the Chicago Board of Trade (CBOT) launched catastrophe (CAT) futures. They can be viewed as offering an alternative to the CatXL treaties discussed above. From the simplest perspective a *futures contract* is an agreement between two parties to make a particular exchange at a particular future date. For example, a future contract made on June 30 could call for the purchasing agent to pay \$400, the *futures price*, on September 30 in exchange for an ounce of gold delivered on September 30. The last few lines were taken from Duffie [189]. If in the example above we would change "an ounce of gold" to "an insurer's loss ratio", then we would come close to the definition of CAT futures. The main problem in CBOT's product design was the construction of the underlying, i.e. the equivalent of the gold price say. For that reason, pools of insurance companies were created, allowing for a broad data base on losses within the home–owners market. Companies were mainly pooled on a geographical basis. From such a pool the industry's loss ratio (losses over premiums) was constructed. This stochastic process defined the *underlying* on which various *derivatives* (like futures and options) can be constructed. If the time to maturity is T say, then the settlement value $V(T)$ of the CAT futures was put at

$$V(T) = \$25\,000 \times \min\left(\frac{S_p(T)}{P_p(T)}, 2\right), \tag{8.111}$$

where $S_p(T)$ stands for the pool's loss process and $P_p(T)$ the (deterministic) premiums covering the losses over the period $[0, T]$. The exact construction of $S_p(T)$, i.e. of the losses, is more involved as a clear distinction has to be made between actual versus reserved losses and between date of occurrence versus settlement date of the claims. Typically, the first three months (event quarter) of the contract would define the claim occurrence period, the next three months (runoff quarter) were added to allow for claim settlement. By the end of these 6 months (reporting period) one would hope that a high percentage (80–90%) of the claims were indeed settled. The value $V(T)$ would then be made available in a first interim report shortly after the end of the reporting period. The final report for this particular future would be published during the 4th month after the reporting period. For further details we refer the reader to the various publications now available on these products. See for instance CBOT [105, 106]. A very readable introduction is Albrecht, König and Schradin [6]. In order to see how a particular home–owner insurer can use these futures as a hedging instrument, denote by $S_i(t)$, respectively $P_i(t)$, the insurer's loss process, respectively premium function. By way of example, suppose the futures are at present trading at a loss ratio of 0.6, i.e. a 60% loss ratio for the pool. The insurer interested in having a loss ratio of 60 % at maturity T can achieve this by buying now $n_i = P_i(T)/25\,000$ futures at the quoted loss ratio of 0.6. For simplicity, we assume that all losses have been settled by the end of the reporting period. It is not difficult to adjust the argument when we have a $100\,(1-\delta)\%$ settlement quota, say. The insurer's "wealth" at maturity T then becomes:

$$\begin{aligned}
&P_i(T) - S_i(T) + \text{"gain or losses from futures transactions"}\\
&= P_i(T) - S_i(T) + \left[n_i \times 25\,000 \times \min\left(\frac{S_p(T)}{P_p(T)}, 2\right) - n_i \times 25\,000 \times 0.6\right]\\
&= P_i(T)\left[(1-0.6) + \left(\min\left(\frac{S_p(T)}{P_p(T)}, 2\right) - \frac{S_i(T)}{P_i(T)}\right)\right]\\
&= P_i(T)\,(0.4 + \Delta_{i,p}(T))\,.
\end{aligned}$$

If $\Delta_{i,p}(T) = 0$, i.e. the insurer's loss ratio corresponds exactly to the pool's ratio, then his/her loss ratio $S_i(T)/P_i(T)$ at maturity is exactly 60%. Depending on the value of $\Delta_{i,p}(T)$, more or less futures will have to be bought in order to achieve the required hedge.

Though novel in construction, the contracts were not really a success. Various reasons can be given for this:

- time difference: quarterly futures based on a three months claim period versus standard year–contracts in the reinsurance industry,
- information on prices came very slowly and was incomplete,
- danger of adverse selection and moral hazard,
- who constitutes the secondary market, i.e. who sells these futures?

As CAT futures come close to being a so–called beta–zero asset, i.e. nearly uncorrelated with other assets, they should be the Holy Grail of any portfolio manager. That they were not perceived like that was mainly due to the very slow flow of information on the futures prices. Rather than the futures themselves, options traded better, especially the call option spreads where insurers would (or could) appear on both sides of the transaction, i.e. as buyer as well as seller. The CBOT has reacted to the market's criticism and launched its new generation of so–called PCS options which are designed to counter the most obvious defects of the CAT futures. For a discussion of these options see CBOT [107]. Schradin [569] offers a very readable introduction.

Before we turn to the mathematical analysis of some of the products discussed above, consider the yearly data on California loss ratios (in %) for earthquake insurance given in Table 8.7.2.

1971	17.4	1977	0.7	1983	2.9	1989	129.8
1972	0	1978	1.5	1984	5.0	1990	47.0
1973	0.6	1979	2.2	1985	1.3	1991	17.2
1974	3.4	1980	9.2	1986	9.3	1992	12.8
1975	0	1981	0.9	1987	22.8	1993	3.2
1976	0	1982	0	1988	11.5		

Table 8.7.2 *Yearly loss ratios (in %) for earthquake insurance in California.*

On the basis of these data,

How would one predict the loss ratio for 1994?

The answer came indeed one year later

1994	2272.7(!)

The event that had happened in 1994 was the Northridge earthquake. These data from the California Department of Insurance (Insurance Information Institute), are taken from Jaffee and Russell [353]. In the latter paper an excellent discussion on insurance of catastrophic events is to be found. At this point we would like to repeat a statement made in the Reader Guidelines. "Though not providing the risk manager with the final product he or she can use for monitoring risk on a global scale, we will provide that manager with stochastic methodology needed for the construction of various components of such a global tool." The next section should be viewed with this statement in mind.

8.7.2 Probabilistic Analysis

In this section we investigate some of the standard reinsurance treaties using the techniques from extreme value theory and fluctuation theory for random walks provided in this book. Throughout the individual claim sizes X_n are iid non–negative, non–degenerate with common df F, independent of the number $N(t)$ of claims occurring up to time t. The latter rvs are supposed to be Poisson–distributed, but they need not necessarily constitute a Poisson process. The total claim amount of an insurance portfolio is then given by the process

$$S(t) = \sum_{i=1}^{N(t)} X_i\,, \quad t \geq 0\,.$$

Throughout we suppose that $\mu = EX_1$ exists and we also write

$$X = X_1\,, \quad \mu(t) = ES(t) = \mu\, EN(t)\,, \quad t \geq 0\,.$$

We are particularly interested in heavy–tailed dfs F which are more realistic models for claims in the context of reinsurance.

Example 8.7.3 (CAT futures)
Recall from (8.111) that the settlement value $V(T)$ of the CAT futures at maturity T equals

$$\begin{aligned} V(T) &= \$25\,000 \times \min\left(\frac{S(T)}{P(T)}, 2\right) \\ &= \$25\,000 \times \left(\frac{S(T)}{P(T)} - \max\left(\frac{S(T)}{P(T)} - 2, 0\right)\right). \end{aligned}$$

For notational convenience we have dropped the suffix p referring to the pool. The last equality represents $V(T)$ as a long position in the pool's loss ratio and a short position in a European call written on the loss ratio with strike 2 and maturity T. Hence *in principle* one should be able to price these futures using the standard theory of no–arbitrage pricing, i.e. *the value* $V(t)$ at time t of the contingent claim $V(T)$ equals

$$V(t) = E^Q\left[e^{-r(T-t)}\, V(T) \Big| \mathcal{F}_t\right], \quad 0 \leq t \leq T\,,$$

where r stands for the risk–free interest rate, and $(\mathcal{F}_t)_{t\in[0,T]}$ is an increasing family of σ–algebras (filtration) such that $\mathcal{F}_t$ describes the information available up to time t. The measure Q appearing mysteriously as an index denotes the equivalent martingale measure which, since the paper by Harrison and Pliska [318], enters all pricing models in finance. To be a bit more precise we assume, as is usual in stochastic finance, that the underlying process is defined on a probability space

$$\left(\Omega, \mathcal{F}, (\mathcal{F}_t)_{0\leq t\leq T}, P\right),$$

and the process $(S(t))$ is assumed to be adapted to $(\mathcal{F}_t)$, i.e. for all t, $S(t)$ is $\mathcal{F}_t$–measurable. The measure Q is equivalent to P so that $(S(t))$ becomes a Q–martingale. Within this framework, we could consider the pricing problem as being solved. In the case of CAT futures, *this is however far from the truth*! This whole set–up does work well for instance if $(S(t))$ follows a geometric Brownian motion. We know however that the underlying claim process is usually modelled by a compound Poisson, compound mixed Poisson or even compound doubly stochastic (or Cox) process. In these cases, mainly due to the random jumps corresponding to the individual claim sizes, the market based on $(S(t))$ is *incomplete* and therefore allows for infinitely many equivalent martingale measures: which one to choose? The interested reader wanting to learn more about this should consult Delbaen and Haezendonck [176] for general (re)insurance contracts and Meister [454] more in particular for the CAT futures. From a more pragmatic, and indeed actuarial point of view it is definitely worthwhile to calculate the distributional properties of $V(T)$ under the so–called physical measure P. This is exactly what is done below.

In general, $P(t)$ can be taken as a loaded version of the mean value $\mu(t)$; thus

$$P(t) = c\,\mu(t)\,, \quad t \geq 0\,,$$

for some constant $c > 1$, but we require only $c > 0$. For evaluating the futures contract under the physical measure P it is of particular interest to determine

$$E(V(T)) = E\left[\$25.000 \times \left(\frac{S(T)}{P(T)} - \max\left(\frac{S(T)}{P(T)} - 2, 0\right)\right)\right].$$

Since

$$E\left(\frac{S(T)}{P(T)}\right) = \frac{\mu(T)}{c\mu(T)} = \frac{1}{c}\,,$$

it remains to calculate

$$E\max\left(\frac{S(T)}{P(T)} - 2, 0\right) = E\left(\frac{S(T)}{P(T)} - 2\right)^{+}.$$

It is one objective of this section to give an asymptotic expression for this value, but also for the variance of $V(T)$. For this reason we will exploit a large deviation result for $S(t)$ as provided by Theorem 8.6.4. Since T is in general "small" (for instance three months $\approx$ 90 days), but $N(T)$ is "large" it makes sense to speak about ***high density data*** and to generalise the Cramér–Lundberg model in so far as to require only that $EN(t)$ becomes large with increasing time. □

Example 8.7.4 (Reinsurance treaties of random walk type)
In this example we assume throughout that $(S(t))$ is given by the Cramér–Lundberg model driven by a homogeneous Poisson process $(N(t))$ with constant intensity $\lambda > 0$.

Three common types of reinsurance treaties are the following:

Proportional reinsurance. This is a common form of reinsurance for claims of "moderate" size. Here simply a fraction $p \in (0,1)$ of each claim (hence the pth fraction of the whole portfolio) is covered by the reinsurer. Thus the reinsurer pays for the amount $R_1(t) = pS(t)$ whatever the size of the claims.

Stop–loss reinsurance. The reinsurer covers losses in the portfolio exceeding a well defined limit K, the so–called *ceding company's retention level.* This means that the reinsurer pays for $R_2(t) = (S(t) - K)^+$. This type of reinsurance is useful for protecting the company against insolvency due to excessive claims on the coverage.

Excess–of–loss reinsurance. The reinsurance company pays for all individual losses in excess of some limit D, i.e. it covers $R_3(t) = \sum_{i=1}^{N(t)} (X_i - D)^+$. The limit D has various names in the different branches of insurance. In life insurance, it is called the *ceding company's retention level.* In non–life insurance, where the size of loss is unknown in advance, D is called *deductible.* The reinsurer may in reality not insure the whole risk exceeding some limit D but rather buy a layer of reinsurance corresponding to coverage of claims in the interval $(D_1, D_2]$. This can be done directly or by itself obtaining reinsurance from another reinsurer. The typical example of the CatXL was discussed in Section 8.7.1; see Figure 8.7.1 for an example with perhaps $D_1 = 2$ million and $D_2 = 6$ million Swiss francs.

It is an important question to value the losses R_1, R_2, R_3 by probabilistic means. For example, it is of interest to estimate the probabilities

$$
\begin{aligned}
& P\left(R_1(t) > x\right) \\
& \quad = \; P\left(S(t) - \mu(t) > p^{-1}x - \mu(t)\right), \qquad (8.112) \\
& P\left(R_2(t) > x\right) \\
& \quad = \; P\left((S(t) - K)^+ > x\right) = P(S(t) - K > x), \\
& \quad = \; P(S(t) - \mu(t) > x + K - \mu(t)) \qquad (8.113) \\
& P\left(R_3(t) > x\right) \\
& \quad = \; P\left(\sum_{i=1}^{N(t)} (X_i - D)^+ > x\right) \qquad (8.114)
\end{aligned}
$$

$$= P\left(\sum_{i=1}^{N(t)} (X_i - D)^+ - \lambda E(X-D)^+ t > x - \lambda E(X-D)^+ t\right).$$

A mathematical study of these probabilities is important especially for large values of x. Typically, x and K in (8.113) depend on t and indeed may be of the same order as $\mu(t)$.

We could apply the CLT to estimate the probabilities (8.112)–(8.114), given $\text{var}(X) < \infty$. For example, Theorem 2.5.16 with

$$x_t(c) = p\left(\mu(t) + c\sqrt{\lambda t\,(\text{var}(X) + \mu^2)}\right), \quad c \in \mathbb{R},$$

yields

$$P\left(R_1(t) > x_t(c)\right) \to \overline{\Phi}(c),$$

where Φ denotes the standard normal df. Thus the CLT provides an answer only in a relatively small x_t–band of the order $\sqrt{t}$ around the mean value $\mu(t)$. If x is of the critical order t, i.e. of the same order as the mean value $\mu(t)$, or even larger, large deviation results are the appropriate tools.

In the context of stop–loss reinsurance it is also of interest to study the quantities

$$E\left(\frac{S(t)}{P(t)} - K\right)^+ \quad \text{and} \quad \text{var}\left(\frac{S(t)}{P(t)} - K\right)^+ \tag{8.115}$$

for a fixed positive constant K and with the premium income $P(t) = c\mu(t)$ for some constant $c > 0$. Notice that $S(t)/P(t)$ is just the loss ratio at time t which is compared with a fixed limit K. Probabilities of large deviations will also help to evaluate quantities like (8.115). □

The main tool for dealing with the problems mentioned in Examples 8.7.3 and 8.7.4 is the large deviation result of Theorem 8.6.4. Under the assumption $\overline{F} \in \mathcal{R}_{-\alpha}$ for some $\alpha > 1$ the relation

$$P(S(t) - \mu(t) > y) \sim EN(t)\overline{F}(y) = EN(t)P(X > y) \tag{8.116}$$

holds uniformly for $y \geq \gamma\mu(t)$ for every positive $\gamma > 0$, provided $EN(t) \to \infty$ as $t \to t_0 \in (0, \infty]$. This formula immediately yields approximations to the probabilities (8.112)–(8.114) when x is of the same order as $\mu(t)$. For example, consider the situation of (8.112) with $y_t(c) = c\mu(t)$ for some $c > p$. Then

$$P\left(R_1(t) > y_t(c)\right) \sim \lambda t \overline{F}\left(\left(p^{-1}c - 1\right)\mu(t)\right),$$

i.e. $P(R_1(t) > y_t(c)) \in \mathcal{R}_{-\alpha+1}$ and therefore this probability is not negligible even for large t.

Both quantities in (8.115) can be rewritten in a such way that asymptotic analysis becomes fairly straightforward. Thus

$$\begin{aligned} & E\left(\frac{S(t)}{c\mu(t)} - K\right)^+ \\ &= E\left(\frac{S(t)}{c\mu(t)} - K\right) I_{\{S(t)/(c\mu(t))-K>0\}} \\ &= \int_0^\infty P\left(\frac{S(t)}{c\mu(t)} - K > x\right) dx \\ &= \frac{1}{c\mu(t)} \int_{(Kc-1)\mu(t)}^\infty P\left(S(t) - \mu(t) > x\right) dx\,, \end{aligned} \tag{8.117}$$

where c, K are positive constants such that

$$\gamma = Kc - 1 > 0\,.$$

Also,

$$\begin{aligned} & \operatorname{var}\left(\frac{S(t)}{c\mu(t)} - K\right)^+ \\ &= \operatorname{var}\left(\frac{S(t)}{c\mu(t)} - K\right) I_{\{S(t)/(c\mu(t))-K>0\}} \\ &= \int_0^\infty P\left(\frac{S(t)}{c\mu(t)} - K > \sqrt{x}\right) dx - \left(\int_0^\infty P\left(\frac{S(t)}{c\mu(t)} - K > x\right) dx\right)^2 \\ &= \frac{2}{c\mu(t)} \int_{\gamma\mu(t)}^\infty \left(\frac{x}{c\mu(t)} - \frac{\gamma}{c}\right) P\left(S(t) - \mu(t) > x\right) dx \qquad (8.118) \\ & \quad - \left(\frac{1}{c\mu(t)} \int_{\gamma\mu(t)}^\infty P\left(S(t) - \mu(t) > x\right) dx\right)^2 . \end{aligned}$$

This solves the corresponding problems for the expectation and the variance of the futures price (see Example 8.7.3) and of the stop–loss reinsurance treaty (see Example 8.7.4). Since relation (8.116) holds uniformly for $x \geq \gamma\mu(t)$, given $EN(t) \to \infty$, a straightforward analytic argument applied to (8.117) and (8.118) yields

$$E\left(\frac{S(t)}{c\mu(t)} - K\right)^+ \sim \frac{1}{c\mu} \int_{\gamma\mu EN(t)}^\infty \overline{F}(y)\, dy\,,$$

given $\alpha > 1$, and that

$$\begin{aligned}
&\operatorname{var}\left(\frac{S(t)}{c\mu(t)}-K\right)^{+}\\
&\sim \frac{2}{c\mu}\int_{\gamma\mu EN(t)}^{\infty}\left(\frac{x}{c\mu EN(t)}-\frac{\gamma}{c}\right)\overline{F}(x)\,dx-\left(\frac{1}{c\mu}\int_{\gamma\mu EN(t)}^{\infty}\overline{F}(x)\,dx\right)^{2}\\
&\sim \frac{2}{c\mu}\int_{\gamma\mu EN(t)}^{\infty}\left(\frac{x}{c\mu EN(t)}-\frac{\gamma}{c}\right)\overline{F}(x)\,dx\,,
\end{aligned}$$

given $\alpha>2$. Notice that the conditions $\alpha>1$ and $\alpha>2$ guarantee the existence of the expectation and of the variance of $(S(t)/c\mu(t)-K)^{+}$, respectively. Using Karamata's theorem (Theorem A3.6) we obtain the following approximations:

$$E\left(\frac{S(t)}{c\mu(t)}-K\right)^{+} \sim \frac{\gamma EN(t)}{c(\alpha-1)}\overline{F}(\gamma\mu EN(t))\,, \tag{8.119}$$

$$\operatorname{var}\left(\frac{S(t)}{c\mu(t)}-K\right)^{+} \sim \frac{2\gamma^{2}EN(t)}{c^{2}(\alpha-2)(\alpha-1)}\overline{F}(\gamma\mu EN(t))\,. \tag{8.120}$$

Example 8.7.5 (Insurance futures, continuation of Example 8.7.3)
In the context of insurance futures it may be of interest to consider a high density model for $(S(t))$: over the fixed period of time to maturity T (or better said until the end of the event period) many claims may enter into the pool so that $EN(T)$ will be large. The latter can be modelled by Poisson rvs $N(t)$ such that $EN(t)\to\infty$ as $t\to T$. For every $t\in[0,T]$, consider

$$V(t)=\$25\,000\times\min\left(\frac{S(t)}{c\mu(t)},2\right).$$

Clearly, the notation $V(t)$ above should not be confused with the no–arbitrage value as defined in Example 8.7.3. From the relations (8.119) and (8.120) one can then derive the following asymptotic expressions for $EV(t)$ and $\operatorname{var}(V(t))$: assume $\gamma=2K-1=2c-1>0$ and set

$$\widetilde{V}(t)=V(t)/\$25.000=\frac{S(t)}{c\mu(t)}-\left(\frac{S(t)}{c\mu(t)}-2\right)^{+}.$$

If $\alpha>1$ then

$$E(\widetilde{V}(t))=\frac{1}{c}\left(1-(1+o(1))\,\frac{\gamma EN(t)}{\alpha-1}\,\overline{F}(\gamma\mu EN(t))\right). \tag{8.121}$$

If $\alpha>2$ then

$$\operatorname{var}(\widetilde{V}(t))=\frac{1}{c^{2}}\left(\frac{EX^{2}}{\mu^{2}EN(t)}-(1+o(1))\frac{2\gamma^{2}EN(t)}{(\alpha-2)}\overline{F}(\gamma\mu EN(t))\right). \tag{8.122}$$

The evaluation of (8.121) with the help of (8.119) does not cause difficulties. Next we derive (8.122). Observe that by (8.119) and (8.120),

$$
\begin{aligned}
&\operatorname{var}(\widetilde{V}(t))\\
&= E\widetilde{V}^2(t) - (E\widetilde{V}(t))^2\\
&= E\left(\frac{S(t)}{c\mu(t)}\right)^2 + E\left(\left(\frac{S(t)}{c\mu(t)}-2\right)^+\right)^2 - 2E\frac{S(t)}{c\mu(t)}\left(\frac{S(t)}{c\mu(t)}-2\right)^+\\
&\qquad -(E\widetilde{V}(t))^2\\
&= E\left(\frac{S(t)}{c\mu(t)}\right)^2 - E\left(\left(\frac{S(t)}{c\mu(t)}-2\right)^+\right)^2 - 4E\left(\frac{S(t)}{c\mu(t)}-2\right)^+ - (E\widetilde{V}(t))^2\\
&= \operatorname{var}\left(\frac{S(t)}{c\mu(t)}\right) - \operatorname{var}\left(\frac{S(t)}{c\mu(t)}-2\right)^+\\
&\qquad -2\left(E\left(\frac{S(t)}{c\mu(t)}-2\right)^+\right)^2 - \frac{2}{c}\gamma E\left(\frac{S(t)}{c\mu(t)}-2\right)^+\\
&= \frac{EX^2}{c^2\mu^2 EN(t)} - (1+o(1))\frac{2\gamma^2 EN(t)}{c^2(\alpha-2)}\overline{F}(\gamma\mu EN(t))\,. \qquad \square
\end{aligned}
$$

Whereas large deviations seem to be the right instrument for dealing with the treaties and futures of Examples 8.7.3 and 8.7.4, extreme value theory is very much involved in handling the following problems related to reinsurance. In an insurance portfolio we consider the iid claims $X_1, \ldots, X_{N(t)}$ which occur up to time t. We study the randomly indexed ordered sample

$$X_{N(t),N(t)} \le \cdots \le X_{1,N(t)}\,.$$

Recall that $(N(t))$ is independent of (X_n), and throughout we will assume that $(N(t))$ constitutes a homogeneous Poisson process with constant intensity $\lambda > 0$.

Example 8.7.6 (Distribution of the number of claims in a layer and of the kth largest claim)
We will touch on two important questions in reinsurance. The first one is

How many claims can occur in a layer $(D_1, D_2]$ or (D_1, ∞) up to time t?

This means we are interested in the quantity

$$B_t(A) = \sum_{i=1}^{N(t)} I_{\{X_i \in A\}}$$

for some Borel set A. Conditional on $N(t)$, $B_t(A)$ is binomially distributed with success probability $F(A) = P(X \in A)$. Hence

$$\begin{aligned} P\left(B_t(A) = l\right) &= \sum_{k=0}^{\infty} P\left(B_t(A) = l \,|\, N(t) = k\right) P(N(t) = k) \\ &= e^{-\lambda t} \sum_{k=l}^{\infty} \left(\binom{k}{l} (F(A))^l \left(\overline{F}(A)\right)^{k-l} \right) \frac{(\lambda t)^k}{k!} . \end{aligned} \tag{8.123}$$

This solves our first problem. Indeed, depending on the type of layer, we can estimate $F(A)$ and $\overline{F}(A)$ for instance with methods discussed in Chapter 6. However, if we assume that the limits of the layer increase with time we can apply a Poisson approximation to these probabilities. For example, assume that the layer boundaries form a sequence (D_n) such that $n\overline{F}(D_n) \to \tau$ for some $\tau \in \mathbb{R}_+$. Then by Theorem 5.3.4,

$$\sum_{i=1}^{N(n)} I_{\{X_i > D_n\}} \xrightarrow{d} Poi(\tau\lambda) .$$

In particular,

$$P\left(\sum_{i=1}^{N(n)} I_{\{X_i > D_n\}} = l \right) \to e^{-\tau\lambda} \frac{(\tau\lambda)^l}{l!} .$$

Next we ask

What do we know about the size of the largest claims?

In Proposition 4.1.2 we learnt about the distribution of the kth largest order statistic $X_{k,n}$ in a sample of n iid rvs, namely

$$P\left(X_{k,n} \le x\right) = \sum_{r=0}^{k-1} \binom{n}{r} \overline{F}^r(x) \, F^{n-r}(x) . \tag{8.124}$$

Again conditioning on $N(t)$ we get an analogous formula for $X_{k,N(t)}$:

$$\begin{aligned} P\left(X_{k,N(t)} \le x\right) &= \sum_{l=0}^{\infty} P\left(X_{k,l} \le x \,|\, N(t) = l\right) P(N(t) = l) \\ &= e^{-\lambda t} \sum_{l=k}^{\infty} P\left(X_{k,l} \le x \,|\, N(t) = l\right) \frac{(\lambda t)^l}{l!} \end{aligned}$$

$$= e^{-\lambda t} \sum_{l=k}^{\infty} \left(\sum_{r=0}^{k-1} \binom{l}{r} \overline{F}^r(x) F^{l-r}(x) \right) \frac{(\lambda t)^l}{l!}$$

$$= e^{-\lambda t} \sum_{r=0}^{k-1} \frac{(\lambda \overline{F}(x) t)^r}{r!} \sum_{l=r}^{\infty} \frac{(\lambda t F(x))^{l-r}}{(l-r)!}$$

$$= e^{-\lambda t \overline{F}(x)} \sum_{r=0}^{k-1} \frac{\left(\lambda \overline{F}(x) t\right)^r}{r!} . \qquad (8.125)$$

In comparing formulae (8.124) and (8.125), we see that (8.124) is the probability that a binomial rv with parameters $\overline{F}(x)$ and n does not exceed k, whereas (8.125) is the probability that a Poisson rv with parameter $\lambda t \overline{F}(x)$ does not exceed k.

Formula (8.125) can also be generalised for a finite vector of upper order statistics. Exact calculations, though feasible, quickly become tedious. An asymptotic estimate may therefore be useful. We apply the results of Section 4.3. Since $N(t)/t \stackrel{\text{a.s.}}{\to} \lambda$ for the homogeneous Poisson process $(N(t))$ we are in the framework of Theorem 4.3.2. Therefore assume that $F \in \text{MDA}(H)$ for an extreme value distribution H, i.e. there exist $c_n > 0$ and $d_n \in \mathbb{R}$ such that

$$c_n^{-1} (M_n - d_n) \stackrel{d}{\to} H , \qquad (8.126)$$

where $M_n = \max(X_1, \ldots, X_n)$. Then, for every $k \geq 1$,

$$P\left(c_n^{-1}\left(X_{k,N(n)} - d_n\right) \leq x\right) \to \Gamma_k\left(-\ln H^{\lambda}(x)\right) , \quad x \in \mathbb{R},$$

where Γ_k denotes the incomplete Gamma function

$$\Gamma_k(x) = \frac{1}{(k-1)!} \int_x^{\infty} e^{-t} t^{k-1} \, dt , \quad x \geq 0 .$$

The following approximation for the df of $X_{k,N(n)}$ is obtained:

$$P\left(X_{k,N(n)} \leq u\right) \approx \Gamma_k \left(-\ln H^{\lambda} \left(\frac{u - d_n}{c_n} \right) \right) . \qquad \square$$

In order to exemplify further the usefulness of extreme value theory as presented in the previous chapters, we consider some more reinsurance treaties which are defined via the upper order statistics of a random sample.

Example 8.7.7 (Reinsurance treaties of extreme value type)

Largest claims reinsurance. At the time when the contract is underwritten (i.e. at $t = 0$) the reinsurance company guarantees that the k largest claims in the time frame $[0, t]$ will be covered. For example, the company will cover the 10 largest annual claims in a portfolio over a period of 5 years, say. This means that one has to study the quantity

$$R_4(t) = \sum_{i=1}^{k} X_{i,N(t)}$$

either for a fixed k or for a k which grows sufficiently slowly with t.

ECOMOR reinsurance (Excédent du coût moyen relatif). This form of a treaty can be considered as an excess–of–loss reinsurance (see Example 8.7.4) with a random deductible which is determined by the kth largest claim in the portfolio. This means that the reinsurer covers the claim amount

$$R_5(t) = \sum_{i=1}^{N(t)} \left(X_{i,N(t)} - X_{k,N(t)}\right)^+ = \sum_{i=1}^{k-1} X_{i,N(t)} - (k-1)X_{k,N(t)}$$

for a fixed number $k \geq 2$. The link to extreme value theory is again immediate. Moreover, $(k-1)/R_5$ looks very much like Hill's estimator for the index of a regularly varying tail; see Section 6.4.2.

The quantities $R_4(t)$ and $R_5(t)$ are functions of the k upper order statistics in a randomly indexed sample, a theory for which was given in Section 4.3. Thus we can calculate the limit distribution of R_5 for every fixed k: assume that (8.126) is satisfied for appropriate constants c_n, d_n and an extreme value distribution H. From Theorem 4.3.4 we know that

$$\left(c_n^{-1}\left(X_{i,N(n)} - d_n\right)\right)_{i=1,\ldots,k} \xrightarrow{d} (Y_\lambda^{(i)})_{i=1,\ldots,k}\,,$$

where $(Y_\lambda^{(1)},\ldots,Y_\lambda^{(k)})$ denotes the k–dimensional extremal variate corresponding to the extreme value distribution H^λ. Arguments as in Section 4.2, see for instance the proof of Corollary 4.2.11, yield

$$\begin{aligned} c_n^{-1} R_5(n) &= c_n^{-1}\left(\sum_{i=1}^{k-1} X_{i,N(n)} - (k-1)X_{k,N(n)}\right) \\ &\xrightarrow{d} \sum_{i=1}^{k-1} i\left(Y_\lambda^{(i)} - Y_\lambda^{(i+1)}\right) \end{aligned}$$

for $k \geq 2$. Now suppose that $F \in \mathrm{MDA}(\Lambda)$ where $\Lambda(x) = \exp\{-\exp\{-x\}\}$ denotes the Gumbel distribution. Calculation shows that

$$(Y_\lambda^{(1)},\ldots,Y_\lambda^{(k)}) \stackrel{d}{=} (Y_1^{(1)} + \ln\lambda,\ldots,Y_1^{(k)} + \ln\lambda)\,. \qquad (8.127)$$

Hence

$$
\begin{aligned}
c_n^{-1} R_5(n) &\xrightarrow{d} \sum_{i=1}^{k-1} i\left(Y_1^{(i)} - Y_1^{(i+1)}\right) \\
&\stackrel{d}{=} \sum_{i=1}^{k-1} E_i \qquad (8.128)
\end{aligned}
$$

for iid standard exponential rvs E_i, where (8.128) follows from Corollary 4.2.11. Hence the limit in (8.128) has a $\Gamma(k-1,1)$ distribution.

Such a nice formula does not exist for $F \in \mathrm{MDA}(\Phi_\alpha)$ where for some $\alpha > 0$, $\Phi_\alpha(x) = \exp\{-x^{-\alpha}\}$ denotes the Frechét distribution. However, a straightforward calculation shows that the following relation holds

$$
\left(Y_\lambda^{(1)}, \ldots, Y_\lambda^{(k)}\right) \stackrel{d}{=} \left(\lambda^{1/\alpha} Y_1^{(1)}, \ldots, \lambda^{1/\alpha} Y_1^{(k)}\right),
$$

so that the joint density of $(Y_1^{(1)}, \ldots, Y_1^{(k)})$ can be used to derive the limit distribution of $R_5(n)$. This, however, will in general lead to complicated numerical integration problems.

The same remark also applies to the limit distribution of the quantities $R_4(n)$, where for every fixed $k \geq 1$

$$
\begin{aligned}
c_n^{-1}\left(R_4(n) - k d_n\right) &= c_n^{-1} \sum_{i=1}^{k} \left(X_{i,N(n)} - d_n\right) \\
&\xrightarrow{d} \sum_{i=1}^{k} Y_\lambda^{(i)} . \qquad (8.129)
\end{aligned}
$$

In the case $F \in \mathrm{MDA}(\Lambda)$ we can give an explicit, though complicated formula for the limit distribution (8.129). Recall from Example 4.2.10 that

$$
\left(E_{i,n} - \ln n\right)_{i=1,\ldots,k} \xrightarrow{d} \left(Y^{(i)}\right)_{i=1,\ldots,k}, \qquad n \to \infty,
$$

where $(E_{i,n})$ denote the order statistics of a sample of size n from iid standard exponential rvs E_i. From Example 4.1.10 we also know that

$$
\left(E_{i,n}\right)_{i=1,\ldots,n} \stackrel{d}{=} \left(\sum_{j=i}^{n} j^{-1} E_j\right)_{i=1,\ldots,n} .
$$

Hence

$$
\begin{aligned}
\sum_{i=1}^{k}(E_{i,n}-\ln n) &\stackrel{d}{=} \sum_{i=1}^{k}\sum_{j=i}^{n} j^{-1}E_j - k\ln n \\
&= \sum_{j=1}^{k} E_j - k\ln k + k\sum_{j=k+1}^{n} j^{-1}E_j - k(\ln n - \ln k) \\
&= \sum_{j=1}^{k} E_j - k\ln k + k\sum_{j=k+1}^{n} j^{-1}(E_j - 1) + o(1) \\
&\stackrel{d}{\to} \sum_{j=1}^{k} E_j - k\ln k + k\sum_{j=k+1}^{\infty} j^{-1}(E_j - 1) \\
&\stackrel{d}{=} \sum_{i=1}^{k} Y_\lambda^{(i)} .
\end{aligned}
$$

The infinite series on the rhs converges since

$$
\operatorname{var}\left(\sum_{i=k+1}^{\infty} j^{-1}(E_j-1)\right) = \sum_{i=k+1}^{\infty} j^{-2} \sim k^{-1}, \quad k \to \infty .
$$

Now, recalling (8.127), we finally obtain the formula

$$
\begin{aligned}
\sum_{i=1}^{k} Y_\lambda^{(i)} &\stackrel{d}{=} \sum_{i=1}^{k} Y^{(i)} + k\ln\lambda \\
&\stackrel{d}{=} \sum_{j=1}^{k} E_j - k\ln(k/\lambda) + k\sum_{j=k+1}^{\infty} j^{-1}(E_j-1) .
\end{aligned}
$$

For small k the distribution of this limit rv can be derived by simulations of iid standard exponential rvs. For larger k an asymptotic theory seems appropriate. □

Notes and Comments

In the large deviation approach to insurance futures and reinsurance treaties we have closely followed Klüppelberg and Mikosch [398]. Kremer [408] gives the representation of reinsurance treaties in terms of order statistics. Teugels [620] covers this topic in a set of lecture notes. Beirlant and Teugels [55], see also the references therein, give some asymptotic theory for the quantities R_5 related to ECOMOR treaties. They assume that the number of order statistics k increases as $t \to \infty$ and that F is either in the maximum domain of attraction of the Fréchet or of the Gumbel law. Notice that some asymptotic results for R_5 can already be derived from the theory of Hill estimation as provided in Section 6.4.2. Various authors have contributed to the pricing of CAT futures: an early discussion is Cox and Schwebach [137]. A model based on integrated geometric Brownian motion with or without a Poisson component with fixed jump sizes was proposed by Cummins and Geman [149, 150]. Because of the geometric Brownian motion assumption, the latter papers use the valuation theory of Asian options. The precise model assumptions (i.e. fixed claim sizes) render the model complete and hence allow for unique no–arbitrage pricing. An approach based on marked point processes is discussed in Aase [1]. In Aase and Ødegaard [2] various models are empirically tested. Meister [454] discusses in detail the equivalent martingale construction for the underlying risk processes; he derives various pricing formulae within a utility and general equilibrium framework. A summary of the latter work is to be found in Embrechts and Meister [215]. See also Buhr and Carrière [101] for a related approach. Chichilnisky [113] discusses the important issue of hedging and Chichilnisky and Heal [114] offer a new related financial instrument. Various papers on the subject of securitisation of insurance risk are published in Cox [136]. The general issue of comparing and contrasting actuarial versus financial pricing of insurance is summarized in Embrechts [203]. This paper also contains various references for further reading. An interesting paper to start with concerning reinsurance in arbitrage–free markets is Sondermann [601]. Delbaen and Haezendonck [176] give the relevant martingale theory in order to embed premium calculation principles for risk processes in a no–arbitrage framework. The more actuarial approach to pricing in finance is beautifully summarized in Gerber and Shiu [258]. The latter paper singles out the so–called Esscher premium principle when it comes to arbitrage pricing of products in the intersection of insurance and finance. We encountered the Esscher transform in our discussion of the Cramér-Lundberg theorem, see Theorem 1.2.2, and also in our analysis of the path and claim leading to ruin, see Section 8.3.2. The Esscher transform appeared as an exponentially tilted df. The notion can be generalised to processes and indeed turns out to be useful in a much wider context. For a generalisation to conditional Esscher transforms for semi-martingales and their applications to finance see for instance Bühlmann et al. [99] . The discrete time case is treated more in detail in Bühlmann et al. [100].

In the near future we will see a large increase in the number as well as diversity of (re)insurance products based on ideas coming from finance (the CAT futures are such an example). Further examples from the latter family are the so–called catastrophe–linked bonds for which the payout is contingent on the occurrence of specific catastrophic events. For instance, one buys today a bond at \$80 say; if during the next two years no well–defined catastrophe occurs, then the bond repays at \$100, if one catastrophe occurs, only \$85, and in the case of two, \$65 say. A similar bond can be constructed where the repayment value is contingent on the size of specific catastrophic losses. It is clear that the pricing of such and similar products very much depends on our understanding of the modelling of the underlying extremal events. Extreme value theory will definitely offer a key set of relevant tools.

8.8 Stable Processes

In Section 2.4 we learnt about a particular stable process, the α–stable motion. It occurs in a natural way as the weak limit of (properly normalised and centred) partial sum processes $(\sum_{i=1}^{[nt]} X_i)_{t\geq 0}$ for iid rvs X_i. In the special case when X_1 has a finite variance, Brownian motion $(B_t)_{t\geq 0}$ is the limit process. The central role of Brownian motion and, more generally, of Gaussian processes in probability theory is uncontested. Thus they find applications not only in martingale theory and stochastic analysis, but also in insurance and stochastic finance. For a financial engineer, it is to an increasing extent more important to know about Brownian motion and Itô's lemma than to wear a dark suit and a tie.

Geometric Brownian motion $(\exp\{ct + \sigma B_t\})_{t\geq 0}$ for constants c, σ is believed to be an elementary model for returns. However, a glance at any real financial data set makes it clear that geometric Brownian motion is a very poor approximation to reality. It does not explain changing volatility, or jumps. Therefore, attempts have been made to move away from this simple model. For example, infinitely divisible processes (Brownian motion is one of them) are under discussion (see for instance Barndorff–Nielsen [46, 47] and Eberlein and Keller [197]). α–Stable motion is also infinitely divisible. Apart from any drift, it is a pure jump process and its marginal distributions have an infinite variance. As such it is a candidate for modelling real phenomena with erratic behaviour. α–Stable processes allow for generalisations and extensions in many ways. They are mathematical models as attractive as the Gaussian processes. This has been proved convincingly in the recent books by Janicki and Weron [354] and Samorodnitsky and Taqqu [565].

It is our intention now to give a short introduction to the topic of stable processes. We do this for the following two reasons: 1. We believe that stable processes constitute an important class of stochastic processes with the potential for wide applications in modelling extremal events. 2. They are not very familiar (if not even unknown) to the applied worker. Even in circles where they are known there is a lot of suspicion about the infinite variance of these processes, which is considered something extraordinary. (It took 70 years before the role of Brownian motion was fully recognised in finance.) Books like Mittnik and Rachev [465] will certainly contribute to wider applications of stable processes in finance and insurance.

8.8.1 Stable Random Vectors

By virtue of Kolmogorov's consistency theorem (see for instance Billingsley [69]), the distribution of a stochastic process $(X_t)_{t\in T}$ is determined by its self–consistent family of finite–dimensional distributions, i.e. the distributions of the random vectors

$$(X_{t_1},\ldots,X_{t_d})\,,\quad t_1,\ldots,t_d\in T\,,\quad d\geq 1\,. \tag{8.130}$$

This is also the point to start with. We restrict ourselves to symmetric processes which means that the distribution of the vector (8.130) does not change when the latter is multiplied by -1. We use the abbreviation $s\alpha s$ (symmetric α–stable) and assume that $\alpha<2$; the case $\alpha=2$ corresponds to the Gaussian processes.

Recall (see Section 2.2) that an $s\alpha s$ rv X has chf

$$Ee^{itX}=e^{-c|t|^\alpha}\,,\quad t\in\mathbb{R}\,.$$

for some $c>0$ (a scaling parameter) and some $\alpha\in(0,2]$. For completeness we will also admit the parameter choice $c=0$ which corresponds to a "degenerate stable distribution". For iid symmetric X,X_1,X_2 this is equivalent to

$$a_1X_1+a_2X_2\stackrel{d}{=}X\left(|a_1|^\alpha+|a_2|^\alpha\right)^{1/\alpha}\,,\quad a_1,a_2\in\mathbb{R}\,.$$

Stable random vectors are defined in a similar fashion:

Definition 8.8.1 ($s\alpha s$ random vector)
The random vector $\mathbf{X}$ (*the distribution* F *of* $\mathbf{X}$) *with values in* $\mathbb{R}^d$ *is* $s\alpha s$ *for some* $\alpha\in(0,2)$ *if it has chf*

$$Ee^{i(\mathbf{t},\mathbf{X})}=\exp\left\{-\int_{S^{d-1}}|(\mathbf{t},\mathbf{y})|^\alpha dm_s(\mathbf{y})\right\}\,,\quad \mathbf{t}\in\mathbb{R}^d\,. \tag{8.131}$$

Here $(\cdot,\cdot)$ *denotes the usual scalar product in* $\mathbb{R}^d$ *and* m_s *is a symmetric (i.e.* $m_s(A)=m_s(-A)$*) finite measure on the Borel sets of the unit sphere*

$$S^{d-1}=\left\{\mathbf{s}=(s_1,\ldots,s_d):\|\mathbf{s}\|=\sqrt{s_1^2+\cdots+s_d^2}=1\right\}$$

of $\mathbb{R}^d$. *It is called the* spectral measure *of the random vector* $\mathbf{X}$ (*of its distribution* F) *and* (8.131) *is the corresponding* spectral representation. □

Remarks. 1) The spectral measure m_s in (8.131) is unique on S^{d-1} for $\alpha < 2$.

2) The spectral representation (8.131) yields immediately the following: let $\mathbf{X}$, $\mathbf{X_1}$, $\mathbf{X_2}$ be iid $s\alpha s$ in $\mathbb{R}^d$ and $a_1, a_2 \in \mathbb{R}$. Then

$$\begin{aligned} Ee^{i(\mathbf{t},a_1\mathbf{X_1}+a_2\mathbf{X_2})} &= Ee^{i(\mathbf{t},a_1\mathbf{X})}\,Ee^{i(\mathbf{t},a_2\mathbf{X})} \\ &= \exp\left\{-\left(|a_1|^\alpha + |a_2|^\alpha\right)\int_{S^{d-1}} |(\mathbf{t},\mathbf{y})|^\alpha dm_s(y)\right\} \\ &= E\exp\left\{i\left(\mathbf{t},\mathbf{X}\left(|a_1|^\alpha + |a_2|^\alpha\right)^{1/\alpha}\right)\right\}, \qquad \mathbf{t} \in \mathbb{R}^d. \end{aligned}$$

Hence

$$a_1\mathbf{X_1} + a_2\mathbf{X_2} \stackrel{d}{=} \mathbf{X}\left(|a_1|^\alpha + |a_2|^\alpha\right)^{1/\alpha}, \quad a_1, a_2 \in \mathbb{R}.$$

For symmetric $\mathbf{X}$, the latter can be shown to be equivalent to the defining spectral representation (8.131) for an $s\alpha s$ random vector. □

Example 8.8.2 (Independent $s\alpha s$ rvs constitute an $s\alpha s$ random vector) Assume $\mathbf{X} = (X_1, \ldots, X_d)$ is a vector of independent $s\alpha s$ rvs:

$$Ee^{itX_j} = \exp\{-c_j|t|^\alpha\}, \quad t \in \mathbb{R}, \quad j = 1, \ldots, d.$$

Then, for every $\mathbf{t} = (t_1, \ldots, t_d) \in \mathbb{R}^d$,

$$Ee^{i(\mathbf{t},\mathbf{X})} = \prod_{j=1}^d Ee^{it_jX_j} = \prod_{j=1}^d \exp\{-c_j\,|t_j|^\alpha\} = \exp\left\{-\sum_{j=1}^d c_j\,|t_j|^\alpha\right\}.$$

Define the jth unit vector $\mathbf{e_j} = (e_{ji})_{i=1,\ldots,d}$ by

$$e_{ji} = \begin{cases} 1 & i = j, \\ 0 & i \neq j, \end{cases} \qquad i = 1, \ldots, d.$$

Then

$$\begin{aligned} Ee^{i(\mathbf{t},\mathbf{X})} &= \exp\left\{-\sum_{j=1}^d |(\mathbf{e_j},\mathbf{t})|^\alpha\, c_j\right\} \\ &= \exp\left\{\int_{S^{d-1}} |(\mathbf{y},\mathbf{t})|^\alpha d\widetilde{m}_s(\mathbf{y})\right\}, \end{aligned}$$

where $\widetilde{m}_s$ is the symmetrised version of the discrete measure

$$m_s = \sum_{j=1}^d c_j\varepsilon_{\mathbf{e_j}}$$

($\varepsilon_{\mathbf{x}}$ denotes Dirac measure concentrated in x) on the unit sphere S^{d-1} of $\mathbb{R}^d$. Conversely, it can easily be seen that an $s\alpha s$ random vector in $\mathbb{R}^d$ with spectral measure m_s concentrated on the set of vectors $\{\mathbf{e_1}, \ldots, \mathbf{e_d}, -\mathbf{e_1}, \ldots, -\mathbf{e_d}\}$ has necessarily independent $s\alpha s$ components. □

Example 8.8.3 (Subgaussian stable vector)
Let A be an $\alpha/2$–stable positive rv. It is well known (see Samorodnitsky and Taqqu [565], Proposition 1.2.12) that A has Laplace transform

$$Ee^{-sA} = \exp\{-cs^{\alpha/2}\}, \quad s \geq 0,$$

for some $c > 0$. In the following we assume wlog that $c = 1$. Recall that a normal $N(0, 2\sigma^2)$ rv N has chf

$$Ee^{itN} = \exp\{-\sigma^2 t^2\}, \quad t \in \mathbb{R}.$$

Assume A and N are independent and define

$$X = A^{1/2} N. \tag{8.132}$$

Then

$$\begin{aligned} Ee^{itX} &= E\left(E\left(e^{itA^{1/2}N} \middle| A\right)\right) = E\exp\left\{-A\sigma^2 t^2\right\} \\ &= \exp\{-|\sigma|^\alpha |t|^\alpha\}, \quad t \in \mathbb{R}, \end{aligned}$$

i.e. X is $s\alpha s$, and every $s\alpha s$ rv X has representation (8.132). This is the motivation for the following multivariate generalisation: assume $\mathbf{N} = (N_1, \ldots, N_d)$ is mean–zero Gaussian in $\mathbb{R}^d$ given by its chf

$$Ee^{i(\mathbf{t},\mathbf{N})} = \exp\left\{-\mathbf{t}^\top R\mathbf{t}\right\}, \quad \mathbf{t} \in \mathbb{R}^d,$$

where the vectors $\mathbf{t}$ are understood as columns and $\top$ stands for transpose. Moreover,

$$2R = 2\left(r_{ij}\right)_{i,j=1,\ldots,d} = \left(\operatorname{cov}\left(N_i, N_j\right)\right)_{i,j=1,\ldots,d}$$

is the covariance matrix of $\mathbf{N}$. If $\mathbf{N}$ and A are independent then $\mathbf{X} = A^{1/2}\mathbf{N}$ is called *subgaussian*. It has chf

$$\begin{aligned} Ee^{i(\mathbf{t},\mathbf{X})} &= Ee^{iA^{1/2}(\mathbf{t},\mathbf{N})} \\ &= E\left(E\left(e^{iA^{1/2}(\mathbf{t},\mathbf{N})} \middle| A\right)\right) \\ &= E\exp\left\{-A\mathbf{t}^\top R\mathbf{t}\right\} \\ &= \exp\left\{-\left|\mathbf{t}^\top R\mathbf{t}\right|^{\alpha/2}\right\} \\ &= \exp\left\{-\left|\sum_{i,j=1}^d t_i t_j r_{ij}\right|^{\alpha/2}\right\}, \quad \mathbf{t} \in \mathbb{R}^d, \end{aligned}$$

which can be shown to be the chf of an $s\alpha s$ random vector (it suffices to prove that every linear combination of components of a subgaussian vector is $s\alpha s$; see Proposition 1.3.1 in Samorodnitsky and Taqqu [565]; this follows from the chf above). If $\mathbf{N}$ has iid $N(0, 2\sigma^2)$ components the chf of X is particularly simple:

$$\begin{aligned} Ee^{i(\mathbf{t},\mathbf{X})} &= \exp\left\{-|\sigma|^\alpha \left|\sum_{j=1}^d t_j^2\right|^{\alpha/2}\right\} \\ &= \exp\left\{-|\sigma|^\alpha \|\mathbf{t}\|^\alpha\right\}, \quad \mathbf{t} \in \mathbb{R}^d . \end{aligned} \tag{8.133}$$

For example, let $d = 2$, $\alpha = 1$ in (8.133). Then we obtain

$$Ee^{i(\mathbf{t},\mathbf{X})} = \exp\left\{-|\sigma|\sqrt{t_1^2 + t_2^2}\right\}, \quad \mathbf{t} \in \mathbb{R}^2 ,$$

which is the chf of the *two-dimensional isotropic Cauchy law.* It has density

$$f_{\mathbf{X}}(x_1, x_2) = \frac{|\sigma|}{2\pi\left(x_1^2 + x_2^2 + \sigma^2\right)^{3/2}}, \quad x_1, x_2 \in \mathbb{R}.$$

The meaning of the word "isotropic" will become clear by the arguments given below.

The $s\alpha s$ distribution with spectral representation (8.133) corresponds to a spectral measure m_s which, up to a constant multiple, is Lebesgue measure on S^{d-1}. We verify this for $d = 2$. Write $\mathbf{t}$ in polar coordinates:

$$\mathbf{t} = \begin{pmatrix} t_1 \\ t_2 \end{pmatrix} = r\begin{pmatrix} \cos\phi_t \\ \sin\phi_t \end{pmatrix}, \quad r = \sqrt{t_1^2 + t_2^2} .$$

Then we have

$$\begin{aligned} r^\alpha \int_0^{2\pi} |\cos\phi|^\alpha \, d\phi &= r^\alpha \int_0^{2\pi} |\cos(\phi - \phi_t)|^\alpha \, d\phi \\ &= r^\alpha \int_0^{2\pi} |\cos\phi_t \cos\phi + \sin\phi_t \sin\phi|^\alpha \, d\phi \\ &= \int_0^{2\pi} \left|\left(\begin{pmatrix} r\cos\phi_t \\ r\sin\phi_t \end{pmatrix}, \begin{pmatrix} \cos\phi \\ \sin\phi \end{pmatrix}\right)\right|^\alpha d\phi \\ &= \int_0^{2\pi} \left|\left(\begin{pmatrix} t_1 \\ t_2 \end{pmatrix}, \begin{pmatrix} \cos\phi \\ \sin\phi \end{pmatrix}\right)\right|^\alpha d\phi . \end{aligned}$$

In a similar way, one can proceed in $\mathbb{R}^d$, using spherical coordinates.

Notice that the distribution of a subgaussian $\mathbf{X}$ is invariant under rotations (isotropic): let O be an orthogonal matrix ($O^\top = O^{-1}$). The vector $O\mathbf{X}$ is the vector $\mathbf{X}$ after a rotation in $\mathbb{R}^d$. Then

$$Ee^{i(\mathbf{t},\mathbf{X})} = \exp\{-|\sigma|^\alpha \|\mathbf{t}\|^\alpha\} = \exp\left\{-|\sigma|^\alpha \left\|O^\top \mathbf{t}\right\|^\alpha\right\} = Ee^{i(\mathbf{t},O\mathbf{X})}, \quad \mathbf{t} \in \mathbb{R}^d,$$

since

$$\|\mathbf{t}\|^2 = \mathbf{t}^\top \mathbf{t} = \mathbf{t}^\top \left(OO^\top\right) \mathbf{t} = \left(O^\top \mathbf{t}\right)^\top \left(O^\top \mathbf{t}\right) = \left\|O^\top \mathbf{t}\right\|^2 . \qquad \square$$

8.8.2 Symmetric Stable Processes

Now we are in the position to define an *sαs* process:

Definition 8.8.4 (*sαs* process)
A stochastic process $(X_t)_{t\in T}$ *is called sαs if all its finite-dimensional distributions are sαs, i.e. the random vectors* (8.130) *are sαs in the sense of Definition 8.8.1.* $\square$

From the definition of an *sαs* process $(X_t)_{t\in T}$ it follows that all linear combinations

$$\sum_{i=1}^d a_i X_{t_i}, \quad (a_i)_{i=1,\dots,d} \in \mathbb{R}^d, \quad (t_i)_{i=1,\dots,d} \in T^d, \quad d \geq 1,$$

are *sαs*. The converse is also true as it follows from Theorem 3.1.2 in Samorodnitsky and Taqqu [565].

Example 8.8.5 (*sαs* motion)
In Section 2.4 we learnt about α–stable motion as a process with independent, stationary α–stable increments. *sαs* motion is a special α–stable process with symmetric α–stable finite–dimensional distributions: it is a process $(X_t)_{t\geq 0}$ satisfying the conditions

- $X_0 = 0$ a.s.
- (X_t) has independent increments: $X_{t_2} - X_{t_1}, \dots, X_{t_d} - X_{t_{d-1}}$ are independent for every choice of $0 \leq t_1 < \cdots < t_d < \infty$ and $d \geq 1$.
- $X_t - X_s \stackrel{d}{=} X_{t-s}$, $t > s$, and X_t is *sαs* for every $t > 0$.

Notice that for $t_1 < \cdots < t_d$, by virtue of the stationary, independent increments,

$$\begin{aligned}
&(X_{t_1}, \dots, X_{t_d}) \\
&\stackrel{d}{=} \left(Z_1 t_1^{1/\alpha}, Z_1 t_1^{1/\alpha} + Z_2 (t_2 - t_1)^{1/\alpha}, \dots,\right. \\
&\qquad \left. Z_1 t_1^{1/\alpha} + Z_2 (t_2 - t_1)^{1/\alpha} + \cdots + Z_d (t_d - t_{d-1})^{1/\alpha}\right)
\end{aligned}$$

for iid $s\alpha s$ Z_i such that $Z_1 \stackrel{d}{=} X_1$. In particular, for all $c > 0$

$$(X_{ct_1}, \ldots, X_{ct_d}) \stackrel{d}{=} c^{1/\alpha} (X_{t_1}, \ldots, X_{t_d}) .$$

This property is called *$1/\alpha$–self–similarity* of the process (X_t) (see Section 8.9 for more information on self–similarity). Notice that Brownian motion is 1/2–self–similar and the Cauchy (i.e. $s1s$) motion is 1–self–similar. □

8.8.3 Stable Integrals

Many stable processes of interest have a stochastic–integral representation with respect to a stable random measure. This is similar to the Gaussian case. As for Gaussian processes, there exist different ways of defining stable integrals. We prefer here a constructive approach to the topic.

First we introduce the notion of an α–stable random measure with respect to which we will integrate. It is in general not a signed measure since it can have infinite variation. We again restrict ourselves to the symmetric ($s\alpha s$) case.

Definition 8.8.6 (α–Stable random measure)
Let $[E, \mathcal{E}, m_c]$ be a measure space, i.e. $\mathcal{E}$ is a σ–algebra of subsets of E and m_c is a measure on $\mathcal{E}$. A set function M on $\mathcal{E}$ is called an $s\alpha s$ random measure with control measure m_c *if the following conditions are satisfied:*

(a) *For every $A \in \mathcal{E}$ with $m_c(A) < \infty$, $M(A)$ is an $s\alpha s$ rv with chf*

$$Ee^{itM(A)} = \exp\{-m_c(A)|t|^\alpha\} , \quad t \in \mathbb{R} .$$

(b) *For disjoint $A_1, \ldots, A_d \in \mathcal{E}$ with $\sum_{i=1}^d m_c(A_i) < \infty$, the rvs $M(A_1), \ldots, M(A_d)$ are independent.*
(c) *For disjoint $A_1, A_2, \ldots \in \mathcal{E}$ with $\sum_{i=1}^\infty m_c(A_i) < \infty$ the relation*

$$M\left(\bigcup_{i=1}^\infty A_i\right) = \sum_{i=1}^\infty M(A_i) \quad \text{a.s.}$$

holds. □

Remarks. 1) For our purposes, E will be a subset of the real line equipped with the corresponding σ–algebra of the Borel sets.

2) Motivated by the defining properties (b) and (c), M is also called an *independently scattered σ–additive set function.*

3) The existence of an $s\alpha s$ random measure has to be proved. One way is to apply Kolmogorov's consistency theorem (see for instance Billingsley [69])

since M can be considered as an $s\alpha s$ process indexed by the sets $A \in \mathcal{E}$ with finite m_c–measure. It is easy to see that the rhs in (c) is a.s. convergent. As a series of independent terms it converges a.s. if and only if it converges in distribution (see Dudley [187], Section 9.7.1 on p. 251). And it converges weakly to the distribution with chf

$$\begin{aligned}\prod_{k=1}^{\infty} E\exp\{itM(A_k)\} &= \prod_{k=1}^{\infty} \exp\{-m_c(A_k)|t|^{\alpha}\} \\ &= \exp\left\{-|t|^{\alpha}\sum_{k=1}^{\infty} m_c(A_k)\right\}.\end{aligned}$$

□

Example 8.8.7 ($s\alpha s$ motion as an $s\alpha s$ random measure)
Assume that M is an $s\alpha s$ random measure on $[[0,\infty),\sigma([0,\infty)),m_c]$ where $\sigma([0,\infty))$ denotes the σ–algebra of the Borel sets in $[0,\infty)$ and m_c is Lebesgue control measure. Define

$$X_t = M([0,t]), \quad 0 \le t < \infty.$$

By definition of an $s\alpha s$ random measure, $X_t \stackrel{d}{=} t^{1/\alpha}X_1$ and $(X_t)_{t\ge 0}$ has stationary, independent increments. Thus $(X_t)_{t\ge 0}$ is an $s\alpha s$ motion. □

Now we want to construct an $s\alpha s$ integral. We start with a simple function as integrand:

$$f_n = \sum_{j=1}^{n} c_j I_{A_j}, \tag{8.134}$$

where $A_1,\ldots,A_n \in \mathcal{E}$ is a disjoint partition of E and $c_1,\ldots,c_n$ are arbitrary real numbers. Then immediately

$$\int_E |f_n(x)|^{\alpha}\, dm_c(x) = \sum_{j=1}^{n} |c_j|^{\alpha}\, m_c(A_j).$$

We require that the rhs is finite, i.e. $f_n \in L^{\alpha}[E,\mathcal{E},m_c]$. For such a function we define the stochastic integral

$$I(f_n) = \int_E f_n(x)\, dM(x) = \sum_{j=1}^{n} c_j M(A_j).$$

Using the independence of the $M(A_j)$, the chf of $I(f_n)$ is easily evaluated:

$$
\begin{aligned}
&Ee^{itI(f_n)} \\
&= \prod_{j=1}^{n} Ee^{itc_j M(A_j)} \\
&= \exp\left\{-|t|^\alpha \sum_{j=1}^{n} |c_j|^\alpha m_c(A_j)\right\} \\
&= \exp\left\{-|t|^\alpha \int_E |f_n(x)|^\alpha \, dm_c(x)\right\}. \qquad (8.135)
\end{aligned}
$$

Hence $I(f_n)$ is $s\alpha s$, and therefore the notion $s\alpha s$ or α–stable integral is justifed.

Every measurable real–valued function f on $[E, \mathcal{E}]$ can be approximated by simple functions f_n as defined in (8.134) which, in addition, satisfy the conditions $|f_n| \le |f|$ and $f_n \to f$ m_c–a.s. Samorodnitsky and Taqqu [565], p. 122, give a particular construction for f_n. If $f \in L^\alpha[E, \mathcal{E}, m_c]$ then also $f_n \in L^\alpha[E, \mathcal{E}, m_c]$ for all n, and dominated convergence yields

$$
\int_E |f_n(x) - f_m(x)|^\alpha \, dm_c(x) \to 0, \quad n, m \to \infty, \qquad (8.136)
$$

$$
\int_E |f_n(x) - f(x)|^\alpha \, dm_c(x) \to 0, \quad n \to \infty.
$$

Since it is always possible to find a disjoint partition (A_j) of E jointly for f_n and f_m, we can write, for $n > m$,

$$
f_k = \sum_j c_j^{(k)} I_{A_j}, \quad k = n, m.
$$

Hence $I(f_n - f_m) = I(f_n) - I(f_m)$ and by (8.135) and (8.136) we may conclude that

$$
Ee^{it(I(f_n) - I(f_m))} = \exp\left\{-|t|^\alpha \int_E |f_n(x) - f_m(x)|^\alpha \, dm_c(x)\right\} \to 1, \quad t \in \mathbb{R}.
$$

This proves that $(I(f_n))$ is a Cauchy sequence with respect to convergence in probability and hence it converges in probability to a limit rv which we denote by

$$
I(f) = \int_E f(x) \, dM(x)
$$

and which we call *the stochastic integral of f with respect to M* or an *$s\alpha s$ integral.* The limit $I(f)$ is independent of the choice of the approximating simple functions (f_n). (If there is another approximating sequence (g_n), one can construct a joint sequence (h_n) from elements of (f_n) and (g_n) which is again Cauchy and has a unique limit $I(f)$.)

In the following we give some simple properties of the stochastic integrals $\int_E f(x)\,dM(x)$.

Proposition 8.8.8 (Elementary properties of $s\alpha s$ integrals)
Assume $f, f_1, f_2, \ldots \in L^\alpha[E, \mathcal{E}, m_c]$.

(a) $I(f)$ *is* $s\alpha s$ *with chf*

$$Ee^{itI(f)} = \exp\left\{-|t|^\alpha \int_E |f(x)|^\alpha \, dm_c(x)\right\}, \quad t \in \mathbb{R}.$$

(b) $I(f)$ *is linear, i.e. for every* $a_1, a_2 \in \mathbb{R}$,

$$\int_E (a_1 f_1(x) + a_2 f_2(x))\, dM(x) = a_1 \int_E f_1(x)\, dM(x) + a_2 \int_E f_2(x)\, dM(x).$$

(c) *The relation* $I(f_n) \xrightarrow{P} I(f)$ *holds if and only if*

$$\int_E |f_n(x) - f(x)|^\alpha \, dm_c(x) \to 0.$$

Proof. From (8.135) we know that (a) holds for a simple function f_n. For general f, (a) follows from (8.135) and from the continuity theorem for chfs since $I(f_n) \xrightarrow{P} I(f)$ for the simple functions f_n used in the definition of $I(f)$.
(b) is immediate for simple functions f_1, f_2. For general f_1, f_2 let $f_1^{(n)}$, $f_2^{(n)}$ be simple functions as used for the definition of $I(f_1)$, $I(f_2)$. Then $I(f_i^{(n)}) \xrightarrow{P} I(f_i)$, $i = 1, 2$, and (b) follows by first applying (b) to $f_i^{(n)}$, $i = 1, 2$, and then by passing to the limit as $n \to \infty$.
(c) Notice that

$$Ee^{it(I(f_n) - I(f))} = \exp\left\{-|t|^\alpha \int_E |f_n(x) - f(x)|^\alpha \, dm_c(x)\right\}, \quad t \in \mathbb{R}.$$

A necessary and sufficient condition for $I(f_n) \xrightarrow{P} I(f)$ is that $E\exp\{it(I(f) - I(f_n))\} \to 1$ for all t, but this means that $\int_E |f_n(x) - f(x)|^\alpha \, dm_c(x) \to 0$. □

Proposition 8.8.9 *For every* $f_1, \ldots, f_d \in L^\alpha[E, \mathcal{E}, m_c]$, *the random vector* $\mathbf{J_d} = (I(f_1), \ldots, I(f_d))$ *is* $s\alpha s$.

Proof. We have the chf

$$
\begin{aligned}
Ee^{i(\mathbf{t},\mathbf{J}_d)} &= E\exp\left\{i\sum_{j=1}^{d} t_j I(f_j)\right\} \\
&= \exp\left\{-\int_E \left|\sum_{j=1}^{d} t_j f_j(x)\right|^{\alpha} dm_c(x)\right\} \\
&= \exp\left\{-\int_{E+} \left|\frac{\sum_{j=1}^{d} t_j f_j(x)}{\left(\sum_{j=1}^{d} f_j^2(x)\right)^{1/2}}\right|^{\alpha} \left(\sum_{j=1}^{d} f_j^2(x)\right)^{\alpha/2} dm_c(x)\right\},
\end{aligned}
$$

where $E^+ = \{x \in E : \sum_{j=1}^{d} f_j^2(x) > 0\}$. Introducing the new coordinates

$$
g_j = g_j(x) = \frac{f_j(x)}{\left(\sum_{j=1}^{d} f_j^2(x)\right)^{1/2}}, \quad j = 1, \ldots, d,
$$

on S^{d-1} we can write

$$
Ee^{i(\mathbf{t},\mathbf{J}_d)} = \exp\left\{-\int_{S^{d-1}} \left|\sum_{j=1}^{d} t_j g_j\right|^{\alpha} d\widetilde{m}(\mathbf{g})\right\},
$$

where

$$
\widetilde{m}(A) = \int_{\mathbf{g}^{-1}(A)} \left(\sum_{j=1}^{d} f_j^2(x)\right)^{\alpha/2} dm_c(x). \qquad \square
$$

Proposition 8.8.10 *The rvs $I(f_1)$ and $I(f_2)$ are independent if and only if*

$$
f_1(x) f_2(x) = 0 \quad m_c - \text{a.e.} \tag{8.137}
$$

Sketch of the proof. Using chfs, one has to show that

$$
\begin{aligned}
Ee^{itI(f_1+f_2)} &= \exp\left\{-|t|^{\alpha} \int_E |f_1(x) + f_2(x)|^{\alpha}\, dm_c(x)\right\} \\
&= \exp\left\{-|t|^{\alpha} \left(\int_E |f_1(x)|^{\alpha}\, dm_c(x) + \int_E |f_2(x)|^{\alpha}\, dm_c(x)\right)\right\} \\
&= Ee^{itI(f_1)} Ee^{itI(f_2)}, \quad t \in \mathbb{R}.
\end{aligned}
$$

This means that one has to prove that

$$
\int_E |f_1(x) + f_2(x)|^{\alpha}\, dm_c(x) = \int_E |f_1(x)|^{\alpha}\, dm_c(x) + \int_E |f_2(x)|^{\alpha}\, dm_c(x). \tag{8.138}
$$

If f_1 and f_2 have disjoint supports as assumed by (8.137), (8.138) is immediate. For the converse we refer to Samorodnitsky and Taqqu [565], Theorem 3.5.3. □

Remark. 4) It should be mentioned that the above construction of an $s\alpha s$–stable integral also works in the Gaussian case when $\alpha = 2$. But notice that then $E\exp\{itI(f)\} = \exp\{-t^2 \int_E f^2(x)dm_c(x)\}$, i.e. $I(f)$ is $N(0, 2\int_E f^2(x)dm_c(x))$. Thus one can define a Gaussian stochastic integral with respect to a $s2s$ random measure via only the variances of the rvs $I(f)$, which is standard in that case. □

8.8.4 Examples

In this section we consider some more examples of stable processes. In most cases we will make use of the representation as $s\alpha s$ integrals with respect to an $s\alpha s$ random measure. For the reader who is not interested in the construction of such an integral as given in Section 8.8.3, it is convenient to think of $\int_E f(x)\,dM(x)$ for some $E \subset \mathbb{R}$ as a discrete sum of type $\sum_j f(x_{j-1})M((x_{j-1}, x_j])$ for a finite partition $x_1 < x_2 < \cdots < x_n$ of elements of E, where the rvs $M((x_{j-1}, x_j])$ are independent $s\alpha s$ with chfs

$$E\exp\{itM((x_{j-1}, x_j])\} = \exp\left\{-|t|^\alpha \left(m_c\left(x_j\right) - m_c\left(x_{j-1}\right)\right)\right\}$$

for a non–decreasing, right–continuous function m_c on $\mathbb{R}$.

Example 8.8.11 ($s\alpha s$ motion)
Let

$$X_t = \int_0^\infty I_{(-\infty,t]}(x)\,dM(x) = \int_0^t dM(x)\,, \quad t \geq 0\,,$$

where M is an $s\alpha s$ random measure on $\mathbb{R}$ with Lebesgue control measure m_c. Thus

$$Ee^{it(X_{s_2}-X_{s_1})} = \exp\left\{-|t|^\alpha (s_2 - s_1)\right\}\,, \quad s_1 < s_2\,, \quad t \in \mathbb{R}\,. \tag{8.139}$$

By definition of an $s\alpha s$ random measure, the increments

$$X_{s_i} - X_{s_{i-1}} = \int_{s_{i-1}}^{s_i} dM(x)\,, \quad 0 \leq s_1 < \cdots < s_d < \infty$$

are independent and, in view of (8.139), also stationary.

In a similar way, we can introduce $s\alpha s$ *noise*:

$$Y_t = \int_t^{t+1} dM(x) = M(t+1) - M(t)\,, \quad t \in \mathbb{Z}\,,$$

i.e. $(Y_t)_{t\in\mathbb{Z}}$ are iid $s\alpha s$ with chf $Ee^{isY_1} = e^{-|s|^\alpha}$, $s \in \mathbb{R}$. □

Example 8.8.12 ($s\alpha s$ linear process or moving average process)
In Chapter 7 we considered the linear or infinite moving average process

$$X_t = \sum_{j=-\infty}^{\infty} \psi_{t-j} Z_j = \sum_{j=-\infty}^{\infty} \psi_j Z_{t-j}\,, \quad t \in \mathbb{Z}\,, \tag{8.140}$$

where $(Z_t)_{t\in\mathbb{Z}}$ are iid $s\alpha s$ innovations or noise variables and $(\psi_j)_{j\in\mathbb{Z}}$ are real coefficients satisfying a certain summability condition. By the 3–series theorem (for instance Petrov [495], p. 205),

$$\sum_{j=-\infty}^{\infty} |\psi_j|^\alpha < \infty \tag{8.141}$$

can be seen to be a necessary and sufficient condition for the a.s. convergence of the series (8.140). The latter condition is for instance satisfied for every causal, invertible ARMA process driven by (Z_t) (see for instance Brockwell and Davis [92] or Example 7.1.1). Notice that X_t can be written as a stochastic integral

$$X_t = \int_{-\infty}^{\infty} \psi(t-x)\, dM(x)\,, \quad t \in \mathbb{R}, \tag{8.142}$$

where

$$\psi(y) = \psi_j\,, \quad y \in (j, j+1]\,, \quad j \in \mathbb{Z}\,,$$

and M is an $s\alpha s$ random measure with Lebesgue control measure m_c on $\mathbb{R}$. Condition (8.141) means that

$$\int_{\mathbb{R}} |\psi(x)|^\alpha\, dx < \infty\,, \tag{8.143}$$

which is needed for a proper definition of the integral (8.142). More generally, any $s\alpha s$ process $(X_t)_{t\in\mathbb{R}}$ defined by (8.142) with a function ψ satisfying (8.143) is called a *moving average process*. It is a strictly stationary process (see Appendix A2.1) since

$$\begin{aligned} & E \exp\left\{ i \sum_{j=1}^{d} s_j X_{t_j+h} \right\} \\ = \ & \exp\left\{ - \int_{\mathbb{R}} \left| \sum_{j=1}^{d} s_j \psi(t_j + h - x) \right|^\alpha dx \right\} \\ = \ & \exp\left\{ - \int_{\mathbb{R}} \left| \sum_{j=1}^{d} s_j \psi(t_j - x) \right|^\alpha dx \right\} \end{aligned}$$

$$= \quad E \exp\left\{ i \sum_{j=1}^{d} s_j X_{t_j} \right\}, \qquad \mathbf{s}, \mathbf{t} \in \mathbb{R}^d, \quad h \in \mathbb{R}.$$

□

Example 8.8.13 (*sαs* Ornstein–Uhlenbeck process)
Autoregressive processes are among the most popular stationary processes in time series analysis. A causal, invertible autoregressive process of order 1 (AR(1) process) is defined via the difference equations

$$X_t = \phi X_{t-1} + Z_t, \quad t \in \mathbb{Z},$$

for some $\phi \in (-1,1)$. Assume that $(Z_t)_{t \in \mathbb{Z}}$ is iid *sαs* noise. Then (X_t) has the series representation

$$X_t = \sum_{j=0}^{\infty} \phi^j Z_{t-j} = \sum_{j=-\infty}^{t} \phi^{t-j} Z_j, \quad t \in \mathbb{Z},$$

which is a moving average process in the sense of Example 8.8.12. A continuous time version of an AR(1) process can be defined via the integral representation

$$X_t = \int_{-\infty}^{t} e^{-\lambda(t-x)} \, dM(x), \quad t \in \mathbb{R}, \tag{8.144}$$

for a positive constant $\lambda > 0$. In view of the discussion in Example 8.8.12, this is a well defined *sαs* moving average process with

$$\psi(x) = e^{-\lambda x} I_{[0,\infty)}(x), \quad x \in \mathbb{R}.$$

The stationary process (8.144) is called an *sαs Ornstein–Uhlenbeck process.* If M stands for Brownian motion and the integral in (8.144) is interpreted in the Itô sense, (X_t) defines one of the important Gaussian processes, an Ornstein–Uhlenbeck process. □

Example 8.8.14 (*sαs* bridge)
Let $(B_t)_{t \in [0,1]}$ denote Brownian motion. Recall that the Gaussian process $X_t = B_t - tB_1$ on $[0,1]$ defines a Brownian bridge (see for instance Revuz and Yor [541]). It is the weak limit of the uniform empirical process (for instance Billingsley [69], Pollard [504] or Shorack and Wellner [579]), but it is also closely related to the weak limit of integrated periodogram processes (see for instance Anderson [17], Grenander and Rosenblatt [285] or Klüppelberg and Mikosch [395]) in time series analysis. See also the Notes and Comments in Section 7.4.

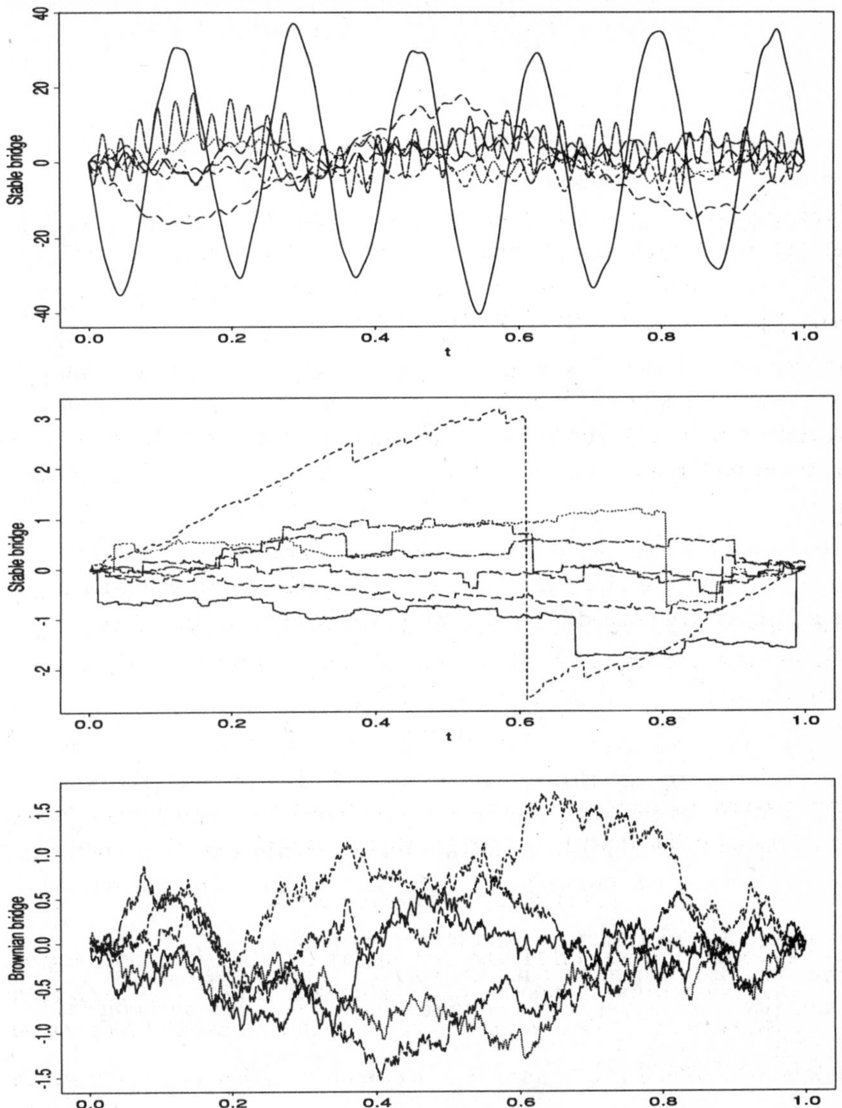

Figure 8.8.15 *Visualisation of the "bridge" processes* (Y_t) *given by the series* (8.145) *(top) and* (X_t) *by the formula* (8.146) *(middle): seven sample paths for* $\alpha = 1.2$*. The sample paths of* (Y_t) *are basically slightly perturbated sine curves; the regular sine shape of some of the paths is because of the dominating influence of a very large value* Z_i*. On the other hand,* (X_t) *inherits the jump process character of* 1.2*–stable motion. For comparison, five realisations of the Brownian bridge (bottom).*

An alternative definition of a Brownian bridge is given by the infinite series

$$Y_t = \frac{\sqrt{2}}{\pi} \sum_{k=1}^{\infty} \frac{\sin(k\pi t)}{k} Z_k \,, \quad 0 \le t \le 1 \,, \tag{8.145}$$

for iid standard normal rvs Z_t. This is a particular case of the ***Lévy–Ciesielski representation of a Brownian bridge*** (see for instance Hida [325]).

For the definition of an α–stable bridge we may proceed in similar ways: let $(\xi_t)_{t\in[0,1]}$ be an $s\alpha s$ motion. Then

$$X_t = \xi_t - t\xi_1 \,, \quad t \in [0,1] \,, \tag{8.146}$$

is a natural definition of an α–stable bridge. Since (ξ_t) is a pure jump process, (X_t) inherits this property and $X_0 = X_1 = 0$ a.s. Except for the latter fact, the Brownian bridge and an $s\alpha s$ bridge for $\alpha < 2$ do not have very much in common. For example, the Brownian bridge has a.s. continuous sample paths.

Alternatively, we can define an α–stable bridge via the series (8.145) with Z_t iid $s\alpha s$ rvs. For $\alpha \in (1,2)$ this process is well defined, but it is not for $\alpha \le 1$. Indeed, if it were defined for $\alpha \le 1$, we would have that

$$Y_t \stackrel{d}{=} \frac{\sqrt{2}}{\pi} Z_1 \left(\sum_{k=1}^{\infty} \left| \frac{\sin(k\pi t)}{k} \right|^{\alpha} \right)^{1/\alpha}$$

but the series on the rhs diverges in general. The process (Y_t) occurs in a natural way as the weak limit of a certain integrated periodogram process (see Klüppelberg and Mikosch [395]). It has a.s. continuous sample paths and $Y_0 = Y_1 = 0$ a.s., two properties which it shares with the Brownian bridge. □

Example 8.8.16 (Fractional $s\alpha s$ processes and noises)
Fractional Brownian motion and fractional Brownian noise have gained some popularity as processes with a close relation to so–called *long–range dependence* or *long memory processes.*

Standard *fractional Brownian motion* has representation as a stochastic integral

$$B_t^{(H)} = c(H) \int_{\mathbb{R}} \left[\left((t-x)^+\right)^{H-1/2} - ((-x)^+)^{H-1/2} \right] dM(x) \,, \quad t \in \mathbb{R},$$

where M is Brownian motion on $\mathbb{R}$ and the integral has to be interpreted in the Itô sense. The paramter H is taken from the interval $(0,1)$. Alternatively, a Gaussian mean–zero process on $\mathbb{R}$ with covariance function

$$\operatorname{cov}\left(B_{t_1}^{(H)}, B_{t_2}^{(H)}\right) = \frac{1}{2} \left\{ |t_1|^{2H} + |t_2|^{2H} - |t_1 - t_2|^{2H} \right\} \operatorname{var}\left(B_1^{(H)}\right)$$

is called fractional Brownian motion. For $H = 1/2$ we get Brownian motion. Using for instance the covariance function we can easily check that the scaling property

$$\left(B_{ct_1}^{(H)}, \ldots, B_{ct_d}^{(H)}\right) \stackrel{d}{=} c^H \left(B_{t_1}^{(H)}, \ldots, B_{t_d}^{(H)}\right)$$

holds for every choice of real (t_i), $c > 0$ and $d \geq 1$, i.e. fractional Brownian motion is an *H–self–similar process* (see Section 8.9). Moreover, fractional Brownian motion is the only non–degenerate Gaussian self–similar process with stationary increments.

If we define the increment process

$$Z_t^{(H)} = B_{t+1}^{(H)} - B_t^{(H)}, \quad t \in \mathbb{Z},$$

we thus obtain a stationary Gaussian process, called *fractional Brownian noise.* For $H = 1/2$, $(Z_t^{(H)})$ is iid Gaussian noise. For $1/2 < H < 1$,

$$\operatorname{cov}(Z_0, Z_t) \sim ct^{2(H-1)}, \quad t \to \infty,$$

for some constant $c > 0$. Thus the sequence $(\operatorname{cov}(Z_0, Z_t))_{t \geq 0}$ is not summable; it decreases to zero very slowly and is therefore referred to as indicating *long memory* or *long–range dependence.* This is in contrast for instance to Gaussian ARMA processes where the covariance function decreases to zero exponentially fast. Another interesting feature which has stimulated recent research is the fact that Gaussian fractional noise with $H \neq 1/2$ does not satisfy the classical CLT. Indeed, for $H \neq 1/2$ the normalisation n^H is required for weak convergence of the centred cumulative sums. This was first oberserved by Hurst [349] (therefore the symbol H for the "Hurst coefficient") who discovered empirically that cumulative yearly flows of the river Nile have a magnitude of order $n^{0.7}$ instead of the expected (from the CLT) $n^{1/2}$. Mandelbrot (see for instance [438, 439]) popularised the idea of long memory and proposed fractional Brownian motion for modelling it. In the frequency domain of time series analysis (see Section 7.1) long–range dependence is characterised by the fact that the spectral density of $(Z_t^{(H)})$ for $1/2 < H < 1$ has a singularity (a spike) at zero. This is in contrast to classical time series where the spectral density is usually a smooth function (for instance for ARMA processes). Recall from Theorem 7.1.4 the spectral representation of a stationary sequence. From it we may conclude the following: a singularity of the spectral density implies that the underlying time series has (random) cycles of arbitrary length. Mandelbrot called the latter behaviour "Joseph effect" since the biblical Joseph was able to predict the long sequence of seven good and bad harvests in the ancient Egypt.

One can also define "long–range dependence" in the *sαs* case, but certainly not via the covariance function because the second moments of the marginal distributions do not exist. An example of an *sαs* H–self–similar process is given by the integrals

$$X_t^{(H)} = \int_{\mathbb{R}} \left[\left((t-x)^+\right)^{H-1/\alpha} - \left((-x)^+\right)^{H-1/\alpha} \right] dM(x) \,, \quad t \in \mathbb{R} \,,$$

where M is an *sαs* random measure with Lebesgue control measure and $0 < H < 1$. For $H = 1/\alpha$ the process is *formally* interpreted as *sαs* motion. The *sαs* process so defined is called *linear fractional stable motion.* It is a process with stationary increments, but for *sαs* processes the subclass of self–similar processes does not consist only of these fractional motions. The corresponding *fractional sαs noise* can be defined as

$$Y_t^{(H)} = X_{t+1}^{(H)} - X_t^{(H)} \,, \quad t \in \mathbb{Z} \,.$$

For $H = 1/\alpha$ it is *formally* interpreted as *sαs* noise, whereas for $H \in (1/\alpha, 1)$ and $1 < \alpha < 2$, in analogy to fractal Brownian noise, it is considered as a process with "long–range dependence". Mandelbrot also coined the word "Noah effect" for extremal events and related heavy-tailed phenomena. The biblical Noah survived an enormous flood during which he and his family stayed for a very long time on a boat. In this sense, α–stable fractional noise is a process that enjoys both the Joseph and the Noah effect. □

Example 8.8.17 (*sαs* harmonisable processes)
Every nice (i.e. continuous in probability) stationary Gaussian process has a representation of the form

$$X_t = \int_{\mathbb{R}} e^{itx} \, dM(x) \,, \tag{8.147}$$

where M is a (complex–valued) Gaussian random measure. A process of the form (8.147) is called *harmonisable.* The stochastic integrals (8.147) with an *sαs* random measure M do not represent the whole class of *sαs* stationary processes (see Rosinski [556]). Nevertheless, the class of harmonisable *sαs* processes deserves some attention. An *sαs* harmonisable, stationary process is usually defined as the real part of a stochastic integral $\int_{\mathbb{R}} e^{itx} \, dM(x)$ for an *sαs* complex–valued random measure. To avoid this difficulty we just give here the chf of a finite vector of values of such a process $(X_t)_{t\in\mathbb{R}}$ (the totality of these chfs determines the finite–dimensional distributions of the process):

$$E \exp\left\{ i \sum_{j=1}^{d} s_j X_{t_j} \right\}$$

$$= \exp\left\{-\int_{\mathbb{R}}\left|\sum_{j=1}^{d} e^{it_j x} s_j\right|^{\alpha} dm_c(x)\right\} \tag{8.148}$$

$$= \exp\left\{-\int_{\mathbb{R}}\left|\left(\sum_{j=1}^{d} s_j\cos(t_j x)\right)^2 + \left(\sum_{j=1}^{d} s_j\sin(t_j x)\right)^2\right|^{\alpha/2} dm_c(x)\right\}$$

$$= \exp\left\{-\int_{\mathbb{R}}\left|\sum_{j,k=1}^{d} s_j s_k\cos((t_j - t_k)x)\right|^{\alpha} dm_c(x)\right\}, \quad \mathbf{t}, \mathbf{s} \in \mathbb{R}^d .$$

where m_c is a finite measure on $\mathbb{R}$. This can formally be interpreted as the chf of the vector

$$\left(\int_{\mathbb{R}} e^{it_1 x}\, dM(x), \ldots, \int_{\mathbb{R}} e^{it_d x}\, dM(x)\right)$$

with respect to an $s\alpha s$ random measure M with control measure m_c. The stationarity of the process (X_t) is easily seen since we can replace $e^{it_j x}$ in (8.148) by $e^{i(t_j+h)x}$, $j = 1, \ldots, d$ for every real h.

The $s\alpha s$ harmonisable, stationary processes have gained a certain popularity in the theory of stochastic processes since they allow one to introduce a pseudo spectral distribution whose theory very much parallels the corresponding L^2 spectral analysis; see also the Notes and Comments in Section 7.2. □

Notes and Comments

In the lattter sections we have briefly introduced $s\alpha s$ random vectors, processes and integrals and considered some particular examples. From the computer graphs of the corresponding sample paths it is immediate that stable processes are models for real phenomena with large fluctuations and big jumps. This makes them attractive for modelling extremal events.

The marginal distributions of an $s\alpha s$ process are $s\alpha s$, hence $E|X_t|^\alpha = \infty$ (except for the cases $X_t = 0$ a.s.), in particular, (X_t) is an infinite variance process with heavy tails of the form $P(X_t > x) \sim c_t x^{-\alpha}$, $x \to \infty$ (see Section 2.2), for some constant c_t. This tail behaviour is preserved under several operations acting on an $s\alpha s$ process $(X_t)_{t\in T}$.

For example, it is immediate from the definition of an $s\alpha s$ vector via its chf that any linear combination $\sum_{i=1}^{d} a_i X_{t_i}$ for $\mathbf{t} \in T^d$ and $\mathbf{a} \in \mathbb{R}^d$ is $s\alpha s$. A similar statement can be made for an infinite series $\sum_{i=1}^{\infty} a_i X_{t_i}$, provided that the latter series is well defined.

Products of independent $s\alpha s$ rvs have also been studied and extensions of these results exist for quadratic and multilinear forms in independent $s\alpha s$ rvs. The latter results are also used to define multiple stochastic intergrals with respect to stable random measures. For an account of that theory see Kwapień and Woyczyński [411] and the literature cited therein. Heavy tails are also typical for all such structures derived from $s\alpha s$ processes. For example, if X, Y are iid $s\alpha s$ then

$$P(XY > x) \sim c x^{-\alpha} \ln x \,, \quad x \to \infty \,,$$

(see Rosinski and Woyczyński [557]) which shows that XY is still in the domain of attraction of an α–stable law (see Section 2.2).

It is surprising that the order statistics of an $s\alpha s$ random vector have the same tail behaviour as any of its components. To be more precise, assume that

$$(X_1, \ldots, X_n) \stackrel{d}{=} \left(\int_E f_1 \, dM(x), \ldots, \int_E f_n dM(x) \right) \tag{8.149}$$

where $f_1, \ldots, f_n \in L^\alpha[E, \mathcal{E}, m_c]$ and M is an $s\alpha s$ random measure with control measure m_c. Such an integral representation always exists for any $s\alpha s$ random vector and appropriate functions (f_i); see Theorem 3.5.6 in Samorodnitsky and Taqqu [565]. The following is due to Samorodnitsky [563]:

Theorem 8.8.18 (Tail behaviour of the order statistics of an $s\alpha s$ sample)
Let $X_{n,n} \le \cdots \le X_{1,n}$ denote the order statistics of the $s\alpha s$ vector $(X_1, \ldots, X_n)$ with integral representation (8.149) *and let $|X|_{n,n} \le \cdots \le |X|_{1,n}$ be the order statistics of $(|X_1|, \ldots, |X_n|)$. Then the relations*

$$\begin{aligned} P(X_{k,n} > x) & \sim & x^{-\alpha} \frac{1}{2} C_\alpha \int_E \left((h_k^+)^\alpha + (h_k^-)^\alpha \right) dm_c \,, \quad x \to \infty \,, \\ P(|X|_{k,n} > x) & \sim & x^{-\alpha} C_\alpha \int_E h_k^\alpha dm_c \,, \quad x \to \infty \,, \end{aligned}$$

hold, where C_α is a positive constant, $h_k(x)$ is the kth largest among the $|f_i(x)|$, and $h_k^\pm(x)$ the kth largest among the $f_i^\pm(x)$, for $i = 1, \ldots, n$. □

A recent overview of results for suprema of α–stable continuous time processes is given in Samorodnitsky and Taqqu [565].

The latter book is an encyclopaedic source on stable and self–similar processes, containing a wealth of theoretical results. The monograph by Janicki and Weron [354] is devoted to stable processes and simulations of their sample paths. They also treat problems in the simulation of stable processes and the numerical solution of stochastic differential equations driven by stable processes. Mittnik and Rachev [466] consider applications of stable models in finance (their notion of "stability" is much wider than the one introduced in Section 2.2).

Discussion about the importance of stable distributions and processes for applications is still going on. Theoretically, they are much better understood than 20 or even 15 years ago, as the above mentioned monographs show, but they are nevertheless quite difficult objects to deal with. In particular, the dependence of the stable laws on several parameters and the particularly unpleasant behaviour of statistical estimators of α (see the discussion in Chapter 6) have scared many people away from working with stable processes. On the other hand, there is an increasing interest in such processes, as the great variety of the published literature shows (see for instance not only the list of references in the books mentioned above, but also the references in Chapter 7). Stable processes seem to be the right model wherever a real phenomenon with very large fluctuations occurs. Thus it is not surprising that stable processes have been applied in the financial context. However, multivariate stable distributions have also found their way into the modelling of earthquakes (see Kagan [362], Kagan and Vere–Jones [364]), and related insurance questions (see Kagan [363]).

Long memory processes such as the various forms of fractional noise (see Example 8.8.16) are by now well studied. There seems to be empirical evidence that certain financial time series (for instance exchange rates) exhibit long memory; see for instance Cheung [112] for some empirical studies, and the references cited therein. An introduction to long memory processes (in particular fractional ARIMA processes) is provided in Brockwell and Davis [92], Section 13.2. Samorodnitsky and Taqqu [565] touch on the problem of infinite variance long memory processes. Beran [61] deals with statistical estimation techniques for data exhibiting long memory.

8.9 Self–Similarity

A "self–similar" structure is one that looks roughly the same on a small or on a large scale. For example, share prices of stock when plotted against time have very much the same shape on a yearly, monthly, weekly, yes even on a daily basis. The same happens if we study a snow flake under the microscope and change the scale; we will always see a similar structure. Different parts of a tree (branches, foliage) look self–similar. Segments of coastline of a particular country repeat their patterns on a smaller scale. So we will not be surprised to discover the "Italian boot" in Norway or anywhere else hundreds of times, but on a much smaller scale and always with slight deviations from the original shape. Networks of tributaries of a big river are believed to be self–similar. This example also highlights the possible use of self–similarity: if we knew the right scaling factor we could, for example, extrapolate from the water flow through a given territory to the water flow through any other part of the network. If we knew about the development of a price on a small time scale, we could use self–similarity to extrapolate ("predict") by re–scaling to a longer or shorter interval of time. That is actually what we want to understand under this notion: changes of scale in "time" determine in a specific way a change of scale in "space", for instance the values of a price. From these examples we can also see what the limits of the method are: in reality it does not make sense to extrapolate into "too long" or "too short" intervals of "time" or "space". In other words, one has to be careful about the scaling factor in order to avoid results which are physically impossible.

In the context of stochastic processes, self–similarity has a very precise meaning. It describes scale invariance of the underlying distribution of the process, rather than of its sample paths:

Definition 8.9.1 (Self–similar stochastic process)
The real-valued stochastic process $(X_t)_{t\in T}$ with index set $T \in \{\mathbb{R}, \mathbb{R}_+, [0,\infty)\}$ is said to be self–similar *with index $H > 0$ (H–ss) if its finite–dimensional distributions satisfy the relation*

$$(X_{at_1}, \ldots, X_{at_m}) \stackrel{d}{=} a^H (X_{t_1}, \ldots, X_{t_m}) \tag{8.150}$$

for any choice of values $t_1, \ldots, t_m \in T, a > 0$. □

Remarks. 1) Notice that (8.150) indeed means "scale–invariance" of the finite–dimensional distributions of (X_t). It does not imply this property for the sample paths. Therefore, pictures trying to explain self–similarity by zooming in and out on one sample path, are by definition misleading.

2) If we interpret t as "time" and X_t as "space" then (8.150) tells us that every change of time scale $a > 0$ corresponds to a change of space scale a^H. The bigger H, the more dramatic the change of the space coordinate.

3) Self–similarity is convenient for simulations: a sample path of (X_t) on $[0,1]$ multiplied by a^H and a re–scaling of the time axis by a immediately provide a sample path on $[0,a]$ for any $a > 0$.

4) Notice that

$$X_{a0} \stackrel{d}{=} a^H X_0 \,, \quad a > 0 \,.$$

Hence $X_0 = 0$ a.s. □

Example 8.9.2 (Gaussian self–similar process)
Consider a Gaussian self–similar process $(X_t)_{t\geq 0}$, i.e. its finite–dimensional

distributions are Gaussian and satisfy (8.150). Gaussian processes are uniquely determined via their mean value function

$$m(t) = EX_t$$

and their covariance function

$$C(s,t) = \operatorname{cov}(X_s, X_t) .$$

If (X_t) is H–ss, then

$$m(t) = t^H EX_1$$

and

$$\begin{aligned} C(s,t) &= E\left[(X_t - m(t))(X_s - m(s))\right] \\ &= E\left[\left(X_{s(t/s)} - m(s(t/s))\right)(X_s - m(s))\right] \\ &= E\left[s^H \left(X_{t/s} - m(t/s)\right) s^H (X_1 - m(1))\right] \\ &= s^{2H} C(t/s, 1), \quad s \neq 0 . \end{aligned}$$

Conversely, every H–ss Gaussian process $(X_t)_{t\geq 0}$ is determined by a power function $m(t) = c\,t^H$ for a constant c and by an H–homogeneous non-negative definite function $C(s,t) = s^{2H} C(t/s, 1)$, $s \neq 0$. □

In the sense of Definition 8.9.1 we have already discussed two typical examples of self–similar processes, namely Brownian motion and α–stable motion; see Section 2.4 and Example 8.8.5. They occurred as weak limits of normalised and centred iid sum processes. A general result of Lamperti [413] states that every self–similar process can be obtained as a weak limit of certain normalised sum processes.

Example 8.9.3 (Brownian motion)
We consider Brownian motion $(B_t)_{t\geq 0}$; see Definition 2.4.1. For a Gaussian process with independent, stationary increments (see Definition 2.4.6) we immediately have the relation

$$(B_{at_1}, \ldots, B_{at_m}) \stackrel{d}{=} a^{1/2} (B_{t_1}, \ldots, B_{t_m}) .$$

Thus Brownian motion is 1/2–ss. Alternatively, 1/2–self–similarity of Brownian motion follows from Example 8.9.2 since

$$m(t) = 0\, t^H = 0 , \quad C(s,t) = \min(s,t) = s \min(1, t/s) , s \neq 0 ,$$

hence $H = 1/2$.

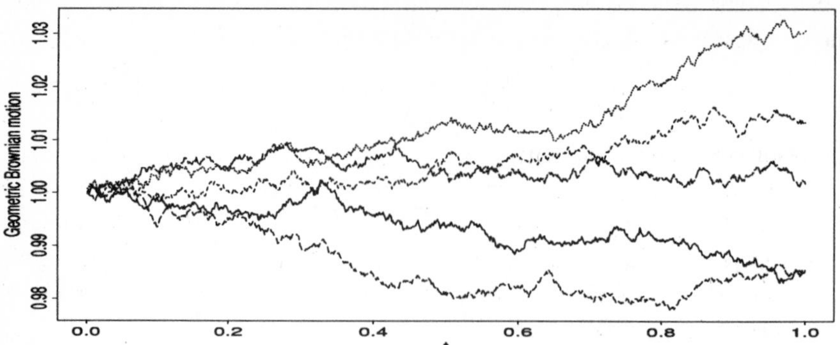

Figure 8.9.4 *Visualisation of geometric Brownian motion: five sample paths of the process* $X_t = \exp\{0.01B_t\}$.

Brownian motion is an important stochastic process in the context of insurance and finance. In insurance, it is used as an approximation to the total claim amount process (see Section 2.5) and in order to calculate the probability of ruin in finite time (the so–called "diffusion approximation"); see Grandell [282] for an extensive discussion, and also Example 2.5.18. In finance, increments of Brownian motion are employed as a surrogate for the "noise" of the market. In his famous paper, Bachelier [35] proposed Brownian motion as an appropriate model for prices. More recently, the Black–Scholes–Merton model for returns (see Duffie [190]) is based on the Itô stochastic differential equation (SDE)

$$dX_t = cX_t\,dt + \sigma X_t\,dB_t\,, \quad t \geq 0\,, \tag{8.151}$$

with constants c and $\sigma > 0$. This simple SDE has the unique strong solution

$$X_t = X_0 \exp\left\{\left(c - \frac{1}{2}\sigma^2\right)t + \sigma B_t\right\}, \quad t \geq 0\,,$$

which process is called *geometric Brownian motion*. It is not self–similar, but the underlying 1/2–self–similarity of Brownian motion creates similar patterns of returns or prices on a large or small scale. As already discussed in Section 7.6, relative returns in the time interval $[(n-1)\Delta, n\Delta]$ are roughly of the form

$$\begin{aligned} \Delta_n X &= \frac{X_{n\Delta} - X_{(n-1)\Delta}}{X_{(n-1)\Delta}} \\ &\approx \ln\left(1 + \frac{X_{n\Delta} - X_{(n-1)\Delta}}{X_{(n-1)\Delta}}\right) \end{aligned}$$

$$
\begin{aligned}
&= \ln X_{n\Delta} - \ln X_{(n-1)\Delta} \\
&= \left(c - \frac{1}{2}\sigma^2\right)\Delta + \sigma\left(B_{n\Delta} - B_{(n-1)\Delta}\right), \quad n \geq 1 .
\end{aligned}
$$

Hence observations of $\Delta_n X$ should roughly look like Gaussian white noise. This, as we have already stressed on various occasions, is never the case for real data. There are too many large fluctuations in the time series $(\Delta_n X)$, and therefore the model (8.151) has to be treated with care. In particular, it is not reasonable to use the self–similarity of (B_t) to get information ("predict") (X_t) in too short or too long time intervals.

Nevertheless, Brownian motion as limit process is an unavoidable tool in insurance and finance when it is understood as a mathematical model with the natural limitations that are often obvious from the context. □

Example 8.9.5 (α–stable motion)
In Section 2.4 and in Example 8.8.5 we introduced α–stable motion $(\xi_t)_{t\geq 0}$ as a process with independent, stationary α–stable increments for some $\alpha \leq 2$. For $\alpha = 2$, (ξ_t) is just Brownian motion, a process with a.s. continuous sample paths. For $\alpha < 2$, (ξ_t) is a pure jump process. As a consequence of Lemma 2.4.8, the finite–dimensional distributions of an α–stable motion satisfy the relation

$$
(\xi_{at_1}, \ldots, \xi_{at_m}) \stackrel{d}{=} a^{1/\alpha}(\xi_{t_1}, \ldots, \xi_{t_m}) .
$$

Thus (ξ_t) is $1/\alpha$–ss.

We established in Section 2.4 that (ξ_t) is a limit process of an iid sum process where in the case $\alpha < 2$ each summand has infinite variance. Hence α–stable motion can be taken as an approximation to the total claim amount process when large claims are present in the portfolio; see for instance Furrer, Michna and Weron [247]. In finance, certain attempts have been made to replace Brownian motion in the SDE (8.151) by an α–stable motion; see Rachev and Samorodnitsky [521] and Mittnik and Rachev [466]. It is believed that, to some extent, such a model would explain the large jumps which evidently occur in prices and which are caused by dramatic political or economic events. Rachev and Samorodnitsky [521] provide a Black–Scholes type pricing formula for this situation which, however, seems difficult to evaluate for practical purposes. A general theory of stochastic integration with respect to α–stable motion (see Kwapień and Woyczyński [411], Samorodnitsky and Taqqu [565]) and with respect to much more general processes exists (see for instance Chung and Williams [121], Protter [514]). For example, the simple SDE without drift

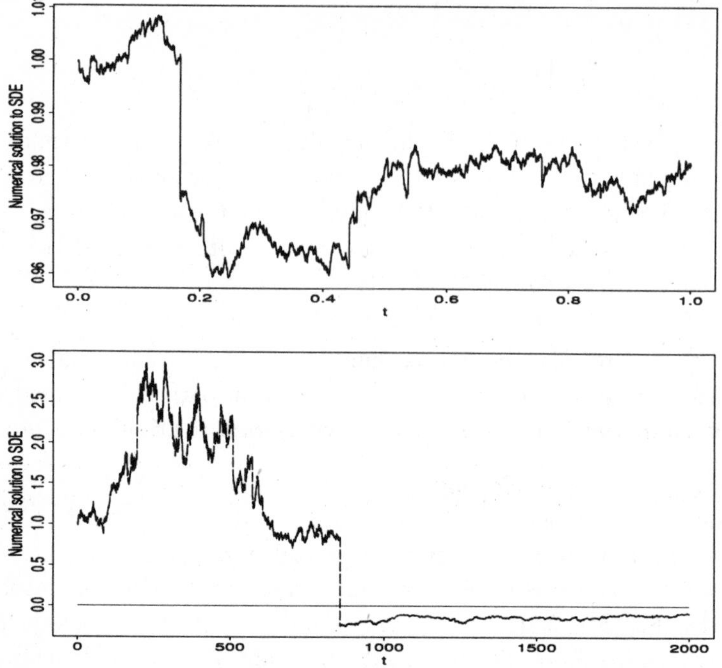

Figure 8.9.6 *Numerical solution to the SDE* $dX_t = 0.018X_{t-}d\xi_t$, $X_0 = 1$ *with driving* $s1.8s$ *process* (ξ_t). *The top figure clearly visualises the jumps in the sample paths of* (X_t) *which is a consequence of the jumps of the driving process. The bottom figure indicates that* X_t *may become negative when time goes by. This is in contrast to geometric Brownian motion; see Figure 8.9.4.*

$$dX_t = cX_t dt + \sigma X_{t-}\, d\xi_t\,, \quad \sigma > 0\,, c \in \mathbb{R}, \quad t \in [0,T]\,, \tag{8.152}$$

has solution

$$\begin{aligned} X_t &= X_0 e^{ct} \prod_{0 \le s \le t} (1 + \sigma \Delta\xi_s) \qquad (8.153) \\ &= X_0 e^{ct} \lim_{\delta \downarrow 0} \prod_{0 \le s \le t, |\Delta\xi_s| > \delta} (1 + \sigma\Delta\xi_s)\,, \quad t \in [0,T]\,. \end{aligned}$$

Here $\Delta\xi_s = \xi_s - \xi_{s-}$ for a cadlag version of the process (ξ_s). Almost all sample paths of (ξ_s) have countably many jumps so that the product in (8.153) is meaningful. The solution (8.153) can be defined as a limit in probability; see Protter [514]. But it can also be interpreted pathwise as a so-called product integral; see Dudley and Norvaiša [188]. This implies in particular that

$$X_t = X_0 e^{ct} \lim_{n\to\infty} \prod_{i=1}^{n} \left(1+\sigma(\xi_{iT/n} - \xi_{(i-1)T/n})\right) .$$

This formula is convenient for numerical simulations of the solution to the SDE (8.152). Relation (8.153) and Figure 8.9.6 make clear that, in comparison with the Brownian case, solutions to SDE driven by α-stable motions are of a completely different nature. In particular, since $\Delta\xi_s$ can be both negative or positive, the "price process" in (8.153) can also assume negative values which may not be very desirable. It is proved in Mikosch and Norvaiša [463] that the process (X_t) in (8.153) with $X_0 \neq 0$ a.s. changes its sign infinitely often. The instants of time when the change happens form the points of a Poisson process. □

In what follows we give a simple characterisation of self-similar processes via stationarity. For that reason, recall the definition of a strictly stationary process from Section 4.4 or Appendix A2.1. The following result is due to Lamperti [413].

Theorem 8.9.7 (Relation of H-self-similarity to stationarity)
If $(X_t)_{t>0}$ is H-ss then

$$Y_t = e^{-Ht} X_{e^t} , \quad t \in \mathbb{R},$$

is stationary. Conversely, if $(Y_t)_{t\in\mathbb{R}}$ is stationary, then

$$X_t = t^H Y_{\ln t} , \quad t > 0 ,$$

is H-ss.

Proof. Let (X_t) be H-ss. Then any linear combination of the Y_{t_i+h} satisfies

$$\begin{aligned} \sum_{i=1}^{m} a_i Y_{t_i+h} &= \sum_{i=1}^{m} a_i e^{-t_i H} e^{-Hh} X_{\exp\{t_i+h\}} \\ &\stackrel{d}{=} \sum_{i=1}^{m} a_i e^{-t_i H} X_{\exp\{t_i\}} \\ &= \sum_{i=1}^{m} a_i Y_{t_i} \end{aligned}$$

for any real a_i, t_i and h. Thus, by employing the Cramér–Wold device (see Appendix A2.5), we find that the finite-dimensional distributions of (Y_t) are invariant under shifts of time, hence (Y_t) must be stationary.
Now if $(Y_t)_{t\in\mathbb{R}}$ is stationary then, for all positive a and t_i,

$$\begin{aligned}
\sum_{i=1}^m a_i X_{at_i} &= \sum_{i=1}^m a_i (at_i)^H Y_{\ln(at_i)} \\
&= a^H \sum_{i=1}^m a_i t_i^H Y_{\ln a + \ln t_i} \\
&\stackrel{d}{=} a^H \sum_{i=1}^m a_i t_i^H Y_{\ln t_i} \\
&= a^H \sum_{i=1}^m a_i X_{t_i} \,,
\end{aligned}$$

and an application of the Cramér–Wold device completes the proof that $(X_t)_{t>0}$ is H–ss. □

This theorem makes clear that self–similarity is very closely related to stationarity; a logarithmic time transform translates shift invariance of the stationary process into scale invariance of the self–similar process.

Example 8.9.8 (Ornstein–Uhlenbeck process)
From Example 8.9.3 we know that Brownian motion is 1/2–ss. From Theorem 8.9.7 we conclude that

$$Y_t = e^{-0.5t} B_{\exp\{t\}} \,, \quad t \in \mathbb{R} \,,$$

is a stationary process. It is clearly a Gaussian process, as a scaled and time–transformed Brownian motion. A check for stationarity in the Gaussian case requires only the constancy of the mean value function $m(t) = EY_t$ and shift invariance of the covariance function $C(s,t) = E[(Y_s - m(s))(Y_t - m(t))]$. Indeed,

$$\begin{aligned}
m(t) &= 0 \,, \\
C(s,t) &= e^{-0.5(t+s)} \operatorname{cov}\left(B_{\exp\{s\}}, B_{\exp\{t\}}\right) \\
&= e^{-0.5(t+s)} \min\left(e^s, e^t\right) \\
&= e^{-0.5|t-s|} \,.
\end{aligned}$$

The process $(Y_t)_{t\in\mathbb{R}}$ is one of the well–studied Gaussian processes. It is called an *Ornstein–Uhlenbeck process.* It is characterised as a mean–zero Gaussian process with covariance function C given above. Alternatively, an Ornstein–Uhlenbeck process is the solution to an Itô SDE with additive Brownian noise. Also notice that we can define an Ornstein–Uhlenbeck process via stochastic integrals; see Example 8.8.13.

Analogously, if (ξ_t) is $s\alpha s$ then

$$Y_t = e^{-t/\alpha} \xi_{\exp\{t\}}, \quad t \in \mathbb{R},$$

defines a stationary $1/\alpha$–ss process, an $s\alpha s$ Ornstein–Uhlenbeck process. Its representation via stochastic integrals was mentioned in Example 8.8.13. □

Brownian motion and α–stable motion are two examples of self–similar processes with completely different sample path behaviour. Though the sample paths of Brownian motion oscillate wildly, they are a.s. continuous, whereas almost every sample path of α–stable motion, for $\alpha < 2$, has countably many jumps. Moreover, H–ss processes with $H \in (0,1)$ do not have differentiable sample paths; the proof below is taken from Mandelbrot and van Ness [439].

Proposition 8.9.9 (Non–differentiability of self–similar processes.) *Suppose (X_t) is H–ss with $H \in (0,1)$. Then for every fixed t_0,*

$$\limsup_{t \downarrow t_0} \frac{|X_t - X_{t_0}|}{t - t_0} = \infty,$$

i.e. sample paths of H–ss processes are nowhere differentiable with probability 1.

Proof. Without loss of generality we choose $t_0 = 0$. Let (t_n) be a sequence such that $t_n \downarrow 0$. Then, by H–self–similarity,

$$\begin{aligned} P\left(\lim_{n\to\infty} \sup_{0\le s\le t_n} \left|\frac{X_s}{s}\right| > x\right) &= \lim_{n\to\infty} P\left(\sup_{0\le s\le t_n} \left|\frac{X_s}{s}\right| > x\right) \\ &\ge \limsup_{n\to\infty} P\left(\left|\frac{X_{t_n}}{t_n}\right| > x\right) \\ &= \limsup_{n\to\infty} P\left(t_n^{H-1}|X_1| > x\right) \\ &= 1, \quad x > 0. \end{aligned}$$

Hence, with probability 1, $\lim_{n\to\infty} X_{t_n}/t_n$ does not exist for any sequence $t_n \downarrow 0$. □

H–ss processes with stationary increments (H–*sssi* processes) are of particular interest. They satisfy the relation

$$(X_{t_1+h} - X_h, \ldots, X_{t_m+h} - X_h) \stackrel{d}{=} (X_{t_1}, \ldots, X_{t_m})$$

for all possible choices of $h, t_1, \ldots, t_m$. For example, Brownian motion and α–stable motion, by definition, have stationary (and independent) increments; see Definitions 2.4.1 and 2.4.7. The class of the Gaussian H–sssi processes is relatively small. For the following result see Samorodnitsky and Taqqu [565], Lemma 7.2.1.

Proposition 8.9.10 *Suppose $(X_t)_{t\in\mathbb{R}}$ is a finite variance H–sssi process. Then*

$$0 < H \leq 1\,, \qquad \operatorname{cov}(X_s, X_t) = \frac{1}{2}\left\{|s|^{2H} + |t|^{2H} - |s-t|^{2H}\right\} \operatorname{var}(X_1)\,.$$

Moreover,

$$\begin{aligned} EX_t &= 0 & &\text{if } 0 < H < 1\,, \\ X_t &= tX_1 \text{ a.s.} & &\text{if } H = 1\,. \end{aligned}$$

□

The finite–dimensional distributions of Gaussian H–sssi processes are uniquely determined by their mean–value and covariance functions. Hence every H–sssi Gaussian process has necessarily the mean value and covariance function as given by Proposition 8.9.10, and these functions determine the only possible Gaussian H–sssi processes. A Gaussian H–sssi process is called *fractional Brownian motion*; see also Example 8.8.16. A critical discussion on the use of fractional Brownian motion in mathematical finance is given by Rogers [545]. In particular, if $H = 1/2$ and $\operatorname{var}(X_1) = 1$ then we obtain a standard Brownian motion.

There exist non–Gaussian finite variance H–sssi processes; see Taqqu [612, 614] or Major [434]. Among the infinite variance H–sssi processes are the α–stable ones with $\alpha < 2$ of particular interest; see also Example 8.8.16. In contrast to the Gaussian case, the property of H–sssi does not typically determine the finite–dimensional distributions.

Notes and Comments

Self–similarity as given by Definition 8.9.1 is due to Mandelbrot; see for instance Mandelbrot and van Ness [439]; Lamperti [413] uses a slightly more general notion. An extensive and mathematically rigorous discussion (which is more the exeption than the rule) has been provided by Samorodnitsky and Taqqu [565]; see also Taqqu [613, 614]. A more philosophical treatment of the topic can be found in many articles by Mandelbrot, we refer to his monograph [437]. There exist many entertaining and educational books on self–similarity and related topics; see for instance Schroeder [570].

Appendix

In the following we will provide some basic tools which are used throughout the book. All rvs and stochastic processes are assumed to be defined on a common probability space $[\Omega, \mathcal{F}, P]$.

We commence with elementary results on the convergence of rvs and of probability distributions.

A1 Modes of Convergence

The following theory can be found for instance in Billingsley [69], Feller [235], Karr [372] or Loève [427].

We introduce the main modes of convergence for a sequence of rvs $A, A_1, A_2, \ldots$. The corresponding definitions for processes can be derived by the subsequence principle. This means, for example, the following: let $(A_t)_{t\geq 0}$ be a real–valued stochastic process. Then $A_t \to A$ (in some mode of convergence) as $t \to \infty$ means that $A_{t_n} \to A$ (in the same mode) as $n \to \infty$ for every sequence $(t_n)_{n\geq 1}$ of real numbers such that $t_n \uparrow \infty$.

A1.1 Convergence in Distribution

Definition A1.1 *We say that* (A_n) converges in distribution *or* converges weakly to the rv A ($A_n \stackrel{d}{\to} A$) *if for all bounded, continuous functions* f *the relation*

$$Ef(A_n) \to Ef(A), \quad n \to \infty,$$

holds. □

Although weak convergence is convergence of the underlying probability measures we prefer the notation $A_n \xrightarrow{d} A$ instead of symbols for corresponding measures. However, for convenience we sometimes write $A_n \xrightarrow{d} F_A$ where F_A is the distribution or probability measure of A. We use the same symbol both for the distribution and for the df of a rv.

Weak convergence can be described by the dfs F_{A_n} and F_A of A_n and A, respectively: $A_n \xrightarrow{d} A$ holds if and only if for all continuity points y of the df F_A the relation

$$F_{A_n}(y) \to F_A(y), \quad n \to \infty, \tag{A.1}$$

is satisfied. Moreover, if F_A is continuous then (A.1) can even be strenghtened to uniform convergence:

$$\sup_x |F_{A_n}(x) - F_A(x)| \to 0, \quad n \to \infty.$$

Weak convergence is metrizable, i.e. there exists a metric $\rho(G, H)$ which is defined on the space of all dfs or distributions on $[\Omega, \mathcal{F}]$ such that $A_n \xrightarrow{d} A$ if and only if $\rho(F_{A_n}, F_A) \to 0$. A well-known metric for this purpose is the Lévy metric:

$$\rho_L(F, G) = \inf\{\varepsilon > 0 : \forall x \in \mathbb{R}, \quad F(x - \varepsilon) - \varepsilon \le G(x) \le F(x + \varepsilon) + \varepsilon\}.$$

A1.2 Convergence in Probability

Definition A1.2 *We say that* (A_n) converges in probability to the rv A $(A_n \xrightarrow{P} A)$ *if for all positive* ϵ *the relation*

$$P(|A_n - A| > \epsilon) \to 0, \quad n \to \infty,$$

holds. □

Convergence in probability implies convergence in distribution. The converse is true if and only if $A = a$ a.s. for some constant a.

The relation $A_n \xrightarrow{P} \infty$ has to be interpreted as $1/A_n \xrightarrow{P} 0$.

Convergence in probability is metrizable. An appropriate metric is given by

$$\rho(X, Y) = E\frac{|X - Y|}{1 + |X - Y|}.$$

Hence $A_n \xrightarrow{P} A$ if and only if

$$E\frac{|A_n - A|}{1 + |A_n - A|} \to 0, \quad n \to \infty.$$

In a metric space S the convergence $a_n \to a$ for elements $a, a_1, a_2, \ldots$ is equivalent to the *subsequence principle*: every subsequence (a_{n_k}) contains a subsequence $(a_{n_{k_j}})$ which converges to a. Hence $A_n \stackrel{P}{\to} A$ if and only if every subsequence (A_{n_k}) contains a subsequence $(A_{n_{k_j}})$ which converges in probability to A.

A1.3 Almost Sure Convergence

Definition A1.3 *We say that* (A_n) converges almost surely (a.s.) *or* with probability 1 to the rv A ($A_n \stackrel{\text{a.s.}}{\to} A$) *if for P–almost all* $\omega \in \Omega$ *the relation*

$$A_n(\omega) \to A(\omega), \quad n \to \infty,$$

holds. □

This means that

$$P(A_n \to A) = P(\{\omega : A_n(\omega) \to A(\omega)\}) = 1.$$

Convergence with probability 1 is equivalent to the relation

$$\sup_{k \geq n} |A_k - A| \stackrel{P}{\to} 0.$$

Hence convergence with probability 1 implies convergence in probability, hence convergence in distribution.

The relation $A_n \stackrel{\text{a.s.}}{\to} \infty$ has to be interpreted as $1/A_n \stackrel{\text{a.s.}}{\to} 0$.

There exist sequences of rvs (A_n) such that $A_n \stackrel{P}{\to} A$ but not $A_n \stackrel{\text{a.s.}}{\to} A$. For example, the necessary and sufficient conditions for the WLLN and the SLLN are different; see Section 2.1. However, if $A_n \stackrel{P}{\to} A$ then $A_{n_k} \stackrel{\text{a.s.}}{\to} A$ for a subsequence (n_k).

Recalling the *subsequence principle for convergence in probability* this means that $A_n \stackrel{P}{\to} A$ if and only if every subsequence (A_{n_k}) contains a subsequence $(A_{n_{k_j}})$ such that $A_{n_{k_j}} \stackrel{\text{a.s.}}{\to} A$. Hence *a.s. convergence is not metrizable.*

A1.4 L^p–Convergence

Definition A1.4 *Let* $p > 0$. *We say that* (A_n) converges in L^p *or* in pth mean to A ($A_n \stackrel{L^p}{\to} A$) *if* $E|A_n|^p < \infty$ *and* $E|A|^p < \infty$ *and*

$$E|A_n - A|^p \to 0, \quad n \to \infty.$$ □

By Markov's inequality, $P(|A_n - A| > \epsilon) \leq \epsilon^{-p} E|A_n - A|^p$ for positive p and ϵ. Thus $A_n \stackrel{L^p}{\to} A$ implies that $A_n \stackrel{P}{\to} A$. The converse is in general not true.

Convergence in L^p is metrizable by the metric

$$\rho(X, Y) = (E|X - Y|^p)^{1/\max(1,p)}.$$

A1.5 Convergence to Types

For two rvs X and Y (and, more generally, for two random elements X, Y assuming values in a measurable space) we write

$$X \stackrel{d}{=} Y$$

if X, Y have the same distribution.

We say that X and Y (and the corresponding distributions and dfs) *belong to the same type* or *are of the same type* if there exist constants $a \in \mathbb{R}$ and $b > 0$ such that

$$X \stackrel{d}{=} bY + a\,.$$

The following result is a particularly important tool for weak convergence. It tells us that the limit law of a sequence of rvs is uniquely determined up to changes of location and scale. For a proof see for instance Gnedenko and Kolmogorov [267], Petrov [495, 496] or Resnick [530].

Theorem A1.5 (Convergence to types theorem)
Let $A, B, A_1, A_2, \ldots$ be rvs and $b_n > 0$, $\beta_n > 0$ and a_n, $\alpha_n \in \mathbb{R}$ be constants. Suppose that

$$b_n^{-1}\left(A_n - a_n\right) \stackrel{d}{\to} A\,.$$

Then the relation

$$\beta_n^{-1}\left(A_n - \alpha_n\right) \stackrel{d}{\to} B \qquad \text{(A.2)}$$

holds if and only if

$$\lim_{n\to\infty} b_n/\beta_n = b \in [0,\infty)\,, \qquad \lim_{n\to\infty} \left(a_n - \alpha_n\right)/\beta_n = a \in \mathbb{R}\,. \qquad \text{(A.3)}$$

If (A.2) *holds then $B \stackrel{d}{=} bA + a$ and a, b are the unique constants for which this holds.*

When (A.2) *holds, A is non–degenerate if and only if $b > 0$, and then A and B belong to the same type.* □

It is immediate from (A.3) that the constants a_n and b_n are uniquely determined only up to the asymptotic relation (A.3).

A1.6 Convergence of Generalised Inverse Functions

For a non–decreasing function h on $\mathbb{R}$ we define the *generalised inverse of h* as

$$h^{\leftarrow}(q) = \inf\{x \in \mathbb{R} : h(x) \geq q\}\,.$$

We use the convention that $\inf \emptyset = \infty$. Then $h^{\leftarrow}$ is left–continuous.

Proposition A1.6 (Properties of a generalised inverse function)

If h is right–continuous, then the following properties hold.

(a) $h(x) \geq q \iff h^{\leftarrow}(q) \leq x$.
(b) $h(x) < q \iff h^{\leftarrow}(q) > x$.
(c) $h(x_1) < q \leq h(x_2) \iff x_1 < h^{\leftarrow}(q) \leq x_2$.
(d) $h(h^{\leftarrow}(q)) \geq q$ *for all* $t \in [0,1]$, *with equality for* h *continuous.*
(e) $h^{\leftarrow}(h(x)) \leq x$ *for all* $x \in \mathbb{R}$, *with equality for* h *increasing.*
(f) *h is continuous* $\iff$ *$h^{\leftarrow}$ is increasing.*
(g) *h is increasing* $\iff$ *$h^{\leftarrow}$ is continuous.*
(h) *If X is a rv with df h, then* $P(h^{\leftarrow}(h(X)) \neq X) = 0$. □

Proposition A1.7 (Convergence of generalised inverse functions)
Let $h, h_1, h_2, \ldots$ be non–decreasing functions such that $\lim_{n\to\infty} h_n(x) = h(x)$ *for every continuity point of h. Then* $\lim_{n\to\infty} h_n^{\leftarrow}(y) = h^{\leftarrow}(y)$ *for every continuity point y of $h^{\leftarrow}$.* □

Proofs of these results and more theory on generalised inverse functions can be found in Resnick [530], Section 0.2.

A2 Weak Convergence in Metric Spaces

The following theory on weak convergence in metric spaces can be found in Billingsley [69] or Pollard [504]. More details about stochastic processes are given in any textbook on the topic; see for instance Gikhman and Skorokhod [262] or Resnick [531].

We deal with weak convergence in general metric spaces. For applications we are mainly interested in four particular types of spaces: $\mathbb{R}^d$, $\mathbb{C}[0,1]$, $\mathbb{D}[0,1]$, $\mathbb{D}(0,\infty)$ and M_p. The theory for the spaces $\mathbb{C}[a,b]$, $\mathbb{D}[a,b]$ for $a < b$ is completely analogous to $\mathbb{C}[0,1]$, $\mathbb{D}[0,1]$ and therefore omitted. The space $\mathbb{D}(0,\infty)$ must be treated with care; see Pollard [504] and Lindvall [423]. The spaces $\mathbb{C}$ and $\mathbb{D}$ are appropriate in order to deal with weak convergence of stochastic processes. The space M_p is needed to define weak convergence of point processes.

A2.1 Preliminaries about Stochastic Processes

Recall that a *stochastic process* $(Y_t)_{t\in T}$ is a family of rvs with indices taken from the set T. For our purposes, T is usually a finite or infinite continuous interval on the real axis or a discrete set such as $\mathbb{Z}$ or $\mathbb{N}$. Thus a stochastic process $(Y_t(\omega))$ is a function of two variables: of the "time" $t \in T$ and the *random outcome* ω. A stochastic process is commonly supposed to be jointly measurable as a function of ω and t. For fixed ω, $(Y_t(\omega))_{t\in T}$ is called a *realisation*, a *trajectory* or a *sample path* of the process. For fixed t, $Y_t(\omega)$ is just a random variable. For any finite set $\{t_1, \ldots, t_d\} \subset T$, the probability measure of the random vector

$$(Y_{t_1}, \ldots, Y_{t_d})$$

is called a *finite–dimensional distribution* of the process. The totality of the finite–dimensional distributions of a process determines under general conditions (as given by *Kolmogorov's consistency theorem*; see for instance Billingsley [69], Appendix II) the *distribution of the process,* i.e. the probabilities

$$P\left(\{\omega : (Y_t(\omega))_{t\in T} \in B\}\right)$$

for appropriate sets B of functions defined on T. In most cases of interest treated in this book, Y assumes values in a metric function space, and B belongs to the σ–algebra generated by the open sets. This is the *σ–algebra of the Borel sets.*

A stochastic process is called *Gaussian* if all its finite–dimensional distributions are *multivariate normal.* The distribution of a d–dimensional non–degenerate multivariate normal vector $\mathbf{X}$ with covariance matrix Σ and mean μ is given by its *density*; see for instance Tong [626],

$$\left((2\pi)^d \det \Sigma\right)^{-1/2} \exp\left\{-\frac{1}{2}(\mathbf{x}-\mu)^T \Sigma^{-1} (\mathbf{x}-\mu)\right\}, \quad \mathbf{x} \in \mathbb{R}^d .$$

The distribution of a Gaussian stochastic process is determined by its (normal) finite–dimensional distributions. Hence it is determined by its mean–value and covariance function. In particular, Brownian motion (B_t) as introduced in Section 2.4 is uniquely determined via its covariance function $\operatorname{cov}(B_s, B_t) = \min(s,t)$ and by the fact that $EB_t = 0$ for all t.

A stochastic process is called *α–stable* for some $\alpha \in (0,2]$ if all its finite–dimensional distributions are α–stable. The notions of a Gaussian and of a 2–stable process are identical. For a definition of the finite–dimensional distributions we restrict ourselves to a d–dimensional *symmetric α–stable* ($s\alpha s$) process (Y_t) (see Definition 8.8.1): the random vector $\mathbf{Y}_d = (Y_{t_1}, \ldots, Y_{t_d})$ is $s\alpha s$ (i.e. a finite–dimensional distribution of the $s\alpha s$ process (Y_t)) for some $\alpha < 2$ if it has chf

$$E\exp\{i\,(\mathbf{x}, \mathbf{Y}_d)\} = \exp\left\{-\int_{S^{d-1}} |\,(\mathbf{x}, \mathbf{y})\,|^\alpha \, dm_s(\mathbf{y})\right\}, \quad \mathbf{x} \in \mathbb{R}^d ,$$

where S^{d-1} denotes the unit sphere in $\mathbb{R}^d$ (with respect to Euclidean metric) and m_s is a symmetric measure on the Borel sets of S^{d-1}. For $\alpha < 2$ the measure m_s uniquely determines the distribution of $\mathbf{Y}_d$. We refer to Section 8.8 for more information on stable processes.

A process $(Y_t)_{t\in T}$, $T = \mathbb{Z}$ or $T = \mathbb{R}$, is *stationary in the wide sense* or simply *stationary* if its mean value function is a constant and if the covariance function $\operatorname{cov}(Y_s, Y_t)$ is only a function of $|t-s|$. It is called *strictly stationary* if its finite–dimensional distributions are invariant under shifts of time, i.e. the finite–dimensional distributions for any instants of time $\{t_1, \dots, t_d\} \in T$ are the same as for $\{t_1 + h, \dots, t_d + h\} \in T$ for any h.

A2.2 The Spaces $\mathbb{C}\,[0,1]$ and $\mathbb{D}\,[0,1]$

The spaces $\mathbb{C}\,[0,1]$ and $\mathbb{D}\,[0,1]$ are appropriate function spaces for weak convergence of stochastic processes. We commence with the space $\mathbb{C}\,[0,1]$ of continuous, real–valued functions on $[0,1]$. It is always supposed that $\mathbb{C}\,[0,1]$ is equipped with the *supremum norm* (*sup–norm*): for $x \in \mathbb{C}\,[0,1]$,

$$\|x\| = \sup_{0\le t\le 1} |x(t)|\,.$$

It generates the *uniform metric* on $\mathbb{C}\,[0,1]$. We also assume that $\mathbb{C}\,[0,1]$ is endowed with the σ–algebra of the *Borel sets* which is generated by the open sets in $\mathbb{C}\,[0,1]$.

Example A2.1 Let $X_1, \dots, X_n$ be iid rvs with mean μ and (finite) variance $\sigma^2 > 0$. Set

$$S_0 = 0, \quad S_k = X_1 + \cdots + X_k\,, \quad k = 1, \dots, n\,. \tag{A.4}$$

Then the process $(S_n(t))_{0\le t\le 1}$ defined by

$$S_n(t) = \begin{cases} \dfrac{1}{\sigma\sqrt{n}}\,(S_k - \mu\, k) & \text{if} \quad t = k/n\,, k = 0, \dots, n\,, \\ \text{linearly interpolated} & \text{otherwise,} \end{cases}$$

has continuous sample paths. It is a process which is fundamental for the FCLT (see Section 2.4) since $S_n(\cdot)$ converges weakly to Brownian motion. The sample paths of the limit process are a.s. continuous, i.e. Brownian motion assumes values in $\mathbb{C}\,[0,1]$. □

The space $\mathbb{D}\,[0,1]$ consists of the *cadlag* (continue à droite, limites à gauche) functions on $[0,1]$, i.e. all real–valued functions x on $[0,1]$ such that

(*a*) $\lim_{t\uparrow t_0} x(t)$ exists for every $t_0 \in (0,1]$, i.e. x has limits from the left.

(b) $\lim_{t \downarrow t_0} x(t) = x(t_0)$ for every $t_0 \in [0,1)$, i.e. x is continuous from the right.

In particular, $\mathbb{C}[0,1] \subset \mathbb{D}[0,1]$.

Example A2.2 Let $X_1, \ldots, X_n$ be iid rvs with mean μ and (finite) variance $\sigma^2 > 0$. We define $S_0, \ldots, S_n$ as in (A.4). Then the process

$$\widetilde{S}_n(t) = \frac{1}{\sigma\sqrt{n}} \left(S_{[nt]} - \mu\,[nt]\right), \quad 0 \le t \le 1,$$

has cadlag sample paths on $[0,1]$. It is a fundamental process for the FCLT (see Section 2.4); it converges weakly to Brownian motion on $[0,1]$. In contrast to $\widetilde{S}_n(\cdot)$, the limit process has continuous sample paths with probability 1.□

The space $\mathbb{D}[0,1]$ can be equipped with different metrics in order to metrize weak convergence of stochastic processes with cadlag sample paths. Skorokhod [584] introduced some of these metrics. Therefore the space $\mathbb{D}$ is also referred to as *Skorokhod space.*

Let Λ be the class of functions $h : [0,1] \to [0,1]$ that are continuous and increasing and such that $h(0) = 0$ and $h(1) = 1$. Note that these conditions imply that h is a bicontinuous bijection from $[0,1]$ onto $[0,1]$.

Definition A2.3 (J_1–convergence)
The functions $x_n \in \mathbb{D}[0,1]$ converge to $x \in \mathbb{D}[0,1]$ in the J_1–sense *if for every n there exists $h_n \in \Lambda$ such that*

$$\sup_{0 \le t \le 1} |h_n(t) - t| \to 0, \quad n \to \infty, \tag{A.5}$$

and

$$\sup_{0 \le t \le 1} |x_n(t) - x(h_n(t))| \to 0, \quad n \to \infty. \tag{A.6}$$

□

Note that conditions (A.5) and (A.6) reduce to uniform convergence when we can choose $h_n(t) = t$ for every n.

Condition (A.5) means that we are allowed to make a transformation h_n of "real time" t on the interval $[0,1]$ but these transformations are asymptotically "negligible". Condition (A.6) measures the "distance in space" between x in the "transformed time" $h_n(t)$ and x_n in the "real time" t. Thus the notion of uniform convergence is weakened by allowing for small perturbations at the "time scale".

The space $\mathbb{D}[0,1]$ can be equipped with a metric $d(x,y)$ which metrizes J_1–convergence. Then $\mathbb{D}[0,1]$ is a separable metric space. For example, we can choose the *Skorokhod metric*

$$d_{0,1}(x,y) = \inf_{h\in\Lambda} \max\left\{ \sup_{0\le t\le 1} |h(t)-t| \,,\, \sup_{0\le t\le 1} |x(t)-y(h(t))| \right\} . \tag{A.7}$$

However, this metric does not ensure completeness, i.e. Cauchy sequences do not necessarily converge. Fortunately, there exists an equivalent metric d' on $\mathbb{D}\,[0,1]$ (i.e. there exists a bicontinuous bijection between $(\mathbb{D}\,[0,1],d)$ and $(\mathbb{D}\,[0,1],d'))$ which makes it into a complete metric space. In particular, any of these two metrics generates the same open sets such that the σ–algebras of the Borel sets (which are generated by the open sets) are identical. In this book we refer to d' as the J_1*–metric.*

A2.3 The Skorokhod Space $\mathbb{D}\,(0,\infty)$

For practical purposes (for instance when dealing with weak convergence in extreme value theory) it is important to work on the whole positive real line and not on a finite interval. This calls for the definition of the space $\mathbb{D}\,(0,\infty)$ of cadlag functions on $(0,\infty)$ (cadlag is defined in the same way as in Section A2.2). Convergence in $\mathbb{D}\,(0,\infty)$ is not an easy matter but it can fortunately be relaxed to convergence in the spaces $\mathbb{D}\,[a,b]$ equipped with the Skorokod metrics $d_{a,b}$ (see (A.7) for the definition of $d_{0,1}$). We refer to Lindval [423], Whitt [640], Pollard [504] or Resnick [530] for a detailed treatment.

We construct a metric $d(x,y)$ on $\mathbb{D}\,(0,\infty)$ such that

$$d\,(x_n,x_0) \to 0$$

for $x_0,x_1,\ldots \in \mathbb{D}\,(0,\infty)$ if and only if for any continuity points $0<a<b<\infty$ of x_0

$$d_{a,b}\,(r_{a,b}x_n, r_{a,b}x_0) \to 0\,,$$

where $r_{a,b}x$ denotes the restriction of the cadlag function x on $(0,\infty)$ to the interval $[a,b]$. Such a metric is given by

$$d(x,y) = \int_{s=0}^{1}\int_{t=1}^{\infty} e^{-t}\left(d_{s,t}\left(r_{s,t}x, r_{s,t}y\right)\wedge 1\right)\,dt\,ds\,. \tag{A.8}$$

Equipped with this metric, $\mathbb{D}\,(0,\infty)$ is a complete metric space.

A2.4 Weak Convergence

In Section A1.1 we already introduced the notion of weak convergence for sequences of ordinary rvs. We will slightly modify that definition to define weak convergence in a general metric space.

Definition A2.4 (Weak convergence in a metric space)
Let K be a space with metric ρ and suppose that K is endowed with the σ–algebra generated by the open subsets with respect to ρ. Moreover, let $A, A_1, A_2, \ldots$ be random elements assuming values in K. The sequence (A_n) converges weakly to A $(A_n \xrightarrow{d} A)$ *if for every bounded, continuous, real–valued function f on K the relation*

$$Ef(A_n) \to Ef(A), \quad n \to \infty, \tag{A.9}$$

holds. □

(*a*) If we specify $K = \mathbb{R}$ then we have weak convergence as introduced in Section A1.1.
(*b*) If we set $K = \mathbb{C}[0,1]$ and equip K with the sup–norm then we obtain weak convergence of stochastic processes with a.s. continuous sample paths on $[0,1]$.
(*c*) If we define $K = \mathbb{D}[0,1]$ for K equipped with the J_1–metric then we obtain weak convergence of stochastic processes with cadlag sample paths on $[0,1]$.
(*d*) If we specify $K = \mathbb{D}(0,\infty)$ for K equipped with the metric $d(x,y)$ in (A.8) we obtain convergence of stochastic processes with cadlag sample paths on $(0,\infty)$. By construction of the metric $d(x,y)$ one can show that weak convergence in $\mathbb{D}(0,\infty)$ is equivalent to weak convergence of the restrictions of the stochastic processes to any compact interval $[a,b]$, i.e. to convergence in $\mathbb{D}[a,b]$ for any $a < b$.

By (A.9), the crucial difference between weak convergence in $\mathbb{C}[0,1]$ and in $\mathbb{D}[0,1]$ is determined by the bounded, real–valued functions f on $[0,1]$ that are continuous with respect to the uniform or with respect to the J_1–metric, respectively. Note that these are different classes of functions!

We mention an elegant version of weak convergence in $\mathbb{D}[0,1]$ under the assumptions that the limiting process has continuous sample paths with probability 1:

Theorem A2.5 *Suppose that the stochastic processes $Y_1, Y_2, \ldots$ on $[0,1]$ have cadlag sample paths, i.e. they assume values in $\mathbb{D}[0,1]$, and that the stochastic process Y has continuous sample paths on $[0,1]$ with probability 1, i.e. it assumes values in $\mathbb{C}[0,1]$. Assume that $Y_n \xrightarrow{d} Y$ in $\mathbb{D}[0,1]$ equipped with the J_1–metric and with the σ–algebra generated by the open sets. Then $Y_n \xrightarrow{d} Y$ in $\mathbb{D}[0,1]$ equipped with the sup–norm and with the σ–algebra generated by the open balls.* □

This theorem applies, for example, if the limit process is Brownian motion on $[0,1]$. This theorem suggests that the σ–algebra of the Borel sets with respect to the J_1–metric and the σ–algebra generated by the open balls with respect to the uniform metric are different for $\mathbb{D}$. This is unfortunately true, and one needs then a slight modification of the definiton of weak convergence (A.9) as well. The interested reader is referred to Chapter V in Pollard [504].

A2.5 The Continuous Mapping Theorem

The importance of weak convergence in metric spaces is increased by the so–called *continuous mapping theorem.* It means, roughly speaking, that weak convergence of a sequence of random elements in a metric space is preserved under continuous mappings:

Theorem A2.6 (Continuous mapping theorem)
Let h be a mapping from the metric space K to the metric space K' (both equipped with the corresponding σ–algebras of Borel sets generated by the open subsets). Suppose that $A_n \xrightarrow{d} A$ in K and denote by P_A the distribution of A. Then $h(A_n) \xrightarrow{d} h(A)$ in the space K', provided that the set of discontinuities of h has P_A–measure zero. □

Example A2.7 (Slutsky arguments)
Let (A_n), (B_n) be two sequences of rvs such that $(A_n, B_n) \xrightarrow{d} (A, B)$. Then Theorem A2.6 ensures that $A_n + B_n \xrightarrow{d} A + B$, $A_n B_n \xrightarrow{d} AB$ etc. Moreover, if $A_n \xrightarrow{P} a$ for constant a and h is continuous at a then $h(A_n) \xrightarrow{P} h(a)$.
The continuous mapping theorem for rational functions is called *Slutsky's theorem.* It states, in particular, that, if $A_n \xrightarrow{d} A$ and $a_n \to a$, $b_n \to b$ for constants a, b, a_n, b_n, then $a_n A_n + b_n \xrightarrow{d} aA + b$, and if $A_n \xrightarrow{d} A$, $a_n \to a$ and $B_n \xrightarrow{P} 0$, then $a_n A_n + B_n \xrightarrow{d} aA$. □

Example A2.8 (Cramér–Wold device)
Let $(A_n^{(i)})$, $i = 1, \ldots, k$, be sequences of real–valued rvs such that

$$(A_n^{(1)}, \ldots, A_n^{(k)}) \xrightarrow{d} (A^{(1)}, \ldots, A^{(k)})$$

for rvs $A^{(1)}, \ldots, A^{(k)}$. Then the continuous mapping theorem implies that

$$c_1 A_n^{(1)} + \cdots + c_k A_n^{(k)} \xrightarrow{d} c_1 A^{(1)} + \cdots + c_k A^{(k)}$$

for any real numbers $c_1, \ldots, c_k$. The converse is also true as a consequence of the continuity theorem for chfs. This rule is also called the *Cramér–Wold device.* □

Example A2.9 The continuous mapping theorem has important consequences for the convergence of stochastic processes: suppose that $Y, Y_1, Y_2, \ldots$ are stochastic processes on $[0,1]$ which assume values in $\mathbb{C}[0,1]$ or $\mathbb{D}[0,1]$ equipped with the natural metrics. Then $Y_n \xrightarrow{d} Y$ implies *convergence of the finite–dimensional distributions,* i.e. for any d and any $0 \le t_1 < \cdots < t_d \le 1$,

$$(Y_n(t_1), \ldots, Y_n(t_d)) \xrightarrow{d} (Y(t_1), \ldots, Y(t_d)) ,$$

provided that

$$P(Y \text{ is continuous at } t_1, \ldots, t_d) = 1 .$$

In particular, weak convergence of the sum processes in Examples A2.1 and A2.2 to Brownian motion implies convergence of the corresponding finite–dimensional distributions to those of Brownian motion. We mention that *convergence of the finite–dimensional distributions is in general not sufficient for the weak convergence of stochastic processes.* Indeed, one still needs a *tighness* argument for the converging processes. This ensures that during the whole limiting process the probability mass of the stochastic processes does not disappear from "good" (i.e. compact) sets.
An application of the continuous mapping theorem with the continuous supremum and infimum functionals yields the following:

$$\sup_{0\le t\le 1} Y_n(t) \xrightarrow{d} \sup_{0\le t\le 1} Y(t) , \qquad \inf_{0\le t\le 1} Y_n(t) \xrightarrow{d} \inf_{0\le t\le 1} Y(t) .$$

Analogous relations hold with Y_n and Y replaced by $|Y_n|$ and $|Y|$.
If Y_n and Y are cadlag on $(0,\infty)$ then $\inf_{t>0} Y_n(t)$ does not in general converge weakly to $\inf_{t>0} Y(t)$, because the infimum functional on $(0,\infty)$ is not continuous. For example,

$$x_n(t) = \begin{cases} 0 & \text{if } t < n , \\ -1 & \text{if } t \ge n , \end{cases}$$

is an element of $\mathbb{D}(0,\infty)$ and $d(x_n, x_0) \to 0$ for $x_0 = 0$. However,

$$\inf_{t>0} x_n(t) = -1 \not\to \inf_{t>0} x_0(t) = 0 .$$

□

A2.6 Weak Convergence of Point Processes

Recall from Section 5.1.1 the definition of a point process with state space E. Throughout, E is a subset of a compactified finite–dimensional Euclidean space equipped with the corresponding Borel σ–algebra $\mathcal{E}$. A point process N assumes point measures as values. The space of all point measures on E is denoted by $M_p(E)$. It is equipped with the σ–algebra $\mathcal{M}_p(E)$ which consists of all sets of the form $\{m \in M_p(E) : m(A) \in B\}$ for $A \in \mathcal{E}$ and any Borel set $B \subset [0, \infty]$, i.e. it is the smallest σ–algebra making the maps $m \to m(A)$ measurable for all $A \in \mathcal{E}$.

We equip $M_p(E)$ with an appropriate (the *vague*) metric so that it becomes a complete metric space. This will then allow us to define weak convergence in $M_p(E)$ in the sense of Definition A2.4.

Definition A2.10 (Definition of vague convergence)
Let $\mu, \mu_1, \mu_2, \ldots$ be measures in $M_p(E)$. The sequence of measures (μ_n) converges vaguely *to the measure μ (we write $\mu_n \stackrel{v}{\to} \mu$) if*

$$\int_E g(x)\, \mu_n(dx) \to \int_E g(x)\, \mu(dx)$$

for all $g \in C_K^+(E)$, the set of all continuous, non–negative functions g with compact support. □

Recall that the real–valued function g on E has compact support if there exists a compact set $K \subset E$ such that $g(x) = 0$ on K^c, the complement of K.

Note that this kind of convergence is very similar to the notion of weak convergence of probability measures on metric spaces. However, if the μ_n are probability measures on $\mathcal{E}$ and $\mu_n \stackrel{v}{\to} \mu$ then it is not guaranteed that the limit measure μ is a probability measure on $\mathcal{E}$ or that there is weak convergence at all.

Example A2.11 Suppose $E = \mathbb{R}$. Choose $\mu_n = \varepsilon_n$ (ε_n is the Dirac measure concentrated at n) and $\mu = 0$, the null measure. We see immediately that (μ_n) does not converge weakly at all. On the other hand, for any $g \in C_K^+(\mathbb{R})$

$$\int_{\mathbb{R}} g(x)\, \mu_n(dx) = g(n) = 0 = \int_{\mathbb{R}} g(x)\, \mu(dx)$$

for sufficiently large n, i.e. $\mu_n \stackrel{v}{\to} \mu$. □

For practical purposes, the following criterion is of great value (see Proposition 3.12 in Resnick [530]). Recall that a set in a metric space is relatively compact if it has compact closure.

Proposition A2.12 (Criterion for vague convergence)
The following are equivalent:

(a) $\mu_n \stackrel{v}{\to} \mu$ *as* $n \to \infty$.
(b) *For every relatively compact set $B \in \mathcal{E}$ such that $\mu(\partial B) = 0$ the relation*

$$\lim_{n\to\infty} \mu_n(B) = \mu(B)$$

holds (∂B is the boundary of B). □

Remark. Proposition A2.12 is our main tool for checking vague convergence. In particular, if $E \subset \mathbb{R}$ then it suffices for $\mu_n \xrightarrow{v} \mu$ to show that $\mu_n(a,b] \to \mu(a,b]$ for all intervals $(a,b]$ with a,b not being atoms of μ. See also Remark 4 in Section 5.2. □

There exists a metric $d_v(\mu,\nu)$ which metrizes vague convergence in $M_p(E)$ and which makes $M_p(E)$ a complete, separable metric space. We avoid defining this metric because it is not essential for us.

Now one can define *weak convergence of point processes* in the sense of Definition A2.4, but it is then not clear what it actually means. In Section 5.2 $N_n \xrightarrow{d} N$ is defined as convergence of the finite–dimensional distributions for all P_N–stochastic continuity sets. This is an intuitive definition, and we want to understand weak convergence of point processes in this sense. In the case of point processes, weak convergence is indeed equivalent to convergence of the finite–dimensional distributions.

For a rigorous treatment of weak convergence of point processes we refer to Daley and Vere–Jones [153], Kallenberg [365], Matthes, Kerstan and Mecke [447] or Resnick [530].

A3 Regular Variation and Subexponentiality

A3.1 Basic Results on Regular Variation

Asymptotic estimates are omnipresent in insurance mathematics and mathematical finance. In many cases transforms (Laplace, Fourier, Mellin) play a crucial role. This opens the door to classical Abel–Tauber theory. Starting with the pioneering work by Karamata [367] and imported into probability theory mainly through Feller [235], the theory of regular variation has now obtained the status of "standard knowledge" for any probabilist or statistician. Below we summarise some of the main results, relevant for our applications. Everything we discuss, and indeed much more, is to be found in the encyclopaedic volume on the subject by Bingham, Goldie and Teugels [72].

Definition A3.1 (Regular variation in Karamata's sense)

(a) *A positive, Lebesgue measurable function L on $(0,\infty)$ is* slowly varying *at ∞ (we write $L \in \mathcal{R}_0$) if*

$$\lim_{x\to\infty} \frac{L(tx)}{L(x)} = 1\,, \quad t>0\,. \tag{A.10}$$

(b) *A positive, Lebesgue measurable function h on $(0,\infty)$ is* regularly varying *at ∞ of index $\alpha \in \mathbb{R}$ (we write $h \in \mathcal{R}_\alpha$) if*

$$\lim_{x\to\infty} \frac{h(tx)}{h(x)} = t^\alpha\,, \quad t > 0\,. \tag{A.11}$$

□

Remarks. 1) Under (a) and (b) above, we have defined regular variation at infinity, i.e. for $x \to \infty$. Similarly one can define regular variation at zero replacing $x \to \infty$ by $x \to 0$, or at any positive a. Indeed, regular variation of h at $a > 0$ is defined as regular variation at (infinity) for the function $h_a(x) = h(a - x^{-1})$. For example, in Theorem 3.3.12 the maximum domain of attraction of the Weibull distribution is characterised via regular variation of the tails at a finite point. In cases where the distinction is important, we shall speak about *regular variation* at a. Whenever from the context the meaning is clear, we shall just refer to *regular variation*.

2) Condition (A.11) can be relaxed in various ways, the most important specifying only that the limit exists and is positive, rather than that it has the functional form t^α. Indeed, if we suppose that in (A.11) the limit exists for all $t > 0$ and equals $\chi(t)$ say, then it immediately follows that $\chi(st) = \chi(s)\chi(t)$ and hence $\chi(t) = t^\alpha$ for some $\alpha \in \mathbb{R}$.

3) Typical examples of slowly varying functions are positive constants or functions converging to a positive constant, logarithms and iterated logarithms. For instance for all real α the functions

$$x^\alpha\,, x^\alpha \ln(1+x)\,,\ (x\ln(1+x))^\alpha\,,\ x^\alpha \ln(\ln(e+x))$$

are regularly varying at ∞ with index α. The following examples are not regularly varying

$$2 + \sin x\,, \quad e^{[\ln(1+x)]}\,,$$

where $[\cdot]$ stands for integer part. In Theorem A3.3 below we give a general representation of regularly varying functions. It is perhaps interesting to note that a slowly varying function L may exhibit infinite oscillation in that it can happen that

$$\liminf_{x\to\infty} L(x) = 0 \quad \text{and} \quad \limsup_{x\to\infty} L(x) = \infty\,.$$

An example is given by

$$L(x) = \exp\left\{(\ln(1+x))^{1/2}\cos\left((\ln(1+x))^{1/2}\right)\right\}\,.$$ □

An important result is the fact that convergence in (A.10) is *uniform on each compact subset of* $(0, \infty)$.

Theorem A3.2 (Uniform convergence theorem for regularly varying functions)
If $h \in \mathcal{R}_\alpha$ (in the case $\alpha > 0$, assuming h bounded on each interval $(0,x]$, $x > 0$), then for $0 < a \le b < \infty$,

$$\lim_{x\to\infty} \frac{h(tx)}{h(x)} = t^\alpha\,, \quad \text{uniformly in } t$$

(a) *on each $[a,b]$ if $\alpha = 0$,*
(b) *on each $(0,b]$ if $\alpha > 0$,*
(c) *on each $[a,\infty)$ if $\alpha < 0$.* □

A further important result concerns the representation of regularly varying functions.

Theorem A3.3 (Representation theorem for regularly varying functions)
If $h \in \mathcal{R}_\alpha$ for some $\alpha \in \mathbb{R}$, then

$$h(x) = c(x) \exp\left\{\int_z^x \frac{\delta(u)}{u}\,du\right\}, \quad x \ge z\,, \tag{A.12}$$

for some $z > 0$ where c and δ are measurable functions, $c(x) \to c_0 \in (0,\infty)$, $\delta(x) \to \alpha$ as $x \to \infty$. The converse implication also holds. □

An immediate consequence from (A.12) is

Corollary A3.4 *If $h \in \mathcal{R}_\alpha$ for some $\alpha \ne 0$, then as $x \to \infty$,*

$$h(x) \to \begin{cases} \infty & \text{if} \quad \alpha > 0\,, \\ 0 & \text{if} \quad \alpha < 0\,. \end{cases}$$

□

In applications the following question is of importance.

Suppose $h \in \mathcal{R}_\alpha$. Can one find a smooth function $h_1 \in \mathcal{R}_\alpha$ so that $h(x) \sim h_1(x)$ as $x \to \infty$?

In the representation (A.12) we have a certain flexibility in constructing the functions c and δ. By taking the function c for instance constant, we already have a (partial) positive answer to the above question. Much more can however be obtained as can be seen from the following result by Adamovič; see Bingham et al. [72], Proposition 1.3.4.

Proposition A3.5 (Smooth versions of slow variation)
Suppose $L \in \mathcal{R}_0$, then there exists $L_1 \in C^\infty$ (the space of infinitely differentiable functions) so that $L(x) \sim L_1(x)$ as $x \to \infty$. If L is eventually monotone, so is L_1. □

The following result of Karamata is often applicable. It essentially says that integrals of regularly varying functions are again regularly varying, or more precisely, one can take the slowly varying function out of the integral.

Theorem A3.6 (Karamata's theorem)
Let $L \in \mathcal{R}_0$ be locally bounded in $[x_0, \infty)$ for some $x_0 \geq 0$. Then

(a) for $\alpha > -1$,

$$\int_{x_0}^{x} t^{\alpha} L(t)\, dt \sim (\alpha+1)^{-1} x^{\alpha+1} L(x)\,, \quad x \to \infty\,,$$

(b) for $\alpha < -1$,

$$\int_{x}^{\infty} t^{\alpha} L(t)\, dt \sim -(\alpha+1)^{-1} x^{\alpha+1} L(x)\,, \quad x \to \infty\,.$$

□

Remarks. 4) The result remains true for $\alpha = -1$ in the sense that then

$$\frac{1}{L(x)} \int_{x_0}^{x} \frac{L(t)}{t} dt \to \infty\,, \quad x \to \infty\,,$$

and $\int_{x_0}^{x} (L(t)/t) dt \in \mathcal{R}_0$. If $\int_{x_0}^{\infty} (L(t)/t) dt < \infty$ then

$$\frac{1}{L(x)} \int_{x}^{\infty} \frac{L(t)}{t} dt \to \infty\,, \quad x \to \infty\,,$$

and $\int_{x}^{\infty} (L(t)/t) dt \in \mathcal{R}_0$.

5) The conclusions of Karamata's theorem can alternatively be formulated as follows. Supppose $h \in \mathcal{R}_\alpha$ for some $\alpha \in \mathbb{R}$ and h is locally bounded on $[x_0, \infty)$ for some $x_0 \geq 0$. Then

(a') for $\alpha > -1$,

$$\lim_{x\to\infty} \frac{\int_{x_0}^{x} h(t)\, dt}{x h(x)} = \frac{1}{\alpha+1}\,,$$

(b') for $\alpha < -1$,

$$\lim_{x\to\infty} \frac{\int_{x}^{\infty} h(t)\, dt}{x h(x)} = -\frac{1}{\alpha+1}\,.$$

Whenever $\alpha \neq -1$ and the limit relations in either (a') or (b') hold for some positive function h, locally bounded on some interval $[x_0, \infty)$, $x_0 \geq 0$, then $h \in \mathcal{R}_\alpha$. □

The following result is crucial for the differentiation of regularly varying functions.

Theorem A3.7 (Monotone density theorem)
Let $U(x) = \int_0^x u(y)\,dy$ (or $\int_x^\infty u(y)\,dy$) where u is ultimately monotone (i.e. u is monotone on (z,∞) for some $z > 0$). If

$$U(x) \sim cx^\alpha L(x)\,, \quad x \to \infty\,,$$

with $c \ge 0$, $\alpha \in \mathbb{R}$ and $L \in \mathcal{R}_0$, then

$$u(x) \sim c\alpha x^{\alpha-1} L(x)\,, \quad x \to \infty\,.$$

For $c = 0$ the above relations are interpreted as $U(x) = o(x^\alpha L(x))$ and $u(x) = o(x^{\alpha-1} L(x))$. □

For applications to probability theory conditions of the type $\overline{F} \in \mathcal{R}_{-\alpha}$ for $\alpha \ge 0$, where F is a df, are common; see for instance Chapters 2 and 3. Below we have summarised some of the results which are useful for our purposes. For proofs and further references see Bingham et al. [72].

Proposition A3.8 (Regular variation for tails of dfs)
Suppose F is a df with $F(x) < 1$ for all $x \ge 0$.

(a) *If the sequences (a_n) and (x_n) satisfy $a_n/a_{n+1} \to 1$, $x_n \to \infty$, and if for some real function g and all λ from a dense subset of $(0,\infty)$,*

$$\lim_{n\to\infty} a_n \overline{F}(\lambda x_n) = g(\lambda) \in (0,\infty)\,,$$

then $g(\lambda) = \lambda^{-\alpha}$ for some $\alpha \ge 0$ and $\overline{F}$ is regularly varying.

(b) *Suppose F is absolutely continuous with density f such that for some $\alpha > 0$, $\lim_{x\to\infty} x\, f(x)/\overline{F}(x) = \alpha$. Then $f \in \mathcal{R}_{-1-\alpha}$ and consequently $\overline{F} \in \mathcal{R}_{-\alpha}$.*

(c) *Suppose $f \in \mathcal{R}_{-1-\alpha}$ for some $\alpha > 0$. Then $\lim_{x\to\infty} x\, f(x)/\overline{F}(x) = \alpha$. The latter statement also holds if $\overline{F} \in \mathcal{R}_{-\alpha}$ for some $\alpha > 0$ and the density f is ultimately monotone.*

(d) *Suppose X is a non-negative rv with distribution tail $\overline{F} \in \mathcal{R}_{-\alpha}$ for some $\alpha > 0$. Then*

$$\begin{aligned} EX^\beta &< \infty \quad \text{if } \beta < \alpha\,,\\ EX^\beta &= \infty \quad \text{if } \beta > \alpha\,. \end{aligned}$$

(e) *Suppose $\overline{F} \in \mathcal{R}_{-\alpha}$ for some $\alpha > 0$, $\beta \ge \alpha$. Then*

$$\lim_{x\to\infty} \frac{x^\beta\, \overline{F}(x)}{\int_0^x y^\beta\, dF(y)} = \frac{\beta-\alpha}{\alpha}\,.$$

The converse also holds in the case that $\beta > \alpha$. If $\beta = \alpha$ one can only conclude that $\overline{F}(x) = o(x^{-\alpha} L(x))$ for some $L \in \mathcal{R}_0$.

(f) *The following are equivalent:*

(1) $\int_0^x y^2\,dF(y) \in \mathcal{R}_0$,

(2) $\overline{F}(x) = o\left(x^{-2}\int_0^x y^2\,dF(y)\right), \quad x \to \infty.$ □

Remark. 6) The statements (b), (c), (e) and (f) above are special cases of the general version of Karamata's theorem and the monotone density theorem. For more general formulations of (e) and (f) see Bingham et al. [72], p. 331. Relations (e) and (f) are important in the analysis of the domain of attraction of stable laws; see for instance Corollary 2.2.9. □

The applicability of regular variation is further enhanced by *Karamata's Tauberian theorem* for Laplace–Stieltjes transforms.

Theorem A3.9 (Karamata's Tauberian theorem)
Let U be a non-decreasing, right-continuous function defined on $[0,\infty)$. If $L \in \mathcal{R}_0$, $c \geq 0$, $\alpha \geq 0$, then the following are equivalent:

(a) $U(x) \sim c x^\alpha L(x)/\Gamma(1+\alpha)\,, \quad x \to \infty,$

(b) $\widehat{u}(s) = \int_0^\infty e^{-sx}\,dU(x) \sim c s^{-\alpha} L(1/s)\,, \quad s \downarrow 0.$

When $c = 0$, (a) *is to be interpreted as* $U(x) = o(x^\alpha L(x))$ *as* $x \to \infty$; *similarly for* (b). □

This is a remarkable result in that not only the power coefficient α is preserved after taking Laplace–Stieltjes transforms but even the slowly varying function L. From either (a) or (b) in the case $c > 0$, it follows that

(c) $U(x) \sim \widehat{u}(1/x)/\Gamma(1+\alpha)\,, \quad x \to \infty\,.$

A surprising result is that the converse (i.e. (c) implies (a) and (b)) also holds. This so–called *Mercerian theorem* is discussed in Bingham et al. [72], p. 274. Various extensions of the above result exist; see for instance Bingham et al. [72], Theorems 1.7.6 and 8.1.6.

Corollary A3.10 *Suppose F is a df with Laplace–Stieltjes transform $\widehat{f}$. For $0 \leq \alpha < 1$ and $L \in \mathcal{R}_0$, the following are equivalent:*

(a) $1 - \widehat{f}(s) \sim s^\alpha L(1/s)\,, \quad s \downarrow 0,$

(b) $\overline{F}(x) \sim (1/\Gamma(1-\alpha))x^{-\alpha} L(x)\,, \quad x \to \infty.$ □

For $\alpha \geq 1$ see Theorem 8.1.6 in Bingham et al. [72].

So far we have considered regular variation of positive functions on $(0,\infty)$. Various extensions exist including

- regular variation on $\mathbb{R}$ or $\mathbb{R}^k$,
- regularly varying sequences,
- rapid variation with index $+\infty$ and $-\infty$.

Of these possible generalisations we treat only those which are explicitly needed in the book.

Definition A3.11 (Rapid variation)
A positive, Lebesgue measurable function h on $(0,\infty)$ *is* rapidly varying with index $-\infty$ (*we write* $h \in \mathcal{R}_{-\infty}$) *if*

$$\lim_{x\to\infty} \frac{h(tx)}{h(x)} = \begin{cases} 0 & \text{if } t>1\,, \\ \infty & \text{if } 0<t<1\,. \end{cases}$$

□

An example of a function $h \in \mathcal{R}_{-\infty}$ is $h(x) = e^{-x}$. In the following theorem we have summarised some of the main properties of $\mathcal{R}_{-\infty}$.

Theorem A3.12 (Properties of functions of rapid variation)

(*a*) *Suppose* $h \in \mathcal{R}_{-\infty}$ *is non–increasing, then for some* $z > 0$ *and all* $\alpha \in \mathbb{R}$

$$\int_z^\infty t^\alpha\, h(t)\, dt < \infty\,,$$

and

$$\lim_{x\to\infty} \frac{x^{\alpha+1} h(x)}{\int_x^\infty t^\alpha\, h(t)\, dt} = \infty\,. \tag{A.13}$$

If for some $\alpha \in \mathbb{R}$, $\int_1^\infty t^\alpha h(t)\, dt < \infty$ *and* (A.13) *holds, then* $h \in \mathcal{R}_{-\infty}$.

(*b*) *If* $h \in \mathcal{R}_{-\infty}$, *then there exist functions c and* δ *such that* $c(x) \to c_0 \in (0,\infty)$, $\delta(x) \to -\infty$ *as* $x \to \infty$ *and for some* $z > 0$,

$$h(x) = c(x) \exp\left\{\int_z^x \frac{\delta(u)}{u}\, du\right\}, \quad x \ge z\,. \tag{A.14}$$

The converse also holds.

Proof. See de Haan [292], Theorems 1.3.1 and 1.2.2. □

Remarks. 7) Suppose $\overline{F} \in \mathcal{R}_{-\infty}$. It then follows from Theorem A3.12(a) that all power moments of F are finite and

$$\lim_{x\to\infty} \frac{\int_x^\infty \overline{F}(t)\, dt}{x\overline{F}(x)} = 0\,.$$

The latter limit relationship characterises $\overline{F} \in \mathcal{R}_{-\infty}$.

8) The classes $\mathcal{R}_{-\alpha}$, $\alpha > 0$, play a key role in Chapters 2 and 3. For instance, the condition $\overline{F} \in \mathcal{R}_{-\alpha}$ characterises the maximum domain of attraction of the Fréchet distribution Φ_α; see Theorem 3.3.7. The class $\mathcal{R}_{-\infty}$ enters through the characterisation of the maximum domain of attraction of the

Gumbel distribution; see for instance Proposition 3.3.24. In the latter case the representation (A.14) is rewritten as

$$\overline{F}(x) = c(x) \exp\left\{ \int_z^x \frac{du}{a(u)} \right\},$$

where $a'(x) \to 0$ (a' being the density of a, assumed to exist); see Theorem 3.3.26 for details. □

Definition A3.13 (Regularly varying sequences)
A sequence (c_n) *of positive numbers is* regularly varying of index $\alpha \in \mathbb{R}$ *if*

$$\lim_{n\to\infty} \frac{c_{[tn]}}{c_n} = t^\alpha, \quad t > 0. \qquad \square$$

Whenever (c_n) is regularly varying with index α, then $c(x) = c_{[x]}$ belongs to $\mathcal{R}_\alpha$. Through this property, most of the results of $\mathcal{R}_\alpha$ carry over to the sequence case. For details see Bingham et al. [72], Section 1.9.

Notes and Comments

The reader interested in the above results and further aspects of regular variation should consult Bingham et al. [72] where also various generalisations to dfs on $\mathbb{R}$ or $\mathbb{R}^k$, together with higher–order theories, are discussed. Further interesting texts are de Haan [292], Geluk and de Haan [252], Resnick [530] and Seneta [575]. For the analysis of domain of attraction conditions in extreme value theory especially the work by de Haan has been of great importance.

A3.2 Properties of Subexponential Distributions

In Section 1.3 we have introduced the class $\mathcal{S}$ of subexponential distributions, i.e. $F \in \mathcal{S}$ if F has support $(0, \infty)$ and

$$\lim_{x\to\infty} \frac{\overline{F^{n*}}(x)}{\overline{F}(x)} = n, \quad n \geq 2. \tag{A.15}$$

Remark. Though subexponentiality is basically a condition on the df of non–negative rvs, occasionally we need a version for real–valued rvs. A df G with support on $(-\infty, \infty)$ will be called *subexponential on* $\mathbb{R}$ if there exists a subexponential df F such that $\overline{F}(x) \sim \overline{G}(x)$ as $x \to \infty$. For an example where this condition is needed see Section A3.3. □

In Lemma 1.3.4 we proved that, for $F \in \mathcal{S}$, it suffices to check (A.15) for $n = 2$ or indeed

$$\limsup_{x\to\infty} \frac{\overline{F^{2*}}(x)}{\overline{F}(x)} \le 2\,. \tag{A.16}$$

A slight generalisation of (A.16) is contained in the next result.

Lemma A3.14 (A sufficient condition for subexponentiality)
If there is an integer $n \ge 2$ such that

$$\limsup_{x\to\infty} \frac{\overline{F^{n*}}(x)}{\overline{F}(x)} \le n\,,$$

then $F \in \mathcal{S}$.

Proof. Observe that for arbitrary $m \ge 1$,

$$\begin{aligned} \overline{F^{(m+1)*}}(x) &= \overline{F^{m*}}(x) + \int_0^x \overline{F}(x-t)\,dF^{m*}(t) \\ &\ge \overline{F^{m*}}(x) + \overline{F}(x)\;F^{m*}(x)\,. \end{aligned}$$

Dividing by $\overline{F}(x)$ and taking limsup,

$$\limsup_{x\to\infty} \overline{F^{(m+1)*}}(x)/\overline{F}(x) \ge \limsup_{x\to\infty} \overline{F^{m*}}(x)/\overline{F}(x) + 1\,.$$

Therefore, under the hypothesis of the lemma, $\limsup_{x\to\infty} \overline{F^{2*}}(x)/\overline{F}(x) \le 2$, and an appeal to (A.16) completes the proof. □

The following lemma is useful towards proving $\mathcal{S}$–membership.

Lemma A3.15 (Closure of $\mathcal{S}$ under tail–equivalence)
Suppose F and G are dfs on $(0,\infty)$. If $F \in \mathcal{S}$ and

$$\lim_{x\to\infty} \frac{\overline{G}(x)}{\overline{F}(x)} = c \in (0,\infty)\,,$$

then $G \in \mathcal{S}$.

Proof. Suppose $v > 0$ fixed and $x > 2v$, and X, Y independent rvs with df G. Then

$$\begin{aligned} \{X+Y>x\} &= \{X \le v, X+Y>x\} \cup \{Y \le v, X+Y>x\} \\ &\quad \cup \{v < X \le x-v, X+Y>x\} \cup \{Y>v, X>x-v\}\,, \end{aligned}$$

where the above events are disjoint, hence

$$\begin{aligned}\frac{\overline{G^{2*}}(x)}{\overline{G}(x)} &= 2\int_0^v \frac{\overline{G}(x-y)}{\overline{G}(x)}\,dG(y) + \int_v^{x-v} \frac{\overline{G}(x-y)}{\overline{G}(x)}\,dG(y) \\ &\quad + \frac{\overline{G}(x-v)}{\overline{G}(x)}\,\overline{G}(v) \qquad\qquad (A.17)\\ &= I_1(x,v) + I_2(x,v) + I_3(x,v)\,.\end{aligned}$$

By Lemma 1.3.5(a), $I_1(x,v) \to 2G(v)$, $I_3(x,v) \to \overline{G}(v)$ as $x \to \infty$. For $\varepsilon > 0$ there exists $x_0 = x_0(\varepsilon)$ such that for all $x \geq x_0$,

$$c - \varepsilon \leq \overline{G}(x)/\overline{F}(x) \leq c + \varepsilon\,.$$

Therefore for $v \leq y \leq x - v$, v large enough,

$$\overline{G}(x-y)/\overline{F}(x-y) \leq c + \varepsilon \quad \text{and} \quad \overline{F}(x)/\overline{G}(x) \leq (c-\varepsilon)^{-1}.$$

Consequently,

$$\begin{aligned} & I_2(x,v) \\ &\leq \frac{c+\varepsilon}{c-\varepsilon}\int_v^{x-v} \frac{\overline{F}(x-y)}{\overline{F}(x)}\,dG(y) \\ &= \frac{c+\varepsilon}{c-\varepsilon}\left(\frac{\overline{F}(x-v)}{\overline{F}(x)}\overline{G}(v) - \frac{\overline{F}(v)}{\overline{F}(x)}\overline{G}(x-v) + \int_v^{x-v} \frac{\overline{G}(x-t)}{\overline{F}(x)}\,d\overline{F}(t)\right) \\ &\leq \frac{c+\varepsilon}{c-\varepsilon}\left(\frac{\overline{F}(x-v)}{\overline{F}(x)}\,\overline{G}(v) - \overline{F}(v)\frac{\overline{G}(x-v)}{\overline{G}(x)}\frac{\overline{G}(x)}{\overline{F}(x)}\right. \\ &\qquad \left. + (c+\varepsilon)\int_v^{x-v} \frac{\overline{F}(x-t)}{\overline{F}(x)}\,dF(t)\right) \\ &\to -\frac{c+\varepsilon}{c-\varepsilon}\left(\overline{F}(v)\,c - \overline{G}(v) - (c+\varepsilon)\,o_v(1)\right), \qquad x \to \infty\,,\end{aligned}$$

where $o_v(1)$ means $\lim_{v\to\infty} o_v(1) = 0$, the latter being a consequence of $F \in \mathcal{S}$. Therefore

$$\begin{aligned}\limsup_{x\to\infty} \frac{\overline{G^{2*}}(x)}{\overline{G}(x)} &\leq 2G(v) - \frac{c+\varepsilon}{c-\varepsilon}\left(c\overline{F}(v) - \overline{G}(v) - (c+\varepsilon)\,o_v(1)\right) + \overline{G}(v) \\ &\to 2\,, \qquad v \to \infty\,,\end{aligned}$$

hence $G \in \mathcal{S}$, because of (A.16). □

An interesting result yielding a complete answer to $\mathcal{S}$–membership for absolutely continuous F with density f and *hazard rate* $q(x) = f(x)/\overline{F}(x)$ eventually decreasing to 0 is given in Pitman [501].

Proposition A3.16 (A characterisation theorem for $\mathcal{S}$)
Suppose F is absolutely continuous with density f and hazard rate $q(x)$ eventually decreasing to 0. Then

(a) *$F \in \mathcal{S}$ if and only if*

$$\lim_{x\to\infty} \int_0^x e^{y\,q(x)} f(y)\,dy = 1\,. \tag{A.18}$$

(b) *If the function $x \mapsto \exp\{x\,q(x)\}\,f(x)$ is integrable on $[0,\infty)$ then $F \in \mathcal{S}$.*

Proof. Denote by $Q(x) = \int_0^x q(y)\,dy = -\ln \overline{F}(x)$ the associated hazard function. Then

$$\begin{aligned}\frac{\overline{F^{2*}}(x)}{\overline{F}(x)} - 1 &= \int_0^x \frac{\overline{F}(x-y)}{\overline{F}(x)}\,dF(y)\\ &= \int_0^x e^{Q(x)-Q(x-y)-Q(y)}\,q(y)\,dy\,.\end{aligned}$$

Note that with this notation,

$$\int_0^x e^{y\,q(x)}\,f(y)\,dy = \int_0^x e^{y\,q(x)-Q(y)}\,q(y)\,dy\,.$$

If q is not monotone over the whole range $[0,\infty)$, there is an equivalent Q_0 (i.e. F_0 with $\overline{F}_0(x) \sim \overline{F}(x)$) with a monotone derivative q_0. In view of Lemma A3.15 we may therefore assume that q is decreasing over the whole range. Therefore

$$\begin{aligned}\frac{\overline{F^{2*}}(x)}{\overline{F}(x)} - 1 &\geq \int_0^x e^{y\,q(x)-Q(y)}\,q(y)\,dy\\ &\geq \int_0^x e^{-Q(y)}\,q(y)\,dy = F(x)\,.\end{aligned}$$

From this it follows that the condition (A.18) is necessary for $F \in \mathcal{S}$.
In order to prove sufficiency, suppose that the condition (A.18) holds. After splitting the integral below over $[0,x]$ into two integrals over $[0,x/2]$ and $(x/2,x]$ and making a substitution in the second integral, we obtain

$$\begin{aligned}&\int_0^x e^{Q(x)-Q(x-y)-Q(y)}\,q(y)\,dy\\ &= \int_0^{x/2} e^{Q(x)-Q(x-y)-Q(y)}\,q(y)\,dy\\ &\quad + \int_0^{x/2} e^{Q(x)-Q(x-y)-Q(y)}\,q(x-y)\,dy\\ &= I_1(x) + I_2(x)\,.\end{aligned}$$

It follows by the same monotonicity argument as above that $I_1(x) \geq F(x/2)$. Moreover, for $y \leq x/2$ and therefore $x - y \geq x/2$,

$$Q(x) - Q(x-y) \leq y\, q(x-y) \leq y\, q(x/2)\,.$$

Therefore

$$F(x/2) \leq I_1(x) \leq \int_0^{x/2} e^{y\, q(x/2) - Q(y)}\, q(y)\, dy\,,$$

and (A.18) implies that

$$\lim_{x\to\infty} I_1(x) = 1\,. \tag{A.19}$$

The integrand in $I_1(x)$ converges pointwise to $f(y) = \exp\{-Q(y)\}q(y)$. Thus we can reformulate (A.19) as "the integrand of $I_1(x)$ converges in f–mean to 1". The integrand in $I_2(x)$ converges pointwise to 0, it is however everywhere bounded by the integrand of $I_1(x)$. From this and an application of Pratt's lemma (see Pratt [509]), it follows that $\lim_{x\to\infty} I_2(x) = 0$. Consequently,

$$\lim_{x\to\infty} \frac{\overline{F^{2*}}(x)}{\overline{F}(x)} - 1 = 1\,,$$

i.e. $F \in \mathcal{S}$, proving sufficiency of (A.18) and hence assertion (a).

(b) The assertion follows immediately from Lebesgue's dominated convergence theorem, since $q(x) \leq q(y)$ for $y \leq x$. □

Applications of this result are given in Example 1.4.7.

The following situation often occurs: suppose that in some probability model there is an input rv with df F and an output rv with df G. Assume further that F and G are linked through some model dependent functional relationship $G = T(F)$ say. In various cases one can show a result of the following type.

The following are equivalent	
(a) $\overline{F}$ is regularly varying,	
(b) $\overline{G}$ is regularly varying.	(A.20)
Moreover, either (a) or (b) implies	
(c) $\lim_{x\to\infty} \overline{G}(x)/\overline{F}(x) = c \in (0,\infty)$.	

A key question is now

How far can one enlarge the class of regularly varying functions in (A.20), *keeping the implications in force?*

In several fairly general situations the answer is that one can replace regular variation by subexponentiality, and that this is best possible, in that (c) becomes *equivalent* to (a) and (b). This type of questions is known under the name of *Mercerian theorems*; see Bingham et al. [72] for a detailed discussion.

Example A3.17 (Regular variation for the compound Poisson df)
With applications to insurance in mind, we restrict the functional T above to the compound Poisson type, i.e. assume that for some $\lambda > 0$,

$$G(x) = e^{-\lambda} \sum_{k=0}^{\infty} \frac{\lambda^k}{k!} F^{k*}(x) , \quad x \geq 0 , \tag{A.21}$$

which is the df of $\sum_{i=1}^{N} X_i$, where the X_i are iid with df F, independent of the $Poi(\lambda)$ rv N. Taking Laplace–Stieltjes transforms in (A.21), one obtains

$$\widehat{g}(s) = e^{-\lambda(1-\widehat{f}(s))} , \quad s \geq 0 ,$$

and hence

$$1 - \widehat{g}(s) \sim \lambda(1 - \widehat{f}(s)) , \quad s \downarrow 0 . \tag{A.22}$$

In order to distinguish between regular variation at infinity and at zero, we denote the former by $\mathcal{R}^{\infty}$ and the latter by $\mathcal{R}^{0}$. Applying Corollary A3.10 and (A.22), one obtains for $0 \leq \alpha < 1$,

$$\begin{aligned}
\overline{G} \in \mathcal{R}^{\infty}_{-\alpha} \quad &\Leftrightarrow \quad 1 - \widehat{g} \in \mathcal{R}^{0}_{\alpha} \quad \Rightarrow \lim_{x\to\infty} \frac{1 - \widehat{g}(1/x)}{\overline{G}(x)} = \Gamma(1-\alpha) \\
&\Leftrightarrow \quad 1 - \widehat{f} \in \mathcal{R}^{0}_{\alpha} \quad \Rightarrow \lim_{x\to\infty} \frac{1 - \widehat{g}(1/x)}{1 - \widehat{f}(1/x)} = \lambda \\
&\Leftrightarrow \quad \overline{F} \in \mathcal{R}^{\infty}_{-\alpha} \quad \Rightarrow \lim_{x\to\infty} \frac{1 - \widehat{f}(1/x)}{\overline{F}(x)} = \Gamma(1-\alpha) .
\end{aligned}$$

So finally we obtain,

$$\boxed{\overline{F} \in \mathcal{R}^{\infty}_{-\alpha} \Leftrightarrow \overline{G} \in \mathcal{R}^{\infty}_{-\alpha} \Rightarrow \lim_{x\to\infty} \frac{\overline{G}(x)}{\overline{F}(x)} = \lambda .}$$

In order to go from $0 \leq \alpha < 1$ to arbitrary α, one uses Theorem 8.1.6 in Bingham et al. [72]. We leave the details to the reader. □

An answer to the question following (A.20) now crucially depends on the following result due to Goldie; see Embrechts et al. [207].

Proposition A3.18 (Convolution root closure of $\mathcal{S}$)
If $F^{n*} \in \mathcal{S}$ for some positive integer n, then $F \in \mathcal{S}$.

Proof. Since $F^{n*} \in \mathcal{S}$ we know by Lemma 1.3.5(a),

$$\lim_{x\to\infty} \frac{\overline{F^{n*}}(x-t)}{\overline{F^{n*}}(x)} = 1\,, \tag{A.23}$$

uniformly on compact t–sets. Fix $A > 0$. We have for $x > A$

$$1 + \left(\int_0^A + \int_A^x\right) \frac{\overline{F^{n*}}(x-t)}{\overline{F^{n*}}(x)}\, dF^{n*}(t) = \frac{\overline{F^{2n*}}(x)}{\overline{F^{n*}}(x)} \to 2\,, \quad x \to \infty\,.$$

By dominated convergence and (A.23) the first integral converges to $F^{n*}(A)$, so that

$$\lim_{x\to\infty} \int_A^x \frac{\overline{F^{n*}}(x-t)}{\overline{F^{n*}}(x)}\, dF^{n*}(t) = \overline{F^{n*}}(A). \tag{A.24}$$

Fix $u > 0$ so that $F^{(n-1)*}(u) > 0$. Then for $x > u$,

$$\begin{aligned} \overline{F^{n*}}(x) &= \overline{F}(x) + \left(\int_0^u + \int_u^x\right) \overline{F^{(n-1)*}}(x-t)\, dF(t) \\ &= \overline{F}(x) + J_1(x) + J_2(x)\,. \end{aligned} \tag{A.25}$$

We show that given $\varepsilon > 0$, there exists $x_0 = x_0(u,\varepsilon)$ such that

$$\frac{J_1(x)}{\overline{F^{n*}}(x)} \le 1 - n^{-1} + \frac{1}{2}\,\varepsilon\,, \qquad x \ge x_0 \tag{A.26}$$

$$\frac{J_2(x)}{\overline{F^{n*}}(x)} \le \frac{\overline{F^{n*}}(u)}{F^{(n-1)*}(u)} + \frac{1}{2}\,\varepsilon\,, \qquad x \ge x_0\,. \tag{A.27}$$

For (A.26), note that

$$\begin{aligned} \frac{J_1(x)}{\overline{F^{n*}}(x)} &\le \frac{\overline{F^{(n-1)*}}(x-u)}{\overline{F^{n*}}(x)}\, F(u) \\ &\le \frac{\overline{F^{(n-1)*}}(x-u)}{\overline{F^{(n-1)n*}}(x-u)}\, \frac{\overline{F^{(n-1)n*}}(x-u)}{\overline{F^{n*}}(x-u)}\, \frac{\overline{F^{n*}}(x-u)}{\overline{F^{n*}}(x)}\,. \end{aligned}$$

On the rhs, the first quotient has limsup at most $1/n$ by Remark 3 after Lemma 1.3.4, while the second quotient converges to $n-1$ because $F^{n*} \in \mathcal{S}$. Finally, the third quotient converges to 1 by (A.23). Thus (A.26) follows. For (A.27), on some probability space, let S_{n-1}, X_n and S_n' be independent rvs with dfs $F^{(n-1)*}$, F and F^{n*} respectively, and let $S_n = S_{n-1} + X_n$. Then

$$\begin{aligned}
F^{(n-1)*}(u)\, J_2(x) &\le F^{(n-1)*}(u) \int_u^x \overline{F^{n*}}(x-t)\, dF(t) \\
&= P\,(S_{n-1} \le u)\; P\,(u < X_n \le x < X_n + S_n') \\
&= P\,(S_{n-1} \le u < X_n \le x < X_n + S_n') \\
&\le P\,(u < S_n \le x + u\,,\; S_n + S_n' > x) \\
&= P\,(u < S_n \le x < S_n + S_n') + P\,(x < S_n \le x + u)
\end{aligned}$$

so that

$$\frac{F^{(n-1)*}(u)\, J_2(x)}{\overline{F^{n*}}(x)} \le \int_u^x \frac{\overline{F^{n*}}(x-t)}{\overline{F^{n*}}(x)}\, dF^{n*}(t) + \frac{\overline{F^{n*}}(x) - \overline{F^{n*}}(x+u)}{\overline{F^{n*}}(x)}\,.$$

The rhs converges to $\overline{F^{n*}}(u)$ using (A.23) and (A.24). Then (A.27) follows. From (A.25)–(A.27) we find

$$\liminf_{x\to\infty} \frac{\overline{F}(x)}{\overline{F^{n*}}(x)} \ge 1 - \left(1 - n^{-1}\right) - \frac{\overline{F^{n*}}(u)}{F^{(n-1)*}(u)}\,.$$

The rhs converges to $1/n$ as $u \to \infty$. Thus $\limsup_{x\to\infty} \overline{F^{n*}}(x)/\overline{F}(x) \le n$. Hence $F \in \mathcal{S}$, by Lemma A3.14. □

We are now in the position to prove the following key result.

Theorem A3.19 (Subexponentiality and compound Poisson dfs)
Let G, F and λ be related by (A.21). *Then the following assertions are equivalent:*

(a) $G \in \mathcal{S}$,

(b) $F \in \mathcal{S}$,

(c) $\lim_{x\to\infty} \overline{G}(x)/\overline{F}(x) = \lambda$.

Proof. (b) $\Rightarrow$ [(a) and (c)]. This follows as in the proof of Theorem 1.3.6 from Lemma 1.3.5, dominated convergence and Lemma A3.15.
(a) $\Rightarrow$ (b). Assume in the first part of the proof that $0 < \lambda < \ln 2$. Consider the df

$$R(x) = \left(e^{\lambda} - 1\right)^{-1} \sum_{k=1}^{\infty} \frac{\lambda^k}{k!}\, F^{k*}(x)\,, \quad x \ge 0\,.$$

Taking Laplace–Stieltjes transforms, for all $s \ge 0$,

$$\widehat{r}(s) = \frac{e^{\lambda \widehat{f}(s)} - 1}{e^{\lambda} - 1}\,,$$

so

$$\lambda\widehat{f}(s) = \ln\left(1 - \left(1 - e^{\lambda}\right)\widehat{r}(s)\right),$$

and since $e^{\lambda} - 1 < 1$ we have

$$\lambda\widehat{f}(s) = -\sum_{k=1}^{\infty} \frac{\left(1 - e^{\lambda}\right)^k}{k}\,\widehat{r}^{\,k}(s).$$

Inversion yields for all $x \geq 0$,

$$\frac{\lambda\overline{F}(x)}{\overline{R}(x)} = -\sum_{k=1}^{\infty} \frac{\left(1 - e^{\lambda}\right)^k}{k}\,\frac{\overline{R^{k*}}(x)}{\overline{R}(x)}. \tag{A.28}$$

Moreover, $e^{\lambda}\overline{G}(x) = (e^{\lambda} - 1)\overline{R}(x)$, so $G \in \mathcal{S}$ implies $R \in \mathcal{S}$. Choose $\varepsilon > 0$ such that $(e^{\lambda} - 1)(1+\varepsilon) < 1$. By Lemma 1.3.5(c), there exists $K = K(\varepsilon) < \infty$ such that $\overline{R^{n*}}(x)/\overline{R}(x) \leq K(1+\varepsilon)^n$ for all $x \geq 0$ and for all n. Hence, using dominated convergence in (A.28), we have

$$\lim_{x\to\infty} \frac{\overline{F}(x)}{\overline{R}(x)} = \frac{e^{\lambda} - 1}{\lambda e^{\lambda}},$$

so $F \in \mathcal{S}$ by Lemma A3.15.
Consider now $\lambda > 0$ arbitrary, hence there exists an integer $\ell = \ell(\lambda) > 0$ such that $\lambda/\ell < \ln 2$. Define for $x \geq 0$,

$$H(x) = e^{-\lambda/\ell} \sum_{k=0}^{\infty} \frac{(\lambda/\ell)^k}{k!}\, F^{k*}(x),$$

so

$$\widehat{h}(s) = \left(\exp\left\{-\lambda\left(1 - \widehat{f}(s)\right)\right\}\right)^{1/\ell} = \widehat{g}^{\,1/\ell}(s), \qquad s \geq 0.$$

This implies $G = H^{\ell*} \in \mathcal{S}$, giving $H \in \mathcal{S}$ by virtue of Proposition A3.18. The first part of the proof (since $\lambda/\ell < \ln 2$) gives $F \in \mathcal{S}$.
(c) $\Rightarrow$ (b). For all $x \geq 0$,

$$\overline{F^{2*}}(x) = \left(2/\lambda^2\right) e^{\lambda} \left(\overline{G}(x) - e^{-\lambda} \sum_{k\neq 2} \frac{\lambda^k}{k!}\, \overline{F^{k*}}(x)\right).$$

Dividing both sides by $\overline{F}(x)$, we obtain

$$\limsup_{x\to\infty} \overline{F^{2*}}(x)/\overline{F}(x) \leq 2,$$

so $F \in \mathcal{S}$ by (A.16). □

The same techniques as used in the above proof lead to the following more general result.

Theorem A3.20 (Subexponentiality and compound dfs)
Suppose (p_n) *is a probability measure on* $\mathbb{N}_0$ *such that for some* $\varepsilon > 0$,

$$\sum_{n=0}^{\infty} p_n (1+\varepsilon)^n < \infty \,,$$

and set

$$G(x) = \sum_{n=0}^{\infty} p_n \, F^{n*}(x) \,, \quad x \geq 0 \,.$$

(*a*) *If* $F \in \mathcal{S}$, *then* $G \in \mathcal{S}$, *and*

$$\lim_{x\to\infty} \frac{\overline{G}(x)}{\overline{F}(x)} = \sum_{n=1}^{\infty} n \, p_n \,. \tag{A.29}$$

(*b*) *Conversely, if* (A.29) *holds and there exists* $\ell \geq 2$ *such that* $p_\ell > 0$, *then* $F \in \mathcal{S}$.

Proof. (a) The same argument as given in the proof of Theorem 1.3.6 applies.
(b) For ℓ fixed,

$$p_\ell \, \frac{\overline{F^{\ell*}}(x)}{\overline{F}(x)} \leq \frac{\overline{G}(x)}{\overline{F}(x)} - \sum_{k=0}^{\infty} p_k \, \frac{\overline{F^k}(x)}{\overline{F}(x)} + p_\ell \, \frac{\overline{F^\ell}(x)}{\overline{F}(x)} \,.$$

Hence by (A.29) and Fatou's lemma, $p_\ell \limsup_{x\to\infty} \overline{F^{\ell*}}(x)/\overline{F}(x) \leq p_\ell \ell$, giving $F \in \mathcal{S}$ because of $p_\ell > 0$ and Lemma A3.14. □

Theorem A3.19 generalises Theorem 1.3.6 because any compound geometric distribution

$$G(x) = (1-\alpha) \sum_{n=0}^{\infty} \alpha^n \, F^{n*}(x) \,, \qquad 0 < \alpha < 1 \,, \tag{A.30}$$

can be written as a compound Poisson distribution. Indeed, using Laplace–Stieltjes transforms, $\widehat{g}(s) = (1-\alpha)/(1-\alpha\widehat{f}(s))$, which is of the compound Poisson form $\widehat{g}(s) = \exp\{-\lambda(1-\widehat{h}(s))\}$, where $\lambda = \ln(1/(1-\alpha))$ and $\widehat{h}(s) = -\lambda^{-1}\ln(1-\alpha\widehat{f}(s))$. In distributional form, this becomes

$$G(x) = e^{-\lambda} \sum_{n=0}^{\infty} \frac{\lambda^n}{n!} \, H^{n*}(x) \,,$$

where

$$H(x) = -(\ln(1-\alpha))^{-1} \sum_{n=1}^{\infty} \frac{\alpha^n}{n} \, F^{n*}(x) \,.$$

From these relationships, we easily obtain:

Corollary A3.21 (Subexponentiality and compound geometric dfs)
Suppose $0 < \alpha < 1$, *and* F *and* G *are related as in* (A.30), *then the following assertions are equivalent:*

(a) $F \in \mathcal{S}$,

(b) $G \in \mathcal{S}$,

(c) $\lim_{x\to\infty} \overline{G}(x)/\overline{F}(x) = \alpha/(1-\alpha)$. □

Notes and Comments

The compound Poisson model (A.21) leads to one of the key examples of the so–called *infinitely divisible distributions.* A df F is infinitely divisible if for all $n \in \mathbb{N}$, there exists a df H_n such that $F = H_n^{n*}$. The Laplace–Stieltjes transform of an infinitely divisible df F on $[0,\infty)$ can be expressed (see Feller [235], p. 450) as

$$\widehat{f}(s) = \exp\left\{-as - \int_0^\infty \left(1-e^{-sx}\right) d\nu(x)\right\}, \qquad s \ge 0,$$

where $a \ge 0$ is constant and ν is a Borel measure on $(0,\infty)$ for which $\nu(1,\infty) < \infty$ and $\int_0^1 x\, d\nu(x) < \infty$. This is the so–called Lévy–Khinchin representation theorem. In Embrechts et al. [207] the following generalisation of Theorem 1.3.8 is proved.

Theorem A3.22 (Infinite divisibility and $\mathcal{S}$)
For F *infinitely divisible on* $(0,\infty)$ *with Lévy–Khinchin measure* ν, *the following are equivalent:*

(a) $F \in \mathcal{S}$,

(b) $\nu(1,x]/\nu(1,\infty) \in \mathcal{S}$,

(c) $\lim_{x\to\infty} \overline{F}(x)/\nu(x,\infty) = 1$. □

A key step in the proof of the above result concerns the following question.

If F, $G \in \mathcal{S}$, *does it always follow that the convolution* $F * G \in \mathcal{S}$*?*

The rather surprising answer to this question is

In general, NO!

The latter was shown by Leslie [422]. Necessary and sufficient conditions for convolution closure are given in Embrechts and Goldie [204]. The main result needed in the proof of Theorem A3.22 is the following (Embrechts et al. [207], Proposition 1).

Lemma A3.23 (Convolution in $\mathcal{S}$)
Let $H = F * G$ be the convolution of two dfs on $(0,\infty)$. If $G \in \mathcal{S}$ and $\overline{F}(x) = o(\overline{G}(x))$ as $x \to \infty$, then $H \in \mathcal{S}$. □

With respect to asymptotic properties of convolution tails, the papers by Cline [123, 124] offer a useful source of information.

In the discrete case, the following classes of dfs have been found to yield various interesting results.

Definition A3.24 (Discrete subexponentiality)
Suppose (p_n) defines a probability measure on $\mathbb{N}_0$, let $p_n^{2*} = \sum_{k=0}^{n} p_{n-k}p_k$ be the two–fold convolution of (p_n). Then $(p_n) \in \mathrm{SD}$, if

(a) $\lim_{n\to\infty} p_n^{2*}/p_n = 2 < \infty$,

(b) $\lim_{n\to\infty} p_{n-1}/p_n = 1$.

The class SD *is the class of* discrete subexponential sequences. □

Suppose (p_n) is infinitely divisible and $\widehat{p}(r) = \sum_{k=0}^{\infty} p_k r^k$ its generating function. Then by the Lévy–Khinchin representation theorem

$$\widehat{p}(z) = \exp\left\{-\lambda\left(1 - \sum_{j=1}^{\infty} \alpha_j\, z^j\right)\right\},$$

for some $\lambda > 0$ and a probability measure (α_j). The following result is the discrete analogue to Theorem A3.22, or for that matter Theorem A3.19 (Embrechts and Hawkes [210]).

Theorem A3.25 (Discrete infinite divisibility and subexponentiality)
The following three statements are equivalent:

(a) $(p_n) \in \mathrm{SD}$,

(b) $(\alpha_n) \in \mathrm{SD}$,

(c) $\lim_{n\to\infty} p_n/\alpha_n = \lambda$ *and* $\lim_{n\to\infty} \alpha_n/\alpha_{n+1} = 1$. □

See the above paper for applications and further references.

A recent survey paper on subexponentiality is Goldie and Klüppelberg [274].

A3.3 The Tail Behaviour of Weighted Sums of Heavy–Tailed Random Variables

In this section we study the tail of the weighted sum

$$X = \sum_{j=-\infty}^{\infty} \psi_j Z_j$$

for iid Z_j with common df F and real coefficients ψ_j. The latter are chosen in such a way that X is well defined as an a.s. converging series. The behaviour of the tail $P(X > x)$ as $x \to \infty$ is crucial for the extremal behaviour of linear processes; see Section 5.5.

Throughout we consider heavy–tailed dfs F. Assume first that F satisfies the tail balance condition

$$\overline{F}(x) \sim p x^{-\alpha} L(x)\,, \quad F(-x) = q x^{-\alpha} L(x)\,, \quad x \to \infty\,, \tag{A.31}$$

for some $\alpha > 0$, a slowly varying function L and $p \in (0,1]$, $q = 1-p$. We also suppose that

$$\sum_{j=-\infty}^{\infty} |\psi_j|^{\delta} < \infty \quad \text{for some } 0 < \delta < \min(\alpha, 1)\,. \tag{A.32}$$

Lemma A3.26 *Assume that conditions (A.31) and (A.32) are satisfied. Then as $x \to \infty$,*

$$P(X > x) \sim x^{-\alpha} L(x) \sum_{j=-\infty}^{\infty} |\psi_j|^{\alpha} \left(p I_{\{\psi_j > 0\}} + q I_{\{\psi_j < 0\}} \right) . \tag{A.33}$$

Notice that (A.33) implies that every rv Z_j contributes to the tail $P(X > x)$. This contribution depends on the size and the sign of the weight ψ_j.

Proof. We combine the arguments of Feller [235], p. 278, (cf. Lemma 1.3.1) and Resnick [530], Lemma 4.24. First notice that (A.32) implies the absolute a.s. convergence of the infinite series X. Write

$$X_m = \sum_{|j| \le m} \psi_j Z_j\,, \quad X'_m = X - X_m\,.$$

Observe that, for $m \ge 1$ and $\varepsilon \in (0,1)$,

$$\begin{aligned} & P\left(X_m > x(1+\varepsilon)\right) - P\left(\sum_{|j|>m} |\psi_j Z_j| \ge \varepsilon x \right) \\ \le\ & P\left(X_m > x(1+\varepsilon)\right) - P\left(X'_m \le -\varepsilon x\right) \end{aligned} \tag{A.34}$$

$$\begin{aligned}
&\leq P\left(X_m > x(1+\varepsilon)\,,\ X'_m > -\varepsilon x\right)\\
&\leq P\left(X > x\right)\\
&\leq P\left(X_m > x(1-\varepsilon)\right) + P\left(X'_m > \varepsilon x\right)\\
&\leq P\left(X_m > x(1-\varepsilon)\right) + P\Big(\sum_{|j|>m} |\psi_j Z_j| > \varepsilon x\Big)\,. \qquad \text{(A.35)}
\end{aligned}$$

Lemma 4.24 in Resnick [530] implies that

$$\lim_{m\to\infty}\lim_{x\to\infty} P\Big(\sum_{|j|>m} |\psi_j Z_j| \geq \varepsilon x\Big)\Big/P(|Z|>x) = \lim_{m\to\infty}\sum_{|j|>m}|\psi_j|^\alpha = 0\,.$$

The latter relation and (A.34), (A.35) show that it suffices to prove, for every $m \geq 1$, that

$$P\left(X_m > x\right) \sim x^{-\alpha}L(x)\sum_{|j|\leq m}|\psi_j|^\alpha\left(pI_{\{\psi_j>0\}} + qI_{\{\psi_j<0\}}\right)\,.$$

We restrict ourselves to two summands $\psi_1 Z_1$ and $\psi_2 Z_2$ with non–zero ψ_1, ψ_2 to demonstrate the method; the general case can be proved analogously. As in the proof of Lemma 1.3.1 we have for $\delta \in (0, 1/2)$ that

$$\begin{aligned}
&\{\psi_1 Z_1 + \psi_2 Z_2 > x\}\\
&\quad \subset\ \{\psi_1 Z_1 > (1-\delta)x\} \cup \{\psi_2 Z_2 > (1-\delta)x\} \cup \{\psi_1 Z_1 > \delta x\,,\ \psi_2 Z_2 > \delta x\}\,.
\end{aligned}$$

Hence, by regular variation of the tails,

$$\begin{aligned}
&\limsup_{x\to\infty} P\left(\psi_1 Z_1 + \psi_2 Z_2 > x\right)/P(|Z|>x)\\
&\qquad \leq\ \sum_{i=1}^{2}|\psi_i|^\alpha\left(pI_{\{\psi_i>0\}} + qI_{\{\psi_i<0\}}\right)(1-\delta)^{-\alpha}\,. \qquad \text{(A.36)}
\end{aligned}$$

Moreover,

$$\begin{aligned}
&\{\psi_1 Z_1 + \psi_2 Z_2 > x\}\\
&\quad \supset\ \{\psi_1 Z_1 > x(1+\delta)\,,\ |\psi_2 Z_2| \leq \delta x\} \cup \{\psi_2 Z_2 > x(1+\delta)\,,\ |\psi_1 Z_1| \leq \delta x\}\,.
\end{aligned}$$

Hence

$$\begin{aligned}
&\liminf_{x\to\infty} P\left(\psi_1 Z_1 + \psi_2 Z_2 > x\right)/P(|Z|>x)\\
&\qquad \geq\ \sum_{i=1}^{2}|\psi_i|^\alpha\left(pI_{\{\psi_i>0\}} + qI_{\{\psi_i<0\}}\right)(1+\delta)^{-\alpha}\,. \qquad \text{(A.37)}
\end{aligned}$$

Letting $\delta \to 0$ in (A.36) and (A.37), we see that the proof is complete. □

Next we consider $F \in \mathrm{MDA}(\Lambda) \cap \mathcal{S}$, where $F \in \mathcal{S}$ is understood as subexponentiality on $\mathbb{R}$; see the Remark in Appendix A3.2. Then $\overline{F}$ is rapidly varying; see Lemma 3.3.24. This means that

$$\lim_{x\to\infty} \frac{\overline{F}(xt)}{\overline{F}(x)} = \begin{cases} 0 & \text{if} \quad t > 1\,, \\ \infty & \text{if} \quad 0 < t < 1\,. \end{cases}$$

In contrast to regularly varying tails we may expect in this case that the size of the weights ψ_j has much more influence on the tail behaviour of X. As before we assume a tail balance condition of the form

$$\overline{F}(x) \sim p\, P(|Z| > x)\,, \quad F(-x) \sim q\, P\left(|Z| > x\right)\,, \quad x \to \infty\,, \tag{A.38}$$

for some $p \in (0,1]$, $q = 1 - p$. Let ψ_j be real numbers such that

$$\sum_{j=-\infty}^{\infty} |\psi_j|^\delta < \infty \quad \text{for some } \delta \in (0,1) \text{ and } \max_j |\psi_j| = 1\,. \tag{A.39}$$

Define

$$k^+ = \mathrm{card}\,\{j : \psi_j = 1\}\,, \quad k^- = \mathrm{card}\,\{j : \psi_j = -1\}\,.$$

The following is proved in Davis and Resnick [163] for more general classes of dfs in $\mathrm{MDA}(\Lambda)$.

Lemma A3.27 *Assume* $F \in \mathrm{MDA}(\Lambda) \cap \mathcal{S}$ *and that the conditions* (A.38) *and* (A.39) *are satisfied. Then*

$$P(X > x) \sim \left(p\, k^+ + q\, k^-\right) P(|Z| > x)\,, \quad x \to \infty\,.$$

A comparison with Lemma A3.26 shows that in the subexponential case in $\mathrm{MDA}(\Lambda)$ only the summands $\psi_j Z_j$ with $\psi_j = \pm 1$ (i.e. the largest coefficients) have an influence on the tail $P(X > x)$ as $x \to \infty$. This is in sharp contrast to the regularly varying case, where every $\psi_j Z_j$ makes some contribution.

Sketch of the proof. First assume that Z is non–negative. The same decomposition of X as in the proof of Lemma A3.26 yields $X = X_m + X_m'$ for every $m \ge 1$. Choose $m \ge 1$ such that $\sum_{|j|>m} |\psi_j|^\delta < 1$. We first show that

$$\lim_{x\to\infty} \frac{P(X_m' > x)}{\overline{F}(x)} = 0\,. \tag{A.40}$$

Take $x > 0$, then

$$
\begin{aligned}
P\left(X_m' > x\right) &\le P\left(\sum_{|j|>m} |\psi_j Z_j| > x\right) \\
&\le P\left(\sum_{|j|>m} |\psi_j Z_j| > \sum_{|j|>m} |\psi_j|\, |\psi_j|^{\delta-1}\, x\right) \\
&\le P\left(\bigcup_{|j|>m} \left\{|Z_j| > |\psi_j|^{\delta-1}\, x\right\}\right) \\
&\le \sum_{|j|>m} P\left(|Z| > |\psi_j|^{\delta-1}\, x\right).
\end{aligned}
$$

It follows that

$$
\frac{P\left(X_m' > x\right)}{\overline{F}(x)} \le \sum_{|j|>m} \frac{P\left(|Z| > |\psi_j|^{\delta-1}\, x\right)}{\overline{F}(x)}. \tag{A.41}
$$

In view of the rapid variation of $\overline{F}$ the relation

$$
\lim_{x\to\infty} \frac{P\left(|Z| > |\psi_j|^{\delta-1}\, x\right)}{\overline{F}(x)} = 0
$$

holds for every fixed $|j| > m$. Thus (A.40) follows if we may interchange $\sum_{|j|>m}$ and $\lim_{x\to\infty}$ in (A.41). This is possible by the following argument. Recall the Balkema–de Haan representation of $F \in \mathrm{MDA}(\Lambda)$ from Theorem 3.3.26:

$$
\overline{F}(x) = c(x) \exp\left\{-\int_z^x \frac{1}{a(t)}\, dt\right\}, \quad z < x < \infty,
$$

where $c(x)$ is a measurable function satisfying $c(x) \to c > 0$ and $a(x)$ is a positive, absolutely continuous function with density $a'(x)$ and $\lim_{x\uparrow\infty} a'(x) = 0$. Using this representation, it is shown in Proposition 1.1 of Davis and Resnick [163] that for any $\varepsilon > 0$ and sufficiently large x,

$$
\frac{\overline{F}(x/t)}{\overline{F}(x)} \le (1+\varepsilon)\left(\frac{a(x)}{x}\right)^{1/\varepsilon}\left(\frac{t}{\varepsilon(1-t)}\right)^{1/\varepsilon}, \quad 0 < t < 1,
$$

Hence, for some constant $c_0 > 0$,

$$
\frac{P\left(|Z| > |\psi_j|^{\delta-1}\, x\right)}{\overline{F}(x)} \le c_0\, |\psi_j|^{(1-\delta)/\varepsilon} \left(\frac{a(x)}{x}\right)^{1/\varepsilon}.
$$

Now choose ε such that $(1-\delta)/\varepsilon \geq \delta$, implying that $\sum_{|j|>m} |\psi_j|^{(1-\delta)/\varepsilon} < \infty$. Applying in (A.41) the dominated convergence theorem and the fact that $\lim_{x\to\infty} x^{-1}a(x) = \lim_{x\to\infty} a'(x) = 0$, we have proved that (A.40) holds.

Now we want to make use of a variant of Theorem 2.7 in Embrechts and Goldie [205]; see also Theorem 1 in Cline [123].

Lemma A3.28 *Let $Y_1, \ldots, Y_m$ be independent rvs and $F \in \mathcal{S}$. Assume*

$$\lim_{x\to\infty} \frac{P(Y_i > x)}{\overline{F}(x)} = a_i \in [0,\infty]\,, \quad i = 1,\ldots,m\,.$$

Then

$$\lim_{x\to\infty} \frac{P\left(\sum_{i=1}^m Y_i\right)}{\overline{F}(x)} = \sum_{i=1}^m a_i\,. \qquad \square$$

Since $\overline{F}$ is rapidly varying we have for $|j| \leq m$, in view of the tail balance condition (A.38), that

$$\lim_{x\to\infty} \frac{P(\psi_j Z > x)}{\overline{F}(x)} = \begin{cases} 1 & \text{if} \quad \psi_j = 1\,, \\ 0 & \text{if} \quad |\psi_j| < 1\,, \\ qp^{-1} & \text{if} \quad \psi_j = -1\,. \end{cases}$$

Now an application of Lemma A3.28 together with (A.40) proves that

$$\lim_{x\to\infty} \frac{P(X > x)}{\overline{F}(x)} = \lim_{x\to\infty} \frac{P(X_m + X_m' > x)}{\overline{F}(x)} = k^+ + qp^{-1}k^-\,. \qquad \square$$

A4 Some Renewal Theory

Recall the set–up of Section 2.5.2 where for a given iid sequence of positive rvs $Y, Y_1, Y_2, \ldots$ we defined the *renewal counting process*

$$N(t) = \text{card}\,\{n \geq 1 : Y_1 + \cdots + Y_n \leq t\}\,, \quad t \geq 0\,.$$

Often $N(t)$ is interpreted as the number of *renewals* before time t in a replacement experiment where the individual interarrival times Y_k are iid with non–defective (proper) df F such that $F(0) < 1$ and $EY_1 = 1/\lambda \leq \infty$. The function $V(t) = EN(t)$ is called the *renewal function.* It gives the expected number of renewals in the interval $[0,t]$. Proposition 2.5.12 yields $V(t)/t \to \lambda = 1/EY_1$ as $t \to \infty$. This is known as the *elementary renewal theorem.* The following basic facts concerning V are easy to prove. Before we formulate the result, we would like to warn the reader that one has to be careful as regards the interpretation of Lebesgue–Stieltjes integrals over finite intervals. This means one has to care about possible jumps of the df, with respect to which one integrates, at the endpoints of the integration interval. In this section, all such integrals are taken over closed intervals. For an extensive discussion on this point and its consequences for renewal theory we refer to Section 3.2 in Resnick [531].

Proposition A4.1 (Basic properties of the renewal function)

(a) *V is a right-continuous, non-decreasing function,*
(b) *$V(t) = \sum_{n=1}^{\infty} F^{n*}(t) < \infty$ for all $t \geq 0$,*
(c) *$V(t)$ satisfies the* renewal equation

$$V(t) = F(t) + \int_0^t V(t-x)\, dF(x)\,, \quad t \geq 0\,. \qquad \square$$

If $(N(t))$ is a homogeneous Poisson process, then for all $h \geq 0$, $V(t+h) - V(t) = \lambda h$. Concerning the general case covered by the elementary renewal theorem, one may ask:

Does the approximation $V(t+h) - V(t) \approx \lambda h$ hold for large t?

The answer to this question is YES and forms the content of a fundamental theorem of applied probability. Before its formulation, recall that a rv X (respectively its df F) is said to be *lattice* if there exist $a, d \geq 0$ such that $\sum_{n=0}^{\infty} P(X = a + nd) = 1$. The largest d having this property is said to be the *maximal step* of X (respectively F).

Theorem A4.2 (Blackwell's renewal theorem)

(a) *If F is non-lattice, then for all $h > 0$,*

$$\lim_{t \to \infty} (V(t+h) - V(t)) = \lambda h\,.$$

(b) *If F is lattice with maximal step d, then the same limit relation holds, provided h is a multiple of d.* $\square$

Blackwell's theorem can be restated as a weak convergence result. Recall from Appendix A2 the basic notions on weak convergence. To any renewal function $V(t)$ one may associate a *renewal measure* $V(\cdot)$ via $V(a,b] = V(b) - V(a)$, $a, b > 0$. $V(\cdot)$ may be extended to all Borel sets A on $[0, \infty)$. In this set-up, Theorem A4.2(a) tells us that $V(t + A)$ converges to $\lambda|A|$ as $t \to \infty$, where $|\cdot|$ stands for Lebesgue measure. We may therefore expect to have a result which approximates the integral of a nice function with respect to V by the integral of that function with respect to $\lambda|\cdot|$. This is indeed true! It turns out that the property of "nice function" is slightly technical; the right notion is that of *direct Riemann integrability* as discussed for instance in Alsmeyer [8], p. 69, Asmussen [27], p. 118, or Resnick [531], Section 3.10.1. A sufficient condition for a function g to be directly Riemann integrable is that $g \geq 0$, g is non–increasing and Riemann integrable in the ususal sense, i.e. $\int_0^\infty g(t)\, dt < \infty$. Also differences of such functions are directly Riemann integrable. The following result is now *key* to a multitude of applications.

Theorem A4.3 (Smith's key renewal theorem)
Let $\lambda^{-1} = EY_1 < \infty$ and $V(t) = EN(t)$ be the renewal function associated with a non–lattice df F.

(a) If h is directly Riemann integrable, then

$$\lim_{t\to\infty} \int_0^t h(t-x)\, dV(x) = \lambda \int_0^\infty h(x)\, dx\,.$$

(b) Consider a renewal–type equation of the form

$$g(t) = h(t) + \int_0^t g(t-x)\, dF(x)\,, \quad t \geq 0\,,$$

where h is locally bounded. Then the unique solution is given by

$$g(t) = \int_0^t h(t-x)\, dV(x)\,, \quad t \geq 0\,.$$

Moreover, if h is directly Riemann integrable, then

$$\lim_{t\to\infty} g(t) = \lambda \int_0^\infty h(x)\, dx\,.$$

□

Remarks. 1) Theorem A4.3 was basic for proving the Cramér–Lundberg estimate for eventual ruin; see Theorem 1.3.6.

2) Theorem A4.3 remains true if h is integrable and bounded on $[0,\infty)$ with $\lim_{x\to\infty} h(x) = 0$ and F is an absolutely continuous df. This version was also used for proving that the marginal distribution tail of a stationary ARCH(1) process is Pareto–like; see Theorem 8.4.12. For a more general result along these lines see Alsmeyer [8], Theorem 3.1.7.

3) A renewal process for which the interarrival times have a *defective df* F, i.e. $F(\infty) = \alpha < 1$, is called *transient* or *terminating.* We encountered an example of such a process in Section 8.3.1. In this case, $V(\infty)$ is a geometrically distributed rv with parameter α, hence the renewal process terminates with probability 1.

4) Renewal processes are closely linked to so–called *regenerative processes.* For such a stochastic process $(X_t)_{t\geq 0}$, there exists a renewal process of times

where the process X regenerates (probabilistically restarts itself) so that the future after one of these *regeneration times* looks probabilistically exactly as if the process just started at time 0. For a discussion of such processes we refer to Resnick [531], Section 3.7.1. These ideas were also used in Section 8.3.1.□

Notes and Comments

A summary of renewal theory is to be found in Feller [235]. More recent accounts are Alsmeyer [8], Asmussen [27], Resnick [531] and Ross [558]. We especially found Resnick [531] very informative. Blackwell's renewal theorem has various proofs, ranging from purely analytic (using Tauberian theory) to more probabilistic ones (such as Feller's approach using ladder epochs and variables). An appealing approach is by so-called *coupling techniques.* The main idea consists of proving the theorem first for an easy renewal function V' (the obvious candidate being the stationary renewal measure) and then showing that V and V' are close in some sense. This kind of technique is becoming increasingly popular in applied probability; a recent survey is given by Lindvall [424]. For a discussion on renewal theory in the infinite mean case start with Bingham et al. [72], Section 8.6.

Generalisations of Theorem A4.3 to weighted renewal measures of the type $\sum_{n=1}^{\infty} a_n F^{n*}$ together with various applications to risk theory are given in Embrechts, Maejima and Omey [212, 213] and Embrechts, Maejima and Teugels [214].

References

Each reference is followed, in square brackets, by a list of the page numbers where this reference is cited.

1. Aase, K.K. (1999) An equilibrium model of catastrophic insurance futures contracts. *The Geneva Papers of Risk and Insurance Theory*. To appear. *[520]*
2. Aase, K.K. and Ødegaard, B.A. (1996) Empirical tests of models of catastrophe insurance futures. Preprint, University of Bergen. A first version of this paper has appeared at the Wharton Financial Institutions Center, Working Paper Series #96–18. *[520]*
3. Abramowitz, M. and Stegun, I.A. (1972) *Handbook of Mathematical Functions*. Dover Publications, New York. *[184]*
4. Adler, R.J. (1990) *An Introduction to Continuity, Extrema, and Related Topics for General Gaussian Processes*. Hayward, California: Institute of Mathematical Statistics. *[119, 120]*
5. Aebi, M., Embrechts, P. and Mikosch, T. (1992) A large claim index. *Mitteilungen SVVM*, 143–156. *[433, 438]*
6. Albrecht, P., König, A. and Schradin, H.D. (1994) Katastrophenversicherungstermingeschäfte: Grundlagen und Anwendungen im Risikomanagement von Versicherungsunternehmen. *Zeitschrift für Versicherungswissenschaft* **83**, 633-682. *[506]*
7. Aldous, D. (1989) *Probability Approximations via the Poisson Clumping Heuristic*. Springer, New York. *[118–119, 204]*
8. Alsmeyer, G. (1991) *Erneuerungstheorie*. Teubner, Stuttgart. *[472, 589–590]*
9. An, H.Z. and Chen, Z.G. (1982) On convergence of LAD estimates in autoregression with infinite variance. *J. Multivariate Anal.* **12**, 335–345. *[402]*
10. Andersen, P.K., Borgan, Ø., Gill, R.D. and Keiding, N. (1993) *Statistical Models Based on Counting Processes*. Springer, New York. *[310]*
11. Anderson, C.W. (1970) Extreme value theory for a class of discrete distributions with applications to some stochastic process. *J. Appl. Probab.* **7**, 99–113. *[119]*
12. Anderson, C.W. (1980) Local limit theorems for the maxima of discrete random variables. *Math. Proc. Cambridge Philos. Soc.* **88**, 161–165. *[119]*
13. Anderson, C.W. (1984) Large deviations of extremes. In: Tiago de Oliveira, J. (Ed.) *Statistical Extremes and Applications*, pp. 81–89. Reidel, Dordrecht. *[344]*

14. Anderson, C.W. and Turkman, K.F. (1991) The joint limiting distribution of sums and maxima of stationary sequences. *J. Appl. Probab.* **28**, 33–44. *[439]*
15. Anderson, C.W. and Turkman, K.F. (1991) Sums and maxima in stationary sequences. *J. Appl. Probab.* **28**, 715–716. *[439]*
16. Anderson, T.W. (1971) *The Statistical Analysis of Time Series.* Wiley, New York. *[378]*
17. Anderson, T.W. (1993) Goodness of fit tests for spectral distributions. *Ann. Statist.* **21**, 830–847. *[392, 393, 534]*
18. ANEX (1996) A statistical ANalysis program with emphasis on EXtreme values. Designed by Beirlant, J., Van Acker, L. and Vynckier, P. KU Leuven. *[370]*
19. Araujo, A. and Giné, E. (1980) *The Central Limit Theorem for Real and Banach Valued Random Variables.* Wiley, New York. *[80]*
20. Arnold, B.C., Balakrishnan, N. and Nagaraja, H.N. (1992) *A First Course in Order Statistics.* Wiley, New York. *[119, 186, 195]*
21. Arov, D.Z. and Bobrov, A.A. (1960) The extreme terms of a sample and their role in the sum of independent random variables. *Theory Probab. Appl.* **5**, 377–396. *[436, 438]*
22. Arratia, R., Goldstein, L. and Gordon, L. (1990) Poisson approximation and the Chen–Stein method (with discussion). *Statist. Science* **5**, 403–434. *[204]*
23. Arratia, R. and Waterman, M.S. (1985) An Erdös–Rényi law with shifts. *Adv. in Math.* **55**, 13–23. *[497]*
24. Asmussen, S. (1982) Conditioned limit theorems relating a random walk to its associate, with applications to risk reserve processes and the GI/G/1 queue. *Adv. Appl. Probab.* **14**, 143–170. *[444–454]*
25. Asmussen, S. (1984) Approximations for the probability of ruin within finite time. *Scand. Actuar. J.*, 31–57. *[448]*
26. Asmussen, S. (1985) Conjugate processes and the simulation of ruin problems. *Stoch. Proc. Appl.* **20**, 213–229. *[454]*
27. Asmussen, S. (1987) *Applied Probability and Queues.* Wiley, Chichester. *[27, 111, 441, 589, 590]*
28. Asmussen, S. (1998) *Ruin Probabilities.* World Scientific, Singapore. To appear. *[27, 110, 441–444]*
29. Asmussen, S. (1998) Subexponential asymptotics for stochastic processes: extremal behaviour, stationary distributions and first passage probabilities. *Ann. Appl. Probab.* **8**, 354–374. *[57]*
30. Asmussen, S. and Binswanger, K. (1997) Simulation of ruin probabilities for subexponential claims. *ASTIN Bulletin* **27**, 297–318. *[57, 454]*
31. Asmussen, S., Fløe Henriksen, L. and Klüppelberg, C. (1994) Large claims approximations for risk processes in a Markovian environment. *Stoch. Proc. Appl.* **54**, 29–43. *[56, 454]*
32. Asmussen, S. and Højgaard, B. (1996) Ruin probability approximations for Markov–modulated risk processes with heavy tails. *Theory Random Proc.* **2**, 96–107. *[454]*

33. Asmussen, S. and Klüppelberg, C. (1996) Large deviation results for subexponential tails, with applications to insurance risk. *Stoch. Proc. Appl.* **64**, 103–125. *[444 -454, 502]*
34. Athreya, K.B. and Ney, P.E. (1972) *Branching Processes.* Springer, Berlin. *[48]*
35. Bachelier, L. (1900) Théorie de la spéculation. *Ann. Sci. École Norm. Sup.* **III–17**, 21–86. Translated in: Cootner, P.H. (Ed.) (1964) *The Random Character of Stock Market Prices,* pp. 17–78. MIT Press, Cambridge, Mass. *[8, 544]*
36. Bahadur, R. and Rao, R.R. (1960) On deviations of the sample mean. *Ann. Math. Statist.* **31**, 1015–1027. *[499]*
37. Balkema, A.A. and Haan, L. de (1974) Residual lifetime at great age. *Ann. Probab.* **2**, 792–804. *[168]*
38. Balkema, A.A. and Haan, L. de (1990) A convergence rate in extreme–value theory. *J. Appl. Probab.* **27**, 577–585. *[151]*
39. Ball, C.A. (Ed.) (1993) A review of stochastic volatility models with application to option pricing. In: *Financial Markets, Institutions and Instruments* vol. 2, pp. 55–71. Blackwell, Cambridge. *[408]*
40. Ball, C.A. and Roma, A. (1994) Stochastic volatility option pricing. *J. Financial Quant. Anal.* **29**, 589–607. *[411]*
41. Ballerini, R. and Resnick, S.I. (1987) Records in the presence of a linear trend. *Adv. Appl. Probab.* **19**, 801–828. *[316]*
42. Barakat, H.M. and El–Shandidy, M.A. (1990) On the limit distribution of the extremes of a random number of independent random variables. *J. Statist. Plann. Inference* **26**, 353–361. *[209]*
43. Barbour, A.D. , Holst, L. and Janson, S. (1992) *Poisson Approximation.* Oxford University Press, New York. *[204, 483]*
44. Barndorff–Nielsen, O.E. (1961) On the rate of growth of the partial maxima of a sequence of independent identically distributed random variables. *Math. Scand.* **9**, 383–394. *[170, 179, 392]*
45. Barndorff–Nielsen, O.E. (1963) On the limit behaviour of extreme order statistics. *Ann. Math. Statist.* **34**, 992–1002. *[170, 179, 392]*
46. Barndorff–Nielsen, O.E. (1997) Normal inverse Gaussian distributions and stochastic volatility modelling. *Scand. J. Statist.* **24**, 1–14. *[411, 521]*
47. Barndorff–Nielsen, O.E. (1998) Processes of normal inverse Gaussian type. *Finance and Stochastics* **2**, 41–68. *[411, 521]*
48. Barndorff–Nielsen, O.E. and Cox, D.R. (1990) *Asymptotic Techniques for Use in Statistics.* Chapman and Hall, London. *[87]*
49. Barndorff–Nielsen, O.E. and Schmidli, H. (1995) Saddlepoint approximations for the probability of ruin in finite time. *Scand. Actuar. J.*, 169–186. *[454]*
50. Barnett, V. (1975) Probability plotting methods and order statistics. *Applied Statistics* **24**, 95–108. *[293]*
51. Barnett, V. and Lewis, T. (1978) *Outliers in Statistics.* Wiley, New York. *[2]*
52. Barnett, V. and Turkman, K.F. (Eds.) (1993) *Statistics for the Environment.* Wiley, New York. *[290]*

53. Baxter, M. and Rennie, A. (1996) *Financial Calculus. An Introduction to Derivative Pricing.* Cambridge University Press, Cambridge. *[482]*
54. Beard, R.E., Pentikäinen, T. and Pesonen, R. (1984) *Risk Theory,* 3rd edition. Chapman and Hall, London. *[27]*
55. Beirlant, J. and Teugels, J.L. (1992) Limit distributions for compounded sums of extreme order statistics. *J. Appl. Probab.* **29**, 557–574. *[519]*
56. Beirlant, J. and Teugels, J.L. (1992) Modeling large claims in non–life insurance. *Insurance: Math. Econom.* **11**, 17–29. *[345]*
57. Beirlant, J., Teugels, J.L. and Vynckier, P. (1996) *Practical Analysis of Extreme Values.* Leuven University Press, Leuven. *[119, 168, 296, 310, 344, 345, 351]*
58. Beirlant, J. and Willekens, E. (1990) Rapid variation with remainder and rates of convergence. *Adv. Appl. Probab.* **22**, 787–801. *[151]*
59. Benabbou, Z. and Partrat, C. (1994) "Grand" sinistres et lois mélanges, I. In: *Proceedings XXV ASTIN Colloquium, Cannes,* pp. 368–387. *[54]*
60. Benktander, G. (1970) Schadenverteilung nach Grösse in der Nicht–Leben–Versicherung. *Mitteilungen SVVM,* 263–283. *[296]*
61. Beran, J. (1994) *Statistics for Long-Memory Processes.* Chapman and Hall, New York. *[541]*
62. Berman, S.M. (1992) *Sojourns and Extremes of Stochastic Processes.* Wadsworth, Belmont, California. *[119, 120]*
63. Bertoin, J. (1996) *Lévy Processes.* Cambridge University Press, Cambridge. *[96]*
64. Bhansali, R.J. (1984) Order determination for processes with infinite variance. In: Franke, J., Härdle, W. and Martin, D. (Eds.) *Robust and Nonlinear Time Series Analysis,* pp. 17–25. Springer, New York. *[403]*
65. Bhansali, R.J. (1988) Consistent order determination for processes with infinite variance. *J. Roy. Statist. Soc. Ser. B* **50**, 46–60. *[403]*
66. Bhansali, R.J. (1997) Robustness of the autoregressive spectral estimate for linear processes with infinite variance. *J. Time Series Anal.* **18**, 213–229 *[392]*
67. Bibby, B.M. and Sørensen, M. (1995) Martingale estimation functions for discretely observed diffusion processes. *Bernoulli* **1**, 17–39. *[411]*
68. Billingsley, P. (1965) *Ergodic Theory and Information.* Wiley, New York. *[63, 68, 378]*
69. Billingsley, P. (1968) *Convergence of Probability Measures.* Wiley, New York. *[92, 96, 109, 235, 522, 527, 534, 551–556]*
70. Billingsley, P. (1995) *Probability and Measure,* 3rd edition. Wiley, New York. *[62]*
71. Bingham, N.H. and Maejima, M. (1985) Summability methods and almost–sure convergence. *Z. Wahrscheinlichkeitstheorie verw. Gebiete* **68**, 383–392. *[69]*
72. Bingham, N.H., Goldie, C.M. and Teugels, J.L. (1987) *Regular Variation.* Cambridge University Press, Cambridge. *[35, 57, 80, 93, 129, 132, 145, 190–191, 236, 564–576, 590]*

73. Bingham, N.H. and Teugels, J.L. (1981) Conditions implying domains of attraction. In: *Proceedings 6th Conference Probab. Theory, Brasov*, pp. 23–34. Ed. Acad. Republ. Soc. Rom., Bucharest. *[438]*
74. Binswanger, K. (1997) *Rare Events in Insurance.* PhD thesis, ETH Zürich. *[57]*
75. Binswanger, K. and Embrechts, P. (1994) Longest runs in coin tossing. *Insurance: Math. Econom.* **15**, 139–149. *[497]*
76. Blattberg, R.C. and Gonedes, N.J. (1974) A comparison of the stable and the Student distributions as statistical models for stock prices. *J. Busin. Univ. Chicago* **47**, 244–280. *[405]*
77. Bollerslev, T. (1986) Generalized autoregressive conditional heteroskedasticity. *J. Econometrics* **31**, 307–327. *[462]*
78. Bollerslev, T., Chou, R.Y. and Kroner, K.F. (1992) ARCH models in finance: A review of the theory and evidence. *J. Econometrics*, **52**, 5–59. *[412]*
79. Boos, D.D. (1984) Using extreme value theory to estimate large percentiles. *Technometrics* **26**, 33–39. *[344]*
80. Borkovec, M. (1995) *Extremales Verhalten von stark oszillierenden Zeitreihen.* Diplomarbeit, ETH Zürich. *[480]*
81. Borkovec, M. and Klüppelberg, C. (1998) Extremal behaviour of diffusion models in finance. *Extremes* **1**, 47–80. *[411]*
82. Borodin, A.N. and Salminen, P. (1996) *Handbook of Brownian Motion – Facts and Formulae.* Birkhäuser, Basel. *[96]*
83. Borovkov, A.A. (1976) *Stochastic Processes in Queueing.* Springer, New York. *[48]*
84. Borovkov, A.A. (1984) *Asymptotic Methods in Queueing Theory.* Wiley, Chichester. *[48]*
85. Bowers, N.L., Gerber, H.U., Hickman, J.C., Jones, D.A. and Nesbitt, C.J. (1987) *Actuarial Mathematics.* The Society of Actuaries, Schaumburg, Illinois. *[27]*
86. Box, G.E.P. and Jenkins, G.M. (1970) *Time Series Analysis: Forecasting and Control.* Holden–Day, San Francisco. *[378, 398]*
87. Brandt, A., Franken, P. and Lisek, B. (1990) *Stationary Stochastic Models.* Wiley, Chichester. *[454]*
88. Brealey, R.A. and Myers, S.C. (1996) *Principles of Corporate Finance,* 5th edition. McGraw–Hill, New York. *[456]*
89. Breiman, L. (1965) On some limit theorems similar to the arc–sine law. *Theory Probab. Appl.* **10**, 323–331. *[438]*
90. Breiman, L. (1968) *Probability.* Addison–Wesley, Reading, Mass. *[67, 250]*
91. Brillinger, D.R. (1981) *Time Series Analysis: Data Analysis and Theory.* Holt, Rinehart & Winston, New York. *[378]*
92. Brockwell, P.J. and Davis, R.A. (1991) *Time Series: Theory and Methods,* 2nd edition. Springer, New York. *[217, 218, 316, 371–403, 533, 541]*
93. Brockwell, P.J. and Davis, R.A. (1996) *Introduction to Time Series and Forecasting.* Springer, New York. *[371, 378]*

94. Broniatowski, M. (1990) On the estimation of the Weibull tail coefficient. Preprint. *[345]*
95. Buchwalder, M., Chevallier, E. and Klüppelberg, C. (1993) Approximation methods for the total claimsize distributions – an algorithmic and graphical presentation. *Mitteilungen SVVM*, 187–227. *[49]*
96. Bucklew, J.A. (1991) *Large Deviation Techniques in Decision, Simulation, and Estimation.* Wiley, New York. *[87, 499, 502]*
97. Bühlmann, H. (1970) *Mathematical Methods in Risk Theory.* Springer, Berlin. *[27]*
98. Bühlmann, H. (1989) Tendencies of development in risk theory. In: *Centennial Celebration of the Actuarial Profession in North America*, vol. 2, pp. 499–521. Society of Actuaries, Schaumburg, Illinois. *[23]*
99. Bühlmann, H., Delbaen, F., Embrechts, P. and Shiryaev, A. (1997) No-arbitrage, change of measure and conditional Esscher transforms. *CWI Quarterly* **9**, 291–317. *[520]*
100. Bühlmann, H., Delbaen, F., Embrechts, P. and Shiryaev, A. (1998) On Esscher transforms in discrete finance models. *ASTIN Bulletin* **28**, 171–186. *[520]*
101. Buhr, K.A. and Carrière, J.F. (1995) Valuation of catastrophe insurance futures using conditional expectations of Poisson and Cox processes. *Actuarial Research Clearing House* **1**, 273–298. *[520]*
102. Buishand, T.A. (1989) Statistics of extremes in climatology. *Statist. Neerlandica* **43**, 1–30. *[324, 425]*
103. Cambanis, S. and Soltani, A.R. (1984) Prediction of stable processes: Spectral and moving average representations. *Z. Wahrscheinlichkeitstheorie verw. Gebiete* **66**, 593–612. *[381]*
104. Castillo, E. (1988) *Extreme Value Theory in Engineering.* Academic Press, San Diego. *[293, 310]*
105. CBOT (1995) *The Chicago Board of Trade, Catastrophe Insurance: Background Report.* *[506]*
106. CBOT (1995) *The Chicago Board of Trade, Catastrophe Insurance: Reference Guide.* *[506]*
107. CBOT (1995) *The Chicago Board of Trade, PCS Options: A User's Guide.* *[507]*
108. Chambers, J.M. (1977) *Computational Methods for Data Analysis.* Wiley, New York. *[293]*
109. Chambers, J.M., Cleveland, W.S., Kleiner, B. and Tukey, P.A. (1983) *Graphical Methods for Data Analysis.* Wadsworth, Belmont Ca., Duxbury Press, Boston. *[290]*
110. Chambers, J.M., Mallows, C.L. and Struck, B.W. (1976) A method for simulating stable random variables. *J. Amer. Statist. Assoc.* **71**, 340–344. *[398, 412]*
111. Chatfield, C. (1984) *The Analysis of Time Series: An Introduction,* 3rd edition. Chapman and Hall, London. *[378]*

112. Cheung Y.–W. (1993) Long memory in foreign–exchange rates. *J. Busin. Econ. Statist.* **11**, 93–101. *[541]*
113. Chichilnisky, G. (1996) Markets with endogenous uncertainty. Theory and policy. *Theory and Decision* **41**, 99–131. *[520]*
114. Chichilnisky, G. and Heal, G.M. (1993) Global environmental risks. *J. Economic Perspectives* **7**, 65–86. *[520]*
115. Chistyakov, V.P. (1964) A theorem on sums of independent positive random variables and its applications to branching random processes. *Theory Probab. Appl.* **9**, 640–648. *[42, 48]*
116. Chover, J., Ney, P. and Wainger, S. (1972) Functions of probability measures. *J. Analyse Math.* **26**, 255–302. *[48]*
117. Chow, T.L. and Teugels, J.L. (1979) The sum and the maximum of i.i.d. random variables. In: Mandl, P. and Huskova, M. (Eds.) *Proceedings Second Prague Symp. Asymptotic Statistics*, pp. 81–92. *[438, 439]*
118. Chow, Y.S. and Teicher, H. (1978) *Probability Theory. Independence, Interchangeability, Martingales*, 2nd edition. Springer, Berlin. *[67]*
119. Christoph, G. and Wolf, W. (1992) *Convergence Theorems with a Stable Limit Law.* Akademie–Verlag, Berlin. *[81, 83, 85]*
120. Chung, K.L. (1974) *A Course in Probability Theory,* 2nd edition. Academic Press, New York. *[27]*
121. Chung, K.L. and Williams, R.J. (1990) *Introduction to Stochastic Integration,* 2nd edition. Birkhäuser, Boston. *[410, 546]*
122. Cleveland, W.S. (1993) *Visualizing Data.* Hobart Press, Summit, N.J. *[290]*
123. Cline, D.B.H. (1986) Convolution tails, product tails and domains of attraction. *Probab. Theory Related Fields* **72**, 529–557. *[52, 55, 582]*
124. Cline, D.B.H. (1987) Convolutions of distributions with exponential and subexponential tails. *J. Austral. Math. Soc. Ser. A* **43**, 347–365. *[45, 582]*
125. Cline, D.B.H. and Brockwell, P.J. (1985) Linear prediction of ARMA processes with infinite variance. *Stoch. Proc. Appl.* **19**, 281–296. *[381]*
126. Cline, D.B.H. and Hsing, T. (1991) Large deviation probabilities for sums and maxima of random variables with heavy or subexponential tails. Preprint, Texas A&M University. *[501]*
127. Cohen, J. (1989) On the compound Poisson limit for high level exceedances. *J. Appl. Probab.* **26**, 458–465. *[425]*
128. Cohen, J.P. (1982) The penultimate form of approximation to normal extremes. *Adv. Appl. Probab.* **14**, 324–339. *[151]*
129. Cohen, J.P. (1982) Convergence rates for the ultimate and penultimate approximations in extreme–value theory. *Adv. Appl. Probab.* **14**, 833–854. *[152]*
130. Cohen, J.W. (1992) *Analysis of Random Walks.* Studies in Probability, Optimization and Statistics, vol. 2. IOS Press, Amsterdam. *[70]*
131. Coles, S.G. (1993) Regional modelling of extreme storms via max–stable processes. *J. Roy. Statist. Soc. Ser. B* **55**, 797–816. *[425]*

132. Conti, B. (1992) Familien von Verteilungsfunktionen zur Beschreibung von Grossschäden: Anforderungen eines Praktikers. Winterthur–Versicherungen, Winterthur. *[56, 166]*
133. Cox, D.R. and Hinkley, D.V. (1974) *Theoretical Statistics.* Chapman and Hall, London. *[318]*
134. Cox, D.R. and Isham, V. (1980) *Point Processes.* Chapman and Hall, London. *[46, 231]*
135. Cox, J. R. and Rubinstein, M. (1985) *Options Markets.* Prentice Hall, Englewood Cliffs. *[482]*
136. Cox, S.H. (Ed.) (1997) *Securitization of Insurance Risk: The 1995 Bowles Symposium.* SOA Monograph M-F-I97-1. Society of Actuaries, Schaumburg, Illinois. *[520]*
137. Cox, S.H. and Schwebach, R.G. (1992) Insurance futures and hedging insurance price risk. *J. Risk Insurance* **59**, 575–600. *[520]*
138. Cramér, H. (1930) *On the Mathematical Theory of Risk.* Skandia Jubilee Volume, Stockholm. Reprinted in: Martin–Löf, A. (Ed.) Cramér, H. (1994) *Collected Works.* Springer, Berlin. *[442]*
139. Cramér, H. (1938) Sur un nouveau théorème–limite de la théorie des probabilités. *Actual. Sci. Indust.* **736**, 5–23. Reprinted in: Martin–Löf, A. (Ed.) Cramér, H. (1994) *Collected Works.* Springer, Berlin. *[85]*
140. Cramér, H. (1954) *The Elements of Probability Theory and Some of Its Applications.* Almqvist och Wiksell, Uppsala, and Wiley, New York. Reprint: Ed. Huntington: Krieger 1973. *[27, 454]*
141. Cramér, H. (1994) *Collected Works*, vol. I. Edited by Martin–Löf, A. Springer, Berlin. *[27]*
142. Cramér, H. (1994) *Collected Works*, vol. II. Edited by Martin–Löf, A. Springer, Berlin. *[27]*
143. Crowder, M.J., Kimber, A.C., Smith, R.L. and Sweeting, T.J. (1991) *Statistical Analysis of Reliability Data.* Chapman and Hall, London. *[310]*
144. Csörgő, M., Csörgő, S. and Horváth, L. (1986) *An Asymptotic Theory for Empirical Reliability and Concentration Processes.* Lecture Notes in Statistics **33**. Springer, Berlin. *[433]*
145. Csörgő, M. and Révész, P. (1981) *Strong Approximations in Probability and Statistics.* Academic Press, New York. *[69]*
146. Csörgő, M. and Steinebach, J. (1981) Improved Erdös–Rényi and strong approximation laws for increments of partial sums. *Ann. Probab.* **9**, 988–996. *[498]*
147. Csörgő, S., Deheuvels, P. and Mason, D.M. (1985) Kernel estimates of the tail index of a distribution. *Ann. Statist.* **13**, 1050–1077. *[345]*
148. Csörgő, S. and Mason, D.M. (1985) Central limit theorems for sums of extreme values. *Math. Proc. Cambridge Philos. Soc.* **98**, 547–558. *[344]*
149. Cummins, J.D. and Geman, H. (1994) An Asian option approach to valuation of insurance futures contracts. *Review Futures Markets* **13**, 517–557. *[520]*

150. Cummins, J.D. and Geman, H. (1995) Pricing catastrophe futures and call spreads: an arbitrage approach. *J. Fixed Income* **4**, 46–57. *[520]*
151. Dahlhaus, R. (1988) Empirical spectral processes and their applications to time series analysis. *Stoch. Proc. Appl.* **30**, 69–83. *[392]*
152. Dahlhaus, R. (1989) Efficient parameter estimation for self–similar processes. *Ann. Statist.* **17**, 1749–1766. *[399]*
153. Daley, D. and Vere–Jones, D. (1988) *An Introduction to the Theory of Point Processes.* Springer, Berlin. *[225, 231–236, 564]*
154. Danielson, J. and Vries, C.G. de (1997) Tail index and quantile estimation with very high frequency data. *J. Empirical Finance* **4**, 241–257. *[289, 344, 351]*
155. Darling, D.A. (1952) The role of the maximum term in the sum of independent random variables. *Trans. Amer. Math. Soc.* **73**, 95–107. *[438]*
156. David, H.A. (1981) *Order Statistics*, 2nd edition. Wiley, New York. *[195, 293]*
157. Davis, R.A. (1996) Gauss–Newton and M–estimation for ARMA processes with infinite variance. *Stoch. Proc. Appl.* **63**, 75–95. *[403]*
158. Davis, R.A., Knight, K. and Liu, J. (1992) M–estimation for autoregressions with infinite variance. *Stoch. Proc. Appl.* **40**, 145–180. *[402]*
159. Davis, R.A. and Resnick, S.I. (1984) Tail estimates motivated by extreme value theory. *Ann. Statist.* **12**, 1467–1487. *[344, 352]*
160. Davis, R.A. and Resnick, S.I. (1985) Limit theory for moving averages of random variables with regularly varying tail probabilities. *Ann. Probab.* **13**, 179–195. *[265, 281, 385]*
161. Davis, R.A. and Resnick, S.I. (1985) More limit theory for the sample correlation function of moving averages. *Stoch. Proc. Appl.* **20**, 257–279. *[281, 385]*
162. Davis, R.A. and Resnick, S.I. (1986) Limit theory for the sample covariance and correlation functions of moving averages. *Ann. Statist.* **14**, 533–558. *[281, 385–386, 399]*
163. Davis, R.A. and Resnick, S.I. (1988) Extremes of moving averages of random variables from the domain of attraction of the double exponential distribution. *Stoch. Proc. Appl.* **30**, 41–68. *[278–282, 585–586]*
164. Davis, R.A. and Resnick, S.I. (1996) Limit theory for bilinear processes with heavy tailed noise. *Ann. Appl. Probab.* **6**, 1191–1210. *[386]*
165. Davison, A.C. (1984) Modelling excesses over high thresholds, with an application. In: Tiego de Oliveira, J. (Ed.) *Statistical Extremes and Applications*, pp. 461–482. Reidel, Dordrecht. *[366]*
166. Davison, A.C. and Smith, R.L. (1990) Models for exceedances over high thresholds (with discussion). *J. Roy. Statist. Soc. Ser. B* **52**, 393–442. *[356, 366, 425]*
167. Daykin, C.D., Pentikäinen, T. and Pesonen, M. (1994) *Practical Risk Theory for Actuaries.* Chapman and Hall, London. *[27]*
168. Deheuvels, P. (1985) On the Erdös–Rényi theorem for random fields and sequences and its relationships with the theory of runs and spacings. *Z. Wahrscheinlichkeitstheorie verw. Gebiete* **70**, 91–115. *[489, 498]*

169. Deheuvels, P. and Devroye, L. (1987) Limit laws of Erdös–Rényi–Shepp type. *Ann. Probab.* **15**, 1363–1386. *[498]*
170. Deheuvels, P., Häusler, E. and Mason, D.M. (1988) Almost sure convergence of the Hill estimator. *Math. Proc. Cambridge Philos. Soc.* **104**, 371–381. *[193, 337]*
171. Deheuvels, P. and Steinebach, J. (1987) Exact convergence rates in strong approximation laws for large increments of partial sums. *Probab. Theory Related Fields* **76**, 369–393. *[498]*
172. Deheuvels, P. and Steinebach, J. (1989) Sharp rates for increments of renewal processes. *Ann. Probab.* **17**, 700–722. *[498]*
173. Dekkers, A.L.M. and Haan, L. de (1989) On the estimation of the extreme-value index and large quantile estimation. *Ann. Statist.* **17**, 1795–1832. *[328–329, 349–351]*
174. Dekkers, A.L.M. and Haan, L. de (1993) Optimal choice of sample fraction in extreme–value estimation. *J. Multivariate Anal.* **47**, 173–195. *[160]*
175. Dekkers, A.L.M., Einmahl, J.H.J. and Haan, L. de (1989) A moment estimator for the index of an extreme–value distribution. *Ann. Statist.* **17**, 1833–1855. *[339]*
176. Delbaen, F. and Haezendonck, J.M. (1989) A martingale approach to premium calculation principles in an arbitrage free market. *Insurance: Math. Economics* **8**, 269–277. *[509, 520]*
177. Dembo, A. and Zeitouni, O. (1998) *Large Deviations Techniques and Applications,* 2nd edition. Springer, New York. *[87, 502]*
178. Deuschel, J.–D. and Stroock, D.W. (1989) *Large Deviations.* Academic Press, Boston. *[87, 502]*
179. De Vylder, F. and Goovaerts, M. (1984) Bounds for classical ruin probabilities. *Insurance: Math. Econom.* **3**, 121–131. *[57]*
180. Dickson, D.C.M. (1995) A review on Panjer's recursion formula and its applications. *British Actuarial J.* **1**, 107–124. *[49]*
181. Diebold, F.X. (1988) *Empirical Modeling of Exchange Rate Dynamics.* Springer, Berlin. *[412]*
182. Diebolt, J. and Guégan, D. (1993) Tail behaviour of the stationary density of general non–linear autoregressive processes of order 1. *J. Appl. Probab.* **30**, 315–329. *[474]*
183. Dijk, V. and Haan, L. de (1992) On the estimation of the exceedance probability of a high level. In: Sen, P.K. and Salama, I.A. (Eds.) *Order Statistics and Nonparametrics: Theory and Applications,* pp. 79–92, Elsevier, Amsterdam. *[348]*
184. Djehiche, B. (1993) A large deviation estimate for ruin probabilities. *Scand. Actuar. J.*, 42–59. *[454, 503]*
185. Doney, R.A. (1997) One–sided local large deviation and renewal theorems in the case of infinite mean. *Probab. Theory Related Fields* **107**, 451–465. *[502]*
186. Drees, H. (1996) Refined Pickands estimators of the extreme value index. *Ann. Stat.* **23**, 2059–2080. *[344]*

187. Dudley, R.M. (1989) *Real Analysis and Probability.* Wadsworth and Brooks/Cole, Pacific Grove, California. *[528, 538]*

188. Dudley, R.M. and Norvaiša, R. (1996) *Product Integrals and p–Variation.* Unpublished book manuscript. *[547]*

189. Duffie, D. (1989) *Futures Markets.* Prentice Hall, Englewood Cliffs. *[505]*

190. Duffie, D. (1992) *Dynamic Asset Pricing Theory.* Princeton University Press, Princeton. *[544]*

191. Dufresne, D. (1991) The distribution of a perpetuity, with applications to risk theory and pension funding. *Scand. Actuar. J.*, 39–79. *[456]*

192. Dufresne, D. (1994) *Mathématiques des Caisses de Retraite.* Editions Supremum, Montréal. *[456]*

193. Dufresne, F. and Gerber, H.U. (1988) The surplus immediately before and at ruin, and the amount of claim causing ruin. *Insurance: Math. Econom.* **7**, 193–199. *[442]*

194. DuMouchel, W.H. (1973) On the asymptotic normality of the maximum–likelihood estimate when sampling from a stable distribution. *Ann. Statist.* **1**, 948–957. *[345]*

195. DuMouchel, W.H. (1983) Estimating the stable index α in order to measure tail thickness: a critique. *Ann. Statist.* **11**, 1019–1031. *[345]*

196. Dzhaparidze, K.O. (1986) *Parameter Estimation and Hypothesis Testing in Spectral Analysis of Stationary Time Series.* Springer, Berlin. *[392]*

197. Eberlein, E. and Keller, U. (1995) Hyperbolic distributions in finance. *Bernoulli* **1**, 281–299. *[407, 411, 521]*

198. Einmahl, J.H.J. (1990) The empirical distribution function as a tail estimator. *Statist. Neerlandica* **44**, 79–82. *[352]*

199. Eisenberg, L. and Jarrow, R. (1994) Option pricing with random volatilities in complete markets. *Review Quant. Fin. Account.* **4**, 5–17. *[411]*

200. Ellis, R.S. (1985) *Entropy, Large Deviations and Statistical Mechanics.* Springer, Berlin. *[87]*

201. Embrechts, P. (1983) A property of the generalized inverse Gaussian distribution with some applications. *J. Appl. Probab.* **20**, 537–544. *[36, 57]*

202. Embrechts, P. (1995) Risk theory of the second and third kind. *Scand. Actuar. J.*, 35–43. *[27]*

203. Embrechts, P. (1998) Actuarial versus financial pricing of insurance. Wharton Financial Institutions Center, Working Paper Series #96–17. Russian version in: *Surveys in Applied and Industrial Mathematics* **5**, 6–22. *[520]*

204. Embrechts, P. and Goldie, C.M. (1980) On closure and factorization theorems for subexponential and related distributions. *J. Austral. Math. Soc. Ser. A* **29**, 243–256. *[51, 581]*

205. Embrechts, P. and Goldie, C.M. (1982) On convolution tails. *Stoch. Proc. Appl.* **13**, 263–278. *[296, 587]*

206. Embrechts, P. and Goldie, C.M. (1994) Perpetuities and random equations. In: Mandl, P., Huskova, M. (Eds.) Asymptotic Statistics. *Proceedings of the*

5th Prague Symposium, September 4–9, 1993, pp. 75–86. Physica–Verlag, Heidelberg. *[454–457]*

207. Embrechts, P., Goldie, C.M. and Veraverbeke, N. (1979) Subexponentiality and infinite divisibility. *Z. Wahrscheinlichkeitstheorie verw. Gebiete* **49**, 335–347. *[576, 581]*
208. Embrechts, P., Grandell, J. and Schmidli, H. (1993) Finite–time Lundberg inequalities in the Cox case. *Scand. Actuar. J.*, 17–41. *[36]*
209. Embrechts, P., Grübel, R. and Pitts, S.M. (1993) Some applications of the fast Fourier transform algorithm in insurance mathematics. *Statist. Neerlandica* **47**, 59–75. *[49]*
210. Embrechts, P. and Hawkes, J. (1982) A limit theorem for the tails of discrete infinitely divisible laws with applications to fluctuation theory. *J. Austral. Math. Soc. Ser. A* **32**, 412–422. *[582]*
211. Embrechts, P. and Klüppelberg, C. (1993) Some aspects of insurance mathematics. *Theory Probab. Appl.* **38**, 374–416. *[27]*
212. Embrechts, P., Maejima, M. and Omey, E. (1984) A renewal theorem of Blackwell type. *Ann. Probab.* **12**, 561–570. *[590]*
213. Embrechts, P., Maejima, M. and Omey, E. (1985) Some limit theorems for generalized renewal measures. *J. London Math. Soc.* (*2*) **31**, 184–192. *[590]*
214. Embrechts, P., Maejima, M. and Teugels, J. (1985) Asymptotic behaviour of compound distributions. *ASTIN Bulletin* **15**, 45–48. *[49, 590]*
215. Embrechts, P. and Meister, S. (1997) Pricing insurance derivatives, the case of CAT–futures. In: Cox, S.H. (Ed.) *Securitization of Insurance Risk: The 1995 Bowles Symposium.* SOA Monograph M-F-I97-1. Society of Actuaries, Schaumburg, Illinois *[520]*
216. Embrechts, P. and Omey, E. (1984) A property of longtailed distributions. *J. Appl. Probab.* **21**, 80–87. *[50]*
217. Embrechts, P., Samorodnitsky, G., Dacorogna, M.M. and Müller, U.A. (1998) How heavy are the tails of a stationary HARCH(k) process? A study of the moments. In: Karatzas, I., Rajput, B.S. and Taqqu, M.S. (Eds.) *Stochastic Processes and Related Topics: In Memory of Stamatis Cambanis 1943–1995* pp. 69–102. Birkhäuser, Boston. *[412]*
218. Embrechts, P. and Veraverbeke, N. (1982) Estimates for the probability of ruin with special emphasis on the possibility of large claims. *Insurance: Math. Econom.* **1**, 55–72. *[27, 35–36, 49]*
219. Engle, R.F. (1982) Autoregressive conditional heteroscedastic models with estimates of the variance of United Kingdom inflation. *Econometrica* **50**, 987–1007. *[462]*
220. Erdös, P. and Rényi, A. (1970) On a new law of large numbers. *J. Analyse Math.* **22**, 103–111. *[486, 497]*
221. Erdös, P. and Révész, P. (1975) On the length of the longest head run. In: Csiszár, I. and Elias, P. (Eds.) *Topics in Information Theory*, pp. 219–228. Colloq. Math. Soc. János Bolyai **16**. North Holland, Amsterdam. *[489]*

222. Erickson, K.B. (1976) Recurrence sets of normed random walk in $\mathbb{R}^d$. *Ann. Probab.* **4**, 802–828. *[70]*

223. Erickson, K.B. (1977) Slowing down d–dimensional random walks. *Ann. Probab.* **5**, 645–651. *[70]*

224. Falk, M. (1995) On testing the extreme value index via the POT method. *Ann. Statist.* **23**, 2013–2035. *[344, 366]*

225. Falk, M., Hüsler, J. and Reiss, R.D. (1994) *Laws of Small Numbers: Extremes and Rare Events.* Birkhäuser, Basel. *[119, 263, 290, 348, 367, 370]*

226. Fama, E.F. (1965) The behaviour of stock market prices. *J. Busin. Univ. Chicago* **38**, 34–105. *[405]*

227. Fama, E.F. (1976) *Foundations of Finance.* Blackwell, Oxford. *[406]*

228. Fama, E.F. and Roll, R. (1968) Some properties of symmetric stable distributions. *J. Amer. Statist. Assoc.* **63**, 817–836. *[405]*

229. Feigin, P. and Resnick, S.I. (1992) Estimation for autoregressive processes with positive innovations. *Commun. Statist. Stochastic Models* **8**, 479–498. *[402]*

230. Feigin, P. and Resnick, S.I. (1994) Limit distributions for linear programming time series estimators. *Stoch. Proc. Appl.* **51**, 135–166. *[402]*

231. Feigin, P. and Resnick, S.I. (1996) Pitfalls of fitting autoregressive models for heavy–tailed time series. *Extremes.* To appear. *[316]*

232. Feilmeier, M. and Bertram, J. (1987) *Anwendung Numerischer Methoden in der Risikotheorie.* Verlag Versicherungswirtschaft e.V., Karlsruhe. *[49]*

233. Feller, W. (1946) A limit theorem for random variables with infinite moments. *Amer. J. Math.* **68**, 257–262. *[68]*

234. Feller, W. (1968) *An Introduction to Probability Theory and Its Applications I,* 3rd edition. Wiley, New York. *[51]*

235. Feller, W. (1971) *An Introduction to Probability Theory and Its Applications II.* Wiley, New York. *[26–37, 67–69, 80–81, 441–445, 551, 564, 581–583, 590]*

236. Ferguson, T.S. (1996) *A Course in Large Sample Theory.* Chapman and Hall, London. *[80]*

237. Feuerverger, A. (1990) An efficiency result for the empirical characteristic function in stationary time-series models. *Canad. J. Stat.* **18**, 155-161. *[345]*

238. Feuerverger, A. and McDunnough, P. (1981) On efficient inference in symmetric stable laws and processes. In: Csörgő, M., Dawson, D.A., Rao, J.N.K. and Saleh, A.K.Md.E. (Eds.) *Statistics and Related Topics.* North Holland, Amsterdam. *[345]*

239. Field, C.A. and Ronchetti, E. (1990) *Small Sample Asymptotics.* IMS Lecture Notes Series **13**. American Mathematical Society, Providence, R.I. *[85, 87]*

240. Fisher, R.A. and Tippett, L.H.C. (1928) Limiting forms of the frequency distribution of the largest or smallest member of a sample. *Proc. Cambridge Philos. Soc.* **24**, 180–190. *[119, 125, 151]*

241. Florens-Zmirou, D. (1993) On estimating the diffusion coefficient from discrete observations. *J. Appl. Prob.* **30**, 790–804. *[411]*

242. Föllmer, H. and Schweizer, M. (1993) A microeconomic approach to diffusion models for stock prices. *Math. Finance* **3**, 1–23. *[10]*
243. Foster, F.G. and Stuart, A. (1954) Distribution–free tests in time–series based on the breaking of records. *J. Roy. Statist. Soc. Ser. B* **16**, 1–22. *[310]*
244. Fox, R. and Taqqu, M.S. (1986) Large sample properties of parameter estimates for strongly dependent stationary Gaussian time series. *Ann. Statist.* **14**, 517–532. *[399]*
245. Frey, R. (1997) Derivative asset analysis in models with level–dependent and stochastic volatility. *CWI Quarterly* **10**, 1-34. *[411]*
246. Fuller, W.A. (1976) *Introduction to Statistical Time Series.* Wiley, New York. *[378]*
247. Furrer, H.J., Michna, Z. and Weron, A. (1997) Stable Lévy motion approximation in collective risk theory. *Insurance: Math. Econom.* **20**, 97–114. *[110, 545]*
248. Furrer, H.J. and Schmidli, H. (1994) Exponential inequalities for ruin probabilities of risk processes perturbed by diffusion. *Insurance: Math. Econom.* **15**, 23–36. *[56]*
249. Galambos, J. (1987) *Asymptotic Theory of Extreme Order Statistics*, 2nd edition. Krieger, Malabar, Florida. *[119, 169, 170, 179, 209]*
250. Gantert, N. (1996) Large deviations for a heavy–tailed mixing sequence. Preprint, Technische Universität Berlin. *[502]*
251. Garman, M.B. and Klass, M.J. (1980) On the estimation of security price volatilities from historical data. *J. Busin. Univ. Chicago* **53**, 67–78. *[411]*
252. Geluk, J.L. and Haan, L. de (1987) *Regular Variation, Extensions and Tauberian Theorems.* CWI Tract **40**, Amsterdam. *[341, 571]*
253. Geman, H. and Yor, M. (1993) Bessel processes, Asian options and perpetuities. *Math. Finance* **3**, 349–375. *[456]*
254. Genon-Catalot, V. and Jacod, J. (1993) On the estimation of the diffusion coefficient for multidimensional diffusions. *Ann. Inst. H. Poincaré (Probab. Statist.)* **29**, 119–151. *[411]*
255. Gerber, H.U. (1973) Martingales in risk theory. *Mitteilungen SVVM* **73**, 205–216. *[34]*
256. Gerber, H.U. (1979) *An Introduction to Mathematical Risk Theory.* Huebner Foundation Monographs 8, distributed by Richard D. Irwin Inc., Homewood Illinois. *[27, 34]*
257. Gerber, H.U. (1990) *Life Insurance Mathematics.* Springer, Berlin. *[456]*
258. Gerber, H.U. and Shiu, E.S.W. (1994) Option pricing by Esscher transforms. With discussion. *Trans. Soc. Actuar.* **46**, 99–191. *[520]*
259. Gerontidis, I. and Smith, R.L. (1982) Monte Carlo generation of order statistics from general distributions. *Applied Statistics* **31**, 238–243. *[189]*
260. Ghose, D. and Kroner, K.F. (1995) The relationship between GARCH and symmetric stable processes: Finding the source of fat tails in financial data. *J. Empirical Finance* **2**, 225–251. *[405]*

261. Ghysels, E., Harvey, A. and Renault, E. (1996) Stochastic volatility. In: Maddala, G.S. and Rao, C.R. (Eds.) *Statistical Methods of Finance. Handbook of Statistics*, vol. 14, pp. 119-191. North-Holland, Amsterdam. *[411]*
262. Gikhman, I.I. and Skorokhod, A.V. (1975) *The Theory of Stochastic Processes II.* Springer, Berlin. *[96, 555]*
263. Giraitis, L. and Surgailis, D. (1990) A central limit theorem for quadratic forms in strongly dependent linear variables and its application to asymptotic normality of Whittle's estimate. *Probab. Theory Related Fields* **86**, 87–104. *[399]*
264. Glick, N. (1978) Breaking records and breaking boards. *Amer. Math. Monthly* **85**, 2–26. *[308, 310]*
265. Gnanadesikan, R. (1980) Graphical methods for internal comparisons in ANOVA and MANOVA. In: Krishnaiah, P.R. (Ed.) *Handbook of Statistics*, vol. 1, pp. 133–177. North Holland, Amsterdam. *[293]*
266. Gnedenko, B.V. (1943) Sur la distribution limité du terme d'une série aléatoire. *Ann. Math.* **44**, 423–453. *[125]*
267. Gnedenko, B.V. and Kolmogorov, A.N. (1954) *Limit Theorems for Sums of Independent Random Variables.* Addison–Wesley, Cambridge, Mass. *[80, 81, 554]*
268. Gnedenko, B.V. and Korolev, V.Yu. (1996) *Random Summation.* CRC Press, Boca Raton. *[111]*
269. Göing, A. (1995) Estimation in financial models. Risklab, ETH Zürich. *[411]*
270. Goldie, C.M. (1977) Convergence theorems for empirical Lorenz curves and their inverses. *Adv. Appl. Prob.* **9**, 765–791. *[433]*
271. Goldie, C.M. (1978) Subexponential distributions and dominated–variation tails. *J. Appl. Probab.* **15**, 440–442. *[51]*
272. Goldie, C.M. (1991) Implicit renewal theory and tails of solutions of random equations. *Ann. Appl. Probab.* **1**, 126–166. *[455, 467]*
273. Goldie, C.M. and Grübel, R. (1996) Perpetuities with thin tails. *Adv. Appl. Probab.* **28**, 463–480. *[456, 480]*
274. Goldie, C.M. and Klüppelberg, C. (1998) Subexponential distributions. In: Adler, R., Feldman, R. and Taqqu, M.S. (Eds.) *A Practical Guide to Heavy Tails: Statistical Techniques for Analysing Heavy–Tailed Distributions*, pp. 435–459. Birkhäuser, Boston. *[48, 582]*
275. Goldie, C.M. and Maller, R.A. (1996) A point process approach to almost sure behaviour of record values and order statistics. *Adv. Appl. Probab.* **28**, 426–462. *[179]*
276. Goldie, C.M. and Resnick, S. (1988) Distributions that are both subexponential and in the domain of attraction of an extreme–value distribution. *Adv. Appl. Probab.* **20**, 706–718. *[149, 150]*
277. Goldie, C.M. and Resnick, S.I. (1989) Records in a partially ordered set. *Ann. Probab.* **17**, 678–699. *[310]*

278. Goldie, C.M. and Smith, R.L. (1987) Slow variation with remainder: a survey of the theory and its applications. *Quart. J. Math. Oxford (2)* **38**, 45–71. *[151, 341]*

279. Goovaerts, M., De Vylder, F. and Haezendonk, J. (1984) *Insurance Premiums.* North Holland, Amsterdam. *[25]*

280. Gordon, L., Schilling, M.F. and Waterman, M.S. (1986) An extreme value theory for long head runs. *Z. Wahrscheinlichkeitstheorie verw. Gebiete* **72**, 279–287. *[119, 489, 496, 497]*

281. Grandell, J. (1977) A class of approximations of ruin probabilities. *Scand. Actuar. J. Suppl.*, 38–52. *[109, 454]*

282. Grandell, J. (1991) *Aspects of Risk Theory.* Springer, Berlin. *[27–29, 36, 96, 109–111, 544]*

283. Grandell, J. (1991) Finite time ruin probabilities and martingales. *Informatica* **2**, 3–32. *[34]*

284. Grandell, J. (1997) *Mixed Poisson Processes.* Chapman and Hall, London. *[48, 111, 208]*

285. Grenander, U. and Rosenblatt, M. (1984) *Statistical Analysis of Stationary Time Series,* 2nd edition. Chelsea Publishing Co., New York. *[392–393, 534]*

286. Groenendijk, P.A., Lucas, A.L. and Vries, C.G. de (1995) A note on the relationship between GARCH and symmetric stable processes. *J. Empirical Finance* **2**, 253–264. *[405]*

287. Grübel, R. (1991) G/G/1 via FFT. *Applied Statistics* **40**, 355–365. *[481]*

288. Guibas, L.J. and Odlyzko, A.M. (1980) Long repetitive patterns in random sequences. *Z. Wahrscheinlichkeitstheorie verw. Geb.* **53**, 241–262. *[489, 497]*

289. Guillaume, D.M., Dacarogna, M.M., Davé, R.R., Müller, U.A., Olsen, R.B. and Pictet, O.V. (1997) From the bird's eye to the microscope. A survey of new stylized facts of the intra–daily foreign exchange markets. *Finance and Stochastics* **1**, 95–129. *[330]*

290. Gumbel, E.J. (1958) *Statistics of Extremes.* Columbia University Press, New York. *[119, 290]*

291. Gut, A. (1988) *Stopped Random Walk. Limit Theorems and Applications.* Springer, New York. *[104–111]*

292. Haan, L. de (1970) *On Regular Variation and Its Application to Weak Convergence of Sample Extremes.* CWI Tract **32**, Amsterdam. *[125, 132, 144–145, 571]*

293. Haan, L. de (1984) Slow variation and the characterization of domains of attraction. In: Tiago de Oliveira, J. (Ed.) *Statistical Extremes and Applications,* pp. 31–48. Reidel, Dordrecht. *[159–160, 350]*

294. Haan, L. de (1990) Fighting the arch–enemy with mathematics. *Statist. Neerlandica* **44**, 45-68. *[288–289, 330]*

295. Haan, L. de (1994) Extreme value statistics. In: Galambos et al. (Eds.) *Extreme Value Theory and Applications*, pp. 93–122. *[335, 352]*

296. Haan, L. de and Karandikar, R.L. (1989) Embedding a stochastic difference equation into a continuous–time process. *Stoch. Proc. Appl.* **32**, 225–235. *[411]*

297. Haan, L. de and Peng, L. (1998) Comparison of tail index estimators. *Statist. Neerlandica* **52**, 60-70. *[341–342]*

298. Haan, L. de and Peng, L. (1998) Exact rates of convergence to a stable law. Preprint, Erasmus University Rotterdam. *[83]*

299. Haan, L. de and Resnick, S.I. (1977) Limit theory for multivariate sample extremes. *Z. Wahrscheinlichkeitstheorie verw. Gebiete* **40**, 317–337. *[236]*

300. Haan, L. de and Resnick, S.I. (1979) Derivatives of regularly varying functions in $\mathbb{R}^d$ and domains of attraction of stable distributions. *Stoch. Proc. Appl.* **8**, 349–355. *[236]*

301. Haan, L. de and Resnick, S.I. (1996) Second order regular variation and rates of convergence in extreme value theory. *Ann. Probab.* **24**, 97–124. *[151]*

302. Haan, L. de and Resnick, S.I. (1998) On asymptotic normality of the Hill estimator. *Commun. Statist. Stochastic Models* **14**, 849–866. *[341]*

303. Haan, L. de, Resnick, S.I., Rootzén, H. and Vries, C.G. de (1989) Extremal behaviour of solutions to a stochastic difference equation with applications to ARCH processes. *Stoch. Proc. Appl.* **32**, 213–224. *[455, 474–480]*

304. Hagerman, R.L. (1978) More evidence on the distribution of security returns. *J. Finance* **23**, 1213–1221. *[405]*

305. Hahn, M.G., Mason D.M. and Weiner, D.C. (Eds.) (1991) *Sums, Trimmed Sums and Extremes.* Birkhäuser, Boston. *[439]*

306. Hahn, M.G. and Weiner, D.C. (1991) On joint estimation of an exponent of regular variation and an asymmetry parameter for tail distribution. In: Hahn, M.G., Mason, D.M., and Weiner, D.C. (Eds.) *Sums, Trimmed Sums and Extremes*, pp. 111–134. Birkhäuser, Basel. *[344]*

307. Hall, P.G. (1979) On the rate of convergence of normal extremes. *J. Appl. Probab.* **16**, 433–439. *[151]*

308. Hall, P.G. (1981) On the rate of convergence to a stable law. *J. London Math. Soc.* **23**, 179–192. *[83]*

309. Hall, P.G. (1982) On some simple estimates of an exponent of regular variation. *J. Roy. Statist. Soc. Ser. B* **44**, 37–42. *[341]*

310. Hall, P.G. (1990) Using the bootstrap to estimate mean squared error and select smoothing parameter in nonparametric problems. *J. Multivariate Anal.* **32**, 177–203. *[344]*

311. Hall, P.G. (1992) *The Bootstrap and Edgeworth Expansion.* Springer, New York. *[85]*

312. Hall, P.G. and Heyde, C.C. (1980) *Martingale Limit Theory and Its Applications.* Academic Press, New York. *[67, 69, 96]*

313. Hall, W.J. and Wellner, J.A. (1979) The rate of convergence in law of the maximum of an exponential sample. *Statist. Neerlandica* **33**, 151–154. *[150]*

314. Hamilton, J.D. (1994) *Time Series Analysis.* Princeton University Press, Princeton. *[378]*

315. Hannan, E.J. (1970) *Multiple Time Series.* Wiley, New York. *[378, 399]*

316. Hannan, E.J. (1973) The asymptotic theory of linear time series. *J. Appl. Probab.* **10**, 130–145. *[399]*

317. Hannan, E.J. and Kanter, M. (1977) Autoregressive processes with infinite variance. *J. Appl. Probab.* **14**, 411–415. *[399]*
318. Harrison, J. and Pliska, S. (1981) Martingales and stochastic integrals in the theory of continuous trading. *Stoch. Proc. Appl.* **11**, 215–260. *[508]*
319. Harvey, A.C. (1987) *Forecasting, Structural Time Series Models and the Kalman Filter.* Cambridge University Press, Cambridge, UK. *[462]*
320. Häusler, E. and Teugels, J.L. (1985) On asymptotic normality of Hill's estimator for the exponent of regular variation. *Ann. Statist.* **13**, 743–756. *[341, 344]*
321. Heyde, C.C. (1967) A contribution to the theory of large deviations for sums of independent random variables. *Z. Wahrscheinlichkeitstheorie verw. Gebiete* **7**, 303–308. *[87, 501]*
322. Heyde, C.C. (1967) On large deviation problems for sums of random variables not attracted to the normal law. *Ann. Math. Statist.* **38**, 1575–1578. *[87, 501]*
323. Heyde, C.C. (1969) A note concerning behaviour of iterated logarithm type. *Proc. Amer. Math. Soc.* **23**, 85–90. *[69, 501]*
324. Heyde, C.C. and Gay, R. (1993) Smoothed periodogram asymptotics and estimation for processes and fields with possible long–range dependence. *Stoch. Proc. Appl.* **45**, 169–182. *[399]*
325. Hida, T. (1980) *Brownian Motion.* Springer, New York. *[96, 536]*
326. Hill, B.M. (1975) A simple general approach to inference about the tail of a distribution. *Ann. Statist.* **3**, 1163–1174. *[190, 331, 347–348]*
327. Hipp, C. (1996) The delta–method for actuarial statistics. *Scand. Actuar. J.*, 79–94. *[433]*
328. Hipp, C. and Michel, R. (1990) *Risikotheorie: Stochastische Modelle und Statistische Methoden.* Verlag Versicherungswirtschaft e.V., Karlsruhe. *[49]*
329. Hoffman–Jørgensen, J. (1993) Stable densities. *Theory Probab. Appl.* **38**, 350–355. *[74]*
330. Hogg, R.V. and Klugman, S.A. (1984) *Loss Distributions.* Wiley, New York. *[33–36, 56, 168, 310, 330]*
331. Hooghiemstra, G. and Meester, L.E. (1996) Computing the extremal index of special Markov chains and queues. *Stoch. Proc. Appl.* **65**, 171–185. *[481]*
332. Höpfner, R. (1997) Two comments on parameter estimation in stable processes. *Math. Methods Statist.* **6**, 125–134. *[345]*
333. Höpfner, R. (1997) On tail parameter estimation in certain point process models. *J. Statist. Plann. Inference.* **60**, 169–187. *[345]*
334. Höpfner, R. and Jacod, J. (1994) Some remarks on the joint estimation of index and scale parameter for stable processes. In: Mandl, P., Huskova, M. (Eds.) *Asymptotic Statistics. Proceedings 5th Prague Symp. 1993*, pp. 273–284. Physica, Heidelberg. *[345]*
335. Hosking, J.R.M. (1984) Testing whether the shape parameter is zero in the generalized extreme–value distribution. *Biometrika* **71**, 367–374. *[324]*
336. Hosking, J.R.M. (1985) Maximum–likelihood estimation of the parameter of the generalised extreme–value distribution. *Applied Statistics* **34**, 301–310. *[318]*

337. Hosking, J.R.M. and Wallis, J.R. (1987) Parameter and quantile estimation for the generalized Pareto distribution. *Technometrics* **29**, 339–349. *[358]*
338. Hosking, J.R.M., Wallis, J.R. and Wood, E.F. (1985) Estimation of the generalized extreme–value distribution by the method of probability–weighted moments. *Technometrics* **27**, 251–261. *[321–323]*
339. Hsing, T. (1988) On the extreme order statistics for a stationary sequence. *Stoch. Proc. Appl.* **29**, 155–169. *[425]*
340. Hsing, T. (1991) Estimating the parameters of rare events. *Stoch. Proc. Appl.* **37**, 117–139. *[425]*
341. Hsing, T. (1991) On tail index estimation using dependent data. *Ann. Statist.* **19**, 1547–1569. *[336, 425]*
342. Hsing, T. (1993) On some estimates based on sample behavior near high level excursions. *Probab. Theory Related Fields* **95**, 331–356. *[402, 425]*
343. Hsing, T. (1995) Limit theorems for stable processes with application to spectral density estimation. *Stoch. Proc. Appl.* **57**, 39–71. *[393]*
344. Hsing, T. (1995) A note on the asymptotic independence of the sum and maximum of strongly mixing stationary random variables. *Ann. Probab.* **23**, 938–947. *[439]*
345. Hsing T., Hüsler, J. and Leadbetter, M.R. (1988) On the exceedance point process for a stationary sequence. *Probab. Theory Related Fields* **78**, 97–112. *[421, 425]*
346. Hull, J. (1997) *Options, Futures and Other Derivatives,* 3rd edition. Prentice Hall, Englewood Cliffs. *[482]*
347. Hull, J. and White, A. (1987) The pricing of options on assets with stochastic volatility. *J. Finance* **42**, 281–300. *[411]*
348. Hull, J. and White, A. (1988) An analysis of the bias in option pricing caused by a stochastic volatility. *Adv. Futures Options Res.* **3**, 29–61. *[411]*
349. Hurst, H. (1951) Long–term storage capacity of reservoirs. *Trans. Amer. Soc. Civil Engrs.* **116**, 778–808. *[537]*
350. Ibragimov, I.A. and Linnik, Yu.V. (1971) *Independent and Stationary Sequences of Random Variables.* Wolters–Noordhoff, Groningen. *[74–85]*
351. Iglehart, D.L. (1969) Diffusion approximations in collective risk theory. *J. Appl. Probab.* **6**, 285–292. *[96, 109, 454]*
352. Jacod, J. and Shiryaev, A.N. (1987) *Limit Theorems for Stochastic Processes.* Springer, Berlin. *[96, 231]*
353. Jaffee, D.M. and Russell, T. (1996) Catastrophe insurance, capital markets, and uninsurable risks. In: Cummins, J.D. (Ed.) *Proceedings of the Conference on Risk Management in Insurance Firms.* Financial Institutions Center, The Wharton School, Philadelphia. *[507]*
354. Janicki, A. and Weron, A. (1993) *Simulation and Chaotic Behaviour of α–Stable Stochastic Processes.* Marcel Dekker, New York. *[74, 81, 96, 521, 540]*
355. Jenkinson, A.F. (1969) Statistics of extremes. In: *Estimation of Maximum Flood,* pp. 183–227. Technical Note 98. World Meteorological Organization, Geneva. *[318]*

356. Jensen, J.L. (1995) *Saddlepoint Approximations.* Oxford University Press, Oxford. *[49, 85, 87]*
357. Joe, H. (1987) Estimation of quantiles of the maximum of N observations. *Biometrika* **74**, 347–354. *[152]*
358. Johnson, N.L. and Kotz, S. (1970) *Distributions in Statistics. Continuous Univariate Distribution I.* Wiley, New York. *[33]*
359. Johnson, N.L. and Kotz, S. (1970) *Distributions in Statistics. Continuous Univariate Distribution II.* Wiley, New York. *[33]*
360. Johnson, N.L. and Kotz, S. (1970) *Distributions in Statistics. Continuous Multivariate Distribution.* Wiley, New York. *[33]*
361. Johnson, N.L., Kotz, S. and Kemp, A.W. (1992) *Univariate Discrete Distributions,* 2nd edition. Wiley, New York. *[304]*
362. Kagan, Y.Y. (1994) Observational evidence for earthquakes as a non–linear dynamic process. *Physica D* **77**, 160–192. *[541]*
363. Kagan, Y.Y. (1997) Earthquake size distribution and earthquake insurance. *Commun. Statist. Stochastic Models* **13**, 775–797. *[541]*
364. Kagan, Y.Y. and Vere–Jones, D. (1996) Problems in the modelling and statistical analysis of earthquakes. In: Heyde, C.C., Prokhorov, Yu.V., Pyke, R. and Rachev, S.T. (Eds.) *Proceedings of the Athens International Conference on Applied Probability and Time Series,* vol. 1: Applied Probability, pp. 398–425. Springer, Berlin. *[541]*
365. Kallenberg, O. (1983) *Random Measures,* 3rd edition. Akademie–Verlag, Berlin. *[221, 231, 235, 564]*
366. Kallsen, J. and Taqqu, M.S. (1998) Option pricing in ARCH–type models. *Math. Finance* **8**, 13-26. *[480]*
367. Karamata, J. (1933) Sur un mode de croissance régulière. Théorèmes fondamentaux. *Bull. Soc. Math. France* **61**, 55–62. *[564]*
368. Karatzas, I. and Shreve, S.E. (1988) *Brownian Motion and Stochastic Calculus.* Springer, Berlin. *[96, 410]*
369. Karatzas, I. and Shreve, S.E. (1998) *Methods of Mathematical Finance.* Springer, Heidelberg. *[410, 482]*
370. Karlin, S. and Ost, F. (1988) Maximal length of common words among random letter sequences. *Ann. Probab.* **16**, 535–563. *[497]*
371. Karlin, S. and Taylor, H.M. (1981) *A Second Course in Stochastic Processes.* Academic Press, New York. *[27]*
372. Karr, A.F. (1986) *Point Processes and Their Statistical Inference.* Marcel Dekker, New York. *[231, 551]*
373. Karr, A.F. (1993) *Probability.* Springer, New York. *[206]*
374. Keller, B. and Klüppelberg, C. (1991) Statistical estimation of large claim distributions. *Mitteilungen SVVM,* 203–216. *[345]*
375. Kendall, M.G. and Stuart, A. (1966) *The Advanced Theory of Statistics, Volume 3. Design and Analysis of Time Series.* Charles Griffin, London. *[316]*
376. Kennedy, J. (1994) Understanding the Wiener–Hopf factorization for the simple random walk. *J. Appl. Probab.* **31**, 561–563. *[27]*

377. Kessler, M. (1997) Estimation of an ergodic diffusion from discrete observations. *Scand. J. Statist.* **24**, 211–229. *[411]*
378. Kesten, H. (1970) The limit points of a normalized random walk. *Ann. Math. Statist.* **41**, 1173–1205. *[69, 70]*
379. Kesten, H. (1973) Random difference equations and renewal theory for products of random matrices. *Acta Math.* **131**, 207–248. *[412, 454–457, 467, 473]*
380. Kesten, H. and Maller, R.A. (1994) Infinite limits and infinite limit points of random walks and trimmed sums. *Ann. Probab.* **22**, 1473–1513. *[439]*
381. Klass, M.J. (1976) Toward a universal law of the iterated logarithm. Part 1. *Z. Wahrscheinlichkeitstheorie verw. Gebiete* **36**, 165–178. *[69]*
382. Klass, M.J. (1977) Toward a universal law of the iterated logarithm. Part 2. *Z. Wahrscheinlichkeitstheorie verw. Gebiete* **39**, 151–166. *[69]*
383. Klass, M.J. (1984) The minimal growth rate of partial maxima. *Ann. Probab.* **12**, 380–389. *[170, 173, 179]*
384. Klass, M.J. (1985) The Robbins–Siegmund series criterion for partial maxima. *Ann. Probab.* **13**, 1369–1370. *[170, 173, 179]*
385. Kloeden, P.E. and Platen, E. (1992) *Numerical Solution of Stochastic Differential Equations.* Springer, Berlin. *[96, 410]*
386. Kloeden, P.E., Platen, E. and Schurz, H. (1994) *Numerical Solution of SDE through Computer Experiments.* Springer, Berlin. *[96, 410]*
387. Klüppelberg, C. (1988) Subexponential distributions and integrated tails. *J. Appl. Probab.* **25**, 132–141. *[50, 55]*
388. Klüppelberg, C. (1989) Estimation of ruin probabilities by means of hazard rates. *Insurance: Math. Econom.* **8**, 279–285. *[55]*
389. Klüppelberg, C. (1989) Subexponential distributions and characterisations of related classes. *Probab. Theory Related Fields* **82**, 259–269. *[57]*
390. Klüppelberg, C. and Mikosch, T. (1993) Spectral estimates and stable processes. *Stoch. Proc. Appl.* **47**, 323–344. *[392]*
391. Klüppelberg, C. and Mikosch, T. (1994) Some limit theory for the self–normalised periodogram of stable processes. *Scand. J. Statist.* **21**, 485–491. *[392–393]*
392. Klüppelberg, C. and Mikosch, T. (1995) Explosive shot noise processes with applications to risk retention. *Bernoulli* **1**, 125–147. *[111]*
393. Klüppelberg, C. and Mikosch, T. (1995) Modelling delay in claim settlement. *Scand. Actuar. J.*, 154–168. *[111]*
394. Klüppelberg, C. and Mikosch, T. (1996) Self–normalised and randomly centred spectral estimates. In: Heyde, C.C., Prokhorov, Yu.V., Pyke, R. and Rachev, S.T. (Eds.) *Proceedings of the Athens International Conference on Applied Probability and Time Series,* vol. 2: Time Series, pp. 259–271. Springer, Berlin. *[380]*
395. Klüppelberg, C. and Mikosch, T. (1996) The integrated periodogram for stable processes. *Ann. Statist.* **24**, 1855–1879. *[392, 534–536]*
396. Klüppelberg, C. and Mikosch, T. (1996) Gaussian limit fields for the integrated periodogram. *Ann. Appl. Probab.* **6**, 969–991. *[392]*

397. Klüppelberg, C. and Mikosch, T. (1996) Parameter estimation for a misspecified ARMA process with infinite variance innovations. *J. Math. Sciences.* **78**, 60–65. *[385, 392]*
398. Klüppelberg, C. and Mikosch, T. (1997) Large deviations of heavy–tailed random sums with applications to insurance and finance. *J. Appl. Probab.* **34**. To appear. *[502–503, 519]*
399. Klüppelberg, C. and Stadtmüller, U. (1996) Ruin probabilities in the presence of heavy–tails and interest rates. *Scand. Actuar. J.* To appear. *[57]*
400. Klüppelberg, C. and Villaseñor, J.A. (1993) Estimation of distribution tails – a semiparametric approach. *Blätter der DGVM* **21**, 213–235. *[345]*
401. Knight, K. (1989) Consistency of Akaike's information criterion for autoregressive processes. *Ann. Statist.* **17**, 824–840. *[403]*
402. Kokoszka, P. and Mikosch, T. (1997) The integrated periodogram for long–memory processes with finite or infinite variance. *Stoch. Proc. Appl.* **66**, 55-78. *[392]*
403. Kokoszka, P. and Taqqu, M.S. (1996) Parameter estimation for infinite variance fractional ARIMA. *Ann. Statist.* **24**, 1880–1913. *[402]*
404. Korolev, V.Yu. (1994) Convergence of random sequences with independent random indices I. *Theory Probab. Appl.* **39**, 282-297. *[209]*
405. Kotulski, M. (1995) Asymptotic distributions of the continuous–time random walks: a probabilistic approach. *J. Statist. Phys.* **81**, 777–792. *[109]*
406. Koutrouvelis, I.A. (1980) Regression–type estimation of the parameters of stable laws. *J. Amer. Statist. Assoc.* **75**, 918–928. *[345]*
407. Kratz, M. and Resnick, S.I. (1996) The QQ-estimator and heavy tails. *Commun. Statist. Stochastic Models* **12**, 699–724. *[310]*
408. Kremer, E. (1985) *Einführung in die Versicherungsmathematik.* Vandenhoeck and Ruprecht, Göttingen. *[519]*
409. Kremer, E. (1986) Simple formulas for the premiums of the LCR and ECOMOR treaties under exponential claim sizes. *Blätter der DGVM*, **XVII**, 237–243. *[366]*
410. Küchler, U., Neumann, K., Sørensen, M. and Streller, A. (1994) Stock returns and hyperbolic distributions. Discussion paper 23. Humboldt Universität Berlin. *[411]*
411. Kwapień, S. and Woyczyński, W.A. (1992) *Random Series and Stochastic Integrals: Single and Multiple.* Birkhäuser, Boston. *[81, 96, 540, 545]*
412. Lamberton, D. and Lapeyre, B. (1991) *Introduction au calcul stochastique appliqué à la finance.* Edition Marketing S.A., Ellipses, SMAI, Paris. English translation by Rabeau, N. and Mantion, F. (1996) *Introduction to Stochastic Calculus Applied to Finance.* Chapman and Hall. *[91]*
413. Lamperti, J.W. (1962) Semi–stable stochastic processes. *Trans. Amer. Math. Soc.* **104**, 62–78. *[543, 547, 550]*
414. Lawless, J.F. (1983) Statistical methods in reliability. *Technometrics* **25**, 305–355. *[324]*

415. Lawless, J.F. (1983) *Statistical Models and Methods for Lifetime Data.* Wiley, New York. *[324]*

416. Leadbetter, M.R. (1983) Extremes and local dependence of stationary sequences. *Z. Wahrscheinlichkeitstheorie verw. Gebiete* **65**, 291–306. *[417, 425]*

417. Leadbetter, M.R. (1991) On a basis for 'peaks over threshold' modeling. *Statist. Probab. Letters* **12**, 357-362. *[167, 366, 417]*

418. Leadbetter, M.R., Lindgren, G. and Rootzén, H. (1983) *Extremes and Related Properties of Random Sequences and Processes.* Springer, Berlin. *[117–120, 151, 211-218, 235-247, 263, 282, 417-418]*

419. Leadbetter, M.R. and Rootzén, H. (1988) Extremal theory for stochastic processes. *Ann. Probab.* **16**, 431–478. *[218, 282, 425]*

420. Lehmann, E.H. (1983) *Theory of Point Estimation,* Wiley, New York. *[318]*

421. Lerche, H.R. (1986) *Boundary Crossing of Brownian Motion.* Lecture Notes in Statistics **40**. Springer, Berlin. *[110]*

422. Leslie, J. (1989) On the non–closure under convolution of the subexponential family. *J. Appl. Probab.* **26**, 58–66. *[581]*

423. Lindvall, T. (1973) Weak convergence of probability measures and random functions in the function space $\mathbb{D}[0,\infty)$. *J. Appl. Probab.* **10**, 109–121. *[555, 559]*

424. Lindvall, T. (1992) *Lectures on the Coupling Method.* Wiley, New York. *[590]*

425. Lo, G.S. (1986) Asymptotic behavior of Hill's estimate and applications. *J. Appl. Probab.* **23**, 922–936. *[344]*

426. Lockhart, R.A. and Stephens, M.A. (1994) Estimation and tests of fit for the three–parameter Weibull distribution. *J. Roy. Statist. Soc. Ser. B* **56**, 491–500. *[324]*

427. Loéve, M. (1978) *Probability Theory II,* 4th edition. Springer, Berlin. *[67, 80, 551]*

428. Longin, F. (1996) The Asymptotic Distribution of Extreme Stock Market Returns. *J. Busin.* **63**, 383–408. *[290, 330, 386, 406]*

429. Loretan, M. and Phillips, C.B. (1994) Testing the covariance stationarity of heavy–tailed time series: An overview of the theory with applications to several financial instruments. *J. Empirical Finance* **1**, 211–248. *[330, 406, 386]*

430. Loynes, R.M. (1965) Extreme values in uniformly mixing stationary stochastic processes. Ann. Math. Statist. **36**, 993–999. *[425]*

431. Lundberg, F. (1903) *Approximerad framställning av sannolikhetsfunktionen. Återförsäkring av kollektivrisker.* Akad. Afhandling. Almqvist och Wiksell, Uppsala. *[8, 22]*

432. Macleod, A.J. (1989) ASR76 – A remark on algorithm AS215: Maximum-likelihood estimation of the parameters of the generalised extreme–value distribution. *Applied Statistics* **38**, 198–199. *[318]*

433. Maejima, M. (1987) Some limit theorems for summability methods of i.i.d. random variables. In: Kalashnikov, V.V., Penkov, B., Zolotarev, V.M. (Eds.) *Stability Problems for Stochastic Models.* Proceedings 9th Intern. Seminar Varna

(Bulgaria), pp. 57–68. Lecture Notes in Mathematics **1233**. Springer, Berlin. *[69]*

434. Major, P. (1981) *Multiple Wiener–Itô Integrals.* Lecture Notes in Mathematics **849**. Springer, Berlin. *[550]*
435. Maller, R.A. and Resnick, S.I. (1984) Limiting behaviour of sums and the term of maximum modulus. *Proc. London Math. Soc.* (3) **49**, 385–422. *[436]*
436. Mandelbrot, B. (1963) The variation of certain speculative prices. *J. Busin. Univ. Chicago* **36**, 394–419. *[345, 405]*
437. Mandelbrot, B.B. (1963) New methods in statistical economics. *J. Political Economy,* **71**, 421–440. *[550]*
438. Mandelbrot, B.B. (1966) Forecasts of future prices, unbiased markets and "martingale" models. *J. Busin.*, **39**, 242–255. *[537]*
439. Mandelbrot, B. and van Ness, J.V. (1968) Fractional Brownian motions, fractional noises and applications. *SIAM Rev.* **10**, 422–437. *[537, 549–550]*
440. Mann, N.R. (1984) Statistical estimation of parameters of the Weibull and Fréchet distribution. In: Tiago de Oliveira, J. (Ed.) *Statistical Extremes and Applications*, pp. 81–89. Reidel, Dordrecht. *[324]*
441. Martikainen, A.I. (1983) One–sided versions of strong limit theorems. *Theory Probab. Appl.* **28**, 48–64. *[69]*
442. Martikainen, A.I. (1985) On the one–sided law of the iterated logarithm. *Theory Probab. Appl.* **30**, 736–749. *[69]*
443. Martin–Löf, A. (1983) Entropy estimates for ruin probabilities. In: Gut, A. and Holst, J. (Eds.) *Probability and Statistics, Essays in Honor of C.–G. Esséen.*, pp. 29–39. Department of Mathematical Statistics, Uppsala University. *[454]*
444. Martin–Löf, A. (1986) Entropy, a useful concept in risk theory. *Scand. Actuar. J.*, 223–235. *[454, 503]*
445. Mason, D.M. (1982) Laws of large numbers for sums of extreme values. *Ann. Probab.* **10**, 756–764. *[193, 336]*
446. Masry, E. and Cambanis, S. (1984) Spectral density estimation for stationary stable processes. *Stoch. Proc. Appl.* **18**, 1–31. *[393]*
447. Matthes, K., Kerstan, J. and Mecke, J. (1978) *Infinitely Divisible Point Processes.* Wiley, Chichester. *[231, 235, 564]*
448. Matthews, R. (1996) Far out forecasting. In: *The New Scientist* **2051**, October 12, 1996, pp. 36–40. *[VII]*
449. McCullagh, P. and Nelder, J.A. (1983) *Generalized Linear Models.* Chapman and Hall, London. *[46]*
450. McCulloch, J.H. (1986) Simple consistent estimators of stable distribution parameters. *Commun. Statist. Simulation Comput.* **15**, 1109–1136. *[345]*
451. McCulloch, J.H. and Panton, D.B. (1997) Precise tabulation of the maximally–skewed stable distributions and densities. *Comput. Statist. Data Anal.* **23**, 307–320. *[81]*
452. McNeil, A. (1997) Estimating the tails of loss severity distributions using extreme value theory. *ASTIN Bulletin* **27**, 117–137. *[367]*
453. Medhi, J. (1992) *Statistical Methods.* Wiley Eastern, New Delhi. *[407]*

454. Meister, S. (1995) *Contributions to the Mathematics of Catastrophe Insurance Futures.* Diplomarbeit, ETH Zürich. *[509, 520]*
455. Melino, A. (1994) Estimation of continuous–time models in finance. In: Sims, C.A. (Ed.) *Advances in Econometrics. Proceedings of the 6th World Congress,* vol. II, pp. 314–351. Econometric Society, CUP, Cambridge. *[411]*
456. Mijnheer, J.L. (1975) *Sample Path Properties of Stable Processes.* CWI Tract **59**, Amsterdam. *[80, 96]*
457. Mikosch, T. (1997) Periodogram estimates from heavy–tailed data. In: Adler, R., Feldman, R. and Taqqu, M.S. (Eds.) *A Practical Guide to Heavy Tails: Statistical Techniques for Analysing Heavy–Tailed Distributions*, pp. 241–257. Birkhäuser, Boston. *[392]*
458. Mikosch, T., Gadrich, T., Klüppelberg, C. and Adler, R.J. (1995) Parameter estimation for ARMA models with infinite variance innovations. *Ann. Statist.* **23**, 305–326. *[399]*
459. Mikosch, T. and Norvaiša, R. (1987) Limit theorems for summability methods of independent random variables I. *Lith. Math. J.* **27**, 142–155. *[69]*
460. Mikosch, T. and Norvaiša, R. (1987) Limit theorems for summability methods of independent random variables II. *Lith. Math. J.* **27**, 303–326. *[69]*
461. Mikosch, T. and Norvaiša, R. (1987) Strong laws of large numbers for fields of Banach space valued random variables. *Probab. Theory Related Fields* **74**, 241–253. *[69]*
462. Mikosch, T. and Norvaiša, R. (1997) Uniform convergence of the empirical spectral distribution function. *Stoch. Proc. Appl.* **70**, 85–114. *[392]*
463. Mikosch, T. and Norvaiša, R. (1999) Stochastic integral equations without probability. *Bernoulli.* To appear. *[547]*
464. Mills, T.C. (1993) *The Econometric Modelling of Financial Time Series.* Cambridge University Press, Cambridge. *[462]*
465. Mittnik, S. and Rachev, S.T. (1993) Modeling asset returns with alternative stable distributions. *Econom. Rev.* **12**, 261–396. *[345, 412, 522]*
466. Mittnik, S. and Rachev, S.T. (1999) *Financial Modeling and Option Pricing with Alternative Stable Models.* Series in Financial Economics and Quantitative Analysis. Wiley, New York. To appear. *[345, 412, 540, 545]*
467. Mood, A.M., Graybill, F.A. and Boes, D.C. (1974) *Introduction to the Theory of Statistics,* 3rd edition. McGraw–Hill, Kogakusha, Tokyo. *[407]*
468. Mori, T. and Oodaira, H. (1976) A functional law of the iterated logarithm for sample sequences. *Yokohama Math. J.* **24**, 35–49. *[263]*
469. Müller, U.A., Dacorogna, M.M., Davé, R.D., Olsen, R.B., Pictet, O.V. and von Weizsäcker, J.E. (1997) Volatilities of different time resolutions – analyzing the dynamics of market components. *J. Empirical Finance* **4**, 214–240. *[412, 480]*
470. Müller, U.A., Dacorogna, M.M., Olsen, R.B., Pictet, O.V., Schwarz, M. and Morgenegg, C. (1990) Statistical study of foreign exchange rates, empirical evidence of a price change scaling law, and intraday analysis. *J. Banking Finance* **14**, 1189–1208. *[289]*

471. Müller, U.A., Dacorogna, M.M. and Pictet, O.V. (1996) Heavy tails in high–frequency financial data. In: Adler, R., Feldman, R. and Taqqu, M.S. (Eds.) *A Practical Guide to Heavy Tails: Statistical Techniques for Analysing Heavy-Tailed Distributions*, pp. 55–77. *[386, 406, 412]*
472. Nagaev, A.V. (1969) Limit theorems for large deviations where Cramér's conditions are violated (in Russian). *Izv. Akad. Nauk UzSSR Ser. Fiz.-Mat. Nauk* **6**, 17–22. *[500]*
473. Nagaev, A.V. (1969) Integral limit theorems for large deviations when Cramér's condition is not fulfilled I,II. *Theory Probab. Appl.* **14**, 51–64 and 193–208. *[500]*
474. Nagaev, S.V. (1979) Large deviations of sums of independent random variables. *Ann. Probab.* **7**, 745–789. *[500, 502]*
475. Nagaraja, H.N. (1988) Record values and related statistics – a review. *Commun. Statist. Theory Methods* **17**, 2223–2238. *[310]*
476. Nelson, D.B. (1990) ARCH models as diffusion approximations. *J. Econometrics* **45**, 7–38. *[411]*
477. NERC (Natural Environment Research Council) (1975) *Flood Studies Report, 5 Volumes.* Whitefriars Press, London. *[284]*
478. Nevzorov, V.B. (1987) Records. *Theory Probab. Appl.* **32**, 201–228. *[310]*
479. Newell, G.F. (1964) Asymptotic extremes for m–dependent random variables. *Ann. Math. Statist.* **35**, 1322–1325. *[425]*
480. Nolan, J. (1996) An algorithm for evaluating stable densities in Zolotarev's (M) parameterization. *Math. and Comp. Modelling.* To appear. *[81]*
481. Nolan, J. (1997) Numerical approximations of stable densities and distribution functions. *Commun. Statist. Stochastic Models* **13**, 759-774. *[81]*
482. North, M. (1980) Time dependent stochastic models of floods. *J. Hyd. Div. ASCE*, **106**, 649–655. *[366]*
483. O'Brien, G.L. (1974) Limit theorems for the maximum term of a stationary process. *Ann. Probab.* **2**, 540–545. *[425]*
484. O'Brien, G.L. (1980) A limit theorem for sample maxima and heavy branches in Galton-Watson trees. *J. Appl. Probab.* **17**, 539–545. *[436]*
485. O'Brien, G.L. (1987) Extreme values for stationary and Markov sequences. *Ann. Probab.* **15**, 281–291. *[422]*
486. Omey, E. and Willekens, E. (1986) Second order behaviour of the tail of a subordinated probability distribution. *Stoch. Proc. Appl.* **21**, 339–353. *[57]*
487. Omey, E. and Willekens, E. (1987) Second order behaviour of distributions subordinate to a distribution with finite mean. *Commun. Statist. Stochastic Models* **3**, 311–342. *[57]*
488. Pakes, A.G. (1975) On the tails of waiting time distributions. *J. Appl. Probab.* **12**, 555–564. *[48]*
489. Panjer, H.H. and Willmot, G. (1992) *Insurance Risk Models.* Society of Actuaries, Schaumburg, Illinois. *[48, 49, 111]*

490. Pedersen, A.R. (1995) A new approach to maximum likelihood estimation for stochastic differential equations based on discrete observations. *Scand. J. Statist.* **22**, 55–71. *[411]*

491. Perfekt, R. (1993) *On Extreme Value Theory for Stationary Markov Chains.* PhD thesis, University of Lund. *[425, 481]*

492. Perry, P.R. (1983) More evidence on the nature of the distribution of security returns. *J. Financial Quant. Anal.* **18**, 211–221. *[405]*

493. Petrov, V.V. (1954) The generalisation of the Cramér limit theorem (in Russian). *Uspekhi Mat. Nauk* **9**, 195–202. Translated in: *Selected Transl. Math. Statist. Probab.* **6** (1966), pp. 1–8. American Mathematical Society, Providence, R.I. *[499]*

494. Petrov, V.V. (1965) On the probabilities of large deviations of the sums of independent random variables. *Theory Probab. Appl.* **10**, 287–298. *[499]*

495. Petrov, V.V. (1975) *Sums of Independent Random Variables.* Springer, Berlin. *[67–69, 81–87, 173, 378, 499, 533, 554]*

496. Petrov, V.V. (1995) *Limit Theorems of Probability Theory.* Oxford University Press, Oxford. *[67–69, 81–87, 103, 173, 378]*

497. Pfeifer, D. (1989) *Einführung in die Extremwertstatistik.* Teubner, Stuttgart. *[119, 310]*

498. Pickands, J. III (1975) Statistical inference using extreme order statistics. *Ann. Statist.* **3**, 119–131. *[166–168, 328–330, 366]*

499. Pictet, O.V., Dacorogna, M.M. and Müller, U.A. (1997) Hill, bootstrap and jackknife estimators for heavy tails. In: Adler, R., Feldman, R. and Taqqu, M.S. (Eds.) *A Practical Guide to Heavy Tails: Statistical Techniques for Analysing Heavy-Tailed Distributions*, pp. 283–310. Birkhäuser, Boston. *[344]*

500. Pinelis, I.F. (1985) On the asymptotic equivalence of probabilities of large deviations for sums and maxima of independent random variables (in Russian). In: *Limit Theorems in Probability Theory*, pp. 144–173. Trudy Inst. Math. **5**. Nauka, Novosibirsk. *[501, 502]*

501. Pitman, E.J.G. (1980) Subexponential distribution functions. *J. Austral. Math. Soc. Ser. A* **29**, 337–347. *[51, 573]*

502. Pitts, S.M. (1994) Nonparametric estimation of the stationary waiting time distribution function for the GI/G/1 queue. *Ann. Statist.* **22**, 1428–1446. *[49]*

503. Pitts, S.M. (1994) Nonparametric estimation of compound distributions with applications in insurance. *Ann. Inst. Statist. Math.* **46**, 537–555. *[49]*

504. Pollard, D. (1984) *Convergence of Stochastic Processes.* Springer, Berlin. *[96, 195, 236, 433, 534, 555–561]*

505. Prabhu, N.U. (1965) *Stochastic Processes.* Macmillan, New York. *[27]*

506. Prabhu, N.U. (1973) Recent research on the ruin problem of collective risk theory. In: Prekopa, A. (Ed.) *Inventory Control and Water Storage.* North Holland, Amsterdam. *[27]*

507. Prabhu, N.U. (1980) *Stochastic Processes, Queues, Insurance Risk, and Dams.* Springer, New York. *[27]*

508. Prabhu, N.U. (1987) A class of ruin problems. In: MacNeil, I.B. and Umphrey, G.J. (Eds.) *Actuarial Science,* pp. 63–78. Reidel, Dordrecht. *[27]*
509. Pratt, J.W. (1960) On interchanging limits and integrals. *Ann. Math. Statist.* **31**, 74–77. *[433, 575]*
510. Prescott, P. and Walden, A.T. (1980) Maximum–likelihood estimation of the parameters of the three–parameter generalized extreme–value distribution. *Biometrika* **67**, 723–724. *[318]*
511. Prescott, P. and Walden, A.T. (1983) Maximum–likelihood estimation of the parameters of the three–parameter generalized extreme–value distribution from censored samples. *J. Statist. Comput. Simulation* **16**, 241–250. *[318]*
512. Priestley, M.B. (1981) *Spectral Analysis and Time Series, vols. I and II.* Academic Press, New York. *[378, 392–393]*
513. Proceedings of the 1st International Conference on High Frequency Data in Finance (1995) Preprint. Olsen & Associates, Zürich. Some papers published in these Proceedings have also appeared in *J. Empirical Finance* **4**, issues 2–3 (1997). *[480]*
514. Protter, P. (1992) *Stochastic Integration and Differential Equations.* Springer, Berlin. *[410, 546–547]*
515. Pruitt, W.E. (1981) The growth of random walks and Lévy processes. *Ann. Probab.* **9**, 948–956. *[70]*
516. Pruitt, W.E. (1987) The contribution to the sum of the summand of maximum modulus. *Ann. Probab.* **15**, 885–896. *[438]*
517. Pruitt, W.E. (1990) The rate of escape of a random walk. *Ann. Probab.* **18**, 1417–1461. *[70]*
518. Pyke, R. (1965) Spacings. *J. Roy. Statist. Soc. Ser. B* **7**, 395–449. *[196]*
519. Pyke, R. (1970) Spacings revisited. *Proceedings 6th Berkeley Symposium Mathematical Statistics Probability,* vol. **I**, pp. 417–427. University of California Press, Berkeley. *[196]*
520. Rachev, S.T. (1991) *Probability Metrics and the Stability of Stochastic Models.* Wiley, Chichester. *[83]*
521. Rachev, S.T. and Samorodnitsky, G. (1993) Option pricing formula for speculative prices modelled by subordinated processes. *Pliska* (Studia Mathematica Bulgarica) **19**, 175–190. *[545]*
522. Ramlau–Hansen, H. (1988) A solvency study in non–life insurance. Part 1: Analysis of fire, windstorm, and glass claims. *Scand. Actuar. J.*, 3–34. *[34, 54]*
523. Ramlau–Hansen, H. (1988) A solvency study in non–life insurance. Part 2: Solvency margin requirements. *Scand. Actuar. J.*, 35–60. *[34, 54]*
524. Reinhard, J.M. (1984) On a class of semi–Markov risk models obtained as classical risk models in a Markovian environment. *ASTIN Bulletin* **14**, 23–43. *[57]*
525. Reiss, R.D. (1987) Estimating the tail index of the claim size distribution. *Blätter der DGVM* **18**, 21–25. *[366]*
526. Reiss, R.D. (1989) *Approximate Distributions of Order Statistics.* Springer, New York. *[119, 152, 184–200, 290]*

527. Reiss, R.D. (1993) *A Course on Point Processes.* Springer, New York. *[231, 263]*

528. Rényi, A. (1970) *Probability Theory.* Akadémiai Kiadó, Budapest. *[486]*

529. Resnick, S.I. (1986) Point processes, regular variation and weak convergence. *Adv. Appl. Probab.* **18**, 66–138. *[96, 231–235, 263, 439]*

530. Resnick, S.I. (1987) *Extreme Values, Regular Variation, and Point Processes.* Springer, New York. *[119–122, 132–151, 220, 232–236, 249–265, 281, 310, 441, 554–584]*

531. Resnick, S.I. (1992) *Adventures in Stochastic Processes.* Birkhäuser, Boston. *[27, 231, 442, 555, 588–590]*

532. Resnick, S.I. (1997) Heavy tail modeling and teletraffic data. *Ann. Statist.* **25**, 1805-1848. *[289]*

533. Resnick, S.I. (1997) Why non–linearities can ruin the heavy tailed modeller's day. In: Adler, R., Feldman, R. and Taqqu, M.S. (Eds.) *A Practical Guide to Heavy Tails: Statistical Techniques for Analysing Heavy–Tailed Distributions,* pp. 219–239. Birkhäuser, Boston. *[386]*

534. Resnick, S.I. (1997) Discussion of the Danish data on large fire insurance losses. *ASTIN Bulletin* **27**, 139-151. *[367]*

535. Resnick, S.I. and Stărică, C. (1995) Consistency of Hill's estimator for dependent data. *J. Appl. Probab.* **32**, 239–167. *[270, 337]*

536. Resnick, S.I. and Stărică, C. (1997) Smoothing the Hill estimator. *Adv. Appl. Probab.* **29**, 271-293. *[339]*

537. Resnick, S.I. and Stărică, C. (1998) Asymptotic behaviour of Hill's estimator for autoregressive data. Preprint, Cornell University. *[270, 337, 406]*

538. Révész, P. (1967) *The Laws of Large Numbers.* Akadémiai Kiadó, Budapest. *[68]*

539. Révész, P. (1970) Strong theorems on coin tossing. In: *Proceedings Intern. Congress of Mathematicians,* Helsinki, pp. 749–754. *[481]*

540. Révész, P. (1990) *Random Walk in Random and Non–Random Environment.* World Scientific, Singapore. *[70, 483]*

541. Revuz, D. and Yor, M. (1991) *Continuous Martingales and Brownian Motion.* Springer, Berlin. *[96, 534]*

542. Ripley, B.R. (1987) *Stochastic Simulation.* Wiley, New York. *[189]*

543. RiskMetrics (1995) Technical document. JP Morgan, New York. *[287, 370]*

544. Robbins, H. and Siegmund, D. (1970) On the law of the iterated logarithm for maxima and minima. *Proceedings 6th Berkeley Symposium Mathematical Statistics Probability,* vol. **III**, pp. 51–70. University of California Press, Berkeley. *[179]*

545. Rogers, L.C.G. (1997) Arbitrage from fractional Brownian motion. *Math. Finance.* **7**, 95–105. *[550]*

546. Rogers, L.C.G. and Satchell, S.E. (1991) Estimating variance from high, low and closing prices. *Ann. Appl. Probab.* **1**, 504–512. *[411]*

547. Rogers, L.C.G. and Williams, D. *Diffusions, Markov Processes, and Martingales.* Volume 1 (1994) Foundations, 2nd edition. Volume 2 (1987) Itô Calculus. Wiley, New York. *[410]*

548. Rogozin, B.A. (1969) A remark on a theorem due to W. Feller. *Theory Probab. Appl.* **14**, 529. *[27]*

549. Rootzén, H. (1978) Extremes of moving averages of stable processes. *Ann. Probab.* **6**, 847–869. *[282]*

550. Rootzén, H. (1986) Extreme value theory for moving average processes. *Ann. Probab.* **14**, 612–652. *[282]*

551. Rootzén, H. (1988) Maxima and exceedances of stationary Markov chains. *Adv. Appl. Probab.* **20**, 371–390. *[425, 474]*

552. Rootzén, H., Leadbetter, M.R. and de Haan, L. (1998) On the distribution of tail array sums for strongly mixing stationary sequences. *Ann. Appl. Probab.* **8**, 868–885. *[336]*

553. Rootzén, H. and Tajvidi, N. (1997) Extreme value statistics and wind storm losses: a case study. *Scand. Actuar. J.*, 70–94. *[289, 296, 430]*

554. Rosbjerg, D. (1983) *Partial Duration Series in Water Resources.* PhD thesis, Technical University of Lyngby, Denmark. *[310]*

555. Rosenblatt, M. (1985) *Stationary Sequences and Random Fields.* Birkhäuser, Boston. *[378]*

556. Rosinski, J. (1995) On the structure of stationary stable processes. *Ann. Probab.* **23**, 1163–1187. *[381]*

557. Rosinski, J. and Woyczyński, W.A. (1987) Multilinear forms in Pareto–like random variables and product random measures. *Coll. Math.* **51**, 303–313. *[540]*

558. Ross, S.M. (1983) *Stochastic Processes.* Wiley, New York. *[590]*

559. Rozovski, L.V. (1994) Probabilities of large deviations on the whole axis. *Theory Probab. Appl.* **38**, 53–79. *[501, 502]*

560. Rydberg, T.H. (1997) The normal inverse Gaussian Lévy process: simulation and approximation. *Commun. Statist. Stochastic Models* **13**, 887–910. *[411]*

561. Rydberg, T.H. (1999) Generalized hyperbolic diffusions with applications towards finance. *Math. Finance* **9**. To appear. *[411]*

562. Rytgaard, M. (1996) Simulation experiments on the mean residual life function $m(x)$. In: *Proceedings of the XXVII ASTIN Colloquium, Copenhagen, Denmark,* vol. 1, pp. 59–81. *[296, 298]*

563. Samorodnitsky, G. (1988) Extrema of skewed stable processes. *Stoch. Proc. Appl.* **30**, 17–39. *[540]*

564. Samorodnitsky, G. (1996) A class of shot noise models for financial applications. In: Heyde, C.C., Prokhorov, Yu.V., Pyke, R. and Rachev, S.T. (Eds.) *Proceedings of the Athens International Conference on Applied Probability and Time Series,* vol. 1: Applied Probability, pp. 332–353. Springer, Berlin. *[412]*

565. Samorodnitsky, G. and Taqqu, M.S. (1994) *Stable Non–Gaussian Random Processes. Stochastic Models with Infinite Variance.* Chapman and Hall, London. *[81, 96, 380, 521–550]*

566. Sato, K. (1995) *Lévy Processes on the Euclidean Spaces.* Lecture Notes. University of Zürich. *[96]*
567. Schmidli, H. (1992) *A General Insurance Risk Model.* PhD thesis, ETH Zürich. *[110]*
568. Schnieper, R. (1993) Praktische Erfahrungen mit Grossschadenverteilungen. *Mitteilungen SVVM*, 149–171. *[54, 56]*
569. Schradin, H.D. (1996) PCS catastrophe insurance options – a new instrument for managing catastrophe risk. In: Albrecht, P. (Ed.) *Aktuarielle Ansätze für Finanz-Risiken.* Beiträge zum 6. Internationalen AFIR-Colloquium, Nürnberg, 1.–3. Oktober 1996, Band II., pp. 1519–1537. Verlag Versicherungswirtschaft e.V., Karlsruhe. *[507]*
570. Schroeder, M. (1990) *Fractals, Chaos, Power Laws.* Freeman, New York. *[550]*
571. Scott, L. (1987) Option pricing when the variance changes randomly: theory, estimation and an application. *J. Financial Quant. Anal.* **22**, 419–438. *[411]*
572. Seal, H.L. (1978) *Survival Probabilities. The Goal of Risk Theory.* Wiley, Chichester. *[46]*
573. Seal, H.L. (1983) Numerical probabilities of ruin when expected claim numbers are large. *Mitteilungen SVVM*, 89–104. *[46, 53]*
574. Segerdahl, C.O. (1955) When does ruin occur in the collective theory of risk? *Skand. Aktuar Tidskr.*, 22–36. *[454]*
575. Seneta, E. (1976) *Functions of Regular Variation.* Lecture Notes in Mathematics **506**. Springer, New York. *[55, 571]*
576. Serfling, R.J. (1980) *Approximation Theorems of Mathematical Statistics.* Wiley, New York. *[80]*
577. Serfozo, R. (1982) Functional limit theorems for extreme values of arrays of independent random variables. *Ann. Probab.* **10**, 172–177. *[263]*
578. Shephard, N. (1996) Statistical aspects of ARCH and stochastic volatility. In: Cox, D.R., Hinkley, D.V. and Barndorff-Nielsen, O.E. (Eds.) *Likelihood, Time Series with Econometric and other Applications.* Chapman and Hall, London. *[412, 454, 462]*
579. Shorack, G.R. and Wellner, J.A. (1986) *Empirical Processes with Applications to Statistics.* Wiley, New York. *[83, 195, 292, 296, 534]*
580. Siegmund, D. (1979) The time until ruin in collective risk theory. *Mitteilungen SVVM*, 157–166. *[454]*
581. Sigma (1995) Non–proportional reinsurance of losses due to natural disasters in 1995: Prices down despite insufficient cover. *Sigma publication No 6*, Swiss Re, Zürich. *[503]*
582. Sigma (1996) Natural catastophes and major losses in 1995: decrease compound to previous year but continually high level of losses since 1989. *Sigma publication No 2*, Swiss Re, Zürich. *[3–7]*
583. Sin, C.A. (1998) Complications with stochastic volatility models. *Adv. Appl. Probab.* **30**, 256–268. *[411]*
584. Skorokhod, A.V. (1957) Limit theorems for stochastic processes. *Theory Probab. Appl.* **1**, 261–290. *[558]*

585. Slud, E. and Hoesman, C. (1989) Moderate and large deviation probabilities in actuarial risk theory. *Adv. Appl. Probab.* **21**, 725–741. *[454, 503]*

586. Smirnov, N.V. (1952) Limit distributions for the terms of a variational series. *Trans. Amer. Math. Soc.* (*1*) **11**, 82–143. *[438]*

587. Smith, R.L. (1982) Uniform rates of convergence in extreme–value theory. *Adv. Appl. Probab.* **14**, 600–622. *[151, 341]*

588. Smith, R.L. (1985) Statistics of extreme values. *Bull. ISI*, LI (Book 4), Paper 26.1 *[356]*

589. Smith, R.L. (1985) Maximum likelihood estimation in a class of non–regular cases. *Biometrika* **72**, 67–90. *[318, 324]*

590. Smith, R.L. (1986) Extreme value theory based on the r largest annual events. *J. Hydrology* **86**, 27–43. *[321]*

591. Smith, R.L. (1987) Estimating tails of probability distributions. *Ann. Statist.* **15**, 1174–1207. *[168, 356–366]*

592. Smith, R.L. (1988) A counterexample concerning the extremal index. *Adv. Appl. Probab.* **20**, 681–683. *[422]*

593. Smith, R.L. (1988) Forecasting records by maximum likelihood. *J. Amer. Statist. Assoc.* **83**, 331–338. *[310, 316]*

594. Smith, R.L. (1989) Extreme value analysis of environmental time series: An application to trend detection in ground–level ozone. *Statist. Science* **4**, 367–393. *[289, 367]*

595. Smith, R.L. (1990) Extreme value theory. In: Ledermann, W. (Chief Ed.) *Handbook of Applicable Mathematics, Supplement*, pp. 437–472. Wiley, Chichester. *[167, 319–323, 356]*

596. Smith, R.L. and Naylor, J.C. (1987) A comparison of maximum likelihood and Bayesian estimators for the three–parameter Weibull distribution. *Applied Statistics* **36**, 358–369. *[324]*

597. Smith, R.L. and Shively, T.S. (1995) A point process approach to modeling trends in tropospheric ozone. *Atmosp. Environm.* **29**, 3489–3499. *[367]*

598. Smith, R.L., Tawn, J.A. and Coles, S.G. (1997) Markov chain models for threshold exceedances. *Biometrika* **84**, 249–268. *[425]*

599. Smith, R.L. and Weissman, I. (1987) Large deviations of tail estimators based on the Pareto approximation. *J. Appl. Probab.* **24**, 619–630. *[344]*

600. Smith, R.L. and Weissman, I. (1994) Estimating the extremal index. *J. Roy. Statist. Soc. Ser. B* **56**, 515–528. *[425]*

601. Sondermann, D. (1991) Reinsurance in arbitrage free markets. *Insurance: Math. Econom.* **10**, 191–202. *[520]*

602. Sørensen, M. (1997) Estimating functions for discretely observed diffusions: a review. In Basawa, I.V., Godambe, V.P. and Taylor, R.L. (Eds.): *Selected Proceedings of the Symposium on Estimating Functions.* IMS Lecture Notes – Monograph Series, vol. 32, 305–325. *[411]*

603. Spector, P. (1994) *An Introduction to S and S–PLUS.* Duxbury Press, Belmont. *[370]*

604. Spitzer, F. (1976) *Principles of Random Walk,* 2nd edition. Springer, Berlin. *[70]*
605. Stam, A.J. (1977) Regular variation in $\mathbb{R}_+^d$ and the Abel–Tauber theorem. Preprint, University of Groningen. *[236]*
606. Stein, E. and Stein, C. (1991) Stock price distributions with stochastic volatility: an analytic approach. *Rev. Fin. Studies* **4**, 727–752. *[411]*
607. Steinebach, J. (1986) Improved Erdös–Rényi and strong approximation laws for increments of renewal processes. *Ann. Probab.* **14**, 547–559. *[498]*
608. Stout, W.F. (1974) *Almost Sure Convergence.* Academic Press, New York. *[68–70]*
609. Straub, E. (1989) *Non-Life Insurance Mathematics.* Springer, Berlin. *[27]*
610. Székely, G. and Tusnády, G. (1979) Generalized Fibonacci numbers, and the number of "pure heads". *Matemaikai Lapok* **27**, 147–151. *[483]*
611. Tajvidi, N. (1996) *Characterisation and Some Statistical Aspects of Univariate and Multivariate Generalised Pareto Distributions.* PhD Thesis, Department of Mathematics, University of Gothenburg. *[366]*
612. Taqqu, M.S. (1981) Self–similar processes and related ultraviolet and infrared catastrophes. In: *Random Fields: Rigorous Results in Statistical Mechanics and Quantum Field Theory,* pp. 1027–1096. Colloq. Math. Soc. János Bolyai **27**. North Holland, Amsterdam. *[550]*
613. Taqqu, M.S. (1986) A bibliographical guide to self–similar processes and long–range dependence. In: Eberlein, E. and Taqqu, M.S. (Eds.) *Dependence in Probability and Statistics,* pp. 137–162. Birkhäuser, Boston. *[550]*
614. Taqqu, M.S. (1988) Self–similar processes. In: Kotz, S. and Johnson, N.L. (Eds.) *Encyclopaedia of Statistical Sciences* **8**, pp. 352–357. Wiley, New York. *[550]*
615. Tawn, J.A. (1988) An extreme–value theory model for dependent observations. *J. Hydrology* **101**, 227–250. *[425]*
616. Taylor, S.J. (1986) *Modelling Financial Time Series.* Wiley, Chichester. *[403–405, 412, 462]*
617. Teugels, J.L. (1975) The class of subexponential distributions. *Ann. Probab.* **3**, 1001–1011. *[48]*
618. Teugels, J.L. (1981) Remarks on large claims. *Bull. Inst. Intern. Statist.* **49**, 1490–1500. *[366]*
619. Teugels, J.L. (1984) Extreme values in insurance mathematics. In: Tiago de Oliveira, J. (Ed.) *Statistical Extremes and Applications. Proceedings of the International Conference at Vimeiro,* pp. 252–259. Reidel, Dordrecht. *[366]*
620. Teugels, J.L. (1985) *Selected Topics in Insurance Mathematics.* University of Leuven, Belgium. *[519]*
621. Thépaut, A. (1950) Une nouvelle forme de réassurance. Le traité d'excédent du coût moyen relatif (Ecomor). *Bulletin trimesteriel de l'Institut des Actuaires Français* **49**, 273. *[287]*
622. Thorin, O. (1970) Some remarks on the ruin problem in the case the epochs of claims form a renewal process. *Skand. Aktuar. Tidskr.*, 29–50. *[36]*

623. Thorin, O. (1974) On the asymptotic behaviour of the ruin probability for an infinite period when the epochs of claims form a renewal process. *Scand. Actuar. J.*, 81–99. *[48]*

624. Thorin, O. (1982) Probability of ruin. *Scand. Actuar. J.*, 65–102. *[36]*

625. Thorin, O. and Wikstad, N. (1977) Calculation of ruin probabilities when the claim distribution is lognormal. *ASTIN Bulletin* **9**, 231–246. *[48]*

626. Tong, Y.L. (1990) *The Multivariate Normal Distribution.* Springer, Berlin. *[556]*

627. Tufts, E.R. (1983) *The Visual Display of Quantitative Information.* Graphics Press, Cheshire. *[290]*

628. van der Vaart, A.W. and Wellner, J.A. (1996) *Weak Convergence and Empirical Processes.* Springer, New York. *[195]*

629. Venables, W.N. and Ripley, B.D. (1994) *Modern Applied Statistics with S–Plus.* Springer, New York. *[370]*

630. Veraverbeke, N. (1993) Asymptotic estimates for the probability of ruin in a Poisson model with diffusion. *Insurance: Math. Econom.* **13**, 57–62. *[56]*

631. Vervaat, W. (1979) On a stochastic difference equation and a representation of non–negative infinitely divisible random variables. *Adv. Appl. Probab.* **11**, 750–783. *[454, 457]*

632. Vinogradov, V. (1994) *Refined Large Deviation Limit Theorems.* Longman, Harlow (England). *[501, 502]*

633. von Bahr, B. (1975) Asymptotic ruin probability when exponential moments do not exist. *Scand. Actuar. J.*, 6–10. *[49]*

634. Vries, C.G. de (1991) On the relation between GARCH and stable processes. *J. Econometrics* **48**, 313–324. *[412]*

635. Wei, X. (1996) Asymptotically efficient estimation of the index of regular variation. Ann. Statist. **23**, 2036–2058. *[345]*

636. Weissman, I. (1975) On location and scale functions of a class of limiting processes with application to extreme–value theory. *Ann. Probab.* **3**, 178–181. *[125]*

637. Weissman, I. (1978) Estimation of parameters and large quantiles based on the k largest observations. *J. Amer. Statist. Assoc.* **73**, 812–815. *[321, 348]*

638. Weissman, I. (1994) On the extremal index of stationary sequences. Preprint, Technion, Israel. *[418, 425]*

639. Weissman, I. and Cohen, U. (1995) The extremal index and clustering of high values for derived stationary sequences. *J. Appl. Probab.* **32**, 972–981. *[420, 425]*

640. Whitt, D. (1980) Some useful functions for functional limit theorems. *Math. Oper. Res.* **5**, 67–85. *[559]*

641. Whittle, P. (1951) *Hypothesis Testing in Time Series Analysis.* Almqvist och Wicksel, Uppsala. *[393, 399]*

642. Widder, V. (1966) *The Laplace Transform.* Princeton University Press, Princeton. *[35]*

643. Williams, D. (1991) *Probability with Martingales.* Cambridge University Press, Cambridge, UK. *[485]*

644. Yang, M.C.K. (1975) On the distribution of the inter–record times in an increasing population. *J. Appl. Probab.* **12**, 148–154. *[316]*

645. Zolotarev, V.M. (1986) *One–Dimensional Stable Distributions.* American Mathematical Society, Providence, R.I. *[74, 80, 85]*

646. Zolotarev, V.M. (1994) On the representation of densities of stable laws by special functions. *Theory Probab. Appl.* **39**, 354–362. *[74]*

Index

List of Abbreviations and Symbols

We have tried as much as possible to use uniquely defined abbreviations and symbols. In various cases, however, symbols can have different meanings in different sections. The list below gives the most typical usage. Commonly used mathematical symbols are not explained.

Abbreviation or symbol	Explanation	Page
α	characteristic exponent of a stable distribution,	72
	index of a regularly varying function	565
AR(p)	autoregressive process of order p	372
ARCH(p)	autoregressive conditionally heteroscedastic process of order p	409
ARMA(p,q)	autoregressive–moving average process of order (p,q)	372
ARMACH	autoregressive–moving average conditionally heteroscedastic process	409
a.s.	almost sure, almost surely, with probability 1	
B, B_t	Brownian motion	88
$\mathbb{C}$	function space of continuous functions	557
$\mathbb{C}[0,1]$, $\mathbb{C}[a,b]$	space of continuous function on $[0,1]$ or $[a,b]$	557
card(A)	cardinal number of the set A	
chf	characteristic function	
$C_K^+(E)$	set of all continuous, non–negative functions on E with compact support	232
CLT	central limit theorem	74
corr(X,Y)	correlation between the rvs X and Y	
cov(X,Y)	covariance between the rvs X and Y	
$\mathcal{D}$	class of dominatedly varying functions	49
$\mathbb{D}$	cadlag function space (Skorokhod space)	
$\mathbb{D}[0,1]$, $\mathbb{D}(0,\infty)$	space of cadlag functions on $[0,1]$ or $(0,\infty)$	557
DA	domain of attraction	74
DA(α)	domain of attraction of an α–stable law	75
DA(G_α)	domain of attraction of G_α	74
df	distribution function	

DNA	domain of normal attraction	80
DNA(α)	domain of normal attraction of an α–stable law	80
DNA(G_α)	domain of normal attraction of G_α	80
E_i	exponential random variable	
$e(u)$	mean excess function	160
ε_x	Dirac measure	220
$Exp(\lambda)$	exponential distribution with parameter λ: $F(x) = 1 - e^{-\lambda x}\,, \quad x \geq 0$	
F	df and distribution of a rv	
F_A	df and distribution of the rv or random element A	
F_I	integrated tail distribution: $F_I(x) = \mu^{-1} \int_0^x \overline{F}(y)dy$	28
F_n	empirical (sample) df	182
F_u	excess df	160
$F^{\leftarrow}(p)$	p–quantile of F,	130
	quantile function of F	130
$F_n^{\leftarrow}(p)$	empirical p–quantile	183
$\overline{F}$	tail of the df F: $\overline{F} = 1 - F$	
F^{n*}	n–fold convolution of the df or distribution F	
$\widehat{f}_X$	Laplace–Stieltjes transform of the rv X: $\widehat{f}_X(s) = Ee^{-sX}$	28
G_α	distribution or df of an α–stable distribution	74
Γ	gamma function : $\Gamma(x) = \int_0^\infty t^{x-1}e^{-t}dt$	
$\Gamma(a,b)$	gamma distribution with parameters a and b: $f(x) = b^a(\Gamma(a))^{-1}x^{a-1}e^{-bx}\,, \quad x \geq 0$	
Γ_a	incomplete gamma function: $\Gamma_a(x) = (\Gamma(a))^{-1} \int_x^\infty e^{-t}t^{a-1}dt\,, \quad x \geq 0$	
$\gamma(h)$	autocovariance at lag h	374
GARCH	generalised autoregressive–moving average conditionally heteroscedastic process	409
GEV	generalised extreme value distribution	152
GPD	generalised Pareto distribution	162

$G_{\xi,\beta}$	generalised Pareto distribution with shape parameter ξ and scale parameter β	162
H	extreme value distribution	124
$H_\xi, H_{\xi;\mu,\psi}$	generalised extreme value distribution with shape parameter ξ, location parameter μ and scale parameter ψ	152
H_θ	generalised extreme value distribution	316
I_A	indicator function of the set (event) A	
iid	independent, identically distributed	
$I_{n,A}$	periodogram of the sequence (A_n)	386
$\mathcal{L}$	a class of heavy–tailed distributions	49
λ	intensity parameter of Poisson process	
Λ	Gumbel distribution: $\Lambda(x) = \exp\{-e^{-x}\}$	121
lhs	left–hand side	
LIL	law of the iterated logarithm	67
LLN	law of large numbers	60
$\ln x$	logarithm with basis e	
$\ln^+ x$	$\ln^+ x = \max(\ln x, 0)$	
$L(x)$	slowly varying function	564
MA(q)	moving average process of order q	372
MDA	maximum domain of attraction	128
MDA(H)	MDA of the extreme value distribution H	128
med(X)	median of X	
M_n	maximum of $X_1, \ldots, X_n$	
$M_p(E)$	space of point measures on E	220
μ	expectation of a rv, mean measure of a PRM	227
μ_A	expectation of the rv A	
$\mathbb{N}$	set of the positive integers	
$\mathbb{N}_0$	set of the non–negative integers	
$N, N(t), N_t$	integer–valued rv or point process, renewal counting process	97

$N(\mu,\sigma^2), N(\mu,\Sigma)$	Gaussian (normal) distribution with mean μ and variance σ^2 or covariance matrix Σ	
$N(0,1)$	standard normal distribution	
$o(1)$	$a(x)=o(b(x))$ as $x\to x_0$ means that $\lim_{x\to x_0} a(x)/b(x)=0$	
$O(1)$	$a(x)=O(b(x))$ as $x\to x_0$ means that $\limsup_{x\to x_0} \lvert a(x)/b(x)\rvert<\infty$	
ω	$\omega\in\Omega$ random outcome	
$[\Omega,\mathcal{F},P]$	probability space	
$\phi_X(t)$	chf of the rv X: $\phi_X(t)=Ee^{itX}$	
Φ	standard normal distribution or df	
Φ_α	Frechet distribution: $\Phi_\alpha(x)=\exp\{-x^{-\alpha}\}$, $\alpha>0$	121
$Poi(\lambda)$	Poisson distribution with parameter λ: $p(n)=e^{-\lambda}\lambda^n/n!\,,\quad n\in\mathbb{N}_0$	
PRM	Poisson random measure	227
PRM(μ)	Poisson random measure with mean measure μ	227
ψ_j	coefficient of a linear process	372
$\psi(u)$	ruin probability in infinite time	24
$\psi(u,T)$	ruin probability in finite time T	24
$\psi(z)$	transfer function of linear process	376
$\lvert\psi(z)\rvert^2$	power transfer function	376
Ψ_α	Weibull distribution: $\Psi_\alpha(x)=\exp\{-(-x)^\alpha\}$, $\alpha>0$	121
Ψ_N	Laplace functional of the point process N	225
$\mathcal{R}$	class of the dfs with regularly varying right tail	49
$\mathcal{R}_\alpha$	class of the regularly varying functions with index α	565
$\mathcal{R}_{-\infty}$	class of the rapidly varying functions	570
$\mathbb{R}, \mathbb{R}^1$	real line	
$\mathbb{R}_+$	$\mathbb{R}_+=(0,\infty)$	
$\mathbb{R}^d$	d–dimensional Euclidean space	
ρ	safety loading	26

$\rho(h)$	autocorrelation at lag h	374
rhs	right–hand side	
rv	random variable	
$\mathcal{S}$	class of the subexponential distributions	39
$\mathcal{S}(\gamma)$	extension of the class $\mathcal{S}$	57
$s\alpha s$	symmetric α–stable	72
σ^2	variance of a rv	
σ_A^2	variance of the rv A	
sign(a)	sign of the real number a	
SLLN	strong law of large numbers	61
S_n	cumulative sum of $X_1, \ldots, X_n$	60
ss	self–similar	542
$S(t)$	random sum,	96
	total claim amount	23
u	initial capital	23
u, u_n	threshold value in extreme value theory	
$U(a,b)$	uniform random variable on (a,b)	
$U(t)$	risk process	23
	$U(t) = F^{\leftarrow}(1 - t^{-1})$	158
var(X)	variance of the rv X	
WLLN	weak law of large numbers	60
wlog	without loss of generality	
ξ	shape parameter of the GEV and GPD	152
x_F	right endpoint of the df F	115
$X_{1,n}$	maximum of $X_1, \ldots, X_n$	182
$X_{k,n}$	kth largest order statistic	182
$\overline{X}_n$	sample mean	60
$\mathbb{Z}$	set of the integers	
$\sim$	$a(x) \sim b(x)$ as $x \to x_0$ means that $\lim_{x \to x_0} a(x)/b(x) = 1$	
	$a(x) \sim 0$ means $a(x) = o(1)$	
$\approx$	$a(x) \approx b(x)$ as $x \to x_0$ means that $a(x)$ is approximately (roughly) of the same order as $b(x)$ as $x \to x_0$. It is only used in a heuristic sense.	

$\asymp$	$a(x) \asymp b(x)$ as $x \to x_0$ means that $0 < \liminf_{x\to x_0} a(x)/b(x) \le \limsup_{x\to x_0} a(x)/b(x) < \infty$	
$*$	convolution	
$\\|\cdot\\|$	$\\|x\\|$ norm of x	
$[\cdot]$	$[x]$ integer part of x	
$\{\cdot\}$	$\{x\}$ fractional part of x	
x^+	positive part of a number: $x^+ = \max(0,x)$	
$\vert\cdot\vert$	Lebesgue measure of a set A: $\vert A\vert$	
$\bigvee_{i\in I} A_i$	$\bigvee_{i\in I} A_i = \sup_{i\in I} A_i$	
$\bigwedge_{i\in I} A_i$	$\bigwedge_{i\in I} A_i = \inf_{i\in I} A_i$	
∂B	boundary of the set B	
$\overline{B}$	closure of the set B	
B^c	complement of the set B	
$\stackrel{\text{a.s.}}{\to}$	$A_n \stackrel{\text{a.s.}}{\to} A$: a.s. convergence	553
$\stackrel{d}{\to}$	$A_n \stackrel{d}{\to} A$: convergence in distribution	551
$\stackrel{J_1}{\to}$	$A_n \stackrel{J_1}{\to} A$: J_1–convergence	558
$\stackrel{L^p}{\to}$	$A_n \stackrel{L^p}{\to} A$: convergence in L^p	553
$\stackrel{P}{\to}$	$A_n \stackrel{P}{\to} A$: convergence in probability	552
$\stackrel{v}{\to}$	$\mu_n \stackrel{v}{\to} \mu$: vague convergence	563
$\stackrel{d}{=}$	$A \stackrel{d}{=} B$: A and B have the same distribution	

For a measure ν on $\mathbb{R}^d$ and intervals $(a,b]$, $[a,b]$ etc. we write $\nu(a,b] = \nu((a,b])$, $\nu[a,b] = \nu([a,b])$ etc.